ENVIRONMENTAL
ENGINEERING
AND SANITATION

ENVIRONMENTAL SCIENCE AND TECHNOLOGY

A Wiley-Interscience Series of Texts and Monographs

Edited by ROBERT L. METCALF, *University of Illinois*
JAMES N. PITTS, Jr., *University of California*

ENVIRONMENTAL ENGINEERING AND SANITATION

SECOND EDITION

JOSEPH A. SALVATO, Jr.

Sanitary and Public Health Engineer

Wiley-Interscience

A DIVISION OF JOHN WILEY & SONS, INC.
NEW YORK   LONDON   SYDNEY   TORONTO

Library of Congress Cataloging in Publication Data

Salvato, Joseph A.
  Environmental engineering and sanitation.

  (Environmental science and technology)
  1958 ed. published under title: Environmental sanitation.
  Includes bibliographies.
  1. Sanitation.   2. Environmental engineering.
I. Title.

RA565.S3 1972   620.8   79-39724
ISBN 0-471-75077-8

Printed in the United States of America.

10 9 8 7 6 5

SERIES PREFACE

Environmental Sciences and Technology

The Environmental Sciences and Technology Series of Monographs, Textbooks, and Advances is devoted to the study of the quality of the environment and to the technology of its conservation. Environmental science therefore relates to the chemical, physical, and biological changes in the environment through contamination or modification, to the physical nature and biological behavior of air, water, soil, food, and waste as they are affected by man's agricultural, industrial, and social activities, and to the application of science and technology to the control and improvement of environmental quality.

The deterioration of environmental quality, which began when man first collected into villages and utilized fire, has existed as a serious problem since the industrial revolution. In the last half of the twentieth century, under the ever-increasing impacts of exponentially increasing population and of industrializing society, environmental contamination of air, water, soil, and food has become a threat to the continued existence of many plant and animal communities of the ecosystem and may ultimately threaten the very survival of the human race.

It seems clear that if we are to preserve for future generations some semblance of the biological order of the world of the past and hope to improve on the deteriorating standards of urban public health, environmental science and technology must quickly come to play a dominant role in designing our social and industrial structure for tomorrow. Scientifically rigorous criteria of environmental quality must be developed. Based in part on these criteria, realistic standards must be established and our technological progress must be tailored to meet them. It is obvious that civilization will continue to require increasing amounts of fuel, transportation, industrial chemicals, fertilizers, pesticides, and countless other products and that it will continue to produce waste products of all descriptions. What is urgently needed is a total systems approach to modern civilization through which the pooled talents of scientists and engineers, in cooperation with social scientists and the medical profession, can be focused on the development of order and equilibrium to the presently disparate segments of the human environment. Most of the skills and tools that are needed are already in existence. Surely a technology that has created such manifold environmental problems is also capable of solving them. It is our hope that this Series in Environmental Sciences and Technology will not only serve to make this challenge more explicit to the established professional but that it also will help to stimulate the student toward the career opportunities in this vital area.

Robert L. Metcalf
James N. Pitts, Jr.

PREFACE

Workers in environmental health who have had experience with environmental sanitation and engineering problems have noted the need for a book that is comprehensive in its scope and more directly applicable to conditions actually encountered in practice.

Many standard texts adequately cover the specialized aspects of environmental engineering and sanitation; but little detailed information is available in one volume dealing with the urban, suburban, and rural community.

In this text emphasis is placed on the practical application of sanitary science and engineering theory and principles to environmental control. This is necessary if available knowledge is to benefit man in his lifetime. In addition, and in deliberate contrast to complement other texts, empirical formulas, rules of thumb, and good practice are applied when possible to illustrate the "best" possible solution under the particular circumstances. It is recognized that this may be hazardous in some instances when blindly applied by the practitioner. It is sincerely hoped, however, that individual ingenuity and investigation will not be stifled by such practicality, but will be challenged and stimulated.

A special effort was made to include design, construction, maintenance, and operation details as they relate to plants and structures. Examples and drawings are used freely to help in the understanding and use of the subject matter.

Since the field is a very broad one, the following subjects are specifically covered in this new revised edition:

1. Control of communicable and certain noninfectious diseases.
2. Environmental Engineering Planning.
3. Water Supply.
4. Wastewater Treatment and Disposal.
5. Solid Waste Management.
6. Air Pollution Control.

7. Radiation Uses and Protection.
8. Food Protection.
9. Recreation Areas and Temporary Residences.
10. Vector and Weed Control and Pesticide Use.
11. Housing and the Residential Environment.
12. Administration.

The control of diseases is discussed to emphasize the importance of the proper application of sanitary principles in disease prevention. This is particularly important in areas of the world where communicable and related diseases have not yet been brought under control; what can happen in the more advanced countries when basic sanitary safeguards are relaxed is also discussed. Without this important subject, this text would be as incomplete as one dealing with preventive medicine or planning that failed to discuss water supply, sewage disposal, solid waste management, air pollution control, and other phases of environmental engineering and sanitation.

Teachers and students of civil, chemical, mechanical, environmental, and sanitary engineering, as well as public health engineering, environmental health, and the environmental sciences will find much of direct value in this text. Others too will find the content especially useful. These would include the health officer, professional sanitarian, social scientist, ecologist, biologist, conservationist, public health nurse, health educator, environmental health technician, and sanitary inspectors of towns, villages, cities, counties, states, and federal governments both in the United States and abroad. City and county engineers and managers, consulting engineers, architects, planners, equipment manufacturers and installers, contractors, farm extension personnel, and institution, resort, and camp directors can all benefit from the contents. The many environmentalists who are interpreting and applying the principles of sanitary science and environmental engineering, sanitation, and hygiene to both the advanced and the developing areas of the world will find the material in this text particularly helpful in accomplishing their objectives.

In the preparation of the first edition (1958) the following people were especially helpful: Col. William A. Hardenbergh, Professor William C. Gibson, Dr. Wendell R. Ames, Arthur Handley, Professor John E. Kiker, Gerald A. Fleet, Professor Nicholas A. Milone, and Richard M. McLaughlin. The assistance they gave is again acknowledged.

Four new chapters have been added to this revised edition, thus expanding *Environmental Sanitation* to *Environmental Engineering and Sanitation*. Most of the material in the original chapters has been reorganized, consolidated, updated, and rewritten. The new chapters cover environmental engineering planning, solid waste management, air pollution control, and radiation uses and protection. For chapter review and suggestions on air pollution control I am grateful to Alexander Rihm, Director of Air Pollution Control, Harry Hovey, Associate Director, and staff members of the New York State Department of Environmental Conservation; and for the chapter on radiation uses and protection to Sherwood Davies, Director, Bureau of Radiological Health, New York State Deparment of Health. In addition, I am appreciative of the general professional encouragement and stimulation received from Meredith H. Thompson, Assistant Commissioner, Division of Sanitary Engineering, New York State Department of Health.

As in the first edition, I am especially indebted to my wife Hazel for the typing involved in an undertaking such as this and for her patience and understanding in helping to make possible this new revised edition.

JOSEPH A. SALVATO, JR.

Troy, New York
November 1971

CONTENTS

Types of planning—Comprehensive community planning, comprehensive functional regional planning, definitive, or project, planning.

The process of comprehensive community planning—Statement of goals and objectives; basic studies, mapping, and data analysis; plan preparation, plan implementation, public information and community action, re-evaluation and continual planning; conclusion.

Regional planning for environmental engineering controls—General, content of a regional planning report, project study, comprehensive solid waste study, comprehensive wastewater study, comprehensive water supply study, comprehensive environmental engineering and health planning, financing.

Environmental factors in site selection and planning—Desirable features; topography; geology, soil, and drainage; utilities; meteorology; location; resources; animal and plant life; improvements needed; site planning.

Fringe and rural-area housing developments—Growth of suburbs and rural areas, facilities and services needed, causes and prevention of haphazard development, cooperative effort needed, subdivision planning, design of water and sewerage service for subdivisions.

Introduction, travel of sewage pollution through the ground, disease transmission, water cycle and its characteristics, water quality, sanitary survey and water sampling, interpretation of water analyses, bacterial examination, physical examination, chemical examination, microscopic examination, quantity.

Source and protection of water supply—General, dug well, bored well, driven and jetted wells, drilled well, well contamination—cause and removal, spring, infiltration gallery, cistern, desalting, surface water.

Treatment of water—design and operation control—Treatment required; chlorination; gas chlorinator, testing for residual chlorine, chlorine treatment for operation control; distribution system contamination; plain sedimentation; microstraining; coagulation and settling; filtration; slow sand filter; rapid sand filter; pressure sand filter; diatomaceous earth filter; taste and odor control; control of microorganisms; aquatic weed control; other taste and odor causes and controls; methods to remove or reduce objectionable tastes and odors; hydrogen sulfide, sources and removal; iron and manganese, occurrence and removal; corrosion cause and control; water softening; fluoridation.

Water system design principles—Water quantity, design period, watershed runoff and reservoir design, intakes and screens, pumping, distribution storage requirements, peak demand estimates, distribution system design standards, the American Insurance Association's Water System Design Standards, cross-connection control, hydropneumatic systems.

Pumps—Displacement pump, centrifugal pump, jet pump, air-lift pump, hydraulic ram, pump and well protection, pump power and drive, automatic pump control.

Examples—Design of small water systems, design of a camp water system, cost estimates.

Cleaning and disinfection—Wells and springs, pipelines, storage reservoirs and tanks.

Emergency water supply and treatment—Boiling, chlorination, iodine, filtration, bottled and tank-truck water.

4. Wastewater Treatment and Disposal

Disease hazard, definitions, stream pollution and recovery.

Small water-borne sewage disposal systems—General soil conditions, soil percolation test, estimate of sewage flow, house sewer and plumbing, grease trap, septic tank, care of septic tank and tile field, division of flow to soil absorption system, subsurface soil absorption systems, tile field system, leaching or seepage pit, cesspool, dry well.

Small sewage disposal systems for tight soils—Evapotranspiration, transvap systems, modification of conventional tile field system, waste stabilization pond, aerobic sewage treatment unit, sand filter, sand filter design, dosing tank, lift station.

Sewage works design—small treatment plants—Preparation of plans and reports, design details, chlorination, trickling filter, extended aeration, waste stabilization pond, low-cost sanitation, inspection during construction, operation control.

Typical designs of small plants—Design for a small community, children's camp design, town highway building design, elementary school design, design for a subdivision, toll road service-area design.

Sewage works design—large systems—General, plans and report, sewers, inverted siphons, pumping stations, sewage treatment, hydraulic overloading of sewers and treatment plant, cost of sewerage and treatment.

Excreta disposal—Privies and latrines.

Industrial wastes—Industrial wastes surveys, milk wastes, poultry wastes, canning wastes, packing-house wastes, laundry wastes.

Composition, storage, and collection—General; composition, weight, and volume; storage; can washing; apartment house and institution compactors, macerators, and pneumatic tubes; collection; transfer station.

Treatment and disposal of solid wastes—Open dump, hog feeding, grinding, disposal at sea, garbage reduction, composting, incineration, sanitary landfill, pyrolization, high-temperature incineration, wet oxidation, size reduction, high-density compaction, disposal of animal wastes.

Incineration—General; nonincinerable waste and residue; site selection, plant layout, and building design; incinerator design; types of furnaces; control of incineration; on-site incineration.

Sanitary landfill—Introduction, sanitary landfill planning and design, sanitary landfill methods, equipment for disposal by sanitary landfill, operation and supervision, summary of recommended operating practices, modified sanitary landfill.

The problem and its effects—Health effects; economic effects; effects on plants; effects on animals; aesthetic, climatic, and related effects.

Sources and types of air pollution—Man-made sources, natural sources, types of air pollutants.

Sampling and measurement—Measurement of materials' degradation, smoke measurement, particulate sampling, gas sampling.

Environmental factors—Meteorology, topography.

Air pollution surveys—Inventory, air sampling, basic studies and analyses.

Ambient air quality standards—Federal standards, a state classification system.

Controls—Source control, control equipment, collectors and separators, filters, electrical precipitators, scrubbers, afterburners, other collectors, dilution by stack height, planning and zoning.

Program and enforcement—General, organization and staffing, regulation.

architectural details; washing facilities, toilets, and locker rooms; water supply; liquid and solid waste disposal.

Food handling, quality, and storage—Food handling, food inspection, microbiological standards, dry food storage.

Milk source, transportation, processing, and control tests—Milk quality, dairy farm sanitation, barn, pen stabling, pipeline milker, milkhouse, bulk cooling and storage, transportation, pasteurization, cooler, bottle washer, cooler and boiler capacity, quality control.

Milk program administration—Certified industry inspection, cooperative state—Public Health Service program for certification of interstate milk shippers, official local program supervision and inspection, official state surveillance and program evaluation.

Hospital infant formula.

Restaurants, slaughterhouses, poultry dressing plants, and other food establishments—Basic requirements, inspection and inspection forms, shellfish, typical kitchen floor plans.

Design details—General design guides; lighting; ventilation; uniform design standards; space requirements; refrigeration; refrigeration design; cleansing; hand dishwashing, floor plans, and designs; machine dishwashing, floor plans, and designs; hot water for general utility purposes.

Beach and pool standards and regulations—Health considerations, regulations and standards, accident prevention.

Swimming pool types and design—Recirculating swimming pool, fill-and-draw pool, flow-through pool, partly artificial pool, summary of pool design, a small pool design.

Swimming pool operation—Bacterial, chemical, and physical water quality; disinfection; control of pH; clarity; swimming pool water temperature; testing for free available chlorine and pH; algae control; prevention of ringworm infections; personnel; pool regulations.

Swimming pool maintenance—Recirculating pump, hair catcher, filters, chemical feed equipment, pool structure.

Wading pools—Types.

Bathing beaches—Disinfection and water quality, control of algae, control of aquatic weeds, control of swimmer's itch.

Temporary residences—Operation of temporary residences, camps and resort hotels, compliance guide, travel trailer parks, migrant labor camps, mass gatherings.

Pesticides—Pesticide use, pesticide groupings, pesticide disposal.

The housefly—Spread of diseases, fly control.

Mosquito control—Disease hazard, life cycle and characteristics, control program, domestic mosquito control.

Control of miscellaneous arthropods—bedbugs, ticks, chiggers, lice, fleas, roaches, ants, punkies and sand flies, blackflies, termites.

Control of rats and mice—Rat control program, rat control techniques, rat control at open dumps, control of rats in sewers, control of rats in open areas, control of mice, rat control by use of chemosterilants.

Pigeon control—The health hazard, control measures.

Control of poison ivy, poison oak, and poison sumac—Eradication, remedies.

Ragweed and noxious-weed control—Reasons for control, hay fever symptoms and pollen sampling, problem, a control program, eradication, equipment for chemical treatment, chemical dosages, control of aquatic weeds.

Substandard housing and its effects—Growth of the problem; health, economic, and social effects.

Appraisal of quality of living—APHA appraisal method; census data; health, economic, and social factors; planning; environmental sanitation and hygiene indices; other survey methods; APHA criteria; basic health principles of housing and its environment; minimum standards housing ordinance; summary of APHA–PHS recommended housing maintenance and occupancy ordinance.

Housing program—Approach, components for a good housing program, outline of a housing program, solutions to the problem, selection of work areas, enforcement program, enforcement procedures.

Housing form paragraphs for letters.

Plumbing—Plumbing code, backflow prevention, indirect waste piping, plumbing details, other.

Ventilation—Spread of respiratory diseases, thermal requirements, space ventilation and heating, toilet ventilation, venting of heating units.

Mobile home parks.

Institution sanitation—Definition, institutions as small communities, hospitals and nursing homes, schools, other institutions.

Organization—Introduction, state and local programs, manpower, salaries and salary surveys.

Environmental control program planning—Statement of goals and objectives; data collection, studies, and analyses; environmental control program plan; administration.

Program supervision—Evaluation, reporting, time sheet, coding, record keeping and data processing, statistical report, workload and program supervision, what is an inspection?, frequency of inspection, inspection supervision, production, efficiency, performance.

Enforcement—Philosophy, performance objectives and specification standards, correspondence, compliance guides, in-service training, education, persuasion and motivation, legal action.

Emergency sanitation.

ENVIRONMENTAL
ENGINEERING
AND SANITATION

INTRODUCTION

In many parts of the world, simple survival or prevention of disease and poisoning are still serious concerns. In other areas, maintenance of an environment that is suited to man's efficient performance and to the preservation of comfort and enjoyment of living are the goals for the future. These levels of life and progress can be the basis for action programs in environmental health.[1]

As urbanization increases, man's impact on the environment, and the impact of the environment on man, must be controlled to protect the human and natural resources essential to life and at the same time enhance man's well-being. The environment encompasses "the sum of all external influences and conditions affecting life and development of an organism (including man)." The achievement of an environment to enhance man's well-being requires the application of environmental science and engineering principles. This means "the control of all those factors in man's physical environment which exercise or may exercise a deleterious effect on his physical development, health, and survival"[2] and "the application of engineering principles to the control, modification or adaptation of the physical, chemical, and biological factors of the environment in the interest of man's health, comfort, and social well-being,"[3] with consideration of the impact of the control measures applied.

The chapters that follow attempt to point out major areas of concern to be attacked by a host of disciplines working in close harmony. Effec-

---

[1] Frank M. Stead, "Levels in Environmental Health," *Am. J. Public Health*, **50**, No. 3, 312 (March 1960); *WHO Tech. Rep. Ser.* No. 77, 9 (1954).

[2] WHO Expert Committee on Environmental Sanitation, *WHO Tech. Rep. Ser.*, No. 10 (1950), 5.

[3] "The Education of Engineers in Environmental Health," *WHO Tech. Rep. Ser.*, No. 376 (1967), 6.

tive control or management of these conditions will help achieve the highest possible quality of environment and living, keeping in mind the physical, social,[4] and economic factors involved and their interdependence.

[4] Includes political, cultural, educational, biological, medical, and public health. Social health is the "ability to live in harmony with other people of other kinds, with other traditions, with other religions and with other social systems throughout the world." (WHO)

1

CONTROL OF COMMUNICABLE AND CERTAIN NONINFECTIOUS DISEASES

A communicable disease is "an illness due to a specific infectious agent or its toxic product which arises through transmission of that agent or its products from a reservoir to a susceptible host—either directly, as from an infected person or animal, or indirectly, through the agency of an intermediate plant or animal host, vector, or the inanimate environment."[1] Illness may be caused by bacteria, bacterial toxins, viruses, protozoa, spirochetes, parasitic worms (helminths), poisonous plants and animals, chemical poisons, fungi, rickettsiae, certain yeasts, and molds. In this text the diseases are grouped into respiratory, water- and food-borne, insect- and rodent-borne, and miscellaneous diseases. Noninfectious diseases would include mercury poisoning, lead poisoning, nutritional deficiency diseases, carbon monoxide poisoning, and illnesses associated with air pollution.

Definitions

Certain terms with which one should become familiar are frequently used in the discussion of communicable diseases. The definitions given in *Control of Communicable Diseases in Man*[2] are considered most authoritative and should be used as a reference and source of more detailed information. Some of the more common definitions are given here.

[1] *Control of Communicable Diseases in Man,* 11th ed., American Public Health Association, New York, 1970.
[2] Ibid.

3

CARRIER—A carrier is an infected person (or animal) that harbors a specific infectious agent in the absence of discernible clinical disease and serves as a potential source of infection for man. The carrier state may occur in an individual with an infection inapparent throughout its course (commonly known as *healthy carrier*), or during the incubation period, convalescence, and postconvalescence of an individual with a clinically recognizable disease (commonly known as *incubatory carrier* or *convalescent carrier*). Under either circumstance the carrier state may be of short or long duration (*temporary carrier* or *chronic carrier*).

CONTACT—A person or animal that has been in such association with an infected person or animal or a contaminated environment as to have had opportunity to acquire the infection.

CONTAMINATION—The presence of an infectious agent on a body surface; also on or in clothes, bedding, toys, surgical instruments or dressings, or other inanimate articles or substances including water, milk, and food. Contamination is distinct from *pollution*, which implies the presence of offensive but not necessarily infectious matter in the environment.

DISINFECTION—Killing of infectious agents outside the body by chemical or physical means directly applied.

a. *Concurrent disinfection* is the application of disinfective measures as soon as possible after the discharge of infectious material from the body of an infected person, or after the soiling of articles with such infectious discharges, all personal contact with such discharges or articles being prevented prior to such disinfection.

b. *Terminal disinfection* is application of disinfective measures after the patient has been removed by death or to a hospital, or has ceased to be a source of infection, or after isolation practices have been discontinued. Terminal disinfection is rarely practiced; terminal cleaning generally suffices, along with airing and sunning of rooms, furniture, and bedding. It is necessary only for diseases spread by indirect contact; steam sterilization of bedding is desirable after smallpox.

DISINFESTATION—Any physical or chemical process serving to destroy undesired small animal forms, particularly arthropods or rodents, present upon the person, the clothing, or in the environment of an individual, or on domestic animals. Disinfestation includes delousing for infestation with *Pediculus humanus humanus*, the body louse. Synonyms include the term *disinsection* when insects only are involved.

ENDEMIC—The habitual presence of a disease or infectious agent within a given geographic area; may also refer to the usual prevalence of a given disease within such area. *Hyperendemic* expresses a persistent intense transmission, usually applied to malaria.

EPIDEMIC—The occurrence in a community or region of a group of illnesses (or an outbreak) of similar nature, clearly in excess of normal expectancy and derived from a common or a propagated source. The number of cases indicating presence of an epidemic will vary according to the infectious agent, size and type of population exposed, previous experience or lack of exposure to the disease, and time and place of occurrence; epidemicity is thus relative to usual frequency of the disease in the same area, among the specified population, at the same season of the year. A single case of a communicable disease long absent from a population (as smallpox in Boston) or first invasion by a disease not previously recognized in that area (as American trypanosomiasis in Arizona) is to be considered sufficient evidence of an epidemic to require immediate reporting and full field investigation.

TRANSMISSION OF INFECTIOUS AGENTS—Any mechanism by which a susceptible human host is exposed to an infectious agent. These mechanisms are

a. Direct transmission: Direct and essentially immediate transfer of infectious agents (other than from an arthropod in which the organism has undergone essential multiplication or development) to a receptive portal of entry by which infection of man may take place. This may be by touching, as in kissing or sexual intercourse (direct contact); or by the direct projection of droplet spray onto the conjunctivae, or onto the mucous membranes of the nose or mouth during sneezing, coughing, spitting, singing, or talking (usually not possible over a distance greater than 3 ft) (droplet spread); or, as in the systemic mycoses, by direct exposure of susceptible tissue to soil, compost, or decaying vegetable matter that contains the agent and where it normally leads a saprophytic existence.

b. Indirect transmission:

    (1) VEHICLE-BORNE—Contaminated materials or objects such as toys, handkerchiefs, soiled clothes, bedding, surgical instruments, or dressings (indirect contact); water, food, milk, biological products including serum and plasma, or any substance serving as an intermediate means by which an infectious agent is transported and introduced into a susceptible host through a suitable portal of entry. The agent may or may not have multiplied or developed in or on the vehicle before being introduced into man.

    (2) VECTOR-BORNE—(*a*) *Mechanical:* Includes simple mechanical carriage by a crawling or flying insect through soiling of its feet or proboscis, or by passage of organisms through its gastrointestinal tract. This does not require multiplication or development of the

organism. (*b*) *Biological: Propagation* (multiplication), cyclic development, or a combination of these (cyclopropagation) is required before the arthropod can transmit the infective form of the agent to man. An incubation period (extrinsic) is required following infection before the arthropod becomes *infective.* Transmission may be by saliva during biting, or by regurgitation or deposition on the skin of agents capable of penetrating subsequently through the bite wound or through an area of trauma following scratching or rubbing. This is transmission by an infected nonvertebrate host and must be differentiated for epidemiological purposes from simple mechanical carriage by a vector in the role of a vehicle. An arthropod in either role is termed a *vector.*

(3) AIRBORNE—The dissemination of microbial aerosols with carriage to a suitable portal of entry, usually the respiratory tract. Microbial aerosols are suspensions in air of particles consisting partially or wholly of microorganisms. Particles in the 1 to 5 micron range are quite easily drawn into the lungs and retained there. They may remain suspended in the air for long periods of time, some retaining and others losing infectivity or virulence. Not considered as airborne are droplets and other large particles, which promptly settle out (see a. Direct transmission, above); the following are airborne, their mode of transmission indirect:

(*a*) *Droplet nuclei:* Usually the small residues which result from evaporation of droplets emitted by an infected host (see above). Droplet nuclei also may be created purposely by a variety of atomizing devices, or accidentally, in microbiology laboratories or in abattoirs, rendering plants, autopsy rooms, etc. They usually remain suspended in the air for long periods of time.

(*b*) *Dust:* The small particles of widely varying size which may arise from contaminated floors, clothes, bedding, other articles; or from soil (usually fungus spores separated from dry soil by wind or mechanical stirring).

HOST—A man or other living animal, including birds and arthropods, affording under natural conditions subsistence of lodgment to an infectious agent. Some protozoa and helminths pass successive stages in alternate hosts of different species. Hosts in which the parasite attains maturity or passes its sexual stage are *primary* or *definitive hosts;* those in which the parasite is in a larval or asexual state are *secondary* or *intermediate hosts.*

INCUBATION PERIOD—The time interval between exposure to an infectious agent and appearance of the first sign or symptom of the disease in question.

RESERVOIR OF INFECTIOUS AGENTS—Any human beings, animals, arthropods, plants, soil, or inanimate matter in which an infectious agent normally lives and multiples and on which it depends primarily for survival, reproducing itself in such manner that it can be transmitted to a susceptible host.

SOURCE OF INFECTION—The person, animal, object or substance from which an infectious agent passes immediately to a host. Source of infection should be clearly distinguished from source of contamination, such as overflow of a septic tank contaminating a water supply, or an infected cook contaminating a salad.

SUSCEPTIBLE—A person or animal presumably not possessing sufficient resistance against a particular pathogenic agent and for that reason liable to contract a disease if or when exposed to the disease agent.

Disease Control

Communicable and certain noninfectious diseases may be controlled or prevented by taking steps to regulate the "source," "mode of transmission," or the "susceptibility" of persons. This is shown in Figure 1–1 and is frequently pictured as a three-link chain. Although the diseases can be brought under control by eliminating one of the links, it is far better to direct one's attack simultaneously toward all three links and erect "barriers" or "dams" where possible. Sometimes it is only practical to partially break one link in the chain. Phelps called this the "principle of multiple barriers." It recognizes as axiomatic the fact that "all human efforts, no matter how well conceived or conscientiously applied, are imperfect and fallible."* Therefore the number and type of barriers should be determined by the practicality and cost of providing the protections, the benefits to be derived, and the probable cost if the barriers are not provided. Cost is used in the sense not only of dollars but also in terms of human misery, loss of productiveness, ability to enjoy life, and loss of life. Here is a real opportunity for applying professional judgment to the problems at hand to obtain the maximum return for the effort expended.

Communicable and certain noninfectious diseases can usually be regulated or brought under control. A health department having a complete and competent staff to prevent or control diseases that affect individuals

* Earle B. Phelps, *Public Health Engineering,* John Wiley & Sons, Inc., p. 347, New York, 1948.

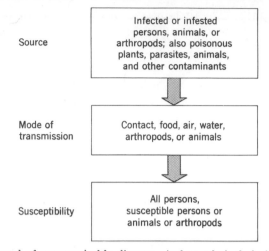

Source	Infected or infested persons, animals, or arthropods; also poisonous plants, parasites, animals, and other contaminants
Mode of transmission	Contact, food, air, water, arthropods, or animals
Susceptibility	All persons, susceptible persons or animals or arthropods

Figure 1–1 Spread of communicable diseases. Arthropods include insects, arachnids, crustaceans, and myriapods (invertebrate animals with jointed legs and a segmented body).

and animals is usually established for this purpose. The preventive and control measures conducted by a health department might include supervision of water supply, wastewater, and solid wastes; housing and the residential environment; milk and food; stream pollution; recreation areas including swimming pools and beaches; occupational health and accident prevention; insects and rodents; rural and resort sanitation; air pollution; noise; radiological health; hospitals, nursing homes, and other institutions; medical clinics, maternal and child health services, school health, dental clinics, nutrition, medical rehabilitation; medical care; disease control, including immunizations, cancer, heart disease, tuberculosis, and venereal diseases; vital statistics; health education; epidemiology; nursing services. Personnel, fiscal, and public relations support functions would be carried out by the office of business management. In some states, certain environmental and medical activities are combined with the activities of other agencies and vice versa.

Control of Source

General sources of disease agents are noted in Figure 1–1. Eliminating the source by isolation, treatment, destruction, or control of the source is possible but not always practicable. An individual frequently is not aware that he is being exposed to a potential source of disease, particularly when it is an insidious and cumulative source that is not clearly understood, such as certain chemicals in the air, water, and food.

Control of Mode of Transmission

The means whereby specific agents may become the vehicle for the transmission of disease are numerous. Prevention of disease requires the continual application of control measures and elimination of the human element to the extent feasible as discussed in this text.

Control of Susceptibles

There are many diseases to which all persons are considered to be generally susceptible. Among these are measles, impetigo contagiosa, the common cold, ascariasis, chickenpox, amebic dysentery, bacillary dysentery, cholera, malaria, trichinosis, and typhoid fever. Then there are other diseases such as influenza, meningococcus meningitis, pneumonia, human brucellosis (undulant fever), and food-borne illnesses to which some people apparently have an immunity or resistance.

In order to reduce, when possible, the number of persons who may be susceptible to a disease at any one time, certain fundamental principles are followed to improve the general health of the public. This may be accomplished by instructions in personal hygiene, immunization, and conserving or improving the general resistance of individuals to disease by a balanced diet and good food, fresh air, moderate exercise, sufficient sleep, rest, and the avoidance of fatigue and exposure.

Immunization can be carried out by the injection of vaccines, toxoids, or other immunizing substances for the prevention or lessening of the severity of specific diseases. Smallpox, typhoid, and paratyphoid fevers, poliomyelitis, and tetanus are some of the diseases against which all in the armed forces are routinely immunized. Typhoid immunization is about 70 to 90 percent effective, depending on degree of exposure.[3] Children are generally immunized against diphtheria, tetanus, whooping cough, poliomyelitis, measles, mumps, rubella (German measles), and smallpox.

Typical Epidemic Control

Outbreaks of illnesses such as influenza, measles, smallpox, poliomyelitis, and other diseases still occur. At such times the people become apprehensive and look to the health department for guidance, assurance, and information to quiet their fears.

An example of the form health department assistance can take is illustrated in the precautions released June 1, 1951, by the Illinois Health

[3] B. Cyjetanovic and K. Uemura, "The Present Status of Field and Laboratory Studies of Typhoid and Paratyphoid Vaccine," *WHO Bull.*, No. 32, 29–36 (1965).

Messenger for the control of poliomyelitis. These recommendations are quoted here, even though the disease can now be controlled, for the principles are generally applicable to other epidemics.

General Precautions during Outbreaks

1. The Illinois Department of Public Health will inform physicians and the general public as to the prevalence or increase in incidence of the disease.
2. *Early diagnosis* is extremely important. Common early signs of polio are headache, nausea, vomiting, muscle soreness or stiffness, stiff neck, fever, nasal voice, and difficulty in swallowing, with regurgitation of liquids through the nose. Some of these symptoms may be present in several other diseases, but in the polio season they must be regarded with suspicion.
3. *All children with any of these symptoms should be isolated in bed, pending diagnosis.* Early medical care is extremely important.
4. Avoid undue fatigue and exertion during the polio season.
5. *Avoid unnecessary travel and visiting in areas where polio is known to be prevalent.*
6. Pay special attention to practice of good personal hygiene and sanitation:
a. Wash hands before eating.
b. Keep flies and other insects from food.
c. Cover mouth and nose when sneezing or coughing.

Surgical Procedures

Nose, throat, or dental operations, unless required as an emergency, should not be done in the presence of an increased incidence of poliomyelitis in the community.

General Sanitation, Including Fly Control

1. Although there has been no positive evidence presented for spread of polio-myelitis by water, sewage, food, or insects, certain facts derived from research indicate that they might be involved in the spread.
  a. Water—Drinking water supplies can become contaminated by sewage con-taining poliomyelitis virus. Although no outbreaks have been conclusively traced to drinking water supplies, only water from an assuredly safe source should be used to prevent any possible hazards that might exist.
  b. Sewage—Poliomyelitis virus can be found for considerable periods of time in bowel discharges of infected persons and carriers and in sewage con-taining such bowel discharges. Proper collection and disposal facilities for human wastes are essential to eliminate the potential hazard of transmis-sion through this means.
  c. Food—The infection of experimental animals by their eating of foods de-liberately contaminated with poliomyelitis virus has been demonstrated in the laboratory, but no satisfactory evidence has ever been presented to incriminate food or milk in human outbreaks. Proper handling and pre-

paration of food and pasteurization of milk supplies should reduce the potential hazard from this source.

d. Insects—Of all the insects studied, only blowflies and houseflies have shown the presence of the poliomyelitis virus. This indicates that these flies might transmit poliomyelitis. It does not show how frequently this might happen; it does not exclude other means of transmission; nor does it indicate how important fly transmission might be in comparison with other means of transmission.

2. Fly eradication is an extremely important activity in maintaining proper sanitation in every community.

3. Attempts to eradicate flies by spraying of effective insecticides have not shown any special effect on the incidence of polio in areas where it has been tried. Airplane spraying is not considered a practical and effective means in reducing the number of flies in a city. The best way to control flies and thus prevent them from spreading any disease is to eliminate fly breeding places. Eradicate flies by:

a. Proper spreading or spraying of manure to destroy fly breeding places.

b. Proper storage, collection, and disposal of garbage and other organic waste.

c. Construction of all privies with fly- and rodent-proof pits.

Proper sanitation should be supplemented by using effective insecticides around garbage cans, manure piles, privies, etc. Use effective insecticide spray around houses or porches or paint on screen to kill adult flies.

Swimming Pools

1. Unsatisfactorily constructed or operated swimming pools should be closed whether or not there is poliomyelitis in the community.

2. On the basis of available scientific information, the State Department of Public Health has no reason to expect that closure of properly equipped and operated swimming pools will have any effect on the occurrence of occasional cases of poliomyelitis in communities.

3. In communities where a case of poliomyelitis has been associated with the use of a swimming pool, that pool and its recirculation equipment should be drained and thoroughly cleaned. (The State Department of Public Health should be consulted about specific cleansing procedures.) After the cleaning job is accomplished, the pool is ready for reopening.

4. Excessive exertion and fatigue should be avoided in the use of the pool.

5. Swimming in creeks, ponds, and other natural waters should be prohibited if there is any possibility of contamination by sewage or too many bathers.

Summer Camps

Summer camps present a special problem. The continued operation of such camps is contingent on adequate sanitation, the extent of crowding in quarters, the prevalence of the disease in the community, and the availability of medical supervision. Full information is available from the Illinois Department of Public Health to camp operators and should be requested by the latter.

1. Children should not be admitted from areas where outbreaks of the disease are occurring.

2. Children who are direct contacts to cases of polio should not be admitted.

3. The retention of children in camps where poliomyelitis exists has not been shown to increase the risk of illness with polio. Furthermore, return of infected children to their homes may introduce the infection to that community if it is not already infected. Similarly, there will be no introduction of new contacts to the camp and supervized curtailment of activity will be carried out, a situation unduplicated in the home. This retention is predicated upon adequate medical supervision.

4. If poliomyelitis occurs in a camp it is advisable that children and staff remain there (with the exception of the patient, who may be removed with consent of the proper health authorities). If they do remain:

  a. Provide daily medical inspection for all children for two weeks from occurrence of last case.
  b. Curtail activity on a supervised basis to prevent overexertion.
  c. Isolate all children with fever or any suspicious signs or symptoms.
  d. Do not admit new children.

Schools

1. Public and private schools should not be closed during an outbreak of poliomyelitis, nor their opening delayed except under extenuating circumstances and then only upon recommendation of the Illinois Department of Health.

2. Children in school are restricted in activity and subject to scrutiny for any signs of illness. Such children would immediately be excluded and parents urged to seek medical attention.

3. Closing of schools leads to unorganized, unrestricted, and excessive neighborhood play. Symptoms of illness under such circumstances frequently remain unobserved until greater spread of the infection has occurred.

4. If poliomyelitis occurs or is suspected in a school:

  a. Any child affected should immediately be sent home with advice to the parents to seek medical aid, and the health authority notified.
  b. Classroom contacts should be inspected daily for any signs or symptoms of illness and excluded if these are found.

Hospitals

1. There is no reason for exclusion of poliomyelitis cases from general hospitals if isolation is exercised—rather, such admissions are necessary because of the need for adequate medical care of the patient.

2. Patients should be isolated individually, or with other cases of poliomyelitis in wards.

3. Suspect cases should be segregated from known cases until the diagnosis is established.

4. The importation of cases to hospitals in a community where poliomyelitis

is not prevalent has not been demonstrated to affect the incidence of the disease in the hospital community.

Recreational Facilities

1. Properly operated facilities for recreation should not be closed during outbreaks of poliomyelitis.

2. Supervized play is usually more conducive to restriction of physical activities in the face of an outbreak.

3. Playground supervisors should regulate activities so that overexertion and fatigue are avoided.

RESPIRATORY DISEASES

Definition

The respiratory diseases are a large group of diseases spread by discharges from the mouth, nose, throat, or lungs of an infected individual. The disease-producing organisms are spread by coughing, sneezing, talking, spitting, by dust, and by direct contact as in kissing, eating contaminated food, using contaminated eating and drinking utensils or common towels, drinking glasses, and toys. Included are the insidious diseases associated with the contaminants in polluted air.

Group

A list of respiratory diseases and their incubation periods are shown in Tables 1–1 and 1–2. Many are transmitted in ways other than through the respiratory tract. Scarlet fever, streptococcal sore throat, and diphtheria, for example, may also be spread by contaminated milk, particularly raw milk. Infectious hepatitis may be carried by sewage-contaminated water, uncooked clams and oysters, milk, sliced meats, salads, and bakery products. Smallpox, chickenpox, mumps, infectious mononucleosis, meningococcus meningitis, and others may also be transmitted by contact with infected persons. Eye irritations and emphysema are associated with air pollution.

Control of Source

When the source of a respiratory disease is an infected individual, control would logically start with him. The individual should be taught the importance of personal hygiene and cleanliness, particularly when ill, to prevent the spread of disease. Such things as avoiding spitting, covering up a cough or sneeze with paper tissue, and staying away from people while ill are some of the simple yet important precautions that

TABLE 1-1   RESPIRATORY DISEASES

Disease	Communicability, (in days)*	Incubation Period, (in days)	Disease	Communicability, (in days)*	Incubation Period, (in days)
Chickenpox (v)	−1 to +6	14 to 21	Pertussis (b) (whooping cough)	−7 to +21	7 to 10
Common cold (v)	−1 to +5	$\frac{1}{2}$ to 3			
Diphtheria (b)	14	2 to 5	Plague, pneumonic (b)	In illness	3 to 4
German measles (v)	−7 to +4	14 to 21	Plague, bubonic (b)	−	2 to 6
Influenza (v)	3	1 to 3	Pneumonia (v)	†	1 to 3
Measles (v)	−4 to +5	10	Smallpox (v)	14 to 21	7 to 16
Meningococcal meningitis (b)	†	2 to 10	Scarlet fever and streptococcal sore throat (b)	†	1 to 3
Mumps (v)	−7 to +9	12 to 26			

(b) bacteria; (v) virus.
* Period from onset of symptoms.
† Meningococci usually disappear within 24 hr after appropriate chemotherapeutic treatment. Pneumococcus eliminated within 3 days after penicillin treatment. Streptococcus transmission eliminated within 24 hr after penicillin treatment.

are not always followed. Every effort should be made to detect and treat the carriers and promptly hospitalize the seriously ill. Identification of the reservoir or agent of disease and its control or elimination should be the goal. When this is impractical or not completely effective, attention should be given to control of the mode of transmission and to the susceptible persons, animals, or arthropods.

TABLE 1-2   RESPIRATORY DISEASES, OTHER

Disease	Communicability, (in days)*	Incubation Period, (in days)	Disease	Communicability, (in days)*	Incubation Period, (in days)
Coccidioidomycosis (f)	Open lesions	10 to 21	Poliomyelitis (v)	21 to 42	3 to 21
			Psittacosis (v)	In illness	4 to 15
Infectious hepatitis (v)	−3 to +7	10 to 50	Q Fever (r)	−	14 to 21
Infectious mononucleosis(?)	−	14 to 42	Vincent's infection (b) (s)	−	Unknown
			Tuberculosis (b)	Extended	28 to 42

(f) fungus; (v) virus; (?) unknown; (s) spirochete; (b) bacteria; (r) rickettsias, airborne.
Note: For greater details see *The Control of Communicable Diseases in Man,* American Public Health Association, 11th Ed., 1970.

WATER- AND FOOD-BORNE DISEASES

General

The disease agents spread by water and food not only incapacitate large groups of people, but sometimes result in serious disability and death. The diseases are caused by the disregard of known fundamental sanitary principles and hence are in most cases preventable. In some instances, as among the very young, the very old, and those who are critically ill with some other illness, the added strain of a water- or food-borne illness might be more than the body can overcome.

The water- and food-borne diseases are sometimes referred to as the intestinal or filth diseases because they are frequently transmitted by food or water contaminated with feces. Included as food-borne diseases are those caused by poisonous plants and animals used for food, toxins produced by bacteria, and foods accidentally contaminated with chemical poisons. They are usually, but not always, characterized by diarrhea, vomiting, nausea, or fever. Symptoms may appear in susceptible persons within a few minutes, several hours, several days, or longer periods, depending on the type and quantity of deleterious material swallowed and the resistance of the individual.

Water may be polluted at its source by excreta or sewage, which is almost certain to contain pathogenic microorganisms and cause illness by draining into an improperly treated surface or ground water supply. Food may also be contaminated by unclean food handlers who can inoculate the food with infected excreta, pus, respiratory drippings, or other infectious discharges by careless or dirty personal habits. In addition, food can be contaminated by flies carrying the causative organisms of such diseases as typhoid, dysentery, or gastroenteritis from an open privy or overflowing cesspool to the kitchen. The role that the fly plays in disease transmission is treated separately in Chapter 10. Briefly then, the intestinal diseases can be transmitted by feces, fingers, flies, food, and water.

Reservoir or Source of Disease Agents

In view of the fact that the water- and food-borne diseases result in discomfort, disability, and even death, a better understanding of their source, method of transmission, control, and prevention is desirable. A grouping of a number of these diseases is given in the folded insert, Figure 1–2.

Gastroenteritis is a vague disease that has also been listed. The term is often used to designate a water- or food-borne disease for which the causative agent has not been determined. As the term implies, it is an inflammation of the stomach and intestines, with resultant diarrhea and extreme discomfort. The occurrence of a large number of diarrheal cases indicates that there has been a breakdown in the sanitary control of water or food and may be followed by cases of typhoid fever, hepatitis, dysentery, or other illness. Much remains to be learned about this broad catchall classification. There are undoubtedly many bacterial toxins, bacteria, viruses, chemicals, etc., that are not suspected or that are not discovered by available laboratory methods.

The diseases listed in Figure 1–2 under bacterial toxins are also known as bacterial intoxications or food poisoning, to distinguish them from food infections. In bacterial intoxications, certain strains of staphylococci multiply under favorable conditions and produce a toxin that is stable at boiling temperature. The consumption of food containing sufficient toxin, therefore, even after heating, may cause food poisoning. The botulinus bacillus in improperly bottled or low-acid canned food will also produce a toxin, but this poison is destroyed by boiling or thorough cooking. There is also danger of cooking large masses of meat on the outside but leaving the interior of the food remaining underdone, thereby permitting survival of spores introduced in handling or intrinsically present that can germinate and cause *Clostridium perfringens* (*C. welchii*) food poisoning. Such cooking gives a false sense of safety.

The headings in Figure 1–2 are defined briefly. Bacteria are single-celled plantlike microscopic organisms. Viruses are organisms that pass through filters that retain bacteria. They can be seen only with an electron microscope, grow only inside living cells, and possess other characteristics to distinguish them from microorganisms such as bacteria and protozoa. Rickettsias resemble bacteria in shape, are Gram-negative, nonmotile, and difficult to stain with ordinary dyes. They are considered to be related and intermediate between bacteria and viruses. Spirochetes are a family of spiral organisms without a nucleus that multiply by transverse division. Protozoa are single-cell animals that reproduce by fission and have other special characteristics. Helminths include intestinal worms or wormlike parasites. Poisonous plants contain toxic substances that, when consumed by man or other animals, may cause illness or even death. Poisonous animals include fish whose meat is poisonous when eaten in a fresh and sometimes cooked state. Poisonous meat is not to be confused with decomposed food. The toxic substance in some poisonous fish meat appears to be heat stable. Chemical poisons are certain elements or compounds, usually metallic, that may cause illness or death when consumed by man or animals in food and water. Fungi are small plantlike organisms that are found in soil, rotting vegetation,

Antimony poisoning	Antimony.	Gray-enameled cooking utensils.	Foods cooked in cheap enameled pan.	Vomiting, paralysis of arms,	Few minutes to an hour.	Avoid purchase of poor quality gray-enameled, or use of chipped enamel, utensils.
Arsenic poisoning	Arsenic.	Arsenic compounds.	Arsenic-contaminated food or water.	Vomiting, diarrhea, painful tenesmus. (A cumulative poison.)	10 min and longer.	Keep arsenic sprays, etc., locked, wash fruits, vegetables. >0.05 mg/l.
Cadmium poisoning	Cadmium.	Cadmium-plated utensils.	Acid food prepared in cadmium utensils.	Nausea, vomiting, cramps, and diarrhea.	15 to 30 min.	Watch for cadmium-plated utensils, racks, and destroy. Inform mfr.
Cyanide poisoning	Cyanide, sodium.	Cyanide silver polish.	Cyanide-polished silver.	Dizziness, giddiness, dyspnea, palpitation, and unconsciousness.	Rapid.	Select silver polish of known composition. Prohibit sale of poisonous polish.
Fluoride or sodium fluoride poisoning	Fluoride or sodium fluoride.	Roach powder.	Sodium fluoride taken for baking powder, soda, flour.	Acute poisoning, vomiting, abdominal pains, convulsions; paresis of eye, face, finger muscles, and lower extremities; diarrhea.	Few minutes to 2 hr.	Keep roach powder under lock and key; mark "Poison"; color the powder, apply with care.
Lead poisoning	Lead.	Lead pipe, sprays, oxides, and utensils. Lead-base paints.	Lead-contaminated food or acid drinks; toys, fumes, paints.	Abdominal pain, vomiting, and diarrhea. (A cumulative poison.)	30 minutes and longer.	Do not use lead pipe; Pb <0.1 mg/l. Wash fruits. Label paints. Avoid using unglazed pottery.
Mercury poisoning	Mercury.	Mercury, bichloride of mercury.	Mercury-contaminated food.	Astringent metallic taste, salivation, great thirst, vomiting abdominal pain, watery diarrhea.	2 to 30 min.	Keep mercuric compound under lock and key.
Methyl chloride poisoning	Methyl chloride.	Refrigerant, methyl chloride.	Food stored in refrigerator having leaking unit.	Progressive drowsiness, stupor, weakness, nausea, vomiting, pain in abdomen, convulsions.	Variable.	Use nontoxic refrigerant, such as Freon, water, brine, dry ice.
Selenium poisoning	Selenium.	Selenium-bearing vegetation.	Wheat from soil containing selenium.	Gastrointestinal, nervous, and mental disorders, dermatitis in sunlight.	Variable.	Avoid semiarid selenium-bearing soil for growing of wheat, or water with more than 0.05 mg/l Se.
Zinc poisoning	Zinc.	Galvanized iron.	Acid food made in galvanized iron pots and utensils.	Pain in mouth, throat, and abdomen followed by diarrhea.	Variable.	Do not use galvanized utensils in preparation of food or drink, or water with >15.0 mg/l zinc.

			foods. Fly.	spots on trunk, other symptoms.	3 weeks for enteric fever.	
	hirschfeldii C.					
BACTERIA						
Streptococcus food poisoning	Alpha-streptococci Beta-hemolytic S., S. fecalis, viridans.	Human mouth, nose, throat, respiratory tract.	Contaminated meats, milk, croquettes, cheese, dressings.	Nausea, sometimes vomiting, colicky pains, and diarrhea.	2 to 18 hr, average 12 hr.	Similar preventive and control measures as in Staphylococcus. Pasteurize milk and milk products.
Shigellosis (bacillary dysentery)	Genus, Shigella, i.e., Flexner, Sonne, Shiga, and others.	Bowel discharges of carriers and infected persons.	Contaminated water or foods, milk and milk products. Fly.	Acute onset with diarrhea, fever, tenesmus, and frequent stools containing blood and mucus.	1 to 7 days, usually less than 4 days.	Food, water, sewage sanitation as in typhoid. Pasteurize milk (boil for infants); control flies; supervise carriers.
Cholera†	Cholera, vibrio, Vibrio comma.	Bowel discharges, vomitus; carriers.	Contaminated water, raw foods. Fly.	Diarrhea, rice-water stools, vomiting, thirst, pain, coma.	A few hours to 5 days, usually 3 days.	Similar to typhoid. Immunize, quarantine; isolate patients.
Melioidosis†	Pseudomonas pseudomallei.	Rats, guinea pigs, cats, rabbits, dogs, horses.	Contact with or ingestion of contaminated excreta, soil or water	Acute diarrhea, vomiting, high fever, delirium, mania.	Less than 2 days or longer.	Destroy rats; protect food; thoroughly cook food; control biting insects; personal hygiene.
Brucellosis (undulant fever)	Brucella melitensis-goat, Br. abortus-cow, Br. suis-pig.	The tissues, blood, milk, urine, infected animal.	Raw milk from infected cows or goats, also contact with infected animals.	Insidious onset, irregular fever, sweating, chills, pains in joints and muscles.	5 to 21 days or longer.	Pasteurize all milk; eliminate infected animals. Handle infected carcasses with care.
Streptococcal sore throat	Hemolytic Streptococci.	Nose, throat, mouth secretions.	Contaminated milk or milk products.	Sore throat and fever, sudden in onset, vomiting.	1 to 3 days.	Pasteurize all milk. Inspect contacts. Exclude carriers.
Diphtheria	Corynebacterium diphtheriae.	Respiratory tract, patient, carrier.	Contact and milk or milk products.	Acute febrile infection of tonsils, throat, and nose.	2 to 5 days or longer.	Pasteurize milk, disinfect utensils. Inspect contacts, immunize.
Tuberculosis	Mycobacterium tuberculosis (hominis and bovis).	Respiratory tract of man. Rarely cattle.	Contact, also eating and drinking utensils, food and milk.	Cough, fever, fatigue, pleurisy.	4 to 6 weeks.	Pasteurize milk, eradicate TB from cattle. X-ray. Control contacts and infected persons. Selective use of BCG.
Tularemia	Pasteurella tularensis (Bacterium tularense).	Rodent, rabbit, horsefly, wood tick, dog, fox, hog.	Meat of infected rabbit, contam. water, handling wild animals.	Sudden onset with pains and fever, prostration.	1 to 10 days, average of 3.	Thorough cooking of meat of wild rabbits. Purify drinking water. Use rubber gloves (care in dressing wild rodents).

	Disease	Microorganisms	Probably man and animals	water, food, including milk, air.			
		Microorganisms unknown.		water, food, including milk, air.	Diarrhea, nausea, vomiting, cramps, possibly fever.	Variable, 8 to 12 hr average.	Environmental sanitation, education, personal hygiene.
RICKETTSIAS	Q Fever	Coriella burneti.	Dairy cattle, sheep, goats.	Slaughterhouse, dairy employees, handling infected cattle; raw cow and goat milk.	Heavy perspiration and chills, headache, malaise.	2 to 3 weeks.	Pasteurization of milk and dairy products, elimination of infected animal reservoir, cleanliness in slaughterhouses and dairies. Pasteurize at 145°F for 30 min or 161°F for 15 sec.
VIRUSES	Choriomeningitis, lymphocytic	L. choriomeningitis virus.	House-mice urine, feces, secretions.	Contaminated food.	Fever, grippe. Severe headache, stiff neck, vomiting, somnolence.	8 to 13 days, 15 to 21 days.	Eliminate or reduce mice. General cleanliness, sanitation.
VIRUSES	Infectious hepatitis	Viruses unknown.	Discharges of infected persons.	Water, food, milk, oysters. Contacts.	Fever, nausea, loss of appetite; possibly vomiting, fatigue, headache, jaundice.	10 to 50 days, average 30 to 35 days.	Sanitary sewage disposal, food sanitation, personal hygiene; coag. and filter water supply, and +0.6 mg/l free Cl_2.
PROTOZOA	Amebiasis (amebic dysentery)	Endamoeba histolytica.	Bowel discharges of carrier, and infected person, possibly also rat.	Cysts, contaminated water, foods, raw vegetables and fruits. Flies, cockroaches.	Insidious and undetermined onset, diarrhea or constipation, or neither; loss of appetite, abdominal discomfort; blood, mucus in stool.	5 days or longer, average 3 to 4 weeks.	Same as Bacillary. Boil water or coag. set. filter through diatomite 5 gpm/ft², Cl_2. Usual Cl_2 and high rate filtration no; 100% effective. Slow sand filtration +Cl_2.
SPIROCHETES	Leptospirosis (Weil's disease) (Spirochetosis icterohemorrhagic)	Leptospira icterohaemorrhagiae, L. hebdomadis, L. canicola, L. pomona, other.	Urine and feces of rats, swine, dogs, cats, mice, foxes, sheep.	Food, water, soil contaminated with excreta or urine of infected animal. Contact.	Fever, rigors, headaches, nausea, muscular pains, vomiting, thirst. Prostration, jaundice.	4 to 19 days. Average 9 to 10 days.	Destroy rats; protect food; avoid polluted water; treat abrasion of hands and arms. Disinfect utensils, treat infected dogs.
	Trichinosis (Trichiniasis)	Trichinella spiralis.	Pigs, bears, wild boars, rats, foxes, wolves.	Infected pork and pork products, bear and wild boar meat.	Nausea, vomiting, diarrhea, muscle pain, swelling of face and eyelids, difficulty in swallowing.	9 days, varies 2 to 28 days.	Thoroughly cook pork (150°F). pork products, bear and wild boar meat; destroy rats. Feed hogs boiled garbage or discor-

Poisoning	Poison	Source	Form in food	Symptoms	Time	Prevention
Ergotism‖	Ergot, a parasitic fungus (*Claviceps purpurea*).	Fungus of rye and occasionally other grains.	Ergot fungus contaminated meal or bread.	Gangrene involving extremities, fingers, and toes; or weakness and drowsiness, headache, giddiness, painful cramps in limbs.	Gradual, after prolonged use of diseased rye in food.	Do not use discolored or spoiled grain. Fungus grows in the grain; meal is grayish, possibly with violet-colored specks.
Rhubarb poisoning	Probably oxalic acid.	Rhubarb.	Rhubarb leaves.	Intermittent cramplike pains, vomiting, convulsions, coma.	2 to 12 hr.	Do not use rhubarb leaves for food.
Mushroom poisoning	Phalloidine and other alkaloids; also other poisons in mushroom.	Mushrooms—Amanita phalloides and other Amanita.	Poisonous mushrooms (Amanita phalloides, Amanita muscaria, others).	Severe abdominal pain, intense thirst, retching, vomiting, and profuse watery evacuations.	6 to 15 hr or 15 min to 6 hr with muscaria.	Do not eat wild mushrooms; warn others. Amanita are very poisonous, both when raw or cooked.
Favism‡	Poison from Vicia fava bean, pollen.	Plant and bean *Vicia fava*.	The bean when eaten raw, fava pollen.	Acute febrile anemia with jaundice, passage of blood in urine.	1 to 24 hr.	Avoid eating fava, particularly when green, or inhalation of pollen.
Fish poisoning	Poison in fish, ovaries and testes, roe (heat-stable).	Fish. Pike, carp, sturgeon roe in breeding season.	Fish—Tedrodon, Meletta, Clupea, Pickerel eggs, muki-muki.	Painful cramps, dyspnoea, cold sweats, dilated pupils, difficulty in swallowing and breathing.	30 min to 2 hr or longer.	Avoid eating roes during breeding season. Heed local warnings concerning edible fish.
Shellfish poisoning	Poison (*Gonyaulax catenella*) in food of mussels and clams.	Dinoflagellates—plankton, food of clams and mussels.	Mussels and clams.	Respiratory paralysis. In milder form, trembling about lips to loss of control of the extremities and neck.	5 to 30 min and longer.	Obtain shellfish from certified dealers and from approved areas. Toxicity reduced 70–90% by steaming or frying.
Snakeroot poisoning	Trematol in snakeroot (*Eupatorium urticaefolium*).	White snakeroot jimmy weed.	Milk from cows pastured on snakeroot.	Weakness or prostration, vomiting, severe constipation & pain, thirst; temperature normal.	Variable. Repeated use of the milk.	Prevent cows from pasturing in wooded areas where snakeroot exists.
Potato poisoning	*Solanum tuberosum*; other S.	Sprouted green potatoes.	Possibly green sprouted potatoes.	Vomiting, diarrhea, headache, abdominal pains, prostration.	Few hours.	Do not use sprouts or peel of sprouted green potatoes.
Water-hemlock poisoning	Cicutoxin or resin from hemlock (*Cicuta maculata*).	Water hemlock.	Leaves and roots of water hemlock.	Nausea, vomiting, convulsions, pain in stomach, diarrhea.	1 to 2 hr.	Do not eat roots, leaves, or flowers of water hemlock.

DISEASE	SPECIFIC AGENT	RESERVOIR	COMMON VEHICLE	SYMPTOMS—IN BRIEF	ONSET*	PREVENTION AND CONTROL
Paragonimiasis (lung flukes)	*Paragonimus ringeri, P. westermani, P. kellicotti.*	Respiratory and intes. tract of man, cat, dog, pig, rat, wolf.	Infected water, fresh-water crabs or crayfish.	Chronic cough, clubbed fingers, dull pains, and diarrhea.	Variable.	Boil drinking water in endemic areas; thoroughly cook fresh water crabs and crayfish.†
Clonorchiasis‡ (liver flukes)	*C. sinensis, Opisthorchis felineus.*	Liver of man, cats, dogs, pigs.	Infected fresh water fish.	Chronic diarrhea, night blindness.	Variable.	Boil drinking water in endemic areas; thoroughly cook fish.†
Fascioliasis (sheep liver flukes)	*Fasciola hepatica.*	Live of sheep.	Sheep liver eaten raw.	Irregular fever, pain, diarrhea.	Several months.	Thoroughly cook sheep liver.†
Trichuriasis (whipworm)	*Trichuris trichiura.*	Large intestine of man.	Contaminated food, soil.	No special symptoms, possibly stomach pain.	Long and indefinite.	Sanitation. Boil water, cook good, properly dispose feces.†
Oxyuriasis (pinworm, threadworm or enterobiasis)	*Oxyuris vermicularis,* or *Enterobius vermicularis.*	Large intestine of man, particularly children.	Fingers. Ova-laden dust. Contaminated food, water; sewage. Clothing, bedding.	Nasal itching, anal itching, diarrhea.	3 to 6 weeks; months.	Wash hands after defecation, keep fingernails short, sleep in cotton drawers. Sanitation.
Fasciolopsiasis‡ (intestinal flukes)	*Fasciolopsis buski.*	Small intestine of man, dog, pig.	Raw fresh-water plants; water, food.	Stomach pain, diarrhea, greenish stools, constipation. Edema.	6 to 8 weeks.	Cook or dip in boiling water roots of lotus, bamboo, water chestnut, caltrop.
Dwarf tapeworm	*Hymenolepis nana.*	Man and rodents.	Food contaminated with ova; direct contact.	Diarrhea or stomach pain, irritation of intestine.	1 month.	Sanitary excreta disposal, personal hygiene, food sanitation, rodent control; treat cases.

Continued on reverse side

HELMINTHS	Schistosomiasis (Bilharziasis)‡ (blood flukes)	*Schistosoma haematobium, S. mansoni, S. japonicum.*	Venous circulation of man; urine, feces.	Cercariae infected drinking and bathing water.	Dysenteric, pulmonary, and abdominal symptoms. Rigors. itching on skin, dermatitis.	4 to 6 months or longer.	continue feeding. Store meat 20 days at 5°F or 36 hr at −27°F. Avoid infected water; coag., set 1 hr, sand filt. 3 gpm/ft² or 5 with diatomite, Cl_2 1 mg/l; boil water; 10 mg/l $CuSO_4$, and impound water 48 hr, Cl_2. (Also 10 mg/l sodium or copper pentachlorophenate.) Slow sand filt. + Cl_2.
	Ascariasis (intestinal roundworm)	*Ascaris lumbricoides.*	Small intestine of man, gorilla, ape.	Contaminated food, water; sewage.	Worm in stool, abdominal pain, skin rash, protuberant abdomen, nausea, large appetite.	About 2 months.	Personal hygiene, sanitation.† Boil drinking water in endemic areas. Sanitary excreta disposal.
	Echinococcosis (Hydatidosis)	*Echinococcus granulosus,* dog tapeworm.	Dog, sheep, wolf, dingo, swine, horse, monkey.	Contaminated food and drink; hand to mouth; contact with infected dogs.	Cysts in tissues—liver, lung, kidney, pelvis. May give no symptoms. May cause death.	Variable, months to several years.	Keep dogs out of abattoir and do not feed raw meat. Mass treatment of dogs. Educate children and adults in the dangers of close association with dogs.
	Taeniasis (pork tapeworm) (beef tapeworm)	*Taenia solium* (pork tapeworm), *T. saginata* (beef tapeworm).	Man, cattle, pig, buffalo, possibly rat, mouse.	Infected meats eaten raw. Food contaminated with feces of man, rats, or mice.	Abdominal pain, diarrhea, convulsions, insomnia, excessive appetite.	8 to 10 weeks.	Thoroughly cook meat, control flies, properly dispose of excreta, food-handler hygiene. Use only inspected meat. Store meat 6 days at 15°F or lower.
	Fish Tapeworm (broad tapeworm)	*Diphyllobothrium latum,* other.	Man, frog, dog, cat, bear.	Infected fresh-water fish eaten raw.	Abdominal pain, loss of weight, weakness, anemia.	3 to 6 weeks.	Thoroughly cook fish, roe, (caviar). Proper excreta disposal.

Continued on reverse side

Figure 1-2 Characteristics and control of water- and food-borne diseases.

DISEASE	SPECIFIC AGENT	RESERVOIR	COMMON VEHICLE	SYMPTOMS—IN BRIEF	ONSET*	PREVENTION AND CONTROL
Botulism food poisoning	*Clostridium botulinum* and *C. parabotulinum* that produce toxin.	Soil, dust, fruits, vegetables, foods, mud.	Improperly processed canned and bottled foods containing the toxin.	Gastrointestinal pain, diarrhea or constipation, prostration, difficulty in swallowing, double vision, difficulty in respiration.	2 hr to 8 days, generally 12 to 36 hr.	Boil home canned nonacid food 5 min; thoroughly cook meats, fish, and dried foods held over. Do not taste suspected food!
Staphylococcus food poisoning	Staphylococci that produce enterotoxin, *Staphylococcus albus, S. aureus* (toxin is stable at boiling temperature).	Skin, mucous membranes, pus, dust, air, sputum, and throat.	Contaminated custard pastries, cooked or processed meats, poultry, dairy products, hollandaise sauce, salads, milk.	Acute nausea, vomiting, and prostration; diarrhea, abdominal cramps. Usually explosive in nature followed by rapid recovery of those afflicted.	1 to 6 hr or longer, average of 2 to 4 hr.	Refrigerate prepared food in shallow containers at a temperature below 45°F immediately upon cooling. Reuse leftover food within 4 hr. Avoid handling food. Educate food handlers in personal hygiene and sanitation.
Salmonellosis (Salmonella infection)	*Salmonella typhimurium, S. newport, S. enteritidis, S. monterideo,* other.	Hogs, cattle, and other livestock, poultry, pets, eggs, carriers, powdered eggs.	Contaminated sliced cooked meat, salads, infected meats, warmed over foods, milk, milk products.	Abdominal pain, diarrhea, chills, fever; vomiting and nausea. Diarrhea usually persists several days.	6 to 48 hr, usually 12 to 24 hr.	Protected storage of food. Thoroughly cook food. Eliminate rodents, pets, and carriers. Similar measures as in Staphylococcus. Poultry and meat sanitation.
Typhoid fever	Typhoid bacillus, *Salmonella typhosa.*	Feces and urine of typhoid carrier or patient.	Contaminated water, milk and milk products, shellfish, and foods. Fly.	General infection characterized by continued fever, usually rose spots on the trunk, diarrheal disturbances.	Average 14 days, usually 7 to 21 days.	Protect and purify water supply; pasteurize milk and milk products; sanitary sewage disposal; educate food handlers; food, fly, shellfish control; supervise carriers; immunize. Personal hygiene.
Paratyphoid fever	*Salmonella paratyphi* A, *S. schott-*	Feces and urine of carrier or	Contaminated water, milk and milk prod-	General infection characterized by continued fever, diarrheal	1 to 10 days for gastro-enteritis; 1 to	Similar preventive and control measures as in typhoid fever and salmonellosis.

BACTERIAL TOXINS§

Methemo-globinemia	Nitrate nitrogen.	Ground water; shallow dug wells, also drilled wells.	Drinking water from private wells.	Vomiting, diarrhea, and cyanosis in infants.	2 to 3 days.	Use water with less than 45 mg/l NO₃ for drinking water and in infant formula. Properly develop and locate wells.
Sodium nitrite poisoning	Sodium nitrite.	Impure sodium nitrate and nitrite.	Sodium nitrate taken for salt.	Dizziness, weakness, stomach cramps, diarrhea, vomiting, blue skin.	5 to 30 min.	Use USP sodium nitrate in curing meat. Nitrite is poisonous, keep locked.
Copper poisoning	Copper.	Copper pipes and utensils.	Carbonated beverages and acid foods in prolonged contact with copper.	Vomiting, weakness.	1 hr or less.	Do not prepare or store acid foods or liquids or carbonated beverages in copper containers. Cu should not exceed 0.3 mg/l. Prevent CO_2 backflow into copper lines.

* Incubation period. † Take same precautions with drinking, culinary, and bathing water as in Schistosomiasis. ‡ Does not originate in the U.S. § A food poisoning receiving more attention is caused by *Clostridium perfringens* (*C. welchii*). The organism is a very prevalent one in soil; it is a spore-former; certain strains are resistant to heat and withstand normal cooking. It is spread by poultry, meats and meat products. The symptoms include abdominal pains, nausea, some vomiting and diarrhea; the illness is explosive, with rapid recovery. The incubation period is 6 to 22 hours with an average of 15 hours. To prevent the illness, cool prepared foods rapidly and refrigerate immediately. Do not allow foods to cool slowly for several hours or overnight. ‖ Many other fungi that produce toxin are associated with food and feedstuffs. The mycotoxins cause illness in humans and animals. More needs to be known. For more information see F. L. Bryan, *Diseases Transmitted by Foods*, 58 pp., USDHEW, PHS, Atlanta, 1971, and *Procedure for the Investigation of Foodborne Disease Outbreaks*, International Association of Milk, Food, and Environmental Sanitarians, Shelbyville, Ind., 1966.

Figure 1–2 Characteristics and control of water- and food-borne diseases. This figure represents a summary of information selected from: 1. G. M. Dack, Food Poisoning, 251 pp., University of Chicago Press, 1956. 2. C. E. Dolman, "Bacterial Food Poisoning," 46 pp., *Canad. Pub. Health J. Assoc.*, 1943. 3. V. A. Getting, "Epidemiologic Aspects of Food-Borne Disease," 75 pp., *New Eng. J. Med.*, 1943. 4. F. A. Korff, "Food Establishment Sanitation in a Municipality," *Am. J. Pub. Health* **32**, 740 (1952). 5. P. Manson-Bahr, *Synopsis of Tropical Medicine*, 224 pp., Williams & Wilkins Co., Baltimore, 1943. 6. New York State Department of Health, *Health News*. 7. Miscellaneous military and civilian texts and reports. 8. R. P. Strong, *Stitt's Diagnosis, Prevention and Treatment of Tropical Diseases*, 2 vols., Blakiston Co., Philadelphia, 1942. 9. "*The Control of Communicable Diseases in Man*," Am. Pub. Health Assoc., New York, 1970. (Sept. 1944, Revised May 1945, 1946, 1952, 1971. Copyright 1946, Joseph A. Salvato, Jr., MCE.) More complete characteristics, preventive and control measures, and modes of transmission, other than food and water have been omitted for brevity as has been the statement "epidemiological study," and "education of the public" opposite each disease under the heading "Prevention and Control." Milk and milk products are considered foods. Under "Specific Agent" and "Common Vehicle" above, only the more common agents are listed.

Figure 1–2 Characteristics and control of water- and food-borne diseases.

and bird excreta. Systemic fungal diseases are transmitted through the soil, vegetation, or excreta by contact or ingestion.

Man is in intimate contact with the sources of the water- and food-borne diseases at all times. Man, rodent, livestock, dogs and cats, soil, certain plants, fish, insecticides, and metals are sources, or reservoirs, of the deleterious substances causing illness. By knowing more about these sources of infection or poisoning, it is possible to take precautions to prevent the harmful substances harbored by these sources from gaining access to food and drink. This is not easily done. Practically speaking, it is not yet possible to remove man, for example, from the kitchen where food is prepared. The entrance into the kitchen of cats and dogs, unauthorized persons, mice and rats, and persons who are ill can, however, be prevented or controlled so as not to be hazardous.

The mouth, nose, throat, or respiratory tract of man may be the reservoir of organisms that directly or indirectly cause a large group of illnesses. Staphylococci that produce enterotoxin are also found on the skin, on the mucous membranes, in pus, feces, dust, air, and in insanitary food processing plants. They are the principal cause of boils, pimples, and other skin infections and are particularly abundant in the nose and throat of a person with a cold. It is no surprise, therefore, that staphylococcus food poisoning is one of the most common food-borne diseases. Scrupulous cleanliness in food processing plants and in the kitchen and among food handlers is essential if contamination of food with salmonellas, staphylococci, and other organisms is to be reduced or prevented.

The feces or urine of infected persons may be the reservoir of a broad group of intestinal diseases. A certain percentage of the population is always infected with one or more of these diseases. For example, the incidence of amebic dysentery in the native populations of tropical countries varies between 10 and 25 percent and may be as high as 60 percent; shigellosis may be higher. Surveys made in the United States have shown that 5 to 10 percent of the population are carriers of *Endamoeba histolytica*.[4] Stoll has ventured to hazard a guess of the incidence of helminthic infections in the world.[5] He estimates that at least 500 million persons harbor ascarids, 400 million hookworms, and 100 million other worms. This would indicate that if a person was a carrier of only one infection, one-half of the world population would be infected. Actually, if a person

[4] L. T. Coggeshall, "Current and Postwar Problems Associated with the Human Protozoan Diseases," *Ann. N.Y. Acad. Sci.,* **44,** 198 (1943).
[5] Norman R. Stoll, "Changed Viewpoints on Helminthic Disease: World War I vs. World WAR II," *Ann. N.Y. Acad. Sci.,* **44,** 207–209 (1943).

is ill with a helminthic disease, he probably is infected with more than one parasite, since the conditions conducive to one infection would admit additional species to be present. A study made at a missionary college in east central China showed that 49 percent of the students harbored parasitic worms, and a survey made in an elementary school in New Jersey showed that 23 percent of the children were infected. Stoll states further that 25 percent of the American population are carriers of ascarids, hookworm, or tapeworms. In 1970 Lease reported on the study of the day care and elementary school programs in four counties in South Carolina involving 884 children. He found that 22.5 percent of the Negro children harbored Ascaris intestinal roundworms and, of the 52 white children in the group, 13.5 percent had worms. Central sewerage and water supply was lacking. The rural infection rate was higher and the infected rural child had twice the number of worms as the infected city child.[6] Since parasitic infection plus poor diet or other causes may result in serious debility and perhaps death, preventive measures, including better sanitation, are essential.

Intestinal diseases occur more frequently where there are low standards of hygiene and sanitation and where there is poverty and ignorance. Contamination of food and drink either directly with excreta or indirectly by flies that have had contact with infected excreta is usually necessary for disease transmission. A high incidence of these diseases in developed areas of the world is an indication of lack of or faulty sanitation and shows a disregard of fundamental hygienic practices.

The U.S. Department of Agriculture has indicated that between 1 and 2 percent of all the swine examined had living trichinae. In 1948 the incidence of trichinosis in grain-fed hogs was 0.95 percent and in garbage-fed hogs 5.7 percent.[7] A 1967 report shows the rate reduced to 0.12 percent and 2.2 percent respectively.[8]

Salami, cervelat, mettwurst, and Italian-style ham may be considered acceptable when stamped "U.S. inspected for wholesomeness," but this is no guarantee that the product is absolutely safe. This is particularly true of raw meat and poultry, which frequently contain salmonellas and other pathogenic organisms, even though stamped inspected. Such raw products require hygienic handling and adequate cooking. Uncooked summer sausage (ground fresh pork, beef, and seasoning plus light smoking) should be avoided. The incidence of adult infection in the United

[6] E. John Lease, "Study Finds Carolina Children Afflicted by Intestinal Parasites," *New York Times*, March 28, 1970, p. 30.

[7] *Public Health Reports*, **63**, No. 15, 478–488 (April 9, 1948).

[8] W. J. Zimmermann, "The Incidence of *Trichinella spiralis* in Humans of Iowa," *Public Health Reports*, **82**, 127–130 (1967).

States is estimated at 4 percent or less.[9] Swine are a reservoir of the organisms causing

1. Fascioliasis (intestinal fluke) and fasciolopsiasis
2. Paragonimiasis (lung fluke)
3. Salmonella infection (salmonellosis)
4. Taeniasis (pork or beef tapeworm) and cysticercosis
5. Trichinosis (trichiniasis)
6. Trichuriasis (whipworm)
7. Tularemia
8. Brucellosis (undulant fever)

The spread of all these diseases can be prevented by thoroughly cooking all foods. Using only inspected meats, prohibiting the feeding of uncooked garbage or offal to hogs, and practicing good sanitation will also help. Storage of pork 20 days at 5°F or 36 hours at −27°F is believed adequate to kill trichina larvae. Swine and other livestock are also reservoirs of infection for clonorchiasis, echinococcosis, leptospirosis, lymphocytic choriomeningitis, schistosomiasis, and toxoplasmosis.

The excreta or urine of rats or mice contain the causative organisms of

1. Choriomeningitis, lymphocytic
2. Leptospirosis (Weil's disease)
3. Salmonella infection (salmonellosis)

Because of the practical difficulty of permanently eliminating all mice and rats, the threat of contaminating food with the organisms causing the foregoing diseases is ever present. This emphasizes the necessity of keeping all food covered or protected. In addition, rats are frequently infested with fleas that, if infected, can spread murine typhus. Sodoku and Haverhill fevers transmitted by the bite of infected rats are also to be guarded against. Sodoku is not common in the United States.

Dust, eggs, poultry, pigs, sheep, cattle and other livestock, rabbits, cats, and dogs may harbor salmonella and other causative organisms. Shelled eggs and egg powder may also contain salmonella. Salmonella food infection is common. Poisonous plants, fish, and chemicals should also be guarded against.

Food Decomposition

When fresh foods are allowed to stand at room temperature, they begin to deteriorate. Enzymes develop in decomposing food or are introduced by bacteria into the food. The changes in the food, brought about

[9] Harry Most, "Trichinellosis in the United States," *JAMA*, **193**, No. 11, 871–878, (September 13, 1965).

by the action of enzymes, cause.the food to change in its composition. Such outside favorable factors as oxygen, sunlight, warmth, and moisture cause an acceleration of the rate of decomposition. This shows up in the unpleasing appearance of the food, by the loss of freshness, by the color, and by the odor. Food that has been permitted to decompose loses much of its nutritive value. Oxidation causes reduction of the vitamin content and breaking down of the fats, then the proteins, to form hydrogen sulfide, ammonia, and other products of decay. Ptomaine poisoning is a misnomer for food poisoning or food infection. Ptomaines are basic products of decay formed as the result of the action of bacteria on nitrogenous matter such as meat. Ptomaines in and of themselves are not considered dangerous. Contamination, which almost always accompanies putrefaction, is dangerous. In certain instances, the bacterial activity will produce a toxin which even ordinary cooking will not destroy.

The Vehicle or Means by which Water- and Food-Borne Diseases are Spread

The means by which the water- and food-borne disease agents are transmitted to individuals include water, milk, milk products, and other foods. These vehicles are not, however, the only methods by which diseases are spread. Some, as previously discussed, are also spread through the air, some by contact with persons who are ill, some by insects, and others by hands or articles soiled with infectious discharges. This discussion will cover the role of water, milk, milk products, and food as bearers or carriers of disease-producing organisms and poisonous substances.

In 1940 Frank[10] stated that there were probably five to ten times as many outbreaks, cases, and deaths resulting from faulty sanitation as were reported in *Public Health Reports* for 1938. The number of water-borne outbreaks have been reported periodically.[11] In the United States they averaged 45 per year for the years 1938–1940, 39 for 1941–1945, 23 for 1946–1950, 8 for 1951–1955, and 7 for 1956–1960.[12]

---

[10] C. Leslie Frank, "Disease Outbreaks Resulting from Faulty Environmental Sanitation," *Public Health Reports,* 55, No. 31, 1373–1383 (August 2, 1940).
[11] A. Wolman and A. E. Gorman, "Water-borne Typhoid Fever Still a Menace," *Am. J. Public Health,* 21, No. 2, 115–129 (February 1931); *Water-Borne Outbreaks in the United States and Canada 1930–36 and Their Significance,* Eighth Annual Yearbook, American Public Health Assoc., 1937–1938, p. 142. Gainey and Lord, *Microbiology of Water and Sewage,* Prentice-Hall, Englewood Cliffs, N.J., 1952.
[12] S. R. Weibel, F. R. Dixon, R. B. Weidner, and L. J. McCabe, "Waterborne-Disease Outbreaks, 1946–60," *J. Am. Water Works Assoc.,* 56, 947–958 (August 1964).

Weibel et al. studied the incidence of water-borne disease in the United States for 1946–1960.[13] They reported 22 outbreaks (10%) with 826 cases due to use of untreated surface water; 95 outbreaks (42%) with 8811 cases due to untreated ground water; 3 outbreaks (1%) with 189 cases due to contamination of reservoir or cistern; 35 outbreaks (15%) with 10,770 cases due to inadequate control of treatment; 38 outbreaks (17%) with 3344 cases due to contamination of distribution system; 7 outbreaks (3%) with 1194 cases due to contamination of collection or conduit system; and 28 outbreaks (12%) with 850 cases due to miscellaneous causes, for a total of 228 outbreaks with 25,984 cases.

Wolman and Gorman showed that the greatest number of the water-borne diseases occurred among population groups of 1000 and under and among groups from 1000 to 5000, that is, predominately in the rural communities.[14] Weibel reported the greatest number of outbreaks and cases in communities of 10,000 population or less. The need for emphasis on water control and sewage disposal at small existing and new communities, as well as at institutions, resorts, and rural places, is apparent. These figures show what can happen when there is a breakdown in the water-distribution system, treatment plant, or water source protection. See also Chapter 3.

Drinking water contaminated with sewage is the principal cause of water-borne diseases. The diseases that usually come to mind in this connection are typhoid fever, paratyphoid fever, dysentery, gastroenteritis, and more recently infectious hepatitis. However, because of the supervision given public water supplies and control over a lessening number of typhoid carriers, the incidence of typhoid fever has been reduced to a low residual level. Occasional outbreaks, due mostly to carriers, remind us that the disease is still a potential danger.

In 1940 some 35,000 cases of gastroenteritis and six cases of typhoid fever resulted when about five million gallons of untreated, grossly polluted Genesee River water was accidentally pumped into the Rochester, New York, public water-supply distribution system. A valved cross-connection between the public water supply and the polluted Genesee River fire-fighting supply had been unintentionally opened. In order to maintain the proper high pressure in the fire supply, the fire pumps were placed in operation and hence river water entered the potable public water-supply system. The check valve was also inoperative.

At Manteno State Hospital in Illinois, 453 cases of typhoid fever

[13] Ibid.
[14] Op. cit.

were reported resulting in 60 deaths in 1939.[15] It was demonstrated by dye and salt tests that sewage from the leaking vitrified clay tile hospital sewer line passing within a few feet of the drilled well-water supply seeped into the well. The hospital water supply consisted of four wells drilled in creviced limestone. The state sanitary engineer had previously called the hospital administrator's attention to the dangerously close well location to the sewer and made several very strong recommendations over a period of eight years; but his warning went unheeded until after the outbreak. Indictment was brought against three officials, but only the director of the Department of Public Welfare was brought to trial. Although the county court found the director guilty of omission of duty, the Illinois Supreme Court later reversed the decision.

An explosive epidemic of infectious hepatitis in Delhi, India, started during the first week of December 1955 and lasted about six weeks. A sample survey showed about 29,300 cases of jaundice in a total population of 1,700,000.[16] No undue incidence of typhoid or dysentery occurred. Water was treated in a conventional rapid sand filtration plant; but raw water may have contained as much as 50 percent sewage. Inadequate chlorination (combined chlorine), apathetic operation control, and poor administration apparently contributed to the cause of the outbreak, although the treated water was reported to be well clarified and bacteriologically satisfactory.[17]

An outbreak of gastroenteritis in Riverside, California, affected an estimated 18,000 persons in a population of 130,000. Epidemiological investigation showed that all patients were carriers of *Salmonella typhimurium*, serological type B and phage type II. The water supply was implicated. There was no evidence of coliform bacteria in the distribution system, although 5 of 75 water samples were found positive for *S. typhimurium*, type B, phage II. The cause was not found in spite of an extensive investigation.[18]

Raw or improperly pasteurized milk, poor milk-handling practices, and carriers have been the principal causes of the milk-borne diseases. Contaminated food and improper refrigeration were the usual cause of the food-borne diseases. Surgeon General Leonard A. Scheele of the U.S.

[15] Illinois Department of Public Health, "A Report on a Typhoid Fever Epidemic at Manteno State Hospital in 1939," *J. Am. Water Works Assoc.*, 38, 1315–1316 (November 1946).

[16] The authorities estimated the total number of infections at 1,000,000.

[17] Joseph M. Dennis, "1955–56 Infectious Hepatitis Epidemic in Delhi, India," *J. Am. Water Works Assoc.*, 51, 1288–1287 (October 1959).

[18] Everett C. Ross and Howard L. Creason, "The Riverside Epidemic," *Water and Sewage Works*, 113, 128–132 (April 1966).

Public Health Service, in commenting about the 1949 summary of illnesses, pointed out that the outbreaks and cases implicating food, other than milk and milk products, represented 10 to 15 percent of the probable actual number.

Fuchs states that from 1923 to 1937, inclusive, 639 milk-borne outbreaks were reported, involving 25,863 cases and 709 deaths.[19] In other words there was an average of 43 outbreaks involving 1724 cases and 47 deaths reported each year. Between 1938 and 1956 an average of 24 milk outbreaks per year, with 980 cases and 5 deaths, were reported to the U.S. Public Health Service. Between 1957 and 1960 the outbreaks averaged 9, and the cases, 151 per year. There were no deaths reported between 1949 and 1960. It is to be noted that the number of outbreaks, cases, and deaths have decreased. This is because of better equipment and more effective control over the pasteurization of milk and milk products. Evidence from the past shows what can happen if controls are relaxed.

Whereas milk-borne diseases have been brought under control, food-borne illnesses remain unnecessarily high. Between 1938 and 1956 there were reported 4647 outbreaks, 179,773 cases, and 439 deaths. In 1967, 273 outbreaks were reported, with 22,171 cases and 15 deaths. The bacteria associated with food-borne illnesses in recent years (1967–1968) are salmonella, *Clostridium perfringens*, and Staphylococcus. Banquets accounted for over half of the illnesses reported; schools and restaurants made up most of the rest. The largest number of outbreaks occurred in the home.[20]

The Control and Prevention of Water- and Food-Borne Diseases

Many health departments, particularly on a local level, are placing greater emphasis on food protection at food-processing establishments, catering places, schools, restaurants, institutions, and the home, and on the training of food management and staff personnel. In addition, an educated and observant public, coupled with a food quality laboratory surveillance system and program evaluation, can help greatly in making health department programs more effective.

When there is carelessness or negligence in the sanitary handling, protection, processing, refrigeration, or storage of food, in the treatment, distribution, or protection of the water supply, or in personal hygiene and general sanitation, a disease outbreak is almost certain to occur.

[19] A. W. Fuchs, address before Philadelphia College of Physicians, February 5, 1940.
[20] William E. Woodward et al., "Foodborne Disease Surveillance in the United States, 1966 and 1967," *Am. J. Public Health*, **60**, 130–137 (January 1970).

The application of known and well-established sanitary principles has been effective in reducing such diseases to a minimum. See Figure 1–3. Refrigeration, proper cooking, and sanitation are most important. These are particularly pertinent to prepared frozen dinners.

Most of the above items are discussed in the text in greater detail under the appropriate headings. For ready reference to a summary of the characteristics and control of water- and food-borne diseases, the reader is referred to the folded insert, Figure 1–2. Although extensive, it should not, however, be accepted as being complete or final, but should be used as a starting point for further study.

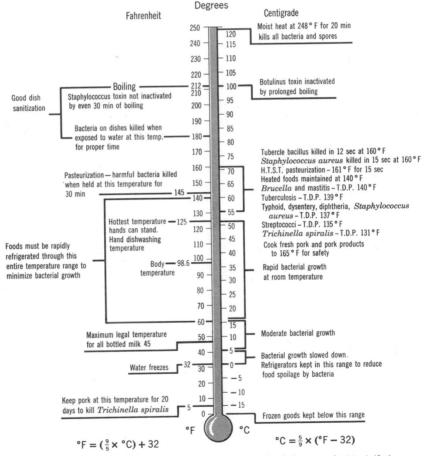

$$°F = (\tfrac{9}{5} \times °C) + 32 \qquad\qquad °C = \tfrac{5}{9} \times (°F - 32)$$

Note: T.D.P. = thermal death point temperature that will completely destroy a test culture in the presence of moisture in 10 min.
H.T.S.T. = high temperature, short time.

Figure 1–3   Food sanitation temperature chart.

INSECT-BORNE DISEASES AND ZOONOSES

General

The diseases transmitted by arthropods, commonly known as insect-borne diseases, are those diseases that are usually transmitted by biting insects from man to man or from animal to man. The ordinary housefly, which acts as a mechanical carrier of many disease agents, is discussed separately in Chapter 10.

Insect-Borne Diseases

A list of insect-borne diseases together with their important reservoirs is given in Tables 1-3, 1-4, and 1-5. The list is not complete but includes some of the common as well as less known diseases. Tick- or flea-borne diseases may be spread directly by the bite of the tick or flea and indirectly by crushing the insect into the wound made by the bite. Usuaally mosquitoes, lice, ticks, and other blood-sucking insects spread disease from man to man, or animal to man, by biting a person or animal carrying the disease-causing organisms. By taking blood containing the disease-producing organisms, the insect is in a position to transmit the disease when biting another person or animal.

TABLE 1-3   SOME EXOTIC INSECT-BORNE DISEASES (NOT NORMALLY FOUND IN THE UNITED STATES)

Disease	Incubation Period	Reservoir	Vector
Bartonellosis	16 to 22 days	Man	Sandflies (*Phlebotomus*)
Leishmaniasis	Weeks to months	Man, animals	Sandflies (*Phlebotomus*)
Loa loa	Years	Man	Chrysops-blood-sucking fly
Onchocerciasis	3 months or more	Man	Blackflies (genus *Simulium*)
Phlebotomus fever	3 to 4 days	Fly, man	Sandfly (*Phlebotomus*)
Relapsing fever	5 to 15 days	Man, tick, rodents	Lice, crushed in wound; ticks
Trench fever	7 to 30 days	Man	Lice, crushed in wound (*Pediculus humanus*)

Source: *Control of Communicable Diseases in Man,* op. cit.

Since many diseases are known by more than one name, other nomenclature is given to avoid confusion. As time goes on and more information is assembled, there will undoubtedly be greater standardization of ter-

TABLE 1-4 CHARACTERISTICS OF SOME INSECT-BORNE DISEASES

Disease	Etiologic Agent	Reservoir	Transmission	Incubation Period	Control*
Endemic typhus (Murine) (Flea-borne)†	Rickettsia typhi (R. mouseri) also possibly Ctenocephalides felis	Infected rodents. Rattus rattus and Rattus norvegicus. Also fleas, possibly opossums.	Bite or feces of rat flea Xenopsylla cheopis. Also possibly ingestion or inhalation of dust contaminated with flea feces or urine.	7 to 14 days, commonly 12	First elimination of rat flea by insecticide applied to rat runs, burrows, and harborages, then rat control. Spray kennels, beds, floor cracks.
Epidemic typhus (Louse-borne)	Rickettsia prowazeki	Infected persons and infected lice.	Crushing infected body lice Pediculus humanus or feces into bite, abrasions or eyes. Possibly louse feces in dust.	7 to 14 days, commonly 12	Insecticidal treatment of clothing and bedding; personal hygiene, bathing, elimination of overcrowding. Immunization. Delousing of individuals in outbreaks.
Bubonic plague	Pasteurella pestis, plague bacillus	Wild rodents and infected fleas.	Bite of infective flea Xenopsylla cheopis, occasionally bedbug and human flea. Pneumonic plague spread person-to-person.	2 to 6 days	Immunization. Surveys in endemic areas. Chemical destruction of flea. Community sanitation; rat control. (Plague in wild rodents called sylvatic plague.)
Q fever	Coxiella burneti (Rickettsia burneti)	Infected wild animals (bandicoots); cattle, sheep, goats, ticks, carcasses of infected animals.	Airborne rickettsias in or near premises contaminated by placental tissues. Raw milk from infected cows, direct contact with infected animals or meats.	2 to 3 weeks	Immunization of persons in close contact with rickettsias or possibly infected animals. Pasteurization of all milk at 145°F for 30 min or 161°F for 15 sec.

*Investigation and survey usually precede preventive and control measures. See also Chapter 10.
†"The association of seropositive opossums with human cases of murine (endemic) typhus in Southern California and the heavy infestation of the animals with Ctenocephalides felis fleas, which readily bite man, suggest that opossums and their ectoparasites are responsible for some of the sporadic cases of typhus in man." William H. Adams, Richard W. Emmons, and Joe E. Brooks, "The Changing Ecology of Murine (Endemic) Typhus in Southern California," Am. J. of Tropical Medicine & Hygiene (March 1970), pp. 311–318.

TABLE 1-4   CHARACTERISTICS OF SOME INSECT-BORNE DISEASES (cont'd)

Disease	Etiologic Agent	Reservoir	Transmission	Incubation Period	Control*
Rocky Mountain spotted fever	*Rickettsia rickettsii*	Infected ticks, dog tick, wood tick, Lone Star tick.	Bite of infected tick or crushed tick blood or feces in scratch or wound.	3 to 10 days	Avoid tick-infested areas and crushing tick in removal, clear harborages, insecticides.
Colorado tick fever	Colorado tick fever virus	Infected ticks and small animals,	Bite of infected tick, *Dermacentor andersoni,*	4 to 5 days	See Rocky Mountain spotted fever.
Tularemia	*Francisella tularensis (Pasteurella tularensis)*	Wild animals, rabbits, muskrats; also wood ticks.	Bite of infected flies or ticks, handling infected animals. Ingestion of contaminated water or insufficiently cooked rabbit meat.	1 to 10 days, usually 3	Avoid bites of ticks, flies. Use rubber gloves in dressing wild animals, avoid contaminated water; thoroughly cook rabbit meat.
Rickettsial-pox	*Rickettsia akari*	Infected house mice; possibly mites.	Bite of infective rodent mite.	10 to 24 days	Mouse and mite control. Application of miticides to infested areas—incinerators.
Scabies	*Sarcoptes scabiei;* a mite	Persons harboring itch mite.	Contact with persons harboring mite and use of infested garments or bedding. Also during sexual contact.	Several days or weeks	Personal hygiene, bathing, clean laundry. Exclude children from school until treated.
Trypanosomiasis, American	*Trypanosoma cruzi*	Infected persons, dogs, cats, wood rats, opossums.	Fecal material of infected insect vectors, cone-nosed bugs in eye, nose, wounds in skin.	5 to 14 days	Screening and rat-proofing of dwellings; destruction of vectors by insecticides and of infected domestic animals.
Scrub typhus	*Rickettsia tsutsugamushi*	Infected larval mites, wild rodents.	Bite of infected larval mites.	10 to 12 days	Eliminate rodents and mites; use repellents, clear brush.
Trypanosomiasis, African	*Trypanosoma gambiense*	Man, wild game, and cattle.	Bite of infected tsetse fly.	2 to 3 weeks	Fly control, treatment of population, clear brush, education in prevention.

Source: *Control of Communicable Diseases in Man,* op. cit.
*Investigation and survey usually precede preventive and control measures.

TABLE 1-5 MOSQUITO-BORNE DISEASES

Disease	Etiologic Agent	Reservoir	Transmission	Incubation Period	Control
Dengue or Breakbone fever*	Viruses of dengue fever	Infected vector mosquitoes, man, and possibly animals, the monkey.	Bite of infected *Aedes aegypti, Aedes albopictus, Aedes scutellaris* complex.	3 to 15 days, commonly 5 to 6 days	Eliminate *Aedes* vectors and breeding places. Screen rooms; use mosquito repellents.
Encephalitis, arthropod-borne viral	Virus of Eastern Equine, Western, St. Louis, Venezuelan Equine, Japanese B, Murray Valley, West Nile, and others.	Possibly wild and domestic birds and infected mosquito. Ring-necked pheasant. Rodents, bats, reptiles.	Bite of infected *Culex tarsalis,* also *Culiseta melanura* suspected. *Culex tritaeniorhynchus, Culex pipens, Culex pipens-quinquefasciatus.* Possibly *Aedes, Psorophora* and *Mansonia.*	Usually 5 to 15 days	Destruction of larvae and breeding places of *Culex* vectors. Space spraying, screening of rooms, use mosquito bed-nets where disease present. Avoid exposure during biting hours or use repellents. Public education on control of disease. Vaccination of equines.
Filariasis* (Elephantiasis after prolonged exposure)	Nematode worms, *Wuchereria bancrofti,* and *W. malayi*	Blood of infected person bearing microfiliariae, mosquito vector.	Bite of infected mosquito— *Culex fatigans, C. pipens; Aedes polynesiensis* and several species of anopheles.	3 months. Microfilariae do not appear in blood until at least 9 months.	Antimosquito measures. Determine insect vectors, locate breeding places and eliminate. Spray buildings. Educate public in spread and control of disease.
Malaria	*Plasmodium vivax, P. malariae, P. falciparum, P. ovale*	Man and infected mosquito. Found between 45°N. and 45°S. latitude and where average summer temperature is above 70°F or where	Bite of certain species of infected anopheles and injection or transfusion of blood of infected person.	Average of 12 days for *falciparum,* 14 for *vivax,* 30 for *malariae.* Sometimes delayed for 8 to 10 months.	Application of residual insecticide on inside walls and places where anopheles rests. Community spraying until malaria no longer endemic. Screen rooms and use bed-nets in endemic areas. Apply

*Normally not found in United States.

TABLE 1-5 MOSQUITO-BORNE DISEASES (cont'd)

Disease	Etiologic Agent	Reservoir	Transmission	Incubation Period	Control
Malaria (contd.)		the average winter temperature is above 48° F.			repellents to skin and clothing. Eliminate breeding places by drainage and filling, use larvicides—oil and Paris green. Suppressive drugs, treatment, health education.
Rift Valley fever*	Virus of Rift Valley fever	Sheep, cattle, goats, monkeys, rodents.	Probably through bite of infected mosquito or other blood-sucking arthopod. Laboratory infections and in butchering.	Usually 5 to 6 days	Precautions in handling of infected animals. Protection against mosquitoes in endemic areas. Care in laboratory.
Yellow fever*	Virus of yellow fever	Infected mosquitoes, persons, monkeys, marmosets, and probably marsupials.	Bite of infected *Aëdes aegypti.* In S. Africa forest mosquitoes, *H. Spegazzinii* and others. In E. Africa *Aëdes simpsoni, A. africanus,* and others. In forests of S. America by bite of several species of the genus *Haemagogus* and *A. leucocelaenus.*	3 to 6 days	Control of aëdes breeding places in endemic areas. Intensive vaccination in S. and E. Africa. Immunization of all persons exposed because of residence or occupation. In epidemic area spray interior or of all homes, apply larvicide to water containers, mass vaccination, evaluation surveys.

Source: Various sources and *Control of Communicable Diseases in Man,* op. cit.
*Normally not found in United States.

minology. In some cases there is a distinction implied in the different names that are given to very similar diseases. The names by which the same or similar diseases are referred to are presented below.

Bartonellosis includes oroya fever, Carrion's disease, and verruga peruana. Dengue is also called dandy fever, breakbone fever, bouquet, solar, or sellar fever. Endemic typhus, flea-borne typhus, and murine typhus are synonymous. Epidemic typhus is louse-borne typhus, also known as classical, European, and Old-World typhus. Brill's disease is probably epidemic typhus. Plague, black death, bubonic plague, and flea-borne pneumonic plague are the same. Filariasis or mumu, an infestation of *Wuchereria bancrofti,* may after obstruction of the lymph channels cause elephantiasis. Loa loa and loiasis are the same filarial infection. Cutaneous leishmaniasis, espundia, uta, bubas and forest yaws, aleppo, Baghdad or Delhi boil, chiclero ulcer, and oriental sore are synonyms. Visceral leishmaniasis is also known as kala azar. Malaria, marsh miasma, remittent fever, intermittent fever, ague, and jungle fever are synonymous. Blackwater fever is believed to be associated with malaria. Onchocerciasis is also known as blinding filarial disease. Sandfly fever is the same as phlebotomus fever, three-day fever, and pappataci fever. Q fever is also known as nine-mile fever. Febris recurrens, spirochaetosis, spirillum fever, famine fever, and tick fever are terms used to designate relapsing fever. Rocky Mountain spotted fever, tick fever of the Rocky Mountains, tick typhus, black fever, and blue disease are the same. Tsutsugamushi disease, Japanese river fever, scrub typhus, and mite-borne typhus are used synonymously. Trench fever is also known as five-day fever, Meuse fever, Wolhynian fever, and skin fever. Plague-like diseases of rodents, deer-fly fever, and rabbit fever are some of the other terms used when referring to tularemia. Other forms of arthropod-borne infectious encephalitis in the United States are the St. Louis type, the Eastern equine type, and the Western equine type; still other types are known. Nasal myiasis, aural myiasis, ocular myiasis or myiases, cutaneous myiases, and intestinal myiases are different forms of the same disease. Sleeping sickness, South American sleeping sickness, African sleeping sickness, Chagas' disease, Negro lethargy, and trypanosomiasis are similar diseases caused by different species of trypanosomes. Tick-bite fever is also known as Boutonneuse fever, Tobia fever, and Marseilles fever; Kenya typhus and South African tick fever are related. Scabies, "the itch," and the "seven-year itch" are the same disease.

The complete elimination of rodents and arthropods associated with disease has been a practical impossibility. Man, arthropods, and rodents, therefore, offer ready foci for the spread of infection unless controlled.

Zoonoses and Their Spread

The World Health Organization defines zoonoses as "Those diseases and infections which are naturally transmitted between vertebrate animals and man."[21] The Pan American Health Organization lists as the major zoonoses in the Americas encephalitis (arthropod-borne), psittaco-

---

[21] *WHO Tech. Rep. Ser.,* **169,** 6 (1959). *WHO Tech. Rep. Ser.* **378,** 6 (1967) considers definition too wide but recommends no change.

sis, rabies, jungle yellow fever, Q fever, spotted fever (Rocky Mountain, Brazilian, Colombian), typhus fever (murine), leishmaniasis, trypanosomiasis (Chagas' disease), anthrax, brucellosis, leptospirosis, plague, salmonellosis, tuberculosis (bovine), tularemia, hydatidosis, taeniasis (cysticercosis), and trichinosis.[22] It will be recognized that these diseases are also classified with the water-, food-, or insect-borne diseases.

The rodent-borne diseases include rat-bite fever, Haverhill fever, leptospirosis, choriomeningitis, salmonellosis, tularemia, possibly amebiasis or amebic dysentery, rabies, trichinosis (indirectly), and tapeworm. Epidemic typhus, endemic or murine typhus, Rocky Mountain spotted fever, tsutsugamushi disease, and others are sometimes included in this group. Although rodents are reservoirs of these diseases, the diseases themselves are actually spread by the bite of an infected flea, tick, or mite or by the blood or feces of an infected flea or tick on broken skin, as previously discussed.

Sodoku and Haverhill fever are two types of rat-bite fever. The incubation period for both is 3 to 10 days. Contaminated milk has also been involved as the cause of Haverhill fever. The importance of controlling and destroying rats, particulary around premises and barns, is again emphasized.

The causative organism of leptospirosis, also known as Weil's disease, spirochetosis icterohemorrhagic, leptospiral jaundice, spirochaetal jaundice, hemorrhagic jaundice, and icterohemorrhagic spirochetosis, is transmitted by the urine of infected rodents, cattle, dogs, and swine. Direct contact or the consumption of contaminated food or water, or direct contact with waters containing the leptospira, may cause the infection after 4 to 19 days.

The virus causing lymphocytic choriomeningitis is found in the mouth and nasal secretions, urine, and feces of infected house mice. It is probably spread by the consumption of food contaminated by the discharges of an infected mouse. The disease occurs after an incubation period of 8 to 21 days.

Salmonellosis is a food infection also called salmonella infection. In addition to human carriers, rodents, poultry, eggs, and livestock are sources of infection. Food contaminated with the discharges of infected animals, including rats, may cause the disease within 6 to 72 hours of its consumption.

Melioidosis is an uncommon disease in man. It is a disease primarily of rodents and small animals. The rat, cat, dog, horse, rabbit, and guinea

[22] Miscellaneous Publication No. 74, Pan American Health Organization, Washington, D.C., 1963.

pig are reservoir hosts of the disease-producing organisms. The incubation period is less than 10 days.

Tularemia may be transmitted by handling infected rodents with bruised or cut hands, particularly rabbits and muskrats, or by the bite of infected deer flies, ticks, and animals.

Trichinosis is ordinarily spread by the consumption of infected pork, pork products, and less frequently by bear or wild boar meat.

Taeniasis as used includes beef tapeworm, pork tapeworm, cysticercosis, and dwarf tapeworm. The worm is found in beef and pork that, when consumed in the raw or partially cooked state, causes beef tapeworm and pork tapeworm respectively in 1 to 3 months. Cysticercosis is caused by eating the pork tapeworm egg. Dwarf tapeworm is common in man, rats, and mice. It has been estimated that about 1 to 2 percent of the population, especially children, are infected in southern United States.[23] Higher incidences ranging between 10 and 18 percent are reported from India, Portugal, Spain, and Sicily. Eating raw beef or pork is dangerous. Cattle can contract the disease if permitted to pasture in fields upon which human feces containing tapeworm segments or eggs has been spread. Larval forms of the tapeworm develop into cysts that locate in the muscle or meat.

In addition to the human diseases mentioned above, rats may transmit hog cholera, swine erysipelas, fowl tuberculosis, and probably hoof-and-mouth disease to livestock or domestic animals.

Anthrax, also known as woolsorter's disease, malignant pustule, and charbon, is an infectious disease principally of cattle, swine, sheep, and horses that is transmissible to man. Many other animals may be infected. See Table 1–6.

Control of Zoonoses and Insect-Borne Diseases

Rabies is a disease of many domestic and wild animals and biting mammals, including bats. Every sick-looking dog or other animal that becomes unusually friendly or ill-tempered and quarrelsome should be looked on with suspicion. One should not place one's hand in the mouth of a dog, cat, or cow that appears to be choking. The animal may be rabid. A rabid dog may be furious or it may be listless, depending on the form of disease the dog has. A person bitten by a rabid animal, or one suspected of being rabid, should immediately wash and flush the wound thoroughly with soap and warm water, detergent and water, or just plain water and seek immediate medical attention. The physician will notify the health officer or health department of the existence of

[23] Richard P. Strong, *Stitt's Diagnosis, Prevention and Treatment of Tropical Diseases*, 6th ed., Blakiston Co., Philadelphia, 1942, p. 1470.

the suspected rabid animal and take the required action. The animal should be caged or tied up with a strong chain and isolated for 10 days to see if the symptoms appear. A dog or cat bitten by a rabid animal should be confined for 6 months, vaccinated and confined for 3 months, or destroyed.[24] A wild animal, if suspected, should be killed without unnecessary damage to the head, and the brain should be examined for evidence of rabies. The head should be packed in ice and shipped to the nearest equipped health department laboratory.

Vaccination of dogs is an important part of a good control program. In areas where rabies exists, mass immunization of at least 70 percent of the dog and cat population in the county or similar unit within a 2- or 3-week period is indicated. The "chick embryo vaccine" was used to vaccinate 250,000 dogs between 1952 and 1954; only two animals later developed rabies, as compared to 76 known cases in nonvaccinated dogs, and only one canine death was attributed to the vaccine.[25] Where rabies exists, or where it might be introduced, a good program should include vaccination of all dogs at 5 months of age and older. The avianized rabies vaccine (Flury Strain) produces excellent immunity in dogs for at least 3 years.[26] It must not be used in cats. Prophylactic immunization of man with duck embryo and the HEP chicken embryo vaccine is available. See Table 1–6.

To eliminate or reduce the incidence of zoonoses and insect-borne diseases it is necessary to control the environment and reservoirs and the vehicles or vectors. This would include control of water and food, carriers of disease agents, and the protection of persons and domestic animals (immunization) from the disease. Where possible and practical the reservoirs and vectors of disease should be destroyed and the environment made unfavorable for their propagation. Theoretically, the destruction of one link in the chain of infection should be sufficient. Actually, efforts should be exerted simultaneously toward elimination and control of all the links, since complete elimination of one link is rarely possible and protection against many diseases is difficult, if not impossible, even under ideal conditions. Personnel, funds, and equipment available will frequently determine the action taken to secure the maximum results or return on the investment made. The control of insects and rodents is discussed in some detail in Chapter 10.

[24] See WHO's Expert Committee on Rabies, *WHO Tech. Rep.,* 1966.
[25] Donald J. Dean, *The New York State Conservationist,* 9, No. 5, 8–12 (April–May 1955).
[26] H. Koprowski, H. R. Cox, and M. Welsh, "Advances in the Epidemiology and Prevention of Rabies," American Academy of General Practice Meeting, Los Angeles, March 28–31, 1955.

TABLE 1-6 SOME CHARACTERISTICS OF MISCELLANEOUS DISEASES

Disease	Etiologic Agent	Reservoir	Transmission	Incubation Period	Control
Ringworm of scalp (Tinea capitis)	*Microsporum* and *Trichophyton*	Infected dogs, cats, cattle, children.	Contact with contaminated barber clippers, toilet articles or clothing, dogs, cats, cattle, backs of seats in theaters, planes, and railroads.	10 to 14 days	Survey of children with Wood lamp; education re contact with dogs, cats, infected children. Reporting to school and health authorities, treatment of infected children, pets, and farm animals; investigation of source.
Ringworm of body (Tinea corporis)	*Epidermophyton floccosum*, *Microsporum*, *Trichophyton*	Skin lesions of infected man or animal.	Direct contact with infected person or contaminated floors, shower stalls, benches, towels, etc.	10 to 14 days	Sterilization of towels; fungicidal treatment of floors, benches, mats, shower stalls with creosol or equal. Exclusion of infected persons from pools and gyms. Treatment of infected persons, pets, and animals. Cleanliness, sunlight, dryness.
Ringworm of foot (Tinea pedis) (Athlete's foot)	Same as Ringworm of body.	Skin lesions of infected man or animal.	Contact with infected persons, contaminated floors, shower stalls, benches, mats, towels.	10 to 14 days	In addition to above, drying feet and between toes with individual paper towels; use of individual shower sandals, foot powder, and clean sterilized socks. Well-drained floors in bathhouses, pools, etc.
Ringworm of nails (Tinea unguium)	*Epidermophyton* and *Trichophyton*	Skin or nails of infected persons.	Probably from infected feet, contaminated floors.	Unknown	Same as above.

TABLE 1-6 SOME CHARACTERISTICS OF MISCELLANEOUS DISEASES (cont'd)

Disease	Etiologic Agent	Reservoir	Transmission	Incubation Period	Control
Ancylostomiasis (Hookworm disease)	*Necator americanus* and *Ancylostoma duodenale*	Feces of infected persons, soil containing infective larvae.	Larvae hatching from eggs in contaminated soil penetrate foot. Larvae also swallowed. (Cat and dog larvae cause a dermatitis.)	About 6 weeks	Prevention of soil pollution; sanitary privies or sewage disposal systems; wearing of shoes; education in method of spread; treatment of cases; sanitary water supply.
Rabies (Hydrophobia)	Virus of rabies	Infected dogs, foxes, cats, squirrels, cattle, horses, swine, goats, wolves, vampires, and fruit-eating bats.	Bite of rabid animal or its saliva on scratch or wound.	2 to 6 weeks, or up to 6 months.	Detention and observation for 10 days of animal suspected of rabies. Immediate destruction or 6 months, detention of animal bitten by a rabid animal. Vaccination of dogs, dogs on leash, dogs at large confined. Education of public. Avoid killing animal; if necessary, save head intact. Reduce wild life reservoir in cooperation with conservation agencies. Wash wound immediately and obtain medical attention if bitten.
Tetanus	*Clostridium tetani*, tetanus bacillus	Soil, street dust, animal feces containing bacillus.	Entrance of tetanus bacillus in a wound.	4 days to 3 weeks	Immunization with tetanus toxid plus reinforcing dose and booster within 5 years. Allergic persons should carry record of sensitivity. Thorough cleansing of wounds. Safety program.

Note: Abstracted from *Control of Communicable Diseases in Man,* op. cit.

TABLE 1-6 SOME CHARACTERISTICS OF MISCELLANEOUS DISEASES (cont'd)

Disease	Etiologic Agent	Reservoir	Transmission	Incubation Period	Control
Anthrax	*Bacillus anthracis*	Cattle, sheep, goats, horses, swine.	Contaminated hair, wool, hides, shaving brushes, ingestion or contact with infected meats. Inhalation of spores. Flies possibly; laboratory accidents. (Shaving brushes are under USPHS regulations. Bristles soaked 4 hr in a 10% formalin at a temperature of 110°+F destroys anthrax spores.)	7 days, usually less than 4 days.	Isolation and treatment of suspected animals. Postmortem examination by veterinarian of animals suspected of anthrax and deep burial of carcass, blood, and contaminated soil at a depth of at least 6 ft and above ground water, or incineration. Spores survive a long time. Education of employees in rug manufacturing plants, abbatoirs, tanneries, and in personnel hygiene; prompt treatment of abrasions. Treatment of trade wastes.
Trachoma	Virus of trachoma	Tears, secretions, discharges of nasal mucuous membranes of infected persons.	Direct contact with infected persons and towels, fingers, handkerchiefs, clothing soiled with infective discharges.	5 to 12 days	Routine inspections and examinations of schoolchildren. Elimination of common towels and toilet articles and use of sanitary paper towels. Education in personal hygiene and keeping hands out of eyes.
Psittacosis (Ornithosis)	Viruses of psittacosis	Infected parrots, parakeets, love birds, canaries, pigeons, poultry, other birds.	Contact with infected birds, their wastes and surroundings. Virus is airborne.	6 to 15 days	Importation of birds from psittacosis-free areas. Quarantine of pet shops having infected birds until thoroughly cleaned. Education of public to dangers of parrot illnesses.

TABLE 1-6 SOME CHARACTERISTICS OF MISCELLANEOUS DISEASES (cont'd)

Disease	Etiologic Agent	Reservoir	Transmission	Incubation Period	Control
Impetigo contagiosa	Probably streptococci, also staphylococci	Skin lesions of infected person, possibly nose and throat discharges.	Direct contact with infected skin lesions or articles such as towels and pencils soiled by discharges.	Within 5 days, often 2 days	Personal hygiene, avoidance of common use of toilet articles, prompt recognition and treatment of illness. Inspection of children at camps, nurseries, institutions, schools.
Chancroid (Soft chancre)	Hemophilus ducreyi, Ducrey bacillus	Discharges from open lesions and pus from buboes from infected persons.	Prostitution, indiscriminate sexual promiscuity and uncleanliness.	3 to 5 days or longer	Character guidance, health and sex education, premarital and prenatal examinations. Improvement of social and economic conditions; elimination of slums, housing rehabilitation and conservation, new housing, neighborhood renewal. Suppression of commercialized prostitution, personal prophylaxis, facilities for early diagnosis and treatment, public education concerning symptoms, modes of spread, and prevention. Case-finding, patient interview, contact-tracing and serologic examination of special groups known to have a high incidence of venereal disease, with follow-up. Report to local health authority.
Syphilis (Pox, lues)	Treponema pallidum	Exudates from lesions of skin, mucous membrane, body fluids, and secretions from infected persons.	By direct contact in sexual intercourse, kissing, fondling of children.	10 days to 10 weeks, usually 3 weeks.	
Gonorrhea (Clap, dose)	Neisseria gonorrhoeae, gonococcus	Exudate from mucous membranes of infected persons.	Sexual intercourse; infection during birth; careless use of rectal thermometer.	3 to 9 days, sometimes 14 days.	
Lymphogranuloma venereum (Tropical bubo)	Lymphogranuloma venereum	Lesions of rectum and urethra and sinuses and ulcerations of infected persons.	Sexual intercourse. Contact with contaminated articles.	5 to 21 days, usually 7 to 12 days	
Granuloma inguinale (Tropical sore)	Donovania granulomatis	Probably active lesions of infected persons.	Presumably by sexual intercourse.	Unknown, probably 8 days to 12 weeks.	Same as above.

MISCELLANEOUS DISEASES AND ILLNESSES

These are a divergent group, are not all communicable in the usual sense, and are discussed separately below.

Ringworm

This disease includes a group of fungus infections that develop on the surface of the skin and may involve the nails, scalp, body, and feet. See Table 1-6. Practically all persons have or have had one or more of the fungus infections. Thickening and scaling of the skin, raw inflamed areas, cracked skin, and blistering accompanied by severe itching are usual symptoms of dermatophytosis. The microscopic fungi grow best under conditions of warmth and moisture. Floors, mats, benches, and chairs at swimming pools, shower rooms, bathrooms, and clubs present ideal conditions for the incubation and spread of the fungi that come into contact with the bare skin.

The provision of smooth, nonslip, impervious floors constructed to remain well drained and facilitate quick drying is recommended. The direct entrance of sunlight that has not passed through window glass or the provision of ample window area with a southern exposure for adequate ventilation should be given serious consideration in swimming pool, shower, and bathhouse design. This will encourage prompt drying of the floors and destruction of fungi and their spores. Indoor floors and benches can be disinfected with a solution of sodium hypochlorite (500 mg/l available chlorine) or cresol, but the strong odor may discourage their use unless followed by rinsing with clear water. Cresol should not be used where it can be tracked into a pool. Daily scrubbing with a strong detergent in hot water followed by a rinse containing a quaternary ammonium compound (0.1 percent) would seem to be the most practical fungicide treatment.

Simple laundering of stockings or towels is not effective in destroying the pathogenic fungi responsible for athlete's foot; sterilization is necessary. This can be accomplished by boiling cotton socks 15 minutes; air drying in sunlight may also be effective. Shoes frequently become a reservoir of infection, since leather is difficult to sterilize without ruining it. Investigators found that leather shoes placed in a closed container with ethylene oxide maintained at a concentration of 150 mg/l of gas destroyed the fungi in the leather shoes.[27] The gas presents a serious fire and explosion hazard. Placing of shoes in a box containing formalde-

[27] John D. Fulton and Roland B. Mitchell, "Sterilization of Footwear," *U.S. Armed Forces Medical J.*, 3, No. 3, 425–439 (March 1952).

hyde for several hours followed by airing to prevent skin irritation is also suggested.[28]

Hookworm Disease or Ancylostomiasis

Hookworm disease is most common in those areas where the winters are mild and where moist sandy soil is encountered. The parasite develops best at a temperature of around 81°F, although it will live at temperatures between 57 and 100°F. A hard frost will kill the eggs and larvae. See Table 1–6.

Strongyloidiasis is an infection of the nematode *Strongyloides stercoralis*. It is similar to hookworm disease.

Tetanus or Lockjaw

This is a disease caused by contamination of a wound or burn with soil, street dust, or animal excreta containing the tetanus bacteria. The bacillus lives in the intestines of domestic animals. Gardens that are fertilized with manure, barnyards, farm equipment, and pastures are particular sources of danger. The tetanus germ produces a toxin that affects the nervous system, causing spasms, convulsions, and frequently death. It is a spore former that survives in soil and is resistant to ordinary boiling. Deep puncture-type wounds are most dangerous, regardless of whether they are made by clean or rusty objects. See Table 1–6.

OTHER ILLNESSES ASSOCIATED WITH THE ENVIRONMENT AND FOOD

Treatment of the environment is supplementing treatment of the individual, but more effort and knowledge is needed. The total environment is "the most important determinant of health." A review of more than 10 years of research conducted in Buffalo, New York, showed that the overall death rate for people living in heavily polluted areas was twice as high, and the death rates for tuberculosis and stomach cancer three times as high, as the rates in less polluted areas.[29] Dr. Rene Dubos points out that "Many of man's medical problems have their origin in the biological and mental adaptive responses that allowed him earlier in life to cope with environmental threats. All too often the wisdom of the body is a shortsighted wisdom."[30] In reference to air pollution,

---

[28] *Control of Communicable Diseases in Man,* op. cit., p. 70.
[29] Warren Winkelstein, Address at the 164th Annual Convention of the Medical Society of the State of New York, 1969.
[30] *The Fitness of Man's Environment,* Smithsonian Institution Press, Washington, D.C., 1968.

he adds that "while the inflammatory response is protective (adaptive) at the time it occurs, it may, if continuously called into play over long periods of time, result in chronic pathological states, such as emphysema, fibrosis, and otherwise aging phenomena."

It is necessary to determine the total insult to which the body is subjected, that is, the total body burden of all deleterious substances gaining access to the body both singly and in combination (synergistic effects) through the air, food, drink, and skin. Then the effects must be learned and the indicated preventive measures applied.

Human adjustment to environmental pollutants and emotional stresses due to crowding and other factors can result in later disease and misery.[31] Whereas microbiological causes of many diseases have been brought under control in many parts of the world, effects of the presence or absence of certain chemicals in air, water, and food in trace amounts are not yet clearly demonstrated. Their action is often insidious, and effects are extended in time to the point where direct relationship is difficult to prove in view of the many possible intervening and confusing factors. Some illnesses associated with the environment and food are discussed below.

Lead Poisoning

Lead poisoning is most frequently associated with the children living in old substandard housing who eat lead-based paint that peels or flakes from walls, ceilings, and woodwork. It causes mental retardation, blindness, chronic kidney diseases, and other deficiencies. Control is approached through mass screening of ghetto children between the ages of 1 and 6 for excessive lead in the blood (40 μg or more per 100 ml) and treatment of those affected, with priority given to the 1–3 -year-olds; identification of housing having lead-based paint on the walls, woodwork or ceilings; removal or covering of paint containing more than 1 percent lead; prohibition of sale of toys or baby furniture containing lead paint; education of parents, social workers, sanitarians, health guides, and owners of old buildings to the hazard and its control.[32]

Lead in gasoline discharged with auto exhaust in urban areas, water standing and flowing in lead pipe, natural or added lead in food and drink, and lead in certain household products all contribute to the body burden. Serious illnesses and at least one death have been attributed

[31] Rene Dubos, *Man Adapting,* Yale University Press, New Haven, Conn., 1965, p. 279.
[32] For discussion of the problem and of planning and implementing a control program, see *Control of Lead Poisoning in Children,* Bureau of Community Environmental Management, U.S. Public Health Service, Dept. of HEW (December 1970).

to the use of earthenware pottery with improper glaze. Such glaze dissolves in fruit juice, coffee, wine, soda pop and other acid drinks.[33] Most of the glaze applied to earthenware pottery contains lead. When the pottery is not fired long enough at the correct temperature, the glaze will not seal completely and its lead component can be leached or released. The Lead Industries Association and the U.S. Potters Association have recently set up a surveillance program to check on commercial pottery glazes. Those found acceptable will be allowed to carry a USPA label of approval. Lead is a cumulative poison. The maximum Public Health Service limit for lead in drinking water is 0.05 mg/l. The total daily ingestion from all sources over any period of years should not exceed 0.6 mg per day.[34]

Carbon Monoxide Poisoning

Carbon monoxide is odorless, tasteless, and colorless. It is a product of incomplete combustion of carbonaceous fuel. Poisoning is caused by leaks in an automobile exhaust system; running a car indoors or while standing; unvented space or water heater, gas-range oven, or gas-fired floor furnace; clogged chimney or vent; inadequate fresh air for complete combustion; improperly operating gas refrigerator; incomplete combustion of liquefied petroleum gas in recreational and camping units.

Carbon monoxide poisoning is sometimes confused with food poisoning; nausea and vomiting are common to both. In carbon monoxide poisoning additional symptoms include headache, drowsiness, dizziness, flushed complexion, and general weakness; carbon monoxide is found in the blood. Education of the public and physicians by the utilities and the health department, standards for appliances, and housing-code enforcement can reduce deaths from this poisoning.[35]

Mercury Poisoning

Mercury poisoning in man has been associated with the consumption of methylmercury-contaminated fish, shellfish, bread, and pork and, in wildlife, through the consumption of contaminated seed. Fish and shellfish poisoning occurred in Japan in the Minamata River and Bay region and at Niigata between 1953 and 1964. Bread poisoning occurred as result of the use of wheat seed treated with a mercury fungicide to

---

[33] "Family Health," *Family Health Magazine* (February 1971), p. 12.

[34] Maxcy-Rosenau, *Preventive Medicine and Public Health,* Appleton-Century-Crofts, New York, 1965, p. 796.

[35] Eugene L. Lehr, "Carbon Monoxide Poisoning: A Preventable Environmental Hazard," *Am. J. Public Health,* **60,** 289–293 (February 1970).

make bread in West Pakistan in 1961, in Central Iraq in 1960 and 1965, and in Panorama, Guatemala in 1963 and 1964. Pork poisoning took place in Alamagordo, New Mexico, when methylmercury-treated seed was fed to hogs that were eaten. In Sweden, the use of methyl-mercury as a seed fungicide was banned in 1966 in view of the drastic reduction in the wild bird population attributed to treated seed.

It is also reported that crops grown from seed dressed with minimal amounts of methylmercury contain enough mercury to contribute to an accumulation in the food chain reaching man. The discovery of mod-erate amounts of mercury in tuna and most freshwater fish, and rela-tively large amounts in swordfish, by many investigators in 1969 and 1970 tended to further dramatize the problem.[36]

The organic methylmercury and other alkylmercury compounds are highly toxic. Depending on the concentration and intake, it can cause unusual weakness, fatigue, and apathy followed by neurological disor-ders. Numbness around the mouth, loss of side vision, poor coordination in speech and gait, tremors of hands, irritability, and depression are additional symptoms leading possibly to blindness, paralysis, and death. The methylmercury also attacks vital organs such as the liver and kid-ney. It concentrates in the fetus and can cause birth defects.

Methylmercury has an estimated biological half-life of 70 to 74 days in man, depending on such factors as age, size, and metabolism, and is excreted mostly in the feces at the rate of about 1 percent per day. Mercury persists in large fish such as pike from 1 to 2 years.

Elemental mercury is generally stored under water; it vaporizes on exposure to air. Certain compounds of mercury may be absorbed through the skin, gastrointestinal tract, and respiratory system, although elemen-tal mercury and inorganic mercury compounds are not absorbed to any significant extent through the digestive tract because they do not remain in the body.

A maximum allowable steady intake (ADI) of 0.03 mg for a 70-kg (154-lb) man would provide a safety factor of ten. If fish containing 0.5 ppm mercury were eaten daily, the limit of 0.03 mg would be reached by the daily consumption of 60 g (about 2 oz) of fish.[37] The safe level for whole blood would be 2 μg/100 ml and, for hair, 6 ppm.[38]

[36] Goran Löfroth, "Methylmercury," *Redaktionstajänsten,* Natural Science Research Council, Stockholm, Sweden, March 20, 1969.

[37] *Hazards in Mercury,* a Special Report to the Secretary's Pesticide Advisory Committee, EPA, Dept. of HEW, Washington, D.C., November 1970.

[38] Statement of Roger C. Herdman, Deputy Commissioner for Research and Develop-ment, New York State Department of Health, before the Subcommittee on the Environment of the Committee on Commerce of the U.S. Senate, May 20, 1971.

Mercury is ubiquitous in the environment. The sources are both natural and manmade. Natural sources are leachings, erosion, and volatilization of dimethylmercury from mercury-containing geological formations. Manmade sources are waste discharges from chlor-alkali and pulp manufacturing plants, mining, chemical manufacture and formulation, the manufacture of mercury seals and controls, treated seeds, combustion of fossil fuels, fallout, and surface runoff. The mercury ends up in lakes, streams, tidal waters, and in the bottom mud and sludge deposits.

Microorganisms and macroorganisms in water and bottom deposits can transform metallic mercury, inorganic divalent mercury, phenylmercury and alkoxialkylmercury into methylmercury. The methylmercury thus formed, and perhaps other types, in addition to that discharged in wastewaters, are assimilated and accumulated by aquatic and marine life such as plankton, small fish, and large fish. Alkaline waters tend to favor production of the more volatile dimethylmercury, but acid waters are believed to favor retention of the dimethyl form in the bottom deposits. Under anaerobic conditions, the inorganic mercury ions are precipitated to insoluble mercury sulfide in the presence of hydrogen sulfide. The process of methylation will continue as long as organisms are present and have access to mercury.

The form of mercury in fish has been found to be practically all methylmercury, and there are indications that a significant part of the mercury found in eggs and meat is in the form of methylmercury.

The concentration of mercury in fish and other aquatic animals and in wildlife is not unusual. Examination of preserved fish collected in 1927 and in 1939 from Lake Ontario and Lake Champlain in New York State have shown concentrations up to 1.3 ppm mercury (wet basis). Fish from remote ponds, lakes, and reservoirs have shown 0.05 to 0.7 ppm or more mercury, with the larger and older fish showing the higher concentration.

In 1970 the amount of mercury in canned tuna fish averaged 0.32 ppm; in fresh swordfish, 0.93 ppm; in freshwater fish, 0.42 ppm (up to 2.0 and 3.0 ppm in a few large fish such as walleyed pike); and as high as 8 to 23 ppm in fish taken from heavily contaminated waters. The mercury in urban air is generally in the range of 0.02 to 0.2 $\mu g/m^3$; in drinking water, less than 0.001 ppm; in rainwater, about 0.2 to 0.5 ppb ($\mu g/1$); in ocean water, 0.12 ppb; in Lake Superior water, 0.12 ppb; in Lake Erie, water 0.39 ppb.[39] Reports from Sweden and Denmark (1967–1969) indicate a mercury concentration of 3 to 8 ppb (ng/g)

---

[39] "Sources of Mercury to the Environment," Symposium, Mercury in Man's Environment, Royal Society of Canada, February 15, 1971. (Avg. Hg in soil 0.04 ppm.)

in pork chops, 9 to 21 ppb in pig's liver, 2 to 5 ppb in beef filet, 9 to 14 ppb in hen's eggs, and 0.40 to 8.4 ppm in pike.

In view of the potential hazards involved, steps have been taken to provide standards or guidelines for mercury. The maximum allowable concentration for 8-hr occupational exposure has been set at 0.1 mg metallic vapor and inorganic compounds of mercury per cubic meter of air. For organic mercury the threshold limit is 0.01 mg/m^3 of air. The suggested limit for fish is 0.5 ppm and, for shellfish, 0.2 ppm. The proposed standard for drinking water is 0.005 ppm (5 ppb). A standard of 0.05 ppm has been suggested for food.

There is no evidence to show that the mercury in the current daily dietary intake has caused any harm, although apparent health does not indicate possible nondetectable effects on brain cells or other tissues. Nevertheless, from a conservative health standpoint, it has been recommended that pregnant women not eat any freshwater fish until more is known about the effects of methylmercury on fetal brain tissue.[40] The general population should probably not eat more than one freshwater-fish meal per week. Swordfish should not be eaten.

Further research and studies are needed to determine the actual and subtle effects of mercury (as well as cadmium, arsenic, manganese, lead, zinc, and others) on man and his environment.

Illnesses Associated with Air Pollution

The particulate and gaseous irritants in polluted air are reported to increase susceptibility to upper respiratory diseases, such as the common cold, and to aggravate existing illnesses. Diseases mentioned as being *also* associated with air pollution include bronchial asthma (restriction of the smaller airways or bronchioles and increase in mucous secretions), chronic bronchitis (excessive mucous and frequent cough), pulmonary emphysema (shortness of breath), lung cancer, and conjunctivitis (inflammation of the lids and covers of the eyeball).

A direct single cause-and-effect relationship is often difficult to prove because of the many other causative factors and variables usually involved. Nevertheless, the higher morbidity and mortality associated with higher levels of air pollution and reported episodes (Table 6–2) is believed to show a positive relationship.

Certain air contaminants, depending on the body burden, may produce systemic effects. These include arsenic, asbestos, cadmium, beryllium compounds, mercury, manganese compounds, carbon monoxide, fluorides, hydrocarbons, mercaptans, inorganic particulates, lead, radioactive iso-

[40] Statement by Roger C. Herdman, op. cit.

topes, carcinogens, and insecticides. They require attention and are being given consideration in the development of air-quality criteria. See Chapter 6.

Bronchial asthma affects susceptible sensitive individuals exposed to irritant air contaminants and aeroallergens. The aeroallergens include pollens, spores, rusts, and smuts. There also appears to be a good correlation between asthmatic attacks and air pollution levels.

Chronic bronchitis has many contributing factors including a low socioeconomic status, occupational exposure, and population density; smoking is a major factor. Air pollution resulting in smoke, particulates, and sulfur dioxide is an additional factor.

Emphysema mortality rates in United States urban areas are approximately twice the rural rates, indicating an association with air pollution levels. Asthma and bronchitis often precede emphysema.

Lung cancer rates are reported to be higher among urban populations than rural. The dominant factor in lung cancer is smoking. Air pollution plays a small but continuous role.

Some generalized effects of common air pollutants and their possible relationship to the above are of interest. Sulfur dioxide and sulfuric acid in low concentrations irritate the nose and throat. This can cause the membrane lining of the bronchial tubes to become swollen and eroded, with resultant clotting in the small arteries and veins. Acute carbon monoxide poisoning causes a lowered concentration of oxygen in the blood and body tissues. Ozone irritates the air passages, causing chest pain, cough, shortness of breath, and nausea. Ozone can cause severe damage to the lungs. Nitrogen dioxide in high concentrations can result in acute obstruction of the air passages and inflammation of the smaller bronchi. Nitrogen dioxide at low levels causes eye and bronchial irritation. In the presence of strong sunlight, nitrogen dioxide breaks down into nitric oxide and atomic oxygen, which combines with molecular oxygen in air to form ozone. Benzopyrene and related compounds are known to cause some types of cancer under laboratory conditions and are incriminated as carcinogens. Olefins have an injurious effect on certain body cells and are apt to cause eye irritations.[41]

Nutritional Deficiency Diseases

Severe examples of diseases caused by deficiencies in the diet are not common in the United States. There are, however, many people whose diet is slightly deficient in one or more nutrients but who show

[41] *Act Now for Clean Air,* a program to reduce air pollution in New York State, New York State Department of Health, Albany, 1968.

no clinically detectable symptoms for many years, if ever. Most malnutrition takes the form of protein deficiency.[42] Social, cultural, and emotional patterns and ignorance contribute to the problem, regardless of the economic status of individuals. The vague and insidious nature of these diseases is good reason to apply known preventive and control measures continuously. Several of the nutritional diseases are mentioned below.

Scurvy

This is a disease caused by a deficiency of vitamin C or ascorbic acid, which is found in citrus fruit, potatoes, tomatoes, and raw cabbage. Weakness, anemia, spongy and swollen gums that bleed easily, and tender joints are some of the common symptoms.

Pellagra

This disease is due to a deficiency of niacin or tryptophan (amino acid). Niacin is found in eggs, lean meats, liver, whole-grain cereals, milk, leafy green vegetables, fruits, and dried yeast. Recurring redness of the tongue or ulcerations in the mouth are primary symptoms, sometimes followed by digestive disturbances and mental depressions.

Rickets

This is a childhood disease (under 2 years) caused by the absence of vitamin D, which is associated with proper utilization of calcium and phosphorous. Vitamin D is found in liver, milk, butter, eggs, and fish of high body oil content. An inadequate supply of vitamin D in the diet will probably show in knock-knees or bowed legs, crooked arms, soft teeth, pot belly, and faulty bone growth. Sunshine is a good source of vitamin D, as are vitamin D fortified milk and other foods.

Beriberi

A deficiency of thiamin or vitamin B_1, found in whole-grain cereals, dried beans, peas, peanuts, pork, and liver, may cause changes in the nervous system, loss of appetite, and interference with digestion. Change from unpolished to polished rice in the diet can cause the disease in some countries where the diet is not varied.

[42] M. B. Stoch and P. M. Smythe reported that children who were grossly undernourished from birth had smaller heads, lower intelligence quotients, and poorer body coordination than did a control group from the same socioeconomic environment. "The Effect of Undernutrition During Infancy on Subsequent Brain Growth and Intellectual Development," *South African Med. J.* (October 28, 1967).

Ariboflavinosis

This disease is due to a deficiency of riboflavin, known also as vitamin B_2 or G. Riboflavin is found in liver, milk, eggs, dried yeast, enriched white flour, and leafy green vegetables. An inadequate amount of this vitamin may cause greasy scales on the ear, forehead, and other parts of the body, drying of the skin, cracks in the corners of the mouth, and sometimes partial blindness.

Vitamin A Deficiency

This disease causes night blindness and skin changes. Vitamin A is also needed for bone growth. The diet should be adjusted to include foods rich in vitamin A or carotene, such as dry whole milk, cheese, butter, eggs, liver, carrots, dandelions, kale, and sweet potatoes.

Liver Cirrhosis

This disease is caused by malnutrition and probably aggravated by alcohol. A high protein diet plus choline and dried brewer's yeast is indicated.

Iron-Deficiency Anemia

Lack of vitamin B_{12} or folic acid, repeated loss of blood, and increased iron need during pregnancy cause weakness, irritability, brittle fingernails, cuts and sores on the face at the mouth, and other debility. Prevention of blood loss and treatment with iron salts is suggested.

Goiter

This is a thyroid disorder usually caused by deficient iodine content in food and water and its absorption. Iodized or iodated salt should be universally used.

Kwashiorkor

Kwashiorkor is a protein deficiency nutritional disease common among children under about 6 years of age living in underdeveloped areas of the world. Changes in the color and texture of the hair, diarrhea, and scaling sores are some of the clinical signs. A diet rich in animal proteins, including dry skim milk, meat, eggs, fish, and cheese, and vegetables (including soybeans and Incaparina) containing vitamins and amino acids can control the disease. But the people affected must be taught to accept the high-protein foods, to use a safe water supply in reconstituting manufactured foods, and the importance of environmental hygiene to prevent debility due to diarrhea and parasitism and thus reduce susceptibility to kwashiorkor.

INVESTIGATION OF WATER- AND FOOD-BORNE DISEASE OUTBREAKS

General

The promptness with which an investigation is started will largely determine its success. In this way the cause of the disease may be determined and precautions for the prevention of future incidents learned. One food-poisoning outbreak at an institution, camp, or eating place will cause severe repercussions and may result in great financial loss.

Upon learning of the existence of an unusual incidence of a disease, it is the duty of the owner or manager and physician called to immediately notify the health department. They are required to give such cooperation as may be possible to the investigating authority. The health department, in turn, should report to the owner or operator of the establishment and to the U.S. Public Health Service upon completion of the investigation. Such reporting, however, has been sporadic. The Public Health Service compiles statistical reports for the entire country from such information.

A general knowledge of the cause, transmission, characteristics, and control of diseases is necessary in order to conduct an epidemiological investigation. This is usually made under the direction of an epidemiologist if one is available. The background material was discussed earlier in this chapter. Reference to Figure 1–2 will give, in easily accessible form, the characteristics of various water- and food-borne diseases and perhaps a clue as to what might be responsible for a disease outbreak. It is not a simple matter to quickly determine the cause of illness due to water or food; but a preliminary study of the symptoms, incubation periods, food consumed, and sanitary conditions may give a lead and form a basis for the immedite control action to be instituted.

A common method of determining the probable offending food is a tabulation as shown in Figure 1–4, which is made from the illness questionnaire, Figure 1–5. Comparison of the attack rates for each food will usually implicate or absolve a particular food.

A summation can be made in the field to take the place of individual questionnaires when assistance is available. The tabulation horizontal headings would include:

1. Names of persons served; 2. Age; 3. Ill—yes or no; 4. Day and time ill; 5. Incubation period in hours (determined later); 6. Foods served

at suspected meals—previous 12 to 72 hours; 7. Symptoms—nausea, vomiting, diarrhea, blood in stool, fever, thirst, constipation, stomach ache, sweating, sore throat, headache, dizziness, cough, chills, pain in chest, weakness, cramps, other.

A simple bar graph, with hour and days as the horizontal axis and number ill each hour plotted on the vertical axis, can be made from the data. The time between peaks represents the incubation period. The average incubation period is the sum of the incubation periods of those ill (time elapsing between the initial infection and the clinical onset of a disease) divided by the number of ill persons studied.

Sanitary Survey

The sanitary survey should include a study of all environmental factors that might be the cause, or may be contributing to the cause, of the disease outbreak. These should include water supply, food, housing, sewage disposal, bathing, and any other relevant factors. Each should be considered responsible for the illness until definitely ruled out. The folded insert and Chapters 1, 3, 4, 8, 9, 10, and 11 should be referred to for guidance and possible specific contributing causes to an outbreak and their correction. Figure 1–6 will be helpful.

Medical Survey

The medical survey should develop a clinical picture to enable identification of the disease. Typical symptoms, date of onset of the first case, date of onset of last case, range of incubation periods, number of cases, number hospitalized, number of deaths, and number exposed are usually determined by the health officer. To assemble this information and analyze it carefully, a questionnaire should be completed, by trained personnel if possible, on each person available or on a sufficient number to give reliable information. A typical simple, short question sheet is shown in Figure 1–5.

A very important part of the medical survey is examining all the food handlers and assembling a medical history on each one. Frequently a carrier or a careless infected food handler can be found at the bottom of a food-borne outbreak. The importance of animal reservoirs of infection should not be overlooked. The folded insert gives, in condensed form, symptoms of many diseases that, when compared to a typical clinical picture, may suggest the causative organism and the disease.

Samples

The collection of samples for laboratory examinations is a necessary part of the investigation of water- and food-borne disease outbreaks.

FORM APPROVED
BUDGET BUREAU NO. 68-R1034

DEPARTMENT OF HEALTH, EDUCATION, AND WELFARE
PUBLIC HEALTH SERVICE
Health Services and Mental Health Administration
NATIONAL COMMUNICABLE DISEASE CENTER
EPIDEMIOLOGY PROGRAM
ATLANTA, GEORGIA 30333

INVESTIGATION OF A FOODBORNE OUTBREAK

1. Where did the outbreak occur?

   State _____ (1,2)   City or Town _____   County _____

2. Date of outbreak: (Date of onset 1st case)

   _____ (3-8)

3. Indicate actual (a) or estimated (e) numbers:

   Persons exposed _____ (9-11)

   Persons ill _____ (12-14)

   Hospitalized _____ (15-16)

   Fatal cases _____ (17)

4. History of Exposed Persons:

   No. histories obtained _____ (18-20)

   No. persons with symptoms _____ (21-23)

   Nausea _____ (24-26) Diarrhea _____ (33-35)

   Vomiting _____ (27-29) Fever _____ (36-38)

   Cramps _____ (30-32) Other, specify _____ (39)

5. Incubation period (hours):

   Shortest _____ (40-42) Longest _____ (43-45)

   Approx. for majority _____ (46-48)

6. Duration of Illness (hours):

   Shortest _____ (49-51) Longest _____ (52-54)

   Approx. for majority _____ (55-57)

7. Food-specific attack rates: (58)

Food Items Served	Number of persons who ATE specified food				Number who did NOT eat specified food			
	Ill	Not Ill	Total	Percent Ill	Ill	Not Ill	Total	Percent Ill

8. Vehicle responsible (food item incriminated by epidemiological evidence): (59,60)

9. Manner in which incriminated food was marketed: (Check all applicable)

(a) Food Industry (61)
Raw 1
Processed 2
Home Produced
Raw 3
Processed 4

(b) Vending Machine 1 (62)

(c) Not wrapped 1 (63)
Ordinary Wrapping 2
Canned 3
Canned-Vacuum Sealed 4
Other (specify) 5

(d) Room Temperature 1 (64)
Refrigerated 2
Frozen 3
Heated 4

If a commercial product, indicate brand name and lot number

10. Place of Preparation of Contaminated Item: (65)
Restaurant 1
Delicatessen 2
Cafeteria 3
Private Home 4
Caterer 5
Institution:
School 6
Church 7
Camp 8
Other, specify 9

11. Place where eaten: (66)
Restaurant 1
Delicatessen 2
Cafeteria 3
Private Home 4
Picnic 5
Institution:
School 6
Church 7
Camp 8
Other, specify 9

HSM 4.245 (NCDC)
Rev. 3—69

(Over)

Figure 1-4a Investigation of a food-borne outbreak.

LABORATORY FINDINGS (Include Negative Results)

12. Food specimens examined: (67)

Specify by "X" whether food examined was original (eaten at time of outbreak) or check-up (prepared in similar manner but not involved in outbreak)

Item	Orig.	Check up	Findings Qualitative	Quantitative
Example: beef	X		C. perfringens, Hobbs type 10	2×10^6/gm

13. Environmental specimens examined: (68)

Item	Findings
Example: meat grinder	C. perfringens, Hobbs Type 10

14. Specimens from patients examined (stool, vomitus, etc.): (69)

Item	No. Persons	Findings
Example: stool	11	C. perfringens, Hobbs Type 10

15. Specimens from food handlers (stool, lesions, etc.): (70)

16. Factors contributing to outbreak (check all applicable):

Item	Findings
Example: lesion	C. perfringens, Hobbs type 10

	Yes	No
1. Improper storage or holding temperature	1	2 (71)
2. Inadequate cooking	1	2 (72)
3. Contaminated equipment or working surfaces	1	2 (73)
4. Food obtained from unsafe source	1	2 (74)
5. Poor personal hygiene of food handler	1	2 (75)
6. Other, specify	1	2 (76)

17. Etiology: (77, 78)

Pathogen _____ Suspected ____ 1 (79)

Chemical _____ Confirmed ____ 2

Other _____ Unknown ____ 3

18. Remarks: Briefly describe aspects of the investigation not covered above, such as unusual age or sex distribution; unusual circumstances leading to contamination of food, water; epidemic curve; etc. (Attach additional page if necessary)

Name of reporting agency: (80) _____

Investigating official: _____ Date of investigation: _____

NOTE: Epidemic and Laboratory Assistance for the investigation of a foodborne outbreak is available upon request by the State Health Department to the National Communicable Disease Center, Atlanta, Georgia 30033.

HSM 4.245 (NCDC) (Back)
Rev. 3–69

Figure 1-4b Laboratory findings (include negative results).

Please answer the questions asked below to the best of your ability. This information is desired by the health department to determine the cause of the recent sickness and to prevent its recurrence. Leave this sheet, after you have completed it, at the desk on your way out. (If mailed, enclose self-addressed and stamped envelope and request return of completed questionnaire as soon as possible.)

1. Check any of the following conditions that you have had:

Nausea	Fever	Sore throat	Cough	Chills
Vomiting	Constipation	Headache	Pain in chest	Weakness
Diarrhea	Stomach ache	Dizziness	Laryngitis	Cramps
Thirst	Sweating	Paralysis	Bloody stool	Other

2. Were you ill .... Yes .... No.
3. If ill, first became sick on: Date................Hour................A.M./P.M.
4. How long did the sickness last?..
5. Check below (√) the food eaten at each meal and (×) the food not eaten. Answer even though you may not have been ill.

Meal	Tuesday	Wednesday	Thursday
Breakfast	Apple juice, Corn flakes, Oatmeal, Fried eggs, Bread, Coffee, Milk, Water	Orange, Pancakes, Wheaties, Syrup, Coffee, Milk, Water	Grapefruit, Wheatina, Shredded wheat, Boiled egg, Coffee, Milk, Water
Lunch	Baked salmon, Creamed potatoes, Corn, Apple pie, Lemonade, Water	Roast pork, Baked potatoes, Peas, Rice pudding, Milk, Water, Chef salad	Swiss steak, Home fried potatoes, Turnips, Spinach, Chocolate pudding, Orange drink, Milk, Water
Dinner	Gravy, Hamburger steak, Mashed potatoes, Salmon salad, Cookies, Pears, Cocoa, Water	Roast veal, Rice, Beets, Peas, Jello, Coffee, Water	Fruit cup, Meat balls, Spaghetti, String beans, Pickled beets, Sliced pineapple, Tea, Coffee, Milk

6. Did you eat food or drink water outside?............ If so, where and when?
....

7. Name.. Tel. .... Age............ Sex............
8. Remarks (Physician's name, hospital)..
.... Investigator ....

Figure 1–5 Questionnaire for illness from food, milk, or water.

Date.. Investigator..
Name of place.............................. Owner..
Population..................................... Manager..
Onsets—day and hours.................. Incubation period..
Number afflicted.............. Number hospitalized.............. Number deaths..............
Outbreak: explosive.................. gradual.................. undetermined..............
Samples collected..

Underline symptoms most commonly reported:

Diarrhea, constipation, abdominal pains, stomach cramps, muscular cramps, prostration, high temperature, painful straining at stool or in urination, sore throat, chills, thirst, sweating, vomiting, nausea, swelling of face and eyelids, laryngitis, cough, pain in chest, enlarged tonsils or adenoids, pains in joints, eye movement difficult, swallowing difficult, headache, dizziness, other..............

Water

1. Water sources and treatment..............
2. Method of serving water..............
3. Interconnections: toilet..............
   washbasin.............. bath tubs..............
   tubs.............. other..............
4. Recent repairs..............
5. Cross-connections, with other supplies..............
6. Changes in water taste..............
   color.............. odor..............

Milk and food

7. Source of milk (pasteurized)..............
8. Method of handling milk..............
9. Use of leftover foods..............
10. Source of fowl, meats, ice cream, shellfish, pastries..............
11. Food refrigeration and storage..............
12. Food handling and preparation..............
13. Ice source and handling..............
14. Thawing foods protected..............
15. Dressings, sauces, etc...............

Food handlers

16. Recent illness in food handlers..............
17. Hand-washing facilities..............
18. No. pyogenic skin infections..............
19. Personal hygiene..............

Kitchen and dining hall

20. Storage and use of insecticides..............
    rat poison.............. roach powder..............
    water paint.............. silver polish..............
21. Garbage storage and disposal..............
22. Prevalence of rodents and insects....
23. Fly breeding controlled..............
24. Dish cleansing and disinfection..............
25. Premises and equipment clean..............
26. Food service well organized..............

Other

27. Housing overcrowding..............
28. Bathing beach or swimming pool operation, water source..............
29. Medical and nursing care..............
30. Other..............

Remarks (Comment on unsatisfactory items and probable cause, general impressions, etc.):

Figure 1–6 Outbreak investigation field summary.

Isolation of the incriminating organism from the allegedly responsible food or drink, producing the characteristic symptoms in laboratory animals or human volunteers, and then isolating the same organisms from human volunteers and laboratory animals will confirm the field diagnosis and definitely implicate the responsible vehicle. In the early stages of the field investigation it is very difficult to determine just what samples to collect; yet if the samples of food, for example, are not collected at that moment, the chances are the samples will be gone when they are wanted. It is customary, therefore, to routinely collect samples of water from representative places and available samples of all left-over milk and food that had been consumed and place them under seal and refrigeration. Sterile spatulas or spoons boiled for 5 min can be used to collect samples. Sterile wide-mouth water bottles and petri dishes make suitable containers. In all cases, a sterile technique must be used. Since examination of all the food may be unnecessary it is advisable, after studying the questionnaires and accumulated data, to select the suspicious foods for laboratory examination and set aside the remaining food in protected sterile containers under refrigeration at a temperature of less than 40°F for possible future use. Since stool and vomitus are most likely to reveal the offending organism, it is important to collect such samples to assist the laboratory in its search for specific pathogens rather than a whole group of pathogens.

Samples of water should be collected directly from the source, storage tanks, high and low points of the distribution system at times of high and low pressure, kitchen, and taps near drinking fountains for chemical and bacterial examinations. Samples of milk should be collected from unopened and opened bulk milk cans, cartons or containers, and leftover milk in pitchers or other containers for phosphatase test, coliform test, direct microscopic count, standard plate count, presence of streptococcus, and any other specific tests that may be indicated. Samples of food possibly incriminated should be collected from the refrigerator, storeroom, or wherever available. As a last resort samples have been collected from garbage pails or even dumps; however, results on such samples must be interpreted with extreme caution. The folded insert will be of assistance in suggesting the common vehicle, and hence the foods to suspect, by using the incubation periods and symptoms of those ill as a guide. It should be remembered that the time elapsing before symptoms appear is variable and depends on the causative agent and size of dose, the resistance of individuals, the amount and kind of food or drink consumed. For example, an explosive outbreak with a very short incubation period of a few minutes to less than an hour would suggest a chemical poisoning. Antimony, arsenic, cadmium, cyanide, mercury, sodium fluoride, sodium nitrite, or perhaps shellfish poisoning, favism, fish poisoning, and zinc poisoning are possibilities. An explosive outbreak with an incubation period of several hours would suggest botulism or fish, mushroom, potato, rhubarb-leaf, shellfish, chemical, or staphylococcus food poisoning. An incubation period of 6 to 24 hours would suggest botulism, mushroom poisoning, rhubarb poisoning, salmonella infection, or streptococcus food poisoning. An incubation period of 1 to 5 days would suggest ascariasis, botulism, diphtheria, amebic dysentery, bacillary dysentery, leptospirosis, paratyphoid fever, salmonella infection, scarlet fever, streptococcal sore throat, or trichinosis. For other diseases with more extended incubation periods, refer to the cover insert. The laboratory examinations might be biologic, toxicologic, microscopic, or chemical, depending on the symptoms and incubation period.

Samples collected from food handlers may include stool, vomitus, urine, skin, nose, throat, and other membrane specimens, or swabs for biologic or microscopic examinations. Medical and laboratory study to determine the origin of the food contaminant may be carried over an extended period of time, since the absence of disease-producing organisms in three stools is generally required before a person is considered negative, although in screening, one stool specimen will reveal a high percentage of carriers.

Some of those ill with the severest symptoms should be selected for laboratory study to determine, if possible, the organism causing the illness. The samples collected and the examinations made would be indicated by the symptoms and dates of onset displayed by the sufferers as explained above.

When chemical poisoning is suspected, samples of flour, sugar, and other foods that might have been contaminated in transportation, handling, and preparation should be considered. Galvanized or zinc-, cadmium-, and antimony-plated utensils and liquids prepared in such containers should be collected for chemical and toxicologic tests. Also include samples of insecticides, rodenticides, ant powders or sprays, and silver polish.

The sanitary and medical surveys may involve the swimming pool or bathing beach. In that case, samples should be collected at the peak and toward the end of the bathing period for examinations.

It is customary to notify the laboratory in advance that an outbreak has occurred and that specimens will be delivered as soon as possible. All samples should be carefully identified, dated, sealed, and refrigerated. A preliminary report with the samples, including the probable cause, number ill, age spread, symptoms, incubation period, and so on, will greatly assist the laboratory in its work.

Preliminary Recommendations

Before leaving the scene of an investigation, a preliminary study should be made of the data accumulated and temporary control measures instituted based on the findings summarized in Figure 1–6. Instructions should be left by the health officer and sanitary engineer or sanitarian with the owner or manager regarding those who are ill, the housing, water supply, food and milk supply, food handlers, sewage disposal, swimming and bathing place, and any other items that may be indicated.

Epidemiologic Report

Report of an investigation of a water- or food-borne disease outbreak should include the cause, laboratory findings, transmission, incidence, cases by dates of onset, average incubation period and range, typical symptoms, length of illness, age and sex distribution, deaths, secondary attack rate, and recommendations for the prevention and control of the disease. This can often be accomplished by careful study of the results of the sanitary and medical surveys and results of laboratory examinations, including the questionnaires and summaries given in Figures 1–4 and 1–6. Reports should be sent to the state health department and thence to the Public Health Service.

BIBLIOGRAPHY

Bryan, Frank L., *Diseases Transmitted by Foods,* U.S. Public Health Service, Dept. of HEW, Atlanta, Ga., 1971, 58 pp.

Dack, G. M., *Food Poisoning,* University of Chicago Press, 1956, 251 pp.

Dolman, C. E., *Bacterial Food Poisoning,* Canadian Public Health Association, Toronto, 1943, 46 pp.

Getting, V. A., "Epidemiologic Aspects of Food-Borne Diseases," *New Eng. J. Med.* (June 1943), 75 pp. Reprint.

Hilleboe, Herman E., and Larimore, Granville W., *Preventive Medicine,* W. B. Saunders Co., Philadelphia, 1965, 523 pp.

"Joint FAO/WHO Expert Committee on Zoonoses," *WHO Tech. Rep. Ser.,* 378 (1967).

Korff, F. A., "Food Establishment Sanitation in a Municipality," *Am. J. Public Health,* 32, 740 (July 1942).

Manson-Bahr, P., *Synopsis of Tropical Medicine,* Williams & Wilkins Co., Baltimore, 1943, 224 pp.

Sartwell, Philip E., *Maxcy-Rosenau, Preventive Medicine and Public Health,* Appleton-Century-Crofts, New York, 1965, 1070 pp.

"Microbiological Aspects of Food Hygiene," *WHO Tech. Rep. Ser.,* 399 (1968).

Military Sanitation, FM 21–10, Department of the Army Field Manual, U.S. Government Printing Office, Washington, D.C., May 1957, 304 pp.

Procedure for the Investigation of Foodborne Disease Outbreaks, International

Association of Milk, Food, and Environmental Sanitarians, Inc., Shelbyville, Ind., 1966, 36 pp.

Strong, R. P., *Stitt's Diagnosis, Prevention and Treatment of Tropical Diseases,* 2 vols., Blakiston Co., Philadelphia, 1942, 1747 pp.

Taylor, Joan, *Bacterial Food Poisoning,* Royal Society of Health, London, England, 1969, 175 pp.

The Control of Communicable Diseases in Man, 11th ed., American Public Health Association, New York, 1970, 316 pp.

Lewis, K. H. and Cassel, K. Jr., *Botulism,* Proc. of Symposium, Public Health Service Pub. No. 999-FP-1, Cincinnati, Ohio, December 1964, 327 pp.

Bryan, Frank, "What the Sanitarian Should Know About *Clostridium Perfringens* Foodborne Illness," *Jour. Milk and Food Technology,* **32,** 381–389 (October 1969).

Hall, Herbert E., "Current Developments in Detection of Microorganisms in Foods— *Clostridium Perfringens,*" *Jour. Milk and Food Technology,* **32,** 426–430 (November 1969).

Bowmer, Ernest J., "Salmonellae in Food—A Review," *Jour. Milk and Food Technology,* **28,** 74–86 (March 1965).

Food-Borne Infections and Intoxications, A Four-Year Summary Report 1965–1969, Food Protection and Toxicology Center, University of California, Davis, November 1969, 206 pp.

Rieman, Hans, *Food-Borne Infections and Intoxications,* Academic Press, New York, 1969, 698 pp.

An Evaluation of the Salmonella Problem, National Academy of Sciences, Washington, D.C., 1969, 207 pp.

2

ENVIRONMENTAL
ENGINEERING PLANNING

INTRODUCTION

Definitions

Comprehensive planning takes into account the physical, social, economic, ecologic, and related factors of an area and attempts to blend them into a single compatible whole that will support a healthful and efficient society. A *comprehensive plan* has been defined as

"(a) A model of an intended future situation with respect to: (i) specific economic, social, political and administrative activities; (ii) their location within a geographic area; (iii) the resources required; and (iv) the structures, installations and landscape which are to provide the physical expression, and physical environment for, these activities; and (b) A programme of action and predetermined co-ordination of legislative, fiscal, administrative and political measures, formulated with a view to achieving the situation represented by the model."[1]

The term *planning* is defined as the "systematic process by which goals (policies) are established, facts are gathered and analyzed, alternative proposals and programs are considered and compared, resources are measured, priorities are established and recommendations are made for the deployment of resources designed to achieve the established goals."[2]

---

[1] *Report of Stage Three: The Seminar-Conference, Toronto, August 6–16, 1967*, p. 10. This definition modifies that of Ernest Weissman, in *Planning and Urban Design*, Bureau of Municipal Research, Toronto, 1967, p. 40.
[2] *Local and Regional Planning and Development*, Governor's Task Force Report, West Virginia, Regional Review, National Service to Regional Councils, Washington, D.C., January 1970, p. 2.

There is great need for emphasis on the area-wide, metropolitan, and regional approaches to planning and on the presentation of a total integrated and balanced appraisal of the essential elements for healthful living. The latter is difficult to achieve, as it requires a delicate synthesis of competing economic, social, and physical environmental goals. However, a reasonable balance must be sought and these goals can only be approached if realistic objectives are set and *all* the facts are analyzed and presented with equal force. Regional planning should recognize national and local planning. In many instances, no matter how logical and sound the regional planning, the most that can be achieved is local implementation hopefully within the context of the regional plan.

Environmental Engineering Considerations

Some of the essential environmental health, engineering, and sanitation objectives, frequently overlooked in planning, will be identified to show how they can be blended into the usual planning process to achieve more comprehensive community plans. People are demanding, as a fundamental right, not only clean air and pure water and food, but also decent housing, an unpolluted land, freedom from excessive noise, adequate recreation facilities, and housing in communities that provide comfort, privacy, and essential services. The usual single-purpose planning for a highway, a housing development, or a sanitary landfill must be broadened to consider the environmental, economic, and social purposes to be served and the effects produced by the proposed project. The planner must ensure that potential resultant problems are avoided in the planning stages and that projects are so designed as to be pleasing and to enhance the community well-being.

As an example, major highway planning must take into consideration the need for rest areas with ample picnic spaces, toilet facilities, drinking water, noxious weed control, and solid waste storage and disposal facilities; it must also consider traffic counts, landscaping, buffer zones, and the effects of routes, entrances, and exits on adjoining communities. Environmental effects during and after construction must also be taken into consideration: erosion control during construction, travel of sediment, dust, and possible flooding; pollutional effects on water supply intakes, streams, lakes, fish, and wildlife; and the disposal of trees, stumps, and demolition material without causing air pollution. Problems to be resolved in the planning stages are control of noise, vibration and air pollution; dust and weed suppression; control of de-icing chemicals, herbicides, and other pollutants because they might affect nearby water wells, reservoirs, watersheds, and recreational areas, as well as wildlife, vegetation, streams, and lakes.

TYPES OF PLANNING

There are many kinds and levels of planning for the future. They may range from family planning to national planning for survival. The discussion here will deal with (1) "comprehensive[3] community planning," also referred to as "general planning" from which policy decisions can be made; (2) "comprehensive functional planning," also called "preliminary or feasibility studies" dealing with a single facility or service; and (3) "definitive planning" for a specific project, which includes final engineering and architectural reports, plans, contract drawings, and specifications leading to construction. See Figure 2–1.

Comprehensive Community Planning

In this phase, an attempt is made to take an overall look at the total region. If only a part of the total region is involved the area, or community, is studied within the context of the larger region of which it is a part. This is described later as "the process of comprehensive community planning." Report recommendations should lead to policy decisions that establish priorities to implement specific projects, with an estimate of their cost. At this stage cost is a secondary consideration, although cost may become a major factor in the specific planning and implementation phase. A combination of talents is needed to make a proper study. These may include the architect, artist, attorney, ecologist, civil engineer, economist, environmental engineer, geographer, geologist, hydrologist, landscape architect, mathematician, operations research scientist, planner, political scientist, sociologist, soils scientist, systems engineer, and others.

Comprehensive Functional Regional Planning

If the priority project is a new or improved water system (it could be a sewerage system and treatment plant, park and recreation facilities, a solid waste collection and disposal system), then the next step is a functional "comprehensive" engineering planning study to consider in some detail the several ways in which the regional or area-wide service or facility can be provided, together with approximate costs.[4] Such a

---

[3] The term *comprehensive* referred to here and in the discussion that follows is used with considerable reservation because truly comprehensive planning is an ideal that will rarely if ever be achieved.
[4] For a guide to the selection of a consulting engineer see *Consulting Engineer—A Guide for the Engagement of Engineering Services*, No. 45, American Society of Civil Engineers, New York City (1972). The same general principles would apply to selection of an architect or other planning consultant.

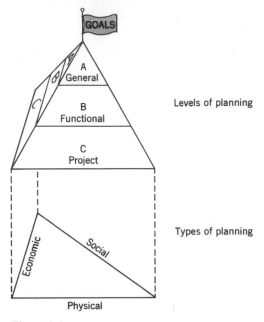

Figure 2-1

A. *General, Overall Policy Planning*—Identification of goals; aspirations and realistic objectives. Establishment of functional priorities.
B. *Functional Planning*—Such as for transportation, water supply, wastewater, recreation, air pollution, solid wastes, or medical care facilities in which *alternative functional solutions are presented,* including the economic, social, and ecological factors, advantages and disadvantages.
C. *Project Planning*—Detailed engineering and architectural specific project plans, specifications, drawings, and contracts for bidding. Implementation follows.
    Construction, Operation and Maintenance—Plan adjustment as constructed; updating and planning for alterations and new construction.

Figure 2-1 Types and levels of planning (applicable to national, state, regional, and local planning).

study is also referred to as a preliminary or feasibility report; it is discussed in greater detail under the heading "Regional Planning for Environmental Engineering Controls." For example, the water system study alternatives for the region might include (1) possible service areas and combinations; (2) a well-water supply with water softening and iron removal; (3) an upland lake or reservoir with multipurpose uses requiring land acquisition, water rights, and a conventional water treatment plant; or (4) a nearby stream requiring preliminary settling and an elaborate water treatment plant to adequately handle the known pollution in the stream.

At this stage no detailed engineering or architectural construction plans are prepared. However, the engineering, political, legal, economic, and social feasibility or acceptance of each alternative is presented together with recommendations, cost estimates, and methods of financing each alternative. The study report should be sufficiently complete and presented so that the officials can make a decision (political) and select one alternative for definitive planning.

Definitive, or Project, Planning

The next step for the example given (a new water system) would be the establishment of a legal entity to administer the project as required by local law, followed by the acquisition of necessary right-of-way and water rights, resolution of any legal constraints, establishment of service districts, approval of bond issues, rate-setting and financing of operation, maintenance, and debt retirement. This step is followed by selecting a consulting engineer (it could be the same engineer who made the preliminary study), preparing plans, specifications, and contract drawings, advertising for bids, and awarding the contract to a contractor. Construction should be under the supervision of the consulting engineer, and a resident engineer responsible to the consultant or municipality should be employed.

The consulting engineer should ensure that the municipality is provided with revised drawings showing the works and location of facilities as constructed and in-place. He would also normally be expected to provide operation manuals and guides, to take responsibility for placing the plant in operation, and, during the first year, to train personnel to take over full operation.

THE PROCESS OF COMPREHENSIVE COMMUNITY PLANNING[5]

For comprehensive community planning, the process includes (1) statement of goals and objectives,[6] (2) basic studies, mapping, and data analysis, (3) plan preparation, (4) plan implementation, (5) public in-

---

[5] Tailored to fit the particular characteristics and needs of the area under study with consideration of social and economic factors.

[6] Sometimes the formulation of goals and objectives is placed second. Goals are the final purpose or aim, the ends to which a design tends. Objectives are the realistically attainable ends. In any case, the goals and objectives should be continually adjusted as the planning study progresses and as the quality of life desired and demanded changes.

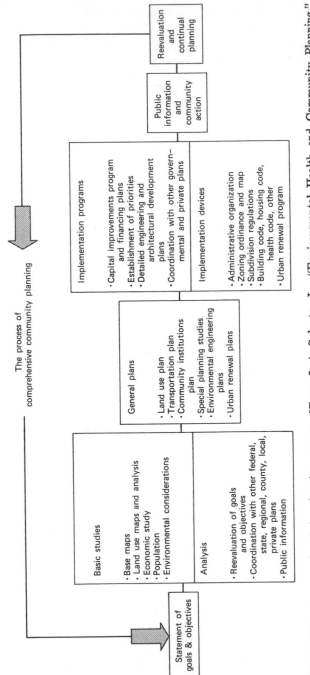

Figure 2-2 An example of the planning process. [From J. A. Salvato, Jr., "Environmental Health and Community Planning," J. Urban Planning and Development Division, ASCE, 94, No. UP 1, 22–30 (August 1968).]

The process of comprehensive community planning

Statement of goals & objectives

Basic studies
· Base maps
· Land use maps and analysis
· Economic study
· Population
· Environmental considerations

Analysis
· Reevaluation of goals and objectives
· Coordination with other federal, state, regional, county, local, private plans
· Public information

General plans
· Land use plan
· Transportation plan
· Community institutions plan
· Special planning studies
· Environmental engineering plans
· Urban renewal plans

Implementation programs
· Capital improvements program and financing plans
· Establishment of priorities
· Detailed engineering and architectural development plans
· Coordination with other governmental and private plans

Implementation devices
· Administrative organization
· Zoning ordinance and map
· Subdivision regulations
· Building code, housing code, health code, other
· Urban renewal program

Public information and community action

Reevaluation and continual planning

formation and community action, and (6) re-evaluation and continual planning. The process is shown in Figure 2–2 and is outlined below.[7]

Statement of Goals and Objectives

Community Aspirations and Environmental Quality. A first step is the preparation of a tentative statement to guide the planning staff that reflects the goals and objectives the community expects will be achieved through the planning process.

The statement should recognize the economic, social, and physical community aspirations, including the environmental health quality goals and objectives. Depending on the need, as confirmed or modified by Step 2, these may be (1) a water supply of satisfactory quality adequate for domestic, industrial, recreational, and fire-fighting purposes; (2) clean air; (3) proper sewage and other wastewater collection and disposal; (4) water pollution abatement; (5) proper solid waste collection and disposal; (6) adequate and safe parks and recreation facilities, including swimming pools or bathing beaches; (7) noise abatement and control; (8) a convenient and acceptable transportation system; (9) elimination of accident hazards; (10) preservation of good housing and residential areas, rehabilitation of sound substandard housing, and construction of new sound housing in a healthful and pleasing environment; (11) elimination of sources and causes of mosquitoes, ticks, blackflies, termites, rats and other vermin; (12) adequate schools; (13) adequate hospitals, nursing homes, and other medical care facilities; (14) adequate gas and electricity; and (15) adequate cultural facilities.

Basic Studies, Mapping, and Data Analysis

Research and Problem Identification. Having tentatively agreed on a statement of goals and objectives, the next step is evaluation of the community. This is done by and includes the following:

1. Mapping. Preparation of a base map, which will be the basis of other detail maps referred to below, and of existing land use maps.
2. Land-Use Analysis. Collection, plotting, and analysis of data on residential, commercial, industrial, public, agricultural, and recreational land uses; blighted and deteriorated structures, inefficient and conflicting land uses, and desirable land uses; areas available for future population growth and industrial development; and availability and adequacy of supporting services.
3. Population and Demographic Studies. Present and future trends, locations,

---

[7] Joseph A. Salvato, Jr., "Environmental Health and Community Planning," *J. of the Urban Planning and Development Division*, ASCE, **94**, No. UP 1, Proc. Paper 6084, 22–30 (August 1968).

and amounts of populations; social composition and characteristics of the population; age distribution and changes.

4. Economic Studies and Proposals. Existing sources of income, future economic base, labor force, markets, industrial opportunities, retail facilities, stability of economy.

5. Transportation Systems. Existing systems, their location and adequacy; effects of air and water pollution; aesthetic, zoning, noise, and vibration controls; population growth; industrial and recreation development of the systems; and modifications or protective features needed in existing and proposed systems.

6. Community Institutions. Description, location, and adequacy of educational, recreational, and cultural facilities; medical, public health, and environmental protection facilities; religious and other institutions; and public buildings such as post offices, fire and police stations, auditoriums and civic centers, public markets, and government offices.

7. Environmental Health and Engineering Considerations. Define and show on base maps and charts both favorable and unfavorable natural and man-made environmental conditions and factors such as meteorology, including wind and solar radiation studies; topography, hydrology, flooding, tidal effects, seismology, and geography including soil drainage, percolation, and bearing characteristics; natural pollution of air, land, and water; background radiation and the flora and fauna of the area; man-made pollution of air, water, and land; noise and vibrations, ionizing radiations, and unsightly conditions; and condition of housing and community facilities and utilities.

8. Coordinate with Other Planning. Federal, state, regional, county, and local agencies and municipalities usually have some plans already completed and certain plans under way or proposed. Private enterprise as represented by individuals, business establishments, industrial plants, and others are simultaneously making plans for renovation, expansion, or relocation and in many instances are in a position to support and implement the official planning. Hence it is extremely important, insofar as possible, to coordinate all the planning and obtain the participation essential to the realization of the planning goals and objectives.

9. Public Information. Acquaint the public, and specifically individual representatives of organizations that may be participating in the implementation of the comprehensive community plans, with the goals, objectives, basic studies, mapping and data analysis, and the results. Encourage and solicit feedback and carefully consider suggestions. Ask for supporting information to clarify suggestions made.

10. Reevaluate Goals and Objectives. Confirm or adjust as indicated by the basic studies, analyses, and feedback.

Plan Preparation

General plans, including area-wide and regional plans, that present alternatives and rough costs to help people understand what is involved and help officials make policy decisions should be prepared. Detailed engineering and architectural drawings and specifications come later, when actual construction is scheduled as determined by implementation (Step 4). The plans should include the following:

1. Land-Use Plan. This type of plan shows the existing uses to be retained and future patterns and areas for residential, commercial, industrial, agricultural,

recreation, open space or buffer, and public purposes. Such plans are based on the findings in Steps 1 and 2 and are integrated with the plans that follow.

2. The Transportation or Circulation Plan. This shows major and minor highways and streets, transit systems, waterways, ports, marinas, airports, service areas, terminals, and parking facilities. The plan should clearly show facilities to be retained, those to be improved or altered, and new facilities proposed.

3. The Community Institutions Plan. This shows location of existing or proposed new, expanded, or remodeled educational and cultural facilities; health, welfare, religious, and other institutions; public buildings and facilities. The plans for these institutions are considered in the light of their adequacy to meet present and future needs based on Steps 1 and 2.

4. Special Planning Studies. These include neighborhood analyses and plans for urban renewal, clearance, rebuilding, rehabilitation, code enforcement, housing conservation, preservation of sites and structures, central business, parking, parks and recreation areas, as well as private enterprise plans and their integration with community plans.

5. The Environmental Engineering Plans. These plans are concerned with the adequacy and needs for water supply, sewerage, solid waste disposal, air and water pollution control, housing and a healthful residential environment, realty subdivision and construction controls, gas, electricity, nuisance control. They are also concerned with the effects of industrial, agricultural, commercial, highway, airport, recreational and power development; storm water, drainage and flood control; forest, open space, soil, estuary and wildlife conservation; more aesthetic structures and public buildings. See also step "7" above.

The plans should take into consideration the environmental factors that need to be improved and those that can be developed to eliminate certain hazardous or annoying conditions while at the same time making a needed or desired improvement. Incompatible or nonconforming structures or uses should be eliminated or adjusted to harmonize with the most desirable and obtainable environment. Additional details on environmental factors and controls are given later in this chapter under "Comprehensive Environmental Engineering and Health Planning" and "Environmental Factors in Site Selection and Planning."

6. Urban Renewal Plans. Federal grants and assistance are available to a municipality that formulates an acceptable "workable program" for community improvement. A workable program is one that "will include an official plan of action . . . for effectively dealing with the problem of urban slums and blight within the community and for the establishment and preservation of a well-planned community with well-organized residential neighborhoods of decent homes and suitable living environment for family life."[8] The workable program requires that the following seven elements be met to be eligible for federal aid. They are basic to the elimination of slums and blight and to the prevention of their spread in any city.

(a) Up-to-date codes and ordinances, including building, electrical, plumbing, housing, health codes and the like, and zoning and subdivision regulations that provide sound standards governing land and building use and occupancy.

(b) A comprehensive community plan, often referred to as a general or master plan and made up of policy statements and plans to guide community growth

---

[8] U.S. Dept. of Housing and Urban Development, *Urban Renewal Fact Sheet,* Government Printing Office, Washington, D.C. 1966 as revised.

and development (as outlined in this chapter). It includes three essential features: a land-use plan; a circulation plan; and a community facilities plan.

(c) Neighborhood Analysis, which is an examination of the physical resources, a pinpointing of deficiencies, a notation of environmental problems of the area under study, and recommendations concerning the steps to be taken to eliminate physical and environmental shortcomings.

(d) Administrative Organization. To insure that codes and ordinances are enforced, the community must demonstrate that it has an effective organizational structure and adequate personnel to carry out these functions.

(e) Financing. The community must demonstrate that it has adequate financing and sound budgetary policies to assure that public improvement projects related to urban renewal are carried out.

(f) Housing for Displaced Persons. Most projects result in temporary or permanent displacement, and relocation assistance must be available to residents requiring it. There is also need to assist businesses that have a relocation problem. Both of these problems must be faced by the community to prevent hardship to those who are displaced.

(g) Citizen Participation. Broad citizen involvement, representative of all segments of community life, is required as part of the democratic process of program formulation.

Plan Implementation

Construction; Problem Correction, Prevention and Control. Plans, to yield a return, must be implemented. Implementation involves political and governmental decisions and public acceptance. This means:

1. Capital Improvement Program and Financing Plans. Project priorities are established. Approximate costs, sources of revenue and financing for five years or longer are determined. Existing and contemplated public and private planning and construction (including industrial, commercial, urban renewal, and slum clearance), local, state and federal assistance programs, the feasibility of area-wide or regional solutions, and the total tax burden are all recognized and coordinated. The program as implemented will largely determine the future quality of the environment and the economic health of the community. Governmental action usually follows adoption of a capital improvement program. The program is reviewed and updated annually.

2. Detailed Engineering, Architectural, and Development Plans. These are plans for specific projects as determined in (1) and may be preceded by feasibility studies. In contrast to general plans, specific drawings, detail plans, and specifications are prepared from which accurate estimates or bids are obtained and construction contracts are let. Included are federal, state, urban renewal, and private construction projects such as a new city hall, school, shopping center, water system, sewerage and treatment plant, marina, sanitary landfill, and housing and park and recreation area development.

3. Regulations, Laws, Codes, and Ordinances. Zoning controls and subdivision regulations are commonly provided. Also considered are flood plain management and provision for long-term land leases and tax deferral for open-space preservation. There is a need for standards and regulations for water supply, sewage disposal, and house connections; air pollution abatement and emission standards; control of noise and vibration nuisances; control of mining and excavations; regulations

for solid waste storage, collection, and disposal. Also needed are a modern building code[9] (including plumbing, fire and safety, electrical, heating, and ventilation regulations), sanitary code,[10] and a housing occupancy and maintenance code.[11]

4. Administrative Organization. The entire process of comprehensive community planning can at best be only of limited value unless provision is made for competent direction, personnel staffing, and administration. Adoption of an official map, land-use plans, an improvement program, codes, ordinances, rules and regulations has little meaning unless implemented by competent personnel who are adequately compensated.

Public Information and Community Action

For a community plan to be effective, it is necessary that public participation and information be involved at all stages, without public interference with the technical planning operations. In keeping the public informed, it is also necessary to stimulate and provide channels for individuals to respond with information and ideas. Public participation can result in additional types of physical improvement and programs that help achieve the public objectives.

Involvement of the public, including community organizations and influential citizens, during the planning process makes possible better appreciation of the community goals and objectives and the problems to be overcome. In this way information presented by the planner has a better chance of reaching more people and in being discussed by individuals, groups, and news media. All this helps to further understanding and gain support for implementing definitive plans and bond issues.

[9] A building code deals primarily with, and contains standards for, the construction or alteration of buildings, structural and fire safety, and prevention of related hazards. The code authorizes plan and specification review functions, inspection for approval of the construction or alteration, and the issuance of permits.

[10] A sanitary code is concerned with environmental sanitation and safety and control and prevention of communicable diseases for protection of the public health, safety, and general welfare. It may set standards and regulations for such matters as disease control; qualifications of personnel including those responsible for the operation of certain facilities; water supply; wastewater disposal; air pollution prevention; solid wastes; food sanitation including milk; radiation; noise; vectors; accident prevention; housing, hospitals, nursing homes, institutions; recreation areas and facilities; schools, camps and resorts; trailer and mobile home parks; bathing beaches, swimming pools; occupational health; emergency sanitation; and other preventive measures that may be required to ensure that the public health is protected.

[11] A housing code is concerned primarily with, and contains minimum standards for, the provision of safe, sanitary, decent dwellings for human habitation. It sets standards for the supplied utilities and facilities, occupancy and maintenance. The code requires inspection to determine compliance and usually is applicable to all dwellings, regardless of when constructed. It is also a tool to obtain housing and community rehabilitation, conservation, and maintenance.

Reevaluation and Continual Planning

Not to be forgotten is the continual need for public information, coordination and administration of planning activities locally on a day-to-day basis, and liaison with state and federal agencies. Studies and analyses in Step 2, the plans developed in Step 3, and the improvement program and the controls in Step 4 must all be kept reasonably current. This is necessary to prevent obsolescence of the comprehensive community plan as well as the community.

Community objectives and goals can be expected to change with time. New and revised land-use concepts, means of transportation, housing needs and design, public desires and aspirations, economic, technological and sociological developments, will require periodic reevaluation of the general plan and revision as indicated. It should be kept in mind that human wants are insatiable. As soon as a goal or objective is approached, another will be sought. This is as it should be, but must be kept in balance and within realistic bounds, which comprehensive planning should help accomplish.

Conclusion

There is an urgent need for more engineering in community planning and more comprehensive planning in engineering, with emphasis on area-wide, metropolitan, and regional approaches. The environmental health, sanitation, and engineering factors that are essential for community survival and growth must not be overlooked in planning and engineering.

Public awareness demands a quality of environment that provides such fundamental needs as pure water, clean air, unpolluted land, pure foods, decent housing, privacy, safe recreation facilities, and open space. A balanced appraisal must be made, and the planning process must blend these goals in its objectives, analyses, plans, and in the capital budgeting.

State and local health and environmental protection departments have vital planning, plan approval, and regulatory responsibilities to assure that the public health and welfare is protected. These responsibilities usually deal with water pollution abatement; wastewater treatment and disposal; safe and adequate water supply; air pollution control; solid waste disposal; x-ray and nuclear facility operations; housing and realty subdivision development, including temporary residences such as trailer parks and resorts; vector control; milk and food protection; medical care facilities; and recreational facilities including, bathing beaches and swimming pools. Responsibilities of the health and environmental protection department extend to the issuance of permits and the continual approval of operational results to protect the public and enhance the home, work, and recreation environment. Hence, they have an important stake in all planning (preventive environmental sanitation and engineering) to assure that the public does not inherit situations that are impossible or costly to correct.

Comprehensive community planning that gives proper attention to the environmental health considerations, followed by phased detail planning and capital budgeting, is one of the most important functions a community can engage in for the immediate and long-term economy and benefit of its people.

The achievement of the World Health Organization health goal requires the control of all those factors in man's physical environment that exercise or may exercise a deleterious effect on his physical, mental, or social well-being. It is necessary, therefore, that single-purpose and general economic, social, and physical planning take into full consideration the environmental factors affected and the facilities and services required for healthful living.

REGIONAL PLANNING FOR ENVIRONMENTAL ENGINEERING CONTROLS

General

Air, water, and land pollution, inefficient transportation facilities, urban and rural blight, and disease do not respect political boundaries. Adequate highways, land-use controls, park and recreation facilities, water, sewers, solid waste disposal, and other services necessary for proper community functions are usually best designed within the context of a regional plan. Such planning, however, must recognize federal and state agency planning and, to the extent feasible, local planning (including planning by private enterprise). The provision of services, however, can be hampered if legal impediments prevent regional solutions. As long as each city, town, or village can obstruct or curtail intermunicipal cooperation, planning for the future cannot be completely effective. Coordination of planning among smaller communities within the context of the applicable county, metropolitan, and regional plan is essential. The smallest practical planning unit for the development and administration of environmental engineering controls appears to be the county.

Local governments normally do not have jurisdiction and hence cannot plan and budget for an entire metropolitan region. The key to the solution of metropolitan problems is the establishment of an area-wide organization that can investigate, plan, and act on an area-wide basis. Unless this is done regional plans that are developed will probably remain on the shelf for a long time.

Although planning the solution of regional problems on a regional basis is generally accepted as being basically sound, it does not necessarily follow that a special organization or authority must be established to carry out the actual construction, operation, and maintenance of the utility or facility. Local experiences and sentiment may dictate that the only way a regional project can be carried out, in whole or in part, is on an individual community basis regardless of the additional costs involved. The important thing is that the particular project or facility be constructed in general conformance with the regional plan and that it be put into service. If this is done, it will be possible at a later date, when the "climate" is propitious, to realize the more efficient and economical consolidated arrangement.

Content of a Regional Planning Report

At the very least it would appear logical to combine into one comprehensive engineering study evaluation of the regional water supply, wastewater, solid wastes, air quality, and perhaps other items. These all have in common the components listed below and are closely interrelated.

The ideal comprehensive regional plan would be one that combines all the project physical planning studies into one tied in with a comprehensive economic and social development plan. In the absence of such a triad of planning, the project or functional regional plans should take into consideration the economic, environmental, and social factors affecting the planning. In addition the effects of comprehensive or overall regional planning should be recognized and steps taken in the project planning to prevent or alleviate potential deleterious side effects.

An outline of a regional or area-wide planning study and report showing the elements that are common to most functional studies follows:

1. Letter of transmittal to the contracting agency.
2. Acknowledgments.
3. Table of contents.
   a. List of tables.
   b. List of figures.
4. Findings, conclusions, and recommendations.
5. Purpose and scope.
6. Background data and analysis, as applicable.
   a. Geography, hydrology, meteorology, geology, and ground water levels.
   b. Population density and characteristics—past, present, future.
   c. Soil characteristics; flora and fauna.

d. Transportation and mobility; adequacy and effects produced—present and future.

e. Residential, industrial, commercial, recreational, agricultural, and institutional development and redevelopment.

f. Land use—present and future; spread of blight and obsolescence; inefficient and desirable land uses.

g. Drainage, water pollution control, and flood control management.

h. Water resources, multiuse planning and development with priority to water supply; environmental impact.

i. Air and water pollution, sewerage, and solid waste management.

j. Public utilities—electricity, gas, oil, heat—and their adequacy.

k. Educational and cultural facilities, size, location, effects.

l. Economic studies—present sources of income, future economic base and balance, labor force, markets, industrial opportunities, retail facilities, stability.

m. Sociological factors—characteristics, knowledge, attitudes, behavior of the people and their expectations.

n. Local government, political organizations, and laws, codes, ordinances.

o. Special problems, previous studies and findings, background data, including tax structure and departmental budgets.

7. Project study. This would be a regional or area-wide in-depth study of one or more projects or functions such as solid waste management, water supply, recreation, vector control, wastewater, or environmental health. Several examples of comprehensive project studies are outlined below.

8. The comprehensive regional plan.

a. Alternative solutions and plans.

b. Economic, social, and ecologic evaluation of alternatives.

c. The recommended regional plan.

d. Site development and reuse plans.

9. Administration and financing.

a. Public information.

b. Administrative arrangements, management, and costs.

c. Financing methods—general obligation bonds, revenue bonds, special assessment bonds; also grants, incentives, federal and state aid.

d. Cost distribution, service charges and rates; capital costs—property, equipment, structures, engineering and legal services; annual costs to repay capital costs, principal and interest, taxes. Regular and special service charges and rates.

e. Legislation, standards, inspection, and enforcement.

f. Evaluation, research and replanning.

10. Appendices.

a. Applicable laws.

b. Special data.

c. Charts, tables, illustrations.

11. Glossary.

12. References.

Project Study

Some examples of specific regional or area-wide comprehensive single-purpose, functional or project studies are outlined below. These expand on item 7, "Project study." A complete study and report would cover items 1 to 12, listed above.

Comprehensive Solid Waste Study (Refer to items 1 to 12, listed above, for complete study and report.)

1. Additional background information and data analysis, including residential, commercial, industrial, and agricultural, solid wastes.

a. Field surveys and investigations.

b. Existing methods and adequacy of collection, treatment, and disposal and their costs.

c. Characteristics of the solid wastes.

d. Quantities, summary tables, and projections.

e. Waste reduction at source, salvage, and reuse.

2. Solid waste collection, including transportation.

a. Present collection routes, restrictions, practices, and costs.

b. Equipment and methods used.

c. Handling of special wastes.

d. Recommended collection systems.

3. Preliminary analyses for solid waste treatment and disposal.

a. Available treatment and disposal methods (advantages and disadvantages)—compaction, shredding, sanitary landfill, incinerator, high-temperature incinerator, pyrolysis, fluidized bed oxidation, bulky waste incinerator, waste heat recovery, composting, garbage grinders.

b. Pretreatment devices—shredders, hammermills, hoggers, compaction—and their applicability.

c. Disposal of special wastes—automobile, water and wastewater treatment plant sludges, scavenger wastes, industrial sludges and slurries, waste oils, toxic and hazardous wastes, rubber tires, agricultural wastes, pesticides, forestry wastes.

d. Treatment and disposal of industrial wastes. This would normally require independent study by the industry when quantities are large or when special treatment problems are involved.

e. Transfer stations, facilities, and equipment.

f. Rail haul; barge haul; other.

g. Alternative solutions and costs.

4. Review of possible solutions.

a. Social and political factors.

b. Existing and potential best land use within 1500 ft of treatment and disposal site and aesthetic considerations.

c. Site development and reuse plans.

d. Special inducements needed.

e. Preliminary public information and education.

Comprehensive Wastewater Study (See "Content of a Regional Planning Report," above.)

1. Additional background information and data analysis.

a. Field surveys and investigations including physical, chemical, biological, and hydrological characteristics of receiving waters.

b. Existing methods of municipal and industrial wastewater collection, treatment, and disposal.

c. Characteristics of municipal wastes and wastewater volumes, strengths, flow rates.

d. Characteristics of industrial wastes, quantities, and amenability to treatment with municipal wastes.

e. Water pollution control requirements; federal, state, and interstate receiving water classifications.

f. Wastewater reduction, reclamation, reuse.

g. Extent of interim and private, on-lot sewage disposal (adequacy, present and future), including soils suitability.

2. Wastewater collection.

a. Existing collection systems (condition and adequacy), including infiltration, surface, and storm-water flows.

b. Existing collection systems with treatment (condition and adequacy), including infiltration, surface, and storm water flows.

c. Areas or districts needing collection systems and construction timetables.

d. Soils, rock, groundwater conditions.

e. Routing and right-of-way.

f. Storm-water separation feasibility, holding tanks, special considerations, local ordinances, and enforcement.

g. Need for storm-water drainage and collection systems.

3. Preliminary analyses for wastewater treatment and disposal.

a. Treatment plant sites, pollution load, degree of treatment required, land requirements including buffer zone, foundation conditions, outfall sewer, hydrologic and oceanographic considerations.

b. Areas served.

c. Trunk lines and pumping stations.

d. Property and easement acquisition problems.

e. Design criteria.

f. Industrial waste flows and pretreatment required, if any.

g. Effect of storm-water flows on receiving waters and need for holding tanks or treatment.

h. Treatment plant and outfall sewer design considerations.

i. Grit, screening, and sludge disposal.

j. Alternative solutions, total costs, and annual charges.

Comprehensive Water Supply Study (See "Content of a Regional Planning Report," above.)

1. Additional background information and data analysis.

a. Field surveys and investigations of distribution system for cross-connections, water pressure, breaks, water usage, storage adequacy.

b. Occurrence of water-borne diseases and complaints.

c. Leak survey of existing system, flow tests, and condition of mains.

d. Existing and future land uses; service areas, domestic and industrial water demands.

e. Areas and number of people served by individual well-water systems, sanitary quality and quantity of water, chemical and physical quality, cost of individual treatment.

f. Recommendations of the American Insurance Association (formerly National Board of Fire Underwriters) and others.

g. Existing fire rates and reductions possible.

2. Alternate sources of water.

a. Chemical, bacteriological, and physical quality ranges.

b. Average, minimum, and safe yields; source development.

c. Storage needed at source and on distribution system.

d. Flow requirements for fires.

e. Service areas, hydraulic analysis, and transmission system needs.

f. Preliminary designs of system and treatment required for taste, odor, turbidity, color, etc.

g. Right-of-way and water rights needed.

h. Preliminary study of total construction and operation costs of each alternate; advantages and disadvantages of each; annual cost to user and how apportioned.

i. Improvements needed on existing system—source, storage, transmission, treatment, distribution system, operation and maintenance, costs.

Comprehensive Environmental Engineering and Health Planning (Additional information under "Content of a Regional Planning Report," above.)

1. Epidemiological survey including mortality, morbidity, births and deaths, age and sex distribution, chronic diseases, and specific incidences of diseases; population distribution of diseases, social, economic, and environmental relationships; respiratory, water-, insect-, and food-borne diseases; animal and animal-related diseases; air-borne and air-related diseases and illnesses; pesticide and other chemical poisonings; health services and their availability; adequacy of data and programs.

2. Public water supply, treatment and distribution including population served, adequacy, operation, quality control, cross-connection control, storage and distribution protection, operator qualifications. For individual systems—population served, special problems, treatment and costs, adequacy, control of well construction. Extension of public water supply based on a comprehensive regional plan, including fire protection, to replace inadequate and unsatisfactory small community water systems and individual well-water supplies in built-up areas. See Chapter 3.

3. Wastewater collection, treatment, and disposal; adequacy of treatment and collection system, population served, operator qualifications, sewer connection control. For individual systems—population served, special problems, control of installations. Water pollution control. Provision of sewerage meeting surface water and groundwater classifications based on a drainage area or regional plan to eliminate pollution by existing discharges, including inadequate sewage and industrial waste treatment plants and septic tank systems. See Chapter 4.

4. Solid waste management—storage, collection, transportation, processing and disposal, adequacy. Salvaging and recycling, including municipal refuse, industrial and agricultural wastes; handling of hazardous wastes and prevention of air, water, and land pollution. Use of solid wastes to accelerate construction of open-space buffer zones and recreation areas. See Chapter 5.

5. Air resources management and air pollution control including sources, air quality and emission standards, topographical and meteorological factors; problems and effects on man, livestock, vegetation and property; regulation and control program. See Chapter 6.

6. Housing and the residential environment—control of new construction, housing conservation and rehabilitation, enforcement of housing occupancy and maintenance code, effectiveness of zoning controls and urban renewal. Realty subdivision develop-

ment and control, also effect of development on the regional surroundings and effect of the region on the subdivision, including the environmental impact of the subdivision. See Chapter 11 and "Fringe and Rural Area Housing Developments" in this chapter.

7. Recreation facilities and open-space planning, including suitability of water quality and adequacy of sewerage, solid waste disposal, water supply, food service, restrooms, safety, and other facilities. See Chapter 9 and "Environmental Factors in Site Selection and Planning" in this chapter.

8. Food protection program—adequacy from source to point of consumption. See Chapter 8.

9. Nuclear energy development; radionuclide and radiation environmental control, including fallout, air, water, food, and land contamination; thermal energy utilization or dissipation and waste disposal; naturally occurring radioactive materials; air, water, plant, and animal surveillance; federal and state control programs; standards; site selection and environmental impact; plant design and operation control; emergency plans. See Chapter 7.

10. Planning for drainage, flood control, and land-use management. Surface water drainage to eliminate localized flooding and mosquito breeding. Development of recreation sites, including artificial lakes, parks, swimming pools, bathing beaches, and marinas.

11. Public health institutions and adequacy of medical care facilities such as hospitals; nursing homes; public health, mental health, and rehabilitation centers; clinics; service agencies. Staffing, budgets, work load.

12. Noise and vibration abatement and control.

13. Noxious weed, insect, rodent, and other vermin control, including disease vectors, nuisance arthropods; regulation, control, and surveillance, including pesticide use for control of aquatic and terrestrial plants and vectors; federal, state, and local programs; effects of water, recreation, housing, and other land resource development. See Chapter 10.

14. Natural and man-made hazards, including slides, earthquakes, brush and forest fires, reservoirs, tides, sand storms, hurricanes, tornadoes, high rainfall, fog and dampness, high winds, gas and high-tension transmission lines, storage and disposal of explosive and flammable substances and other hazardous materials.

15. Aesthetic considerations; wooded and scenic areas, prevailing winds and sunshine.

16. Laws, codes, ordinances, rules, and regulations. See pages 70 and 71.

17. Environmental health and quality protection; adequacy of organization and administration. See Chapter 12.

Financing

Financing for a municipal capital improvement is generally done by revenue bonds or general obligation bonds. Sometimes, as for small projects, funds on hand are used or a special assessment is levied. Other arrangements include a combination of revenue and general obligation bonding; the issuance of mortgage bonds using the physical utility assets as collateral; the creation of a nonprofit corporation or authority with the power to sell bonds; and contract with a private investor to build a structure or facility for lease at an agreed on cost, with ownership reverting to the municipality at the end of a selected time period.

Usually the constitution of a state or other law contains a specific limitation in regard to the amount of debt that a municipality may incur. The debt limit may be set at approximately 5 to 10 percent, and the operating expenses at about 2 percent, of the average full value of real estate. The debt margin established generally does not include bonded indebtedness for schools. In most cases, bond issues for capital improvements require approval of the state fiscal officer to determine if the proposed project is in the public interest and if the cost will be an undue burden on the taxpayer. The ability to pay or per capita financial resources can be expected to vary from one municipality to another. Sometimes indebtedness for an essential revenue-producing service, such as water supply, is excluded from the constitutional limit.

Revenue bonds are repaid from a specific source of revenue, such as water and sewer charges. The service made possible by the bond issue is therefore directly related to the monthly, quarterly, or annual billing for a specific service. The bonds are not backed by the full credit of the municipality. Interest rates can therefore be expected to be somewhat higher than for general obligation bonds.

General obligation bonds are issued by a governmental agency, and their payment is guaranteed by the municipality through its taxing powers. The money thus borrowed is repaid by all of the people in a community, usually as additions to the real property tax. The total amount of general obligation bonds that may be issued by a municipality is generally limited by law. General obligation bonds may also be used to pay off a revenue-producing capital improvement. Approval by the voters in a special referendum may be required.

Combination of revenue and general obligation bonds would generally use revenue to pay off principal and interest, but would also be paid from general tax funds to the extent needed if the revenue is inadequate. This can result in lower interest rates. Other arrangements and cost apportionment are possible.

ENVIRONMENTAL FACTORS IN SITE SELECTION AND PLANNING

Certain general basic information should be known before a suitable site can be selected. One should know the present and future capacity and the total land area desired. The type of establishment to be maintained and the use to which it will be put—whether a new subdivision, shopping plaza, an adult or children's camp, a park and picnic area, resort hotel, dude ranch, marina, trailer park, country club, factory, school, institution, or private home—and the activities and programs to be carried on during the summer, winter, or all year are to be decided on before any property is investigated. In addition one should know the radius in miles or time of travel within which the area must be located; the accessibility to roads, airports, or railroads; availability of utilities; permanency of the project; moneys available; whether lake, river, or stream frontage is necessary; and whether the site need be mountainous, hilly, flat, wooded, or open.

Desirable Features

It probably will not be possible to find a site that will meet all the conditions authorities recommend as being essential. Desirable features include:

1. An adequate groundwater or surface water supply not subject to excessive pollution that can be developed into a satisfactory supply at an accessible and convenient location on the property, if an adequate public water supply is not available.

2. A permeable soil that will readily absorb rainwater and permit the disposal of sewage and other wastewater by conventional subsurface means is most desirable, if not essential, for the smaller establishment where public sewerage is not available. Such soil will contain relatively large amounts of sand and gravel, perhaps in combination with some silt, clay, broken stones, or loam. The underground water should not be closer than 4 ft of the ground surface at any time and there should be a porous earth cover of not less than 4 or 5 ft over impervious subsoil or rock. A suitable receiving stream or land area is needed if a sewage treatment plant is required.

3. Land to be used for housing or other structures must be well above flood- or high-water level. There should be no nearby swamps.

4. Elevated, well-drained, dry land open to the air and sunshine part of the day, on gently sloping, partly wooded hillsides or ridges, should be available for housing and other buildings. The cleared land should have a firm, grass-covered base to prevent erosion and dust. A slope having a southern or eastern exposure protected from strong winds on the north and west is desirable.

5. The area of the property should be large enough to provide privacy, avoid crowding, accommodate a well-rounded program of activities, and allow for future expansion. The property should be accessible by automobile and bus and convenient to airports, superhighways, railroads if needed, and recreation facilities.

An allowance of 1 acre/camper has been suggested as being adequate for children's camps. For elementary schools, 1 acre/100 pupils, and, for high schools, 10 acres plus 1 acre/100 pupils is recommended. The play area should provide 1000 ft^2/child using the area at any one time. For a residential area, ¾ acre of public playgrounds, 1¼ acres of public playfields, and 1 acre of public park land/1000 population are considered minimums. Other suggested standards for parks are given in Table 2–1. Standards (more correctly criteria) should be considered points of departure, to be interpreted in the light of the economic and social structure of the people affected and adjusted accordingly.

6. A satisfactory area should be available for bathing and swimming and other water sports at recreational sites. This may be a clean lake, river, or stream

TABLE 2-1 SUGGESTED STANDARDS FOR PARKS

Type of park	Area of facility	Service radius	Population served
Play or tot lot	2,500 ft^2 minimum	1/8 mi	25 to 75 children
Neighborhood	8 to 15 acres	1/4 mi—high density 3/8 mi—low density	2,000 to 5,000
District	15 to 40 acres	1/2 to 1 mi average	15,000 to 35,000
Community	100 to 500 acres	1 1/2 to 2 1/2 mi	1,000/1/4 acre
Special area (golf course, marina, stadium, parkway)	35 to 175 acres	1 to 1 1/2 mi	—
School			
Elementary	8 to 15 acres	1/4 to 3/8 mi	—
Junior High	10 to 25 acres	3/8 to 1 mi	—
Senior High	25 to 50 acres	1/2 mi minimum	—

Sources: *Suggested Standards for Parks and Recreation,* National Recreation and Park Association. U.S. Dept. of the Interior, *Outdoor Recreation Space Standards,* Government Printing Office, Washington, D.C., 1970.

or an artificial swimming pool. A river or stream should not have a strong current or remain muddy during its period of use. An artificial swimming pool equipped with filtration, recirculation, and chlorination equipment may be substituted to advantage.

7. Noxious plants, poisonous reptiles, harmful insects, excessive dust, steep cliffs, old mine shafts or wells, dangerous rapids, dampness, and fog should be absent. All this is not usually possible to attain; however, the seriousness of each should be considered.

8. A public water supply, sewerage system, and solid waste disposal system, if available and accessible, would be extremely desirable.

9. For residential development, electricity, gas, and telephone service; a sound zoning ordinance and a land use plan that provides for and protects compatible uses; fire protection; and modern building construction and housing codes vigorously enforced by competent people should all be assured.

10. Air pollution, noise, and traffic problems from adjoining areas should not interfere with the proposed use.

Topography

A boundary survey of the property with contours shown at 5-ft intervals, in addition to roads, watercourses, lakes, swamps, woodlands, structures, railroads, power lines, rock outcrops, and any other significant physical features indicated would be of very great value in studying a property. If such a map is not available, a U.S. Geological Survey sheet (scale: 1 in. = $\frac{1}{2}$ mi. or 1 mi, contour interval = 10 ft) that has been blown up or an aerial photograph (approximately 1 in. = 1660 ft) from the U.S. Department of Agriculture or other agency may be used instead for the preliminary study. The outline of the property should be marked on the map, using deed descriptions. Old plot plans of the property that are available and distinctive monuments or other markings that can be found by inspection of the site would add valuable information. A long-time resident or cooperative neighbor may also be of assistance.

With the map as a beginning, one should hike over as much of the area as possible and carefully investigate the property. Be sure to keep complete notes that refer to numbers placed on the topographic map showing beautiful views and other desirable features. Supplementary freehand sketches and rough maps of possible camp, recreation, or building sites with distances paced off will be valuable details.

The bathing area should be sounded and slope of the bottom plotted. Need for cleaning and removal of mud, rocks, and aquatic growths should be noted. The drainage area tributary to a lake or stream to be used for bathing should be determined and the probable minimum contribution or flow computed to ascertain if an ample quantity of water will be available during the dry months of the year.

In most cases the watershed area tributary to a beach on a stream or lake will extend beyond the boundary of the land under consideration. It is important therefore to know what habitation, agriculture, and industry is on the watershed. The probable land usage and pollution of tributary streams should then be determined because persons owning land have the right to reasonable use of their property, including streams flowing through it. This will bring out whether the stream or lake is receiving chemical, bacterial, or physical pollution that would make it dangerous to use for bathing or water supply purposes. In order to obtain this valuable information it is necessary to make a survey on foot of every stream and brook on the watershed. The local health or conservation department sanitary engineer, conservation officer, or sanitarian may be able to give assistance. In addition to quantity and bacterial quality, the water should be relatively clear and slow moving. Study of the stream bottom and float or weir measurements to determine the velocity and quantity of water will give this information.

While a survey is being made, one should look for signs indicating the high-water level of the lake or stream. The topography of the ground, presence of a flood plain, stranded tree trunks and debris, discolorations on rocks and trees, width, depth, and slope of the stream channel, coupled with probable maximum flow, elevation of nearby railroad beds, and type of vegetation growing, may give good clues. Valuable information can be obtained by discussing this point with long-time residents.

Geology, Soil, and Drainage

The soil should be sampled at representative locations to determine its characteristics. Borings should be made to a depth of about 15 ft in order to record variations in the strata penetrated and the elevation of the groundwater level, if encountered, with respect to the ground surface. Borings and posthole and earth auger tests will also indicate the depth to rock and the presence and thickness of clay or hardpan layers that might interfere with proper drainage or foundations for structures.

In addition to borings, soil studies and soil percolation tests should be made in areas that, as indicated by the topography are probably suitable for subsurface sewage disposal, if needed. Explanation of the soil's characteristics and percolation tests and application of the results are discussed in Chapter 4. Public sewerage should be used if possible.

If a proposed development or building is to be located on the side of a long hill or slope, the necessity and feasibility of providing an earth berm or dam, or deep surface water drainage ditches to divert surface water around the site, should be kept in mind. The possibility of earth slides, flooding, erosion, and washout should be given careful study. Buildings or surrounding land on slopes with a greater than 8 percent incline require special engineering study and treatment such as erosion control, drainage, and vegetative cover. The need for and practicability of constructing special foundations, a surface and curtain drain, or a subsurface drainage system to lower the groundwater table, or the need for draining wet and swampy areas (or the feasibility of making an artificial lake) will be apparent from the data accumulated in the topography study.

Utilities

The existence of or need for a water supply system, a sewage disposal system, a solid waste disposal system, roads, electricity or a generator of electricity, gas, oil, coal, and telephones should be studied for they will determine the type of establishment, services, and sanitary facilities that could be provided.

Needless to say if an adequate satisfactory, and safe water supply is not obtainable at a reasonable cost, with or without treatment, the site should be abandoned. This is particularly important to a factory dependent on large volumes of water. See Chapter 3.

The probable cost of a wastewater collection and treatment system should be estimated before any commitments are made. Water classification standards for water pollution abatement will govern the degree of treatment required and hence cost of construction and operation. For a large project, an elaborate wastewater treatment plant may be required. For small establishments a subsurface sewage disposal system may suffice if the soil conditions are satisfactory. An alternative at pioneer-type camps might be the use of privies. See Chapter 4.

Electricity for lighting or for the operation of water pumps, kitchen equipment, refrigerator compressors, and other mechanical equipment is usually taken for granted. If the provision of electricity means the running of long lines, purchase of an electric-generating unit, or gasoline motor-driven equipment, then the first cost, cost of operation, maintenance, and replacement should be estimated. The availability of gas, oil, coal, and telephone service should also be determined.

Roads to the main buildings are needed for access and for bringing in supplies. The distance from the main roads to the property and the length and condition of secondary roads within the property should be determined.

If a power plant or the industrial process will cause air pollution, air pollution control requirements, prevailing winds, temperature, and related factors will have to be studied and cost of treatment devices determined in evaluating the suitability of a proposed site. See Chapter 6. If a plant process will result in the production of large quantities of solid wastes, the treatment and disposal of the residue must also be considered. See Chapter 5.

Meteorology

Slopes having an eastern or southern exposure in the United States are to be preferred for building locations to get the benefit of the morning sun. This possibility can be ascertained by inspection of available topographic maps and by field surveys. Information about the direction of prevailing winds is of value if a summer or winter place is proposed. An indication of the wind direction can be obtained by observing the weathering of objects and the lean of the trees. This information, plus average monthly temperature, humidity, and rainfall data, may be available at local universities, nearby government weather stations, airfields, and at some water or power company offices. Where rainfall data is not available it may be possible to utilize stream flow measurements to judge the general pattern of precipitation in the area.

Location

The relative location of the property can best be appreciated by marking its outline on a recent U.S. Geological Survey Sheet. In this way the distance to railroad and bus stations, first-class roads, shopping centers, neighbors, resort areas, schools, hospitals, and doctors is almost immediately apparent. State and county highway department road plans should also be investigated and proposed roads marked on the topographic plan to see what their probable effect will be.

Resources

To determine the resources on a property (when it is a large tract), one should seek the assistance of a person who is intimately familiar with wildlife, forestry, geology, and engineering. Since this is not always possible, the next best thing would be for several persons having a broad knowledge to make the survey on foot. The location, size, and type of woodland, pasture, rock, sand, and gravel should be carefully noted. The woodland may also serve as building material or firewood; the rock as roadbed or foundation material; the pasture as a recreation area or golf course; the sand and gravel for concrete and cement work or road surfacing with admixtures if needed. The availability of such material near the construction site will result in a considerable saving. Not to be forgotten as resources are water for domestic, recreational, and power purposes and unspoiled scenery.

Animal and Plant Life

Here again, a knowledge of these things is necessary before a worthwhile study can be made. The presence of poison ivy, for example, must be accepted as a potential source of skin irritations; ragweed and other noxious plants must be accepted as sources of hay fever. It is therefore important to mark the location of the infested areas on a topographic map so that attention is directed to their existence in the construction and planning program. The same would apply to the presence of mosquitoes, flies, ticks, chiggers, rodents, poisonous reptiles, and dangerous animals. On the other hand equal emphasis can be placed on the presence of wild flowers, useful reptiles and wild animals, and native trees. A broad knowledge and extensive investigation is needed to properly evaluate the importance of animal and plant life on a property.

Improvements Needed

Inasmuch as the ideal school, institution, industry camp, or housing site is rarely if ever found, the work that needs to be done to make an ideal site should be determined. An undesirable feature may be a low swampy area. This may be filled in if the area is not too large and suitable fill is available on the property; the ground may be drained if the topography or subsoil strata make this possible; the area may be cleared, dammed up, and made into a lake; or the swamp may be retained as a wildlife preserve. Each possibility should be examined, the cost of the work estimated, and the probable value of the improvement appraised. Needed clearing, seeding, or reforestation and the extent of poison ivy, ragweed, thistle, and other objectionable plants should also be considered. If a natural bathing area is not available on a lake or stream on the property, it is advisable to compare the cost of developing and maintaining one with the cost of an artificial swimming pool. If a lake is available, then thought must be given to clearing, the construction of a beach, and perhaps dredging and shore development to keep down heavy aquatic growths and plant life such as algae. The need for new roads inside the property connecting with town, county, or state roads is another consideration. In some cases this may involve blasting, fill, bridges, and special construction, all of which could mean high costs. Other needed physical improvements may be surface water diversion ditches, culverts, groundwater drainage tile, brush clearing, boat docks, parking areas, and the preparation of areas for recreational purposes. Also important is the environmental impact of the proposed land use.

Site Planning

After properties have been explored and studied and a site is selected, the next step is to have prepared, if not already available, a complete large-scale topographic map of the purchased property to a scale of 1 in. equal to 100 ft, with contours at least at 5-ft intervals, incorporating the details already discussed. After this, put down on paper what the ideal place is to consist of. Lay out on the map the approximate location of the proposed establishment, making maximum use of the natural advantages offered by the site. Then, without losing sight of the purpose the property is to serve and the program to be followed, prepare cutouts made to scale to represent plans of present and future buildings, roads, parking spaces, recreation areas, campsites, water supply, sewage and refuse disposal areas, bathing beaches, and other facilities. The preparation of a scale model also has great promotional possibilities. It will help visualize the buildings and their relationship to each other and to roads, streams, lakes, hills, and neighboring communities. Changes can be easily made on paper. This method has been used successfully to help obtain funds for major improvements and new developments.

It is advisable to seek the advice of an engineer, architect, planner, or other consultant trained and experienced in planning work. One should confer with the local planning, health, and building departments to learn of the regulations for the protection of the life, health, and welfare of the people. This is particularly important when a housing development, industrial plant, camp, hotel, shopping center, school, or other public place is contempated. The agency staff may also be in a position to offer valuable assistance. The plan should show, in addition to the items mentioned, such other details that may be required by the departments having jurisdiction to permit their staff to review the plans and pass on them favorably. After approved plans are received, a construction schedule can be established and the work carried on in accordance with a preconceived, carefully thought-out plan. There will then be no looking back at useless, inefficient, wasteful structures; instead, the eventual realization of the "ideal" place by adding to what has already been accomplished will be anticipated.

The importance of obtaining competent professional advice cannot be emphasized too strongly. All too often charitable camps and institutions seek and expect free engineering and architectural services. This is not only unfair to the individuals consulted, but is also unfair to the camp or institution. The consultants would unconsciously try to conclude the planning and the design as rapidly as possible; and the recipient of the service would proceed on the basis of possibly incompletely studied and conceived plans, detailed drawings, specifications, and contract documents. A proper place for free expert advice would be in the selection of the engineer, architect, contractor, and equipment, and in helping to evaluate the bids received for the particular job. The fee for a consultant's services, in comparison to the cost of a project, is relatively small. The reduction of waste caused by incorrect size of buildings, poor planning of equipment and facilities, foundation and structural weaknesses, and bad location or exposure of buildings, as well as savings in the selection of proper equipment for minimum operating costs and assurance that plans and specifications are being complied with, far overshadows the professional fee.

FRINGE AND RURAL-AREA HOUSING DEVELOPMENTS

Growth of Suburbs and Rural Areas

Census data show that the rate of population growth of many cities has been leveling off, but that the population of suburbs has been increasing. In 1966 the Bureau of the Census estimated that 60 million Americans lived in metropolitan-area central cities as compared to 58 million in 1960; 66 million lived in suburbs as compared to almost 59 million in 1960. The more desirable property adjacent to water and sewer service is usually very limited, making it necessary to use more distant areas for development. Public water and sewer extensions and other planning considerations have not kept pace with the land development in many instances. As a result, many dwellings use individual sewage disposal systems. The 1970 census shows continuation of the 1966 trends.

There is a natural urge within a family to have its own home in the suburbs or in the country. Land in the suburbs and rural areas is considered inexpensive and the taxes low. But the cost of constructing a private well and sewage disposal system; the lack of adequate fire, police, street cleaning, and refuse collection service; and the future need for constructing new schools and providing additional sanitary facilities are at first overlooked, as are the increased taxes to provide these services and facilities.

Facilities and Services Needed

For example, every 1000 new people in a community will require:[12]

1. An additional supply of 100,000 to 200,000 gal of water daily; or 35 to 70 million gal/yr, or 300 individual water well systems, many equipped with water conditioners.

2. The collection and disposal of 4000 to 6000 lb of solid wastes daily; or 730 to 1100 tons/yr.

3. Recreational facilities to serve more people with more leisure time.

4. Sewage treatment works to handle 100,000 to 150,000 gpd, or 35 to 53 million gal of sewage/yr containing 170 lb of organic matter (biochemical oxygen demand) per day or 62,000 lb/yr, and 70,000 lb of dry sewage solids/yr or 300 additional septic tanks and appurtenant subsurface disposal facilities.

5. Expenditure of $20,000 to control the sources or air pollution, or $65,000 per year to offset the physical damage caused by the lack of air pollution control.

6. 4.8 new elementary schoolrooms, 3.6 new high-school rooms, and additional teachers.

7. 10.0 or more acres of land for schools, parks, and play areas.

8. 1.8 policemen and 1.5 firemen; also new public service employees in public works, welfare, recreation, health, and administration.

9. More than a mile of new streets.

10. More streets to clean and free of snow and ice, and to drain.

11. Two to four additional hospital beds, three nursing home beds, and appurtenant facilities.

12. 1000 new library books.

13. More automobiles, retail stores, services, commercial and industrial areas, county and state parks, and other private enterprises.

[12] Adapted from *United States Municipal News*, **22**, 16 (August 15, 1955); and "Environmental Health in Community Growth," *Am. J. Public Health*, **53**, No. 5 (May 1963). This is intended to be illustrative and not necessarily typical.

The average home does not yield enough in taxes to pay for the municipal services provided and demanded by the people. Encouragement of industry to locate in selected areas in a town can relieve the tax burden. Desirable types of industries are those that will not cause air, noise, or water pollution or make special demands and have a high assessed valuation.

Industry also contributes to the economic growth of an area through increased personal income, bank deposits, and retail sales by attracting professional, skilled, and semiskilled manpower. This means more households, people including schoolchildren, automobiles, retail stores, and hence jobs.

Causes and Prevention of Haphazard Development

Fringe and rural-area sanitation is an aspect of healthful living that, like the housing problem in cities, requires the combined talents of many private as well as official agencies and individuals to prevent insanitary conditions and promote good health. Those who can contribute to this goal are the developers and builders, consulting engineers and land planners, health department sanitary engineers and sanitarians, local officials, private lending institutions, federal agencies, planning and zoning boards, legislators and attorneys, the press, and an informed home-buying public.

Where controls are lacking, subdivisions are in many cases poorly planned and without utilities, drainage, or streets worthy of the name. Some lots may be under water, on rock shelves or tight clay soil, or poorly drained. Overflowing septic-tank systems and polluted wells are commonplace. Many people "get stuck," and when they seek legal redress they are confronted with a maxim of English and American law, caveat emptor—"let the purchaser beware," that is, let him examine the article he is buying and act on his own judgment and at his own risk. An official agency operating in the public interest cannot adopt such a principle.

The premature development of subdivisions causes an accumulation of tax arrears on vacant lots, thereby shifting the financial burden for governmental expenditures on other properties.[13] The owners of other properties are therefore forced to make up the tax money lost by paying higher taxes or lose their home or place of business, even though they had no part in the land speculation and could not have benefited had the venture been successful.

If the cost and liabilities for subdivision are borne by those who engage

[13] Philip H. Cornick, "Problems Created by Premature Subdivision of Urban Lands in Selected Metropolitan Districts," State of New York Department of Commerce, Bureau of Planning, 112 State Street, Albany, 1938.

in subdivision for profit, rather than by the general public, bona fide developers are aided and irresponsible unscrupulous developers are discouraged. It is therefore proper that reasonable controls be invoked to control land subdivision for the general health and welfare of the people.

The generally accepted methods of preventing subdivision problems and obtaining orderly community growth are education and effective regulation. But laws must be understood and be enforceable; that is, they must be reasonable and fair to the developer and investor, to the purchaser of homes, and to the local community. It must be recognized that the owner of land for sale is primarily desirous of subdividing for profit. The banks and mortgage insurers are concerned that dwellings in a subdivision and in surrounding areas maintain a high value so as to guarantee a continuing satisfactory return on their investments. The homeowner wants to live and bring up his children in a pleasant, healthful environment; and the community is interested in seeing that a subdivision does not turn out to be a liability.

Many problems can be prevented or made less difficult to solve if the conditions surrounding the problem are more fully understood. Giving publicity to existing laws, including explanation of their purpose and how to comply with them, is one approach. Attorneys, lending institutions, realtors, builders and subdividers, local officials and service clubs, land planners and consulting engineers, the home-buying public, and the local press are some of the important groups that can make more effective a subdivision law. The local health department and planning agency, through consultation, advice, leaflets, and letters and by personal contact, can explain the intent of the law and expedite compliance with it. Some excellent bulletins (see references at end of this chapter) are available to the consulting engineer, land planner, builder, and subdivider for their information and guidance.

When considering the many educational measures that can be applied to control or solve subdivision problems, one must not forget the opportunities available to the administrative and enforcing agencies to prevent the problems in the first place. Immediate notification of the subdivider at the first sign of activity, followed by strict surveillance, can prevent many problems from getting out of hand. Informal conferences with interested parties, explanation of the advantages of filing proper plans and public sewer and water supply, and the offer of advice to help solve problems will usually convince the reasonable individual he has little to lose and much to gain by complying with the law. In some instances a hearing before the proper board or commissioner will give the desired result; in rare instances legal action is necessary.

A city, village, or town may also adopt a local sewage and wastes ordinance as part of its sanitary code, building code, plumbing code, zoning, or planning ordinance, depending on which is available and effective. This is done in cooperation with the health department. The individual planning to put in an individual sewerage system is required to make application to a designated department for a permit to build, at which time he also makes application to the health department for approval of a private sewage disposal or treatment system based upon soil percolation tests and a recommended design. The health department, registered architect, or professional engineer makes design recommendations and then inspections during construction. A certificate of compliance is issued by the health department when the system is properly installed. No dwelling or structure is permitted to be occupied or any facility or appurtenance used until the issuance of the health department certification. Potential subdivisions are then brought to the attention of the health department and control measures instituted.

An educational and persuasive tool is reference to legal decisions and opinions to help convince the skeptic. Numerous decisions have been handed down affirming the right of the health department to control subdivision development, to require engineering plans of proposed water supply and sewerage, and to require installation of sewers rather than septic-tank systems.[14]

Health department realty subdivision laws may be enacted on a state level and enforced on both a state and local level, or the laws may be enacted and enforced on a local level, where a competent engineering division is provided in the local health department. The former procedure may be preferred when political and administrative problems are anticipated. Sample laws, regulations, and guidelines are available from many state and local health and planning agencies.

Cooperative Effort Needed

Developers and builders can simplify their sanitation problems by selecting vacant land that is within or contiguous to communities already having water and sewer services. This makes possible the economical extension of these utilities; smaller lots, if desired, than would be required with individual wells and septic-tank leaching systems and hence more lots; and a more desirable development. If the extension of water or sewer lines is not possible, then the developer or builder can in many cases construct a water system or sewers and treatment plant, with the cooperation of local officials. In such cases maintenance

[14] J. A. Salvato, Jr., "Experiences with subdivision regulations," *Public Works,* p. 126 (April 1957).

and operation of the systems should be guaranteed. It is also possible to construct these facilities on a planned step-by-step construction basis so as to reduce or spread out the financial burden of the initial plant cost.

Consulting engineers are sometimes reluctant to take a subdivision job involving the design of water supply and sewerage systems. The time and effort needed to work out satisfactory plans may appear to be too great for the fee that can be reasonably charged. Actually this is where the professional engineers can perform a real public service by wise advice and sound design. Here is an opportunity for the application of imagination and initiative by not only engineers, but also land planners, builders and contractors, investors, and equipment manufacturers in designing and producing less expensive and more efficient sanitary devices as well as a more desirable living environment for population groups involving 50 to 1000 or more persons.

Local officials can encourage and direct proper development by sympathetic cooperation, by assisting in the formation of water and sewer districts, and through the enforcement of planning and zoning ordinances. Enlightened public officials are no more desirous of encouraging insanitary and haphazard construction than the health department. The submergence of sectional jealousies, the formation of drainage-area sanitary districts crossing town, city, or village boundary lines, and annexation, if necessary, are additional aides. In this connection, county and regional planning offer a great deal of promise. Policy agreements regarding certain services can be made among contiguous cities, villages, and townships through a county planning commission or similar agency. Development of certain lands could be encouraged by mutual understanding to extend more complete fire and police protection, sewers and water mains, snow-plowing service, and refuse collection only to areas designated for development. The county real estate division can cooperate by encouraging the purchase of vacant tax-delinquent land in these delineated areas. Developers and builders would be required to do their part by agreeing to certain deed restrictions, improvements, and zoning, and also to share in the cost of constructing water and sewer lines. Participating communities could properly require larger lots outside of areas without water and sewer lines and agree to the review and approval of all subdivision plots by the coordinating agency.

Involved federal agencies and lending institutions are obligated to comply with state and local sanitary code regulations. They usually withhold payments where sanitary facilities do not meet existing legal requirements. Attorneys, realtors, and mortgage and title investigators can protect their clients' interest and help make sounder investments if the adequacy of existing or proposed water supply and sewage disposal facilities at a property are also investigated before investment is recommended.

County, city, and state health department environmental engineers and sanitarians have the responsibility to enforce the state and local public health laws and sanitary codes. They can give assistance and guidance to the developer, builder, and engineer and help interpret the intent and compliance with the laws. Consultation with health department engineers concerning the supply of water and disposal of sewage, and with the highway department concerning drainage of surface water, will expedite the submission and approval of satisfactory subdivision plans. Health department personnel are often called on to talk before service clubs, professional groups, and community associations. This offers opportunities to explain the protection provided by existing laws and the services rendered by the health department to promote a better way of life through a healthful environment.

Subdivision Planning

Haphazard development can be controlled by comprehensive regional planning if broad public powers and regulatory authority are provided. In the absence of such control, public or private acquisition of large potential sites and their preparation for multipurpose uses, including residential development, has merit. Raw land would be planned for residential, recreational, open-space, commercial, industrial, agricultural, or supporting uses. This would then be followed by development and improvement with roads, water supply, drainage, sewers, and other utilities including waste water and solid waste disposal facilities. The land would then be sold to private enterprise for development as *planned*.

In the planning of a housing development careful consideration should be given to the site selection and planning factors discussed earlier in this chapter. The subdivision is dependent on the region and its central community for employment and cultural needs, and the central community is dependent on the surrounding development for human resources and economic survival. Controlled environmental conditions in a newly developed subdivision can only be assured if the neighboring inhabitants live under equally desirable conditions. No matter how remote a new subdivision may be from the urban centers, it is still an extension of the existing environment and is affected by it. The subdivision, in turn, influences the surrounding area environment by the character of its growth and the standards of its facilities and services. This interdependence must always be kept in mind.

Planning and zoning regulations are based on the rights of government to exercise their police power to control the private uses of land for the promotion of the general health, safety, and welfare. A planning board can be empowered to prepare a comprehensive plan to guide the physical development of land and public services within a delineated area. In carrying out this function the planning board may prepare an official map; make investigations, sketches, and reports relating to planning and community development; review and make recommendations for approval of plans of proposed subdivisions; advise concerning capital budget and long-term capital improvement programs; define problems; and conduct research and analyses.

Enforcement tools available to the planning board are subdivision regulation, the zoning code, and guidance of improvement programs relating to parks, major highways, sewers and water supply, including treatment plants, and desirable type of land development and redevelopment.

Land planning is a specialized activity. Frequently the importance of proper design is ignored by the developer or its significance is not realized, yet nothing could be more fundamental in the subdivision of land. A future community being created can be a good place in which to live and work and in which to invest in a home; or it can be an undesirable subdivision right from the start. The subdivision designer should determine if the land is suitable for housing, business, or industry, or should be retained as farmland. The demand for the intended use should govern whether immediate or future development is indicated. Then the designer should fit the proposed project in with the community zoning, major streets and parkways, recreational areas, churches, schools, shopping centers, and so forth, and design the interior elements of the subdivision accordingly. Due regard must be given to topography, roads, lot sizes and shapes, water supply, sewage disposal, gas and electricity, drainage, recreational areas, schools, easements for utilities, trees, and landscaping, as well as the environmental impact of the proposed land use.

Figure 2–3 illustrates the different ways in which a plot of land may be subdivided. An increased number of lots and a reduced cost of public improvements per lot are common results of good design. This of course is advantageous to the developer, the homeowner, and the community.

The legal steps and authority for the formation of a planning board are usually given in state laws. Ordinances creating local planning boards are obtainable from state planning agencies where established. Planning-board suggested subdivision regulations should also be available from this source. Any regulations considered for adoption should be adapted to the local conditions.

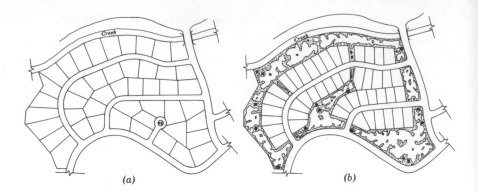

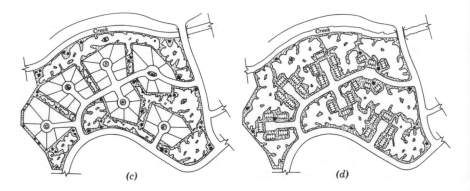

(*a*) Standard subdivision. The area is completely subdivided into 46 lots. No open space and only a few lots have access to the creek.

(*b*) Density zoning. Same number of conventional lots but with half the area in open space.

(*c*) True cluster development. Again 46 lots on 23 acres, but lots are grouped.

(*d*) 110 townhouses on same parcel. Over half the land is untouched, and at 4.7 units per acre, this would be rather crowded if these were single-family units. On clear, flat land, quality townhouse developments can run 10 to 14 units per acre with low cost projects even higher.

Various possibilities exist for subdividing this net area of 23 acres. The one chosen will affect the amount of open space regardless of the number of units.

Figure 2–3 Types of subdivision development. [Reprinted from the article "Trends in Residential Development" by George C. Bestor, *Civil Engineering*, ASCE (September 1969).]

Design of Water and Sewerage Service for Subdivisions

Water and sewerage service for a subdivision can be provided in many different ways. The preferred method is the extension of existing water lines and sewers. If the property is not located within the boundaries of an existing service district, the facilities may become available by annexation to the central municipality, through the formation of an improvement district, or utility corporation if permitted by the existing laws. Where a county or regional service district or authority exists, it would be the controlling agency. If a public water supply or sewerage system is not available, the developer can construct his own facilities, including treatment plants.

The central city fortunate enough to have a water or sewage works has an obligation to serve its suburbs. But it also has the right to maintain sufficient control over the extensions to protect itself and the right to charge for the service what is necessary to obtain a return on its investment. Mathews has discussed the pros and cons of water service extension to the suburbs in some detail.[15] He points out that the obligation to serve suburban consumers on the part of the city is mixed with resentment and resistance, but the usual result is that the service is furnished. Sometimes annexation is made a prerequisite. As to the right to control the distribution system beyond city limits, it is realistically stated that the individual suburban consumer is inclined to blame the city water department for all failures in service, even though the cause may be inadequate distribution facilities in the suburban area. Since inhabitants of the area generally have business and other connections in the city, pressure is brought to bear on the water department to rectify an impossible condition.

The design of small water systems is given in Chapter 3. Experiences in new subdivisions show that peak water demands of six to ten times the average daily consumption rate are not unusual. Lawn-sprinkling demand has made necessary sprinkling controls, metering, or the installation of larger distribution and storage facilities and, in some instances, ground storage and booster stations. As previously stated, every effort should be made to serve a subdivision from an existing public water supply. Such supplies can afford to employ competent personnel and are in the business of supplying water, whereas a subdivider is basically in the business of developing land and does not wish to become involved in operating a public utility.

In general, when it is necessary to develop a central water system

[15] C. K. Mathews, "Water Service Policies for Suburban Areas," *Jour. Am. Water Works Assoc.*, 48, 174–178 (February 1956).

to serve the average subdivision, consideration should first be given to a drilled well-water supply. Infiltration galleries or special shallow wells may also be practical sources of water. Such water systems usually require a minimum of supervision and can be developed to produce a known quantity of water of a satisfactory sanitary quality. Simple chlorination treatment will normally provide the desired factor of safety. Test wells and sampling will indicate the most probable dependable yield and the chemical and bacterial quality of the water. Well logs should be kept.

Where a clean, clear lake supply or stream is available, chlorination and slow sand filtration can provide reliable treatment with daily supervision for the small development. The turbidity of the water to be treated should not exceed 30 ppm. Preliminary settling may be indicated in some cases.

Other more elaborate types of treatment plants, such as rapid sand filters, are not recommended for small water systems unless specially trained operating personnel can be assured. Pressure filters have limitations, as explained in Chapter 3.

The design of small slow sand filter and well-water systems is explained and illustrated in Chapter 3.

Where a public sewerage system is not available or accessible, design of a central sewerage system including treatment should be given careful study. The cost of sewers per dwelling can usually be divided in half, and lot sizes may be made smaller than is necessary when a well and septic-tank system are installed on every lot, thereby making available more lots. In addition such property is more desirable, a better investment, and less likely to cause public health nuisances due to overflowing septic-tank systems.

Various treatment methods have been used to serve subdivisions, depending on the number of persons served, the degree of treatment required, and the type of supervision provided. For the small development of up to 50 homes, a plant consisting of a septic tank, dosing tank, open or covered sand filters, and chlorine contact tank—or a package-type aeration plant, has been found practical. For a housing development of 50 to 300 homes, a plant consisting of an Imhoff tank or primary settling tank, standard rate filter with provision for recirculation, and a secondary settling tank, or a package type plant, followed by chlorination has been preferred. For larger developments, methods that have been used are the standard-rate trickling filter plant; the high-rate trickling filter plant; the activated sludge plant; the aerobic digestion plant; and variations of these processes, followed by oxidation ponds or intermittent sand filters with chlorination. See Chapter 4 for design details.

There are many rural areas that are remote from population centers and public water supplies and sewers. In such cases individual well water and septic-tank sewage disposal systems offer the only practical answer for the immediate future. However, to be acceptable individual sanitary facilities must be carefully designed, constructed, and maintained in accordance with good standards. Where the soil is unsuitable for the disposal of sewage by conventional subsurface means, every effort must be made to prevent the subdivision of land until public sewers, including treatment and drainage if needed, can be provided.

In some situations public water supply is available, leaving only the problem of sewage disposal. In rarer situations public sewers are available, making necessary construction of individual wells. Although the sanitary problems are simplified in such cases, there is still need for careful design, construction, and maintenance of the individual facility that needs to be provided.

When either individual well or septic-tank systems are required, or when both are required, the preparation of a proper subdivision plan makes possible the proper and orderly installation of these facilities on individual lots. When homes are built in accordance with the approved plan, the intent of a subdivision control law is accomplished in the interest of the future homeowners and community. Some health departments have issued practical instructions to designing engineers regarding the preparation and submission of plans for realty subdivisions.[16] Such information simplifies the submission of satisfactory plans and also has a salutary educational effect on the designing engineer who has not had extensive experience in subdivision planning and design.

Details included on a subdivision plan are a location map, topography, typical soil-boring profile, lot sizes, roads, drainage, and location of soil tests and results; typical lot layouts showing the location of the house, well, sewage disposal system, and critical materials and dimensions; sketches showing drilled well development and protection, pump connection and sanitary seal, depth to water and yield, or a community water system; and sewage disposal units, design, and sizes, including septic tank, distribution box, tile field, or leaching pit. Construction details of drilled wells, including pump connections and sanitary seals, are given and illustrated in Chapter 3. Septic-tank, distribution box, tile field, and leaching pit details are given and illustrated in Chapter 4.

There is a danger in the preparation of subdivision plans, as in other engineering and architectural plans, to copy details blindly. Such care-

[16] *Planning the Subdivision as Part of the Total Environment,* Division of General Engineering and Radiological Health, New York State Department of Health, Albany, 1970.

lessness defeats the purpose of employing a consultant or designing engineer and is contrary to professional practice. Subdivision plans involving individual wells and sewage disposal systems must be adapted to the topography and geological formations existing at the particular property. It is well known that no two properties are exactly alike; hence each requires careful study and adaptation of general principles and typical details so that a proper engineering plan results. For example, the soil percolation tests determine the type and size of the required sewage disposal system. The slope of the ground determines the relative locations of the wells and sewage disposal systems on each lot; in most cases these must be established on the plot plan to prevent future interference. The type of well, required minimum depth of casing, and the need for cement grouting of the annular space around the outside of the casing, and the sealing of the bottom of the casing in solid rock vary with each property. These are just a few considerations to bring out the need for the application of engineering training and experience so as to perform a proper professional service.

BIBLIOGRAPHY

Camp Sites and Facilities, Boy Scouts of America, New York, 1950, 90 pp.

The Comprehensive Plan: A Guide for Community Action, State of New York, Office of Planning Coordination, Albany, 1966, 22 pp.

Dickerson, Bruce W., "Selection of Water Supplies for New Manufacturing Operations," *J. Am. Water Works Assoc.,* **62**, 10, 611–615 (October 1970).

Environmental Health Aspects of Metropolitan Planning and Development, *WHO Rep. Ser.,* No. 297 (1965).

"Environmental Health in Community Growth," *Am. J. Public Health,* **53**, No. 5, 802–822 (May 1963).

Environmental Health Planning, U.S. Public Health Service Pub. No. 2120, Dept. of HEW, Washington, D.C., 1971, 139 pp.

Logan, John A., Oppermann, Paul, and Tucker, Norman E., *Environmental Engineering & Metropolitan Planning,* Northwestern University Press, Evanston, Ill., 1962, 265 pp.

Outdoor Recreation Space Standards, Bureau of Outdoor Recreation, Dept. of the Interior, Washington, D.C., April 1967, 67 pp.

Part III, Appraisal of Neighborhood Environment, American Public Health Association, New York, 1950, 132 pp.

McKeever, Ross J., *Community Builders Handbook,* Urban Land Institute Publication, Washington, D.C., 1968, 526 pp.

Salomon, Julian Harris, *Camp Site Development,* Girl Scouts of America, New York, 1959, 160 pp.

Salvato, Joseph A., Jr., "Environmental Health and Community Planning," *J. of the Urban Planning and Development Division,* ASCE, **94**, No. UP 1, Proc. Paper 6084, 23–30 (August 1968).

Salvato, Joseph A., Jr., Smith, Peter J., and Cohn, Morris M., *Planning the Subdivi-*

sion as Part of the Total Environment, New York State Department of Health, Albany, 1970, 77 pp.

Urban Planning Guide, ASCE, Manuals and Reports on Engineering Practice, No. 49, American Society of Civil Engineers, New York, 1969, 299 pp.

Where Not to Build, a Guide for Open Space Planning, Tech. Bull. 1, U.S. Dept. of the Interior, Washington, D.C., April 1968, 160 pp.

Gallion, Arthur B. and Eisner, Simon, *The Urban Pattern,* City Planning and Design, D. Van Nostrand Company, Inc., New York, 1963, 435 pp.

McLean, Mary, *Local Planning Administration,* The International City Managers' Association, Chicago, 1959, 467 pp.

Chapin, Stewart F., *Urban Land Use Planning,* University of Illinois Press, Urbana, 1965, 498 pp.

Anderson, Richard T., *Comprehensive Planning for Environmental Health,* Division of Urban Studies, Cornell University, Ethaca, New York, 1964, 203 pp.

Shomon, Joseph James, *Open Land for Urban America,* an Audubon book published in cooperation with the National Audubon Society by The Johns Hopkins Press, Baltimore, 1971, 171 pp.

Soil, Water, and Suburbia, A Report of the Proceedings of the Conference Sponsored by the United States Department of Agriculture and the United States Department of Housing and Urban Development, June 15 and 16, 1967, Washington, D.C., March 1968, 160 pp.

3

WATER SUPPLY

Introduction

A primary requisite for good health is an adequate supply of water that is of satisfactory sanitary quality. It is also important that the water be attractive and palatable to induce its use; otherwise, water of doubtful quality from a near-by unprotected stream, well, or spring may be used. Where a municipal water supply passes near the property, the owner of the property should be urged to connect to it because such supplies are usually under competent supervision.

When a municipal water supply is not available, the burden of developing a safe water supply rests with the owner of the property. Frequently private supplies are so developed and operated that full protection against dangerous or objectionable pollution is not afforded. Failure to provide satisfactory water supplies in most instances must be charged either to negligence or to ignorance because in the long run it generally costs no more to provide a satisfactory installation that will meet with good health department standards.

A survey made by the U.S. Public Health Service in 1962 points out the number and type of public water supplies in the United States and the populations they served.[1] There were 19,236 public water supplies

---

[1] *Statistical Summary of Municipal Water Facilities in the United States,* U.S. Public Health Service, Pub. No. 1039, Dept. of HEW, Washington, D.C., January 1, 1963.

serving approximately 150 million people;[2] 75 percent were groundwater supplies, 18 percent were surface water supplies, and 7 percent were a combination. Of significance is the finding that 75 million people in communities under 100,000 population were served by 18,873 public water supplies, and 77 million people in communities over 100,000 were served by 399 public water supplies. Also, of the 19,236 supplies, 85 percent served communities of 5000 or less. This information emphasizes the need for giving at least as much attention to small public water supplies as is given to large supplies. In addition, millions of people on vacation in relatively uncontrolled rural environments depend on small water supplies that are often not under close surveillance.

The U.S. Public Health Service completed a study in 1970 covering 969 small to large public surface and groundwater supply systems serving 18.2 million persons (12 percent of the total United States population served by public water supplies) and 84 special systems serving trailer and mobile home parks, institutions, and tourist accommodations.[3] Although the drinking-water supplies in the United States rank among the best in the world, the study showed the need for improvements. Based on the 1962 USPHS Drinking Water Standards it was found that, in 16 percent of the 969 communities surveyed, the water quality exceeded one or more of the mandatory limits established for coliform organisms (120 systems), fluoride (24), and lead (14). It is of interest to note that of the 120 systems that exceeded the coliform standard, 108 served populations of 5000 or less and that 63 of these were located in a state where disinfection was not frequently practiced or was inadequate. An additional 25 percent of the systems exceeded the recommended limits for iron (96 systems), total dissolved solids (95), manganese (90), fluoride (52), sulfate (25), and nitrate (19). The study also showed that 56 percent of the systems were deficient in one or more of the following: source protection, disinfection or control of disinfection, clarification (removal of suspended matter) or control of clarification, and pressure in the distribution system. It was also reported that 90

[2] Leaving about 36 million people dependent on individual water supplies. Cornelius W. Krusé estimates that in 1960 31.4 million people were served by home and farm water systems and 12.6 million people relied on nonpiped water supplies; *Advances in Environmental Sciences Technology*, Vol. 1, edited by James N. Pitts, Jr. and Robert L. Metcalf, Wiley-Intersciences, New York, 1970. A 1965 report shows that about 153 million people in the United States and Puerto Rico were served by public water supplies and 42 million people by individual water supplies; *Estimated Use of Water in the United States, 1965,* Circular 556, Geological Survey, U.S. Dept. of the Interior, 1968.

[3] Leland J. McCabe, James M. Symons, Roger D. Lee, and Gordon G. Robeck, "Survey of Community Water Supply Systems," *J. Am. Water Works Assoc.,* 670–687 (November 1970).

percent of the systems did not have sufficient samples collected for bacteriological surveillance; 56 percent had not been surveyed by the state or local health department within the last 3 years. In 54 percent crossconnection prevention ordinances were lacking; in 89 percent reinspection of existing construction was lacking; in 61 percent the operators had not received any water treatment training; in 77 percent the operators were deficient in training for microbiological work and 46 percent of those who needed chemistry training did not have any. The smaller communities had more water-quality problems and deficiencies than the larger ones, showing the advisability of consolidation and regionalization when feasible.

A safe and adequate water supply for 2 billion people, about two-thirds of the world's population, is still a dream. The availability of any reasonably clean water in the less developed areas of the world just to wash and bathe would go a long way toward the reduction of such scourges as scabies and other skin diseases, yaws and trachoma, and high infant mortality. The lack of safe water makes commonplace high incidences of shigellosis, amebiasis, schistosomiasis, leptospirosis, infectious hepatitis, typhoid, and paratyphoid fever.[4] (See also Chapter 1.) It is believed that the provision of safe water supplies—accompanied by a program of birth control—could vastly improve the living conditions of millions of people in developing countries of the world.[5]

Travel of Sewage Pollution through the Ground

Since the character of soil and rock, quantity of rain, rate of groundwater flow, amount of pollution, bacteria growth media, and other factors beyond control are variable, one cannot say with certainty through what thickness or distance sewage must pass to be purified. Pollution travels a short distance through fine sand or clay; but it will travel indefinite distances through coarse gravel, fissured rock, dried-out cracked clay, or solution channels in limestone.

The U.S. Public Health Service conducted experiments at Fort Caswell, North Carolina, in a sandy soil with groundwater moving slowly through it. The sewage organisms—coliform bacteria—traveled 232 ft, and chemical pollution as indicated by uranin dye traveled 450 ft.[6] The chemical pollution moved in the direction of the groundwater flow largely in the

[4] Abel Wolman, "Water Supply and Environmental Health," *J. Am. Water Works Assoc.*, 746–749 (December 1970).
[5] G. E. Arnold, "Water Supply Projects in Developing Countries," *J. Am. Water Works Assoc.*, 750–753 (December 1970).
[6] C. W. Stiles, H. R. Crohurst, and G. E. Thomson, *Experimental Bacterial and Chemical Pollution of Wells via Ground Water, and the Factors Involved*, U.S. Public Health Service Bull. No. 147, Dept. of HEW, Washington, D.C., June 1927.

upper portion of the groundwater and persisted for 2½ years. The pollution band did not fan out but became narrower as it moved away from the pollution source. It should be noted that in these tests there was a small draft on the experimental wells and that the soil was a sand of 0.14 mm effective size that had a uniformity coefficient of 1.8.

Studies of pollution travel were made by the University of California using twenty-three 6-in. observation wells and a 12-in. gravel-packed recharge well. Diluted primary sewage was pumped through the 12-in. recharge well into a confined aquifer having an average thickness of 4.4 ft approximately 95 ft below ground surface. The aquifer was described as pea gravel and sand having a permeability of 1900 gal/ft²/day. Its average effective size was 0.56 mm and uniformity coefficient, 6.9. The median effective size of the aquifer material from 18 wells was 0.36 mm. The maximum distance of pollution travel was 100 ft in the direction of groundwater flow and 63 ft in other directions. It was found that the travel of pollution was not affected by the groundwater velocity but by the organic mat that built up and filtered out organisms, thereby preventing them from entering the aquifer. The extent of the pollution then regressed as the organisms died away and as pollution was filtered out.[7]

Butler, Orlob, and McGauhey made a study of the literature and reported the results of field studies to obtain more information about the underground travel of harmful bacteria and toxic chemicals.[8] The work of other investigators indicated that pollution from dry pit privies did not extend more than 1 to 5 ft in dry or slightly moist fine soils. However, when pollution was introduced into the underground water, test organisms traveled to wells 50, 69, 80, 112, and 232 ft away. Chemical pollution was observed to travel 300 to 450 ft, although chromate was reported to have traveled 1000 ft in 3 years, and other chemical pollution 3–5 mi. Leachings from a garbage dump in groundwater reached wells 1476 ft away, and a 15-year-old dump continued to pollute wells 2000 ft away. Studies in the Dutch East Indies report the survival of coliform organisms in soil 2 years after contamination and their extension to a depth of 9 to 13 ft, in decreasing numbers, but increasing again as groundwater was approached. The studies of Butler et al. tend to confirm previous reports and have led the authors to conclude "that

[7] *Report on the Investigation of Travel of Pollution,* Pub. No. 11, State Water Pollution Control Board, Sacramento, Calif., 1954.

[8] R. G. Butler, G. T. Orlob, and P. H. McGauhey, "Underground Movement of Bacterial and Chemical Pollutants," *J. Am. Water Works Assoc.,* **46,** 2, 97–111 (February 1954).

the removal of bacteria from liquid percolating through a given depth of soil is inversely proportional to the particle size of the soil."

When pumping from a deep well, the direction of groundwater flow around the well will be toward it. Since the level of the water in the well will probably be 25 to 150 ft more or less below the ground surface, it will exert an attractive influence on groundwater perhaps as far as 400 to 1000 ft away from the well, regardless of the elevation of the top of the well. In other words, distances and elevations of sewage disposal systems must be considered relative to the elevation of the water level in the well while it is being pumped.

A World Health Organization Report reminds us that in nature atmospheric oxygen breaks down accessible organic matter and that topsoil (loam) contains organisms that can effectively oxidize organic matter.[9] However, these benefits are lost if wastes are discharged directly into the groundwater by way of sink holes, pits, or wells.

From the investigations made, it is apparent that the safe distance between a well and a sewage disposal system is dependent on many variables, including chemical, physical, and biological processes.[10] Factors to be considered in arriving at a satisfactory answer include:

1. The amount of sand, clay, and organic (humus) matter, loam in the soil, the soil structure and texture, the effective size and uniformity coefficient, and soil depth largely determine the ability of the soil to remove bacterial pollution deposited in the soil. Whereas the relationship between sand size and bacterial removal has received study in connection with the operation of slow sand water filters, no field relationship is known for soils.

2. The volume, strength, type and dispersion of the polluting material, as well as the distance, elevation, and time for pollution to travel with relation to the groundwater level and flow and soil penetrated, are important. Also important is the volume of water pumped and well drawdown.

3. The well construction, tightness of the pump connection, and sealing of the annular space have a very major bearing on whether a well will be polluted by sewage and surface water.

Considerable judgment is needed to select a proper location for a well. The limiting distances given in Chapter 4 should therefore be used as a guide. Experience has shown them to be reasonable and effective

---

[9] S. Buchan and A. Key, "Pollution of Ground Water in Europe," *WHO Bull.,* **14,** Nos. 5–6, 949–1006 (1956).
[10] A summary of the distances of travel of underground pollution is also given in the Task Group Report, "Underground Waste Disposal and Control," *J. Am. Water Works Assoc.,* **49,** 1334–1341 (October 1957).

in most instances when coupled with proper interpretation of available hydrologic and geologic data and good well construction, location, and protection. See Figure 3–1 for groundwater terms.

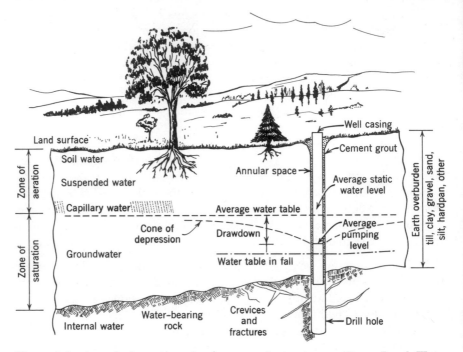

Figure 3–1 *A geologic section showing groundwater terms.* (From *Rural Water Supply,* New York State Dept. of Health, Albany, 1966.)

Disease Transmission

Water, to act as a vehicle for the spread of a specific disease, must be contaminated with the disease organisms from infected persons. All sewage-contaminated waters must be presumed to be potentially danger-ous. Other impurities such as dissolved minerals and heavy concentra-tions of decaying organic matter may also find their way into a water supply, making the water unattractive, harmful, or otherwise unsuitable for domestic use. Diseases that may be spread by water are discussed in Chapter 1 and are listed in the folded insert.

Water Cycle and Its Characteristics

The movement of water can be best illustrated by the hydrologic or water cycle shown in Figure 3–2. Using the clouds and atmospheric

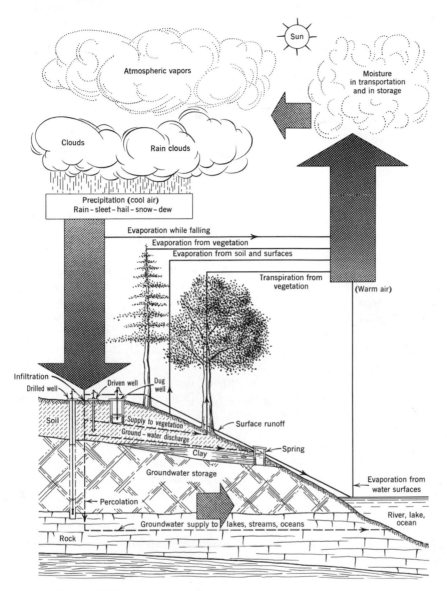

Figure 3–2 The hydrologic or water cycle.

vapors as a starting point, moisture condenses out under the proper conditions to form rain, snow, sleet, hail, frost, fog, or dew. Part of the precipitation is evaporated while falling; some of it reaches vegetation foliage, the ground, and other surfaces. Moisture intercepted by surfaces is evaporated back into the atmosphere. Part of the water reaching the ground surface runs off to streams, lakes, swamps, or oceans from whence it evaporates; part infiltrates into the ground and percolates down to replenish the groundwater storage, which also supplies lakes, streams, and oceans by underground flow. Groundwater in the soil helps to nourish vegetation through the root system. It travels up the plant and comes out as transpiration from the leaf structure and then evaporates into the atmosphere. In its cyclical movement, part of the water is temporarily retained by the earth, plants, and animals to sustain life. The average annual precipitation in the United States is about 30 in., of which 72 percent evaporates from water and land surfaces and transpires from plants, and 28 percent contributes to the groundwater recharge and stream flow.[11] See also Evapotranspiration Chapter 4.

When speaking of water, we are generally concerned primarily with surface water and groundwater, although rain water is also considered. In falling through the atmosphere rain picks up dust particles, plant seeds, bacteria, dissolved gases, ionizing radiation, and chemical substances such as nitrogen, oxygen, carbon dioxide, and ammonia. Hence, rain water is not pure water as one might think; it is, however, very soft. Water in streams, lakes, reservoirs, and swamps is known as surface water. Water reaching the ground and flowing over the surface carries anything it can move or dissolve. This may include waste matter, bacteria, silt, soil, vegetation, and microscopic plants and animals. The water accumulates in streams or lakes. Sewage, industrial wastes, and surface and groundwater will cumulate, contribute to the flow, and be acted upon by natural agencies. Water flowing over the ground may also find its way to lakes or reservoirs where bacteria, suspended matter, and other impurities settle out. On the other hand, microscopic as well as macroscopic plant and animal life grow and die, thereby removing and contributing impurities in the cycle of life.

Part of the water reaching and flowing over the ground seeps down to form the groundwater, also called underground water. In percolating through the ground, water will dissolve materials to an extent dependent on the type and composition of the strata through which the water has passed and the quantity of water. Groundwater will therefore usually

[11] *Hydrology Handbook,* Manual 28, American Society of Civil Engineers, New York, 1949.

contain more dissolved minerals than surface water. The strata penetrated may be unconsolidated, such as sand, clay, and gravel, or consolidated, such as sandstone, granite, and limestone. A brief explanation of the classification and characteristics of formations is given below.

Igneous rocks are those formed by the cooling and hardening of molten rock masses. The rocks are crystalline and contain quartz, feldspar, mica, hornblende, pyroxene, and olivene. Igneous rocks are not usually good sources of water, although basalts are exceptions. Small quantities of water are available in cracks and fissures.

Sedimentary formations are those resulting from the deposition and accumulation of materials weathered and eroded from older rocks by water, ice, or wind and the remains of plants, animals, or material precipitated out of solution. Sand and gravel, clay, silt, chalk, limestone, fossils, gypsum, peat, shale, loess, and sandstone are examples of sedimentary formations. Deposits of sand and gravel generally yield large quantities of water. Sandstones, shales, and certain limestones may yield abundant groundwater, although results may be erratic depending on density, porosity and permeability of the rock.

Metamorphic rocks are produced by the alteration of igneous and sedimentary rocks, generally by means of heat and pressure. Gneisses and schists, quartzites, slates, marble, serpentines, and soapstones are metamorphic rocks. A small quantity of water is available in joints, crevices, and cleavage planes.

Shale and slate have a porosity of about 0.075 gal of water per cubic foot. Sandstone has a porosity of 0.0125 to 0.375 gal of water per cubic foot. Granite has a porosity of 0.0007 to 0.015 gal of water, and sand, 0.625 to 0.712 gal of water per cubic foot. Voids make up as much as 30 percent of the volume in sand and gravel.

Water Quality

The U.S. Public Health Service's drinking water standards dealing with the source, protection, and quality of water supplies, have been developed by a committee consisting of representatives of federal and state organizations, scientific associations, and individual specialists. Drinking water and water supply systems used by railroads, ships, planes, and buses crossing state lines and others subject to federal quarantine regulations must conform to these standards.[12] Bacterial, physical, chemical, and microscopic examination are discussed and interpreted under those respective headings.

[12] *Public Health Service Drinking Water Standards,* U.S. Public Health Service Pub. No. 956, Dept. of HEW, Washington, D.C., 1962. (Being revised in 1972)

Sanitary Survey and Water Sampling

In order to properly interpret water analyses, a sanitary survey of the area draining to the water supply is necessary. The value of the survey is dependent on the training and experience of the person making the investigation. When available, one should seek the advice of the health department sanitary engineer or sanitarian.

If the source of water is a lake, attention would be directed to the location of sewage and waste disposal or treatment systems, bathing areas, storm-water drains, sewer outfalls, swamps, cultivated areas, pastures, and wooded areas in reference to the pump intake; for each would contribute distinctive characteristics to the water. When water is obtained from a stream or creek, all land and habitation above the water supply intake should be investigated. This means inspection of the entire watershed drainage area so that actual and potential sources of pollution can be determined and properly evaluated. All surface water supplies must be considered of doubtful sanitary quality unless given adequate treatment, depending on the type and degree of pollution received.

Groundwater supplies, such as wells or springs, should be investigated with a view toward finding ways whereby the source might be polluted. A complete sanitary survey should include inspection of the drainage area and habitation, local geology and vegetation, nature of soil and rock strata, well logs, evidence of blasting, slope of water table, sources of pollution, and development of the source.

The survey would include the pumping station, treatment plant, and adequacy of each unit; operation records; distribution system; storage facilities; cross-connections with other water supplies; and actual or possible connections with plumbing fixtures that might permit back-siphonage. Where water treatment is provided, the integrity and competence of the person in charge of the plant is an important factor.

Water samples are collected as an adjunct to the sanitary survey and as an aid in measuring the quality of the raw water and effectiveness of treatment given water. Bacterial examinations, chemical and physical analyses, and microscopic examination may be made depending on the problem encountered and the purpose to be served. In any case, all tests should be made by an approved laboratory in accordance with the procedure given in *Standard Methods for the Examination of Water and Wastewater*.[13]

A sanitary technique and a glass or plastic sterile bottle supplied

[13] Published by the American Public Health Association, 1015 Eighteenth Street, N.W., Washington, D.C., 20036; 1971.

by the laboratory for the purpose should be used when collecting a water sample. The hands or faucet must not touch the lip of the bottle or the plug part of the stopper. The sample should be taken from a clean faucet that is not leaking or causing condensation on the outside. Let the water run for about 10 min to get a representative sample. If a sample from a lake or stream is desired, the bottle should be dipped below the surface with a forward sweeping motion so that water coming in contact with the hands will not enter the bottle. When collecting samples of chlorinated water, the sample bottle should contain sodium thiosulfate to dechlorinate the water. It is recommended that all samples be examined promptly after collection and within 6 to 12 hr if possible.

The chemical and physical analyses may be for industrial or sanitary purposes, and the determinations made will be either partial or complete, depending on the information desired. A 2- to 4-liter sample (approximately 2 to 4 qt) collected in a clean bottle supplied by the laboratory, which is thoroughly rinsed with the water to be examined, is usually required. A representative sample should be collected either by dipping about 12 in. below the water surface or from the depth desired, using a special sampling apparatus, creating as little disturbance as possible, or from a faucet at a point nearest the source after the water has been let run for 10 min. In any case, the container should be completely filled and examined by the laboratory within 12 hr; however if the water is not polluted, as much as 3 days may elapse. Selected tests are also made for the routine control and operation of water plants to determine the need for and effectiveness of treatment.

Microscopic examination is used to determine the kinds and quantity of microscopic organisms present. Since microscopic organisms secrete certain oils that cause characteristic tastes and odors, contributed to by the decomposition of dead organisms, the microscopic examination can lead to the prevention of objectionable growths. This is done by eliminating or controlling favorable factors and determining the time to start periodic treatment. Samples for microscopic examination should be collected in clean wide-mouth bottles having a volume of 1 or 2 liters from depths that will yield representative organisms. Some organisms are found relatively close to the surface, whereas others are found at mid-depth or near the bottom, depending on the food, type of organisms, clarity, and temperature of the water.

Interpretation of Water Analyses

As indicated previously, the interpretation of water analyses is based primarily on a sanitary survey of the water supply. A water supply that is coagulated and filtered would be expected to be practically clear,

TABLE 3-1 SOME BACTERIAL AND CHEMICAL ANALYSES

Source of sample	Dug well	Lake	Reservoir	Deep well	Deep well
Time of year	–	April	October	–	–
Treatment	None	Chlor.	None	None	None
Bacteria per ml Agar 35°C. 24 hrs	–	3	–	1	>5000
Coliform MPN per 100 ml	–	<2.2	–	<2.2	2400 or >
Color, units	0	15	30	0	0
Turbidity, units	Trace	Trace	Trace	Trace	5.0
Odor, cold	2 veget.	2 aromatic	1 veget.	1 aromatic	3 disagreeable
Odor, hot	2 veget.	2 aromatic	1 veget.	1 aromatic	3 disagreeable
Iron, mg/l	0.15	0.40	0.40	0.08	0.2
Fluorides, mg/l	<0.05	0.05	–	–	–
Nitrogen as ammonia, free, mg/l	0.002	0.006	0.002	0.022	0.042
Nitrogen as ammonia, albuminoid, mg/l	0.026	0.128	0.138	0.001	0.224
Nitrogen as nitrites, mg/l	0.001	0.001	0.001	0.012	0.030
Nitrogen as nitrates, mg/l	0.44	0.08	0.02	0.02	0.16
Oxygen consumed, mg/l	1.1	2.4	7.6	0.5	16.0
Chlorides, mg/l	17.0	5.4	2.2	9.8	6.6
Hardness (as $CaCO_3$), total, mg/l	132.0	34.0	84.0	168.0	148.0
Alkalinity (as $CaCO_3$), mg/l	94.0	29.0	78.0	150.0	114.0
pH value	7.3	7.6	7.3	7.3	7.5

colorless, and free of iron, whereas the presence of some turbidity, color, and iron in an untreated surface-water supply may be accepted as normal. A summary is given above of the constituents and concentrations considered significant in water examinations. Other compounds and elements not mentioned are also found in water.

Bacterial Examination

The bacterial examination may include the count of total colonies of bacteria developing from measured portions (1 ml) of the water being tested, which have been planted in petri dishes with a suitable culture media (agar) and incubated for 48 hr at 20°C, or for 24 hr at 35°C. The test is of significance when used for comparative purposes under known or controlled conditions.

The bacterial examination of drinking water should always include a quantitative estimation of total organisms of the coliform group, which are *indicative* of fecal contamination. The coliform group includes all of the aerobic and facultative anerobic, gram-negative, non-spore-forming, rod-shaped bacteria that ferment lactose with gas formation within 48 hr at 35°C.[14] This is the presumptive test that can be confirmed and completed by carrying the test further as outlined in *Standard Methods*.

A more rapid method for detecting specific organisms utilizes the membrane filter technique. A special membrane filter that will remove bacteria from water filtered through it is available. By incubation of the filter membrane on special media it is possible to make coliform determinations in about 18 hr, rather than in the 3 to 4 days required by the conventional completed and confirmed test procedure. The isolation of *Salmonella typhosa* and other organisms by cultivation on a selected medium is also possible. Suspended matter, algae, and bacteria in large amounts interfere with the membrane filter procedure.

The fecal coliform test involves incubation at 44.5°C for 24 hr and measures essentially all coliforms in a freshly passed stool of warm-blooded animals. A loop of broth from a positive presumptive tube incubated at 35°C in the total coliform test is transferred to EC medium and incubated at 44.5°C; formation of gas within 24 hr indicates the presence of fecal coliform and hence possibly dangerous contamination. Nonfecal organisms generally do not produce gas at 44.5°C. The test has greatest application in the study of stream pollution, raw water sources, sea waters, sewage treatment plants, and the quality of bathing waters.

The test for fecal streptococci (enterococci) uses special agar media incubated at 35°C for 48 hr. Dark red to pink colonies are counted as fecal streptococci. They are also normally found in the intestinal tract of animals, including man. These organisms may be more resistant to chlorine than coliform and survive longer in some waters.

The analysis for total coliforms would indicate the presence of feces of human and warm-blooded animals, and coliform associated with vegetation, soils, joint and valve packing materials, slimes, swimming pool ropes, pump leathers, sewage, storm-water drainage, surface-water runoff, surface waters, and others. The tests for fecal coliforms and fecal streptococci are helpful in interpreting the significance of surface-water tests and their possible hazard to public health.

In domestic sewage, the fecal coliform concentration is usually at least four times the fecal streptococci and may constitute 30 to 40 percent of the total coliforms. In storm water and in wastes from livestock,

---

[14] This is sometimes also referred to as the *B. coli* group and as the *coli-aerogenes* group. If the membrane filter technique is used, the bacteria produce a dark colony with a metallic sheen within 24 hr on an Endo-type medium containing lactose, instead of fermenting lactose with gas formation in the multiple-tube test.

poultry, animal pets, and rodents the fecal coliform concentration is usually less than 0.4 the fecal streptococci. In streams receiving sewage, fecal coliforms may average 15 to 20 percent of the total coliforms in the stream. The presence of fecal coliform generally indicates fresh and possibly dangerous pollution. The presence of intermediate-aerogenes-cloacae (IAC) subgroups of coliform organisms suggests past pollution or, in a municipal water supply, defects in treatment or in the distribution system.[15]

The presence of any of the coliform organisms in drinking water is a danger sign; it must be carefully interpreted in the light of a sanitary survey and promptly eliminated. There may be some justification for permitting a low coliform density in developing areas of the world where the probability of other causes of intestinal diseases greatly exceeds those caused by water, as determined by epidemiological information. It should also be recognized that a negative coliform test is no assurance that viruses such as infectious hepatitis are absent. In general, a coagulated, filtered, and chlorinated water supply or a properly developed well-water supply should have no coliform organisms. However, a coliform most probable number (MPN) of less than 2.2 (not more than 1 by the membrane filter technique) and a bacteria count of less than 100/100 ml of sample are acceptable. The results of a bacterial examination reflect the quality of the water only at the time of sampling and must be interpreted in the light of the sanitary survey.

Physical Examination (See also pages 154 and 155)

Odor

Odor should be absent or very faint for the water to be acceptable; it should be less than 3 standard units.

Taste

The taste should not be objectionable. Algae, decomposing organic matter, dissolved gases, or industrial wastes may cause tastes and odors. Bone and fish oil and kerosene are particularly objectionable. Phenols in concentrations of 0.2 ppb in combination with chlorine will impart a phenolic or medicinal taste to drinking water.

Turbidity

The public demands a sparklingly clear water. This implies a turbidity less than 1 unit; less than 0.1 unit, which is obtainable when water

---

[15] E. E. Geldreich, *Sanitary Significance of Fecal Coliforms in the Environment,* U.S. Federal Water Pollution Control Administration Pub. No. WP-20-3, Dept. of the Interior, Washington, D.C., November 1966.

is coagulated, settled, and filtered, is practical. Values less than 5 units, however, are acceptable to many people. A turbidity greater than 15 to 25 units would justify treatment of the water. It is a good measure of sedimentation, filtration, or storage efficiency; suspended solids are a more precise measure. The presence of turbidity in well or spring water indicates inadequate protection from surface wash and possible bacterial contamination. In drilled wells turbidity may be caused by improper development, iron, microscopic growths, or failure to cement-grout the annular space. In a reservoir it may call for microscopic examination to determine the need for treatment.

Color

Color should be less than 15 color units, although persons accustomed to clear water will notice a color of only 5 units. Water that has drained through peat bogs, swamps, forests, or decomposing organic matter may contain brownish or reddish stain due to tannates and organic acids dissolved from leaves, bark, and plants. Coagulation, settling, and rapid sand filtration should reduce the color in water to less than 5 units. Slow sand filters should remove about 40 percent of the color.

Macroscopic and Nuisance Organisms

Larvae, crustacea and large numbers of algae or filamentious growths should not be allowed. Tastes, odors, and appearance are affected.

Temperature

The water temperature should preferably be less than 60°F.

Chemical Examination (See also pages 154 and 155)

Hardness

Hardness is due primarily to calcium and magnesium carbonates and bicarbonates (carbonate hardness, which can be removed by heating) and calcium sulfate, calcium chloride, magnesium sulfate, and magnesium chloride (noncarbonate hardness, which cannot be removed by heating). In general water softer than 50 mg/l, as $CaCO_3$, is corrosive, whereas waters harder than about 80 mg/l lead to use of more soap. Desirable hardness values, therefore, should be 50 to 80 mg/l, with 80 to 150 mg/l as passable, and over 150 mg/l as undesirable. Waters high in sulfates (above 600 to 800 mg/l calcium sulfate, 300 mg/l sodium sulfate, or 390 mg/l magnesium sulfate) is laxative to those not accustomed to the water. In addition to being objectionable for laundry and other washing purposes, excessive hardness contributes to the deterioration of fabrics. Satisfactory cleansing of laundry, dishes, and untensils is made difficult or impractical. In boiler and hot-water tanks, the scale resulting from hardness reduces the thermal efficiency and eventually causes restriction of the flow or plugging of the pipe. Calcium chloride when heated becomes acid and pits boiler tubes. There seem to be higher mortality rates from cerebrovascular and cardiovascular diseases in people provided with soft water than in those provided with hard water.

Alkalinity

The alkalinity of water passing through iron distribution systems should be in the range of 30 to 100 mg/l, as $CaCO_3$, to prevent serious corrosion; up to 500 mg/l is acceptable, although this factor must be appraised from the standpoint of pH, hardness, carbon dioxide, and dissolved-oxygen content. Corrosion of iron pipe is prevented by the maintenance of calcium-carbonate stability. Potassium carbonate, potassium bicarbonate, sodium carbonate, and sodium bicarbonate cause alkalinity in natural water. Calcium carbonate, calcium bicarbonate, magnesium carbonate, and magnesium bicarbonate cause hardness as well as alkalinity. Sufficient alkalinity is needed in water to react with added alum to form a floc in water coagulation. Insufficient alkalinity will cause alum to remain in solution.

pH[16]

The pH values of natural water range from about 5.0 to 8.5 and are acceptable except when viewed from the standpoint of corrosion. The pH is a measure of acidity or alkalinity, using a scale of 0.0 to 14.0, with 7.0 being the neutral point. The bactericidal and cysticidal efficiency of free available chlorine as a disinfectant increases with a decrease in pH. The pH determination in water having an alkalinity of less than 20 mg/l by using color indicators is inaccurate; use the electrometric method. The ranges of pH color indicator solutions are: methyl orange, 3.0 to 4.4; bromcresol green, 3.8 to 5.4; methyl red, 4.4 to 6.2; bromcresol purple, 5.2 to 6.8; bromthymol blue, 6.0 to 7.6; phenol red, 6.8 to 8.4; cresol red, 7.2 to 8.8; thymol blue, 8.0 to 9.6; and phenolphthalein, 8.2 to 10.0. Waters containing more than 1.0 mg/l chlorine in any form must be dechlorinated with 1 or 2 drops of ¼ percent sodium thiosulfate before adding the pH indicator solution. This is necessary to prevent the indicator solution from being bleached or decolorized by the chlorine and giving an erroneous reading.

[16] pH is defined as the logarithm of the reciprocal of the hydrogen-ion concentration.

Acidity

Mineral acids should be absent, that is, waters should be alkaline to methyl orange (pH 4.4), but of course may be "acid" to phenolphthalein (pH 8.2).

Carbon Dioxide

The only limitation on carbon dioxide is that pertaining to corrosion. It should be less than 10 mg/l, but when the alkalinity is less than 100 mg/l, the CO_2 concentration should not exceed 5.0 mg/l.

Dissolved Oxygen

Water devoid of dissolved oxygen frequently has a "flat" taste, although many attractive well waters are devoid of oxygen. In general it is preferable for the dissolved oxygen content to exceed 2.5 to 3.0 mg/l to avoid secondary tastes and odors from developing and to support fish life. Game fish require a dissolved oxygen of at least 5.0 mg/l to reproduce, and either die off or migrate when the dissolved oxygen falls below 3.0 mg/l.

Lead

The accepted maximum concentration of lead in potable water is 0.05 mg/l, owing to the fact that this is a cumulative poison. Concentrations exceeding this value occur when corrosive waters are piped through lead pipe; hence copper, red brass, cement-lined iron, or wrought-iron pipe should be used if corrosive waters are piped to buildings. The use of lead pipe to conduct drinking water should be discouraged. Lead, cadmium, zinc, and copper are dissolved by carbonated beverages, which are highly charged with carbon dioxide. Water having a very high or a very low pH will cause lead to go into solution.

Copper

The copper content should be less than 1.0 mg/l. Concentrations of this magnitude are not present in natural waters, but may be due to the corrosion of copper or brass piping; 0.5 mg/l in soft water stains porcelain blue-green. A concentration in excess of 0.2 to 0.3 mg/l will cause off-flavor in coffee and tea; 1 to 1.5 mg/l results in a metallic taste; 1 mg/l may affect film and reacts with soap to produce a green color in water; 0.25 to 1.0 mg/l is toxic to fish. Copper appears to be essential for all forms of life; but excessive amounts, as noted, are toxic to fish.

Zinc

Less than 5.0 mg/l. Zinc is dissolved by surface water and 0.5 mg/l forms a greasy film in water that is boiled. More than 5.0 mg/l causes a metallic taste and 25 to 40 mg/l may cause nausea and vomiting. The ratio of cadmium to zinc may also be of public health importance.

Chlorides

(Of mineral origin.) The permissible chloride content of water depends on the sensitivity of the consumer. Many people notice a brackish taste imparted by 100 mg/l of chlorides, whereas others are satisfied with concentrations as high as 250 mg/l. Irrigation waters should contain less than 200 mg/l. When the chloride is in the form of sodium chloride, use of the water for drinking may be inadvisable for persons who are under medical care for certain forms of heart disease. Hard water softened by an ion exchange unit will increase the concentrations of sodium in the water. See "Sodium" below.

Chlorides (Of Intestinal Origin)

Natural waters remote from the influence of ocean or salt deposits and not influenced by local sources of pollution have a low chloride content—usually less than 4.0 mg/l. Due to the extensive salt deposits in certain parts of the country, it is impractical to assign chloride concentrations that, when exceeded, indicate the presence of sewage, agricultural, or industrial pollution, unless a chloride record over an extended period of time is kept on each water supply. In view of the fact that chlorides are soluble, they will pass through pervious soil and rock for great distances without diminution in concentration, and thus the chloride content must be interpreted with considerable discretion in connection with other constituents in the water. The concentration of chlorides in urine is about 5000 mg/l; in septic-tank effluent, about 80 mg/l; in sewage from a residential community, 50 mg/l.

Residual Chlorine

The concentration of residual chlorine in the water reaching the consumer should be below the taste level. If this cannot be secured without curtailing the effectiveness of chlorination, other supplementary treatment should be provided. See "Chlorination."

Iron

Water should have a soluble iron content of less than 0.1 mg/l to prevent reddish-brown staining of laundry, fountains, and plumbing fixtures. Precipitated ferric hydroxide may cause a slight turbidity in water that can be objectionable and cause clogging of filters and resin bed softeners. In combination with manganese, concentrations in excess of 0.3 mg/l cause complaints. Iron in excess of 1.0 mg/l will cause an unpleasant taste and a false color when orthotolidine is added to chlorinated water for a residual chlorine test. The orthotolidine arsenite test compensates for the false color. A concentration of about 1 mg/l is noticeable in the taste of coffee or tea. Chlorine will precipitate soluble iron. Iron is an essential element for human health.

Manganese

Should be less than 0.1 mg/l to avoid the black-brown staining of plumbing and clothes, although soluble manganese bound to organic matter may be present in considerably higher concentrations without producing difficulties. Manganese in the manganic form, however, will interfere with the orthotolidine test for residual chlorine in chlorinated water, which can be corrected by the use of the orthotolidine arsenite test. Concentrations greater than 0.5 to 1.0 mg/l may give a metallic taste to water.

Sodium

The American Heart Association's 500-mg- and 1000-mg-sodium-per-day diet recommends that distilled water be used if the water supply contains more than 20 mg of sodium/l. The consumption of 2.5 liters of water/day is assumed. Many groundwater supplies and most home-softened (using ion exchange) well waters contain too much sodium for persons on sodium-restricted diets. If the well water is low in sodium (less than 20 mg sodium/l) and the water is softened because of excessive hardness, the house cold-water system can be supplied by a line from the well that bypasses the softener and low-sodium water can be made available at cold-water taps. A laboratory analysis is necessary to determine the exact amount of sodium in water. Persons suffering from heart, kidney, or circulatory illnesses should be guided by their physicians' advice.

Sulfates

Sulfates should not exceed 250 mg/l, although this would have to be modified with zeolite softening where calcium sulfate or gypsum is replaced by an equal concentration of sodium sulfate. Sodium sulfate or Glauber salts in excess of 200 mg/l, magnesium sulfate or Epsom salts in excess of 390 mg/l, and calcium sulfate in excess of 600 to

800 mg/l are laxative to those not accustomed to the water. Magnesium sulfate causes hardness; sodium sulfate causes foaming in steam boilers.

Total Dissolved Solids

The total solid content should be less than 500 mg/l; however, this is based on the industrial uses of public water supplies and not on public health factors. Higher concentrations may cause physiological effects and taste.

Fluorides

Fluorides are found most frequently in groundwaters as a natural constituent. Fluorides in concentrations greater than 3 mg/l cause the teeth of children to become mottled and discolored. Water containing 0.8 to 1.7 mg/l natural or artificially added fluoride is beneficial to children during the period they are developing permanent teeth. The incidence of dental cavities or tooth decay is reduced by about 60 percent. The lower limit is recommended where the annual average of maximum daily air temperature is around 79.3 to 90.5°F, and the upper limit is desirable where the temperature is 50.0 to 53.7°F, since less water is drunk in a colder climate.

Alky Benzene Sulfonate (ABS) or Linear Alkylate Sulfonate (LAS)

ABS may be the active agent in detergent. Household wash water may contain 200 to 1000 mg/l. It will produce foam and cause taste in water supplies when the compound exceeds 0.5 to 1.0 mg/l. ABS in concentrations found in drinking water has not been found to be of health significance, although its very presence indicates the presence of sewage and hence the need for careful water treatment. ABS in drinking water should not exceed 0.05 mg/l. ABS has been replaced by linear alkylate sulfonate (LAS), which can be degraded under aerobic conditions; if not degraded it too will foam. Both ABS and LAS detergents contain phosphates that fertilize plant life in lakes and streams. Decay of plants will use oxygen, leaving less for fish life and sewage oxidation.

Oxygen-Consumed Value

This represents organic matter that is oxidized by potassium permanganate under the test conditions. Pollution significant from a bacteriological examination standpoint is accompanied by so little organic matter as not to significantly raise the oxygen-consumed value. For example, natural waters containing swamp drainage have much higher oxygen-consumed values than water of low original-organic content that are subject to bacterial pollution.

Free Ammonia

Free ammonia represents the first product of the decomposition of organic matter; thus appreciable concentrations of free ammonia usually indicate "fresh pollution" of sanitary significance. The exception is when ammonium sulfate of mineral origin is involved. The following values may be of general significance in appraising free ammonia content: Low—0.015 to 0.03 mg/l; moderate—0.03 to 0.10 mg/l; high—0.10 mg/l or greater. Special care must be exercised to allow for ammonia added if the "chlorine-ammonia" treatment of water is used or if crenothrix organisms are present.

Albuminoid Ammonia

Albuminoid ammonia represents "complex" organic matter and thus would be present in relatively high concentrations in water supporting algae growth, receiving forest drainage, or containing other organic matter. Concentrations of albuminoid ammonia higher than about 0.15 mg/l, therefore, should be appraised in the light of origin of the water and the results of microscopical examination. In general, the following concentrations serve as a guide: Low—less than 0.06 mg/l; moderate—0.06 to 0.15 mg/l; high—0.15 mg/l or greater.

Nitrites

Nitrites represent the first product of the oxidation of free ammonia by biochemical activity. Unpolluted natural waters contain practically no nitrites, so concentrations exceeding the very low value of 0.001 mg/l are of sanitary significance, indicating water subject to pollution that is in the process of change associated with natural purification. The nitrite concentration present is due to the organic matter in the soil through which the water passes.

Nitrates

Nitrates represent the final product of the biochemical oxidation of ammonia. Its presence is probably due to the presence of nitrogenous organic matter of animal and, to some extent, vegetable origin, for only small quantities are naturally present in water. Manure and fertilizer contain large concentrations of nitrates; thus the existence of fertilized fields or cattle feedlots near sources of supply must be carefully considered in appraising the significance of nitrate content. Furthermore, a cesspool may be relatively close to a well and be contributing pollution without a resulting high-nitrate content, because the anaerobic conditions in the cesspool would prevent biochemical oxidation of ammonia to nitrates. In fact, nitrates may be reduced to nitrites under such conditions. In general, however, nitrates disclose the evidence of "previous" pollution of water that has been modified by self-purification processes to a final mineral form. Allowing for these important controlling factors the following ranges in concentration may be used as a guide: Low—less than 0.1 mg/l; moderate—0.1 to 1.0 mg/l; high—greater than 1.0 mg/l.

The presence of more than 45 mg/l nitrates (10 mg/l N) appears to be the cause of methemoglobinemia or "blue babies." Methemoglobinemia is largely a disease confined to infants less than 3 months old. It is caused by the bacterial conversion of the nitrate ion ingested in water, formula, and other food to nitrite. "Nitrite then converts hemoglobin, the blood pigment that carries oxygen from the lungs to the tissues, to methomoglobin. Because the altered pigment no longer can transport oxygen, the physiologic effect is oxygen deprivation, or suffocation."[17] A standard of not more than 45 mg/l nitrates has been used to give a factor of safety, since boiling of water containing nitrates would cause the concentration of nitrates to be increased. A better epidemiological basis for the standard is needed. Addition of the nitrite ion and nitrates ingested through food and air would give a more complete dietary total.

Hydrogen Sulfide

Hydrogen sulfide is most frequently found in groundwaters as a natural constituent and is easily identified by a rotten-egg odor. A concentration of 70 mg/l is an irritant; but 700 mg/l is highly poisonous.

Uranyl Ion

This ion may cause damage to the kidneys. Objectionable taste and color occur at about 10 mg/l. It should be less than 5 mg/l.

Phenols

These should not exceed 0.001 mg/l. In combination with chlorine a medicinal taste is produced in drinking water. The presence of phenols can cause serious problems in the food and beverage industries.

Radioactivity (Gross Beta)

Radioactivity should not exceed 1000 picocurie (micromicrocurie) per liter, in the absence of alpha emitters and strontium 90. Radium 226 should not exceed 3 pCi/l and strontium 90 should not exceed 10 pCi/l.

---

[17] E. F. Winton, R. G. Tardiff, and L. J. McCabe, "Nitrates in Drinking Water," *J. Am. Water Works Assoc.,* **63,** 95 (February 1970).

Carbon-Chloroform Extract (CCE)

CCE may include chlorinated hydrocarbon insecticides, nitrates, nitrobenzenes, aromatic ethers, and many others. Water from uninhabited watersheds usually show CCE concentrations of less than 0.04 mg/l. The taste and odor can be expected to be poor when the concentration of CCE reaches 0.2 mg/l.

Pesticides

Gas chromatography has made possible the routine examination of drinking water for the presence of chlorinated hydrocarbons and other complex chemical compounds. (See pages 154 and 155).

Phosphorous

High phosphorous concentrations together with nitrates and organic carbon are sometimes associated with heavy aquatic plant growth, although other substances in water also have an effect. Uncontaminated waters contain 10 to 30 μg/l total phosphorous, although higher concentrations of phosphorous are also found in "clean" waters. Concentrations associated with nuisances in lakes would not normally cause problems in flowing streams. About 100 μg/l complex phosphates interfere with coagulation.[18]

Mercury

Episodes associated with the consumption of methylmercury-contaminated fish, bread, pork, and seed have called attention to the possible contamination of drinking water. Mercury is found in nature, and concentrations in unpolluted waters are normally less than 1.0 μg/l. A standard of 5.0 μg/l has been proposed for drinking water. See also "Mercury Poisoning" in Chapter 1.

Microscopic Examination

The microscopic organisms (plankton) usually found in drinking water are classified as algae, fungi, protozoa, rotifera, and crustacea. Algae include diatoms, cyanaphyeae or blue-green algae, and chlorophyceae or green algae. Examinations of water sources should be made routinely for control purposes and to determine the need for treatment or other measures. The measure of the concentration of microorganisms present is the "areal standard unit." It represents an area 20 microns square or 400

[18] National Technical Advisory Committee Report "Raw-Water Quality Criteria for Public Water Supplies," *J. Am. Water Works Assoc.*, **61**, 133 (March 1969).

square microns. One micron = 0.001 mm. Microorganisms are reported as the number of areal standard units per milliliter. Protozoa, rotifers, and other animal life are individually counted. Material that cannot be identified is reported as areal standard units of amorphous matter. The apparatus, procedure, and calculation of results and conversion to "Cubic Standard Units" is explained in *Standard Methods*.

1. When more than 300 areal standard units, or organisms, per milliliter are reported, treatment with $CuSO_4$ is indicated to prevent possible trouble with tastes and odors or short filter runs. When more than 500 areal standard units per milliliter are reported, complaints can be expected and the need for immediate action is indicated. A thousand units or more of amorphous matter indicates probable heavy growth of organisms that have died and disintegrated or organic debris from decaying leaves and similar vegetable matter.

2. The presence of *Asterionella, Tabellaria, Synedra, Beggiatoa, Crenothrix, Sphaerotilis natans, Mallomonas, Anabaena, Aphanizomenon, Volvox, Ceratium, Dinobryon, Synura, Uroglenopsis,* and others, some even in small concentrations, may cause tastes and odors that are aggravated where marginal chlorine treatment is used. Free residual chlorination will usually reduce the tastes and odors. More than 25 aeral standard units per milliliter of *Synura, Dinobryon,* or *Uroglena,* or 300 to 700 units of *Asterionella, Dictyosphaerium, Asphanizomenon, Volvox,* or *Ceratium* in chlorinated water will usually cause taste and odor complaints. The appearance of even one areal standard unit of a microorganism may be an indication to start immediate copper sulfate treatment if past experience indicates that trouble can be expected.

3. The blue-green algae, *Anabaena, Microcystis (Polycystis), Nodularia, Gloeotrichia, Coelosphaerium, Nostoc rivulare* and *Aphanizomenon* in large concentrations have been responsible for killing fish and causing illness in horses, sheep, dogs, ducks, chickens, mice, and cattle.[19] Illness in man from this cause has been suspected. Gorham estimated that the oral minimum lethal dose of decomposing toxic *Microcystis* bloom for a 150-lb man is 1 to 2 qt of thick, paintlike suspension, and concluded that toxic waterblooms of blue-green algae in public water supplies are not a significant health hazard. Red tides caused by the dinoflagellates *Gonyaulax monilata* and *Gymno dinium brevis* have been correlated with mass mortality of fish.[20]

Quantity

The quantity of water used for domestic purposes will generally vary directly with the availability of the water, habits of the people, number and type of plumbing fixtures provided, water pressure, air temperature, newness of a community, type of establishments, and other factors. Wherever possible, the actual water consumption under existing or similar circumstances and the number of persons served should be the basis

[19] William Marcus Ingram and G. W. Prescott, "Toxic Fresh-Water Algae," *American Midland Naturalist,* **52,** No. 1 75–87 (July 1954).
[20] Paul R. Gorham, "Toxic Algae as a Public Health Hazard," *J. Am. Water Works Assoc.,* **56,** 1487 (November 1964).

for the design of a water and sewerage system. Special adjustment must be made for industrial use. Municipal water consumption varies widely depending on location, industrial usage, metering, size, economic, social, and other factors. An average is 155 gpcd.

Table 3–2 gives estimates of water consumption at different types of places. Additions should be made for car washing, lawn sprinkling, and miscellaneous uses. If provision is made for fire-fighting requirements, then the quantity of water provided for this purpose to meet fire underwriters' standards will be in addition to that required for normal domestic needs in small communities.

SOURCE AND PROTECTION OF WATER SUPPLY

General

The sources of water supply are divided into two major classifications: groundwater and surface water. To these should be added rainwater and demineralized water. The groundwater supplies include dug, bored, driven and drilled wells, rock and sand or earth springs, and infiltration galleries. The surface-water supplies include lake, reservoir, stream, pond, swamp, river, and creek supplies.

The location of groundwater supplies should take into consideration the tributary area, the probable travel of pollution through the ground, the well construction practices and standards actually followed, and the type of sanitary seal provided at the point where the pump lines pass out of the casing. These factors are explained and illustrated later.[21]

It is sometimes suggested that the top of a well casing terminate below the ground level or in a pit. This is not considered good practice except when the pit can be drained above flood level to the surface by gravity or to a drained basement. Frost-proof sanitary seals with pump lines passing out horizontally from the well casing are generally available. Some are illustrated in Figure 3–6.

In order that the basic data on a new well may be recorded, a form such as the "Well Driller's Log and Report" shown in Figure 3–3 should be completed by the well driller and kept on file by the owner for future reference.

Surface-water supplies are all subject to intermittent pollution and must be treated to make them safe to drink. One never knows when

[21] For more information see *Ground Water Basin Management,* ASCE Manual No. 40, American Society of Civil Engineers, New York, 1961; and *Large-Scale Ground-Water Development* United Nations Water Resources Centre, New York, 1960.

TABLE 3-2 GUIDES FOR WATER USE

Type of Establishment	gpd*
Residential:	
Dwellings and apartments (per bedroom)	150
Temporary quarters:	
Boarding houses	65
Additional (or nonresident boarders)	10
Campsites (per site)	100
Cottages, seasonal	50
Day camps	15 to 20
Hotels	65 to 75
Mobile home parks (per unit)	125 to 150
Motels	50 to 75
Restaurants (toilets and kitchens)	7 to 10
Without public toilet facilities	2½ to 3
With bar or cocktail lounge, additional	2
Summer camps	40 to 50
Public establishments:	
Boarding schools	75 to 100
Day schools	15 to 20
Hospitals (per bed)	250 to 500
Institutions other than hospitals (per bed)	75 to 125
Places of public assembly	3 to 10
Turnpike rest areas	5
Turnpike service areas (10 percent of cars passing)	15 to 20
Amusement and commercial:	
Airports (per passenger)	3 to 5
Country clubs	25
Day workers (per shift)	15 to 35
Drive-in theaters (per car space)	5
Gas station (per vehicle serviced)	10
Milk plant, pasteurization (per 100 lb of milk)	11 to 25
Movie theaters (per seat)	3
Picnic parks with flush toilets	5 to 10
Self-service laundries (per machine)	400
Shopping center (per 1,000 ft² floor area)	250
Stores (per toilet room)	400
Swimming pools and beaches with bathhouses	10
Farming: (per animal)	
Cattle or Steer	12
Milking cow	35
Goat or Sheep	2
Hog	4
Horse or Mule	12
Cleaning milk equipment and tank	2
Cow washer	5 to 10
Liquid manure handling, cow	1 to 3
Poultry (per 100):	
Chickens	5 to 10
Turkeys	10 to 18
Cleaning and sanitizing equipment	4

*Per person unless otherwise stated.

128

TABLE 3-2 (cont'd)

Miscellaneous Water Use Estimates	Water Use
Home:	Gallons
Water closet, tank	4 to 6 per use
Water closet, flush valve 25 psi(pounds per square inch)	30 to 40/min
Washbasin	1½/use
Bathtub	30/use
Shower	25 to 30/use
Dishwashing machine	9½ to 15½/load
Garbage grinder	1 to 2/day
Automatic laundry machine	34 to 57/load
Garden hose:	
5/8 inch, 25-ft head	200/hr
3/4 inch, 1/4 inch nozzle, 25-ft head	300/hr
Lawn sprinkler	120/hr
3,000 square ft lawn, 1 in. per week	1,850/wk
Air conditioner, water-cooled, 3-ton, 8 hr per day	2,880/day

Water demand per dwelling unit:	
Average day	400 gpd
Maximum day	800 gpd
Maximum hourly rate	2,000 gpd
Maximum hourly rate with appreciable lawn watering	2,800 gpd

Home water system: (Minimums)

	Bedrooms			
	2	3	4	5
Pump capacity, gal/hr	250	300	360	450
Pressure tank, gal	42	82	82	120
Service line from pump, diameter in in.†	¾	¾	1	1¼

Other:	Gallons
Fire hose, 1½ in., ½ in. nozzle, 70-ft head	2,400/hr
Drinking fountain, continuous flowing	75/hr
Dishwashing machine, commercial	
Stationary rack type, 15 psi	6 to 9/min
Conveyor type, 15 psi	4 to 6/min
Fire hose, home, 10 gpm at 60 psi for 2 hours, ¾ inch	600/hr
Restaurant, average	35/seat
Restaurant, 24-hr	50/seat
Restaurant, tavern	20/seat
Gas station	500/set of pumps

Developing areas of the world:
One well or tap/200 persons; controlled tap or hydrant — Fordilla type
Average consumption, 5 gal/capita/day at well or tap
Water system design, 30 gal/capita/day (10 gal/capita is common)
Pipe size, 2 in. and preferably larger (1 and 1½ in. common)
Drilled well, cased, 6 to 8 in. diameter
Water system pressure, 20 lb/sq in.
(Keep mechanical equipment to a minimum)

†Service line less than 50 ft long, brass or copper. Use next larger size if iron pipe is used. Use minimum 1¼ in. service with flush valves. Minimum well yield, 5 gal/min.

129

Well at_____ In _____ County of _____
          Name of place   City, village or town

Owner_____ P.O. Address_____

Depth of well_____ Diameter_____ Yield_____ Was well disinfected?_____
              ft   in.   gpm   yes or no

Amt. of casing above ground_____ Below ground_____ Well seal_____
                           in.   ft   cement grout

Draw a well diagram in the space provided below and show the depth and type of casing, the well seal, kind and thickness of formations penetrated, water bearing formations, diameter of drill holes with dotted lines and casing(s) with solid lines.

Well Diagram		Formations Penetrated	Remarks		
Diameter, in.	Depth in ft	Kind, thickness, and if water bearing	Type of well_____ Drilling method_____		
	Grade		Was well dynamited?____		
			Pumping Tests*		
			Details	#1	#2	#3
	25		Static water level, in feet below grade			
	50		Pumping rate in gpm			
	75		Pumping level in feet below grade			
			Duration of test, in hours			
	100		*Water at end of test:*		
	150		Clear ____ Cloudy ____ Turbid ____ Recommended depth of pump in well, ft below grade_____		
	200		*Wells in sand & gravel:* Sand Eff. size_____mm Unif. Coef._____ Length of screen_____ft Diam. of screen_____in.		
	250		Type of screen_____ Screen openings____x____ Comments:		

Show cross-section of well & formations penetrated above. Draw a sketch of the property on the back of this sheet locating the well and sewage disposal systems within 200 ft.

Drilling started_____Completed_____
Well Driller_____
                    Signature

Figure 3–3   Well driller's log and report.

* The specific capacity of a well is the yield at a stabilized drawdown and given pumping rate, expressed as gallons per minute per foot of drawdown. Use chalked tape or known length of air line with pressure gauge. Test run is usually 4 to 8 hr.

the organisms causing typhoid fever, gastroenteritis, infectious hepatitis, or dysentery may be discharged or washed into the water. The extent of the treatment required will depend on the results of a sanitary survey made by a competent sanitary engineer. The minimum required treatment may be simple chlorination or coagulation, sedimentation, filtration, and chlorination. If elaborate treatment is needed it would be best to abandon the idea of using a surface-water supply and resort to a ground-water supply if possible and practical. Where a surface supply must be used a reservoir or a lake that does not receive sewage or industrial pollution and that can be controlled would be preferred to a stream or creek, the pollution of which cannot from a practical standpoint be controlled. There are many situations where there is no practical alternative to the use of streams for water supply. In such cases carefully designed water treatment plants should be provided.

Dug Well

A dug well is one usually excavated by hand, although it may be dug by mechanical equipment. It may be 3 to 6 ft in diameter and 15 to 35 ft deep, depending on where the water-bearing formation or groundwater table is encountered. Wider and deeper wells are less common. Hand pumps over wells or pumplines entering wells should form watertight connections, as shown in Figures 3–4 and 3–5. Since dug wells have a relatively large diameter, they have large storage capacity. The level of the water in dug wells will lower at times of drought and the well may go dry. Dug wells are not usually dependable sources of water supply, particularly where modern plumbing is provided. In some areas, properly developed dug wells provide an adequate and satisfactory water supply.

Bored Well

A bored well is constructed with a hand- or machine-driven auger. Bored wells vary in diameter from a few to 36 in. and in depth from 25 to 60 ft. A casing of concrete pipe, vitrified clay pipe, or metal pipe is usually necessary to prevent the relatively soft formation penetrated from caving into the well. Bored wells have characteristics similar to dug wells in that they have small yields, are easily polluted, and are affected by droughts.

Driven and Jetted Wells

These types of wells consist of a well point with a screen attached, or a screen with the bottom open, which is driven or jetted into a water-

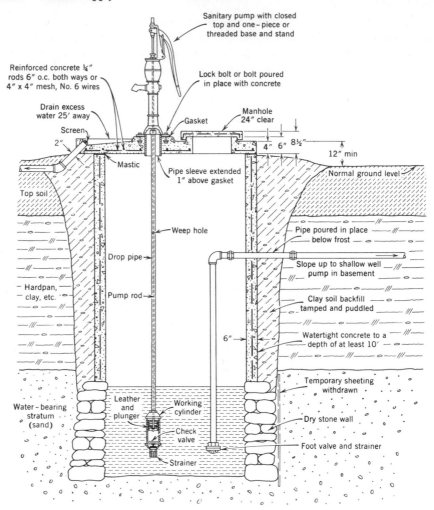

Figure 3–4 A properly developed dug well.

bearing formation found at comparatively shallow depth. A series of pipe lengths are attached to the point or screen as it is forced into position. The driven well is constructed by driving the well point, preferably through at least 10 to 20 ft of casing, with the aid of a maul or sledge, pneumatic tamper, sheet pile driver, drive monkey, hand-operated driver or similar equipment. In many instances the casing is omitted; but then less protection is afforded the driven well, which also serves as the pump suction line. The jetted well is constructed by directing a stream of water at the bottom of the open screen, thereby loosening and flushing the soil up the casing to the surface as the screen is lowered. Driven and jetted wells are commonly between $1\frac{1}{4}$ and 2 in. in diameter and less than 25 ft in depth, although larger and deeper wells can be constructed. In the small-diameter wells, a shallow well hand or mechanical suction pump is connected directly to the well. Large-diameter driven wells make possible installation of the pump cylinder close to or below the water surface in the well at greater depth, in which case the hand or mechanical pump must be located directly over the well. In all cases, however, care must be taken to see that the top of the well is tightly capped, that the concrete pump platform extends 2 or 3 ft around the well pipe or casing, and that the annular space between the well casing and drop pipe, or pipes, is tightly sealed. This is necessary to prevent the entrance of unpurified water or other pollution from close to the surface.

Drilled Well

Studies have shown that, in general, drilled wells are superior to dug or driven wells and springs. There are some exceptions. Drilled wells are less likely to become contaminated and are usually more dependable sources of water. When a well is drilled, a hole is made in the ground with a percussion or rotary drilling machine. Drilled wells are usually 4 to 12 in. in diameter or larger. A steel or wrought-iron casing is lowered as the well is drilled to prevent the hole from caving in and to seal off water of a doubtful quality. Lengths of casing should be threaded and coupled or properly welded. The drill hole must of course be larger than the casing, thereby leaving an irregular space around the outside length of the casing. Unless this space or channel is closed by cement grout or naturally by formations that conform to the casing almost as soon as it is placed, pollution from the surface or from crevices close to the surface, or from polluted formations penetrated, will flow down the side of the casing and into the water source. Water can also move up and down this annular space in an artesian well and as the pumping water level changes.

When the source of water is water-bearing sand and gravel, a gravel wall or gravel-packed well may be constructed. Such a well will usually yield more water than the ordinary drilled well with a screen of the same diameter and with the same drawdown. A slotted or perforated casing in a water-bearing sand will yield only a fraction of the water obtainable by the use of a proper screen. On completion, the well should be overpumped, surged, and tested. Proper surging is an important part of well development. To develop well yield and hydraulic data, a 24-

to 48-hr pump test is desirable. But 4 to 8 hr is more common. Surging or the forcing of water in and out through the screen is done by pumping, use of compressed air, dry ice, or a surge block.

Only water-well casing of clean steel or wrought iron should be used. Used pipe is not satisfactory. Standards for well casing are given in the American Water Works Association's publication *AWWA Standard for Deep Wells,* AWWA A100-66. Doubling the diameter of a casing increases the yield up to only 10 to 12 percent.

Extending the casing at least 5 ft below the pumping water level in the well—or if the well is less than 30 ft deep, 10 ft below the pumping level—will afford an additional measure of protection. In this way the water is drawn from a depth that is less likely to be contaminated. In some sand and gravel areas, extending the casing 5 or 10 ft below the pumping level may shut off the water-bearing sand or gravel. A lesser casing depth would then be indicated, but in no instance should the casing be less than 10 ft, provided sources of pollution are remote and provision is made for chlorination. The recommended depth of casing, cement grouting, and the need for double casing construction or the equivalent is given in Table 3–3.

A vent is necessary on a well because the fluctuation in the water level will cause a change in air pressure above and below atmospheric in a well, resulting in the drawing in of contaminated water from around the pump base over the well or from around the casing if not properly sealed.

It must be remembered that well construction is a very specialized field. Most well drillers are desirous of doing a proper job for they know that a good well is their best advertisement. However, in the absence of a state law dealing with well construction, the enforcement of standards, and the licensing of well drillers, price alone frequently determines the type of well constructed. Individuals proposing to have wells drilled should therefore carefully analyze bids received. Such matters as water quality, well diameter, type and length of casing, minimum well yield, type of pump and sanitary seal where the pump line or lines pass through the casing, provision of a satisfactory well log, method used to seal off undesirable formations and cement grouting of the well, plans to pump the well until clear, and disinfection following construction should all be taken into consideration. See Figures 3–5, 3–6, 3–7, and 3–8. Remember—the cheapest well is not necessarily the best buy.

One of the most common reasons for contamination of wells drilled through rock, clay, or hardpan is failure to seal the well casing properly.

A contaminated well supply causes the homeowner considerable inconvenience and extra expense for it is difficult to seal off contamination after the well is drilled. In some cases the only practical answer is to build a new well.

Proper cement grouting of the space between the drill hole and well casing where the overburden over the water-bearing formation is clay, hardpan, or rock can prevent this common cause of contamination. (Table 3–3).

There are many ways to seal well casings. The best material is neat cement grout.[22] But to be effective the grout must be properly prepared (a proper mixture is $5\frac{1}{2}$ gal clean water to a bag of cement), pumped as one continuous mass, and placed upward from the bottom of the space to be grouted.

The clear annular space around the outside of the casing and the drill hole must be at least $1\frac{1}{2}$ in. on all sides to prevent bridging of the grout. Guides must be welded to the casing.

Cement grouting of a well casing along its entire length of 50 to 100 ft or more is good practice but expensive for the average farm or rural dwelling. An alternative is grouting to at least 20 ft below ground level. This provides adequate protection for most installations, except in limestone formations. It also protects the casing from corrosion.

For a 6-in. diameter well a 10-in. hole is drilled, if 6-in. welded pipe is used to at least 20 ft, or to solid rock if the rock is deeper than 20 ft. If 6-in. coupled pipe is used, a 12-in. hole will be required. From this depth the 6-in. hole is drilled until it reaches a satisfactory water supply. A temporary outer casing, carried down to rock, prevents cave-in until the cement grout is placed.

Upon completion of the well the annular space between the 6-in. casing and temporary casing or drill hole is filled from the bottom up to the grade with cement grout. The temporary pipe is withdrawn as the cement grout is placed—it is not practical to pull the casing after all the grout is in position.

The extra cost of the temporary casing and larger drill hole is small compared to the protection obtained. The casing can be reused as often as needed. In view of this, well drillers who are not equipped should consider adding larger casing and equipment to their apparatus.

A temporary casing or larger drill hole and cement grouting are not required where the entire earth overburden is 40 ft or more of silt or sand and gravel, which immediately close in on the total length of casing to form a seal around the casing; however, this condition is not common.

---

[22] Concrete grout-cement and sand 1:1, with not more than 6 gal water per sack of cement, or puddled clay if approved, are also used.

TABLE 3-3   STANDARDS FOR

Water-bearing formation	Overburden	Oversize Drill Hole for Grout		Well	
		Diameter	Depth†	Cased Portion	Uncased Portion
1. Sand or gravel	Unconsolidated caving material; sand or sand and gravel	None required	None	2" minimum, 5" or more preferred	Does not apply
2. Sand or gravel	Clay, hardpan, silt, or similar material to depth of more than 20'	Casing size plus 4"	Minimum 20'	2" minimum, 5" or more preferred	Does not apply
3. Sand or gravel	Clay, hardpan, silt, or similar material containing layers of sand or gravel within 15' of ground surface	Casing size plus 4"	Minimum 20'	2" minimum, 5" or more preferred	Does not apply
4. Sand or gravel	Creviced or fractured rock, such as limestone, granite, quartzite	Casing size plus 4"	Through rock formation	4" minimum	Does not apply
5. Creviced, shattered or otherwise fractured limestone, granite, quartzite or similar rock types	Unconsolidated caving material, chiefly sand or sand and gravel to a depth of 40' or more and extending at least 2,000' in all directions from the well site	None required	None required	6" minimum	6" preferred
6. Creviced, shattered or otherwise fractured limestone, granite, quartzite or similar rock types	Clay, hardpan, shale, or similar material to a depth of 40' or more and extending at least 2,000' in all directions from well site	Casing size plus 4"	Minimum 20'	6" minimum	6" preferred

Source: *Recommended State Legislation and Regulations,* U.S. Public Health Service,

Note: For wells in creviced, shattered, or otherwise fractured limestone, granite, quartz-in all directions, and no other practical acceptable water supply is available, the well con-

* Requirements for the proper construction of wells vary with the character of subsur-construction details of this Table may be adjusted, as conditions warrant, under the proce-

† In the case of a flowing artesian well, the annular space between the soil and rock and aquifer to the ground surface in accordance with good construction practice.

‡ These diameters shall also be applicable in circumstances where the use of perforated shall be considered as being 2" nominal diameter well screens for purposes of these regula-

§ As used herein, the term "pumping level" shall refer to the lowest elevation of the sur-well contractor, taking into consideration usual seasonal fluctuations in the static water level

THE CONSTRUCTION OF WELLS*

Diameter Well Screen Diameter‡	Minimum Casing Length or Depth†	Liner Diameter in. (if required)	Construction Conditions†	Miscellaneous Requirements
2″ minimum	20′ minimum; but 5′ below pumping level§	2″ minimum		
2″ minimum	5′ below pumping level§	2″ minimum	Upper drill hole shall be kept at least one-third filled with clay slurry while driving permanent casing; after casing is in the permanent position annular space shall be filled with clay slurry or cement grout.	An adequate well screen shall be provided where necessary to permit pumping sand-free water from the well.
2″ minimum	5′ below pumping level§	2″ minimum	Annular space around casing shall be filled with cement grout.	
2″ minimum	5′ below overburden of rock	2″ minimum	Annular space around casing shall be filled with cement grout.	
Does not apply	Through caving overburden	4″ minimum	Casing shall be firmly seated in the rock.	
Does not apply	Through overburden	4″ minimum	Annular space around casing shall be grouted. Casing shall be firmly seated in rock.	

Dept. of HEW, Washington, D.C., July 1965.
zite, or similar rock in which the overburden is less than 40 ft and extends less than 2,000 ft
struction described on Line "7" of this Table is applicable.
face formations, and provisions applicable under all circumstances cannot be fixed. The
dure provided for by the Health Department and in the Note above.
the well casing shall be tightly sealed with cement grout from within 5 ft of the top of the

casing is deemed practicable. Well points commonly designated in the trade as 1¼″ pipe
tions.
face of the water in a well during pumping, determined to the best knowledge of the water
and drawdown level.

TABLE 3–3 STANDARDS FOR THE

| Water-bearing formation | Overburden | Oversize Drill Hole for Grout | | Well | |
		Diameter	Depth†	Cased Portion	Uncased Portion
7. Creviced, shattered or otherwise fractured limestone, granite, quartzite or similar rock	Unconsolidated materials to a depth of less than 40′ and extending at least 2,000′ in all directions	Casing size plus 4″	Minimum 40′	6″ minimum	6″ preferred
8. Sandstone	Any material except creviced rock to a depth of 25′ or more	Casing size plus 4″	15′ into firm sandstone or to 30′ depth, whichever is greater	4″ minimum	4″ preferred
9. Sandstone	Mixed deposits mainly sand and gravel, to a depth of 25′ or more	None required	None required	4″ minimum	4″ preferred
10. Sandstone	Clay, hardpan, or shale to a depth of 25′ or more	Casing size plus 4″	Minimum 20′	4″ minimum	4″ preferred
11. Sandstone	Creviced rock at variable depth	Casing size plus 4″	15′ or more into firm sandstone	6″ minimum	6″ preferred

CONSTRUCTION OF WELLS* (cont'd)

Diameter Well Screen Diameter‡	Minimum Casing Length or Depth†	Liner Diameter in. (if required)	Construction Conditions†	Miscellaneous Requirements
Does not apply	40′ minimum	4″ minimum	Casing shall be firmly seated in rock. Annular space around casing shall be grouted.	If grout is placed through casing pipe and forced into annular space from the bottom of the casing, the oversize drill hole may be only 2″ larger than the casing pipe, Pipe 2″ smaller than the drill hole, and liner pipe 2″ smaller than casing shall be assembled without couplings.
2″ minimum, if well screen required to permit pumping sand-free water from partially cemented sandstone	Same as over-size drill hole or greater	2″ minimum	Annular space around casing shall be grouted. Casing shall be firmly seated in sandstone.	
2″ minimum, if well screen required to permit pumping sand-free water from partially cemented sandstone	Through overburden into firm sandstone	2″ minimum	Casing shall be effectively seated into firm sandstone.	Pipe 2″ smaller than the oversize drill hole, and liner pipe 2″ smaller than casing shall be assembled without couplings.
	Through overburden into dand-stone	2″ minimum	Casing shall be effectively seated into firm sandstone. Oversized drill hole shall be kept at least one-third filled with clay slurry while driving permanent casing; after the casing is in the permanent position, annular space shall be filled with clay slurry or cement grout.	
2″ minimum, if well screen required to permit pumping sand-free water from partially cemented sandstone	15′ into firm sandstone	4″ minimum	Annular space around casing shall be filled with cement grout.	If grout is placed through casing pipe and forced into annular space from the bottom of the casing the oversize drill hole may be only 2″ larger than the casing pipe. Pipe 2″ smaller than the drill hole, and liner pipe 2″ smaller than casing shall be assembled without couplings.

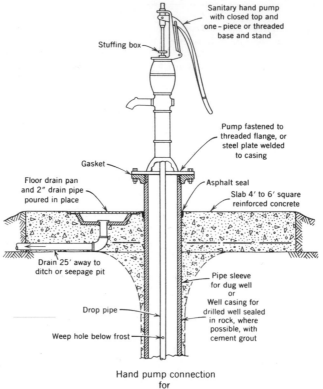

Figure 3–5 Sanitary hand pump and well attachment.

Drilled wells serving public places are usually constructed and cement-grouted as explained in Table 3–3.

In some areas, limestone or shale beneath a shallow overburden represent the only source of water. Acceptance of a well in shale or limestone might be conditioned on an extended observation period to determine the sanitary quality of the water. Continuous chlorination should be required on satisfactory supplies serving the public and recommended to private individuals. However, chlorination should not be relied on to make a heavily contaminated well-water supply satisfactory. Such supplies should be abandoned and filled in with concrete or puddled clay, unless the source of contamination can be eliminated.

Well drillers may have other sealing methods suitable for particular local conidtions, but it is believed that the methods described above utilizing a neat cement grout will give reasonably dependable assurance

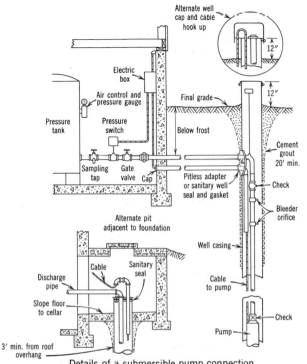

Details of a submersible pump connection

Figure 3–6   Sanitary well caps and seals.

that an effective seal is provided, whereas this cannot be said of some of the other methods used. Driving the casing, a lead packer, drive shoe, rubber sleeves, and similar devices do not provide reliable annular space seals.

Well Contamination—Cause and Removal

Well-water supplies are all too often improperly constructed, protected, or located, with the result that laboratory bacteriological examinations show the water to be contaminated. Under such conditions all water used for drinking or culinary purposes should first be boiled or adequately treated. Boiling will not remove chemical contaminants; treatment may remove some. Abandonment of the well and connection to a public water supply would be the best solution if this is practical. A second alternative would be investigation to find and remove the

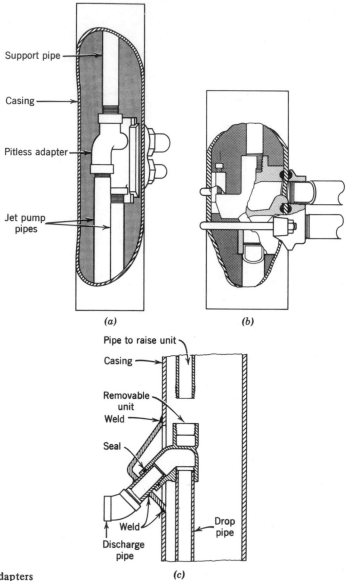

Pitless adapters

Figure 3–6a   Martinson Manufacturing Co., P.O. Box 256, Ramsey, N.J.
Figure 3–6b   Williams Products Co., Joliet, Ill.
Figure 3–6c   Herb Maass Service, 10940 West Congress Street, Milwaukee, Wis.

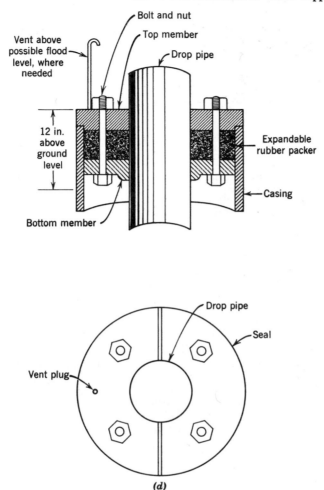

Figure 3–6d   Sanitary expansion well cap.

cause of pollution. A third choice would be a new, properly constructed and located drilled well.

When a well shows the presence of contamination it is usually due to one or more of four probable causes: lack of or improper disinfection of a well following repair or construction; failure to seal the annular space between the drill hole and the outside of the casing; failure to provide a tight sanitary seal at the place where the pump line or lines pass through the casing; sewage pollution of the well through polluted strata or a fissured or channeled formation. On some occasions the casing is found to be only a few feet in length and completely inadequate.

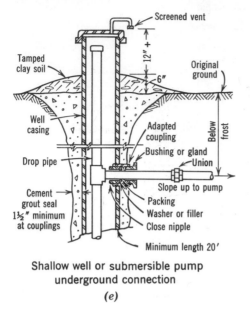

Shallow well or submersible pump
underground connection

(e)

Figure 3–6e Improvised well seal.

If a new well is constructed or if repairs are made to the well, pump, or piping, contamination from the work is probable. Disinfect the well, pump, storage tank, and piping as explained in pages 246 to 251.

If a sewage disposal system is suspected of contamination, a dye such as water soluble sodium or potassium fluorescein or ordinary salt can be used as a tracer. A solution flushed into the disposal system or suspected source may appear in the well water within 12 to 24 hr. It can be detected by sight or taste if a connection exists. Samples should be collected every few hours and set aside for comparison. If the connection is indirect, chemical or fluoroscopic examination for the dye or for chlorides is more sensitive. One part of fluorescein in 10 to 40 million parts of water is visible to the naked eye, and in 10 billion parts if viewed in a long glass tube, or if concentrated in the laboratory. The chlorides in the well before adding salt should, of course, be known. Where chloride determinations are routinely made on water samples, sewage pollution may be apparent without making the salt test. Dye is not decolorized by passage through sand, gravel, or manure; it is slightly decomposed by calcareous soils and entirely decolorized by peaty formations and by free acids, except carbonic acid.[23]

[23] R. B. Dole, "Use of Fluorescein in the Study of Underground Waters," excerpt from *U.S. Geol. Survey, Water-Supply Paper* 160, 73–85 (1906).

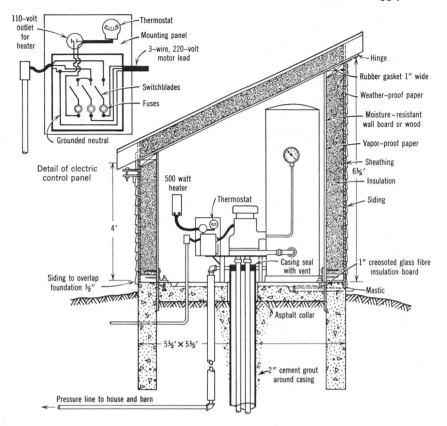

Figure 3–7 Insulated pumphouse (adapted from Minnesota).

If the cause of pollution is an improper underground seal where the pump line or lines pass through the side of the casing, a dye or salt solution or even plain water can be poured around the casing. Samples of the water can be collected for visual or taste test or chemical examination. The seal might also be excavated for inspection. Where the upper part of the casing can be inspected, a mirror or strong light can be used to direct a light beam inside the casing to see if water is entering the well from close to the surface. Sometimes it is possible to hear the water dripping into the well. Inspection of the top of the well will also show if the top of the casing is provided with a sanitary seal and whether the well is subject to flooding. See Figure 3–6.

The path of pollution entry can also be holes in the side of the casing, channels along the length of the casing leading to the well source, crevices

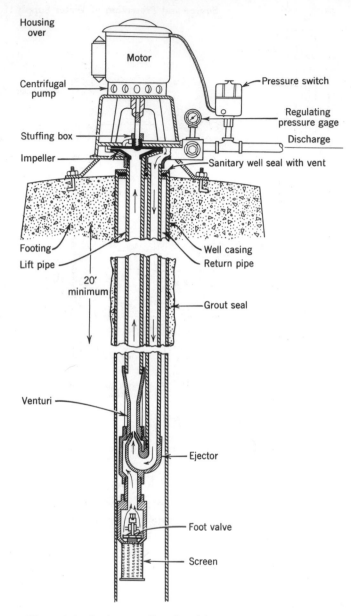

Figure 3–8 Sanitary well seal and jet pump.

or channels connecting surface pollution with the water-bearing stratum, or the annular space around the casing. A solution of dye, salt, or plain water can be used to trace the pollution, as previously explained.

The steps taken to provide a satisfactory water supply would depend on the results of the investigation. If a sanitary seal is needed at the top or side of the casing where the pump lines pass through, then the solution is relatively simple. On the other hand, an unsealed annular space is more difficult to correct. A competent well driller could be engaged to investigate the possibility of grouting the annular space and installing an inner casing or a new casing carefully sealed in solid rock. If the casing is found tight, it would be assumed that pollution is finding its way into the water-bearing stratum through sewage-saturated soil or creviced or channeled rock at a greater depth. It is sometimes possible, but costly, to seal off the polluted stratum and if necessary drill deeper.

Once a stratum is contaminated, it is very difficult to prevent future pollution of the well unless all water from such a stratum is effectively sealed off. Moving the offending sewage-disposal system to a safe distance is possible, but evidence of the pollution may persist for some time. The same general principles apply to dug and driven wells.

Unless all the sources of pollution can be found and removed, it is recommended that the well be abandoned and filled with puddled clay or concrete to prevent the pollution from traveling to other wells. In some special cases, and under controlled conditions, use of a slightly contaminated water supply may be permitted provided approved treatment facilities are installed. Such equipment is expensive and requires constant attention. If a public water supply is not available and a new well is drilled, it should be located and constructed as previously explained.

Spring

Springs are broadly classified as either rock springs or earth springs, depending on the source of water. It is necessary to find the source, properly develop it, and prevent animals from gaining access to the spring area, to obtain a satisfactory water.

Protection and development of a source of water is shown in Figure 3–9. A combination of methods may also be possible under certain ground conditions and would yield a greater supply of water than either alone.

In all cases, the spring should be protected from surface-water pollution by constructing a deep diverting ditch or the equivalent above and around the spring. The spring and collecting basin should have a watertight top, preferably concrete, and water obtained by gravity flow or by means of a properly installed sanitary hand or mechanical pump.

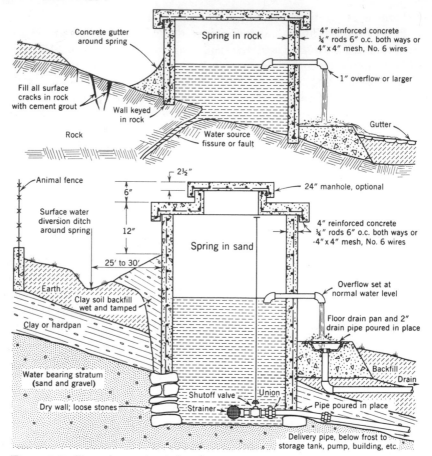

Figure 3-9 Properly constructed springs.

Access or inspection manholes, when provided, should be tightly fitted (as shown) and kept locked. Water from limestone or similar type channeled or fissured rock springs is not purified to any appreciable extent when traveling through the formation and hence may carry pollution from nearby or distant places. Under these circumstances, it is advisable to have periodic bacteriological examinations made and to chlorinate the water. The spring must be developed at the water source.

Infiltration Gallery

An infiltration gallery consists of a system of porous, perforated, or open-joint pipe or other conduit draining to a receiving well. The pipe is surrounded by gravel and located in a porous formation such as sand

and gravel below the water table. The collecting system should be located 20 ft or more from a lake or stream, or under the bed of a stream or lake if installed under expert supervision. It is sometimes found desirable, where possible, to intercept the flow of groundwater to the stream or lake. In such cases a coffer dam, cut-off wall, or puddled clay dam is carefully placed between the collecting conduit and the lake or stream to form an impervious wall. It is not advisable to construct an infiltration gallery unless the water table is relatively stable and the water intercepted is free of pollution. The water-bearing strata should not contain cementing material or yield a very hard water as it may clog the strata or cause incrustation of the pipe, thereby reducing the flow. An infiltration gallery is constructed similar to Figure 3–10. The depth of

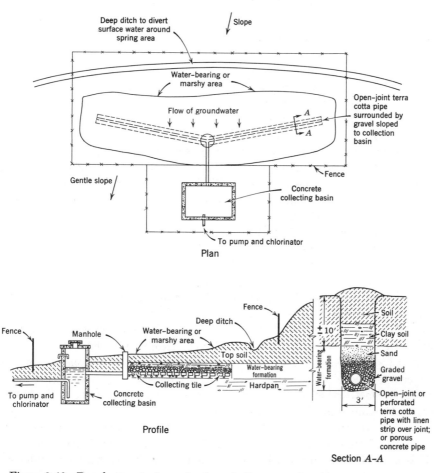

Figure 3–10 Development of a spring in a shallow water-bearing area.

the collecting tile should be about 10 ft below the normal ground level, and below the lowest known water table, to assure a greater and more constant yield. The infiltration area should be protected from pollution by sewage and animals. Water derived from infiltration galleries should be given a minimum of chlorination treatment.

Cistern

A cistern is a watertight tank in which rainwater collected from roof runoff is stored. Where the quantity of groundwater is inadequate or the quality objectionable, and where a municipal water supply is not available, a cistern supply may be acceptable as a limited source of water. Because rainwater is soft, little soap is needed when used for laundry purposes. On the other hand rain will wash dust, dirt, bird droppings, leaves, paint, and other material on the roof into the cistern unless special provision is made to bypass the first rainwater and to filter the water. The bypass may consist of a simple manually or float-operated damper or switch placed in the leader drain. When in one position, all water will be diverted to a float-control tank or to waste away from the building foundation and cistern; when in the other position water will be run into the cistern.

The capacity of the cistern is determined by the size of the roof catchment area, the probable water consumption, the maximum 24-hr rainfall, the average annual rainfall, and maximum length of dry periods. Suggested rainwater cistern sizes are shown in Figure 3–11. The cistern storage capacity given allows for a reserve supply, plus a possible heavy rainfall of $3\frac{1}{2}$ in. in 24 hr. The calculations assume that 25 percent of the precipitation is lost. Weather bureaus, the *World Almanac*, airports, water department, and other agencies give rainfall figures for different parts of the country. Adjustment should therefore be made in the required cistern capacity to fit local conditions. The cistern capacity will be determined largely by the volume of water one wishes to have available for some designated period of time, the total volume of which must be within the limits of the volume of water that the roof catchment area and annual rainfall can safely yield. Monthly average rainfall data can be expected to depart from the true values by 50 percent or more on occasion. The drawing of a mass diagram is a more accurate method of estimating the storage capacity, since it is based on past actual rainfall in a given area.

It is recommended that the cistern water be treated after every rain with a chlorine compound to give a dosage of at least 5 mg/l chlorine. This may be accomplished by adding five times the quantities of chlorine shown in Table 3–4, mixed in 5 gal of water to each 1000 gal of water

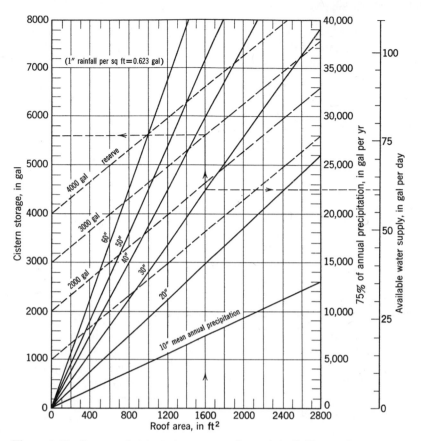

Figure 3–11 Suggested cistern storage capacity and available supply.

TABLE 3-4 QUANTITY AND TYPE OF CHLORINE TO TREAT 1,000 GALLONS OF CLEAN WATER AT RATE OF 1 mg/l

Chlorine Compound	Quantity
High test, 70 percent chlorine	$\frac{1}{5}$ oz or $\frac{1}{4}$ heaping tablespoon
Chlorinated lime, 25 percent chlorine	$\frac{1}{2}$ oz or 1 heaping tablespoon
Sodium hypochlorite, 14 percent chlorine	1 oz
Sodium hypochlorite, 10 percent chlorine	$1\frac{1}{3}$ oz
Bleach, $5\frac{1}{4}$ percent chlorine	$2\frac{3}{5}$ oz

Note: A jigger glass = $1\frac{1}{2}$ liquid ounces. See also pages 251-257.

151

in the cistern. In areas affected by air pollution, fallout on the roof or catchment area will contribute chemical pollutants that may not be neutralized by chlorine treatment.

Example: With a roof area of 1600 ft², in a location where the mean annual precipitation is 30 in. and it is desired to have a reserve supply of 3000 gal, the cistern storage capacity should be about 5600 gal. This should yield an average annual supply of about 62 gal per day.

Desalting

Desalting is resorted to where a satisfactory groundwater, surface water, or rainwater is not available but where seawater or a brackish water is found. Under such circumstances cost is a secondary factor.[24] The conversion of saline water to fresh water is accomplished by distillation using artificial heat or by-product heat, electrodialysis, osmosis, solar distillation, freezing, gas hydration, and other means. Surplus heat from power plants and incinerators can be used to convert seawater or brackish water to fresh water. Distillation is the most common method. Potable water can also be produced from secondary sewage by reverse osmosis and by electrodialysis treatment in a 30-mgd plant at a cost of 20.7 to 22.4 cents/1000 gal.[25] As ordinarily handled, stored, and piped, the water may be subject to contamination; hence the water should be chlorinated before it is distributed.

Surface Water

The quality of surface water depends on the watershed area drained, location and sources of pollution, and the natural agencies of purification, such as sedimentation, sunlight, aeration, nitrification, filtration, and dilution. These are variable and hence cannot be depended on to continuously purify water effectively. In addition increasing urbanization, industrialization, and intensive farming have caused heavy chemical discharges to streams, which are not readily removed by the usual water treatment. Treatment consisting of coagulation, sedimentation, rapid sand filtration, and chlorination has little effect on the contaminants shown in Table 3–5. Because of these factors, heavily polluted surface

---

[24] Cost of production by distillation in a 1-mgd plant is about $1.00 to $1.25 per 1000 gal. A cost of $.50 per 1000 gal in a 10-mgd plant appears feasible.

[25] K. C. Channabasappa, "Reverse osmosis offers useful technique for desalting," *Water and Wastes Engineering/Industrial,* p. A-5 (January 1970).

waters should be avoided as drinking water supplies, if possible, and upland protected water sources should be used and preserved consistent with multipurpose uses in the best public interest.

TREATMENT OF WATER—DESIGN AND OPERATION CONTROL

Treatment Required

As an aid in determining the treatment that should be given water to make it safe to drink, the U.S. Public Health Service has classified waters into several groups.[26] The treatment required by this classification is based on the most probable number (MPN) of coliform bacteria per 100 milliliters (ml) of sample. The classification is summarized in Table 3–6. It needs to be supplemented by chemical, physical, and microscopic examinations. For water to be generally acceptable, other treatment may be required in addition to that necessary for the elimination of disease-producing organisms. People expect the water to be safe to drink, attractive to the senses, soft, nonstaining, and neither scale-forming nor corrosive to the water system. The various treatment processes used to accomplish these results are briefly discussed under the appropriate headings below. In all cases, the water supply must meet the current "Public Health Service Drinking Water Standards." The untrained individual should not attempt to design a water-treatment plant, for life and health will be jeopardized. This is a job for a competent sanitary engineer.

Chlorination[27]

Chlorination is the most common method of destroying the disease-producing organisms that might normally be found in water used for drinking in the United States. The water so treated should be relatively clear and clean with an average monthly MPN of coliform bacteria of not more than 50/100 ml. Clean lake and stream waters and well, spring, and infiltration-gallery supplies not subject to significant pollution can be made of safe sanitary quality by continuous and effective chlorination.

Operation of the chlorinator should be automatic, proportional to the flow of water, and adjusted to the chlorine demand of the water. A

---

[26] *Manual of Recommended Water-Sanitation Practice,* U.S. Public Health Service Bull. No. 296, Dept. of HEW, Washington, D.C., 1946.
[27] See also Table 3–6 and footnotes to table.

TABLE 3-5 SURFACE-WATER CRITERIA
FOR PUBLIC WATER SUPPLIES

Constituent or Characteristic	Permissible Criteria	Desirable Criteria
Physical:		
Color (color units)	75	<10
Odor	†	Virtually absent
Temperature*	do	†
Turbidity	do	Virtually absent
Microbiological:		
Coliform organisms	10,000/100 ml[1]	<100/100 ml[1]
Fecal coliforms	2,000/100 ml[1]	<20/100 ml[1]
Inorganic chemicals:	(mg/l)	(mg/l)
Alkalinity	†	†
Ammonia	0.5 (as N)	<0.01
Arsenic*	0.05	Absent
Barium*	1.0	do
Boron*	1.0	do
Cadmium*	0.01	do
Chloride*	250	<25
Chromium,* hexavalent	0.05	Absent
Copper*	1.0	Virtually absent
Dissolved oxygen	≥4 (monthly mean) ≥3 (individual sample)	Near saturation
Fluoride*	†	†
Hardness*	do	do
Iron (filterable)	0.3	Virtually absent
Lead*	0.05	Absent
Manganese* (filterable)	0.05	do
Nitrates plus nitrites*	10 (as N)	Virtually absent
pH (range)	6.0–8.5	†
Phosphorus*	†	do
Selenium*	0.01	Absent
Silver*	0.05	do
Sulfate*	250	<50
Total dissolved solids* (filterable residue)	500	<200
Uranyl ion*	5	Absent
Zinc*	5	Virtually absent
Organic chemicals:		
Carbon chloroform extract* (CCE)	0.15	<0.04
Cyanide*	0.20	Absent
Methylene blue active substances*	0.5	Virtually absent
Oil and grease*	Virtually absent	Absent
Pesticides:		
Aldrin*	0.017	do
Chlordane*	0.003	do
DDT*	0.042	do

TABLE 3-5 SURFACE-WATER CRITERIA
FOR PUBLIC WATER SUPPLIES (cont'd)

Constituent or Characteristic	Permissible Criteria	Desirable Criteria
Pesticides: (cont'd)		
Dieldrin*	0.017	do
Endrin*	0.001	do
Heptachlor*	0.018	do
Heptachlor epoxide*	0.018	do
Lindane*	0.056	do
Methoxychlor*	0.035	do
Organic phosphates plus carbamates.*	0.1^2	do
Toxaphene*	0.005	do
Herbicides:		
2,4-D plus 2,4,5-T, plus 2,4,5-TP*	0.1	do
Phenols*	0.001	do
Radioactivity:	(pc/l)	(pc/l)
Gross beta*	1,000	<100
Radium-226*	3	<1
Strontium-90*	10	<2

Source: *Water Quality Criteria,* Report of the National Technical Advisory Committee to the Secretary of the Interior, Washington, D.C., April 1968, p. 20.

*The defined treatment process has little effect on this constituent. (Coagulation, sedimentation, rapid sand filtration and chlorination.)

†No consensus on a single numerical value that is applicable throughout the country. See Report and text.

[1] Microbiological limits are monthly arithmetic averages based on an adequate number of samples. Total coliform limit may be relaxed if fecal coliform concentration does not exceed the specified limit.

[2] As parathion in cholinesterase inhibition. It may be necessary to resort to even lower concentrations for some compounds or mixtures. (Permissible levels are based on the recommendations of the Public Health Service Advisory Committee on Use of the PHS Drinking Water Standards.)

spare machine should be on the line. A complete set of spare parts for the equipment will make possible immediate repairs. The chlorinator should provide for the positive injection of chlorine and be selected with due regard to the pumping head and maximum and minimum flow of water to be treated. The point of chlorine application should be selected so as to provide adequate mixing and chlorine contact with the water to be treated before it reaches the first consumer.

Hypochlorinators are generally used to feed relatively small quantities of chlorine as 1 to 5 percent sodium or calcium hypochlorite solution. Positive feed machines are fairly reliable and simple to operate.

TABLE 3-6   A CLASSIFICATION OF WATERS BY CONCENTRATION OF COLIFORM BACTERIA AND TREATMENT REQUIRED TO RENDER THE WATER OF SAFE SANITARY QUALITY*

Group No.	Maximum Permissible Average MPN Coliform Bacteria per Month†	Treatment Required
1	Not more than 10% of all 10-ml or 60% of 100-ml portions positive; not more than 1.0 coliform bacteria/100 ml.	None for protected underground water, but a minimum of chlorination for surface water.
2	Not more than 50/100 ml.	Simple chlorination or equivalent.
3	Not more than 5000/100 ml and this MPN exceeded in not more than 20% of samples.	Rapid sand filtration (including coagulation) or its equivalent plus continuous chlorination.
4	MPN greater than 5000/100 ml in more than 20% of samples and not exceeding 20,000/100 ml in more than 5% of the samples.	Auxiliary treatment such as 30–90 days storage, presettling, prechlorination, or equivalent plus complete filtration and chlorination.
5	MPN exceeds Group No. 4.	Prolonged storage or equivalent to bring within Groups 1–4.

*Physical, inorganic and organic chemicals, and radioactivity concentrations in the raw water and ease of removal by the proposed treatment must also be taken into consideration. See Table 3-5 and *Manual for Evaluating Public Drinking Water Supplies,* Environmental Control Administration, U.S. Public Health Service, Cincinnati, Ohio, 1969.

†Fecal coliforms not to exceed 20% of total coliform organisms. The monthly geometric mean of the MPN for Group 2 may be less than 100 and, for Group 3 and 4, less than 20,000/100 ml with the indicated treatment.

Gas machines feed larger quantities of chlorine and require certain precautions.

Gas Chlorinator

When a dry feed gas chlorinator or a solution feed gas chlorinator is used, the chlorinator and liquid chlorine cylinders should be located in a separate gas-tight room that is mechanically ventilated to provide two air changes per minute, with the exhaust openings at floor level opposite the air inlets. Exhaust ducts must be separate from any other ventilating system of ducts and extend to a height and location that will not endanger personnel or property and ensure adequate dilution. The door to the room should have a glass inspection panel, and a chlorine gas mask or self-contained breathing apparatus approved by the U.S. Bureau of Mines, should be available outside of the chlorinator and chlorine cylinder room.[28] The temperature around the chlorine clyinders should be between 50 and 85°F and cooler than the temperature of the chlorinator room to prevent condensation of chlorine in the line conducting chlorine or in the chlorinator. Cylinders must be stored at a temperature below 140°F.[29] A platform scale is needed for the weighing of chlorine cylinders in use to determine the pounds of chlorine used each day and to anticipate when a new cylinder will be needed. Cylinders should be connected to a manifold so that chlorine may be drawn from several cylinders at a time and so that cylinders can be replaced without interrupting chlorination. A relatively clear source of water of adequate volume and pressure is necessary to prevent clogging of injectors and strainers and to assure proper chlorination at all times. The water pressure to operate a gas chlorinator should be at least 15 psi and about three times the back pressure (water pressure at point of application, plus friction loss in the chlorine solution hose, and difference in elevation between the point of application and chlorinator) against which the chlorine is injected. About 40 to 50 gpd of water is needed per pound of chlorine to be added. Do not draw more than 35 lb of chlorine per day from a 100- or 150-lb cylinder to prevent clogging by chlorine ice. Liquid chlorine comes in 100- and 150-lb cylinders, in 1-ton containers, and in 16- to 90-ton tank cars.

Testing for Residual Chlorine

Chlorination is usually controlled by means of the amperometric titration test and the colorimetric orthotolidine arsenite (OTA) test. Properly made, these tests distinguish between the color produced by interfering substances, free residual chlorine, and combined residual chlorine. Substances that may cause a false color upon the addition of orthotolidine solution include nitrite, ferric and manganic compounds, and possibly organic iron, lignocellulose, and algae. Where free or combined residual chlorine is used, it is the same as free or combined available chlorine. The OTA test must be made according to the procedure given in *Standard Methods for the Examination of Water and Wastewater*.[30]

[28] A chlorine container holding 100 lb of chlorine developing a leak that cannot be repaired can have the chlorine absorbed by 125 lb of caustic soda in 40 gal of water, 300 lb of soda ash in 100 gal of water, or 125 lb of hydrated lime in 125 gal of water continuously agitated. (Chlorine Institute, Inc., New York, N.Y.)
[29] The fusible plugs are designed to soften or melt at a temperature between 158 and 165°F. Chlorinator should have atomatic shutoff if water pressure is lost or if chlorine piping leaks or breaks.
[30] Published by the American Public Health Association, Inc., 1015 Eighteenth Street, N.W., Washington, D.C., 20036; 13th ed., 1971.

Where the interfering substances are not significant, the concentration of free residual chlorine may be closely approximated by adding ortho-tolidine to a sample of chilled water, mixing, and reading the color produced within 5 sec. This is known as the flash test, which has useful practical application with a stable water, provided the OTA test is made occasionally to assure that interfering substances remain absent from the water. If the color, after adding orthotolidine, gets darker on standing 3 min, it indicates the presence of combined chlorine. If combined chlorine is present a slightly higher free residual chlorine is obtained, which is minimized if the temperature of the sample is reduced to 35°F. The flash test, the OTA test and amperometric titration have definite limitations. A new test using leuco crystal violet as an indicator determines free available chlorine with minimal interference from combined chlorine, iron, nitrates and nitrites.[31] For further discussion see Table 3–7.

Chlorine Treatment for Operation Control

To assure that only properly treated water is distributed, it is important to have a competent and trustworthy person in charge of the chlorination plant. He should keep daily records showing the gallons of water treated, the pounds of chlorine or quarts of chlorine solution used and its strength, the gross weight of chlorine cylinders if used, the setting of the chlorinator, the time residual chlorine tests are made, the results of such tests, and any repairs or maintenance, power failure, modifications, or unusual occurrences dealing with the treatment plant or water system. Where large amounts of chlorine are needed, the use of ton containers can effect a saving in cost, as well as less labor and possibly of chlorine gas leakage.

The required chlorine dosage should take into consideration the appearance as well as the quality of a water. Pollution of the source of water, the type of microorganisms likely to be present, the pH of the water, and the temperature and degree of treatment a water receives are all very important.

The chlorine residual that will give effective disinfection of a *clear*

---

[31] A. P. Black and G. P. Whittle, "Determination of Halogen Residuals. Part II. Free and Total Chlorine," *J. Am. Water Works Assoc.* **59**, 607 (May 1967). See also R. J. Lishka and E. F. McFarren, *Water Chlorine (Residual) No. 2*, Report Number 40, Analytical Reference Service, Environmental Protection Agency, Office of Water Programs, Cincinnati, Ohio, 1971. This study "indicated that the best accuracy and precision was obtained by leuco crystal violet and the stabilized neutral orthotolidine (SNORT) procedures, followed by DPD-titrimetric, amperometric titration, DPD-colorimetric and methyl orange. By far the poorest was the orthotolidine-arsenite (OTA) procedure."

TABLE 3-7 CHLORINE RESIDUAL FOR EFFECTIVE DISINFECTION OF DEMAND-FREE WATER

pH	Approximate Percent at 32 to 68°F‡ HOCl	Approximate Percent at 32 to 68°F‡ OCl⁻	Bactericidal Treatment* Free Available Chlorine After 10 min at 32 to 78°F	Bactericidal Treatment* Combined Available Chlorine After 60 min at 32 to 78°F	Cysticidal Treatment Free Available Chlorine After 30 min 36 to 41°F‡	Cysticidal Treatment Free Available Chlorine After 30 min 60°F‡	Cysticidal Treatment Free Available Chlorine After 30 min 78°F‡
5.0	—	—	—	—	—	—	—
6.0	98 to 97	2 to 3	0.2	1.0	7.2	2.3	1.9‖
7.0	83 to 75	17 to 25	0.2	1.5	10.0	3.1	2.5‖
7.2	74 to 62	26 to 38	—	—	—	—	2.6‖
7.3	68 to 57	32 to 43	—	—	—	—	2.8‖
7.4	64 to 52	36 to 48	—	—	—	—	3.0‖
7.5	58 to 47	42 to 53	—	—	14.0‖	4.7	3.2‖
7.6	53 to 42	47 to 58	—	—	—	—	3.5‖
7.7	46 to 37	53 to 64	—	—	16.0‖	6.0	3.8‖
7.8	40 to 32	60 to 68	—	—	—	—	4.2‖
8.0	32 to 23	68 to 77	0.4	1.8	22.0	9.9	5.0‖
9.0	5 to 3	95 to 97	0.8	Reduce pH of water to below 9.0	—	78.0	20.0‖
10.0	0	100	0.8		—	761	170‖

Note 1. Free chlorine = HOCl. Free available chlorine = HOCl + OCl⁻. Combined available chlorine = chlorine bound to nitrogenous matter as chloramine. Only free available chlorine or combined available chlorine is measured by present testing methods; therefore to determine actual free chlorine (HOCl), correct reading by percent shown above. Chlorine residual as the term is generally used is the combined available chlorine and free available chlorine.

2. Viricidal treatment requires a free available chlorine of 0.53 mg/l at pH 7 and 5 mg/l at pH 8.5 in 32°F demand-free water. For water at a temperature of 77 to 82.4°F and pH 7 to 9 a free available chlorine of 0.3 mg/l is adequate.§ At a pH 7 and temperature of 77°F at least 9 mg/l combined available chlorine is needed with 30-min contact time.

3. The above results are based on studies made under laboratory conditions using water free of suspended matter and chlorine demand.

4. In practice, unless otherwise indicated, at least 0.4 to 0.5 mg/l free residual chlorine for 30 min or 2 mg/l combined residual chlorine for 3 hr should be maintained in a clear water before delivery to the consumer. The state health department may require more, dependent on source of raw water and sanitary survey.

* Butterfield, op. cit.

† S. L. Chang, "Studies on *Endamoeba histolytica,*" *War Medicine,* 5, 46 (1944); see also W. Brewster Snow, "Recommended Chlorine Residuals for Military Water Supplies," *J. Am. Water Works Assoc.,* 48, 1510 (December 1956).

‡ *Water Treatment Plant Design,* American Water Works Association, Inc., New York, 1969, pp. 153 and 165; and Edward W. Moore, "Fundamentals of Chlorination of Sewage and Waste," *Water and Sewage Works* (March 1951) 130–136.

§ *Manual For Evaluating Public Drinking Water Supplies,* Environmental Control Administration, U.S. Public Health Service Pub. No. 1820, Cincinnati, Ohio, 1969.

‖ Approximations. All residual chlorine results reported as mg/l. One mg/l hypochlorous acid gives 1.35 mg/l free available chlorine as HOCl and OCl⁻ distributed as noted above. The HOCl component is the markedly superior disinfectant.

water has been studied by Butterfield.[32] The germicidal efficiently of chlorine is primarily dependent on the percent free chlorine that is in the form of hypochlorous acid (HOCl), which in turn is dependent on the pH and temperature of the water, as can be seen in Table 3–7.

In a review of the literature, Greenberg and Kupka concluded that a chlorine dose of at least 20 mg/l with a contact time of 2 hr is needed to adequately disinfect a biologically treated sewage effluent containing tubercle bacilli.[33] Laboratory studies by Kelly and Sanderson indicated that "depending on pH level and temperature, residual chlorine values of greater than 4 ppm, with 15 minutes contact, or contact periods of at least four hours with a residual chlorine value of 0.5 ppm, are necessary to inactivate viruses, and that the recommended standard for disinfection of sewage by chlorine (0.5 ppm residual after fifteen minutes contact) does not destroy viruses."[34] Another study showed that inactivation of partially purified poliomyelitis virus in water required a free residual chlorine after 10 min of 0.05 mg/l at a pH of 6.85 to 7.4. A residual chloramine value of 0.50 to 0.75 mg/l usually inactivated the virus in 2 hr.[35] Destruction of Coxsackie virus required 7 to 46 times as much free chlorine as for *E. Coli*.[36] Infectious hepatitis virus was not inactivated by 1.0 mg/l total chlorine after 30 min, nor by coagulation, settling, and filtration (diatomite); but coagulation, settling, filtration, and chlorination to 1.1 mg/l total and 0.4 mg/l free chlorine was effective.[37] Bush and Isherwood suggest: "The use of activated sludge with abnormally high sludge volume index following by sand filtration may produce the kind of control necessary to stop virus. Chlorination with five-tenths parts per million chlorine residual for an eight hour contact period seems adequate to inactivate Coxsackie virus."[38] The

[32] C. T. Butterfield, "Bactericidal Properties of Chloramines and Free Chlorine in Water," *Public Health Reports,* **63,** 934–940 (1948).

[33] Arnold E. Greenberg and Edward Kupka, "Tuberculosis Transmission by Waste Waters—A Review," *Sewage and Ind. Wastes,* **29,** No. 5, 524–537 (May 1957).

[34] Sally Kelly and Wallace W. Sanderson, "Viruses in Sewage," *Health News,* **36,** 14–17 (June 1959).

[35] Serge G. Lensen et al., "Inactivation of Partially Purified Poliomyelitis Virus in Water by Chlorination, II," *Am. J. Public Health,* **37,** No. 7, 869–874 (July 1947).

[36] N. A. Clarke and P. W. Kabler, "The Inactivation of Purified Coxsackie Virus in Water," *Am. J. Hygiene,* **59,** 119–127 (January 1954).

[37] Greenberg and Kupka, op. cit.

[38] Albert F. Bush and John D. Isherwood, "Virus Removal in Sewage Treatment," *J. Sanitary Engineering Division,* ASCE, Vol. 92, No. SA 1, Proc. Paper 4653, 99–107 (February 1966).

removal of nematodes requires prechlorination to produce 0.4 to 0.5 mg/l after 6 hours' retention followed by settling. The pathogenic fungus *Histoplasma capsulatum* can be expected in surface-water supplies, in treated water stored in open reservoirs, and in improperly protected well-water supplies. Fungicidal action is obtained at a pH of 7.4 and at a water temperature of 26°C with 0.35 mg/l free chlorine after 4 hours' contact and with 1.8 mg/l free chlorine after 35 minutes' contact. Complete rapid sand filter treatment completely removed all viable spores even before chlorination.[39]

Coliform bacteria can be continually found in a chlorinated surface-water supply (turbidity 3.8 to 84 units, iron particles, and microscopic counts up to 2000 units) containing between 0.1 and 0.5 mg/l of free residual chlorine and between 0.7 and 1.0 mg/l total residual chlorine after more than 30 minutes' contact time.[40]

It is evident from available information that the coliform index may give a false sense of security when applied to waters subject to intermittent doses of pollution. The effectiveness of proper disinfection, other conditions being the same, is largely dependent on the freedom of suspended material and organic matter in the water being treated. Treated water having a turbidity of less than 5 Jackson Units (ideally less than 0.1), a pH less than 8, and an HOCl residual of 1 mg/l (0.3 to 0.4 mg/l in practice) after 30 min contact provides an acceptable level of protection.[41]

Free residual chlorination is the addition of sufficient chlorine to yield a free chlorine residual in the water supply in an amount equal to more than 85 percent of the total chlorine present. When the ratio of chlorine to ammonia is 5:1 (by weight), the chlorine residual is all monochloramine; when the ratio reaches 10:1, dichloramine is also formed; when the ratio reaches 15 or 20:1, nitrogen trichloride is formed. Nitrogen trichloride as low as 0.05 mg/l causes an offensive and acrid odor that

---

[39] Dwight F. Metzler, Cassandra Ritter, and Russell L. Culp, "Combined Effect of Water Purification Processes on the Removal of Histoplasma capsulatum from Water," *Am. J. Public Health,* **46**, No. 12, 1571–1575 (December 1956).

[40] Author's personal experience reported in *1956 Annual Report, Division of Environmental Hygiene,* Rensselaer County Department of Health, Troy, N.Y., p. 11. Confirmed by Wallace W. Sanderson and Sally Kelly, New York State Department of Health, Albany, in *Advances in Water Pollution Research Proceedings,* Vol. 2, Int. Conf., London, September 1962, Pergamon Press, London, pp. 536–541.

[41] "Engineering Evaluation of Virus Hazard in Water," *J. Sanitary Engineering Division,* ASCE, Vol. 96, No. SA 1, Proc. Paper 7112, 111–150 (February 1970). (By the Committee on Environmental Quality Management of the Sanitary Engineering Division.)

can be removed by carbon, aeration, exposure to sunlight, or forced ventilation indoors.[42] It is also highly explosive.

In the presence of ammonia, organic matter, and other chlorine-consuming materials, the required chlorine dosage to produce a free residual may be high. The water is then said to have a high chlorine demand. With free residual chlorination, water is bleached, and iron, manganese, and organic matter are coagulated by chlorine and precipitated, particularly when the water is stored in a reservoir or basin for at least 2 hr. Most taste- and odor-producing compounds are destroyed; the reduction of sulfates to taste- and odor-producing sulfides is prevented; and objectionable growths and organisms in the mains are controlled or eliminated, provided a free chlorine residual is maintained in the water. A factor of safety is also provided against accidental pollution of water in the mains.

The reaction of chlorine in water is shown in Figure 3–12.

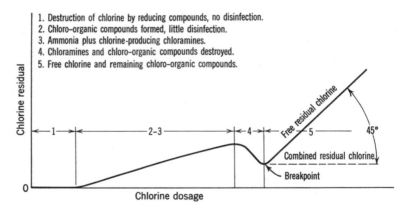

Figure 3–12   The reaction of chlorine in water. (Adapted from *Manual of Instruction for Water Plant Chlorinator Operators,* New York State Dept. of Health, Albany.

Distribution System Contamination

Once a water supply distribution system is contaminated with untreated water, the presence of coliform organisms may persist for an extended period of time. A surface-water supply or an inadequately filtered water supply may admit into a distribution system organic matter, minerals, and sediment, including fungi, algae, macroscopic organ-

[42] George C. White, "Chlorination and Dechlorination: A Scientific and Practical Approach," *J. Am. Water Works Assoc.,* **60,** 540–561 (May 1968).

isms, and microscopic organisms. These settle in the mains or become attached and grow inside the mains when chlorination is marginal or inadequate to destroy them. In some situations iron deposits will intermingle with and harbor the growths. Hence the admission of contaminated water into a distribution system, even for a short time, will have the effect of inoculating the growth media existing inside the mains with coliform organisms. Elimination of the coliform organisms will therefore involve removal of the growth media and harborage material. Bacteriological control of the water supply is lost until the coliform organisms are removed, unless a free chlorine residual of at least 0.2 to 0.4 mg/l is maintained in active parts of a distribution system.

If a positive program of continuous heavy chlorination at the rate of 5 to 10 mg/l, coupled with routine flushing of the main, is maintained, it is possible to eliminate the coliform on the inside surface of the pipes and hence the effects of accidental contamination in 2 to 3 weeks or less. If a weak program of chlorination is followed, with chlorine dosage of less than 5 mg/l, the contamination may persist for 8 or 9 months. The rapidity with which a contaminated distribution system is cleared will depend on many factors: uninterruption of chlorination even momentarily; the chlorine dosage and residual maintained in the entire distribution system; the growths in the mains and degree of pipe incrustation; conscientiousness in flushing the distribution system; the social, economic, and political deterrents; and, mostly, the competency of the responsible individual.

Plain Sedimentation

Plain sedimentation is the quiescent settling or storage of water, such as would take place in a reservoir, lake, or basin, without the aid of chemicals. This natural treatment results in the settling out of suspended solids, reduction of hardness, removal of color due to the action of sunlight, and death of bacteria principally because of the unfavorable temperature, lack of suitable food, and sterilizing effect of sunlight. Certain microscopic organisms, such as protozoa, consume bacteria, thereby aiding in purification of the water. Experiments conducted by Sir Alexander Houston showed that polluted water stored for periods of 5 wk at 32°F, 4 wk at 41°F, 3 wk at 50°F, or 2 wk at 64.4°F effected the elimination of practically all bacteria.[43] This treatment may, under certain conditions, be considered equivalent to filtration. Plain sedimentation, however, has some disadvantages that must be taken into consideration and

---

[43] *Water Quality and Treatment,* 2nd ed. American Water Works Association, New York, 1950, p. 94.

controlled. The growth of microscopic organisms that cause unpleasant tastes and odors is encouraged, and pollution by surface wash, birds, sewage, and industrial wastes may occur unless steps are taken to prevent or reduce these possibilities. Although subsidence permits bacteria to die off, it also permits bacteria to accumulate and grow in reservoir bottom mud under favorable conditions. In addition, iron and manganese may go into solution, carbon dioxide may increase, and hydrogen sulfide may be produced.

Presettling reservoirs are sometimes used to eliminate heavy turbidity or pollution and thus prepare the water for treatment by coagulation, settling, and filtration. Ordinarily, at least two basins are provided to permit one being cleaned while the other is in use. A capacity sufficient to give a retention period of 2 or 3 days is desirable. When heavily polluted water is to be conditioned, provision can be made for preliminary coagulation and prechlorination at the point of entrance of the water into the reservoirs.

Microstraining

Microstraining is a process designed to reduce the suspended solids, including plankton, in a water. The filtering media consist of very finely woven fabrics of stainless steel on a revolving drum. Applications to water supplies are primarily (1) the clarification of relatively clean surface waters low in true color and colloidal turbidity, in which microstraining and disinfection constitute the complete treatment; and (2) the clarification of waters ahead of slow or rapid sand filters and diatomite filters. Removals of the commoner types of algae have been as high as 95 percent. Wash-water consumption may run from 1 to 3 percent of the flow through the unit. Blinding of the fabric rarely occurs but may do so, from inadequate wash-water pressure or the presence of bacterial slimes. Cleansing is readily accomplished with commercial sodium hypochlorite.[44] Small head losses and low maintenance costs may make the microstrainer attractive for small installations.

There are four standard units, the smallest $2\frac{1}{2}$ ft in diameter by 2 ft wide. This has a capacity varying between 50,000 and 250,000 gpd depending on the type and amount of solids in the water and the fabric used. Larger units have capacities in excess of 10 mgd.

Coagulation and Settling

The addition to water of a coagulant such as alum (aluminum sulfate) results in the formation of a flocculent mass, or floc, which enmeshes

[44] George J. Turre, "Use of Micro-strainer Unit at Denver" and George R. Evans, "Discussion," *J. Am. Water Works Assn.*, **51**, 354–362 (March 1959).

microorganisms, suspended particles, and colloidal matter, removing and attracting these materials in settling out. The common coagulants used are alum, "black alum," activated alum, ammonium alum, sodium aluminate, copperas (ferrous sulfate), chlorinated copperas, ferric sulfate, ferric chloride, pulverized limestone, and clays.

To adjust the chemical reaction for improved coagulation, it is sometimes necessary to first add soda ash, hydrated lime, quicklime, or sulfuric acid. Mixing of the coagulant is usually done in two steps. The first step is rapid or flash mix and the second, slow mix. Rapid mix is a violent agitation for 1 or 2 min and may be accomplished by a mechanical agitator, pump impeller and pipe fittings, baffles, hydraulic jump, or other means. Slow mix is accomplished by means of baffles or a mechanical mixer for 20 to 30 min to promote formation of a floc. The coagulated water then flows to the settling or sedimentation basin designed to provide a retention of about 6 hr or an overflow rate of about 500 gpd per square foot of area. Around 80 percent of the turbidity, color, and bacteria are removed by this treatment. It is always recommended that mixing tanks and settling basins be at least two in number to permit cleaning and repairs without interrupting completely the water treatment, even though mechanical cleaning equipment is installed.

For the control of coagulation, jar tests are made in the laboratory to determine the approximate dosage (normally between 10 and 50 mg/l) of chemicals that appear to produce the best results.[45] Then, with this as a guide, the chemical-dosing equipment, dry feed, or solution feed is adjusted to add the desired quantity of chemical proportional to the flow of water treated to give the best results. Standby chemical feed units and alarm devices are necessary to assure continuous treatment.

Zeta-potential is also used to control coagulation. It involves determination of the speed at which particles move through an electric field caused by a direct current passing through the raw water. Best flocculation takes place when the charge approaches zero when a coagulant such as aluminum sulfate, assisted by a polyelectrolyte if necessary, is added.

Another device for the coagulation and settling of water consists of a unit in which the water to be treated is introduced at the bottom and flows upward through a blanket of settled floc. The surface loading (overflow rate) may be as high as 1500 gpd per square foot. The clarified

[45] Charles R. Cox, *Water Supply Control*, Bulletin No. 22, New York State Department of Health, Albany, 1952, pp. 38–53; See also *Standard Methods for the Examination of Water and Wastewater*, op. cit.

water flows off at the top. These basins are referred to as suspended-solids contact clarifiers. A major advantage claimed, where applicable, is reduction of the detention period and hence savings in space.

Filtration

Filters are of the slow sand, rapid sand, and pressure (or vacuum) type. Each has application under various conditions. The primary purpose of filters is to remove suspended materials. Of the filters mentioned, the slow sand filter is recommended for use at small communities and rural places, where adaptable. A rapid sand filter is not recommended because of the rather complicated control required to obtain satisfactory results, unless competent supervision and operation can be assured. The pressure filter, including the diatomaceous earth type, is commonly used for the filtration of industrial water supplies and swimming-pool water; it is not recommended for the treatment of drinking water, except under emergency or carefully controlled conditions.

Filter units, which are attached to faucets, porous stone filters, unglazed porcelain (Pasteur filter), or Berkefield filters may develop hairline cracks. They are unreliable and should not be depended on to remove pathogenic bacteria. Ceramic filters may have limited application under certain conditions.[46] They should be cleaned and sterilized in boiling water once a week.

Slow Sand Filter

A slow sand filter consists of a watertight basin, usually covered, built of concrete. The basin holds a special sand 24 to 48 in. deep, which is supported on a 12-in. layer of graded gravel placed over an underdrain system that may consist of open-joint, porous, or perforated pipe or conduits. The sand should have an effective size of 0.25 to 0.35 mm and a uniformity coefficient of about 1.75, but not exceeding 2.5. Operation of the filter is controlled so that filtration will take place at a rate of 1 to 4 million gal per acre per day, with $2\frac{1}{2}$ million gal as an average rate. This would correspond to a filter rate of 23 to 92 gal/ft^2 of sand area per day or an average rate of 57 gal. A rate of 10 million gal may be used if permitted by the approving authority.

From a practical standpoint, the water that is to be filtered should have low color, less than 30 units, and be relatively clear, with a turbidity of less than 50 units (not clay); otherwise the filter will clog quickly. A plain sedimentation basin ahead of the filter can be used to reduce

[46] Robert Newton Clark, "The Purification of Water on a Small Scale," *Bull. Santé organisation mondiale (Bull. World Health Organization)*, **14**, 820–826 (1956).

turbidity of the water if necessary. A loss of head gauge should be provided on the filter to show the resistance the sand bed offers to the flow of water through it and to show when the filter needs cleaning. This is done by draining the water out of the sand bed and scraping ½ in. of sand with adhering particles off the top of the bed. The sand is replaced when the depth of sand is reduced to about 24 in. A scraper or flat shovel is practical for removing the top layer of clogged sand. Slow sand filters should be constructed in pairs where possible. These filters are easily controlled and produce a consistently satisfactory water.

A well-operated plant will remove 98 to 99.5 percent of the bacteria in the raw water (after a film has formed on the surface of the sand, which will require slow filtration for about 2 weeks). Chlorination of the filtered water is necessary to destroy those bacteria that pass through the filter and to destroy bacteria that grow or enter the storage basin and water system. This type plant will also remove about 25 to 40 percent of the color in the untreated water. Chlorination of the sand filter itself is desirable either continuously or periodically to destroy bacteria that grow within the sand bed, supporting gravel, or underdrain system. Continuous prechlorination at a dosage to provide 0.3 to 0.5 mg/l in the water on top of the filter will not harm the filter film and will increase the length of the filter run.

A slow sand filter suitable for a small rural water supply is shown in Figure 3–13. Details relating to design are given in Table 3–8. The rate of filtration in this filter is controlled by selecting an orifice and filter area that will deliver not more than 50 gal/ft² of filter area per day and thus prevent excessive rates of filtration that would endanger the quality of the treated water. Where competent and trained personnel are available, the rate of flow can be controlled by manipulating a gate valve on the effluent line from each filter, provided a venturi, orifice, or other suitable meter, with indicating and preferably recording instruments, is installed to measure the rate of flow. The valve can then be adjusted to give the desired rate of filtration until the filter needs cleaning. Another practical method of controlling the rate of filtration is by installing a float valve on the filter effluent line as shown in Figure 3–14. The valve is actuated by the water level in a float chamber, which is constructed to maintain a reasonably constant head over an orifice in the float chamber.[47] A special rate-control valve can also be used

[47] A remote float-controlled weighted butterfly valve, with spring-loaded packing glands and stainless steel shafts is described by Thomas M. Riddick, "An Improved Design for Rapid Sand Filter," *J. Am. Water Works Assoc.*, 44, 733–744 (August 1952).

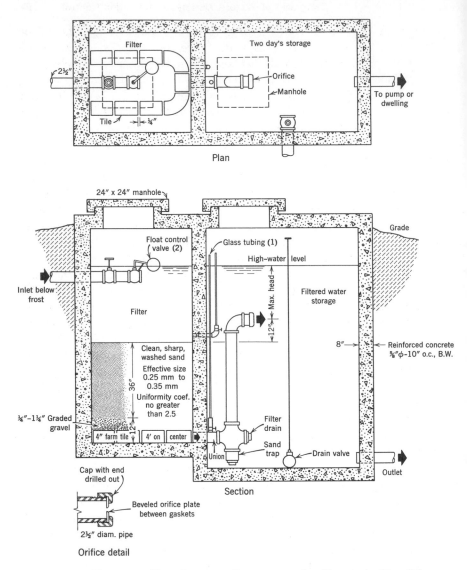

Figure 3–13 Slow sand filter for a small water supply. Notes: 1. The difference in water level between the two glass tubes represents the frictional resistance to the flow of water through the filter. When this difference approaches the maximum head and the flow is inadequate, the filter needs cleaning. To clean, scrape the top ½ in. of sand bed off with a mason's trowel, wash in a pan or barrel, and replace clean sand on bed. 2. Float control valve may be omitted where water on filter can be kept at a desirable level by gravity flow or by an overflow or float switch. 3. Add a meter, venturi, or other flow-measuring device on the inlet to the filter. 4. Rate of flow can also be controlled by maintaining a constant head with a weighted float valve over a triangular weir.

168

TABLE 3-8 FLOWS FROM ORIFICES UNDER VARIOUS HEADS OF WATER

Diameter of Orifice (in in.)

Maximum Flow (in gpd)*

Max. Head, (in ft of water)	$\frac{1}{16}$	$\frac{3}{32}$	$\frac{1}{8}$	$\frac{3}{16}$	$\frac{1}{4}$	$\frac{5}{16}$	$\frac{3}{8}$	$\frac{7}{16}$	$\frac{1}{2}$	$\frac{9}{16}$	$\frac{5}{8}$	$\frac{11}{16}$	$\frac{3}{4}$	$\frac{7}{8}$	1
1	67	149	266	597	1,060	1,660	2,390	3,240	4,240	5,370	6,640	8,010	9,550	13,030	17,000
$1\frac{1}{2}$	82	183	326	732	1,305	2,040	2,930	3,990	5,220	6,580	8,130	9,850	11,700	15,950	20,800
2	96	213	380	852	1,520	2,370	3,410	4,650	6,060	7,680	9,480	11,480	14,300	18,600	24,200
$2\frac{1}{2}$	107	236	421	945	1,680	2,620	3,780	5,150	6,720	8,500	10,500	12,680	15,100	20,600	26,800
3	116	259	462	1,036	1,840	2,880	4,140	5,640	7,380	9,130	11,520	13,920	16,550	22,600	29,500
$3\frac{1}{2}$	126	279	498	1,120	1,990	3,100	4,470	6,060	7,950	10,050	12,420	14,900	17,900	24,300	30,700
4	135	300	534	1,200	2,125	3,330	4,790	6,530	8,520	10,800	13,300	16,100	19,150	26,100	34,000
$4\frac{1}{2}$	145	322	574	1,290	2,290	3,580	5,160	7,020	9,150	11,600	14,350	17,300	20,600	28,000	36,600
5	150	333	594	1,332	2,370	3,700	5,340	7,260	9,480	12,000	14,800	17,950	21,400	29,000	38,700
$5\frac{1}{2}$	157	350	624	1,400	2,490	3,890	5,600	7,620	9,960	12,600	15,550	18,850	22,400	30,500	39,300
6	164	366	650	1,460	2,600	4,070	5,850	7,980	10,400	13,150	16,250	19,900	23,400	31,800	41,500
$6\frac{1}{2}$	169	376	672	1,510	2,680	4,180	6,030	8,220	10,700	13,580	16,750	20,200	24,200	32,800	42,800

Example: To find the size of a filter that will deliver a maximum of 500 gpd: From Table 3-8, a filter with a $\frac{1}{8}$-inch orifice and a head of water of 3' 9" will meet the requirements. Filtering at the rate of 50 gpd/ft² of filter area, the required filter area = $\dfrac{500 \text{ gpd}}{50 \text{ gal}/(\text{ft}^2)(\text{day})} = 10 \text{ ft}^2$. Provide at least 2 days' storage capacity.

*No loss head through sand and gravel or pipe is assumed; flow is based on $Q = C_d V A$, where $V = \sqrt{2gh}$ and C_d = 0.6, with free discharge. (Design filter for twice the desired flow to assure an adequate delivery of water as the frictional resistance in the filter to the flow builds up. Use two or more units in parallel.) Note: The loss of head through a clean filter is about 3 inches, hence add 3 inches to the "maximum head," in the table and sketch, to obtain the indicated flow in practice. A minimum 2 to 3 ft of water over the sand is advised.

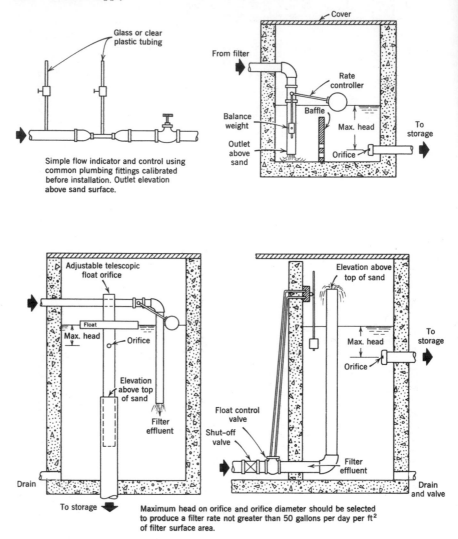

Figure 3–14 Typical devices for the control of the rate of flow or filtration. Plant capacity—50 to 100% greater than average daily demand, with clear well.

if it is accurate within the limits of flow desired. The level of the orifice or filter outlet must be *above* the top of the sand to prevent the developing of a negative head. If a negative head is permitted to develop, the mat on the surface of the sand may be broken and dissolved air in the water may be released in the sand bed, causing the bed to become air-bound. At least 6 in. of water over the sand will minimize possible disturbance of the sand when water from the influent line falls into the filter.

Rapid Sand Filter

A rapid sand gravity filter, also referred to as a mechanical filter, is shown in Figure 3–15. Two important accessories to a rapid sand filter are the loss of head gauge and the rate controller. The loss of head gauge shows the frictional resistance to the flow of water through

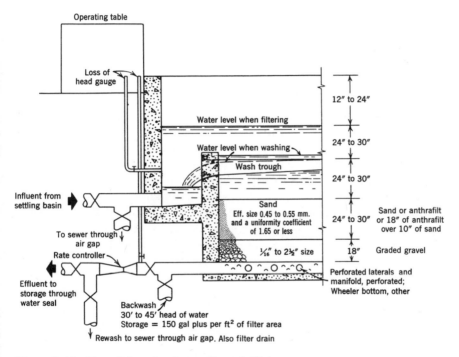

Figure 3–15   Essential parts of a rapid sand filter.

Rate of filtration = $\dfrac{7.48}{\text{minutes for water in filter to fall 1 ft}}$; fill filter with water, shut off influent, open drain.

Backwash time = 15 minutes minimum, until water entering trough is clear

Normal wash water usage = 2 to 2.5% or less of water filtered

Sand expansion = 40 to 50% ▬ 33.6 to 36 in. for 24-in. sand bed
= 25 to 35% for dual media, anthracite and sand

Rate of backwash = $\dfrac{7.48}{\text{minutes for water in filter to rise 1 ft}}$; lower water level to sand, open backwash valve, 15 to 20 gpm/ft² minimum

Orifice area = 0.25 to 0.30% of filter area

the sand, laterals, and orifices. When this reaches about 7 to 9 ft, it indicates that the filter needs to be backwashed. The rate controller is constructed to automatically maintain a uniform predetermined rate of filtration through the filter, usually about 2 gpm/ft², until the filter needs cleaning. Disturbance of filter rate causes breakthrough of filter floc. Filter design and operation should reduce the possible magnitude of filter fluctuations.[48]

A filter rate of 3 to 4 gal may be permitted with skilled operation, if pretreatment can assure water on the filter has a turbidity of less than about 10 units and a coliform concentration of less than 2.2. Sand for the higher rate would have an effective size of 0.5 to 0.7 mm and a uniformity coefficient of 1.5 to 2.0. In a combination anthracite over sand bed, use is made of the known specific gravity of crushed anthracite of 1.5 and the specific gravity of sand of 2.5 plus. The relative weight of sand in water is three times that of anthracite. Fair and Geyer have shown that anthracite grains can be twice as large as sand grains and that after backwashing the sand will settle in place before the anthracite in two separate layers.[49] Combination sand-anthrafilt filters require careful operating attention and use of a filter conditioner to prevent floc penetration. Longer filter runs and less washwater are reported.

Water to be filtered passes downward through the sand and gravel, is collected in the underdrain system, and conducted to the clear water basin or filtered water reservoir, which is usually covered. A rapid sand filter is used for the filtration of polluted (see Table 3–6), turbid, and colored water at a higher rate of filtration than is permissible with a slow sand filter. A flow diagram is shown in Figure 3–16.

Treatment of the raw water by coagulation and settling, and usually chlorination, to remove as much as possible of the pollution is a necessary and important preliminary step in the rapid sand filtration of water. The settled water, in passing to the filter, carries with it some flocculated suspended solids, color, and bacteria. This material forms a mat on top of the sand that aids greatly in the straining and removal of other suspended matter, color, and bacteria; but this also causes rapid clogging of the sand. Special arrangement is therefore made in the design for washing the filter by forcing water backward up through the filter at a rate that will provide a sand expansion of 40 to 50 percent based on the water temperature and sand effective size. For example, with a 0.4 mm effective size sand, a 40 percent sand expansion requires a

[48] J. L. Cleasby, M. W. Williamson, and E. R. Baumann, "Effect of Filtration Rate Changes on Quality," *J. Am. Water Works Assoc.*, **55**, 869–880 (July 1963).

[49] G. M. Fair and J. C. Geyer, *Water Supply and Waste Water Disposal*, John Wiley & Sons, Inc., New York, 1954, p. 677.

Flow diagram

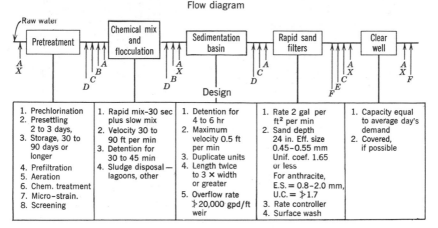

Figure 3–16   Rapid sand filter plant flow diagram.

Possible Chemical Combinations

A   Chlorine. Also bar racks and coarse screens if needed.

B   Coagulant; aluminum sulfate (pH 5.5 to 8.0) ferric sulfate (pH 5.0 to 11.0), ferrous sulfate (pH 8.5 to 11.0), ferric chloride (pH 5.0 to 11.0), sodium aluminate, activated silica, organic chemicals (polyelectrolytes).

C   Alkalinity adjustment; lime, soda ash, or polyphosphate.

D   Activated carbon, potassium permanganate.

E   Dechlorination; sulfur dioxide, sodium sulfite, sodium bisulfite, activated carbon.

F   Fluoridation treatment.

X   Chlorine dioxide.

NOTE: The chlorinator should be selected to prechlorinate surface water at 20 mg/l and postchlorinate at 3 mg/l. Provide for a dose of 3 mg/l plus chlorine demand for groundwater.

wash-water rate rise of 21 in./min with 32°F water and a rise of 33½ in. with water at 70°F.[50] The dirty water is carried off to waste by troughs built in above the sand bed 5 to 6 ft apart. A system of water jets or rakes or a 1½- to 2-in. pressure line at 45 to 75 psi with hose connections should be provided to scour the surface of the sand to assist in loosening and removing the material on the sand. Effective washing of the sand is essential.

When properly operated a filtration plant, including coagulation and settling, can be expected to remove about 98 percent of the bacteria, a great deal of the color, and practically all the suspended solids. Nevertheless, chlorination must be used to assure that the water leaving the

[50] Robert W. Abbett, *American Civil Engineering Practice*, Vol. II, John Wiley & Sons, Inc., New York, 1956.

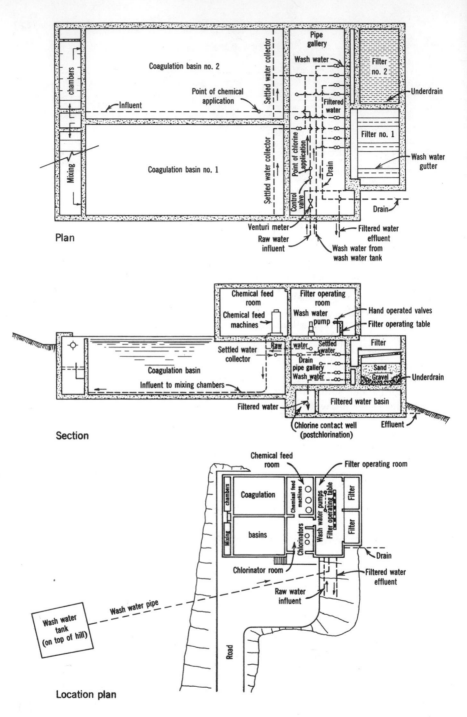

Figure 3–17

plant is safe to drink. Construction of a rapid sand filter should not be attempted unless it is designed and supervised by a competent sanitary engineer. The MPN of coliform organisms in the raw water to be treated should not exceed that listed in Table 3–6 unless the water is brought within the permissible limits by preliminary treatment.

A flow diagram of a typical treatment plant is shown in Figure 3–17.

Pressure Sand Filter

A pressure filter is similar in principle to the rapid sand gravity filter except that it is completely enclosed in a vertical or horizontal cylindrical steel tank through which water under pressure is filtered. The normal filtration rate is 2 gpm/ft^2 of sand. Higher rates are used. Pressure filters are most frequently used in swimming-pool and industrial-plant installations. It is possible to use only one pump to take water from the source or out of the pool (force it through the filter and directly into the plant water system or back into the pool), which is the main advantage of a pressure filter. This is offset by difficulty in introducing chemicals under pressure, inadequate coagulation facilities, and lack of adequate settling. The appearance of the water being filtered and the condition of the sand cannot be seen; the effectiveness of backwashing cannot be observed; the safe rate of filtration may be exceeded; and it is difficult to look inside the filter for the purpose of determining loss of sand or anthrafilt, need for cleaning, replacing of the filter media, and inspection of the wash-water pipes, influent, and effluent arrangements. Because of these disadvantages and weaknesses, a pressure filter is not considered dependable for the treatment of contaminated water to be used for drinking purposes. It may, however, have limited application for small, slightly contaminated water supplies. The coagulant should be well mixed and the water settled in a basin open for inspection before being pumped through a pressure filter. This will require double pumping if the quality of the filter effluent is to be kept under close observation.

Diatomaceous Earth Filter

The pressure filter type consists of a closed steel cylinder inside of which are suspended septa, the filter elements. In the vacuum type the septa are in an open tank under water that is recirculated with a vacuum

Figure 3–17 Flow diagram of typical treatment plant. This plant is compactly arranged and adaptable within a capacity range of 0.25 to 1.0 gpm. Operation is simple as the emphasis is on manual operation with only the essentials in mechanical equipment provided. Design data are described in the text. (From *Water Treatment Plant Design*, American Water Works Association, Inc., New York, 1969, p. 28.)

inside the septa. Normal rates of filtration are 1 to $2\frac{1}{2}$ gpm/ft^2 of element surface. To prepare the filter for use a slurry of filter aid (precoat) of diatomaceous earth is introduced with the water to be treated at a rate of about $1\frac{1}{2}$ oz/ft^2 of filter septum area, and the water is recirculated for at least 1 min before discharge. Then additional filter aid (body coat) is added with the water to maintain the permeability of the filter media. The rate of feed is roughly 2 to 3 mg/l per unit of turbidity in the water. Filter aid comes in different particle sizes. It forms a coating or mat around the outside of each filter element and is more efficient than sand in removing from water suspended matter and such organisms as cysts, which cause amebiasis; cercariae, which cause schistosomiasis; flukes, which cause paragonimiasis and clonorchiasis; and worms, which cause ascariasis and trichuriasis. These organisms, except for amebic cysts and possibly schistosomes, are not common in the United States.

Like the pressure sand filter, the diatomite filter has found greatest practical application in swimming pools and in industrial and military installations. It has a special advantage in the removal of oil from condensate water, since the diatomaceous earth is wasted. It should not be used to treat a public water supply unless pilot plant study results on the water to be treated meet the health department requirements.

A major weakness in the diatomite filter is that failure to add diatomaceous earth to build up the filtering mat, either through ignorance or negligence, will make the filter entirely ineffective and give a false sense of security. In addition, the septa will become clogged and require replacement or removal and chemical cleaning. During filtration, the head loss through the filter increases to 40 or 50 lb/in.^{2}, thereby requiring a pump and motor with a wide range in the head characteristics. The cost of pumping water against this higher head is therefore increased. Diatomite filters cannot be used where pump operation is intermittent, as with a pressure tank installation, for the filter cake will slough off unless sufficient continuous recirculation is provided. A reciprocating pump should not be used.

The filter is backwashed by reversing the flow of the filtered water back through the septum, thereby forcing all the diatomite to fall to the bottom of the filter shell, from which point it is flushed to waste. The filter should not be used to treat raw water with greater than 2400 MPN per 100 ml, 30 JTUs turbidity, or 3000 areal standard units per 100 ml. It does not remove taste- and odor-producing substances. In any case, prechlorination is considered a necessary adjunct to filtration. The diatomite filter must be carefully operated by trained personnel in order to obtain dependable results.

Causes of Tastes and Odors

Tastes and odors in water supplies are caused by oils, minerals, gases, organic matter, and other compounds and elements in the water. Some of the common causes are oils and products of decomposition exuded by algae and some other microorganisms; wastes from gas plants, coke ovens, paper mills, chemical plants, canneries, tanneries, oil refineries, and dairies; high concentrations of iron, manganese, sulfates, and hydrogen sulfide in the water; decaying vegetation such as leaves, weeds, grasses, brush, and moss in the water; and high concentrations of chlorine. The control of taste- and odor-producing substances is best accomplished by eliminating or controlling the source when possible. When this is not possible or practical, study of the origin and type of the tastes and odors should form the basis for needed treatment.

Control of Microorganisms

Microorganisms that cause tastes and odors are for the most part harmless. They are visible under a microscope and include bacteria, protozoa, algae, small worms, small crustacea, and others. *Crenothrix* and *Leptothrix*, also known as iron bacteria, can also be included. Algae utilize carbon dioxide in water and serve as a basic food for fish life. All water is potentially a culture medium for one or more kinds of algae. Bacteria are dormant in water below a temperature of about 48°F (10°C).

Crenothrix and *Leptothrix* are reddish brown, gelatinous, stringy masses that grow in the dark inside distribution systems or wells carrying water devoid of oxygen but containing iron in solution. Control, therefore, may be effected by removal of iron from the water before it enters the distribution system, increase in the concentration of dissolved oxygen in the water above about 2 mg/l, the continuous addition of copper sulfate to provide a dosage of 0.3 to 0.5 mg/l, or the continuous addition of chlorine to provide a free residual chlorine concentration of about 0.3 mg/l, or 0.5 to 1.0 mg/l total chlorine. Chemical treatment of the water will destroy and dislodge growths in the mains, with resulting temporary intensification of objectionable tastes and odors, until all the organisms are flushed out of the water mains. The slime bacteria known as *actinomycetes* are also controlled by this treatment.

High water temperatures, optimum pH values and alkalinities, adequate food such as mineral matter (particularly nitrates, phosphorous, potassium and carbon dioxide), low turbidities, large surface area, shallow depths, and sunlight favor the growth of plankton. Exceptions are diatoms, such as *Asterionella*, which grow also in cold water at considerable depth without the aid of light. Fungi can also grow in the absence of sunlight. Extensive growths of *Anabaena, Oscillaria,* and *Microcystis* resembling pea-green soup are encouraged by calcium and nitrogen. Protozoa such as *Synura* are similar to algae, but they do not need carbon dioxide; they grow in the dark and in cold water. The blue-green algae do not require direct light for their growth; but green algae do. They are found in higher concentrations within about 5 ft of the water surface. Sawyer has indicated that any lake having, at the time of the spring overturn, inorganic phosphorus greater than 0.01 mg/l and inorganic nitrogen greater than 0.3 mg/l can be expected to have major algal blooms.[51]

Inasmuch as the products of decomposition and the oils given off by algae and protozoa cause disagreeable tastes and odors, preventing the growth of these microorganisms will remove the cause of difficulty. Where it is practical to cover storage reservoirs to exclude light, this is the easiest way to prevent the growth of those organisms that require light and cause difficulty. Where this is not possible, copper sulfate or chlorine should be applied to prevent the growth of the microorganisms. A combination of chlorine, ammonia, and copper sulfate has also been used with good results. However, in order that the proper chemical dosage required may be determined, it is advisable to have made microscopic examinations of samples collected at various depths and locations to determine the type, number, and distribution of organisms. This may be supplemented by laboratory tests using the water to be treated and the proposed chemical dose before actual treatment. In New England, diatoms usually appear in the spring, blue-green algae in the summer and fall; then diatoms reappear in fall and winter. The green algae appear between the diatoms and blue-green algae.

In general, the application of about 2½ lb of copper sulfate per million gallons of water treated at intervals of 2 to 4 weeks between April and October, in the temperate zone, will prevent difficulties from most microorganisms. More exact dosages for specific microorganisms are given in Table 3–9. The required copper sulfate dose can be based on the volume of water in the upper 10 ft of a lake or reservoir, as most plankton are found within this depth. Bartsch[52] suggests an arbitrary dosage related to the alkalinity of the water being treated. A copper sulfate dosage of 2¾ lb per million gallons of water in the reservoir

[51] C. N. Sawyer, "Phosphates and Lake Fertilization," *Sewage and Ind. Wastes,* **24,** 768 (June 1952).
[52] A. F. Bartsch, "Practical Methods for Control of Algae and Water Weeds," *Public Health Reports,* **69,** No. 8, 749–757 (August 1954).

is recommended when the methyl orange alkalinity is less than 50 mg/l. When the alkalinity is greater than 50 mg/l, a dosage of 5.4 lb per acre of reservoir surface area is recommended. Higher doses are required for the more resistant organisms. The dose needed should be based on the type of algae making their appearance in the affected areas, as determined by periodic microscopic examinations. See page 180. An inadequate dosage is of very little value and is wasteful. Higher dosages than necessary have caused wholesale fish destruction. For greater accuracy, the copper sulfate dose should be increased by $2\frac{1}{2}$ percent for each degree of temperature above 59°F (15°C), and 2 percent for each 10 mg/l organic matter. Consideration must also be given to the dosage applied to prevent the killing of fish. If copper sulfate is evenly distributed, in the proper concentration, and in accordance with Table 3–10, there should be very little destruction of fish. Fish can withstand higher concentrations of copper sulfate in hard water. If a heavy algal crop has formed and then copper sulfate applied, the decay of algae killed may clog the gills of fish and reduce the supply of oxygen to the point that fish will die of asphyxiation, especially at times of high water temperatures. Tastes and odors are of course also intensified. Blue-green algae may produce a toxin that is lethal to fish. See page 126. Other conditions may also be responsible for destruction of fish. For example, a pH value below 4 to 5 or above 9 to 10; a free ammonia or equivalent of 1.2 to 3 mg/l; an unfavorable water temperature; a carbon dioxide concentration of 100 to 200 mg/l or even less; free chlorine of 0.15 to 0.3 mg/l, chloramine of 0.4 to 0.76 mg/l; 0.5 to 1.0 mg/l hydrogen sulfide and other sulfides; cyanogen; phosphine; sulfur dioxide; and other waste products are all toxic to fish.[53] Lack of food, overproduction, and species survival also result in mass "fish kills."

Copper sulfate may be applied in several ways. The method used usually depends on such things as the size of the reservoir, equipment available, proximity of the microorganisms to the surface, reservoir inlet and outlet arrangement, and time of year. One of the simplest methods of applying copper sulfate is the burlap-bag method. A weighed quantity of crystals (bluestone) is placed in a bag and towed at the desired depth behind a rowboat or preferably motor-driven boat. The copper sulfate is then drawn through the water in accordance with a planned pattern, first in one direction in parallel lanes about 25 ft apart and then at right angles to it so as to thoroughly treat the entire body

---

[53] Peter Doudoroff and Max Katz, "Critical Review of Literature on the Toxicity of Industrial Wastes and their Components to Fish," *Sewage and Ind. Wastes,* **22,** 1432–1458 (November 1950). Trout are usually more sensitive.

TABLE 3-9 DOSAGE OF COPPER SULFATE TO DESTROY MICROORGANISMS, POUNDS PER MILLION GALLONS

Organism	Taste, Odor, Other	Dosage
Diatomaceae	(Usually brown)	
Asterionella	Aromatic, geranium, fishy	1.0 to 1.7
Cyclotella	Faintly aromatic	Use chlorine
Diatoma	Faintly aromatic	—
Fragilaria	Geranium, musty	2.1
Meridon	Aromatic	—
Melosira	Geranium, musty	1.7 to 2.8
Navicula		0.6
Nitzchia		4.2
Stephanodiscus	Geranium, fishy	2.8
Synedra	Earthy, vegetable	3.0 to 4.2
Tabellaria	Aromatic, geranium, fishy	1.0 to 4.2
Chlorophyceae	(Green algae)	
Cladophora	Septic	4.2
Closteriun	Grassy	1.4
Coelastrum		0.4 to 2.8
Conferva		2.1
Desmidium		16.6
Dictyosphaerium	Grassy, nasturtium, fishy	Use chlorine
Draparnaldia		2.8
Entomophora		4.2
Eudorina	Faintly fishy	16.6 to 83.0
Gloeocystis	Offensive	—
Hydrodictyon	Very offensive	0.8
Miscrospora		3.3
Palmella		16.6
Pandorina	Faintly fishy	16.6 to 83.0
Protococcus		Use chlorine
Raphidium		8.3
Scenedesmus	Vegetable, aromatic	8.3
Spirogyra	Grassy	1.0
Staurastrum	Grassy	12.5
Tetrastrum		Use chlorine
Ulothrix	Grassy	1.7
Volvox	Fishy	2.1
Zygnema		4.2
Cyanophyceae	(Blue-green algae)	
Anabaene	Moldy, grassy, vile	1.0
Aphanizomenon	Moldy, grassy, vile	1.0 to 4.2
Clathrocystis	Sweet, grassy, vile	1.0 to 2.1
Coelosphaerium	Sweet, grassy	1.7 to 2.8
Cylindrosphermum	Grassy	1.0
Gloeocopsa	(Red)	2.0
Microcystis	Grassy, septic	1.7
Oscillaria	Grassy, musty	1.7 to 4.2
Rivularia	Moldy, grassy	—
Protozoa		
Bursaria	Irish moss, salt marsh, fishy	—
Ceratium	Fishy, vile (Red-brown)	2.8
Chlamydomonas		4.2 to 8.3
Cryptomonas	Candied violets	4.2
Dinobryon	Aromatic, violets, fishy	1.5
E. histolytica		Use chlorine
(cyst)		5 to 25 mg/l
Euglena		4.2

TABLE 3-9   DOSAGE OF COPPER SULFATE TO DESTROY MICROORGANISMS, POUNDS PER MILLION GALLONS (cont'd)

Organism	Taste, Odor, Other	Dosage
Protozoa (cont'd)		
Glenodinium	Fishy	4.2
Mallomonas	Aromatic, violets, fishy	4.2
Peridinium	Fishy, like clam-shells, bitter taste	4.2 to 16.6
Synura	Cucumber, musk-melon, fishy	0.25
Uroglena	Fishy, oily, codliver oil	0.4 to 1.6
Crustacea		
Cyclops		16.6
Daphnia		16.6
Schizomycetes		
Beggiatoa	Very offensive, decayed	41.5
Cladothrix		1.7
Crenothrix	Very offensive, decayed	2.8 to 4.2
Leptothrix	Medicinal with chlorine	—
Sphaerotilis natans	Very offensive, decayed	3.3
Thiothrix		Use chlorine
Fungi		
Achlya		—
Leptomitus		3.3
Saprolegnia		1.5
Miscellaneous		
Blood worm		Use chlorine
Chara		0.8 to 4.2
Nitella flexilis	Objectionable	0.8 to 1.5
Phaetophyceae	(Brown algae)	—
Potamogeton		2.5 to 6.7
Rhodophyceae	(Red algae)	—
Xantophyceae	(Green algae)	—

Note: Chlorine residual 0.5 to 1.0 mg/l will also control most growths, except *melosira,* cysts of *Endamoeba histolytica, Crustacea,* and *Synura* (2.0 mg/l free).

of water. The rapidity with which the chemical goes into solution may be controlled by regulating the fabric of the bag used, varying the velocity of the boat, using crystals of large or small size, or by combinations of these variables. In another method, a long box is attached vertically to a boat. The bottom vertical side of the box, which is placed below water, has a copper screen opening with an adjustable cover; the top has a hopper provided, into which copper sulfate is added. The rate of solution of the copper sulfate is controlled by raising or lowering the adjustable cover over the screen, thus exposing more or less copper sulfate to the water. Where spraying equipment is available, copper sulfate may be dissolved in a barrel or tank carried in the boat and sprayed on the surface of the water as a ½ or 1 percent solution. Pulverized copper sulfate may be distributed over large reservoirs or lakes by means of a mechanical blower carried on a motor-driven boat. Larger crystals are more effective against algae at lower depths. Where water

TABLE 3-10 DOSAGE OF COPPER SULFATE AND RESIDUAL CHLORINE, WHICH IF EXCEEDED MAY CAUSE FISH KILL

Fish	Copper Sulfate (lb/mil gal)	(mg/l)	Free Chlorine, (in mg/l)	Chloramine, (in mg/l)
Trout	1.2	0.14	0.10 to 0.15	0.4
Carp	2.8	0.33	0.15 to 0.2	0.76 to 1.2
Suckers	2.8	0.33		
Catfish	3.5	0.40		
Pickerel	3.5	0.40		
Goldfish	4.2	0.50	0.25	
Perch	5.5	0.67		
Sunfish	11.1	1.36		0.4
Black bass	16.6	2.0		
Minnows			0.4	0.76 to 1.2
Bullheads				0.4
Trout fry				0.05 to 0.06
Gambusia				0.5 to 1.0

flows into a reservoir, it is possible to add copper sulfate continuously and proportional to the flow, provided fish life is not important. This may be accomplished by means of a commercial chemical feeder, an improvised solution drip feeder, or a perforated box feeder wherein lumps of copper sulfate are placed in the box and the depth of submergence in the water is controlled to give the desired rate of solution. In the winter months when reservoirs are frozen over, copper sulfate may be applied if needed by cutting holes in the ice 20 to 50 ft apart and lowering and raising a bag of copper sulfate through the water several times. If an outboard motor is lowered and rotated for mixing, holes may be 1000 ft apart. Scattering crystals on the ice is also effective in providing a spring dosage when this is practical.

It is possible to control microorganisms in a small reservoir, where chlorine is used for disinfection and water is pumped to a reservoir, by maintaining a free residual chlorine concentration of about 0.3 mg/l in the water. However, chlorine will combine with organic matter and be used up or dissipated by the action of sunlight unless the reservoir is covered and there is a sufficiently rapid turnover of the reservoir water. Where a contact time of 2 hr or more can be provided between the water and disinfectant, the chlorine-ammonia process may be used to advantage. Chlorine may also be added as chloride of lime or in liquid form by methods similar to those used for the application of copper sulfate.

Gnat flies sometimes lay their eggs in reservoirs. The eggs develop into larvae, causing consumer complaints of worms in the water. The best control measure is covering the reservoir or using fine screening to prevent the entrance of gnats.

Aquatic Weed Control

Vegetation that grows and remains below the water surface does not generally cause difficulty. Decaying and emergent aquatic vegetation, as well as decaying leaves, brush, weeds, grasses, and debris in the water can cause tastes and odors in water supplies. The discharge of organic wastes from wastewater treatment plants, storm sewers, and drainage from lawns, pastures, and fertilized fields contains nitrogen and phosphorus, which promote algal and weed growths. The contribution of phosphorus from sewage treatment plants can be relatively small compared to that from surface runoff. Unfortunately little can be done to permanently prevent the entrance of wastes and drainage or destroy growths of rooted plants, although certain chemical, mechanical, and biological methods can provide temporary control.

Reasonably good temporary control of rooted aquatic plants may be obtained by physically removing the growths by dredging or with wire, chain drags, or rakes, and by cutting. Filling of marshy areas and deepening the edges of reservoirs, lakes, and ponds to a depth of 2 ft or more will prevent or reduce plant growths. Weeds that float to the surface should be removed before they decay.

Where it is possible to drain or lower the water level 6 ft to expose the affected areas of the reservoir for about one month, drying of weeds and roots, clearing and removal, direct burning, or oiling of the vegetation, peat, and so forth with a light oil such as kerosene and thorough burning is of great value. Drying out of roots and burning and removal of the ash is effective for a number of years. Flooding 3 ft or more above normal is also effective where possible.

As a last resort, aquatic weeds may be controlled by chemical means. Tastes and odors may result if the water is used for drinking purposes; the chemical may kill fish and persist in the bottom mud, and it may be hazardous to the applicator. The treatment must be repeated annually or more often, and heavy algal blooms may be stimulated, particularly if the plant destroyed is allowed to remain in the water and return its nutrients to the water. Chemical use should be restricted and permitted only after careful review of the toxicity to humans and fish, the hazards involved, and the purpose to be served. Copper sulfate should not be used for the control of aquatic weeds, since the concentration required to destroy the vegetation will assuredly kill any fish present in the water. See also control of Aquatic Weeds in Chapter 9.

Other Causes of Tastes and Odors

Reservoir clearance and drainage reduce algal blooms by removing organic material beneficial to their growth. Organic material, which can cause anaerobic decomposition, odors, tastes, color, and acid conditions in the water is also removed. If topsoil is valuable, its removal may be worthwhile.

Some materials in water cause unpleasant tastes and odors when present in excessive concentrations, although this is not a common source of difficulty. Iron and manganese, for example, may give water a bitter, astringent taste. In some cases sufficient natural salt is present, or saltwater enters to cause a brackish taste in well water. It is not possible to remove the salt in the well water without going to great expense. Elimination of the cause by sealing off the source of the saltwater, groundwater recharge with freshwater, or controlling pump drawdown is sometimes possible. Permissible limits of chemical constituents in water are given on pages 117 to 125 and 154.

Other causes of tastes and odors are sewage and industrial or trade wastes. Sewage would have to be present in very large concentrations to be noticeable in a water supply. If this were the case, the dissolved oxygen in the water receiving the sewage would most probably be used up, with resultant nuisance conditions. On the other hand the billions of bacteria introduced, many of which would cause illness or death if not removed or destroyed before consumption, are the greatest danger in sewage pollution. Trade or industrial wastes introduce in water suspended or colloidal matter, dissolved minerals, phenols, vegetable and animal organic matters, harmful bacteria, poisons, and other materials that produce tastes and odors. Of these, the wastes from steel mills, paper plants, and coal distillation (coke) plants have proved to be the most troublesome in drinking water, particularly in combination with chlorine. Tastes produced have been described as "medicinal," "phenolic," "iodine," "carbolic acid," and "creosote." Concentrations of one part phenol to 500 million parts of water will cause very disagreeable tastes even after the water has traveled 70 miles.[54] The control of these tastes and odors lies in the prevention and reduction of stream pollution through improved plant operation and waste treatment. Chlorine dioxide

[54] *Manual of Water Quality Control,* American Water Works Association, New York, 1940, p. 49.

has been found effective in treating a water supply not too heavily polluted with phenols. The control of stream pollution is a function and responsibility of federal and state agencies, municipalities, and industry. Treatment of water supplies to eliminate or reduce objectionable tastes and odors is discussed separately below.

Sometimes high uncontrolled doses of chlorine produce chlorinous tastes and chlorine odors in water. This may be due to the use of constant feed equipment rather than a chlorinator, which will vary the chlorine dosage proportional to the quantity of water to be treated. In some installations chlorine is added at a point that is too close to the consumers, and in others the dosage of chlorine is marginal or too high, or chlorination treatment is used where coagulation, filtration, and chlorination should be used instead. Where superchlorination is used and high concentrations of chlorine remain in the water, dechlorination with sodium sulphite, sodium thiosulphate, sulphur dioxide, or activated carbon is indicated.

Methods to Remove or Reduce Objectionable Tastes and Odors

Some of the common methods used to remove or reduce objectionable tastes and odors in drinking water supplies, not in order of their effectiveness, are

1. Free residual chlorination or superchlorination.
2. Chlorine-ammonia treatment.
3. Aeration, or forced-draft degasifier.
4. Application of activated carbon.
5. Filtration through granular carbon, or charcoal filters.
6. Coagulation and filtration of water (also using an excess of coagulant).
7. Control of reservoir intake level.
8. Elimination or control of source of trouble. (See pages 177–184)
9. Chlorine dioxide treatment.
10. Ozone treatment.

Free Residual Chlorination

Free residual chlorination will destroy by oxidation most taste- and odor-producing substances and inhibit growths inside water mains. Biochemical corrosion is also prevented in the interior of water mains by destroying the organisms associated with the production of organic acids. The reduction of sulfates to objectionable sulfides is also prevented.

Nitrogen trichloride is formed in water high in organic nitrogen when a high free chlorine residual is maintained and the pH is less than 8.0.

(See page 161). It is an explosive, volatile, oily liquid that is removed by aeration or carbon.

Chlorine-Ammonia Treatment

Chlorine-ammonia treatment in practice is the addition of about three parts of chlorine to one part of ammonia. The ammonia is added a few feet ahead of the chlorine. Chloramines, which react slowly, are formed. Chloramines prevent chlorinous tastes due to the reaction of chlorine with taste-producing substances in water.

Aeration

Aeration is a natural or mechanical process of increasing the contact between water and air for the purpose of improving the chemical and physical characteristics of water. Some waters, such as water from deep lakes and reservoirs in the late summer and winter seasons, cistern water, water from deep wells, and distilled water may have an unpleasant or flat taste due to a deficient dissolved-oxygen content. Aeration will add oxygen to such waters and improve their taste. In some instances the additional oxygen is enough to make the water corrosive. Aerators have other limitations.[55] Free carbon dioxide, hydrogen sulfide, and odors due to volatile oils exuded by algae will also be removed or reduced. Aeration is advantageous in the treatment of water containing dissolved iron and manganese in that oxygen will change or oxidize the dissolved iron and manganese to insoluble ferric and manganic forms that can be removed by settling, contact, and filtration.

Aeration is accomplished by allowing the water to flow in thin sheets over a series of steps, weirs, splash plates, riffles, or waterfalls; by allowing water to drip out of trays, pipes, or pans that have been slotted or perforated with $\frac{1}{8}$ to $\frac{1}{4}$ in. holes; by causing the water to drop through a series of trays containing 6 to 9 in. of coke or broken stones; by means of spray nozzles; by using air-lift pumps; by introducing finely divided air in the water; by permitting water to trickle over 1-in. by 3-in. cypress wood slats with $\frac{1}{2}$ to $\frac{3}{4}$ in. separations in a tank through which air is blown up from the bottom; and by similar means. Coke will become coated and hence useless if the water is not clear. Slat trays are usually 8 to 12 in. apart. Many of these methods are adaptable to small rural water supplies; but care should be taken to protect the water from insects and accidental or willful contamination. Screening of the aerator is desirable to prevent the development of worms.

---

[55] "Aeration of Water," Revision of *Water Quality and Treatment,* Chapter 6 *J. Am. Water Works Assoc.,* **47,** No. 9, 873–885 (September 1955).

Activated Carbon

Activated carbon in the powdered form is used quite generally and removes by adsorption, if a sufficient amount is used, practically all tastes and odors found in water. The powdered carbon may be applied directly to a reservoir as a suspension with the aid of a barrel and boat (as described for copper sulfate), or released slowly from the bag in water near the propeller, but the reservoir should be taken out of service for one or two days, unless the area around the intake can be isolated.

The application of copper sulfate within this time will improve settling of the carbon. Doses vary from 1 to 60 lb or more of carbon to one million gal of water, with 25 lb as an average. In unusual circumstances as much as 1000 lb of carbon/million gallons of water treated may be needed, but cost may make this impractical. Where a filtration plant is provided, carbon is fed by means of a standard chemical dry-feed machine, or as a suspension, to the raw water, coagulation basin, or filters. However, carbon can also be manually applied directly to each filter bed after each wash operation. Activated carbon is also used in reservoirs and settling basins to exclude sunlight causing the growth of algae. This is referred to as "blackout" treatment. The dosage of carbon required can be determined by trial and error and tasting the water, or by a special test known as the "Threshold Odor Test," which is explained in "Standard Methods." If the water is pretreated with chlorine after 15 to 20 minutes, the activated carbon will remove up to about 10 percent of its own weight of chlorine, hence they should *not* be applied together if avoidable. Possible arrangements are activated carbon followed by chlorine, and prechlorination followed by application of carbon to the top of the filters. Careful operation control can make possible prompt detection of taste- and odor-producing compounds reaching the plant and the immediate application of corrective measures.

Charcoal Filters

Granular carbon, or charcoal, filters, either of the open-gravity or closed-pressure type, are also used to remove substances causing tastes and odors in water. The water so treated must be clear, and the filters must be cleaned, reactivated, or replaced when they are no longer effective in removing tastes and odors. Rates of filtration vary from 2 to 4 gpm/ft² of filter area, although rates as high as 10 gpm/ft² are sometimes used. Trays about 4 ft², containing 12 in. of coke, are also used. The trays are stacked about 8 in. apart, and the quantity is determined by the results desired. Granular activated carbon beds are usually about 3 ft thick. They are supported on a few inches of sand. Pressure filters containing sand and activated carbon are often used on small water supplies.

Coagulation

Coagulation of turbidity, color, bacteria, organic matter, and other material in water, followed by settling and then filtration, will also result in the removal of taste- and odor-producing compounds, particularly when activated carbon is included. The use of an excess of coagulants will sometimes result in the production of a better tasting water. See page 172.

Reservoir Intake Control

The quality of reservoir and lake waters varies with the depth, season of the year or temperature, wave action, organisms and food present, condition of the bottom, clarity of the water, and other factors.

Temperature is important in temperate zones. At a temperature of 39.2°F (4°C) water is heaviest, with a specific gravity of 1.0. Therefore, in the fall of the year, the cool air will cause the surface temperature of the water to drop, and when it reaches 39.2°F this water will move to the bottom and set up convection currents, thereby forcing the bottom water up aided by wind action. Then in the winter the water may freeze, and conditions remain static until the spring when the ice melts and the water surface is warmed. A condition is reached when the entire body of water is at a temperature of about 39.2°F, but a slight variation from this temperature, aided by wind action, causes an imbalance, with the bottom colored, turbid water (usually also high in iron, and manganese and nutrient matter) rising and mixing with the upper water. The warm air will cause the temperature of the surface water to rise, and a temporary equilibrium is established, which is upset again with the coming of cold weather. In areas where the temperature does not fall below 39.2°F, and during the warm months of the year, the water will be stratified into three layers: the top mixed zone (epilimnion); the middle transition zone (metalimnion or thermocline), usually the best water quality; and the bottom zone of stagnation (hypolimnion).

A better quality of water can be obtained by drawing from different depth levels. To take advantage of this, provision should be made in deep reservoirs for an intake tower with inlets at different elevations so that the water can be drawn from the most desirable level. Where an artificial reservoir is created by the construction of a dam, it is better to waste surplus water through a blowoff rather than over a spillway; then stagnant bottom water containing decaying organic matter, manganese, iron, and silt can be flushed out.

Chlorine Dioxide Treatment

Chlorine dioxide treatment was developed originally to destroy tastes produced by phenols, but it is also effective against other taste-producing materials. Chlorine dioxide is manufactured at the water plant where it is to be used. Sodium chlorite solution and chlorine water are pumped usually into a glass cylinder where chlorine dioxide is formed and from which it is added to the water being treated, together with the chlorine water. A gas chlorinator is needed to form chlorine water, and for a complete reaction with full production of chlorine dioxide, the pH of the solution in the glass reaction cylinder must be less than 4.0. Where hypochlorinators are used, the chlorine dioxide can be maunfactured by adding hypochlorite solution, a dilute solution of hydrochloric acid, and a solution of sodium chlorite in the glass reaction cylinder so as to maintain a pH of less than 4.0. Three solution feeders are then needed. Cox gives the theoretical ratio of chlorine to sodium chlorite as 1.0 to 2.57 with chlorine water or hypochlorite solution, and sodium chlorite to chlorine dioxide produced as 1.0 to 0.74. Actually, more chlorine is usually needed to drop the pH of the reaction to 4 or less. A chlorine dioxide dosage of 0.2 to 0.3 mg/l will destroy most phenolic taste-producing compounds.

Ozone Treatment

Ozone has been used for many years as a disinfectant (particularly in France) and also as an agent to remove color, taste and odors from water supplies. Although ozone can be produced by electrolysis of perchloric acid and by ultraviolet lamps, the practical method for water treatment is by means of electrical discharges. The ozonized air is injected in a mixing and contact chamber with the water to be treated. Control is by means of the orthotolidine test (leuco crystal violet method also proposed by R. F. Layton and R. N. Kinman, *National Specialty Conference on Disinfection,* American Society of Civil Engineers, New York, 1970, p. 285–305). Ozonation provides no lasting residual in water treated, and is relatively expensive compared to free residual chlorination and chlorine dioxide treatment. It has the advantage however of being more effective in eliminating or controlling difficult color, taste and odor problems not amenable to the other treatment methods mentioned above. Ozone is a powerful oxidizing agent with high disinfection potential over a wide pH and temperature range.

Hydrogen Sulfide, Sources and Removal

Hydrogen sulfide is undesirable in drinking water for aesthetic and economic reasons. Its characteristic "rotten egg" odor is well known;

but the fact that it tends to make water corrosive to iron, steel, stainless steel, copper, and brass is often overlooked. In addition, it should be pointed out that hydrogen sulfide in concentration of 70 mg/l is an irritant and is highly poisonous at 700 mg/l. As little as 0.2 mg/l in water causes bad taste and odor and staining of photographic film.

The sources of hydrogen sulfide are both chemical and biological. Water derived from wells near oil fields or from wells that penetrate shale or sandstone frequently contain hydrogen sulfide. Calcium sulfate, sulfites, and sulfur in water containing little or no oxygen will be reduced to sulfides by anaerobic sulfur bacteria or biochemical action, resulting in liberation of hydrogen sulfide. This is more likely to occur in water at a pH of 5.5 to 8.5, and particularly in water permitted to stand in mains or in water obtained from close to the bottom of deep reservoirs. Organic matter often contains sulfur that, when attacked by sulfur bacteria in the absence of oxygen, will release hydrogen sulfide. Another source of hydrogen sulfide is the decomposition of iron pyrites or iron sulfide.

The addition of 2.0 to 4 mg/l copper sulfate or the maintenance of at least 0.3 mg/l free residual chlorine in water containing sulfate will inhibit biochemical activity and also prevent the formation of sulfides. The removal of H_2S already formed is more difficult, for most complete removal is obtained at a pH of around 4.5. Aeration removes hydrogen sulfide, but this method is not entirely effective; carbon dioxide is also removed, thereby causing an increase in the pH of the water, which reduces the efficiency of removal. Therefore, aeration must be supplemented. Aeration followed by settling and filtration is an effective combination. Chlorination alone can be used without precipitation of sulfur; but large amounts, theoretically 8.4 mg/l chlorine to each milligram per liter of hydrogen sulfide, would be needed. The alkalinity (as $CaCO_3$) of the water is lowered by 1.22 parts for each part of chlorine added. Chlorine in limited amounts, theoretically 2.1 mg/l chlorine for each milligram of hydrogen sulfide, will result in formation of flowers of sulfur, which is a fine colloidal precipitate requiring coagulation and filtration for removal. If the pH of the water is reduced by adding an acid to the water or by adding a sufficient amount of carbon dioxide as flue gas, for example, good hydrogen sulfide removal should be obtained. But pH adjustment to reduce the aggressiveness of the water would be necessary. Another removal combination is aeration, chlorination, and filtration through an activated carbon pressure filter.

Pressure tank aerators, that is, the addition of compressed air to hydropneumatic tanks, can reduce the entrained hydrogen sulfide in a well water from 35 to 85 percent, depending on such factors as the operating pressures and dissolved oxygen in the hydropneumatic tank effluent.[56] The solubility of air in water increases in direct proportion to the absolute pressure. Carbon dioxide is not removed by this treatment. Air in the amount of 0.005 to 0.16 ft^3 per gallon of water and about 15-min detention is recommended, with the higher amount preferred. The air may be introduced through perforated pipe or porous media in the tank bottom, or with the influent water. Unoxidized hydrogen sulfide and excess air in the tank must be bled off. Air relief valves or continuous air bleeders can be used for this purpose. It is believed that oxidation of the hydrogen sulfide through the sulfur stage to alkaline sulfates takes place, since observations show no precipitated sulfur in the tank. Objections to pressure tank aerators are milky water caused by dissolved air, and corrosion. The milky water would cause air binding or upset beds in filters if not removed.

A synthetic resin has been developed that has the property of removing hydrogen sulfide. It can be combined with a resin to remove hardness so that a low-hardness water can be softened and deodorized. The resin is manufactured by Rohm and Haas Company, Philadelphia, Pennsylvania 19104.

Iron and Manganese, Occurrence and Removal

Iron in excess of 0.3 to 0.5 mg/l will stain laundry and plumbing fixtures and cause water to appear rusty. When manganese is predominant the stains will be black. Neither iron nor manganese are harmful in the concentrations found in water. Iron may be present as soluble ferrous bicarbonate in alkaline well or spring waters; as soluble ferrous sulphate in acid drainage waters or waters containing sulfur; as soluble organic iron in colored swamp waters; as suspended insoluble ferric hydroxide formed from iron-bearing well waters, which are subsequently exposed to air; and as product of pipe corrosion producing red water.

Most soils, including gravel, shale, and sandstone rock, contain iron and manganese in addition to other minerals. Decomposing organic matter in water removes the dissolved oxygen usually present in water; then the water dissolves mineral oxides, changing them to soluble compounds. Water containing carbon dioxide or carbonic acid will have the same effect. In the presence of air, however, soluble ferrous bicarbonate will change to ferric iron, which will settle out in the absence of interfering substances. Ferrous iron may be found in the lower levels of deep reservoirs, flooding soils or rock containing iron or its compounds; hence

---

[56] Sidney W. Wells, "Hydrogen Sulfide Problems of Small Water Systems," *J. Am. Water Works Assoc.*, **46**, 2, 160–170 (February 1954).

it is best to draw water from a higher level, but below the upper portion, which supports microscopic growths like algae. This requires the construction and use of multiple-gate intakes, as previously mentioned.

The presence of as little as 0.1 mg/l iron in a water will encourage the growth of such bacteria as *Leptothrix* and *Crenothrix*. Carbon dioxide also favors their growth. These organisms grow in distribution systems and cause complaints. Mains, service lines, meters, and pumps may become plugged by the *Crenothrix* growths. Complaints reporting small gray or brownish flakes, masses of stringy or fluffy growths in water would indicate the presence of iron bacteria. The control of iron bacteria was discussed under "Taste and Odor Control."

Corrosive waters that are relatively free of iron and manganese may attack iron-pipe and house plumbing, particularly hot-water systems, causing discoloration and other difficulties. Such corrosion will cause red water, the control of which is discussed separately.

Iron and manganese can be removed by aeration, sedimentation, the base-exchange process, filtration, the addition of chemicals, and combinations of these methods. The cation exchanger-sodium cycle, if water is clear and unaerated, removes up to 50 mg/l iron; use a manganese filter to remove up to 2 mg/l iron or manganese. A summary of the processes used to remove iron and manganese is given in Table 3–11. Use of a polyphosphate will prevent red water or iron deposit if iron concentration is less than 3 mg/l.

Most of the carbon dioxide in water is removed by aeration; then the iron is oxidized and the insoluble iron is removed by settling or filtration. If organic matter and manganese are also present, the addition of lime or chlorine will assist in changing the iron to an insoluble form and hence simplify its removal.

The open coke-tray aerator is the most common method used to remove iron and manganese. Two or more perforated wooden trays containing about 9 in. of coke are placed in tiers. A 20- to 40-min detention basin is provided beneath the stack of trays, there the heavy precipitate settles out. The lighter precipitate is pumped out with the water to a pressure filter, where it is removed. Although carbon dioxide and hydrogen sulfide are liberated in the coke-tray aerator, when high concentrations of carbon dioxide are present it may be necessary to supplement the treatment by the addition of soda ash, caustic soda, or lime to neutralize the excess carbon dioxide to prevent corrosion of pipe lines.

Open slat-tray aerators operate similar to the coke-tray type but are not as efficient; however, they are easier to clean than the coke tray, and there is no coke to replace. When the trays are enclosed and air under pressure is blown up through the downward falling spray, a compact unit is developed in which the amount of air can be proportioned to the amount of iron to be removed. Theoretically 0.14 mg/l oxygen is required to precipitate 1 mg/l iron. The unit may be placed indoors or outdoors.

Another method for iron removal utilizes a pressure tank with a perforated air distributor near the bottom. Raw water admitted at the bottom of the pressure tank mixes with the compressed air from the distributor and oxidizes the iron present. The water passes to the top of a pressure tank, at which point air is released and automatically bled off. The amount of air injected is proportioned to the iron content by a manually adjusted needle valve ahead of a solenoid valve on the air line.[57]

At a pH of 7.0, 0.6 parts of chlorine removes 1 part iron and 0.9 parts alkalinity. At a pH of 10.0, 1.3 parts of chlorine removes 1 part of manganese and 3.4 parts alkalinity.[58]

Corrosion Cause and Control

Corrosion in water supply usually means the dissolving of pipe in contact with soft water of low alkalinity, containing oxygen. In serious cases water heaters are damaged, the flow of water is reduced, and the water is red or rusty where iron pipe is used. The inside surface of the pipe is dissolved, with consequent weakening or pitting of the pipe and redeposit of iron with reduction of the pipe diameter and hence water flow. Stray electric currents, as from buried defective electric cable, the difference in electrical potential between the water and pipe or tank, dissimilar metals in contact, biochemical changes where iron bacteria such as *Crenothrix* and *Leptothrix* use iron in their growth, high water velocities, and high water temperatures all accelerate corrosion.

Although much remains to be learned concerning the mechanism of corrosion, a simple explanation may aid in its understanding. Water in contact with iron results in the formation of soluble ferrous oxide and hydrogen gas. The ferrous oxide combines with the water and part of the oxygen usually present in water to form ferric hydroxide, which redeposits in other sections of pipe or is carried through with the water. Gaseous hydrogen is attracted to the pipe and forms a protective film if allowed to remain. But gaseous hydrogen will combine with oxygen usually present in an "aggressive" water, thereby removing the protective

[57] H. R. Fosnot, "7 Methods of Iron Removal," *Public Works*, 86, No. 11, 81–83 (November 1955).
[58] Edmund J. Laubusch, "Chlorination of Water," *Water & Sewage Works*, 105, No. 10, 411–417 (October 1958).

TABLE 3–11 PROCESSES OF IRON AND MANGANESE REMOVAL

Treatment Processes	Oxidation Required	Character of Water	Equipment Required	pH Range Required	Chemicals Required	Remarks
Aeration Sedimentation Sand filtration	Yes	Iron alone in absence of appreciable concentrations of organic matter	Aeration, settling basin, sand filter	Over 6.5	None	Easily operated. No chemical control required.
Aeration, contact oxidation, sedimentation, sand filtration	Yes	Iron and manganese loosely bound to organic matter, but no excessive carbon dioxide or organic acids content	Contact aerator of coke, gravel, or crushed pyrolusite, settling basin and sand filter	Over 6.5	None	Double pumping required. Easily controlled.
Aeration, contact filtration	Yes	Iron and manganese bound to organic matter, but no excessive organic acid content	Aerator and filter bed of manganese coated sand, "Birm," crushed pyrolusite ore, or manganese zeolite	Over 6.5 ±	None	Double pumping required unless air compressor, or "sniffler valve" is used to force air into water. Limited air supply adequate. Easily controlled.
Contact filtration	Yes, but not by aeration	Iron and manganese bound to organic matter, but no excessive carbon dioxide or organic acid content	Filter bed of manganese coated sand, "Birm," crushed pyrolusite ore, or manganese zeolite	Over 6.5	Filter bed reactivated or oxidized at intervals with chlorine or sodium permanganate	Single pumping. Aeration not required.
Aeration, chlorination, sedimentation, sand filtration	Yes	Iron and manganese loosely bound to organic matter	Aerator and chlorinator or chlorinator alone, settling basin and sand filter	7.0 to 8.0	Chlorine	Required chlorine dose reduced by previous aeration but chlorination alone permits single pumping.
Aeration, lime treatment, sedimentation, sand filtration	Yes	Iron and manganese in combination with organic matter, and organic acids	Effective aerator, lime feeder mixing basin, settling basin, sand filter	8.5 to 9.6	Lime	pH control required.

194

TABLE 3-11   PROCESSES OF IRON AND MANGANESE REMOVAL (cont'd)

Treatment Processes	Oxidation Required	Character of Water	Equipment Required	pH Range Required	Chemicals Required	Remarks
Aeration, coagulation and lime treatment, sedimentation, sand filtration	Yes	Colored, turbid, surface water containing iron and manganese combined with organic matter	Conventional rapid sand filtration plant	8.5 to 9.6	Lime and ferric chloride or ferric sulfate, or chlorinated copperas, or lime and copperas	Complete laboratory control required.
Zeolite softening	No	Well water *devoid* of oxygen, and containing less than about 1.5 to 2 ppm iron and manganese	Conventional sodium zeolite unit, with manganese zeolite unit or equivalent for treatment of by-passed water	Over 6.5 ±	None, added continuously, but bed is regenerated at intervals with salt solution	Only soluble ferrous and manganous compounds can be removed by base exchange, so aeration or double pumping is not required
Lime treatment, sedimentation, sand filtration	No	Soft well water *devoid* of oxygen containing iron as ferrous bicarbonate	Lime feeder, enclosed mixing and settling tanks and pressure filter	8.0 to 8.5	Lime	Precipitation of iron in absence of oxygen occurs at lower pH than otherwise. Absence of oxygen minimizes or prevents corrosion. Double pumping not required.

Source: Charles R. Cox, *Water Supply Control*, Bull. No. 22, New York State Department of Health, Albany, 1952, pp. 159–160.

hydrogen film and exposing the metal to corrosion. High water velocities also remove the hydrogen film. Another role is played by carbon dioxide. It has the effect of lowering the pH of the water, since more hydrogen ions are formed, which is favorable to corrosion.

The control of corrosion involves the removal of dissolved gases, treatment of the water to make it noncorrosive, the building up of a protective coating inside pipe, the use of resistant pipe materials or coating, cathodic protection, the insulation of dissimilar metals, prevention of electric grounding on water pipe, and control of growths in the mains. Therefore, if the conditions that are responsible for corrosion are recognized and eliminated or controlled, the severity of the problem will be greatly minimized. The particular cause or causes of corrosion should be determined by proper chemical analyses of the water, as well as field inspections and physical tests. The applicable control measures should then be employed.

Cement asbestos, wood-stave, plastic, vitrified clay, and concrete pipe are corrosion resistant. With soft waters, calcium carbonate tends to be removed from new concrete, cement-lined, and asbestos pipe for the first few years. Wood-stave, vitrified clay, and concrete pipe, however, have limited application. Iron and steel pipe are usually lined or coated with cement, tar, paint, or enamel, which resist corrosion. Occasionally the coating spalls off or is imperfect, and isolated corrosion takes place. It should be brought out that even though the distribution system is corrosion resistant, corrosive water should be treated to protect household plumbing systems.

The gases frequently found in water, and which encourage corrosion, are oxygen and carbon dioxide. Where practical, as in the treatment of boiler water or hot water for a building, the oxygen and carbon dioxide can be removed by heating and subjecting the water, in droplets, to a partial vacuum. Some of the oxygen is restored if the water is stored in an open reservoir or storage tank.

Dissolved oxygen can also be removed by passing the water through a tank containing iron chips or filings. Iron is dissolved under such conditions, but it can be removed by filtration. The small amount of oxygen remaining can be treated and removed with a little sodium sulfite. Ferrous sulfate is also used to remove dissolved oxygen.

All except 3 to 5 mg/l carbon dioxide can be removed by aeration, but aeration also increases the dissolved oxygen concentration, which in itself is detrimental. Sprays, cascades, coke trays, diffused air, and zeolite are used to remove most of the carbon dioxide. A filter rate of 25 gpm/ft^2 of coke tray 6 in. thick may reduce the carbon dioxide concentration from 100 to 10 mg/l and increase pH from about

6.0 to 7.0.[59] The carbon dioxide remaining, however, is sufficient to cause serious corrosion in water having an alkalinity caused by calcium carbonate of less than about 100 mg/l. It can be removed where necessary by adding sodium carbonate (soda ash), lime, or sodium hydroxide (caustic soda). With soft waters having an alkalinity greater than 30 mg/l, it is easier to add soda ash or caustic soda in a small water system to eliminate the carbon dioxide and increase the pH and alkalinity of the water. The same effect can be accomplished by filtering the water through broken limestone or marble chips. A well water that has a high concentration of carbon dioxide but no dissolved oxygen can be made noncorrosive by adding an alkali such as sodium carbonate. Soft waters that also have a low carbon dioxide content (3 to 5 mg/l) and alkalinity (20 mg/l) may need a mixture of lime and soda ash to provide both calcium and carbonate for the deposition of a calcium carbonate film.[60]

Sodium hexametaphosphate, tetrasodium pyrophosphate, sodium septaphosphate, sodium silicate—water glass, and lime are used to build up an artificial coating inside of pipe.

Sodium hexametaphosphate dissolves readily and can be added alone or in conjunction with sodium hypochlorite by means of a solution feeder. Concentrated solutions of metaphosphate are corrosive. A dosage of 5 to 10 mg/l is normally used for 4 to 8 wk until the entire distribution system is coated, after which the dosage is maintained at 1 to 2 mg/l. The initial dosage may cause precipitated iron to go into solution, with resultant temporary complaints; but flushing of the distribution system will minimize this problem. Calcium metaphosphate is a similar material, except that it dissolves slowly and can be used to advantage where this property is desirable. Inexpensive and simple pot-type feeders that are particularly suitable for small water supplies are available. Sodium pyrophosphate is similar to sodium hexametaphosphate. All these compounds are reported to coat the interior of the pipe with a film that protects the metal, prevents lime scale and red water trouble, and resists the corrosive action of water. However, heating of water above 140 to 150°F will nullify any beneficial effect.

Sodium silicate in solution is not corrosive to metals and can be easily added to a water supply with any type of chemical feeder to form cal-

[59] Ellsworth L. Filby, "The New Thomas H. Allen Pumping Station and Iron-Removal Plant of Memphis," *Water & Sewage Works*, **99**, No. 4, 133–138 (April 1952).

[60] One grain per gallon (17.1 mg/l) of lime, caustic soda, and soda ash remove, respectively, 9.65 mg/l, 9.55 mg/l, and 7.20 mg/l free CO_2; the alkalinity of the treated water is increased by 23.1 mg/l, 21.4 mg/l, and 16.0 mg/l, respectively.

cium silicate, provided the water contains calcium. Doses vary between 25 and 240 lb/million gal, 70 lb being about average. The recommendations of the manufacturer should be followed in determining the treatment to be used for a particular water.

Adjustment of the pH and alkalinity of a water so that a thin coating is maintained on the inside of piping will prevent its corrosion. Any carbon dioxide in the water must be removed before this can be done, as previously explained. Lime[61] is added to water to increase the alkalinity and pH so as to come within the limits shown on Figure 3–18. The

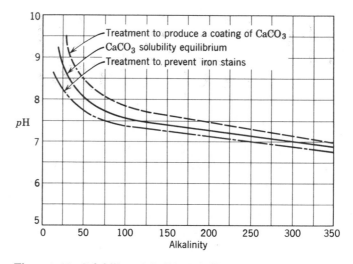

Figure 3–18   Solubility of CaCO₃ at 71°F.

approximate dosage may be determined by the "marble test," but the Enslow stability indicator is a more accurate device. Under these conditions, calcium carbonate is precipitated from the water and deposited on the pipe to form a protective coating, provided a velocity of 1.5 to 3.0 fps is maintained to prevent heavy precipitation near the point of treatment and none at the ends of the distribution system. The addition of 0.5 to 1.5 mg/l metaphosphate will help obtain a more uniform calcite coating throughout the distribution system. The addition of lime must be carefully controlled so as not to exceed a pH of 9.2 to prevent caustic alkalinity being formed. Calcium carbonate is less soluble in hot water than in cold water.

[61] At a pH above 8.3 calcium carbonate is soluble to 13 to 15 mg/l.

The danger of lead or zinc poisoning and off-flavors due to copper can be greatly reduced when corrosive water is conducted through these pipes by simply running the water to waste in the morning. This will flush out most of the metal that has had an opportunity to go into solution while standing during the night.

Biochemical actions, such as the decomposition of organic matter in the absence of oxygen in the dead end of mains, the reduction of sulfates, the biochemical action within tubercles, and the growth of *Crenothrix* and *Leptothrix*, all of which encourage corrosion in mains, can be controlled by the maintenance of at least 0.3 mg/l free residual chlorine in the distribution system. See also pages 177 and 181.

Corrosion caused by electrolysis or stray electric currents can be prevented by making a survey of the piping and removing grounded electrical connections and defective electric cables. Moist soils will permit electric currents to travel long distances. A section of nonconducting pipe in dry soil may confine the current. In the vicinity of powerhouses, this problem is very serious and requires the assistance of the power company involved. Corrosion of water storage tanks can be controlled by providing "cathodic protection," in which a direct current is imposed to make the metal more electronegative. But repainting the metal above the water line is necessary. Special equipment for this purpose is manufactured. Where dissimilar metals are to be joined, a plastic, hard-rubber, or porcelain separating fitting can be used. It must be long enough to prevent the electric charge from jumping the gap. A polyethylene tube around cast-iron pipe will protect it from corrosive soils.

Water Softening

Water softening is the removal from water of minerals causing hardness. For comparative purposes, one grain per gallon of hardness is equal to 17.1 mg/l. Hardness is caused primarily by the presence in water of calcium bicarbonate, magnesium bicarbonate, calcium sufate (gypsum), magnesium sulfate (epsom salts), calcium chloride, and magnesium chloride in solution. In the concentrations usually present these constituents are not harmful in drinking water. The presence of hardness is demonstrated by the use of large quantities of soap in order to make a lather;[62] the presence of a grity or hard curd in laundry or in a basin; the formation of hard chalk deposits on the bottom of pots and

---

[62] With a water hardness of 45 mg/l the annual per capita soap consumption was estimated at 29.23 lb; with 70 mg/l hardness, soap consumption was 32.13 lb; with 298 mg/l hardness, soap consumption was 39.89 lb; and with 555 mg/l, soap consumption was 45.78 lb. (Merrill L. Riehl, *Hoover's Water Supply and Treatment*, National Lime Association, Washington, D.C., April 1957).

inside of piping causing reduction in the flow of water; and the lowered efficiency of heat transfer in boilers, caused by the formation of an insulating scale. Hard water is not suitable for use in boilers, laundries, textile plants, and certain other industrial operations where a zero hardness of water is used, in addition to the normal or partially softened water.

In softening water the lime-soda ash process, zeolite process, and organic resin process are normally used. The soluble bicarbonates and sulfates are removed by converting them to insoluble forms in the lime-soda ash method. The calcium and magnesium are replaced with sodium in the zeolite process, thereby forming sodium compounds in the water that do not cause hardness. With synthetic organic resins, dissolved salts can be almost completely removed. Table 3–12 gives ion exchange values.

Lime-soda ash softening requires the use of lime to convert the soluble bicarbonates of calcium and magnesium to insoluble calcium carbonate and magnesium hydroxide. The calcium and magnesium sulfates are converted to insoluble calcium carbonate and magnesium hydroxide by the additiion of soda ash and lime. A coagulant such as aluminium sulfate (filter alum), ferrous sulfate (copperas), ferric sulfate, or sodium aluminate is usually used to settle the compounds formed. It is sometimes also necessary to add carbon dioxide to cause precipitation of the calcium hydroxide to calcium carbonate. The lime-soda method is not suitable for the softening of small quantities of water because special equipment and technical control are necessary.

The zeolite and synthetic resin softening methods are relatively simple and require little control. For domestic use, only a portion of the hard water is passed through a zeolite softener, since a water of zero hardness is produced by the zeolite filter. The filter effluent is mixed with part of the untreated water to produce a water of about 50 to 80 mg/l hardness. The calcium and magnesium in water to be treated replace the sodium in the zeolite filter media, and the sodium passes through with the treated water. This continues until the sodium is used up, after which the zeolite is regenerated by bringing a solution of common salt in contact with the filter media. Units are available to treat the water supply of a private home or a community. Water having a turbidity of more than 10 mg/l will coat the zeolite grains and reduce the efficiency of a zeolite softener. Iron in the ferric form is also detrimental. The filters are 2 to 6 ft deep and downward flow filters operate at rates between 3 and 5 gpm/ft^2, upward flow filters operate at 4 to 8 gpm/ft^2.

Synthetic resins, in addition to softening water, are also made to remove acids and hydrogen sulfide from water.

Small quantities of water can be softened in batches for laundry pur-

TABLE 3-12 ION EXCHANGE MATERIALS AND THEIR CHARACTERISTICS

Exchange Material	Exchange Capacity, (in grains per ft^3)	Effluent Contains	Regenerate with	Remarks
Natural zeolites	3,000 to 5,000	Sodium bicarbonate, chloride, sulfate	0.37 to 0.45 lb salt per 1,000 grains hardness removed	Ferrous bicarbonate and manganous bicarbonate also removed from well water devoid of oxygen. pH of water
Artificial zeolites	9,000 to 12,000	Sodium bicarbonate, chloride,	0.37 to 0.45 lb salt per 1,000 grains hardness	must be 6.0 to 8.5, moderate turbidity o.k. Use 5 to 10% brine solution. Saturated brine is about 25%.
Carbonaceous zeolites	9,000 to 12,000	Carbon dioxide and acids, sodium chloride˙ and sulfate	0.37 to 0.45 lb salt per 1,000 grains hardness removed	Acid waters may be filtered. CO_2 in effluent removed by aeration, acid by neutralization with bypassed hard water or addition of caustic soda.
Synthetic organic resins	10,000 to 30,000	Carbon dioxide and acids	0.2 to 0.3 lb salt per 1,000 grains hardness removed	Dissolved salts are removed by resins. To remove CO_2 and acids, add soda ash or caustic soda or CO_2 by aeration, and acids by synthetic resin filtration.

Note: One gallon of saturated brine weighs 10 lb and contains 2.5 lb of salt. Hardness is caused by calcium bicarbonate, magnesium bicarbonate, calcium sulfate, magnesium sulfate, and calcium chloride. Natural zeolite is more resistant to waters of low pH than artificial zeolite.

poses by the addition of borax, washing soda, ammonia, or trisodium phosphate. Frequently insufficient contact time is allowed for the chemical reaction to be completed, with resultant unsatisfactory softening.

Fluoridation

Fluorides have been added to public water supplies since about 1943 as an aid to the reduction of tooth decay. The compounds used are sodium fluoride (NaF), sodium silicofloride (Na_2SiF_6), hydrofluosilicic acid (H_2SiF_6), and ammonium silicofluoride ($(NH_4)_2SiF_6$). The average annual per capita cost of fluoridation of a public water supply was estimated to equal 16 cents, with a range of 6 to 33 cents, according to decreasing population served.[63]

WATER SYSTEM DESIGN PRINCIPLES

Water Quantity

The quantity of water upon which to base the design of a water system should be determined in the preliminary planning stages. Future water demand is based on social, economic, and land-use factors, all of which can be expected to change with time. (See Chapter 2.) Population projections are a basic consideration. They are made using arithmetic, geometric, and demographic methods, and with graphical comparisons with the growth of other comparable cities or towns of greater population.[64]

Numerous studies have been made to determine the average per capita water use for water system design. Health departments have design guides, and standard texts give additional information. In any case the characteristics of the community must be carefully studied and appropriate provisions made. A study made in 1965 indicated that the average per capita water use in the United States was 155 gpd, with wide individual variations.[65]

Design Period

The design period is usually determined by the future difficulties to acquire land or replace a structure or pipe line, the cost of money, and the rate of growth of the community or facility served. In general large dams and transmission mains are designed for 50 or more years; wells, filter plants, pumping stations and distribution systems for 25 yr; and water lines less than 12 in. in diameter for the full future life. When interest rates are high or when temporary or short-term use is anticipated, a lesser design period would be in order. Fair et al. suggest that the dividing line is in the vicinity of 3 percent per annum.[66]

[63] William Berner, "Status of Fluoridation in New York State," *J. Am. Water Works Assoc.,* **61**, No. 2, 68–72 (February 1969).
[64] Meyer Zitter, "Population Projections for Local Areas," *Public Works* (June 1957); Gordon M. Fair, John C. Geyer, and Daniel A. Okun, *Water and Wastewater Engineering,* vol. 1, John Wiley & Sons, Inc., New York, 1966.
[65] "Estimated Use of Water in the United States, 1965," Circular **556**, compiled by the Geological Survey, U.S. Dept. of the Interior, 1968. Summarized in *Public Works,* 85–87 (April 1970).
[66] Fair et al., op. cit.

Watershed Runoff and Reservoir Design

Certain basic information, in addition to future water demand, is needed upon which to base the design of water works structures. Long-term rainfall and stream-flow data, as well as groundwater information are available from the Geological Survey, but these seldom apply to small watersheds. Rainfall data for specific areas are also available from local weather stations, airports, and water works. Unit hydrographs, maximum flows, minimum flows, mass diagrams, characteristics of the watershed, rainfall, evaporation losses, percolation and transpiration losses should be considered for design purposes and storage determinations when these are applicable.

Watershed runoff can be estimated in different ways. The rational method for determining the maximum rate of runoff is given by the formula $Q = AIR$. Q is the runoff in ft^3/sec; A is the area of the watershed in acres; R is the rate of rainfall on the watershed in in./hr; and I is the imperviousness ratio, that is, the ratio of water that runs off the watershed to the amount precipitated on it. I will vary from 0.01 to 0.20 for wooded areas; from 0.05 to 0.25 for farms, parks, lawns, and meadows depending on the surface slope and character of the subsoil; from 0.25 to 0.50 for residential semirural areas; from 0.05 to 0.70 for suburban areas; and from 0.70 to 0.95 for urban areas having paved streets, drives, and walks.[67] $R = 360/t + 30$ for maximum storms and $R = 105/t + 15$ for ordinary storms in eastern United States; $R = 7/\sqrt{t}$ for San Francisco; $R = 56/(t + 5)^{.85}$ for New Orleans; and $R = 19/\sqrt{t}$ for St. Louis, in which $t =$ time (duration) of rainfall in min.[68]

Another formula for estimating the average annual runoff is by Vermuelé; it may be written as follows: $F = R - (11 + 0.29R)(0.035T - 0.65)$ in which F is the annual runoff in inches, R is the annual rainfall in inches, and T is the mean annual temperature in degrees Fahrenheit. This formula is reported to be particularly applicable to streams in northern New England and in rough mountainous districts along the Atlantic Coast.[69] For small water systems, it is suggested that design be based on the year of minimum rainfall or about 60 percent of the average.

In any reservoir storage study it is important to take into consideration the probable losses due to evaporation from water surfaces during

[67] Frank W. MacDonald and Adam Mehn, Jr., "Determination of Run-off Coefficients," *Public Works*, 74–76 (November 1963).
[68] Leonard Church Urquhart, *Civil Engineering Handbook*, McGraw-Hill Book Co., New York, 1959.
[69] W. A. Hardenbergh, *Water Supply and Purification*, International Textbook Co., Scranton, Pa., 1953.

the year. This becomes very significant in small systems when the water surfaces exceed 6 to 10 percent of the drainage area.[70] In the North Atlantic states the annual evaporation from land surfaces averages about 40 percent, while that from water surfaces is about 60 percent of the annual rainfall.[71] A more general relationship of monthly transpiration to mean monthly air temperature is given in Figure 4-18 based on studies made by Langbein. The watershed water loss due to land evaporation and plant usage and transpiration is significant and hence must be taken into consideration when determining rainfall minus losses. See Evapotranspiration in Chapter 4.

The minimum stream flow in New England has been estimated to yield 0.2 to 0.4 cfs/mi² of tributary drainage and an annual yield of 750,000 gpd/mi² with a storage of 200 to 250 million gal/mi². New York City reservoirs located in Upstate New York have a dependable yield of about 1 mgd/mi² of drainage area. For design purposes, long-term rainfall and stream flows should be used and a mass diagram constructed similar to Figure 3–19b.

Intakes and Screens

Conditions to be taken into consideration in design of intakes include high- and low-water stages; navigation or allied hazards; floods and storms; floating ice and debris; water velocities, surface and subsurface currents, channel flows, and stratification; location of sanitary, industrial, and storm sewer outlets; and prevailing wind direction.

Small communities cannot afford elaborate intake structures. A submerged intake crib, or one with several branches and upright tee fittings anchored in rock cribs 4 to 10 ft above the bottom, is relatively inexpensive. The inlet fittings should have a coarse strainer of screen with about 1-in. mesh. The total area of the inlets should be at least twice the area of the intake pipe and should provide an inlet velocity less than 0.5 fps. Low-entrance velocities reduce ice troubles and are less likely to draw in fish or debris. Sheet ice over the intake structure also helps avoid anchor ice or frazil ice. If ice-clogging of intakes is anticipated, provision should be made for an emergency intake or for injecting steam, hot water, or compressed air at the intake. Back-flushing is another alternative that may be incorporated in the design. Fine screens at intakes will become clogged, hence they should not be used unless installed

[70] Everett L. MacLeman, "Yield of Impounding Reservoirs," *Water & Sewage Works,* 144–149 (April 1958).
[71] Merriman and Wiggin, *American Civil Engineering Handbook,* John Wiley & Sons, Inc., New York, 1946.

at accessible locations that will make regular cleaning simple. Duplicate stationary screens in the flow channel, with $\frac{1}{8}$- to $\frac{3}{8}$-in. corrosion-resistant mesh can be purchased.

Some engineers have used slotted well screens in place of a submerged crib intake for small supplies. The screen is attached to the end of the intake conduit and mounted on a foundation to keep it off the bottom, and if desired crushed rock or gravel can be dumped over the screen. For example, a 10-ft section of a 24-in. diameter screen made by Edward E. Johnson, Inc., with $\frac{1}{4}$-in. openings is said to be able to handle 12 mgd at an influent velocity of less than 0.5 fps. Attachment to the foundation should be made in such a way that removal for inspection is possible.

In large installations, intakes with multiple-level inlet ports are provided in deep reservoirs, lakes, or streams to make possible selection of the best water when the water quality varies with the season of the year and weather conditions.

For a river intake, the inlet is perpendicular to the flow. It is constructed with vertical slotted channels before and after the bar racks and traveling screens for the placement of stop planks if the structure needs to be dewatered. Bar racks, 1 by 6 in. vertical steel, spaced 2 to 6 in. apart, provided with a rake operated manually or mechanically, keep brush and large debris from entering. This is followed by a continuous slow-moving screen traveling around two drums, one on the bottom of the intake and the other above the operating floor level. The screen is usually a heavy wire mesh with square openings $\frac{3}{8}$ to 1 in.; and it is cleaned by means of water jets inside which spray water through the screen, washing off debris into a wastewater trough. In cold-weather areas heating devices, such as steam jets, are needed to prevent icing and clogging of the racks and screens.

Pumping

When water must be pumped from the source or for transmission, electrically operated pumps should have gasoline or diesel standby units having at least 50 percent of the required capacity. If standby units provide power for pumps supplying chlorinators and similar units, the full 100 percent capacity must be provided where gravity flow of water will continue during the power failure.

The distribution of water usually involves the construction of a pumping station, unless one is fortunate enough to have a satisfactory source of water at an elevation to provide a sufficient flow and water pressure at the point of use by gravity. The size pump selected is based on whether hydropneumatic storage (steel pressure tank), ground level, or elevated storage is to be used; the available storage provided; the yield of the water source; the water usage; and the demand. Actual meter readings should be used, if available, with consideration being given to future plans, periods of low or no usage, and maximum and peak water demands. Metering can reduce water use by 25 percent or more. Average water consumption figures must be carefully interpreted and considered with required fire flows. If the water system is to also provide fire protection, then elevated storage is practically essential, unless ground-level storage with adequate pumps are available.

The capacity of the pump required for a domestic water system with elevated storage is determined by the daily water consumption and volume of the storage tank. Where the topography is suitable the storage tank can of course be located on high ground, although the hydraulic gradient necessary to meet the highest water demand may actually govern. The pump should be of such capacity as to deliver the average daily water demand to the storage tank in 6 to 12 hr. In very small installations the pump chosen may have a capacity to pump in 2 hr all the water used in one day. This may be desirable when the size of the centrifugal pump is increased to 60 gpm or more and the size electric motor to 5 to 10 hp or more, since the efficiencies of these units then approach a maximum. On the other hand larger transmission lines, if not provided, would be required in most cases, which would involve increased cost. Due consideration must also be given to the increased electrical demand and the effects this has. A careful engineering analysis should be made.

Distribution Storage Requirements

Water storage requirements should take into consideration the peak daily water use and the maximum hourly demand, the capacity of the normal and standby pumping equipment, the availability and capacity of auxiliary power, the probable duration of power failure, and the promptness with which repairs can be made.

Water storage is necessary to help meet peak demands, fire requirements, and industrial needs; to maintain relatively uniform water pressures; to eliminate the necessity for continuous pumping; to make possible pumping when the electric rate is low; and to use the most economical pipe sizes. The size and type of storage is determined by which of these specific factors or criteria are to be met. Other things being equal, a large-diameter shallow tank is preferable to a deep tank of the same capacity. It is less expensive to construct, and water pressure fluctuations on the distribution system are less. The cost of storage compared to the decreased cost of pumping, the increased fire protection and lowered fire insurance rate, the greater reliability of water supply, and decreased probability of negative pressures in the distribution system will be factors in making a decision.

In general it is recommended that water storage equal not less than one-half the total daily consumption, with at least one-half the storage in elevated tanks. A preferred minimum storage capacity would be the maximum day usage plus fire requirements, less the daily capacity of the water plant and system for the fire-flow period. In small communities, real estate subdivisions, institutions, camps, and resorts, elevated storage should be equal to at least 1 full day's requirements during hot and dry months when lawn sprinkling is heavy. A 2- or 3-day storage is preferred. The amount of water required during peak hours of the day may equal 15 to 25 percent of the total maximum daily consumption. This amount in elevated storage will meet peak demands, but not fire requirements. Some engineers provide storage equal to 20 to 40 gal per capita or 25 to 50 percent of the total average daily water consumption. A more precise method for computing requirements for elevated storage is to construct a mass diagram. Two examples are shown in Figures 3–19*a* and 3–19*b*. Fire requirements should be taken into consideration.

It is good practice to locate elevated tanks near the area of greatest demand for water and on the side of town opposite which the main enters. Thus peak demands are satisfied with the least pressure loss and smallest main sizes. All distribution reservoirs should be covered; provided with an overflow that will not undermine the footing, foundation, or adjacent structures; and provided with a drain, water-level gage, access manhole with overlapping cover, ladder, and screened air vent.

Water storage tanks are constructed of concrete, steel, wood, or brick masonry. Concrete or brick masonry tanks, preferably reinforced, may be constructed above or partly below ground, except that under all circumstances the manhole covers, vents, and overflows must be well above the normal ground level and the bottom of the tank above groundwater or flood water. Good drainage should be provided around the tank. Tanks located partly below ground must be at a higher level than any sewage disposal systems and not closer than 50 ft. Vents and overflows should be screened and the tanks covered to keep out dust, rain, insects, small animals, and birds. A cover will also prevent the entrance of sunlight, which tends to warm the water and encourage the growth of algae. Manhole covers should be locked and overlap at least 2 in. over a 2- to 6-in. lip around the manhole.

Properly constructed reinforced concrete tanks ordinarily do not require waterproofing. If tanks are built of brick or stone masonry, they

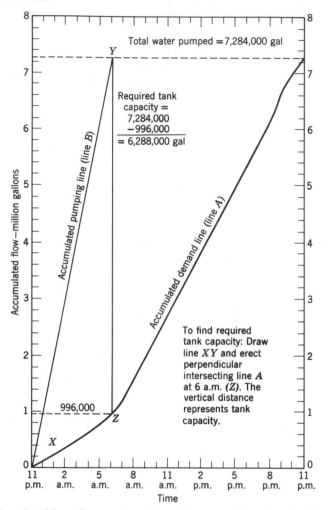

Figure 3–19a (a) Mass diagram for determining capacity of tank when pumping 7 hours, from 11 p.m. to 6 a.m. [From John E. Kiker, Jr., "Design Criteria for Water Distribution Storage," *Public Works*, 102–104 (March 1964).]

should be carefully constructed by experienced craftsmen and only hard, dense material laid with full Portland cement mortar joints used. Two ½-in. coats of 1:3 Portland cement mortar on the inside, with the second coat carefully troweled, should make such tanks watertight. A newly constructed concrete or masonry tank should be allowed to cure for about one month, during which time it should be wetted down frequently. The free lime in the cement can be neutralized by washing the interior

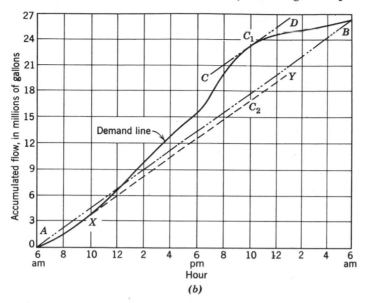

Figure 3-19b .(b) Mass diagram of storage requirements. The cumulative-demand curve is plotted from records or estimates, and the average-demand line, AB, drawn between its extremities. Line CD and XY are drawn parallel to line AB and tangent to the curve at points of greatest divergence from the average. At C_1—the point of maximum divergence—a line is extended down the coordinate to line XY. This line, C_1C_2, represents the required peak-hour storage: in this case, it scales to 6.44 mil gal. [From George G. Schmid, "Peak Demand Storage," *J. Am. Water Works Assoc.*, **48**, 378–386 (April 1956).]

with a weak acid such as a 10 percent muriatic acid solution, or with a solution made up of 4 lb of zinc sulfate per gallon of water.

Wooden elevated storage tanks are constructed of cypress, fir, long-leaf yellow pine, or redwood. They need not be painted or given special treatment; their normal life is 15 to 20 years. Wooden tanks are available with capacities up to 30,000 gal or more. Steel tanks start at 5000 to 25,000 gal; they require maintenance in order to prolong their life. Reinforced prestressed concrete tanks are also constructed.

Steel standpipes, reservoirs, and elevated tanks are made in a variety of sizes and shapes. As normally used, a standpipe is located at some high point to make available most of its contents by gravity flow and at adequate pressure; a reservoir provides mainly storage. The altitude of elevated tanks, standpipes, and reservoirs is usually determined, dependent on topography, to meet special needs and requirements. Elevated tanks rising more than 150 ft above the ground or located within 15,000 ft of a landing area, and in a 50-mi-wide path of civil airways, must meet the requirements of the Civil Aeronautics Administration.

Peak Demand Estimates

The maximum hourly or peak demand flow upon which to base the design of a water distribution system should be determined for each situation. A small residential community, for example, would have characteristics different from a new realty subdivision, central school, or children's camp. Therefore the design flow to determine distribution system capacity should reflect the pattern of living or operation, probable water usage, and demand of that particular type of establishment or community. At the same time consideration should be given to the location of existing and future institutions, industrial areas, suburban or fringe areas, highways, shopping centers, schools, subdivisions, and direction of growth. In this connection, reference to the city, town, or regional comprehensive or master plan, where available, can be very helpful. Larger cities generally have a higher per capita water consumption than smaller cities; but smaller communities have higher percentage peak demand flow than larger communities.

The maximum hourly domestic water consumption for cities above about 50,000 population will vary from 230 to 300 percent of the average annual water consumption; the maximum hourly water demand in smaller cities will probably vary from 300 to 1000 percent of the average annual water consumption. The daily variation is reported to be 150 to 250 percent and the monthly variation 120 to 150 percent of the average annual demand in small cities.[72] Studies in England show that the peak flow is about ten times the average flow in cities of 5000 population.[73] It can be said that the smaller and newer the community, the greater will be the probable variation in water consumption from the average.

Various bases have been used to estimate the probable peak demand at realty subdivisions, camps, apartment buildings, and other places. One assumption for small water plants serving residential communities is to say that, for all practical purposes, almost all water for domestic purposes is used in 12 hr.[74] The maximum hourly rate is taken as twice the maximum daily rate, and the maximum daily rate is $1\frac{1}{2}$ times

[72] Charles P. Hoover, *Water Supply and Treatment,* Bull. 211, National Lime Association, Washington, D.C., 1934. See also page 215.

[73] Institution of Water Engineers, *Manual of British Water Supply Practice,* W. Heffer & Sons, Ltd., Cambridge, England, 1950).

[74] Oscar G. Goldman, "Hydraulics of Hydropneumatic Installation," *J. Am. Water Works Assoc.,* **40,** 144 (February 1948).

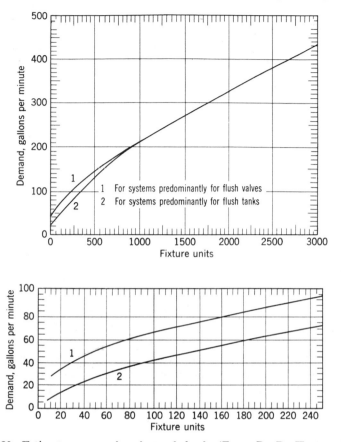

Figure 3–20 Estimate curves for demand load. (From R. B. Hunter, "Water-Distributing Systems for Buildings," National Bureau of Standards Building Materials and Structures, Rep. BMS 79, November 1941.)

the average maximum monthly rate. If the average maximum monthly flow is $1\frac{1}{2}$ times the average monthly annual flow, then the maximum hour's consumption rate is 9 times the average daily flow rate.

Another basis used on Long Island is maximum daily flow rate = 4 times average daily flow rate; maximum 6-hr rate = 8 times average daily flow; and maximum 1-hr rate = $9\frac{1}{2}$ times average daily flow.[75]

A study of small water supply systems in Illinois seems to indicate

[75] *Criteria for the Design of Adequate and Satisfactory Water Supply Facilities in Realty Subdivisions,* Suffolk County Department of Health, Riverhead, New York, 1956.

that the maximum hourly demand rate is 6 times average daily consumption.[76]

An analysis by Wolff and Loos showed that peak water demands varied from 500 to 600 percent over the average day for older suburban neighborhoods with small lots; to 900 percent for neighborhoods with $\frac{1}{4}$- to $\frac{1}{2}$-acre lots; to 1500 percent for new and old neighborhoods with $\frac{1}{3}$- to 3-acre lots.[77] Kuranz,[78] Taylor,[79] and many others[80] have also studied the variations in residential water use.

The results of a composite study of the probable maximum momentary demand are shown in Figure 3–21. It is cautioned, however, that for other than average conditions the required supply should be supplemented as might be appropriate for industries and other special demands.

Peak flows have also been studied at camps, schools, apartment buildings, highway rest areas and other places.

The design of water requirements at toll road and superhighway service areas introduces special considerations that are typical for the installation. It is generally assumed that the sewage flow equals the water flow. In one study of national turnpike and highway restaurant experience, the extreme peak flow was estimated at 1890 gpd/counter seat and 810/table seat; but the peak day was taken as 630 gpd/counter seat and 270 gpd/table seat.[81] In another study of the same problem, the flow was estimated at 350 gpd/counter seat plus 150 gpd/table seat.[82] The flow was 200 percent of the daily average at noon and 160 percent of the daily average at 6 p.m. It was concluded that 10 percent of the cars passing a service area will enter and will require 15 to 20

[76] C. W. Klassen, "Hydropneumatic Storage Facilities," *Technical Release 10-8,* Division of Sanitary Engineering, Illinois Department of Public Health (October 28, 1952).

[77] Jerome B. Wolff and John F. Loos, "Analysis of Peak Demands," *Public Works* (September, 1956).

[78] A. P. Kuranz, "Studies on Maximum Momentary Demand," *J. Am. Water Works Assoc.* (October 1942).

[79] D. R. Taylor, "Design of Main Extensions of Small Size," *Water & Sewage Works* (July 1951).

[80] E. P. Linaweaver, Jr., *Residential Water Use, Report II on Phase Two of the Residential Use Research Project,* Department of Sanitary Engineering and Water Resources, Johns Hopkins University, Baltimore, Md. (June 1965). *Maximum Design Standards for Community Water Supply Systems,* Federal Housing Administration, FHA No. 751, 52 (July 1965). L. P. Linaweaver, Jr., John C. Geyer, Jerome B. Wolff, *A Study of Residential Water Use,* U.S. Dept. of Housing and Urban Development, Washington, D.C., 1967.

[81] Gale G. Dixon and Herbert L. Kaufman, "Turnpike Sewage Treatment Plants," *Sewage and Ind. Wastes* (March 1956).

[82] Clifford Sharp, "Kansas Sewage Treatment Plants," *Public Works* (August 1957).

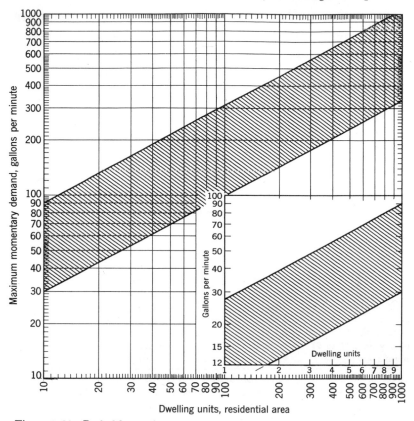

Figure 3-21   Probable maximum momentary water demand.

gal/person. A performance study after one year of operation of the Kansas Turnpike service areas showed that 20 percent of cars passing service areas will enter; there will be 1½ restaurant customers/car; average water usage will be 10 gal/restaurant customer, of which 10 percent is in connection with gasoline service; and plant flows may increase 4 to 5 times in a matter of seconds.[83]

It must be emphasized that actual meter readings from a similar-type establishment or community should be used whenever possible in preference to an estimate. Time spent to obtain this information is a good investment, as each installation has different characteristics. Hence the estimates and procedures mentioned here should be used as a guide to

[83] Mellville Gray, "Sewage from America on Wheels—Designing Turnpike Service Areas Plants to Serve Transient Flows," *Wastes Engineering* (July 1959).

TABLE 3–13   DEMAND WEIGHT OF FIXTURES IN FIXTURE UNITS*

Fixture or Group†	Occupancy	Type of Supply Control	Weight in Fixture Units‡
Water closet	Public	Flush valve	10
Water closet	Public	Flush tank	5
Pedestal urinal	Public	Flush valve	10
Stall or wall urinal	Public	Flush valve	5
Stall or wall urinal	Public	Flush tank	3
Lavatory	Public	Faucet	2
Bathtub	Public	Faucet	4
Shower head	Public	Mixing valve	4
Service sink	Office, etc.	Faucet	3
Kitchen sink	Hotel or restaurant	Faucet	4
Water closet	Private	Flush valve	6
Water closet	Private	Flush tank	3
Lavatory	Private	Faucet	1
Bathtub	Private	Faucet	2
Shower head	Private	Mixing valve	2
Bathroom group	Private	Flush valve for closet	8
Bathroom group	Private	Flush tank for closet	6
Separate shower	Private	Mixing valve	2
Kitchen sink	Private	Faucet	2
Laundry trays (1 to 3)	Private	Faucet	3
Combination fixture	Private	Faucet	3

Source: R. B. Hunter, *Water-Distributing Systems for Buildings,* National Bureau of Standards Building Materials and Structures, Report BMS 79 (November 1941).

*For supply outlets likely to impose continuous demands, estimate continuous supply separately and add to total demand for fixtures.

†For fixtures not listed, weights may be assumed by comparing the fixture to a listed one using water in similar quantities and at similar rates.

‡The given weights are for total demand. For fixtures with both hot and cold water supplies, the weights for maximum separate demands may be taken as three-fourths the listed demand for supply.

supplement specific studies and to aid in the application of informed engineering judgment.

Peak flows for apartment-type buildings can be estimated using the curves developed by Hunter.[84] Figure 3–20 and Tables 3–13 and 3–14 can be used in applying this method. Additions should be made for continuous flows. This method may be used for the design of small water systems, but the peak flows determined will be somewhat high.

[84] R. B. Hunter, *Water-Distributing Systems for Buildings,* Rept. **79,** National Bureau of Standards for Building Materials and Structures (November 1941).

TABLE 3-14   RATE OF FLOW AND REQUIRED PRESSURE DURING FLOW FOR DIFFERENT FIXTURES

Fixture	Flow Pressure* (psi)	Flow Rate (gpm)
Ordinary basin faucet	8	3.0
Self-closing basin faucet	12	2.5
Sink faucet, $\frac{3}{8}$ in.	10	4.5
Sink faucet, $\frac{1}{2}$ in.	5	4.5
Bathtub faucet	5	6.0
Laundry-tub cock, $\frac{1}{2}$ in.	5	5.0
Shower	12	5.0
Ball cock for closet	15	3.0
Flush valve for closet	10 to 20	15 to 40†
Flush valve for urinal	15	15.0
Garden hose, 50 ft and sill cock	30	5.0
Dishwashing machine, commercial	15 to 30	6 to 9

Source: *Report of the Coordinating Committee for a National Plumbing Code,* U. S. Dept. of Commerce, Washington, D.C., 1951.

*Flow pressure is the pressure in the pipe at the entrance to the particular fixture considered.

†Wide range due to variation in design and type of flush-valve closets. (See Chapter 11 for fixture-supply pipe diameters.)

Distribution System Design Standards

So far as possible distribution system design should follow usual good waterworks practice. Mains for residential areas should be designed on the basis of at least 250 gpcd, plus a residual pressure of not less than 30 psi nor more than 125 psi, using the Hazen and Williams coefficient $C = 100$. In small communities and in subdivisions, the peak hourly demand may reach 500 or 1000 percent of the average daily consumption, as explained in the section "Peak Demand Estimates." Fire protection requirements plus lawn-sprinkling and domestic demand will govern. In cities with a population greater than 100,000, the peak hourly demand is greater than the fire demand plus the average domestic demand; it is about equal in cities of 25,000 population. But in smaller communities, the fire demand will greatly exceed the peak domestic hourly demand.[85]

Air-release and vacuum valves are provided where necessary at all high points on the transmission lines, and blowoffs are provided at low drain points. These valves must not discharge to below-ground pits unless provided with a gravity drain to the surface above flood level. So far

[85] John E. Kiker, Jr., "Computing Requirements for Elevated Water Storage," *Public Works* (May 1953).

as possible dead ends should be eliminated or a blowoff provided, and mains should be tied together at least every 600 ft. Lines less than 4 in. in diameter should generally not be considered, except for the smallest system, unless they parallel secondary mains on other streets. Although the design should aim to provide a pressure of 30 psi in the distribution system during peak flow periods, 20 psi minimum may be acceptable. A minimum pressure of 60 to 80 psi is desired in business districts, although 50 psi may be adequate in small villages with one- and two-story buildings.

Valves are spaced between 500 and 750 ft apart and at street intersections. A valve book, at least in triplicate, should show permanent ties for all valves, number of turns to open completely, left- or right-hand turn to open, manufacturer, and dates valves operated. A gate valve should be provided between each hydrant and street main.

Water lines should be laid below frost, about 2 ft above and 10 ft in a horizontal direction from any nearby sewers. Where sewers cross water mains, sewers for a distance of 10 ft on each side should be constructed of pressure pipe laid 18 in. lower and provided with watertight joints. If water and sewer lines must be laid in the same trench, the water main should be offset on an undisturbed earth shelf, with the bottom at least 12 in. above the top of the sewer line. Both lines should be cast iron with mechanical joints or equal construction. It must be recognized that this type of construction is more expensive and requires careful supervision during construction. Mains buried 5 ft are normally protected against freezing and external loads.

The selection of pipe sizes is determined by the required flow of water that will not produce excessive friction loss. Transmission mains for small water systems more than 3 to 4 mi long should not be less than 10 to 12 in. in diameter. Design velocity is kept under 5 fps and head loss under 3 ft/1000 ft. If the water system for a small community is designed for fire flows, the required flow for domestic use will not cause significant head loss. On the other hand, where a water system is designed for domestic supply only, the distribution system pipe sizes selected should not cause excessive loss of head. Velocities may be $1\frac{1}{2}$ to $5\frac{1}{2}$ fps. In any case a special allowance is usually necessary to meet water demands for fire, industrial, and other special purposes.

Design velocities as high as 10 to 15 fps are not unusual, particularly in short runs of pipe. The design of water distribution systems can become very involved and is best handled by a competent sanitary engineer. When a water system is carefully laid out, without dead ends, so as to divide the flow through several pipes, the head loss is greatly reduced. The friction loss in a pipe connected at both ends is about one-quarter the friction loss in the same pipe with a dead end. The friction loss in a pipe from which water is being drawn off uniformly along its length is about one-third the total head loss.

In some communities, where no fire protection is provided, small diameter pipe may be used. In such cases, a 2-in. line should be no more than 300 ft long; a 3-in. line, no more than 600 ft; a 4-in. line, no more than 1200 ft; and a 6-in. line, no more than 2400 ft. If lines are connected at both ends 2- or 3-in. lines should be no longer than 600 ft; 4-in. lines, not more than 2000 ft.

Hudson suggests as a rule of thumb that a 6-in. main can be extended only 500 ft if the average amount of water of 1000 gpm is to be supplied for fire protection, or about 2000 ft if the minimum amount of 500 gpm is to be supplied.[86] An 8-in. line will carry 2.1 times as much water as a 6-in. line for the same loss of head and costs only about $1.50 (1970) more per foot to buy and install. See page 246.

The minimum pipe sizes and rule-of-thumb guides mentioned above are not meant to substitute for distribution system analysis but are intended for checking or rough approximation. Use of the equivalent pipe method, the Hardy Cross method, or one of its modifications should be adequate for the small distribution system.

A water system that provides adequate fire protection is highly recommended where possible. The fire insurance savings should at the very least be compared with the additional cost of increased pipe size, plant capacity, and water storage to provide fire protection. Certainly if the cost of 8-in. pipe is only $1.50 more per ft than 6-in. pipe, the argument for the larger diameter pipe where needed is very persuasive. In any case only pipe and fittings that have a permanent type lining or inner protective surface should be used.

The American Insurance Association's Water System Design Standards

The New York Fire Insurance Rating Organization Engineering Department, now the American Insurance Association, recommends the following design standards for new construction and for long-range improvement of existing systems:

MINIMUM PIPE SIZE
Residential areas: Six-inch and smaller mains should not be installed as dead ends. The gridiron of minor distributors supplying residential districts should consist of mains at least 6-in. in size arranged so that the lengths on the long sides of blocks between intersecting mains do not exceed 600 ft. Where longer lengths are necessary, 8-in. or larger

[86] W. D. Hudson, "Design of Additions to Distribution Systems," *Water and Sewage Works* (July 1959).

mains should be used. (Where initial pressures are high, a satisfactory gridiron may be obtained with longer lengths of 6-in. pipe between intersecting mains.)

In new construction 8-in. pipe should be used where dead ends and poor gridironing are likely to exist for a considerable period or where the layout of the streets and the topography are not well adapted to the above arrangement.

High-value districts: The minimum size should be 8-in. where there are intersecting mains in each street; 12-in. or larger mains should be used on the principal streets and for all long lines that are not connected to other mains at intervals close enough for proper mutual support.

SPACING OF VALVES

The distribution system should be equipped with a sufficient number of valves so located that no single case of accident, breakage, or repair to the pipe system, exclusive of arteries, will necessitate the shutdown of an artery or a length of pipe greater than 500 ft in high-value districts or greater than 800 ft in other sections.

HYDRANT DISTRIBUTION

Distribution of hydrants should be based on the required fire flow and the average area served, not to exceed that given in table.

Fire Flow Required, gpm	Average Area per Hydrant, ft²
1,000 or less	120,000
2,000	110,000
3,000	100,000
4,000	90,000
5,000	85,000
6,000	80,000
9,000	55,000
12,000	40,000

Maximum spacing: Near service limits and in sparsely settled areas, sufficient hydrants should be installed to provide at least one within 600 ft of all buildings.

SIZE AND INSTALLATION OF HYDRANTS

Hydrants should be able to deliver 600 gpm with a friction loss of not more than 5 lb/in.² in the hydrant and a total loss of not more than 5 lb/in.² between the street main and the outlet.

They should have at least two 2½-in. outlets and a large pumper outlet where pumper service is necessary.

They should be of such design that when the barrel is broken off the hydrant will remain closed.

Connection to the street main should be not less than 6 in. in diameter.

A gate valve should be provided on all connections between hydrants and street mains; first attention should be given to providing valves in street main connections on all hydrants installed on supply lines, arteries, and main feeders.

Operating nuts and direction of operation should be standard on all hydrants.

Hydrants should be set so that they are easily accessible to fire department pumpers; they should not be set in depressions, cutouts, or on embankments high above the street; pumper outlets should face directly toward the street; with respect to nearby trees, poles, and fences, there should be adequate clearance for connection of hose lines. Hydrants should be painted a distinguishing color so that they can be quickly spotted at night.

Table 3–13a, "Required Fire Flows, Average City and Residential Districts" summarizes flow rates, the total fire flows, and municipal flows for population groups of 100 to 10,000 for the duration of the indicated fire. The total required fire flow in high-value districts can be estimated from the formula $G = 1020 \sqrt{P}(1.0 - 0.01 \sqrt{P})$, where G is the required amount of water in gpm and P is the population expressed in thousands. The designing engineer is urged to submit his water system plans to the local rating organization of the American Insurance Association for review before final plans and specifications are let out for bids. One must be alert to assure that fire protection programs do not include pumping from polluted sources into public or private water supply mains through hydrant or blowoff connections. Nor should bypasses be constructed around filter plants or provision made for "emergency" raw-water connections to supply water in case of fire.

Cross-Connection Control

A discussion of water system design would not be complete without reference to cross-connection control. It is fundamental that there should be no cross-connection between safe (potable) and unsafe (nonpotable) water supplies. It is the responsibility of the designing engineer, the waterworks official, and the health department to prevent possibilities of pollution of public and private water supplies. Because of the practical difficulties of obtaining and maintaining complete separation of potable

TABLE 3-13a   REQUIRED FIRE FLOWS, AVERAGE CITY AND RESIDENTIAL DISTRICTS*

Population	Flow Rate gpm	Flow Rate mgd	Required Fire Flows for Average City				Required Fire Flows for Residential Districts			
			Duration hours	Fire Flow gal.	Municipal Flow gal.‡	Total Flow gal.§	Duration hours	Fire Flow gal.	Municipal Flow gal.‡	Total Flow gal.§
100	250†	0.36	4	60,000	3,750	63,750	2†	30,000	1,250	31,250
250	500	0.72	4	120,000	9,375	129,375	2†	60,000	3,125	63,125
500	750	1.08	4	180,000	18,750	198,750	2†	90,000	6,250	96,250
1,000	1,000	1.44	4	240,000	37,500	277,500	2†	120,000	12,500	132,500
1,500	1,250	1.80	5	375,000	70,200	445,200	2½†	187,500	23,400	210,900
2,000	1,500	2.16	6	540,000	112,500	652,500	3†	270,000	37,500	307,500
3,000	1,750	2.52	7	735,000	196,875	931,875	3½†	367,500	65,500	333,000
4,000	2,000	2.88	8	960,000	300,000	1,260,000	4	480,000	100,000	580,000
5,000	2,250	3.24	9	1,215,000	422,500	1,637,500	4½	607,500	141,000	748,500
6,000	2,500	3.60	10	1,500,000	562,500	2,062,500	5	750,000	187,500	937,500
10,000	3,000	4.32	10	1,800,000	937,500	2,737,500	5	900,000	312,000	1,212,000

Adapted from Standard Schedule for Grading Cities and Towns of the United States with reference To Their Fire Defenses and Physical Conditions, National Board of Fire Underwriters, New York, 1956, now the American Insurance Association, New York City.

* For residential districts, the required fire flow shall be determined on the basis of structural conditions and congestion of buildings. In districts with about 1/3 the lots in a block built upon having buildings of small area and of low height, at least 500 gpm is required; if the buildings are of larger area or higher, up to 1,000 gpm is required; where districts are more closely built or the buildings consist of high value residences, apartments, tenements, dormitories, or similar structures, 1,500 to 3,000 gpm is required, and in densely built districts with 3-story and higher buildings, up to 6,000 gpm is required.

† In certain states a minimum flow of 500 gpm at 20 psi for at least 4 hours duration is required.

‡ Water usage for fire flow purposes is the maximum day flow (1.5 times the average daily consumption. In this example it is assumed that flow for the average city to be at the rate of 225 gpcd and flow for residential districts to be at the rate of 150 gpcd, both for the hours duration of the fire flow.

§ The total flow may be provided by storage alone or by system capacity plus storage.

and nonpotable water supplies, particularly within a property, the designing engineer would do well to discuss potential problems with state and local health department sanitary engineers. An excellent summary of cross-connection control is given in a Joint Committee Report of the American Water Works Association and the Conference of State Sanitary Engineers[87] and in Public Health Service Publication No. 957.[88]

Hydropneumatic Systems

Hydropneumatic or pressure-tank water systems are suitable for small communities, housing developments, private homes and estates, camps, restaurants, hotels, resorts, country clubs, factories, institutions, and as booster installations. In general only about 10 to 20 percent of the total volume of a pressure tank is actually available. Hydropneumatic tanks are usually made of $\frac{3}{16}$ in. or thicker steel and are available in capacities up to 10,000 or 20,000 gal. Tanks should meet ASME code requirements.

The required size of a pressure tank is determined by peak demand and the capacity of the pump and source. If the pump capacity and water source do not supply enough water for maximum use, ground-level storage and pump should make up the difference, unless the needed water can be provided by elevated storage. See page 129 for suggested home pressure tank sizes. The capacity of a well and pump should be at least 6 to 10 times the average daily water requirement.

A simple and direct method for determining the recommended volume of the pressure storage tank and size pump to provide is given by Figure 3-22. This figure is derived from Boyle's Law and is based on the formula

$$Q = \frac{Qm}{\left(1 - \dfrac{P_1}{P_2}\right)}$$

in which Q is equal to the pressure-tank volume in gal, Qm is equal to 15 minutes' storage at the maximum hourly demand rate, P_1 is the minimum absolute operating pressure (gauge pressure plus 14.7 lb/in.²), and P_2 is the maximum absolute pressure.[89] The pump capacity given on the curve is equal to 125 percent of the maximum hourly demand rate (taken as 6 times the average maximum monthly water consumption

[87] Joint Committee Report, "Use of Backflow Preventers for Cross-Connection Control," *J. Am. Water Works Assoc.*, **50**, 1589-1618 (December 1958). Revised 1970.
[88] *Water Supply and Plumbing Cross-Connections*, U.S. Public Health Service Pub. No. 957, Dept. of HEW, Washington, D.C., 1963.
[89] Joseph A. Salvato, Jr., "The Design of Pressure Tanks for Small Water Systems," *J. Am. Water Works Assoc.*, **41**, 532-536 (June 1949).

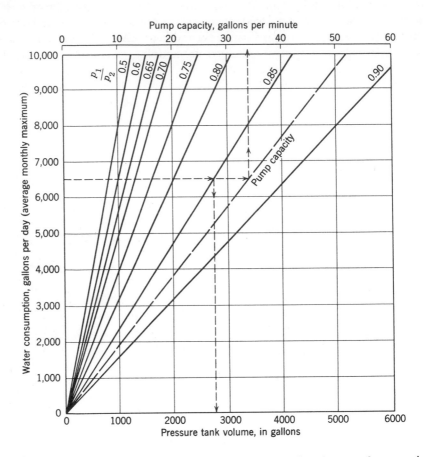

Figure 3–22 Chart for determining pressure storage tank volume and pump size. [From J. A. Salvato, Jr. "The Design of Pressure Tanks for Small Water Systems," *J. Am. Water Works Assoc.* (June 1949).]

rate, with the average maximum monthly rate being $1\frac{1}{2}$ times the average daily rate). The tank is assumed to be just empty when the pressure gauge reads zero. An example is given on page 243.

The water available for distribution is equal to the difference between the dynamic head (friction plus static head) and the tank pressure. Because of the relatively small quantity of water actually available between the usual operating pressures, a higher initial (when the tank is empty) air pressure and range are sometimes maintained in a pressure tank to increase the water available under pressure. When this is done, the escape of air into the distribution system is more likely. Most home pressure tanks come equipped with an automatic air volume control (Figure 3–23), which is set to maintain a definite air-water volume in the pressure tank at previously established water pressures, usually 20 to 40 psi. Air usually needs to be added to replace that absorbed by the water to prevent the tank from becoming waterlogged. Small

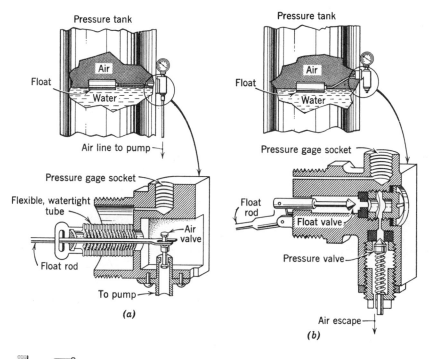

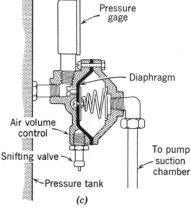

Figure 3–23 Pressure-tank air-volume controls: (a) shallow-well-type for adding air; (b) deep-well-type for air release—used with submersible and piston pumps; (c) diaphragm-type in position when pump is not operating (used mostly with centrifugal pump). [From *Pumps and Plumbing for the Farmstead,* Agricultural Engineering Development Div., TVA (November 1940).]

pressure tanks are also available with a diaphragm inside that separates air contact with the water, thereby minimizing this problem. With deep-well displacement and submersible pumps, an excess of air is usually pumped with the water, causing the pressure tank to become air-bound unless an air-release or needle valve is installed to permit excess air to escape. In large installations an air compressor is needed, and an air-relief valve is installed at the top of the tank.

PUMPS

The pump types commonly used to raise and distribute water are referred to as displacement or reciprocating, rotary, centrifugal, turbine, submersible, jet, and air-lift pumps and hydraulic rams. Other types include the chain pump and screw type.

Displacement Pump

In reciprocating displacement pumps, water is drawn into the pump chamber or cylinder on the suction stroke of the piston or plunger inside the pump chamber and then the water is pushed out on the discharge stroke. This is a simplex or single-acting reciprocating pump. An air chamber Figure 3–24 should be provided on the discharge side of the pump to prevent excessive water hammer caused by the quick-closing flap valves. The air chamber will protect the piping and equipment on the line and will tend to even out the intermittent flow of water. Reciprocating pumps are also of the duplex type wherein water is pumped on both the foward and backward stroke, and of the triplex type in which three pistons pump water. The motive power may be manual, a steam, gas, gasoline, or oil engine, an electric motor, or a windmill. The typical hand pump and deep-well plunger or piston pumps over wells are displacement pumps.

A rotary pump is also a displacement pump, since the water is drawn in and forced out by the revolution of a cam, screw, gear, or vane. It is not used to any great extent to pump water.

Displacement pumps have certain advantages over centrifugal pumps. The quantity of water delivered does not vary with the head against which the pump is operating, but depends on the power of the driving engine or motor. A pressure-relief valve is necessary on the discharge side of the pump to prevent excessive pressure in the line and possible bursting of a pressure tank or water line. They are easily primed and

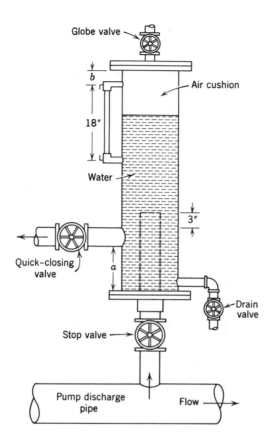

Figure 3–24 Air chamber dimensions for reciprocating pumps. (From *Water Supply and Water Purification*, T.M. 5–295, War Dept., Washington, D.C., 1942, 96 pp.)

Air Chamber Dimensions				
Discharge pipe	Inside diameter of air chamber	Total height	a	b
2″	8″	3′-0″	4″	9″
2½″	8″	3′-6″	4″	12″
3″	10″	4′-0″	5″	15″
4″	10″	5′-0″	6″	21″
5″	12″	6′-0″	6″	27″
6″	16″	7′-0″	6″	33″

operate smoothly under suction lifts as high as 22 ft. Practical suction lifts at different elevations are given in Table 3–15.

TABLE 3-15 ATMOSPHERIC PRESSURE AND PRACTICAL SUCTION LIFT

Elevation Above Sea Level		Atmospheric Pressure		Design Suction Lift (ft)		
(ft)	(mi)	(lb/in.^{2})	(ft of water)	Displacement Pump	Centrifugal Pump	Turbine Pump
0		14.70	33.95	22	15	28
1,320	¼	14.02	32.98	21	14	26
2,640	½	13.33	30.79	20	13	25
3,960	¾	12.66	29.24	18	11	24
5,280	1	12.02	27.76	17	10	22
6,600	1¼	11.42	26.38	16	9	20
7,920	1½	10.88	25.13	15	8	19
10,560	2	9.88	22.82	14	7	18

Note: The possible suction lift will decrease about 2 ft for every 10°F increase in water temperature above 60°F. One lb/in.^{2} = 2.31 ft head of water.

Displacement pumps are flexible and economical. The quantity of water pumped can be increased by increasing the speed of the pump, and the head can vary within wide limits without decreasing the efficiency of the pump. A displacement pump can deliver relatively small quantities of water as high as 800 to 1000 ft. Its maximum capacity is 300 gpm. The overall efficiency of a plunger pump varies from 30 percent for the smaller sizes to 60 to 90 percent for the larger sizes with electric motor drive. It is particularly suited to the pumping of small quantities of water against high heads and can, if necessary, pump air with water.

Centrifugal Pump

Centrifugal pumps are of several types depending on the design of the impeller. Water is admitted into the suction pipe or into the pump casing and rotated in the pump by an impeller inside the pump casing. The energy is converted from velocity head primarily into pressure head. In the submerged turbine-type pump used to pump water out of a well, the centrifugal pump is in the well casing below the water level in the well. If the head against which a centrifugal pump operates is increased beyond that for which it is designed, and the speed remains the same, then the quantity of water delivered will decrease. On the other hand, if the head against which a centrifugal pump operates is less than that for which it is designed, then the quantity of water delivered will be increased. This may cause the load on the motor to be increased and hence the overloading of the electric or other motor, unless the motor selected is large enough to take care of this contingency.

Sometimes two centrifugal pumps are connected in series so that the discharge of the first pump is the suction for the second. Under such an arrangement the capacity of the two pumps together is only equal to the capacity of the first pump, but the head will be the sum of the discharge heads of both pumps. At other times, two pumps may be arranged in parallel so that the suction of each is connected to the same pipe and the discharge of each pump is connected to the same discharge line. In this case the static head will be the same as that of the individual pumps, but the dynamic head, when the two pumps are in operation, will increase because of the greater friction and may exceed the head for which the pumps are designed. It may be possible to force only slightly more water through the same line when using two pumps as when using one pump, depending on the pipe size. Doubling the speed of a centrifugal pump impeller doubles the quantity of water pumped, produces a head four times as great, and requires eight times as much power to drive the pump. In other words the quantity of water pumped varies directly with the speed, the head varies as the square of the speed, and the horsepower as the cube of the speed. It is the usual practice to plot the pump curves for the conditions studied on a graph to anticipate operating results.

The centrifugal pump has no valves or pistons; there is no internal lubrication; it takes up less room and is relatively quiet. A single-stage centrifugal pump is generally used where the suction lift is less than 15 ft and the total head not over 125 to 200 ft. A single-stage centrifugal pump may be used for higher heads, but where this occurs a pump having two or more stages, that is, two or more impellers or pumps in series, should be used. The efficiency of centrifugal pumps varies from about 20 to 85 percent; the higher efficiency can be realized in the pumps with a capacity of 500 gpm or more. The peculiarities of the water system and effect they might produce on pumping cost should be studied from the pump curve characteristics. A typical curve is shown in Figure 3–25.

Centrifugal, turbine, and ejector pumps that are above the pumping water level should have a foot valve on the pump suction line to retain the pump prime. However, foot valves sometimes leak thereby requiring a water connection, or other priming device, or a new check valve on the suction side of the pump. The foot valve should have an area equal to at least twice the suction pipe. It may be omitted where an automatic priming device is provided. In the installation of a centrifugal pump,

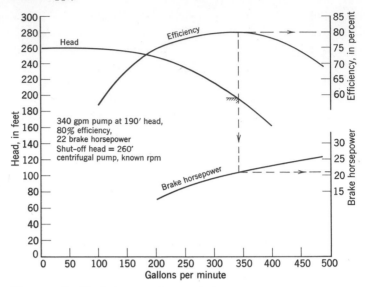

Figure 3–25   Typical pump characteristic curves.

it is customary to install a gate valve on the suction line to the pump and a check valve followed by a gate valve on the pump discharge line near the pump. An air chamber, surge tank, or similar water-hammer suppression should be installed just beyond the check valve, particularly on long pipe lines or when pumping against a high head. Make arrangements for priming a centrifugal pump, unless the suction and pump are under a head of water, and keep the suction line as short as possible. Slope the suction line up toward the pump to prevent air pockets.

Jet Pump

The jet pump is actually the combination of a centrifugal pump and a water ejector down in a well below or near the water level. The pump and motor can be located some distance away from the well, but the pipelines should slope up to the pump about $1\frac{1}{2}$ in. in 20 ft. In this type of pump, part of the water raised is diverted back down into the well through a separate pipe. This pipe has attached to it at the bottom an upturned ejector connected to a discharge riser pipe that is open at the bottom. The water forced down the well passes up through the ejector at high velocity, causing a pressure reduction in the venturi throat, and with it draws up water from the well through the riser or return pipe. A jet pump may be used to raise small quantities of water 90 to 120 ft, but its efficiency is lowered when the lift exceeds

50 ft. Efficiency ranges from 20 to 25 percent. The maximum capacity is 50 gpm. There are no moving parts in the well.

Air-Lift Pump

The air-ejector pump is similar in operation to a water-ejector pump except that air is used instead of water to create a reduced pressure in the venturi throat to raise the water.

In an air-lift pump, compressed air is forced through an air pipe down into a well below the water level and discharged in a finely diffused state at the bottom. The air mixes with the water and in rising carries up with it water in the eductor pipe. The head, and ratio of distances from the point air is injected (A) to the pumping level, and from (A) to the discharge level (times 100), are critical. The maximum capacity is 3000 gpm, head 700 ft, and efficiency 30 to 50 percent.

Hydraulic Ram

A hydraulic ram is a type of pump where the energy of water flowing in a pipe is used to elevate a smaller quantity of water to a higher elevation. An air chamber and weighted check valve are an integral part of a ram. Hydraulic rams are suitable where there is no electricity and the available water supply is adequate to furnish the energy necessary to raise the required quantity of water to the desired level. A battery of rams may be used to deliver larger quantities of water provided the supply of water is ample. Double-acting rams can make use of a nonpotable water to pump a potable water. The minimum flow of water required is 2 to 3 gpm with a fall of 3 ft or more. A ratio of lift to fall of 4 to 1 can give an efficiency of 72 percent, a ratio of 8 to 1 an efficiency of 52 percent, a ratio of 12 to 1 an efficiency of 37 percent, and a ratio of 24 to 1 an efficiency of 4 percent.[90] Rams are known to operate under supply heads up to 100 ft and a lift or delivery heads of 5 to 500 ft. In general, a ram will discharge from $\frac{1}{7}$ to $\frac{1}{10}$ of the water delivered to it. From a practical standpoint, it is found that the pipe conducting water from the source to the ram (known as the drive pipe) should be at least 30 to 40 ft long for the water in the pipe to have adequate momentum or energy to drive the ram. It should not, however, be on a slope greater than about 12 deg with the horizontal. If these conditions cannot be met naturally, it may be possible to do so by providing on the drive pipeline an open stand pipe, so that the pipe beyond it meets the conditions given. The diameter

[90] R. T. Kent, *Mechanical Engineers' Handbook,* Vol. II, John Wiley & Sons, Inc., New York, 1950.

of the delivery pipe is usually about one-half the drive pipe diameter. The following formula may be used to determine the capacity of a ram:

$$Q = \frac{\text{supply to ram} \times \text{power head} \times 960}{\text{pumping head}}$$

where Q = gallons delivered per day.

supply to ram = gallons per minute of water delivered to and used by the ram.

power head = the available supply head of water, in feet or fall.

pumping head = the head pumping against, in feet, or delivery head.

NOTE: This information plus the length of the delivery pipe and the horizontal distance in which the fall occurs are needed by manufacturers to meet specific requirements.

Pump and Well Protection

A power pump located directly over a deep well should have a water-tight well seal at the casing as illustrated in Figures 3–7 and 3–8. An air vent is used on a well that has an appreciable drawdown to compensate for the reduction in air pressure inside the casing, which is caused by a lowering of the water level when the well is pumped. The vent should be carried 18 in. above the floor, and the end should be looped downward and protected with screening. A downward opening sampling tap located at least 12 in. above the floor should be provided on the discharge side of the pump. In all instances, the top of the casing, vent, and motor are located above possible flood level.

The top of the well casing or pump should not be in a pit that cannot be drained to the ground surface by gravity. In most parts of the country, it is best to locate pumps in some type of housing above the ground level and above any high water. Protection from freezing can be provided by installing a thermostatically controlled electric heater in the pump house. Small, well-constructed and insulated pump housings are sometimes not heated but depend on heat from the electric motor and a light bulb to maintain a proper temperature. Some type of ventilation should be provided, however, to prevent the condensation of moisture and destruction of the electric motor and switches. See page 145.

Pump Power and Drive

The power available will usually determine the type of motor or engine used. Electric power, in general, receives first preference, with other sources used for standby or emergency equipment.

Steam power should be considered if pumps are located near existing boilers. The direct-acting steam pump and single, duplex, or triplex displacement pump can be used to advantage under such circumstances. When exhaust steam is available, a steam turbine to drive a centrifugal pump can also be used.

Diesel-oil engines are good, economical pump-driving units when electricity is not dependable or available. They are high in first cost. Diesel engines are constant low-speed units.

Gasoline engines are satisfactory portable or standby pump power units. The first cost is low, but the operating cost is high. Variable speed control and direct connection to a centrifugal pump is common practice. Natural gas, methane, and butane can also be used where these fuels are available.

Electric motor pump drive is the usual practice when this is possible. Residences having low lighting loads are supplied with single-phase current, although this is becoming less common. When the power load may be three or more horsepower, three-phase current is needed. Alternating-current two- and three-phase motors are of three types: the squirrel-cage induction motor, the wound-rotor or slip-ring induction motor, and the synchronous motor. Single-phase motors are the repulsion-induction type having a commutator and brushes; the capacitator or condenser type, which does not have brushes and commutator; and the split-phase type. The repulsion-induction motor is, in general, best for centrifugal pumps requiring ¾ hp or larger. It has good starting torque. The all-purpose capacitor motor is suggested for sizes below ¾ hp. Make sure the electric motor is grounded to the pump. Check electrical code.

The squirrel-cage motor is a constant-speed motor with low starting torque but heavy current demand, low power factor, and high efficiency. Therefore, this type motor is particularly suited where the starting load is large. Larger power lines and transformers are needed, however, with resultant greater power use and operating cost.

The wound-rotor motor is similar to the squirrel-cage motor. The starting torque can be varied from about one-third to three times that of normal, and the speed can be controlled. The cost of a wound-rotor motor is greater than a squirrel-cage motor; but where the pumping head varies, power saving over a long-range period will probably compensate for the greater first cost. Larger transformers and power lines are needed.

The synchronous motor runs at the same frequency as the generator furnishing the power. A synchronous motor is a constant-speed motor even under varying loads; but it needs an exciting generator to start the electric motor. Synchronous motors usually are greater in size than 75 or 100 hp.

An electric motor starting switch is either manually or magnetically operated. Manually operated starters for small motors (less than 1 hp) throw in the full voltage at one time. Overload protection is provided, but undervoltage protection is not. Full-voltage magnetic starters are used on most jobs. Overload and undervoltage control to stop the motor is generally included. Clean starter controls and proper switch heater strips are necessary. Sometimes a reduced voltage starter must be used when the power company cannot permit a full voltage starter or when the power line is too long.

Automatic Pump Control

One of the most common automatic methods of starting and stopping the operation of a pump on a hydropneumatic system is by the use of a pressure switch. This switch is particularly adaptable for pumps driven by electric motors, although it can also be used to break the ignition circuit on a gasoline-engine-driven pump. The switch consists of a diaphragm connected on one side with the pump discharge line and on the other side with a spring-loaded switch. This spring switch makes and breaks the electric contact, thereby operating the motor when the water pressure varies between previously established limits.

Water-level control in a storage tank can be accomplished by means of a simple float switch. Other devices are the float with adjustable contacts and the electronic or resistance probes control and altitude valve. Each has advantages for specific installations.

When the amount of water to be pumped is constant, a time-cycle control can be used. The pumping is controlled by a time setting.

In some installations the pumps are located at some distance from the treatment plant or central control building. Remote supervision can be obtained through controls to start or stop a pump and to report pressure and flow data and faulty operation.

Another type of automatic pump switch is the pressure-flow control. This equipment can be used on ground-level or elevated water storage tanks.

When pumps are located at a considerable distance from a storage tank, and pressure controls are used to operate the pumps, heavy draw-offs may cause large fluctuations in pressure along the line. This will cause sporadic pump starting and stopping. In such cases, and when there are two or more elevated tanks on a water system, altitude valves should be used at the storage tanks. An altitude valve on the supply line to an elevated tank or standpipe is set to close when the tank is full and is set to open when the pressure on the entrance side is less than the pressure on the tank side of the valve. In this way overflowing of the water tank is prevented, even if the float or pressure switch fails to function properly.

EXAMPLES

Design of Small Water Systems

Experiences in new subdivisions show that peak water demands of six to ten times the average daily consumption rate are not unusual. Lawn-sprinkling demand has made necessary sprinkling controls, metering, or the installation of larger distribution and storage facilities and, in some instances, ground storage and booster stations. As previously stated, every effort should be made to serve a subdivision from an existing public water supply. Such supplies can afford to employ competent personnel and are in the business of supplying water, whereas a subdivider is basically in the business of developing land and does not wish to become involved in operating a public utility.

In general, when it is necessary to develop a central water system to serve the average subdivision, consideration should first be given to a drilled well-water supply. Infiltration galleries or special shallow wells may also be practical sources of water. Such water systems usually require a minimum of supervision and can be developed to produce a known quantity of water of a satisfactory sanitary quality. Simple chlorination treatment will normally provide the desired factor of safety. Test wells and sampling will indicate the most probable dependable yield and the chemical and bacterial quality of the water. Well logs should be kept.

Where a clean, clear lake supply or stream is available, chlorination and slow sand filtration can provide reliable treatment with daily supervision for the small development. The turbidity of the water to be treated should not exceed 30 mg/l. Preliminary settling may be indicated in some cases.

Other more elaborate types of treatment plants, such as rapid sand filters, are not recommended for small water systems unless specially trained operating personnel can be assured. Pressure filters have limitations, as explained earlier in this chapter.

The design of small slow sand filter and well-water systems is explained and illustrated in pages 166 to 171, 245, and below.

An example (Figure 3–26) will serve to illustrate the design bases previously discussed. The problem is to determine the probable maximum hourly demand rate for a good summer development consisting of 60 two-bedroom dwellings. The population, at two persons per bedroom, is 240. From Figure 3–21, the demand can vary from 75 to 230 gpm. An average maximum demand would be 152 gpm. Adjustment should be made for local conditions.

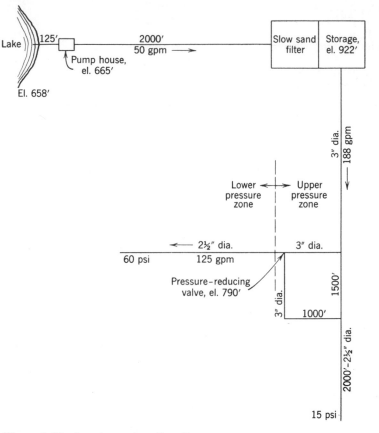

Figure 3-26   A water system flow diagram.

Examples showing calculations to determine pumping head, pump capacity, and motor size follow.

In one instance, assume that water is pumped from a lake at an elevation of 658 ft to a slow sand filter and reservoir at an elevation of 922 ft. See Figure 3–26. The pump house is at an elevation of 665 ft and the intake is 125 ft long. The reservoir is 2000 ft from the pump. All water is automatically chlorinated as it is pumped. The water consumption is 30,000 gpd. With the reservoir at an elevation of 922 ft, a pressure of at least 15 lb/in.² is to be provided at the highest fixture. It is required to find the size of the intake and discharge pipes, the total pumping head, the size pump, and motor. The longest known power failure is 14 hr and repairs can be made locally. Assume that the pump capacity is sufficient to pump 30,000 gal in 10 hr or 50 gpm. Provide one 50-gpm-pump and one 30-gpm-standby, both multistage centrifugal pumps, one to operate at any one time.

From the above, with a flow of 50 gpm, a 2½ in. pipeline to the storage tank is indicated.

The head losses, using Table 3–16 and Table 3–17 are:

Intake, 125 ft of 2½-in. pipe—3.3 (1.25)   =   4.1 ft

For this calculation, assume the entrance loss and loss through pump are negligible. Say that there are four long elbows = 16.8 ft, two globe valves = 140 ft, one check valve = 16 ft, three standard elbows = 18.9 ft, and two 45° elbows = 6 ft. Total equivalent pipe = 198 ft of 2½-in. pipe; head loss =

3.3 (1.98)   =   6.5 ft

Discharge pipe, 2000 ft of 2½-in. − 3.3(20)   =   66.0 ft

Total friction head loss   =   76.6 ft

Suction lift = 7 ft to center line of pump   =   7.0

Static head, difference in elevation = (922 − 665)   = 257.0

Pressure at point of discharge   NONE

Total head   = 340.6 ft

Add for head loss through meters if used (Table 3–18).

If a 3-in. intake and discharge line is used instead of a 2½-in. intake, the total head can be reduced to about 300 ft. The saving thus effected in power consumption would have to be compared with the increased cost of 3-in. pipe over 2½-in. pipe. The additional cost of power would be approximately $76.65 per year, with the unit cost of power at 2 cents per kilowatt-hour. This calculation is shown below using approximate efficiencies.

340-foot head

$$\text{Horsepower to motor} = \frac{\text{gpm} \times \text{total head in ft}}{3960 \times \text{pump efficiency} \times \text{motor efficiency}}$$

$$= \frac{50 \times 340}{3960 \times 0.45 \times 0.83} = 11.5$$

$$11.5 \text{ hp} = 11.5 \times 0.746 \text{ kW} = 8.6 \text{ kW}$$

In 1 hr, at 50 gpm, 50 × 60, or 3000 gal, will be pumped and the power used will be 8.6 kW-hr. To pump 30,000 gal of water will require $8.6 \times \dfrac{30,000}{3000} =$ 86 kW-hr.

If the cost of power is 2 cents per kW-hr, the cost of pumping 30,000 gal of water per day will be 86 × 0.02 = 1.72 or $1.72.

300-foot head

$$\text{Horsepower} = \frac{50 \times 300}{3960 \times 0.45 \times 0.83} = 10.1 \text{ hp} = 10.1 \times 0.746 \text{ kW} = 7.55 \text{ kW.}$$

TABLE 3-16 FRICTION DUE TO WATER FLOWING IN PIPE

Friction Head Loss (ft/100 ft of pipe)
Pipe diameter (in.)

Capacity (gpm)	½	¾	1	1¼	1½	2	2½	3	4	5	6	8	10	12	14
1	2.1	—	—	—	—	—	—	—	—	—	—	—	—	—	—
2	7.4	1.9	—	—	—	—	—	—	—	—	—	—	—	—	—
3	15.8	4.1	1.3	—	—	—	—	—	—	—	—	—	—	—	—
4	27.0	7.0	2.1	0.57	—	—	—	—	—	—	—	—	—	—	—
5	41.0	10.5	3.2	0.84	0.40	—	—	—	—	—	—	—	—	—	—
8	98.0	25.0	7.8	2.0	0.95	—	—	—	—	—	—	—	—	—	—
10	—	38.0	11.7	3.0	1.4	0.50	—	—	—	—	—	—	—	—	—
15	—	80.0	25.0	6.5	3.1	1.1	—	—	—	—	—	—	—	—	—
20	—	136.0	42.0	11.1	5.2	1.8	0.61	—	—	—	—	—	—	—	—
25	—	—	64.0	16.6	7.8	2.7	0.95	0.40	—	—	—	—	—	—	—
30	—	—	89.0	23.5	11.0	3.8	1.3	0.54	—	—	—	—	—	—	—
35	—	—	119.0	31.2	14.7	5.1	1.7	0.75	—	—	—	—	—	—	—
40	—	—	152.0	40.0	18.8	6.6	2.2	0.91	—	—	—	—	—	—	—
50	—	—	—	60.0	28.4	9.9	3.3	1.4	—	—	—	—	—	—	—
60	—	—	—	85.0	39.6	13.9	4.6	1.9	0.47	—	—	—	—	—	—
70	—	—	—	113.0	53.0	18.4	6.2	2.6	0.63	—	—	—	—	—	—
80	—	—	—	—	68.0	23.7	7.9	3.3	0.81	—	—	—	—	—	—
90	—	—	—	—	84.0	29.4	9.8	4.1	1.0	—	—	—	—	—	—
100	—	—	—	—	102.0	35.8	12.0	5.0	1.2	0.41	—	—	—	—	—
125	—	—	—	—	—	54.0	18.2	7.6	1.9	0.64	—	—	—	—	—
150	—	—	—	—	—	76.0	26.0	10.5	2.6	0.87	—	—	—	—	—
175	—	—	—	—	—	102.0	33.8	14.0	3.4	1.2	—	—	—	—	—
200	—	—	—	—	—	129.0	43.1	17.8	4.4	1.5	0.62	—	—	—	—
225	—	—	—	—	—	—	54.0	22.0	5.3	1.8	0.72	—	—	—	—
250	—	—	—	—	—	—	65.0	27.1	6.7	2.2	0.92	—	—	—	—

TABLE 3-16 FRICTION DUE TO WATER FLOWING IN PIPE (cont'd)

Friction Head Loss (ft/100 ft of pipe)

Pipe diameter (in.)

Capacity (gpm)	½	¾	1	1¼	1½	2	2½	3	4	5	6	8	10	12	14
300	—	—	—	—	—	—	92.0	38.0	9.3	3.1	1.3	—	—	—	—
400	—	—	—	—	—	—	—	65.0	16.0	5.4	2.2	—	—	—	—
500	—	—	—	—	—	—	—	98.0	24.0	8.1	3.3	0.83	—	—	—
600	—	—	—	—	—	—	—	—	33.8	11.7	4.7	1.2	—	—	—
700	—	—	—	—	—	—	—	—	45.0	15.2	6.2	1.5	0.52	—	—
800	—	—	—	—	—	—	—	—	57.6	19.4	8.0	2.0	0.67	—	—
900	—	—	—	—	—	—	—	—	71.6	24.2	10.0	2.5	0.83	—	—
1000	—	—	—	—	—	—	—	—	87.0	29.4	12.1	3.0	1.0	0.42	—
1500	—	—	—	—	—	—	—	—	—	62.2	25.6	6.3	2.1	0.88	0.42
2000	—	—	—	—	—	—	—	—	—	—	43.6	10.8	3.6	1.5	0.71
3000	—	—	—	—	—	—	—	—	—	—	—	22.8	7.7	3.2	1.5
4000	—	—	—	—	—	—	—	—	—	—	—	—	13.1	5.4	2.6
5000	—	—	—	—	—	—	—	—	—	—	—	—	19.8	8.2	3.8

Note: If the frictional resistance under certain known conditions is desired, multiply the friction head loss in feet by the following factors:

New cast iron (straight)	0.540	New riveted steel	0.840	Old wrought iron (over 15 yr)	1.51
Cement asbestos	0.540	Cast iron (5 yr old)	0.715	Iron with very rough interior	2.58
New brass, copper	0.540	Steel (10 yr old), C.I. (18 yr old)	1.00	Smooth concrete or masonry	0.615
New cast iron (not straight)	0.615	Cast iron (30 yr old)	1.51	Vitrified sewer	0.840
New wrought iron, wood (smooth)	0.715				

Source: Hazen and William, C = 100, adapted from "Water Supply and Purification," WD FM 5–295 (1942).

TABLE 3-17   FRICTION OF WATER IN FITTINGS

Friction Head Loss as Equivalent Number of ft of Straight Pipe

Pipe Fitting	½	¾	1	1¼	1½	2	2½	3	3½	4	5	6
					Size pipe, in in. (nominal diameter)							
Open gate valve	0.4	0.5	0.6	0.8	0.9	1.2	1.4	1.7	2.0	2.3	2.8	3.5
¾ closed gate valve	40.0	60.0	70.0	100.0	120.0	150.0	170.0	210.0	250.0	280.0	350.0	420.0
Open globe valve	19.0	23.0	29.0	38.0	45.0	58.0	70.0	85.0	112.0	120.0	140.0	170.0
Open angle valve	8.4	12.0	14.0	18.0	22.0	28.0	35.0	42.0	50.0	58.0	70.0	85.0
Standard elbow or through reducing tee	1.7	2.2	2.7	3.5	4.3	5.3	6.3	8.0	9.3	11.0	13.0	16.0
Standard tee	3.4	4.5	5.8	7.8	9.2	12.0	14.0	17.0	19.0	22.0	27.0	33.0
Open swing check	4.3	5.3	6.8	8.9	10.4	13.4	15.9	19.8	24.0	26.0	33.0	39.0
Long elbow or through tee	1.1	1.4	1.7	2.3	2.7	3.5	4.2	5.1	6.0	7.0	8.5	11.0
Elbow 45°	0.75	1.0	1.3	1.6	2.0	2.5	3.0	3.8	4.4	5.0	6.1	7.5
Ordinary entrance	0.9	1.2	1.5	2.0	2.4	3.0	3.7	4.5	5.3	6.0	7.5	9.0

Note: The frictional resistance to flow offered by a meter will vary between that offered by an open angle valve and globe valve of the same size.
See manufacturers for meter and check valve friction losses, also Table 3–18.

TABLE 3-18   HEAD LOSS THROUGH METERS

Flow (gpm)	Head Loss Through Meter (psi) Meter Size (in.)								
	$\frac{5}{8}$	$\frac{3}{4}$	1	$1\frac{1}{4}$	$1\frac{1}{2}$	2	3	4	6
4	1								
6	2								
8	4	1							
10	6	2	1						
15	14	5	2	2					
20	25	9	4	3	1				
30		20	8	7	2	1			
40			15	12	4	2			
50			23	18	6	3			
75					14	5	1		
100					25	10	3	1	
200							10	4	1
300							24	9	2
400								16	4
500								25	6

Source: Adapted from G. Roden, "Sizing and Installation of Service Pipes," *J. Am. Water Works Assoc.*, **38**, No. 5, 637–642 (May 1946).

In 1 hour, 3000 gal will be pumped as before, but the power used will be 7.55 kW-hr. To pump 30,000 gal will require $7.55 \times 10 = 75.5$ kW-hr.

If the cost of power is 2 cents per kW-hr, the cost of pumping 30,000 gal will be $75.5 \times 0.02 = 1.51$ or $1.51.

The additional power cost due to using 2½-in. pipe is $1.72 - 1.51 = 0.21$ per day or $76.65 per year.

At 4 percent interest, i, compounded annually for 25 years, n, $76.65, d set aside each year would equal about $3200, S, as shown below:

$$d = \frac{i}{(1+i)^n - 1} \times S; \qquad 76.65 = \frac{0.04}{(1+0.04)^{25} - 1} \times S$$

$$S = \frac{76.65}{0.024} = \$3,194, \text{ say } \$3,200$$

This assumes that the life of the pipe used is 25 years and the value of money 4 percent. If the extra cost of 3-in. pipe over 2½-in. pipe, plus interest on the difference, minus the saving due to purchasing a smaller motor and lower head pump, is less than $1200 (present worth of $3200), then 3-in. pipe should be used.

The size of electric motor to provide for the 50-gpm pump against a total head of 340 ft is shown above to be 11.5 hp. Since this is a nonstandard size, the next larger size, a 15-hp motor, will be provided.

If a smaller motor is used, it might be overloaded when the pumping head is decreased.

$$\text{Horsepower to the motor drive} = \frac{\text{gpm} \times \text{total head in ft}}{3960 \times \text{pump efficiency} \times \text{motor efficiency}}$$

But the total head loss through 2½-in. pipe when pumping 30 gpm would be:

Intake, 125 ft of 2½-in. pipe	125
Fittings, total equivalent pipe	198
Discharge pipe, 2000 ft of 2½-in.	2000
Total friction head loss	$2323 \times \dfrac{1.3}{100} = $ 32 ft
Suction lift	= 9
Static head	= 255
Total head	= 296 ft

$$\text{Horsepower to the motor drive} = \frac{30 \times 296}{3960 \times 0.35 \times 0.85}$$

= 7.55. Use a 7½-hp electric motor.

Because of the great difference in elevation, 658 ft to 922 ft, it is necessary to divide the distribution system into two zones, so that the maximum pressure in pipes and at fixtures will not be excessive. In this problem, all water is supplied the distribution system at elevation 922 ft. A suitable dividing point would be at elevation 790 ft. All dwellings above this point would have water pressure directly from the reservoir, and all below would be served through a pressure-reducing valve so as to provide not less than 15 lb/in.^{2} at the highest fixture nor more than 60 lb/in.^{2} at the lowest fixture. If one-third of the dwellings are in the upper zone and two-thirds are in the lower zone it can be assumed that the maximum hourly demand rate of flow will be similarly divided.

Assume the total maximum hourly demand rate of flow for an average daily water consumption of 30,000 gpd to be about 188 gpm. Therefore, 63 + 125 gpm can be taken to flow to the upper zone and 125 gpm to the lower zone. If a 3-in. pipe is used for the upper zone, and water is uniformly drawn off, the head loss at a flow of 188 gpm would be about ½ × 17 ft per 100 ft of pipe. And if 2½-in. pipe is used for the lower zone, and water is uniformly drawn off in its length, the head loss at a flow of 125 gpm would be about ⅓ × 18 ft per 100 ft of pipe. If the pipe in either zone is connected to form a loop, thereby eliminating dead ends, the frictional head loss would be further reduced to ¼ of that with a dead end for the portion forming a loop.

In all of the above considerations, actual pump and motor efficiencies obtained from and guaranteed by the manufacturer should be used wherever possible. Their recommendations and installation detail drawings to meet definite requirements should be requested and followed if it is desired to fix performance responsibility.

In another instance, assume that all water is pumped from a deep well through a pressure tank to a distribution system. See Figure 3–27.

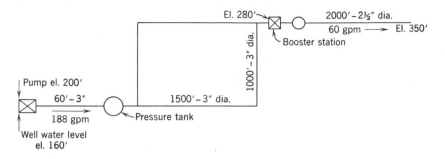

Figure 3–27 A water system flow diagram with booster station.

The lowest water level in the well is at elevation 160, the pump and tank are at elevation 200, and the highest dwelling is at elevation 350. Find the size pump, motor and pressure storage tank, the operating pressures, required well yield, and the size mains to supply a development consisting of 60 two-bedroom dwellings using an average of 30,000 gal of water per day.

Use a deep-well turbine pump. The total pumping head will consist of the sum of the total lift, plus the friction loss in the well drop pipe and connection to the pressure tank, plus the friction loss through the pump and pipe fittings, plus the maximum pressure maintained in the pressure tank. If at the pump, the maximum pressure in the tank is equal to the friction loss in the distribution system plus the static head caused by the difference in elevation between the pump and the highest plumbing fixture, plus the friction loss in the house water system including meter if provided, plus the residual head required at the highest fixture.

With the average water consumption at 30,000 gpd, the maximum hourly demand was assumed to be 188 gpm or less. The recommended pump capacity is taken as 125 percent of the maximum hourly rate, which would be 235 gpm. This assumes that the well can yield 235 gpm, which frequently is not the case. Under such circumstances, the volume of the storage tank can be increased two or three times, and the size of the pump correspondingly decreased to one-half or one-third the original size so as to come within the well yield. Another alternative would be to pump water out of the well, at a rate equal to the safe average yield of the well, into a large ground-level storage tank from which water can be pumped through a pressure tank at a higher rate to meet maximum water demands. This would involve double pumping and hence increased cost. Another arrangement, where possible, would be to pump out of the well directly into the distribution system, which is connected to an elevated storage tank. Although it may not be economical to use a pressure-tank water system, it would be of interest to see just what this would mean.

The total pumping head would be:

Lift from elevation 160 to 200 = 40 ft   40 ft

Figure 3–27 shows a distribution system that forms a rectangle 1000 × 1500 ft with a 2000-ft dead-end line serving one-third of the dwellings taking off at a point diagonally opposite the feed main. The head loss in a line connected at both ends is approximately one-fourth that in a dead-end line. The head loss in one-half the rectangular loop, from which water is uniformly drawn off, is one-third the loss in a line without drawoffs. Therefore, the total head loss in a 3-in. pipeline with a flow of about 188 gpm is equal to

$$\frac{1}{4} \times \frac{1}{3} \times \frac{17}{100} \times 2500 = 35 \text{ ft}$$

and the head loss through a 2000-ft dead-end line, with water being uniformly drawn off, assuming a flow of 60 gpm through 2½-in. pipe is equal to

$$\frac{1}{3} \times \frac{4.6}{100} \times 2000 = 31 \text{ ft}$$

This would make a total of 35 + 31 or 66 ft   66 ft

For a more accurate computation of the head loss in a water distribution grid system by the equivalent pipe, Hardy Cross, or similar method, the reader is referred to standard hydraulic texts. However, the assumptions made here are believed sufficiently accurate for our purpose.

The static head between the pump and the curb of the highest dwelling plus the highest fixture is (350 − 200) + 12 = 162 ft   162 ft

The friction head loss in the house plumbing system (without a meter) is equal to approximately 20 ft   20 ft

The residual head at highest fixture is approximately 20 ft   20 ft

The friction loss in the well drop pipe and connections to the pressure tank and distribution system with a flow of 235 gpm in a total equivalent length of 100 ft of 3-in. pipe is   25 ft

The head loss through the pump and fittings is assumed negligible.   ____

Total pumping head   333 ft

= 145 psi

Because of the high pumping head, and so as not to have excessive pressures in dwellings at low elevations, it will be necessary to divide the distribution system into two parts, with a booster pump and pressure storage tank serving the upper half.

If the booster pump and storage tank are placed at the beginning of the 2000 ft of 2½-in. line, at elevation 280 ft, only one-third of the dwellings need be served from this point. The total pumping head here would be:

Friction loss in 2000 ft of 2½-in. pipe with water withdrawn uniformly along its length and a flow of 60 gpm is

$$\frac{1}{3}\left(2000 \times \frac{4.6}{100}\right) = 31 \text{ ft} \qquad\qquad 31 \text{ ft}$$

The static head between the booster pump and the curb of the highest dwelling plus the highest fixture is

$$(350 - 280) + 12 = 82 \text{ ft} \qquad\qquad 82 \text{ ft}$$

The head loss in the house plumbing is 20 ft	20 ft
The residual pressure at highest fixture is 20 ft	20 ft
Booster pumping station total head	153 ft
	= 67 psi

The total pumping head at the main pumping station at the well would be:

Lift in well = 40 ft	40 ft
Friction loss in distribution system forming loop = 35 ft	35 ft

The static head between the pump and the booster tank, which is also adequate to maintain a 20-ft head at the highest fixture is

$$(280 - 200) = 80 \text{ ft} \qquad\qquad 80 \text{ ft}$$

Friction loss in well drop pipe and connections to the distribution system = 25 ft

	25 ft
Main pumping station, total head	180 ft
	= 78 psi

With an average daily water consumption of 30,000 gal, the average daily maximum demand, on a monthly basis, would be $(30,000 \times 1.5) = 45,000$ gal. The ratio of the absolute maximum and minimum operating pressures at the main pumping station, using a 10-lb differential, would be

$$\frac{78 + 14.7}{88 + 14.7} = \frac{92.7}{102.7} = 0.905, \text{ say } 0.90$$

From Figure 3–22 the pressure tank volume should be about 30,000 gal if 15 min storage is to be provided at the maximum demand rate, or 10,000 gal if 5 min storage is acceptable. The pump capacity, as previously determined, should be 235 gpm. Use a 250-gpm pump. If a 20-lb pressure differential is used, $\dfrac{P_1}{P_2} = 0.82$, and Figure 3–22 indicates a 16,000-gal pressure tank could be used to provide 15 min storage at the probable maximum hourly demand rate of flow.

The booster pumping station would serve one-third of the population, hence the average daily maximum demand on a monthly basis would be 1/3(45,000) or 15,000 gal. The ratio of the absolute maximum and minimum operating pressures at the booster pumping station, using a 10-lb differential, would be

$$\frac{67 + 14.7}{77 + 14.7} = \frac{81.7}{91.7} = 0.89, \text{ say } 0.90$$

From Figure 3–22 the pressure tank should have a volume of about 10,000 gal to provide 15-min storage at times of peak demand. The pump capacity should be 78 gpm. Use a 75-gpm pump. On the other hand, if the operating pressure differential is 20 lb and only 5-min storage at peak demand is desired, the required pressure tank volume would be 1,600 gal.

To determine the size motor required for the main pumping station and booster pumping stations, use the average of the maximum and minimum operating gauge pressures as the pumping head. The size motor for the main pumping station using manufacturer's pump and motor efficiencies is:

$$\frac{250 \text{ gpm} \times (180 + 11\frac{1}{2}) \text{ ft avg. head}}{3960 \times 0.57 \times 0.85} = 24.7$$

Use a 25-hp motor. The size motor for the booster station is:

$$\frac{75 \text{ gpm} \times (153 + 11\frac{1}{2}) \text{ ft avg. head}}{3960 \times 0.50 \times 0.80} = 8$$

Use a 10-hp motor.

In the construction of a pumping hydropneumatic station, provision should be made for standby pump and motive power equipment.

The calculations made on pages 234 to 244 are based on the use of a multistage centrifugal-type pump. Before a final decision is made, the comparison should include the relative merits and cost using a displacement-type pump. Remember that price and efficiency, although important when selecting a pump, are not the only factors to consider. The requirements of the water system and peculiarities should be anticipated and a pump with the desirable characteristics selected.

Design of a Camp Water System

A typical hydraulic analysis and design of a camp water system is shown in Figure 3–28.

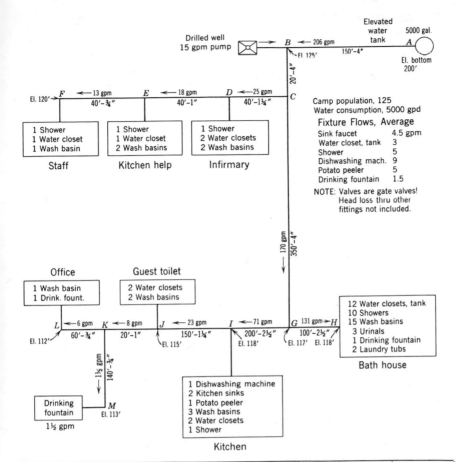

Distance			G.p.m. Flow			Head Available, Ft				Pipe Size, in.	Head Loss		Head, Ft Remaining	Facility Served
From	To	Feet	Max.	%	Probable	Initial	+Fall	−Rise	Total		Ft per 100 ft	Total		
A	C	170	295	70	206	0	75	0	75	4	4.5	7.6	67.4	
C	D	40	50	50	25	67.4	1.6	0	69	1¼	16.6	6.6	62.4	Inf., kitch., staff
D	E	40	30	60	18	62.4	1.0	0	63.4	1	36	14.4	49.0	Kitch., staff
E	F	40	13	100	13	49.0	1.6	0	50.6	¾	70	21	29.6	Staff cabin
C	G	350	245	75	170	67.4	8.0	0	75.4	4	3.4	11.9	63.5	
G	H	100	174	75	131	63.5	0	1	62.5	2½	20	20	42.5	Bath house
G	I	200	71	100	71	63.5	0	1	62.5	2½	6	12	50.5	Kitch, guest, off
I	J	150	23	100	23	50.5	3.0	0	53.5	1¼	15	22	31.5	
J	K	20	8	100	8	31.5	0	0	31.5	1	7.8	1.6	29.9	
K	L	60	6	100	6	29.9	3.0	0	32.9	¾	12	7.2	25.7	Office
K	M	140	1½	100	1½	25.7	2.0	0	27.7	¾	1.0	1.4	26.3	Drink. fount.

Figure 3–28 Typical hydraulic analysis of a camp water system. (See Figure 4–30.)

Cost Estimates

Because of the wide variations in types of water systems and conditions under which constructed, it is impractical to give reliable cost estimates. Some approximations are listed to give a feeling for some of the costs involved.

1. The approximate costs of water pipe, valves, and hydrants, including labor and material, but not including engineering, legal, land, and administrative costs are:

4 in. cast iron pipe, per ft	$2.50 to 3.50
6 in. cast iron pipe, per ft	5.25 to 6.40
8 in. cast iron pipe, per ft	6.80 to 8.00
10 in. cast iron pipe, per ft	10.00 to 12.00
12 in. cast iron pipe, per ft	12.25 to 14.00
16 in. cast iron pipe, per ft	16.00 to 18.00
18 in. cast iron pipe, per ft	18.00 to 20.00
6 in. cement asbestos, per ft	4.80 to 5.60
8 in. cement asbestos, per ft	5.80 to 6.40
12 in. cement asbestos, per ft	8.80 to 10.80
6 in. valve box	112.00 to 136.00
8 in. valve box	160.00 to 180.00
12 in. valve box	320.00 to 350.00
Hydrant	400.00 to 500.00

2. Elevated storage, small capacity—20,000-gal capacity, $13,500 to $15,200; 50,000 gal, $20,000 to $35,800; 100,000 gal, $33,100 to $53,700. Ground-level storage—41,000 gal, $12,100; 50,000 gal, $14,300; 72,000 gal, $15,200; 92,000 gal, $16,900.[91]

For larger installations, standpipe costs may run from $30,000 for 0.15 million gal capacity to $70,000 for 0.5 million gal; $115,000 for 1.0 million gal, and $250,000 for 3.0 million gal. For elevated tanks, cost may run $60,000 for 0.15 million gal capacity; $150,000 for 0.5 million gal; $270,000 for 1.00 million gal; and $650,000 for 3.0 million gal.

3. A complete conventional rapid sand filter plant including roads, landscaping, lagoons, laboratory, and low lift pumps may cost $150,000 for a 0.3-mgd plant; $220,000 for a 0.5-mgd plant; $370,000 for a 1.0-mgd plant; $830,000 for a 3.0-mgd plant; $1,220,000 for a 5.0-mgd plant; $2,000,000 for a 10.0-mgd plant; and $3,400,000 for a 20.0-mgd plant.

CLEANING AND DISINFECTION

Wells and Springs

Wells or springs that have been altered, repaired, newly constructed, flooded, or accidentally polluted should be thoroughly cleaned and disin-

[91] David H. Stoltenberg, "Construction Costs of Rural Water Systems," *Public Works,* 94 (August 1969).

fected after all the work is completed. The sidewalls of the pipe or basin, the interior and exterior surfaces of the new or replaced pump cylinder and drop pipe, and the walls and roof above the water line, where a basin is provided, should be scrubbed clean with a stiff bristled broom or brush and detergent, insofar as possible, and then washed down or thoroughly sprayed with water followed by washing or thorough spraying with a strong chlorine solution. A satisfactory solution for this purpose may be prepared by dissolving 1 oz of 70 percent high-test calcium hypochlorite made into a paste, 3 oz of 25 percent chlorinated lime made into a paste, or 1 pt of 5¼ percent sodium hypochlorite, dissolved in 25 gal of water. The well or spring should be pumped until clear and then be disinfected as explained below.

To disinfect the average well or spring basin, mix 2 qt of 5¼ percent "bleach" such as Clorox or Dazzle in 10 gal water. Pour the solution into the well; start the pump and open all faucets. When chlorine odor is noticeable at the faucets, close each faucet and stop the pump. It will be necessary to open the valve or plug in the top of the pressure tank, where provided, just before pumping is stopped in order to permit the strong chlorine solution to come into contact with the entire inside of the tank. Air must be readmitted and the tank opening closed when pumping is again started. Mix two more qt of bleach in 10 gal water and pour this chlorine solution into the well or spring. Allow the well to stand idle at least 12 to 24 hr; then pump it out to waste, away from grass and shrubbery, through the storage tank and distribution system, if possible, until the odor of chlorine disappears. *It is advisable to return the heavily chlorinated water back into the well, between the casing and drop pipe where applicable, during the first 30 minutes of pumping to wash down and disinfect the inside of the casing insofar as possible.* A day or two after the disinfection, after all the chlorine has been removed, a water sample may be collected for bacterial examination to determine whether all contamination has been removed. If the well is not pumped out, chlorine may persist for a week or longer.

A more precise procedure for the disinfection of a well or spring basin is to base the quantity of disinfectant needed on the volume of water in the well or spring. This computation is simplified by making reference to Table 3–19.

Although a flowing well or spring tends to cleanse itself after a period of time, it is advisable nevertheless to clean and disinfect all wells and springs that have had any work done on them before they are used. Scrub and wash down the basin and equipment. Place twice the amount of calcium hypochlorite indicated by Table 3–19 in a weighted cloth sack or can fitted with a cover. Punch holes in the can and fasten

TABLE 3-19   QUANTITY OF DISINFECTANT REQUIRED TO GIVE A DOSE OF 50 mg/l CHLORINE

Diameter of Well, Spring, of Pipe (in.)	Gallons of Water per ft of Water Depth	Ounces of Disinfectant/10-ft Depth of Water		
		70% Calcium Hypochlorite*	25% Calcium Hypochlorite†	5¼% Sodium Hypochlorite‡
2	0.163	0.02	0.04	0.20
4	0.65	0.06	0.17	0.80
6	1.47	0.14	0.39	1.87
8	2.61	0.25	0.70	3.33
10	4.08	0.39	1.09	5.20
12	5.88	0.56	1.57	7.46
24	23.50	2.24	6.27	30.00
36	52.88	5.02	14.10	66.80
48	94.00	9.00	25.20	120.00
60	149.00	14.00	39.20	187.00
72	211.00	20.20	56.50	269.00
96	376.00	35.70	100.00	476.00

*$Ca(OCl)_2$, also known as high-test calcium hypochlorite. A heaping teaspoonful of calcium hypochlorite holds approximately ½ oz. 1 liquid oz = 615 drops.

†$CaCl(OCl)$.

‡$Na(OCl)$, also known as Bleach, Clorox, Dazzle, etc., can be purchased at most supermarkets, drug, and grocery stores.

a strong string or wire to the can to also secure the cover. Suspend the can near the bottom of the well or spring, moving it around in order to distribute the strong chlorine solution formed throughout the water entering and rising up through the well or spring. Addition of the chlorine solution into a garden hose or plastic tubing extended to the bottom of a well or spring would be more satisfactory.

It should be remembered that disinfection is no assurance that the water entering a well or spring will be free of pollution. The cause for the pollution, if present, should be ascertained and removed. Until this is done, all water used for drinking and culinary purposes should first be boiled.

Pipelines

The disinfection of new or repaired pipelines can be expedited and greatly simplified if special care is exercised in the handling and laying of the pipe during installation. Trenches should be kept dry and a tight fitting plug provided at the end of the line to keep out foreign matter. Lengths of pipe that have soiled interiors should be cleansed and disinfected before being connected up. Each continuous length of main should be separately disinfected with a heavy chlorine dose or other effective disinfecting agent. This can be done by using a portable chlorinator, a hand-operated pump, or a cheap mechanical pump throttled down to inject the chlorine solution at the beginning of the section to be disinfected through a hydrant, corporation cock, or other temporary valved connection. Hypochlorite tablets can also be used to disinfect small systems.

The first step in disinfecting a main is to flush out the line thoroughly by opening a hydrant or drain valve below the section to be treated until the water runs clear. A velocity of at least 3 fps should be obtained. (Use a hydrant flow gauge.) The valve is then partly closed so as to waste water at some known rate. The rate of flow can be estimated with a flow gauge, by running the water into a milk can, barrel, or other container of known capacity and measuring the time to fill it. With the rate of flow known, determine from Table 3–20 the strength

TABLE 3–20   HYPOCHLORITE SOLUTION TO GIVE A DOSE OF 50 mg/l CHLORINE FOR MAIN STERILIZATION

Rate of Water Flow in Pipe-line, in gpm	Quarts of 5¼% Sodium Hypochlorite Made up to 10 gal with Water	Quarts of 14% Sodium Hypochlorite Made up to 10 gal with Water	Pounds of 25% Chlorinated Lime to 10 gal Water	Pounds of 70% Calcium Hypochlorite to 10 gal Water
5	4.6	1.7	2.0	0.7
10	9.1	3.4	4.0	1.4
15	13.7	5.1	6.0	2.1
20	18.3	6.8	8.0	2.9
25	22.8	8.5	10.0	3.5
40	36.6	13.7	16.0	5.7

Add hypochlorite solution at rate of 1 pt in 3 min. The 10-gal solution will last 3 hr if fed at rate of 1 pt in 3 min. Mix about 50 percent more solution than is theoretically indicated to allow for waste.

of chlorine solution to be injected into the main at the established rate of 1 pt in 3 min. The rate of flow of water can, of course, be adjusted and should be kept low for small-diameter pipe. It is a simple matter to approximate the time, in minutes, it would take for the chlorine to reach the open hydrant or valve at the end of the line being treated by dividing the capacity of the main in gal (see Table 3–19) by the rate of flow in gpm. In any case, injection of the strong chlorine solution should be continued at the rate indicated until addition of orthotolidine solution to samples of the water at the end of the main produces a deep red color. The hydrant should then be closed, chlorination treatment stopped, and the water system let stand at least 24 hr. At the end of this time, addition of orthotolidine to a sample of the treated water should show the presence of 25 mg/l residual chlorine. If no residual chlorine is found, the operation should be repeated. Following disinfection, the water main should be thoroughly flushed out with the water to be used and samples collected for bacterial examination for a period of several days. If the laboratory reports the presence of coliform bacteria, the disinfection should be repeated until satisfactory results are received. Where poor installation practices have been followed, it may be necessary to repeat the disinfection three to five times. The water should not be used until all evidence of contamination has been removed as demonstrated by the test for coliform bacteria.

If the pipeline being disinfected is known to have been used to carry polluted water, flush the line thoroughly and double the strength of the chlorine solution injected into the mains. Let the heavily chlorinated water stand in the mains at least 48 hr before flushing it out to waste and proceed as explained in the preceding paragraph. Cleansing of heavily contaminated pipe by the use of detergent, followed by flushing and then disinfection, may prove to be the quickest method.

Where pipe breaks are repaired, flush out the isolated section of pipe and dose the section with 200 mg/l chlorine and try to keep the line out of service at least 2 to 4 hr before flushing out the section and returning it to service. The hypochlorite tablet method can be used to disinfect small systems.

Storage Reservoirs and Tanks

Before disinfecting a reservoir or storage tank it is essential to first remove from the walls (bottom and top) all dirt, scale, and other loose material. The interior should then be flushed out and disinfected by one of the methods explained below.

If it is possible to enter the reservoir or tank, prepare a disinfecting solution by dissolving 1 oz of 70 percent calcium hypochlorite (HTH, Perchloron, Pitt-Chlor, etc.) made into a paste, 3 oz of 25 percent calcium hypochlorite (Chlorinated lime) made into a paste, or 1 pt of 5¼ percent sodium hypochlorite (Bleach, Clorox, Dazzle, etc.) in 25 gal of water. Apply this strong 250 mg/l chlorine solution to the bottom, walls, and top of the storage reservior or tank using pressure-spray equipment. *Make sure the tank is adequately ventilated before entering.* Wear a chlorine gas mask and protective clothing during the work, including goggles. Insist on all safety precautions. The tank or reservoir may be filled an hour after the work is completed.

Another method, for small tanks, is to compute the capacity. Add to the empty tank 10 oz of 70 percent calcium hypochlorite, or 2 lb of 25 percent chlorinated lime completely dissolved, or 4 qt of $5\frac{1}{4}$ percent sodium hypochlorite for each 1000 gal capacity. Fill the tank with water and let it stand for 12 to 24 hr. This will give a 50 mg/l solution. Then drain the water through the distribution system to waste.

A third method for the small tank involves the use of a chlorinator or hand-operated force pump. Admit water to the storage tank at some known rate and add at the same time the chlorine solution indicated in Table 3–20 at a rate of 1 pt in 3 min. Let the tank stand full for 12 to 24 hr and then drain the chlorinated water through the distribution system to waste. Rinse the force pump immediately after use.

It should be remembered, when disinfecting pressure tanks, that it is necessary to open the air-relief or other valve at the highest point so that the air can be released and the tank completely filled with the heavily chlorinated water. Air should be readmitted before pumping is commenced. In all cases, a residual chlorine test should show a distinct residual in the water drained out of the tanks. If no residual can be demonstrated, the disinfection should be repeated.

EMERGENCY WATER SUPPLY AND TREATMENT

Local or state health departments should be consulted when a water emergency arises. Their sanitary engineers are in a position to render valuable, expert advice based on their experience and specialized training.

The treatment to be given a water used for drinking purposes depends primarily on the extent to which the water is polluted and the type of pollution present. This can be determined by making a sanitary survey of the water source to evaluate the significance of the pollution that is finding its way into the water supply. It must be borne in mind that all surface waters, such as from ponds, lakes, streams, and brooks, are almost invariably contaminated and hence must be treated. The degree of treatment required, based on the maximum permissible and average most probable number of coliform bacteria, is given in Table 3–6. However, under emergency conditions it is not practical to wait for the results of bacterial analyses and hence one should be guided by the results of sanitary surveys and such reliable local data as may be available. Using the best information on hand, select the cleanest and most attractive water available and give it the treatment necessary to render it safe. Water passing through inhabited areas is presumed to be polluted with sewage and industrial wastes and must be boiled or given complete treatment, including filtration, to be considered safe to drink.

Boiling

The most reliable and simplest method of purifying small quantities of water, where fuel is available, is to boil the water for one minute. Distillation is also very effective, but special equipment is needed.

Chlorination

Chlorination treatment is a satisfactory method for disinfecting water that is not grossly polluted. It is particularly suitable for the treatment of a relatively clean lake, creek, or well water that is of unknown or questionable quality. Chlorine for use in hand chlorination is available in supermarkets, drugstores, and grocery stores and can be purchased as a powder, liquid, or tablet. Store solutions in the dark.

The powder is a calcium hypochlorite and the liquid a sodium hypochlorite. Both these materials deteriorate with age. The strength of the chlorine powder or liquid is on the container label and is given as a certain percent available chlorine. The quantity of each compound to prepare a stock solution, or the quantity of stock solution to disinfect one gallon of water, is given in Table 3–21. When using the powder, make a paste with a little water, then dissolve the paste in a quart of water. Allow the solution to settle and then use the clear liquid, without shaking. The stock solution loses strength and hence should be made up fresh once a week. It is important to allow the treated water to stand for 30 minutes after the chlorine is added before it is used. Double the chlorine dosage if the water is turbid or colored.

Chlorine-containing tablets suitable for use on camping, hunting, hiking, and fishing trips are available at most drugstores. The tablets contain 4.6 grains of chlorine; they deteriorate with age. Since chloramines are slow-acting disinfectants, the treated water should be allowed to stand at least 60 min before being used.

Homemade chlorinators may be constructed for continuous emergency treatment of a water supply where a relatively large volume of water is needed. Such units require constant observation and supervision as they are not dependable. Figures 3–29 and 3–30 show several arrangements for adding hypochlorite solution. In some parts of the country it may be possible to have a commercial hypochlorinator delivered and installed within a very short time. Some health departments have available a hypochlorinator for emergency use. Communicate with the local

TABLE 3-21   EMERGENCY DISINFECTION OF SMALL
VOLUMES OF WATER

Product	Available Chlorine, %	Stock Solution*	Quantity of Stock Solution to Treat 1 gal of Water†	Quantity of Stock Solution to Treat 1,000 gal of Water†
Zonite	1	Use full strength	30 drops	2 quarts
S.K., 101 Solution	2½	Use full strength	12 drops	1 quart
Clorox, White Sail, Dazzle, Rainbow, Rose-X	5¼	Use full strength	6 drops	1 pint
Sodium Hypochlorite	10	Use full strength	3 drops	½ pint
Sodium Hypochlorite	15	Use full strength	2 drops	¼ pint
Calcium Hypochlorite, "Bleaching Powder," or Chlorinated Lime	25	6 heaping table-spoonfuls (3 oz) to 1 qt of water	one teaspoonful or 75 drops	1 quart
Calcium Hypochlorite	33	4 heaping table-spoonfuls to 1 qt of water	one teaspoonful	1 quart
HTH, Perchloron, Pittchlor	70	2 heaping table-spoonfuls (1 oz) to 1 qt of water	one teaspoonful	1 quart

*One quart contains 135 ordinary teaspoonfuls of water.

†Let stand 30 min before using. To dechlorinate, use sodium thiosulfate in same proportion as chlorine. One jigger = 1½ liquid oz. Chlorine dosage is approximately 5–6 mg/l. (See Chlorination.) (1 liquid oz = 615 drops.) Make sure chlorine solution or powder is fresh; check by making orthotolidine test. Double amount for turbid or colored water.

or state health department for assistance and advice relative to the manufacturers of approved hypochlorinators. When large volumes of water are disinfected, a daily report should be kept showing the gallons of water treated, the amount of chlorine solution used, and the results of hourly residual chlorine tests.

Iodine

Eight drops of 2 percent tincture of iodine may be used to disinfect one qt of water (8 mg/l dose). Allow the water to stand at least 30 min before it is used. (Bromine can also be used to disinfect water, although its use has been restricted to the disinfection of swimming-pool water.) Studies of the usefulness of elemental iodine show it to be a

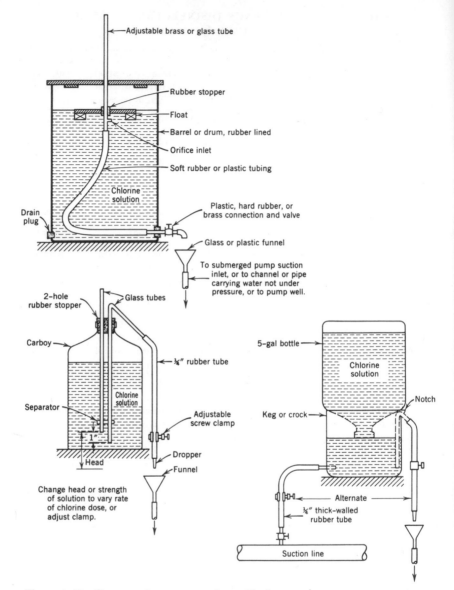

Figure 3–29 Home-made emergency hypochlorinators.

254

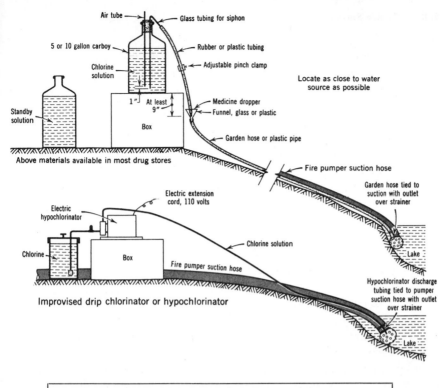

Preparation and Feed of Chlorine Solution		
Chlorine	Quantity to 5 gal water	Rate of feed 500 gpm pumper
Perchloron or HTH, 70% available chlorine	4 pounds*	4 ounces or $\frac{1}{4}$ pint per minute
Chlorinated lime, 24% available chlorine	12 pounds*	4 ounces or $\frac{1}{4}$ pint per minute
Sodium hypochlorite, 14% available chlorine	20 pints	4 ounces or $\frac{1}{4}$ pint per minute

* Make a paste in a jar; add water and mix; let settle for a few min then pour into carboy or other container and make up to 5 gal. Discard white deposit; it has no value. Dosage is 5 mg/l chlorine. Double solution strength if necessary to provide residual of 4 to 5 mg/l.

Figure 3–30 Emergency chlorination for fire supply, under health department supervision.

good disinfectant over a pH range of 3 to 8, even in the presence of contamination.[92] A dosage of 5 to 10 mg/l, with an average of 8 mg/l for most waters, is effective against enteric bacteria, amoebic cysts, cercariae, leptospira, and viruses within 30 min. Tablets that can treat about 1 qt of water may be obtained from the National Supply Service, Boy Scouts of America, large camping supply centers, and from the army, in emergency. These tablets dissolve in less than 1 min and are stable for extended periods of time. They are known as iodine water purification tablets, of which globaline, or tetraglycine hydroperiodide, is preferred. They contain 8.0 mg of active iodine per tablet. The treated water is palatable.

Filtration

Portable pressure filters are available for the treatment of polluted water. These units can produce an acceptable water provided they are carefully operated by trained personnel. Preparation of the untreated water by settling, prechlorination, coagulation, and sedimentation may be necessary, depending on the type and degree of pollution in the raw water. Pressure filters contain special sand, crushed anthracite coal, or diatomaceous earth. Diatomite filtration, or slow sand filtration, should be used where diseases such as amoebic dysentery, ascariasis, schistosomiasis or paragonimiasis are prevalent, in addition of course to chlorination.

Slow sand filters (consisting of barrels or drums) may be improvised in an emergency. Their principles are given in Figure 3–13 and Table 3–8. It is most important to control the rate of filtration so as not to exceed 50 gpd/ft² of filter area and to chlorinate each batch of water filtered as shown in Table 3–21 in order to obtain reliable results.

Bottled and Tank-Truck Water

In an emergency it is sometimes possible to obtain bottled water from an approved source, which is properly handled and distributed. Such water should meet the "Public Health Service Drinking Water Standards" as to source, protection, and bacteriological quality. Water that is transported in tank trucks from an approved source should be batch-chlorinated at the filling point as an added precaution. The tank truck should, of course, be thoroughly cleaned and disinfected before being placed in service. Detergents and steam are sometimes needed, particularly to remove gasoline or other gross pollution. Each tank of water

[92] Shih Lu Chang and J. Carrell Morris, "Elemental Iodine as a Disinfectant for Drinking Water," *Ind. Eng. Chem.*, **45**, 1009 (May 1953).

should be dosed with chlorine at the rate of 1 to 2 mg/l as given in Table 3–4.

Regulations relating to bottled water have been prepared by the State of California Department of Public Health, the New York State Department of Health,[93] and others.

The bottled water industry has shown a large growth in many parts of the country because of the public demand for a more palatable and "pure" water. It is not uncommon to find a wide selection of spring water or mineral water in supermarkets and small stores in almost all parts of the world.

Pasteurization plants and beverage bottling plants have much of the basic equipment needed to package water in paper, plastic, or glass containers. Contamination that can be introduced in processing (filtration through sand and carbon filters) and in packaging (pipelines, storage tanks, fillers) should be counteracted by germicidal treatment of the water just prior to bottling. Ozonation or ultraviolet treatment is sometimes used instead of chlorination. In any case, the source of water, equipment used, and operational practices must meet recognized standards.

BIBLIOGRAPHY

Bennison, E. W., *Ground Water, Its Development, Uses and Conservation*, Edward E. Johnson, Inc., St. Paul, Minn., 1947, 509 pp.

Cox, Charles R., *Operation and Control of Water Treatment Processes*, WHO Mon. Ser., No. 49, Geneva, 1964.

"Engineering Evaluation of Virus Hazard in Water," *J. of the Sanitary Engineering Division*, Proceedings of the American Society of Civil Engineers, Vol. 96, No. SA1, February 1970.

Fair, Gordon M., Geyer, John C., and Okun, Daniel A., *Water and Wastewater Engineering*, Vol. 1, John Wiley & Sons, Inc., New York, 1966.

Individual Water Supply Systems, Recommendations of the Joint Committe on Rural Sanitation, FSA, U.S. Public Health Service, 1950, 61 pp.

Mackenthun, Kenneth M., and Ingram, William M., *Biological Associated Problems in Fresh Water Environments*, Federal Water Pollution Control Administration, U.S. Dept. of the Interior, Washington, D.C., 1967, 287 pp.

Manual of Instruction for Water Treatment Plant Operators, New York State Department of Health, Albany, 1965, 308 pp.

Manual of Recommended Water-Sanitation Practice, FSA, U.S. Public Health Service Bull. No. 296, 1946, 40 pp.

Military Sanitation, Dept. of the Army Field Manual 21-10, May 1957, 304 pp.

Okun, Daniel A., "Alternatives in Water Supply," *J. Am. Water Works Assoc.*, **61**, No. 5, 215–221 (May 1969).

[93] *Public Water Supply Guide for Owners and Operators of Bottled & Bulk Water Facilities*, New York State Department of Health, Albany, January 1971.

Recommended Standards for Water Works, Report of Committee of the Great Lakes–Upper Mississippi River Board of State Sanitary Engineers on Policies for the Review and Approval of Plans and Specifications for Public Water Supplies, 1968, 87 pp.

Standard Methods for the Examination of Water and Wastewater, 13th ed., American Public Health Association, Inc., 1740 Broadway, New York, N.Y., 1971, 874 pp.

Viruses in Waste, Renovated, and Other Waters, edited by Gerald Berg, Federal Water Quality Administration, U.S. Dept. of the Interior, Cincinnati, Ohio, 1969.

Wagner, E. G., and Lanoix, J. N., *Water Supply for Rural Areas and Small Communities,* WHO Mon. Ser., No. 42, Geneva, 1959, 337 pp.

Water Quality Criteria, Federal Water Pollution Control Administration, U.S. Dept. of the Interior, Washington, D.C., April 1, 1968, 234 pp.

Water Treatment Plant Design, American Water Works Association, Inc., New York, N.Y., 1969, 353 pp.

Well Drilling Operations, Technical Manual 5-297, Dept. of the Army and Air Force, September 1965, 249 pp.

Wisconsin Well Construction Code, Division of Well Drilling, Wisconsin State Board of Health, 1951, 56 pp.

Wright, F. B., *Rural Water Supply and Sanitation,* John Wiley & Sons, Inc., New York, 1956, 347 pp.

Manual for Evaluating Public Drinking Water Supplies, Public Health Service Pub. No. 1820, U.S. Dept. of HEW, Bureau of Water Hygiene, Cincinnati, Ohio, 1969, 62 pp.

Water Quality and Treatment, Prepared by The American Water Works Association, Inc., McGraw-Hill Book Company, New York, 1971, 654 pp.

4

WASTEWATER TREATMENT AND DISPOSAL

Disease Hazard

The improper disposal of human excreta and sewage is one of the major factors threatening the health and comfort of individuals in areas where satisfactory sewerage systems are not available. This is so because very large numbers of different disease-producing organisms can be found in the fecal discharges of ill and apparently healthy persons, as explained in Chapter 1. Surveys show that 5 to 10 percent of the population are carriers of *Endamoeba histolytica*, causing amebic dysentery,[1] and 25 percent of the population are carriers of ascarid, hookworm, or tapeworm.[2] Studies in an American city showed that 9.1 percent of the local population harbored *Endamoeba histolytica* and that 23.1 percent harbored parasites.[3] Knowing that other organisms causing various types of diarrhea, bacillary dysentery, infectious hepatitis, salmonella infection, and many other illnesses are found in excreta, it becomes obvious that all sewage should be considered presumptively contaminated, beyond any reasonable doubt, with disease-producing organisms. In addition, it is known that some pathogenic organisms will survive from less

[1] L. T. Coggeshall, "Current and Postwar Problems Associated with the Human Protozoan Diseases," *Ann. N.Y. Acad. Sci.*, **44,** 198 (1943).
[2] Norman R. Stoll, "Changed Viewpoint on Helminthic Disease: World War I vs. World War II," *Ann. N.Y. Acad. Sci.*, **44,** 207–209 (1943).
[3] J. A. Kasper, E. J. Cope, M. Lyon, and M. White, "Report on the Results of Examinations for Intestinal Parasites," *Am. J. Pub. Health,* **40,** 1395–1397 (1950).

than 1 day in peat to more than 2 years in freezing moist soil.[4] Moist soil is favorable and dry soil is unfavorable for survival of the organisms. Therefore the mere exposure of sewage, or its improper disposal, immediately sets the stage for possible disease transmission by direct contact, a vehicle or vector, an individual, an inanimate object such as a child's ball, the housefly, or contaminated water or food.

Awareness of these dangers, coupled with treatment of sewage, provision of potable water, sanitation, and personal hygiene are recognized as being primarily responsible for reducing intestinal and water-borne diseases to their present low level in many parts of the world. Maintenance of the disease barriers and vigorous application of the basic sanitary engineering principles to unsanitary areas is necessary to the enjoyment of a healthful environment. It may appear inconceivable, but there are still many urban areas, as well as suburban and rural areas, in the United States and abroad where the discharge of raw or inadequately treated sewage to roadside ditches and streams is commonplace. See Chapter 1 and Figure 1–2 for additional details on disease control.

Definitions

Before proceeding further it is desirable to define some commonly used terms. *Excreta* is the waste matter eliminated from the body. *Domestic sewage* is the used water from a home or community and includes toilet, bath, laundry, lavatory, and kitchen-sink wastes. Sewage from a community may include some industrial wastes, groundwater, and surface water, hence the more inclusive term *wastewater* is coming into general usage. Normal sewage from a private sewage disposal system contains about 99.8 percent water and 0.2 percent total mineral and organic solids. Domestic sewage contains less than 0.1 percent total solids. The strength of wastewater is commonly expressed in terms of 5-day biochemical oxygen demand (BOD), suspended solids, and chemical oxygen demand (COD).

The *biochemical oxygen demand* of sewage, sewage effluents, polluted waters, industrial wastes, or other wastewaters is the oxygen in parts per million (ppm) or milligrams per liter (mg/l) required during stabilization of the decomposable organic matter by aerobic bacterial action. Complete stabilization requires more than 100 days at 20°C. Incubation for 5 or 20 days is not unusual but as used in this chapter BOD refers to the 5-day test unless otherwise specified.

---

[4] Willem Rudolfs, L. Lloyd Falk, and R. A. Ragotzkie, "Literature Review on the Occurrence and Survival of Enteric, Pathogenic, and Relative Organisms in Soil, Water, Sewage, and Sludges, and on Vegetation," *Sewage & Ind. Wastes,* **22,** 1261 (1950).

Suspended solids are those that are visible and in suspension in water. They are the solids that are retained on the asbestos mat in a Gooch crucible.

The *chemical oxygen demand* is also used, particularly in relation to certain industrial wastes. The COD is the amount of oxygen expressed in ppm or mg/l consumed under specific conditions in the oxidation of organic and oxidizable inorganic material. The test is relatively rapid. It does not oxidize some biodegradable organic pollutants (pyridine, benzene, toluene) but does oxidize some inorganic compounds that are not measured, that is, affected by the BOD analysis.

The *privy*, or one of its modifications, is the common device used when excreta is disposed of without the aid of water. When excreta is disposed of with water, a *water-carriage* sewage-disposal system is used; generally all other domestic liquid wastes are included.

When storm water and domestic sewage enter a sewer it is called a *combined sewer*. If domestic sewage and storm water are collected separately, in a *sanitary sewer* and in a *storm sewer*, the result is a *separate sewer system*. A *sewage works* or *sewer system* is a combination of sewers and appurtenances for the collection, pumping, and transportation of sewage, sometimes called *sewerage*, plus facilities for treatment and disposal of sewage, known as the *sewage treatment plant*.

Stream Pollution and Recovery

Although the 5-day BOD is the best single measure of wastewater or polluted water, organic loading, aquatic organisms when measured with the BOD, the COD where indicated, the dissolved oxygen, and the sanitary survey taken all together are the best indicators of water pollution. Other chemical and bacterial analyses provide additional information.

In a freshly polluted water, during the first stage (5 to 20 days at 20°C), mostly carbonaceous matter is oxidized. This is demonstrated by an immediate increase in the stream BOD and oxygen utilization in the area of pollution discharge, followed by the second, or nitrification, stage in which a lesser but uniform rate of oxygen utilization takes place for an extended period of time. This is accompanied by a related characteristic change in the stream biota, as illustrated in Figure 4–1 for an assumed condition and as discussed below. The amount of dissolved oxygen in a receiving water is the single most important factor determining the waste assimilation capacity of a body of water.

Stream pollution (organic) is apparent along its length by a zone of degradation just below the source, a zone of active decomposition, and, if additional pollution is not added, a zone of recovery. In the

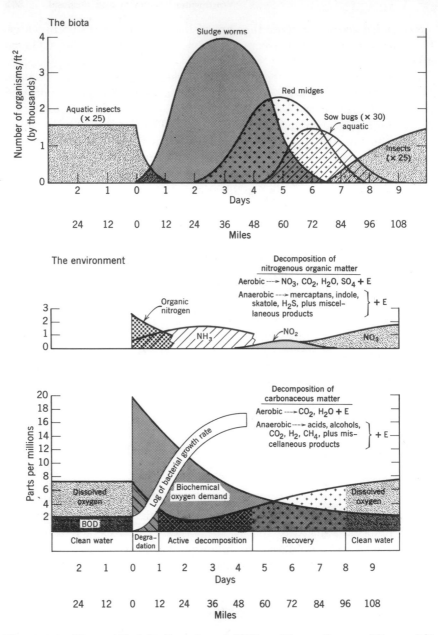

Figure 4–1 [From Alfred F. Bartsch and William Marcus Ingram, "Stream Life and the Pollution Environment," *Public Works* (July 1959).] The assumptions in the hypothetical pollution case under discussion are a stream flow of 100 cfs, a discharge of raw sewage from a community of 40,000, and a water temperature of 25°C, with typical variation of dissolved oxygen and BOD. The biota population

zone of degradation the oxygen in the water is decreased, suspended solids may be increased, and the stream bottom accumulates sludge. The fish life changes from game and food species to coarse. Worms, snails, and other biota associated with pollution increase. In the zone of active decomposition, the dissolved oxygen is further reduced and may approach zero. The water becomes turbid and gives off foul odors. Fish disappear, anaerobes predominate in the bottom mud, and sludge worms become very numerous. In the zone of recovery the process is gradually reversed and the stream returns to normal. The zones mentioned may not be discernible or experienced where sewage has been given adequate treatment before discharge.

SMALL WATER-BORNE SEWAGE DISPOSAL SYSTEMS

The provision of running water in a dwelling or structure immediately introduces the requirement for sanitary removal of the used water. Where public sewage works are available, connection to the sewer will solve a major sanitation problem. Where public or central sewage works are not provided or anticipated, as in predominately rural areas, consideration must be given to the proposed method of collection, removal, treatment, and disposal of sewage. With a suitable soil, the disposal of sewage can be simple, economical, and inoffensive; but careful maintenance is essential for continued satisfactory operation. Where rock or groundwater is close to the surface or the soil is a tight clay, it would be well to investigate some other property.

The common system for sewage treatment and disposal at a private home in a rural area consists of a proper septic tank for the settling and treatment of the sewage and a subsurface leaching system for the disposal of the septic-tank overflow, provided the soil is satisfactory. The soil percolation test and soil characteristics are used as a means for determining soil permeability or the capacity of a soil to absorb settled sewage. This and the quantity of sewage from a dwelling are the bases upon which a subsurface sewage-disposal system is designed. Sand filters, evapotranspiration beds, aeration systems, stabilization ponds or lagoons, recirculating toilets, and various types of privies are used under certain conditions.

---

curve is composed of a series of maxima for individual species, each multiplying and dying off as stream conditions vary. The environment curve shows that with a heavy influx of nitrogen and carbon compounds from sewage, the bacterial growth rate is accelerated and dissolved oxygen is utilized for oxidation of these compounds. As this proceeds, food is "used up" and the BOD declines.

General Soil Considerations

In its broadest sense, soil is made up of decayed or broken-down rock containing varying amounts of organic material such as animal wastes and plants. Destruction of rock is accomplished by water, wind, glacial ice, chemical action, plant life, freezing and thawing, heat, and other forces to form soil. The soil may accumulate in place or may be transported by wind, water, or glacial ice. Soil that accumulates in place is representative of the rock from which it has been derived. Soils may be divided, for reference purposes, into gravel, sand, silt, and clay, and, depending on which is predominant, into sandy loam, gravelly loam, silty loam, loam, clay loam, and clay. Loam is a mixture of gravel, sand, silt, and clay containing decayed plant and animal matter or humus, which would then introduce the term *topsoil*. Clay loam and clay do not drain well and usually are unsuitable for the disposal of sewage by subsurface means.

Generally speaking, if one digs a hole in the earth he will first penetrate topsoil, then loam, then rock fragments, and finally creviced or solid rock. The topsoil may be as much as 2 or 3 ft deep, although on the average it will be about 6 to 8 in. The thickness of the topsoil, loam, rock fragment layers, and rock strata is variable. See Figure 4–2.

The soil percolation test has been used successfully in the design of subsurface sewage disposal systems for many years.[5] A laboratory perco-

[5] John E. Kiker, Jr., "Subsurface Sewage Disposal," *Florida Eng. & Ind. Exp. Sta.*, Bull. **23** (December 1948).

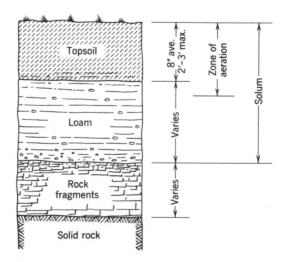

Figure 4–2 Typical earth formation.

lation test using saturated soil cores to determine soil permeability and its correlation with soil characteristics has received much study in connection with soil conservation and usage.[6] The permeability of the soil, or the ability of the soil to absorb and allow water and air to pass through, is related to the depth of the zone of aeration as well as the chemical composition and granular structure of the soil. The microbial population modifies the properties of the soil. A lump of soil with good structure will break apart, with little pressure, along definite cleavage planes. If the color of the soil is yellow, brown, or red, it would indicate that air, and therefore water, pass through; whereas if the soil is greyish or mottled brown and red, it would indicate lack of aeration and therefore a tight soil that is probably unsuitable for subsurface absorption. A greyish soil, however, may be made suitable if drained. Magnesium and calcium tend to keep the soil loose, whereas sodium and potassium have the opposite effect. Sodium hydroxide, a common constituent of so-called septic-tank cleaners, would cause a breakdown of soil structure with resultant smaller pore space and reduced soil permeability.

Soils maps are available for predominately agricultural counties from the Soil Conservation Service of the U.S. Department of Agriculture, although their use is being expanded for other purposes. The latest soil mapping is at a scale of 3.2 in. = 1 mi. Soil types are described giving depth, texture, composition color, drainage, slope, and acidity. The soil survey is done on aerial photograph maps.

Aerobic oxidizing bacteria, that is oxygen-loving bacteria, are found in the zone of aeration. This zone extends through the topsoil and into the upper portion of the subsoil, depending on the soil structure, earthworm population, root penetration, and other factors. The topsoil supports vegetative organisms such as bacteria, fungi, and molds, as well as protozoa, nematodes, earthworms, insects, and larger animals. These organisms have the faculty of reducing complex organic matter to simpler forms through their life processes. Hence septic-tank effluent, for example, which contains material in solution, in colloidal state, and in suspension, when discharged into or close to the topsoil will be acted upon by these organisms and be reduced to "soil," as well as liquids and gases. This is accomplished provided the sewage is not discharged at too rapid a rate or in too great a strength into the earth in the zone of aeration. A waterlogged soil destroys the aerobic organisms, producing anaerobic conditions that tend to preserve the organic matter in septic-

[6] Alfred M. O'Neal, "Soil Characteristics Significant in Evaluating Permeability," *Soil Science*, **67**, No. 5 (May 1949); Alfred M. O'Neal, "A Key for Evaluating Soil Permeability by Means of Certain Field Clues," *Soil Sci. Soc. of Am. Proc.*, **16**, No. 3 (July 1952).

tank effluent, thereby delaying decomposition and increasing mechanical clogging of the liquid-soil interface with organic matter, slimes, and sulfides. To assist in the maintenance of aerobic conditions, subsurface tile fields are usually laid at depths of 24 to 30 in., although depths as great as 36 in. or more are sometimes used. The gravel around the open-joint tile or perforated pipe should extend up into the zone of aeration, usually within 12 to 18 in. of the ground surface.

McGauhey and Winneberger call attention to the recovery infiltrative capacity of trench sidewalls after resting a few hours; the need to have a permeable soil in the first instance; the insulating effect of impermeable lenses or strata to the downward percolation of water; the reduction of percolative capacity of soils containing colloids that swell; and the necessity for the groundwater table to be at a sufficient depth to permit the soil to drain during rest periods rather than water remaining suspended by surface tension and capillary phenomena.[7] The "loss of infiltrative capacity is directly traceable to the organic fraction of sewage which leads to clogging of the soil surface by suspended solids, bacterial growth, and ferrous sulfide precipitation." They report "conclusively that intermittent dosing and draining of the soil system is necessary to the maintenance of optimum infiltration rates." In a subsequent report narrow trenches, 8 to 12 in. in width, and placement of the distributing line as high as possible are advised to provide a maximum effective sidewall surface area.[8] The use of pea gravel adjacent to the trench sidewall permits the clogging organic matter to spread through the gravel fill and thus extend the period of satisfactory operation, particularly when coupled with full loading (siphon chamber) on a continuous basis and maintenance of aerobic conditions in the trench soil, which requires prolonged intermittent resting periods (several months to a year).

It has also been pointed out that most soils will reach a steady-state equilibrium percolation rate of about 0.1 $ft/ft^2/day$ when the soil surface is completely inundated by untreated effluent.

Laboratory study indicates that passage through 40 to 50 cm of an agricultural-type soil is very effective in removing viruses from water.[9]

[7] P. H. McGauhey and John H. Winneberger, "Studies of the Failure of Septic Tank Percolation Systems," *J. Water Pollution Control Federation,* **36,** No. 5, 593–606 (May 1964).

[8] P. H. McGauhey and John H. Winneberger, "Final Report on a Study of Methods of Preventing Failure of Septic-Tank Systems," Sanitary Engineering Research Laboratory, College of Engineering and School of Public Health, University of California, Berkeley, October 31, 1965.

[9] William A. Drewry and Rolf Eliassen, "Virus Movement in Groundwater," *J. Water Pollution Control Fed.,* **40,** No. 8, R257 (August 1968).

Adsorptive capacity is reported to be a very important consideration in the proper design of a septic tank disposal field. Robeck et al. advise that the adsorptive capacity can be obtained with a soil having 0.5 to 1 percent organic matter, which is found in practically all agricultural soil, together with some clay and silt, which add to the adsorptive capacity.[10] However, to obtain the advantages of adsorption, and still have a functioning subsurface absorption system, requires acceptance of a soil with reduced permeability or percolation rate that makes possible the trapping of pathogens and aerobic organisms. This study also emphasizes that a soil with low adsorption or a formation with solution channels, fractures, or fissures will permit pollution to travel long distances without purification. Careful consideration to these factors in the design of subsurface sewage-disposal systems is necessary.

The design of leaching pits and cesspools is based on the ability of the soil found at the depth between 3 and 8 or 10 ft to absorb water. Sometimes pits are made 20 to 25 ft or more in depth, using prefabricated sections, in order to reach permeable soil. It is not known to what extent bacteria, protozoa, and metazoa are active in leaching pits and cesspools. Since relatively large quantities of sewage would be discharged in a small area, designs incorporating cesspools or leaching pits must take into consideration the soil structure, direction and depth of groundwater flow, and the relative location of wells or springs with respect to their possible pollution. Cesspools and leaching pits should be prohibited in shale, course gravel and limestone areas or where groundwater is high, and avoided when shallow wells or springs are in the vicinity unless adequate protecting distances and soils can be assured.

Where the soil is relatively nonpermeable at shallow and deep depths, an artificial sand filter that requires an outlet to a ditch or watercourse, or other treatment and disposal device, is needed in place of a conventional leaching system. Building should preferably be postponed until sewers are available. For greater protection, the filtered sewage is chlorinated.

It is extremely important while in the planning stage, before building construction is started, to consider:

1. The suitability of the soil to absorb settled sewage as determined by soil percolation tests, soil characteristics, and the type and size of the disposal system required. At least 4 to 5 ft of suitable soil should be available over clay, hardpan, rock, or groundwater.

---

[10] Gordon G. Robeck et al., "Factors Influencing the Design and Operation of Soil Systems for Waste Treatment," *J. Water Pollution Control Fed.*, **36**, No. 8, 971 (August 1964).

2. The area of land available for the sewage-disposal system, and its adequacy.

3. The depth of the underground water level, clay or rock, and slope of the ground; for these, determine the elevation of the sewage-disposal units, the house sewer, the house drain, the first-floor level, and location of the lowest plumbing fixture. Sometimes the installation of a sump pump, excavation at a greater depth, or the carting in of earth fill is made necessary because the slope of the ground surface, clay or rock level, and depth of the underground water were not considered in the planning stage.

4. The location of rock outcrops, hills, large trees, storm-water drains, watercourses, adjoining structures, wells, and water lines on the property under consideration as well as on adjoining properties.

5. Surface and underground water drainage, including roof, cellar, and foundation drainage (this drainage must be excluded from the sewage-disposal system to prevent the system becoming overloaded and waterlogged).

6. The average rainfall during the period of use.

The location of sewers and sewage disposal units with respect to water supply should provide the maximum possible protection that is practical. The type of sewage disposal, the slope of the ground surface, the elevation and slope of the underground water level, the type and thickness of the soil formations, the location and construction of well-water sources in the vicinity, and the capacity of pumps used on nearby wells are some of the important things to be taken into consideration. These factors should be carefully evaluated by a trained person before a decision is made and construction is started. The state and county sanitary or public health engineers and sanitarians are trained to give sound advice.

Soil Percolation Test

This test is a measure of soil permeability. Henry Ryon devised the test around 1926, based on his experience in the New York State Department of Health.[11] Ryon obtained his basic data by the investigation of subsurface disposal systems that had failed or were about to fail after 20 years of use. He plotted the results of his tests, covering a wide range of soils in New York State, and developed curves to interpret the results of soil percolation tests. The procedure developed by him has been accepted throughout the world and, except for some refinements, remains as the only rational basis for the design of subsurface sewage disposal systems.

Other investigators have studied the test to determine the effect of the shape of the test hole and the saturation of the soil on the percolation test. It has been found that the shape of the hole has no effect. However, *saturation of the soil is essential* to obtain reproducible results, that

[11] Henry Ryon, *Notes on the Design of Sewage Disposal Works with Special Reference to Small Installations,* New York State Department of Health, Albany, 1928.

is, a constant rate of water drop in the test hole. This was recognized by Ryon, but it is only in recent years that the importance of saturation has been emphasized.

One of the main significant differences of opinion in connection with the soil percolation test is its interpolation to determine the allowable rate of septic settled sewage application per square foot of leaching area. This rate is taken as a percentage of the actual volume of water a test hole accepts in 24 hr. Various investigators have stated that this rate should be 0.45, 2.0, 3.0, 5.0, or 7.0 percent of the actual amount of water absorbed or accepted by the test hole.

The soil percolation test is performed as follows:

1. Dig a hole about 1 ft^2 and to the depth at which it is proposed to lay the drain tile. Scrape the inside of the hole to remove all smooth or cemented patches. A good average depth is 24 to 30 in. About 2 in. of washed gravel should be placed in the bottom of the hole. See Figure 4–3.

2. Pour about 8 in. of water in the hole. If the soil is relatively tight, let the water soak for about 4 hr, adding additional water if necessary, and proceed as explained in Step 3. If the soil is porous, allow the water to seep away. Add 8 in. more of water in the hole and proceed as in Step 3. The test can be expedited by routinely having all test holes filled with water the night before the tests are made to allow ample time for soil swelling and saturation.

3. Measure the rate at which the water surface drops. This can be done by placing a piece of 2- × 4-in. lumber across the hole, being careful to anchor it in a firm position. Then, using a point or line on the 2 × 4 as a guide for the remainder of the test, slide a pointed slat or similar measuring stick straight down until it just touches the water surface. Immediately read the exact time on your watch and draw a horizontal pencil line on the measuring stick using the top of the 2 × 4 as a guide. Repeat the test at one-min intervals, if the water level drops rapidly, or at five- or ten-min intervals if the water level drops slowly. Observe the space between the pencil markings, Keep the depth of water in the hole at 4 to 8 in. When at least three spaces become relatively equal, which may require as long as 3 or 4 hr of presoaking in clayey soils, the test is completed, since equilibrium conditions have been reached for all practical purposes.

4. With the aid of a ruler, measure the space between the *equal* pencil markings and reduce this to minutes for the water level to drop one in. This can be approximated closely by inspection or can be computed.

For example, if the interval between 5-min readings is ⅜ in., the time for the water level to drop one inch (call this x) is calculated as follows:

$$\frac{⅜ \text{ in.}}{5 \text{ min}} = \frac{1 \text{ in.}}{x \text{ min}}, \quad \text{or } \frac{3}{8}x = 5, \quad \text{or}$$
$$3x = 40 \text{ and } x = 13⅓, \text{ say } x = 15 \text{ min.}$$

Make at least two tests for the average lot. Consider the soil characteristics

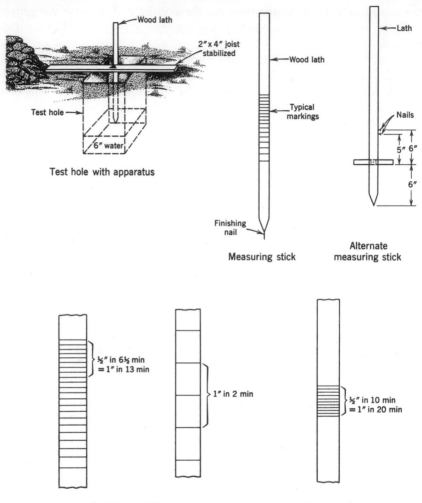

Figure 4-3   The soil percolation test.

and excavate the test holes to a depth of 4 to 5 ft to ensure groundwater, rock, or tighter soil is not encountered. Interpret the results in the light of soil characteristics and soil clues discussed on pages 264 to 267.

5. Use Table 4–1 to determine the allowable rate of settled sewage application in gal per square foot of bottom trench area per day (gpd/ft²) in a tile field. In the example given the rate for a tile field would be 1.3 gpd/ft².

Another example will serve to illustrate how the soil test results are used.

TABLE 4-1   INTERPRETATION OF SOIL PERCOLATION TEST

Time for Water to Fall 1 in. (min)	Allowable Rate of Settled Sewage Application, (gpd/ft²)*	Time for Water to Fall 1 in. (min)	Allowable Rate of Settled Sewage Application, (gpd/ft²)*
1	5.0	12	1.4
2	3.5	15	1.3
3	2.9	20	1.1
4	2.5	25	1.0
5	2.2	30	0.9
6	2.0	35	0.8
7	1.9	40	0.8
8	1.8	50	0.7
9	1.7	60	0.6
10	1.6	Greater than 60	Soil not suitable†

*Based upon *Manual of Septic-Tank Practice,* U.S. Public Health Service Pub. No. 526, Dept. of HEW, Washington, D.C., 1967. Increase leaching area by 20 percent where a garbage grinder is installed and by an additional 40 percent where a home laundry machine is installed. The required length of the tile field may be reduced by 20 percent if 12 in. of gravel is placed under the open joint or perforated distribution lateral, or by 40 percent if 24 in. of gravel is used.

†Consult with health department sanitary engineer or sanitarian. See "Small Sewage Disposal Systems for Tight Soils," in this chapter for difficult situations. Sometimes, where the soil is otherwise suitable, high groundwater at certain times of the year causes subsurface tile field systems to fail. The condition may be alleviated by the installation of subsurface curtain drains to a depth of about 2 ft below the bottom of the tile field trenches to intercept and lower the groundwater table around the field. If gravity drainage is not possible, collect the drainage in a sump and pump the water to a water course or ditch.

Example

Number of bedrooms—3.

Required septic tank—900-gal liquid volume.[12]

Average of soil tests for tile field, from table = 0.9 gpd/ft² (1 in. in 30 min).

Estimated sewage flow at 150 gal per bedroom = 450 gpd.

Required leaching area = $\dfrac{450}{0.9}$, plus 60% for garbage grinder and clothes washer = 800 ft².

Table 4–2 shows suggested trench widths to provide the required leaching area. The required area can be obtained by providing:

$\dfrac{800}{1.0}$ or 800 lineal ft of tile in trenches 12 in. wide, or

$\dfrac{800}{1.5}$ or 533 lineal ft of tile in trenches 18 in. wide, or

⎫ Recommended widths ⎭

[12] Assumes a home laundry machine and a garbage grinder are to be installed.

TABLE 4-2 SUGGESTED TRENCH WIDTH AND SPACING

Result of Soil Percolation Test	Suggested Trench Width*	Effective Area per Lineal Ft	Minimum Spacing of Laterals†
1 in. in 1 min to 1 in. in 15 min	{ 12 in. 24 in.	1 ft² 2 ft²	6 ft 6 ft
1 in. in 16 min to 1 in. in 30 min	{ 24 in. 30 in.	2 ft² 2½ ft²	6 ft 7½ ft
1 in. in 31 min to 1 in. in 60 min	{ 30 in. 36 in.	2½ ft² 3 ft²	7½ ft 9 ft

*Narrow trenches are more efficient.
†A greater than the minimum spacing of laterals is preferable.

$$\frac{800}{2.0}$$ or 400 lineal ft of tile in trenches 24 in. wide, or

$$\frac{800}{2.5}$$ or 320 lineal ft of tile in trenches 30 in. wide ,or

$$\frac{800}{3}$$ or 267 lineal ft of tile in trenches 36 in. wide.

If 12 in. of gravel is placed under the leaching pipe, the required lineal ft of leaching trench may be reduced by 20 percent and with 24 in. of gravel by 40 percent, but the bottom of the trench should be at least two ft above groundwater and tighter soil.

A variation of the soil percolation test is to observe the time for the water level to drop from a depth of 6 to 5 in. Repeat the test, and if the time recorded in the second test is within 10 percent of the time for the first test, use this time to determine the allowable rate of settled sewage application per square foot per day using Table 4-1. If the "times" vary by more than 10 percent, repeat the test until the times for two successive tests do not vary by more than 10 percent. With a tight soil, this method will take somewhat longer to perform than the first method. Some sanitary engineers prefer to continue the test until the times for the water to fall 1 in. are within 3 to 5 percent.

A sufficient number of soil tests should be made that will give information representative of the soil. This will also make possible determination of an average percolation rate that can be used in design. In the case of a single dwelling, at least two soil tests should be made. The test holes should be dug 2 to 3 ft deeper and a soil percolation test made to assure that the character of the soil below the bottom of tile field trenches is similar to the soil tested. Where a small rural real-estate subdivision is under consideration, approximately one hole per acre should be tested and soil borings made. If rock, clay, hardpan, or groundwater is encountered within 4 ft of the ground surface, the property should be considered unsuitable for the disposal of sewage by means of conventional subsurface tile fields. This calls for the exercise of trained engineering judgment. A typical layout is shown in Figure 4–4.

In special cases where tighter soil is encountered just below the bottom of the test hole, interpretation of the results might require adjustment

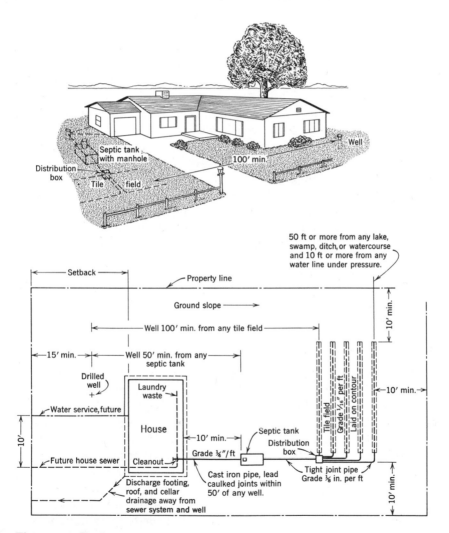

Figure 4–4 Typical private water supply and sewage-disposal layout.

of the allowable sewage application from an apparent, say 2.0 gpd/ft², to 0.5 gal based on the tighter soil.

It should be pointed out that the ill repute of septic-tank leaching systems is due to improperly made and interpreted soil percolation tests, high groundwater, poor construction, or lack of maintenance, and to the use of septic tanks where they were never intended.

Tile fields should not be laid in filled-in ground until it has been thoroughly settled or otherwise stabilized, and then trenches should be dug in the fill. Tests should be made in fill after a six-month settling period and after complete stabilization of the soil. Percolation tests cannot be made in frozen ground.

Where the ground is flat, provision should be made to drain surface water off and around the tile field to prevent the soil from becoming water logged. On steeply sloping ground, a surface-water diversion ditch or berm should be provided above and around the tile field to prevent the tile field from being washed out. This will also prevent silt and mud from washing into the trench during construction and coating of the bottom with a relatively impervious film. It is also important to point out that the undesirable practice of walking in the bottom of trenches causes a compaction of the earth and a reduced percolation capacity. If this happens, rake the bottom to restore the original surface.

Where leaching pits are permitted, the soil test is made in a test hole about 1 ft² at the bottom and at a depth about one-half the proposed *effective depth* of the leaching pit. On completion of the test, the hole should be extended or probed to the proposed full depth to assure that the soil at a greater depth is similar to that tested and that groundwater is not encountered. The test is made in the same manner as explained for a tile field but at a greater depth. The results are interpreted in Table 4–1.

Estimate of Sewage Flow

The sewage flow to be expected from a dwelling or other type of establishment is not constant each day. The day of the week, season of the year, habits of the people, water pressure, type and number of plumbing fixtures, and type of place or business maintained are some of the factors affecting the probable sewage flow. The design cannot be based on the minimum flow but must be based at least on the average maximum. Daily water-meter consumption figures from a similar type of establishment taken over an extended period of time, including weekends and maximum days, would be of value in arriving at a good average maximum daily flow estimate to be used in design. Caution must be used when interpreting quarterly, semiannual, or annual meter readings, as averages derived from these figures will be low unless corrected for vacation periods, weekends and holidays, seasons of the year, and so on. In the absence of actual figures, the per capita or unit estimated water flow given in Table 3–2 may be used as a guide.

Fixture bases of estimating sewage flow assume that all water used finds its way to the sewage disposal or treatment system. Adjustment should be made for lawn watering, car washing, and so forth. In one method, the total number of different types of fixtures is summarized. The sum of each type is multiplied by the usual flow from such a fixture per use or operation. The frequency of use per hour can be estimated for the type of establishment under study. Knowing the number of hours of operation daily, a rough estimate of the probable flow in gpd can be arrived at.

Another fixture basis of estimating sewage flow is given in Table 4–3.

TABLE 4-3   A FIXTURE BASIS OF ESTIMATING SEWAGE FLOW

Type of Fixture	Gallons per Day per Fixture	Gallons per Hour per Fixture	Gallons per Hour per Fixture, Average
	Country Clubs*	Public Parks†	Restaurants‡
Shower	500	150	17
Bathtub	300	–	17
Washbasin	100	–	8½
Water closet	150	36	42 (flush valve)
			21 (flush tank)
Urinal	100	10	21
Faucet	–	15	8½, 21 (hose bib)
Sink	50	–	17 (kitchen)

*John E. Kiker, Jr., "Subsurface Sewage Disposal," *Fla. Eng. and Exp. Sta.,* Bull. No. 23 (December 1948).

†National Park Service.

‡After M. C. Nottingham Companies, California

Although the fixtures refer to country clubs, public parks, and restaurants, they can be applied with modifications to similar types of establishments. The fixture bases of estimating sewage flow are useful in determining the required size drain or sewer line and also as a check on other methods. Fixture unit values in Table 4–4 can also be used with Figure 3–20 for the same purpose.

A third fixture unit basis of estimating sewage flow is that described under water supply, using the probability curves developed by Hunter. These flows are somewhat high, being based on estimated peak discharge.

TABLE 4-4 SANITARY DRAINAGE FIXTURE UNIT VALUES

Fixture or Group	Fixture unit value
Bathroom group consisting of a lavatory, bathtub or shower stall, and a water closet (direct flush valve)	8
Bathroom group consisting of a lavatory, bathtub or shower stall, and a water closet (flush tank)	6
Bathtub with 1½″ trap	2
Bathtub with 2″ trap	3
Bidet with 1½″ trap	3
Combination sink and wash tray with 1½″ trap	3
Combination sink and wash tray with food waste grinder unit (separate 1½″ trap for each unit)	4
Dental unit or cuspidor	1
Dental lavatory	1
Drinking fountain	½
Dishwasher, domestic type	2
Floor drain	1
Kitchen sink, domestic type	2
Kitchen sink, domestic type with food waste grinder unit	3
Lavatory with 1½″ waste plug outlet	1
Lavatory with 1¼″ or 1³/₈″ waste plug outlet	2
Lavatory (barber shop, beauty parlor, or surgeon's)	2
Lavatory, multiple type (wash fountain or wash sink), per each equivalent lavatory unit	2
Laundry tray (1 or 2 compartments)	2
Shower stall	2
Showers (group) per head	3
Sink (surgeon's)	3
Sink (flushing rim type, direct flush valve)	8
Sink (service type with floor outlet trap standard)	3
Sink (service type with P trap)	2
Sink (pot, scullery, or similar type)	4
Urinal (pedestal, syphon jet, blowout type with direct flush valve)	8
Urinal (wall lip type, flush tank)	4
Urinal (stall, washout type, flush tank)	4
Water closet (direct flush valve)	8
Water closet (flush tank)	4
Swimming pools, per each 1000-gal capacity	1
Unlisted fixture, 1¼″ or less fixture drain or trap size	1
Unlisted fixture, 1½″ fixture drain or trap size	2
Unlisted fixture, 2″ fixture drain or trap size	3
Unlisted fixture, 2½″ fixture drain or trap size	4
Unlisted fixture, 3″ fixture drain or trap size	5
Unlisted fixture, 4″ fixture drain or trap size	6

Source: *Plumbing Standards of the State Building Construction Code,* State of New York, Albany, 1958.

Note: Values given are for continuous flow. For a continuous or semicontinuous flow into a drainage system, such as from a pump, pump ejector, or similar device, two fixture units shall be allowed for each gpm/of flow. One fixture unit is equivalent to 7.5 gpm.

An analysis made by Wyly[13] is based on the estimated average discharge. For two-bath houses with automatic dishwasher, clothes washer, and garbage grinder, for a total of 19 fixture units per home, the discharge may be 1.6 gpm for 1 home, 7.6 gpm for 5 homes, 15.2 gpm for 10 homes, 30.4 gpm for 20 homes, 76 gpm for 50 homes, and 152 gpm for 100 homes. See also page 214.

House Sewer and Plumbing

The house or building sewer is that part of the building drainage system carrying sewage that extends from the septic tank or public sewer to a point 3 ft out from the foundation wall. That portion of the drainage system extending from the house sewer horizontally into the structure is the house or building drain. The recommended size of the house or building sewer and drain is given in Table 4–5, although

TABLE 4-5   FIXTURE-UNIT LOAD TO BUILDING DRAINS AND SEWERS

Diameter of Pipe (in.)	Maximum Number of Fixture Units that may be Connected to Any Portion* of the Building Drain or the Building Sewer			
	Fall per ft			
	$\frac{1}{16}$-in.	$\frac{1}{8}$-in.	$\frac{1}{4}$-in.	$\frac{1}{2}$-in.
2	—	—	21	26
2½	—	—	24	31
3	—	20†	27†	36†
4	—	180	216	250
5	—	390	480	575
6	—	700	840	1,000
8	1,400	1,600	1,920	2,300

Source: Adapted from "Report of the Coordinating Committee for a National Plumbing Code," U.S. Dept. of Commerce, Washington, D.C., 1951.

*Includes branches of the building drain.

†Not over 2 water closets.

the local plumbing or building code will govern where one has been adopted. In general, the building sewer should be not less than 4-in. bell and spigot cast-iron pipe with lead-caulked or equal joints. Other

[13] Robert S. Wyly, *Hydraulics of 6- and 8-inch Diameter Sewer Laterals*, Building Research Institute, BRAB, NAS-NRC, 2101 Constitution Ave., Washington, D.C., November 27, 1956.

pipe constructed of durable material and laid with tight joints is also used.

Increasing fittings with smooth joints should be used where the pipe size increases in diameter. Sewer lines should be laid in a straight line to the septic tank where possible. If bends are necessary, use one or two 45° ells, as may be needed, and provide a cleanout. A manhole is sometimes preferable. A cleanout should also be provided at the end of the building drain in the basement and just ahead of the septic tank when the tank is located 30 or more ft from the cleanout on the building drain. A cleanout may be provided on a buried sewer line by installing a "T" fitting in the line with the vertical leg up and connecting to it a section of pipe extending to the ground surface. The fitting, sewer joints, and pipe extension should be encased in 6 in. of concrete. All cleanouts should have tight-fitting brass caps. See Figures 4–5 and 4–6.

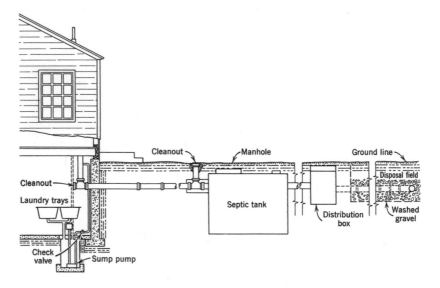

Figure 4–5   Section showing sewage disposal system.

Grease Trap

A grease trap, interceptor, or separator is a unit designed to remove grease and fat from kitchen wastes. Liquid wastes leaving properly designed and maintained units should not cause clogging of pipes nor have a harmful effect on the bacterial and settling action in a septic tank.

Grease traps are of the septic-tank and commercial types. The com-

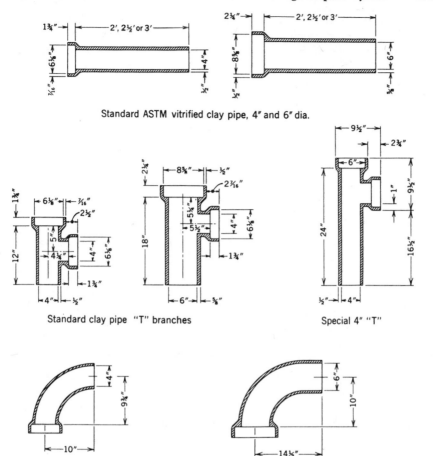

Standard ASTM vitrified clay pipe, 4″ and 6″ dia.

Standard clay pipe "T" branches   Special 4″ "T"

Standard clay pipe long radius elbows, 4″ and 6″ dia.

Figure 4–6   Details of some clay pipe fittings.

mercial types are of questionable value, particularly with the general use of detergents for dishwashing in restaurants. In the septic-tank type, use is made of the cooling and congealing effect obtained when the warm or hot greasy liquid wastes from the kitchen mix with the cooler liquid standing in the tank. There is also a natural tendency, if mixing is not too rapid, for the warmer liquid to rise and the cooler liquid, from which grease has been separated, to settle and be carried out with the food particles. A grease trap is unnecessary in the average private home. The small quantity of fat and grease that does find its way into the kitchen drain is mixed with soaps and detergents and is difficult to separate. In any case, such small quantities would not be harmful when allowed to enter a proper size septic tank. Grease traps of the septic-tank type should be provided, however, in restaurants and similar establishments where the quantity of grease and fats in liquid wastes is likely to be large unless special provision is made in the treatment plant. All grease traps should discharge into the building sewer ahead of the septic tank. Grease traps should be located within 20 or 30 ft from the plumbing fixtures served to prevent congealing and clogging of waste lines.

The septic-tank type of grease trap's capacity is made equal to the maximum volume of water used in a kitchen during a mealtime period. The type of meals served and kitchen equipment used should be taken into consideration. A figure of 2½- or 3-gal capacity per meal served during a mealtime is frequently used. The tank should be located in an accessible place outside the building, using the same precautions as for the septic tank. It should have a tight-fitting cover and be light in weight. Heavy aluminium, cast iron, and steel make satisfactory covers. When covers do not fit tightly, a tar or rubber seal around the cover or a layer of clay soil over the cover may be necessary to eliminate odors.

The construction of several septic-tank-type grease traps is shown in Figure 4–7. Grease traps can be built on the job of concrete or brick or can be prefabricated of metal, concrete, asbestos cement, or terra cotta, with inlet and outlet arrangements as shown in the sketches. Because of the greater capacity of septic-tank-type grease traps they do not require cleaning as frequently as the commercial type. Nevertheless, they must be cleaned so as not to greatly reduce the liquid volume available for cooling of the greasy wastes entering and, of course, to prevent large quantities of grease being carried over to the septic tank. The frequency of cleaning should be determined at each establishment during operation. Cleaning at monthly intervals may be sufficient; but experience should dictate the frequency. One person should be given this responsibility, and a supervisor should check to see that the job is done. Grease removed from a grease trap may be disposed of with the garbage, rendered and sold, or thoroughly buried as explained under sanitary pit privies, depending on local conditions.

Septic Tank

A septic tank is a watertight tank designed to slow down the movement of raw sewage and wastes passing through so that solids can separate or settle out and be broken down by liquefaction and bacterial action. It does not purify the sewage, eliminate odors, or destroy all solid matter.

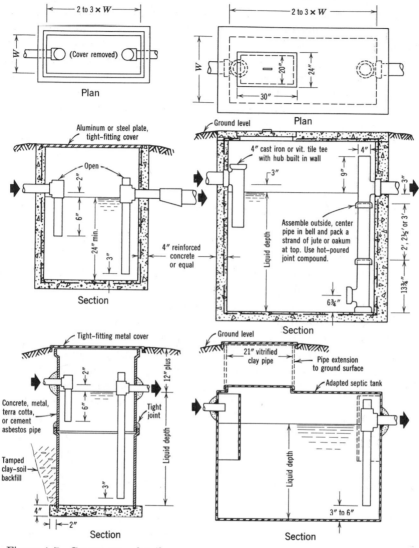

Figure 4–7 Grease trap details.

The septic tank simply conditions the sewage so that it can be disposed of to a subsurface leaching system or to an artificial sand filter without prematurely clogging the system.

Recommended septic tank sizes based on estimated daily flows are given in Tables 4–6 and 4–7. The septic tank should have a liquid volume of not less than 750 gal. When a tank is constructed on the job, its liquid volume can be increased at a nominal extra cost thereby providing capacity for possible future additional flow, garbage grinder, and sludge storage.

The detention time for large septic tanks should not be less than 12 hr. A 24- to 72-hr detention time is recommended. Schools, camps, theaters, factories, and fairgrounds are examples of places where the total or a very large proportion of the daily flow takes place within a few hours. For example if the total daily flow takes place over a period of 6 hr ($\frac{1}{4}$ of 24 hr), the septic tank should have a liquid volume equal to four times the 6-hr flow to provide a detention of 24 hr over the period of actual use.

If the septic tank is to receive ground garbage, its capacity should be increased by at least 50 percent. Some authorities recommend a 30-percent increase.

Septic tanks can be constructed of masonry on the job as explained in Figures 4–8 and 4–9 and Tables 4–6 and 4–7. Precast reinforced concrete and commercial metal tanks are also available. Since some metal tanks have a limited life, it is advisable that their purchase be predicated on their meeting certain minimum specifications. These should include a guaranteed 20-yr life expectancy, not less than 12- or 14-gauge metal thickness, covering both inside and outside with a heavy continuous, protective coating resistant to acids, and a minimum liquid capacity of 750 gal. Metal septic tanks manufactured in accordance with the Department of Commerce Commercial Standard, provided with a minimum 16 to 18-in. manhole, represent a great improvement over the ordinary metal tank.

The depth of septic tanks and ratio of width to length recommended by most health departments are very similar. A liquid depth of 4 ft and a ratio of width of length of 1 to 2 or 3 is common. Depths as shallow as 30 in. and as deep as 6 ft have been found satisfactory. Compartmented tanks are somewhat more efficient. Open-tee inlets and outlets as shown in Figure 4–8 are generally used in small tanks and baffled inlets, and outlets are shown in Figure 4–9 are recommended for the larger tanks.[14] A better distribution of flow and detention is

[14] *Manual of Septic-Tank Practice,* Public Health Service Pub. No. 526 (1967), states that "the outlet device should generally extend to a distance below the surface equal to 40 percent of the liquid depth. For horizontal, cylindrical tanks, this should be reduced to 35 percent." The inlet should penetrate at least 6 in. below the liquid level but not greater than the outlet. The distance between the liquid line and underside of the tank (air space) should be approximately 20 percent of the liquid depth. In horizontal, cylindrical tanks, the area should be 15 percent of the total circle.

TABLE 4-6 WATER SUPPLY AND SEWERAGE SCHEDULE
(USE COMBINATION OF TABLES THAT FIT LOCAL CONDITIONS)

Population		*	Septic Tank—Minimum¶§				Tile Field Laterals§#		
Bedrooms	Persons	Sewage Flow GPD	Length, ft	Width, ft	Depth, ft	Volume, gal	No.	Length, ft	Trench Width, in.
2 or less	4	300	7½	3½	4	750			
3	6	450	8½	3½	4	900			
4	8	600	9	4	4	1,000			
5	10	750	11	4	4	1,250			

Population		†	Leaching Pit System§#			Water Supply—Well, Drilled‡	
Bedrooms	Persons	Service	No. Pits	Size	Depth	Pump Size, gal/hr	Pres. Tank, gal
2 or less	4	¾ in.				250	42
3	6	¾				300	82
4	8	1				360	82
5	10	1¼				450	120

Population		Sand Filter System§‖			Chlorine Contact Tank§			Sump Pump§
Bedrooms	Persons	Length, ft	Width, ft	Area, sq ft	Length, ft	Width, ft	Depth, in.	Float Setting, gal
2 or less	4	21½	12	260	3	2	8	20
3	6	21	18	390	3	2	12	30
3	6	32½	12	390	3	2	12	30
4	8	30	18	520	3	2	16	40
4	8	43	12	520	3	2	16	40
5	10	36	18	650	3	2	20	50
5	10	54	12	650	3	2	20	50

Note: * The design basis is 75 gal per person.
† Use next larger diameter house service line if water is corrosive or hard, if service line is 50 to 100 ft long, if two bathrooms are provided, or flush valve is used for water closet. These pipe sizes are based on the use of brass or copper pipe; use next larger size if iron pipe is proposed.
‡ The minimum dependable well yield should be 3 to 7 gpm.
§ See detail drawings for construction specifications.
‖ Sand filter normally should not be used. Reserve for compelling circumstances to relieve an impending or existing public health hazard.
¶ Includes provision for home garbage grinder and laundry machine.
# Based on the results of soil percolation tests. Discharge all kitchen, bath, and laundry wastes through the septic tank, but *exclude* roof and footing drainage, surface and groundwater, and softener wastes.

TABLE 4–7   SUGGESTED TANK DIMENSIONS

Gallons	Width (ft)	Length (ft)	Depth (ft)
1,000	4	9	4
1,250	4	11	4
1,500	5	10½	4
1,750	5	12	4
2,000	5	14	4
2,250	6	13	4
2,500	6	14½	4
2,750	6	16	4
3,000	6	17	4
3,250	6	15	5
3,500	6	16	5
3,750	6	17	5
4,000	7	16	5
5,000	7	19½	5
6,000	8	20½	5
7,000	8	24	5
8,000	8	23	6
10,000	8	21	8

Concrete Details

1. Concrete for top and bottom 4 in. thick for 2,000-gal tank or smaller and 6 in. for 2,000- to 8,000-gal tank.

2. Concrete for sides and ends 6 in. thick for 6,000-gal tank or smaller and 8 in. for 6,000- to 10,000-gal tank.

3. Reinforce with $\frac{3}{8}$-in. deformed rods 4 in. on center both ways for ordinary loading. Place rods 1 in. above bottom of top slab and 1 in. in from inside of tank for sides, ends, and bottom. Overlap $\frac{3}{8}$-in. rods 15 in. where needed. Adjust steel for local conditions.

4. Concrete mix: 1 bag cement to 2¼ ft³ sand to 3 ft³ gravel with 5 gal water for moist sand.

obtained with the baffle arrangement. A minimum 16-in. manhole over the inlet of a small tank, and a 20- to 24-in. manhole over both the inlet and outlet of a larger tank, constructed with a top slab poured monolithically with the sides, is preferred to a sectional slab top. The sectional slab top can, however, be more easily purchased or constructed on the ground surface with a minimum of form lumber. Joints will require a seal to prevent the entrance of surface water into the tank.

The elevation of the septic tank and the inlet should be selected and established with regard to the landscaping; elevations of sewers that discharge to the septic tank; elevation of dosing tank outlet pipe inverts where used; location available and elevation of the area selected for the disposal or treatment system; and the high-water level of nearby watercourses, to make pumping unnecessary, provided the topography

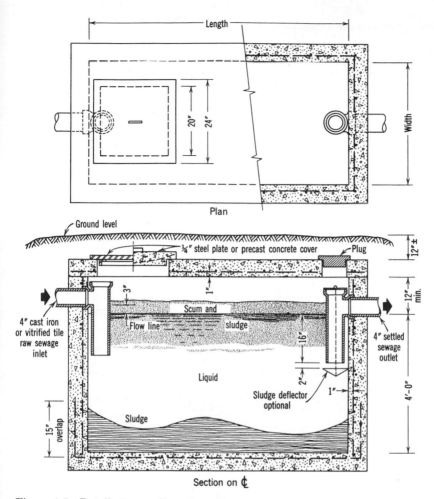

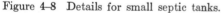

Figure 4–8 Details for small septic tanks.

Recommended construction for small septic tanks.

Top—Reinforced concrete poured 3 to 4 in. thick with two ⅜-in. steel rods per ft of length, or equivalent, and a 20 × 20-in. manhole over inlet, or precast reinforced concrete 1-ft slabs with sealed joints.

Bottom—Reinforced concrete 4 in. thick with reinforcing as in "top" or plain poured concrete 6 in. thick.

Walls—Reinforced concrete poured 4 in. thick with ⅜-in. Steel rods 6 in. on centers both ways, or equivalent; plain poured concrete 6 in. thick; 8-in. brick masonry with 1-in. cement plaster inside finish; or 8-in. stone concrete blocks with 1-in. cement plaster inside finish and cells filled with mortar.

Concrete Mix—One bag of cement to 2¼ ft³ of sand to 3 ft³ of gravel with 5 gal of water for moist sand. Use 1:3 cement mortar for masonry and 1:2 mortar for plaster finish. Gravel or crushed stone and sand shall be clean, hard material. Gravel shall be ¼ to 1½ in. in size; sand from fine to ¼ in.

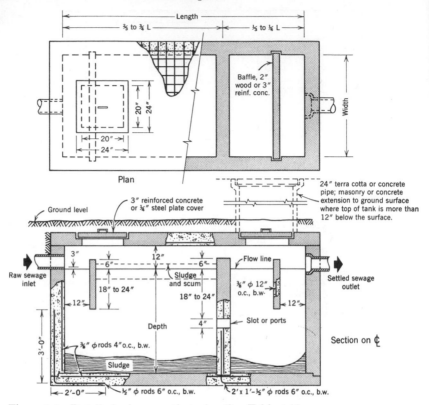

Figure 4–9   Details for large septic tanks. (See Table 4–7)

makes this possible. On the other hand, if pumping is required it should be taken into consideration in the initial design and every advantage taken of this necessity to reduce excavations and shorten and straighten lines. Figure 4–10 gives design information in a form letter.

Care of Septic Tank and Tile Field

Proper maintenance of a septic tank is the best insurance for satisfactory operation of a subsurface sewage disposal or treatment system and prevention of sudden replacement expenses.

A septic tank for a private home will generally require cleaning every 3 to 5 years, but in any case it should be inspected about once a year. When the depth of settled sludge or floating scum approaches the depth given in Table 4–8, the tank needs cleaning. A long pole having a small board about 8 in.² nailed to the bottom to make a plunger, with Turkish toweling wrapped around the lower 18 in. of the pole, can be used to

RENSSELAER COUNTY DEPARTMENT OF HEALTH
DIVISION OF ENVIRONMENTAL HYGIENE

Seventh Ave and State St.   Troy, N. Y.

Re: Private Sewerage System

Dear Sir:

As you know, Mr. .... of this department made an investigation of your property, including soil percolation tests, on .... ,   Our recommendations for an adequate system to serve your .... bedroom house are given on the attached sketches and are checked below.

☐ 1. The design is based upon soil tests indicating that .... gallons of settled sewage can be applied per square foot in a subsurface (tile field system) (leaching pit system) (sand filter system) (tile field system in fill).

☐ 2. The septic tank should have a liquid capacity of .... gallons, and be of an approved type with a manhole over the inlet. It should be followed by a small distribution box. Extension of the septic tank manhole and top of the distribution box close to the surface will make easy inspection and if necessary cleaning of the septic tank, required usually every 2 to 3 years.

☐ 3. The tile field following the septic tank and distribution box is to consist of .... laterals of open joint or perforated pipe laid in washed gravel or crushed stone at least .... feet on centers. Each lateral is to be .... feet long laid in trenches .... inches wide.

☐ 4. The leaching pit system following the septic tank and distribution box is to consist of .... pits. Each pit is to have a depth below the inlet of .... feet and an inside .... of .... feet. If standard 8 inch by 16 inch building blocks are used to construct a pit, then .... blocks will be needed below the inlet for each pit. Coarse gravel shall be placed around the outside wall of the pit.

☐ 5. Suitable gravelly loam fill containing topsoil and having an area of at least .... square feet may be provided in locations where the soil is unsuitable provided at least 12 to 18 inches of natural porous earth exists. Sufficient fill shall be provided to bring the bottom of trenches 2 feet above the clay, rock or ground water. After stabilization of the fill, a tile field is to be constructed with laterals not closer than 20 feet to the edge of the feathered fill. The tile field following the septic tank and distribution box is to consist of .... laterals of open joint or perforated pipe laid .... feet apart. Each lateral is to be .... feet long laid in trenches .... inches wide. Establish grades of all pipes to assure gravity flow of sewage from the house to the tile field without pumping.

☐ 6. The subsurface sand filter following the septic tank and distribution box is to be .... feet long and .... feet wide constructed as shown on the sketches, with a chlorine contact-inspection box on the outlet. The architect, builder and contractor shall establish all grades so as to assure gravity flow from the dwelling through the sewage treatment system to the outlet. Submit a 2-pound sand sample fifteen days before construction for Health Department approval.

The bottom of the tile field trenches and leaching pits must be at least 2 feet above ground water, clay, hardpan, rock, etc. and the bottom of the sand filter above ground water for the system to function properly. Additional important information is given on the back of this sheet and on the attached sketches. Please advise this office when the work is expected to start. We can then arrange to have our representative inspect the construction for compliance with the health department standards and thus help you obtain the best possible job.

HELP MAINTAIN A CLEAN, HEALTHFUL ENVIRONMENT

Figure 4-10 A typical form letter for design of a private sewerage system.

measure the sludge depth and floating-scum thickness. The appearance of particles in the effluent from a septic tank going through a distribution box is also an indication of the need for cleaning. Proper maintenance will prevent solids from being carried over and clogging the treatment or leaching systems. The larger the septic tank above the minimum, the less frequent is the need for cleaning.

Septic tanks can be cleaned by septic-tank cleaning firms. They mix and pump the entire contents out into a tank truck with special equipment. Care must be taken to prevent spillage and consequent pollution of

TABLE 4–8   MAXIMUM ALLOWABLE SLUDGE OR SCUM ACCUMULATION

Typical Septic Tank	Liquid Capacity of Tank (gal)	Liquid Depth of Septic Tank		
		3 ft	4 ft	5 ft
		Bottom of Outlet to Top of Sludge (in.)		
		C	C	C
	500	11	16	21
	600	8	13	18
	750	6	10	13
	900	4	7	10
	1,000	4	6	8

Source: Adapted from *Manual of Septic-Tank Practice*, U.S. Public Health Service Pub. No. 526, Dept. of HEW, Washington, D.C., 1967.

Note: Clean the septic tank before the scum depth = A, that is, within 3 in. of the bottom of the outlet baffle; and before the sludge depth B encroaches on the limiting distance for C.

the surrounding ground. The contents should be emptied into a sanitary sewer system or sludge digestion tank, if permissible, or in a shallow trench or pit at a point 200 ft or more and downgrade from water sources and covered with at least 24 in. of compacted earth, as approved by the department of health. Sludge sticking to the inside of a tank that has just been cleaned would have a seeding effect and assist in renewing the bacterial activity in the septic tank. The septic tank should not be scrubbed clean.

Compounds that are supposed to make the cleaning of tanks unnecessary may actually cause solids to be carried over into the leaching or treatment system with resultant clogging. Some "cleaners" contain sodium hydroxide or potassium hydroxide. Where used, soil clogging is almost certain to be accelerated.[15] It is best to clean a septic tank in the late spring of the year. Bacterial activity will proceed faster in warm weather, thereby effecting quicker adjustment and good operation of the septic tank. A starter, such as yeast, added to a septic tank does not speed up the digestion. The addition of about 6 gal per capita of digested sludge to a new septic tank appears to have a beneficial effect.

[15] A grab sample collected from a septic tank serving a 60-unit trailer camp showed a total solids concentration of 15,058 mg/l one day after a septic-tank cleaner had been added. The effluent from the sand filter following the septic tank showed a total solids content of 1038 mg/l at the same time.

Soap, drain solvents, disinfectants, and similar materials used for household purposes are not harmful unless used in large quantities. Salt or brine from softening units in amounts as little as 1.2 percent retards temporarily the bacterial action in the septic tank and tends to build up in the sludge, but the salt is flushed out when increased digestion causes rising and turnover of the sludge. However, the salt tends to cause soil clogging. Therefore, brine should not discharge to the septic tank.

High weeds, brush, shrubbery, and trees, although consumers of groundwater, should not be permitted to grow over a tile field or sand filter system. It is better to seed the area to grass and build up a lawn. Sunlight is beneficial.

If trees are near the sewage disposal system, difficulty with roots entering poorly joined sewer lines can be anticipated. Lead-caulked cast-iron pipe, a sulphur base or bituminous pipe joint compound, mechanical clay pipe joints, copper rings over joints, and lump copper sulfate in pipe trenches have been found effective in resisting the entrance of roots into pipe joints. Roots will penetrate into the gravel in tile field trenches rather than into the pipe. About 2 to 3 lb of copper sulfate crystals flushed down the toilet bowl once a year will destroy roots the solution comes into contact with, but will not prevent new roots from entering. The application of the chemical should be done at a time, such as late in the evening, when the maximum contact time can be obtained before dilution. Copper sulfate will corrode chrome, iron, and brass, hence it should not be allowed to come into contact with these metals. Cast iron is not affected to any appreciable extent. Some time must elapse before the roots are killed and broken off. Copper sulfate in the recommended dosage will not interfere with operation of the septic tank.[16] The cutting or mechanical removal of roots in sewers tends to increase root growth and size, leading to more problems and sewer repair.

Division of Flow to Soil Absorption System

The overflow from a septic tank should be run to a distribution box to assure equal division of the settled sewage flow to all the leaching pits or laterals comprising the disposal or treatment system. The distribution box should have a removable cover extended to the surface to simplify inspection of the septic-tank effluent and flow distribution to the disposal or treatment system. *Outlets must leave the distribution box at exactly the same level.* A gravel fill or footing under the box,

[16] Don. E. Bloodgood, "Tree Roots, Copper Sulfate, and Septic Tanks," *Water & Sewage Works,* 99, pp. 190–193 (May 1952).

extending below frost, will help keep the box level. A baffle is usually necessary in front of the inlet to break the velocity of the incoming sewage and permit equal distribution to the outlets. Bricks or blocks are very useful for this purpose. Details of distribution boxes are given in Figures 4–11 and 4–13. If the outlets are constructed about 6 in. above the bottom of the distribution box, the liquid collected will have the effect of breaking the incoming velocity of the settled sewage. In any case, it is important to place all outlets at the same level and obtain equal distribution of flow to each of the laterals. A ⅛-in. mesh basket screen over each outlet would prevent large particles of septic-tank sludge from being carried into the tile field. Backup would occur at the box and call attention to the need for cleaning the septic tank without ruining the tile field.

Serial distribution of sewage (Figure 4–12) by the use of tees and elbows is reported to have certain advantages over the use of distribution boxes.[17] It compensates for varying soils and absorptive capacity; it forces full use of a trench before overflow to the next and overcomes the hazard of overflow associated with the parallel system if one trench is overloaded due to uneven flow distribution from an improperly installed or disturbed distribution box.

On steep grades special provision must be made for reducing the velocity of the sewage leaving the septic tank in order to get good distribution to the subsurface tile field or absorption system. Drop manholes are used for this purpose. The flow can be divided approximately in proportion to the length of the absorption system at each manhole. Drop-manhole details are shown in Figure 4–13.

Subsurface Soil Absorption Systems

The conventional subsurface absorption system following the septic tank is the tile field or leaching pit. The cesspool is still used for raw sewage, although generally prohibited, and the dry well is used for the disposal of rainwater, footing, roof and basement floor drainage. Where the soil is not suitable for subsurface disposal, a sand filter, evapotranspiration system, modified tile field system, aeration system, stabilization pond, or some combination may be used.

Tile Field System

The soil percolation test basis for the design of tile field systems was given earlier in this chapter. Design standards and details for tile field systems are shown in Figures 4–4, 4–12, 4–13, 4–14, and in Table 4–9.

---

[17] *Manual of Septic-Tank Practice,* U.S. Public Health Service Pub. No. 526, Dept. of HEW, Washington, D.C., 1967.

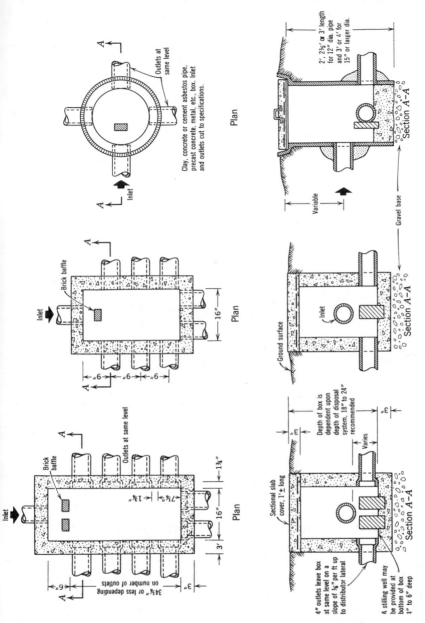

Figure 4-11 Distribution box details. (Bottom of box below frost; level on 12 in. of gravel.)

291

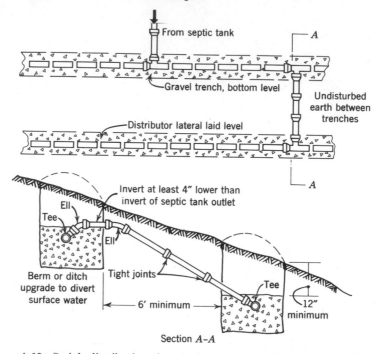

Figure 4–12 Serial distribution for sloping ground. (Adapted from U.S. Public Health Service Pub. No. 526, Dept. of HEW, Washington, D.C., 1967.)

The absorption field laterals should be laid in trenches preferably 24 in. below the ground surface. Where laid at a greater depth, the gravel fill around the open joint or perforated lateral should extend at least to the topsoil and as shown in Figure 4–14. The sunny and open side of a slope is the preferred location for an absorption field, if there is a choice. After settlement and grading, the tile field area should be seeded to grass. Systems in sand or gravel may be laid at depths of 6 to 10 ft provided they are well above groundwater.

When the length of the tile field to provide the required leaching area is 500 to 1000 lineal ft, a siphon should be installed between the septic tank and tile field to distribute the sewage throughout the entire field. If the required length of the tile field is 1000 to 3000 lineal ft, the tile field should be divided into two or four sections with alternating siphons to feed each section, or each two sections when four are provided. Where the tile field required is greater than 3000 lineal ft it is advisable to investigate a secondary treatment process, although larger tile field systems can operate satisfactorily. In some instances flat topography makes it impossible to install siphons and still obtain distribution of

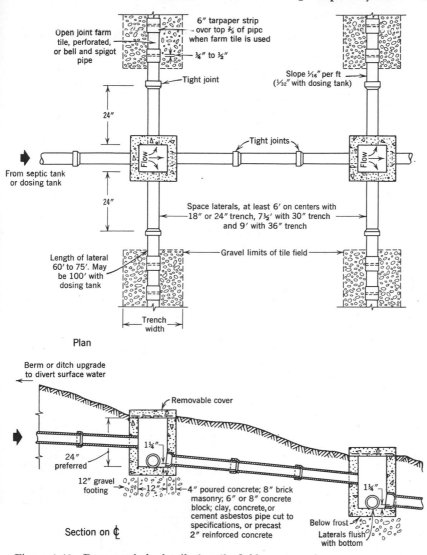

Figure 4–13   Drop-manhole details for tile field on steep slope.

settled sewage by gravity to the tile field. In such cases it would be necessary to install pumps or ejectors; but the design should permit gravity flow to the tile field in case of pump failure, if possible.

Absorption field laterals must be laid on careful grades. The bottom of trenches should be dug on the same grades as the laterals to prevent the sewage running out at one end of a trench or on to the ground

TABLE 4–9   SUGGESTED MINIMUM STANDARDS–

Item	Material	Size
Sewer to septic tank	Cast iron for 10′ from bldg. recommended.	4″ min. dia. recommended.
Septic tank	Concrete or other app'd matrl. Use a 1 : 2¼: 3 mix.	Min. 750 gal 4′ liquid depth, with min. 16″ M.H. over inlet.
Lines to distribution box and disposal system	Cast iron, vit. clay, concrete or composition pipe.	Usually 4″ dia. on small jobs.
Distribution box	Concrete, clay tile, masonry, coated metal, etc.	Min. 12″ × 12″ inside carried to the surface. Baffled.
Tile field†	Clay tile, vit. tile, concrete, composition pipe, laid in washed gravel or crushed stone, ¾″ to 2½″ size, min. 12″ deep.	4″ dia., laid with open joint or perforated pipe. Depth of trench 24″ to 30″.
Sand filter†	Clean sand, all passing ¼″ sieve with effective size of 0.30 to 0.60 mm and uniformity coefficient less than 3.5. Flood bed to settle sand.	Send 2-lb sample to Health Dept. for analysis 15 days before construction.
Leaching or seepage pit†	Concrete block, clay tile, brick, fieldstone, precast.	Round, square, or rectangle.
Chlorine contact-inspection tank	Concrete, concrete block, brick, etc.	2′ × 4′ and 2′ liquid depth recommended.

Note: A slope of $\frac{1}{16}$″ per ft = 6.25″ per 100′ = 0.0052 ft per ft = 0.52 percent.

Note: All parts of disposal and treatment system shall be located above ground water tractor shall establish and verify all grades and check construction. Laundry and kitchen tank by 50 percent if it is proposed to also install a garbage grinder. No softening unit tem. Where local regulations are more restrictive, they govern, if consistent with county

*Water service and sewer lines may be in same trench, if cast-iron sewer with lead-shelf at one side at least 12″ below water service pipe, provided sound sewer pipe is laid be-loads, or vibration.

†*Manual of Septic-Tank Practice,* U.S. Public Health Service Pub. 526 (1967), states that by 40 percent additional where a home laundry machine is also installed. It recommends Serial distribution is also advised.

surface. Laterals for fields of less than 500 ft in total length, without siphons, should be laid on a slope of $\frac{1}{16}$ in./ft or 3 in./50 ft. When siphons are used, the laterals should be laid on a slope of 3 in./100 ft. Absorption fields for steep sloping ground are shown in Figures 4–12 and 4–13, and layouts for level and gently sloping ground are shown in Figure 4–14.

Leaching or Seepage Pit

Leaching pits, also referred to as seepage pits, are used for the disposal of settled sewage where the soil is suitable and a public water supply

SUBSURFACE SEWAGE DISPOSAL SYSTEMS

Grade	Minimum Governing Distances		
	To building or property line	To well or suction line	To water service line
¼" per ft max., $\frac{1}{8}$" per ft min.	5' or more recommended.	25' if cast-iron pipe, otherwise 50'.	10' hor.*
Outlet 2" below inlet.	10'	50'	10'
$\frac{1}{8}$" per ft; but $\frac{1}{16}$" per ft with pump or siphon.	10"	50'	10'
Outlets at same level.	10'	100'	10'
$\frac{1}{16}$" per ft, but $\frac{1}{32}$" per ft with pump or siphon.	10' except when fill is used in which case 20' is required.	100'	10' (25' from any stream; 50' is recommended.)
Laterals laid on slope $\frac{1}{16}$" per ft; but $\frac{1}{32}$" per ft with pump or siphon.	10'	50'	10' (25' from any stream; 50' is recommended.)
Line to pit $\frac{1}{8}$" per ft.	20'	150'; plus in coarse gravel.	20' (50' from any stream).
Outlet 2" below inlet.	10'	50'	10'

and *downgrade* from sources of water supply. The architect, builder, contractor, and subcon-
wastes shall discharge to the septic tank with other sewage. Increase the volume of the septic
wastes, roof or footing drainage, surface water or groundwater shall enter the sewerage sys-
and state regulations.
caulked joints is laid at all points 12" below water service pipe; or sewer may be on dropped
low frost with tight and root-proof joints which is not subject to settlement, superimposed

the leaching area should be increased by 20 percent where a garbage grinder is installed, and
that gravel in tile field extend at least 2" above pipe and 6" below the bottom of the pipe.

is used, or where private well-water supplies are preferably 150 to 200 ft away, at a higher elevation and not likely to be affected. The bottom of the pit should be at least 2 ft and preferably 4 ft above the highest groundwater level and channeled or creviced rock. If this cannot be assured, subsurface absorption fields should be used. In special instances, where suitable soil is found at greater depths, pits can be dug 20 to 25 or more ft deep, using precast perforated wall sections. The soil percolation test is made at mid-depth and at the bottom of the proposed leaching pit and interpreted for design purposes as explained earlier

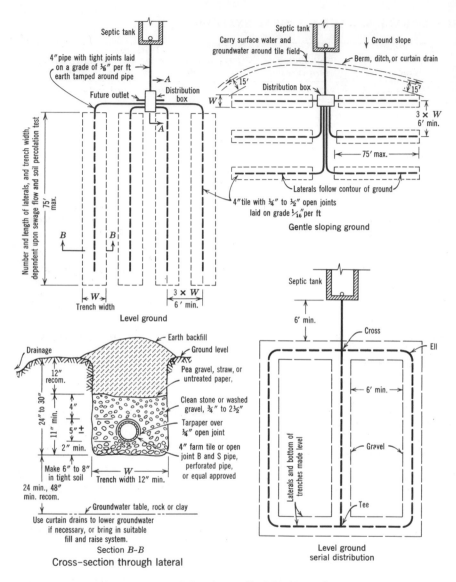

Figure 4–14 Arrangements and details for tile-field disposal systems.

in this chapter and in Table 4–1. The effective leaching area provided by a pit is equal to the sum of the outside vertical wall area below the inlet plus the inside bottom area of the leaching pit wall. Sometimes credit is not given for the pit bottom. A leaching pit may be round, oval, square, or rectangular. The wall below the inlet is dry-wall construction; that is, laid with open joints, without mortar. Field stones, cinder or stone concrete blocks, precast perforated wall sections, or special cesspool blocks are used for the wall construction. Concrete blocks are usually placed with the cell holes horizontal. Crushed stone or coarse gravel should be filled in between the outside of the leaching pit wall and the earth hole. A nomogram to simplify determination of the sizes of circular leaching pits is shown in Figure 4–16. Sketches of leaching pits are given in Figure 4–15 and 4–17.

Cesspool

Cesspools are covered, open-joint walled pits that receive raw sewage. Their use is not recommended where the groundwater serves as a source of water supply. Many health departments prohibit the installation of cesspools where groundwater or creviced and channeled rock is close to the surface. Pollution could travel readily to wells or springs used for water supply. Where cesspools are permitted, they should be located downgrade from sources of water supply and 200 to 500 ft away. Even 500 ft may not be a safe distance in a coarse gravel unless the water-bearing stratum is below the gravel and separated by a thick clay or hardpan stratum. On the other hand lesser distances may be permitted where fine sand and no groundwater is involved. In all cases the bottom of the cesspool should be at least 4 ft above the highest groundwater level.

The construction of a cesspool is the same as a leaching pit, shown in Figure 4–15. In some areas, such as Long Island, cesspools were in common use for many years before cleaning was required. Cleaning the cesspool will not restore it to full use again since the space behind the wall cannot be effectively cleaned. Heavy chlorination may be of value. The cesspool system can be made more efficient under such circumstances by providing a tee outlet, as shown in Figure 4–15, with the overflow discharging to a tile field or leaching pit. Another alternative would be replacement of the cesspool with a septic tank followed by a tile field or leaching pit. The required size of a cesspool can be determined by making soil percolation tests as explained for a leaching pit. No credit is given for the bottom area of a cesspool.

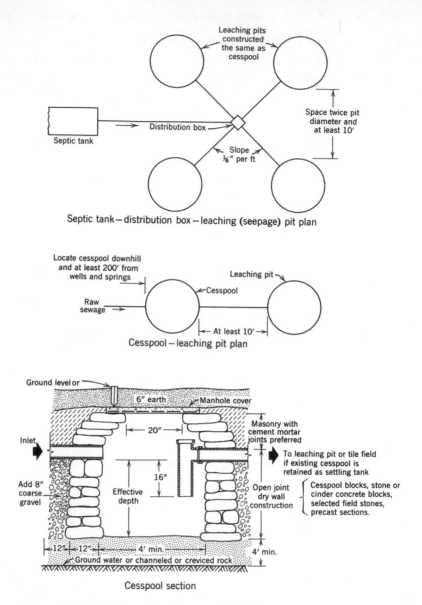

Septic tank – distribution box – leaching (seepage) pit plan

Cesspool – leaching pit plan

Cesspool section

Figure 4–15 Leaching pit and cesspool details.

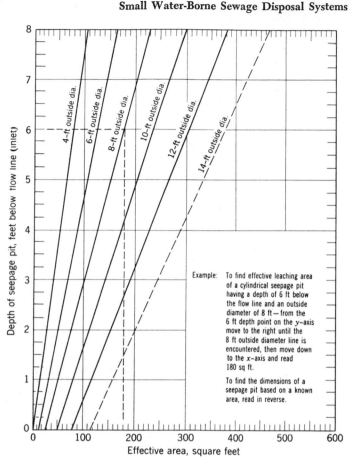

Figure 4–16 Effective areas of round seepage pits preceded by septic tanks or cesspools. (J. A. Salvato, April 1946.)

Dry Well

A dry well is constructed similar to a leaching or seepage pit and with the same care. A dry well is used where the subsoil is relatively porous for the underground disposal of clear rainwater, surface water, or groundwater collected in footing, roof, and basement floor drains and similar devices. Footing, roof, or basement floor drainage should never be discharged to a private sewage disposal or treatment system as the septic-tank and leaching system would be seriously overloaded and cause exposure of the sewage and premature failure of the leaching system. To design the sewerage system for this additional flow is uneconomical and unnecessary. If the soil at a depth of 6 to 10 ft or more is tight

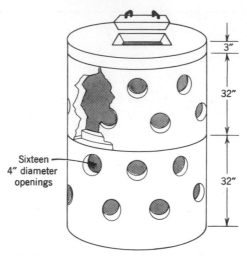

Each section: 32″ high, 4′ inside diameter. (Available in larger sizes.)
　　　　　　　　3″ thick walls; place 8″ coarse gravel all around.
　　　　　　　　250-gal volume.
　　　　　　　　1,100-lb weight.
　　Cover weight: Approximately 400 lb.
NOTE: Manhole can be built up to grade using standard chimney blocks.

Figure 4–17   Precast leaching pit or dry well. (Courtesy the Fort Miller Co., Inc., Fort Miller, N.Y.)

clay, gravel- or stone-filled trenches about 3 ft deep may be found more effective. Dry wells should not be used for the disposal of sewage, household laundry wastes, or kitchen wastes. These wastes should be discharged through a septic tank. In some cases footing and roof drainage is discharged to a nearby watercourse, combined sewer, storm sewer, or roadside ditch, if permitted by local regulations, rather than to dry wells.

Dry wells should be located at least 50 ft from any water well, 20 ft from any leaching portion of a sewage disposal system, and 10 ft or more from building foundations or footings.

SMALL SEWAGE DISPOSAL SYSTEMS FOR TIGHT SOILS

Evapotranspiration

The total length of a subsurface absorption field is determined by the results of soil percolation tests and the quantity of settled sewage

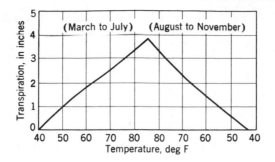

Figure 4–18 Empirical transpiration curve. [From "Computing Runoff from Rainfall and Other Physical Data," ASCE Transactions, **79**, 1094 (1915).] NOTE: The temperature scale refers to mean monthly air temperature and the transpiration scale refers to the corresponding transpiration in inches per month.

to be disposed of. Credit for evaporation from the ground surface and transpiration through vegetation is generally not given, although this is significant during the warm months. See Figure 4–18. Subsurface tile fields that are used only during warm periods might be designed to take into consideration rainfall, transpiration, and evaporation in special instances.

Schwartz and Bendixen reported that vegetation on the surface of subsurface sewage disposal systems doubled the hydraulic longevity.[18]

Phelps states that evaporation from water surfaces varies from about 20 in. per year in Northeastern United States to 90 in. per year in the Imperial Valley, and that evaporation from land areas will be approximately one-third to one-half these values.[19] About two-thirds of the total annual evaporation takes place from April to September. Phelps also gives the following transpiration figures for several types of vegetation for the period in leaf:

Grain and grass crops	9 to 10 inches
Deciduous trees	8 to 12 inches
Small brush	6 to 8 inches
Coniferous trees	4 to 6 inches

[18] Warren A. Schwartz and Thomas W. Bendixen, "Soil Systems For Liquid Waste Treatment and Disposal: Environmental Factors," *J. Water Pollution Control Fed.*, **42**, 624–630 (April 1970).
[19] Earle B. Phelps, *Public Health Engineering*, Vol. 1, John Wiley & Sons, Inc., New York, 1948.

McGauhey gives the following average water recycling distribution:[20]

Evaporation—30 percent
Evapotranspiration—40 percent (from soil mantle)
Surface runoff—20 percent
Groundwater storage—10 percent } Total freshwater resource

Studies in England show that evaporation from a free water surface is 16½ in. of water per year and 14½ in. from bare soil.[21] Evaporation loss from turf or grassland is 0.6 to 0.8 times that from an open water surface.

The quantity of water transpired and evaporated from a cropped area, or the normal loss of water from the soil by evaporation and plant transpiration, is known as the consumptive use. It is expressed in terms of inches of water for any period. The largest proportion of consumptive use is transpiration. Blaney has reported on the studies of a number of investigators giving the consumptive use figures for some crops in California.[22] This information has been used in the development of Table 4–10.

TABLE 4–10   CONSUMPTIVE USE OF CROPS IN CALIFORNIA

Crop	Evaporation (soil) and Transpiration (plant)	Growing Season	Average Annual Transvap*
Alfalfa	5 in./month or 0.83 pt/ft² /day	6 or 7 months	0.35 pt/ft² /day
Truck garden	4 in./month or 0.66 pt/ft² /day	5 months	0.275 pt/ft² /day
Cotton	4 in./month or 0.66 pt/ft² /day	7 months	0.38 pt/ft² /day
Citrus orchard	3 in./month or 0.50 pt/ft² /day	7 months	0.3 pt/ft² /day
Citrus orchard	5 in./month or 0.83 pt/ft² /day	6 months	0.415 pt/ft² /day

*Soil evaporation for 7 to 5 months outside of growing season not included.
†pints per square foot per day = pt/ft² /day

Table 4–11 gives some additional estimates of seasonal consumption of water by crops and vegetation. An example will show the possible use of this data for the design of an evapotranspiration (transvap) system.

With a 150-day (5-month) growing season, using a cover of meadow or lucern grass (15- to 36-in. consumptive use), and 18 in. of precipita-

[20] P. H. McGauhey, *Engineering Management of Water Quality* (New York: Mc-Graw-Hill Book Co., 1968).
[21] *Manual of British Water Supply Practice,* Institute of Water Engineers, W. Heffer & Sons, Ltd., Cambridge, England, 1950.
[22] Harry F. Blaney, "Use of Water by Irrigation Crops in California," *J. Am. Water Works Assoc.,* **43,** No. 3, 189–200 (March 1951).

TABLE 4-11   APPROXIMATE SEASONAL CONSUMPTION OF WATER BY CROPS AND VEGETATION

Growth	in. of Water	Thousands of gal/acre	gal/ft²
Coniferous trees	4 to 9	109 to 245	2.5 to 5.6
Deciduous trees	7 to 10	190 to 270	4.4 to 6.2
Potatoes	7 to 11	190 to 300	4.4 to 6.9
Rye	18 up	490 up	11.2 up
Wheat	20 to 22	540 to 600	12.4 to 13.8
Grapes	6 up	163 up	3.7 up
Alfalfa and clover	2.5 up	68 up	1.6 up
Corn	20 to 75	540 to 2,040	12.4 to 46.8
Oats	28 to 40	760 to 1,090	17.4 to 25
Meadow grass	22 to 60	600 to 1,630	13.8 to 37.4
Lucern grass	26 to 55	706 to 1,500	16.2 to 34.4
Rice	60 to 200	1,630 to 5,400	37.4 to 124

Source: Adapted from Leonard C. Urquhart, *Civil Engineering Handbook*, McGraw-Hill Book Co., New York, January 1950.

tion during the season, of which one-half (9 in.) evaporates and runs off, the water that can be disposed of by

$$\text{evapotranspiration} =$$

$$\frac{(15 - 9) \times 0.623^* \times 8\dagger}{5 \times 30} = 0.2 \text{ pt/ft}^2/\text{day}$$

$$\text{to } \frac{(36 - 9) \times 0.623^* \times 8\dagger}{5 \times 30} = 0.9 \text{ pt/ft}^2/\text{day}$$

This would correspond to an average of 0.08–0.38 pt/ft²/day on an annual basis. A design on this basis, with a gravel bed 24 or more in. deep to provide storage during the nongrowing season, would have application where tight soil exists and groundwater is not too high and there is no practical alternative. A cubic foot of gravel with 30 percent void space would have a capacity of about 2.5 gal/ft³.

Transvap Systems

If clay, hardpan, groundwater, or channeled, creviced, or solid rock is found within and below 4 ft of the ground surface, disposal of sewage by means of a conventional subsurface tile field system is not recommended. For practical purposes clay, hardpan, or solid rock cannot be expected to absorb sewage; and the disposal of sewage directly into

* One in. of water/ft² in gal. † Eight pt in one gal.

the groundwater is to invite failure of the system or, if into channeled or creviced rock, sewage pollution of nearby and distant wells. It may be possible to artificially build up an earth area for sewage disposal *provided at least 12 to 18 in. of natural porous earth exists,* if approved by the health department. The slope or dip of the rock, clay, or hardpan and the limits of the fill must be such as to make improbable seepage of sewage to the ground surface or into creviced rock. The earth gravelly loam fill should be porous and contain topsoil with not less than 20 percent or more than 40 percent clay. See "General Soil Considerations," earlier in this chapter. Fill must be allowed to become stabilized before cutting the trenches and constructing the tile field system. The limits of the fill should extend at least 20 ft beyond the tile field in all directions. A suggested sewage disposal system incorporating fill is illustrated in Figure 4–19. The design is based on a transpiration-evaporation (transvap) rate plus limited percolation, or a transvap rate of 0.5 gpd/ft² of conventional bottom trench area, or $\frac{1}{3}$ pt/ft² of gross area of fill. This principle of design may be used for small motels, private dwellings, and similar establishments *if necessary.*

A system that may be used under more difficult conditions where no topsoil or porous earth exists over clay is shown in Figures 4–20 and 4–21. Here the design is based on a transvap rate of $\frac{1}{4}$ pt/ft² of gross bed area per day. Local climatic conditions and rainfall should be taken into consideration.

The successful operation of a transvap system is largely dependent on the fill-soil composition, the evaporation and transpiration, and the drainage of rainfall off the fill area. Maintenance or restoration of a permeable soil structure and microbial population is essential; otherwise system failure will be a certainty. Curtain drains may be necessary.

Modification of Conventional Tile Field System

Another design basis for a relatively tight soil makes use of the conventional soil percolation test carried to the point of constant rate, beyond the 1 in./60 min. test.

Assume a soil with an actual percolation rate of $\frac{1}{4}$ in./hr. If the rate for a 60 min/in. soil is 0.40 gpd/ft², then for $\frac{1}{4}$ in./hr. soil the rate could be $\frac{1}{4} \times 0.40 = 0.10$ gpd/ft².

Example

Design a subsurface leaching system for a daily flow of 300 gal. The soil test shows $\frac{1}{4}$ in./hr and a permissible settled sewage application of 0.10 gpd/ft².

Required leaching area $= \dfrac{300}{0.10} = 3000$ ft².

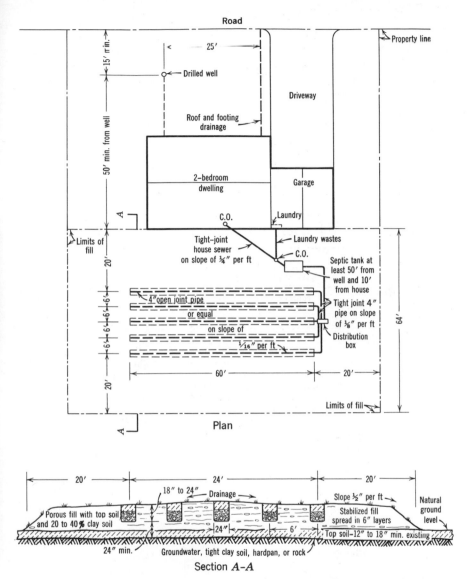

Figure 4–19 Sewage disposal system over clay soil or rock. NOTE: First floor and house sewer elevations must be established to provide gravity flow to sewage disposal system. Design basis: 300 gpd and a transvap-percolation rate of 0.5 gal for trench ft²/day or ±⅓ pt/ft² of gross area of fill (64′ × 100′).

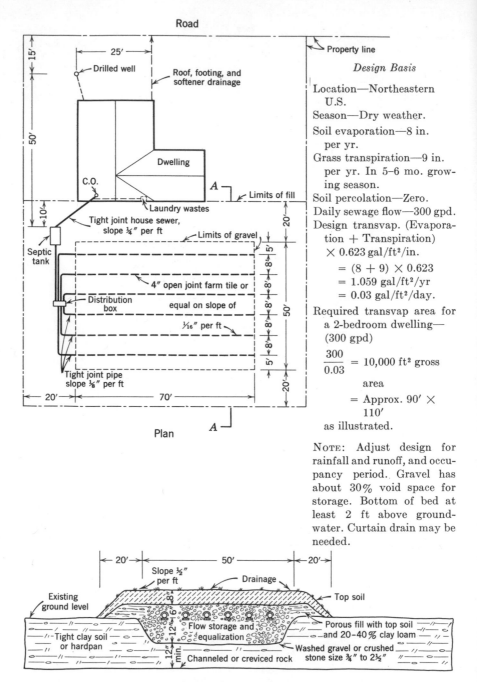

Road

25′

Drilled well

Roof, footing, and softener drainage

15′

50′

Dwelling

C.O.

A

Laundry wastes

Tight joint house sewer, slope ¼″ per ft

Limits of fill

Limits of gravel

Septic tank

4″ open joint farm tile or

equal on slope of

Distribution box

¹⁄₁₆″ per ft

Tight joint pipe slope ⅛″ per ft

10′

20′

70′

20′

5′

8′

8′

8′

8′

8′

5′

50′

20′

A

Plan

Property line

Design Basis

Location—Northeastern U.S.

Season—Dry weather.

Soil evaporation—8 in. per yr.

Grass transpiration—9 in. per yr. In 5–6 mo. growing season.

Soil percolation—Zero.

Daily sewage flow—300 gpd.

Design transvap. (Evaporation + Transpiration) $\times$ 0.623 gal/ft²/in.

$$= (8 + 9) \times 0.623$$
$$= 1.059 \text{ gal/ft}^2/\text{yr}$$
$$= 0.03 \text{ gal/ft}^2/\text{day}.$$

Required transvap area for a 2-bedroom dwelling— (300 gpd)

$$\frac{300}{0.03} = 10{,}000 \text{ ft}^2 \text{ gross}$$

area

$$= \text{Approx. } 90' \times 110'$$

as illustrated.

NOTE: Adjust design for rainfall and runoff, and occupancy period. Gravel has about 30% void space for storage. Bottom of bed at least 2 ft above groundwater. Curtain drain may be needed.

20′

50′

20′

Slope ½″ per ft

Drainage

Existing ground level

Top soil

Tight clay soil or hardpan

12″–6″

12″ min.

Flow storage and equalization

Channeled or creviced rock

Porous fill with top soil and 20–40% clay loam

Washed gravel or crushed stone size ¾″ to 2½″

Section *A–A*

Figure 4–20 Transvap sewage disposal system on clay, for seasonal use only.

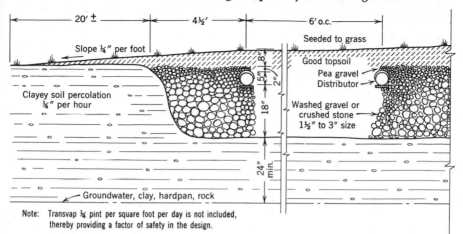

Figure 4–21 Modification of conventional tile field system. NOTE: Curtain drains may be needed to lower groundwater level.

If trenches 36 in. wide with 18 in. gravel underneath lateral distributors are provided, each lineal ft of trench can be expected to provide 5 ft² of leaching area. The required trench $= \dfrac{3000}{5} = 600$ lineal ft, or 8 laterals each 75 ft long, spaced 9 ft on center.

The leaching area can also be provided by a gravel bed 72 × 75 ft, as illustrated in Figure 4–21. This occupies the same land area as the tile field.

Waste Stabilization Pond

In some locations, where tight soil exists and ample property is owned, the waste stabilization pond design principles may be adapted to small installations. Design details are given on page 326.

Aerobic Sewage Treatment Unit

Another type treatment unit that can be used where subsurface absorption systems are not practical is the self-contained aeration unit. Package plants are available. An outlet for the discharge of treated effluent is necessary and approval of the local health department or other regulatory agency is generally required. Disposal to a subsurface tile field system or through a sand filter and chlorination of the effluent may also be required. Routine maintenance and operation of the unit must be assured. Design details are given on pages 325 and 354.

Sand Filter

Sand filters have particular application where conventional subsurface disposal methods, such as tile fields or leaching pits, could not be expected to function satisfactorily and where the cost of constructing any leaching system would be excessive or impractical.

A sand filter is an efficient treatment device. Sewage is distributed over the top of a sand filter bed by means of flooding, tipping buckets, distributing troughs or perforated or open-joint pipe. The sewage is filtered and oxidized in passing through 24 to 30 in. of carefully selected sand. Greater sand depths do not produce significant additional purification. A film containing aerobic and nitrifying organisms forms on the gravel and sand grains of the filter. Bacteria break down the organic matter in sewage. Protozoa feed on bacteria, and metazoa consume bacteria, sludges, and slimes. The annelid worms are reported to be largely responsible for keeping sand filters from clogging.[23]

Typical analyses of sanitary sewage applied to and leaving subsurface sand filters and efficiencies are shown in Tables 4–12 and 4–13. A large reduction in bacteria, protozoa, helminths, turbidity, biochemical oxygen demand, and suspended solids is obtained, in addition to a well-nitrified effluent containing dissolved oxygen. Such effluents would not cause a nuisance in undeveloped areas; but they should be chlorinated if discharged in locations accessible to children or pets because bacteria associated with disease transmission, although greatly reduced, are still present.

Sand Filter Design

The recommended sizes of covered sand filters to serve private homes are shown in Figure 4–22 and 4–23. These systems are designed for a flow of 150 gpd/bedroom and a settled sewage application rate of 1.15 gal per ft^2 of sand filter area per day. It is extremely important to use a proper sand meeting the specifications given in Figure 4–22. The sand grains should be somewhat uniform in size, that is, not graded in size from fine to coarse, as this will surely cause premature clogging of the filter. Assistance and advice of the local or state health department should be sought before a sand is purchased and delivered to the job. Some sand and gravel companies are equipped to make sieve analyses of sand and can assist individuals. The sand filter must be carefully constructed and settled by flooding, with distributor and collector lines laid at exact grade. The use of 1 × 6 in. boards under farm tile distributors, laid on gravel, assist greatly in placing the distributor lines at proper grade. The boards are not needed with long-length perforated pipe. A topsoil

---

TABLE 4-12 TYPICAL SEPTIC TANK AND SUBSURFACE SAND FILTER EFFLUENT

Determination	Sewage Effluent* Septic Tank	Sewage Effluent* Subsurface Sand Filter
Bacteria per ml, Agar, 36°C, 24 hr	76,000,000	127,000
Coliform group MPN	110,000,000	150,000
Color (mg/l)	3.5	2
Turbidity (mg/l)	50	5
Odor†	4.5	1
Suspended matter†	3	1
pH	7.4	7.4
Temperature °C	17	14
BOD, 5-day (mg/l)	140	4
DO (mg/l)	0	5.2
DO saturation (%)	0	52
Nitrogen, total (mg/l)	36	21
Free ammonia	12	0.7
Organic	12	3.4
Nitrites	0.001	0.02
Nitrates	0.12	17
Oxygen consumed (mg/l)	80	20
Chlorides (mg/l)	80	65
Alkalinity (mg/l)	400	300
Total solids (mg/l)	820	810
Susp. solids (mg/l)	101	12

Source: J. A. Salvato, Jr., "Experience with Subsurface Sand Filters," *Sewage and Industrial Wastes,* 27, No. 8, 909–916 (August 1955).

*Median results, using 51 samples from septic tanks and 56 from filters.

†1 = very slight, 2 = slight, 3 = distinct, 4 = decided, 5 = extreme. Normal municipal domestic sewage has an MPN of 50–100 million coliform bacteria per 100 ml.

TABLE 4-13 TYPICAL EFFICIENCIES OF SUBSURFACE FILTERS*

Determination	Percent Reduction
Bacteria per ml, Agar, 36°C, 24 hr	99.5
Coliform group: MPN per 100 ml	99.6
BOD, 5-day (mg/l)	97
Susp. solids (mg/l)	88
Oxygen consumed (mg/l)	75
Total nitrogen (mg/l)	42
Free ammonia	94
Organic	72

Source: J. A. Salvato, Jr., "Experience with Subsurface Sand Filters," *Sewage and Industrial Wastes,* 27, No. 8, 909–916 (August 1955).

*Effluent will contain 5.2 mg/l dissolved oxygen and 17 mg/l nitrates.

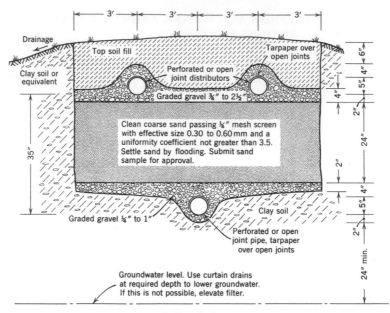

Section *A–A*

Number of Bedrooms	Capacity of Septic Tank	Size of Filter			Alternate Filter Size*		Sump Capacity Between Float Settings†
		Length	Width	Area	Length	Width	
2	750 gal	21½ ft	12 ft	260 ft²	43 ft	6 ft	20 gal
3	900	32½	12	390	65	6	30
4	1000	43	12	520	87	6	40
5	1250	54	12	650	—	—	50

* Use one distributor on top and one underdrain on bottom. † Where needed.

Required size of subsurface sand filter

(b)

NOTE: Architect, builder, and contractors shall determine invert elevations of the house sewer, septic tank outlet, distribution box, sand filter distributor and collector lines, chlorine contact tank, inspection manhole, and outlet sewer, drain, ditch, or watercourse so as to provide gravity flow through the system where possible. Exclude roof and footing drainage. Increase the volume of the septic tank by 50 percent if a garbage grinder is installed, and the area of the filter by 30 to 60 percent if both garbage grinder and home laundry machine are installed.

Figure 4–22 Typical section of a subsurface sand filter.

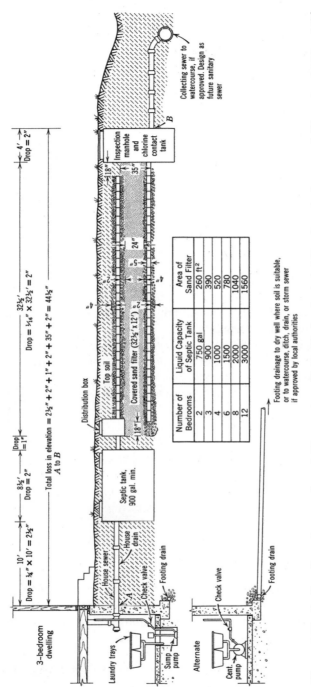

Number of Bedrooms	Liquid Capacity of Septic Tank	Area of Sand Filter
2	750 gal	260 ft²
3	900	390
4	1000	520
6	1500	780
8	2000	1040
12	3000	1560

Figure 4-23   Section through covered sand filter system.   Note: Design basis is 150 gal per bedroom and filter rate of 1.15 gpd/ft².

311

cover preferably not exceeding 8 to 12 in. should be filled in over the gravel covered distributor lines.

Design details for open and covered sand filters are summarized below for easy reference:

Filter rate: Earth or gravel covered—50,000 gpd/acre for settled domestic sewage; 100,000 gpd/acre for temporary summer use if a recommended sand size is used and the effluent does not enter a water supply source. Open filter—75,000 to 100,000 gpd/acre for settled sewage and 200,000 to 400,000 gpd/acre for secondary treated sewage.

Dose: Dosing device recommended when total length of distributor laterals exceeds 300 lineal ft or sand bed area exceeds 1800 ft². Alternating dosing device recommended when length of distributor laterals exceeds 800 lineal ft. Dose should be at least 90 gpm/1000 ft² of filter area at average head.

Volume of dose with distributor laterals = 60 to 75 percent of volume of distributors dosed. Volume on open filter = 1 to 4 in. flooding or 27,150 to 108,600 gal/acre. Design for 1 to 3 doses/day and a minimum 4-hr rest period between doses. (Four-inch pipe holds 0.653 gal/ft.)

Length of lateral distributor, earth or gravel covered: 75 ft or less; 100 ft acceptable with dosing device.

Sand: Depth 24 to 30 in., effective size recommended 0.35 to 0.50 mm, although 0.25 to 0.60 mm is usually acceptable. Uniformity coefficient less than 3.5 recommended, but 4.0 is acceptable. Use clean silica sand passing ¼-in. sieve. A sand analysis is shown in Figure 4–24.

Grade of lateral distributor: $\frac{1}{16}$ in./ft, but $\frac{1}{32}$ in./ft with dosing device.

Sand filters are not always covered with earth, but may be covered only with gravel or may be left entirely uncovered. When covered only with gravel, the distributing lines are laid in the gravel. When a filter is open, distribution of the settled sewage is accomplished by means of troughs laid on the surface, from splash plates in the center or at one or more corners of a filter section, as shown in Figure 4–25, or by means of a sprinkling distributor. The maximum lateral distance of sewage travel from the splash plate to the edge of the sand bed should be 20 ft.

In freezing weather open filters will require greater operation control and maintenance. Scraping the sand before freezing weather into furrows about 8 in. deep with ridges 24 to 48 in. apart will help maintain continuous operation, as ice sheets will form between ridges and help insulate the relatively warm sewage in the furrows. Glass greenhouse covers are very desirable and will help assure continuous operation of the filters; however, they are expensive.

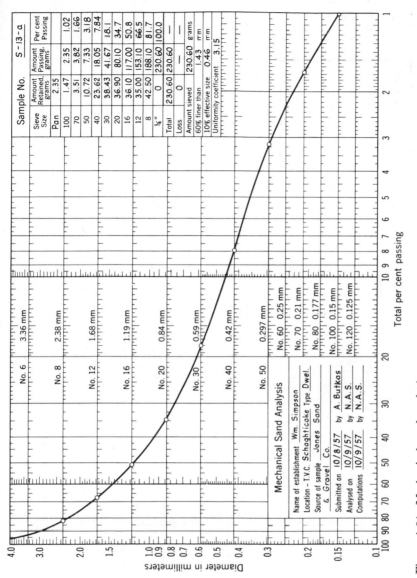

Figure 4-24 Mechanical sand analysis.

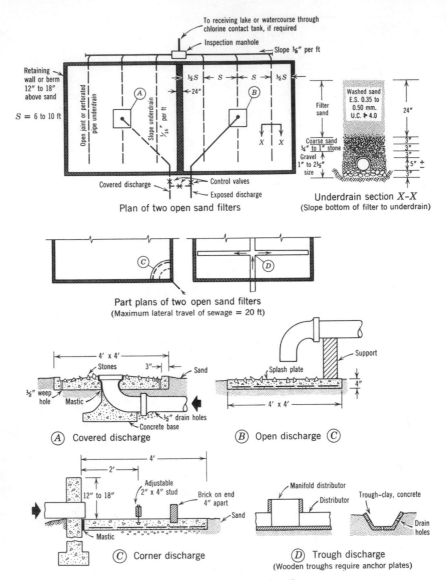

Figure 4–25 Open sand filter and distribution details.

Dosing Tank

The size of a dosing tank is determined by the length of a tile field or area of a sand filter to be dosed at any one time. The volume of the dose should equal 60 to 75 percent of the volume of lines dosed. In general, a siphon or other dosing device should be provided when the total length of tile field exceeds 300 to 500 ft, and when the sand filter distributor laterals exceed 300 lineal ft or the area of the sand bed exceeds 1800 ft^2. Alternating siphons are recommended when the tile field exceeds 1000 ft and when the sand filter distributors exceed 800 lineal ft. Popkin and Bendixen report that "a vastly improved design and operation of soil absorption systems can be obtained through use of once-a-week dosing and/or by use of improved pretreatment without imperiling groundwater quality."[24]

Dosing tanks are usually designed to operate automatically. Dosage is accomplished by a siphon, ejector, tipping bucket, or pump. Hand-operated gates, float valves, and motorized valves are also used. If the available grade does not permit the use of automatic siphons or a similar device, pumps or ejectors will have to be used. In such cases the system should be designed, if possible, to permit gravity flow through the dosing tank or pump well to the filter, in case of pump or power failure, while repairs are being made. If this cannot be done a standby gasoline-engine-driven pump should be provided, in addition to a high-water alarm to warn of pump failure. In larger systems, two siphons or two pumps with proper float setting are installed, each feeding a separate tile field or sand filter section in alternation.

The size siphon or other dosing device that has a minimum discharge rate at least 125 percent greater and preferably twice the probable maximum rate at which settled sewage might enter the dosing tank is selected. This is necessary in order to exceed the rate of flow of the incoming sewage and vent the siphon, making possible its continued automatic operation. It also will prevent overdosing of a tile field system or sand filter and possible flooding. Where open sand beds are dosed, rapid discharge of the sewage gives better distribution over the sand bed. The head or fall available will also determine the size siphon that can be used. Special designs incorporating given drawdowns can be obtained from the manufacturer. The dosing device should be capable of applying the required volume on the filter in less than 10 min. Some design details of the Miller siphon are given in Figure 4–27. Prefabricated metal dosing tanks incorporating single or alternating siphons are

[24] Ronald A. Popkin and Thomas W. Bendixen, "Improved Subsurface Disposal," *J. Water Pollution Control Fed.*, **40**, 8, Part 1, 1513 (August 1968).

also available.[25] The diameter of the carrier line, that is, the line between the dosing tank and the tile field, can be developed with the aid of Figure 4–26, since the maximum discharge rate of the siphon is given in Figure 4–27.

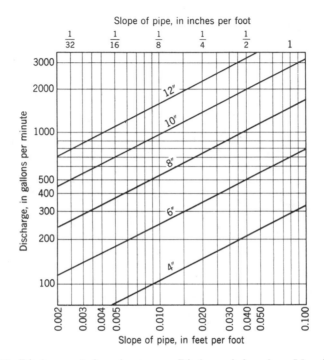

Figure 4–26 Discharge of clay pipe sewers. Discharge is based on Manning formula

$$Q = A \frac{1.486}{n} R^{2/3} S^{1/2}, \text{ with } n = 0.013, \text{ sewer full.}$$

Manufacturers' detail drawings of the specified siphon shown should be obtained and carefully followed during construction and critical elevations staked out. The siphon trap must be filled with water when the system is to be placed in operation. The small vent pipe on the bell must have airtight joints, with the bell perfectly level. Sometimes the floor under the siphon bell is depressed 4 or 6 in. below the remainder of the dosing tank floor with satisfactory results. Another acceptable

[25] San-Equip, Inc., Syracuse, N.Y.; Kaustine Co., Inc., Perry, N.Y.; Pacific Flush-Tank Co., 4211 Ravenswood Ave., Chicago, Ill.

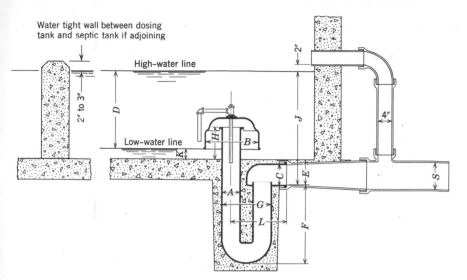

Approximate Dimensions in Inches and Average Weights in Pounds

Diameter of siphon	A	3	4	5	6
Drawing depth	D	13	17	23	30
Diameter of discharge head	C	4	4	6	8
Diameter of bell	B	10	12	15	19
Invert below floor	E	$4\frac{1}{4}$	$5\frac{1}{2}$	$7\frac{1}{2}$	10
Depth of trap	F	13	$14\frac{1}{4}$	23	$30\frac{1}{4}$
Width of trap	G	10	12	14	16
Height above floor	H	$7\frac{1}{4}$	$11\frac{3}{4}$	$9\frac{1}{2}$	11
Invert to discharge $= D + E + K$	J	$20\frac{1}{4}$	$25\frac{1}{2}$	$33\frac{1}{2}$	44
Bottom of bell to floor	K	3	3	3	4
Center of trap to end of discharge ell	L	$8\frac{5}{8}$	$11\frac{3}{4}$	$15\frac{1}{2}$	$17\frac{1}{8}$
Diameter of carrier	S	4	4–6	6–8	8–10
Average discharge rate gpm	—	72	165	328	474
Maximum discharge rate gpm	—	96	227	422	604
Minimum discharge rate gpm	—	48	102	234	340
Shipping weight in pounds	—	60	150	210	300
Detail drawing 1-F	—	373	374.2	375	376

NOTE: Two single siphons of this type set side by side in the same tank will alternate. The draft "D" will be 1 in. to 2 in. less in this case. One foot of 4 in. pipe holds 0.653 gal.

Figure 4–27   Design details of the Miller siphons.

alternative is to move the siphon forward in the dosing tank and install a tee extension on the discharge or carrier line in the dosing tank extending about 2 in. above the high water line. This can take the place of the vent and overflow line passing through the dosing tank wall. Either the open tee or overflow are essential, however, to prevent siphoning out the water seal in the siphon trap and to assure continuous automatic operation of the siphon.

Lift Station

As in the case of the siphon, pump manufacturers' working drawings should be followed when a design calls for the installation of pumps. Typical sump pump details are shown in Figures 4–28 and 4–34. Horizontal pumps are also available and are sometimes preferred; a vacuum pump and air-relief valves permit automatic operation. Pumps should be selected so as to operate for at least 10-min intervals. Compressed

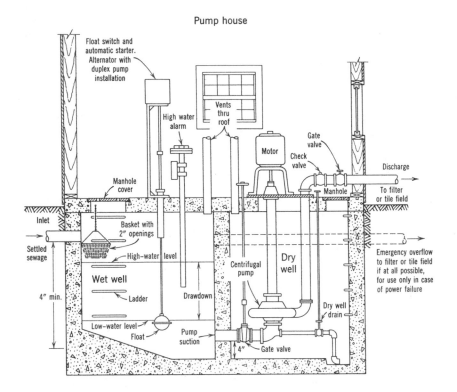

Figure 4–28 Pump sump for dosing sand filter or tile field. When dry well is omitted, the pumps are submerged in the wet well. Sump may be attached to septic tank.

air-operated sewage ejectors may be used to advantage. Obtain the manufacturers' recommendations for the most efficient pump and motor horsepower to meet job requirements. Submersible pumps are also available for pumping small and large quantities of settled sewage. Normally at least two pumps, each being capable of handling the maximum flow, should be provided. See also page 340.

SEWAGE WORKS DESIGN—SMALL TREATMENT PLANTS

Preparation of Plans and Reports

The design should be tailor-made to fit the local conditions and take into consideration probable future additions. Designs for small sewage treatment plants should be prepared by licensed professional engineers experienced in sanitary engineering work and be approved by the regulatory agency involved before construction is started. Designs for typical small installations are sometimes made available through state or county health departments. For small jobs, pencil sketches drawn to scale with dimensions of all units and critical elevations can be prepared for checking by the approving agency. Changes during construction may be very costly. The line of demarcation between large and small sewerage systems is usually an arbitrary one and subject to local interpretation based on the value of the work, its difficultness, the danger to health and safety, and the reasonableness of requiring the services of a professional engineer or registered architect. A practical dividing point might be 25 persons or a sewage flow of 1500 gpd. In some states, when the value of the work exceeds $5000 or $10,000, or danger to health and safety is involved, it is required that plans be submitted by a licensed engineer or architect.

Most states require approval of plans and have rules, regulations, and standards to guide the designing engineer in the preparation of satisfactory plans. The review of preliminary drawings with the local sanitary engineer official will usually expedite approval of final plans. State laws require that a permit be obtained. Permits are issued if the treatment proposed will meet established surface or underground water standards. A permit can be revoked if operation and results are unsatisfactory.

Plans submitted for approval should give all information necessary for review of the design and for construction of the disposal or treatment system as designed. The more complete the plans, the closer bids received for the job will be. Plans should include construction details, engineer's report, an application or statement of information, specifications, and a topographic map showing the location of the disposal plant and outlet, if any. The first step, therefore, when a large sewerage system is to be constructed, is to engage the services of an experienced licensed professional engineer to prepare complete plans for approval by the agency having jurisdiction and for construction of the system. The plans and engineer's report should include the following information.

1. A plot plan giving the boundaries of the property, drawn to a scale of 20 to 100 ft to the in., showing contour lines or critical elevations, streams, swamps, lakes, rock outcrops, wooded areas, structures, roads, play areas, and other significant features. Existing and proposed structures are also shown.

2. Location of existing and proposed sewer and water lines and other utilities, sources of water supply and storage facilities, grease traps, manholes, septic tanks and other disposal or treatment units, and outfall sewers if any.

3. Ground elevations and sizes, materials, slopes, and invert elevations of existing and new sewer lines leaving buildings and entering and leaving manholes, grease traps, septic tanks, siphons, trickling filters, settling tanks, aeration tanks, sludge tanks, drying beds, pump pits, distribution boxes, distributing laterals, sand filters, inspection manholes, chlorine contact tanks, and also the flood level of receiving streams. A profile through the sewerage system is very valuable.

4. Population served and present and future capacity. Where food or drink is served, the seating capacity, number of sales, and meals served should be taken into consideration.

5. Number of different plumbing fixtures including dishwashing machines, potato peelers, laundry machines, and continuously running and automatic flushing devices. Commercial and industrial devices should also be shown.

6. Location, depth, and results of soil percolation tests, with findings of probing 2 or 3 ft deeper and description of the type of soil to a depth of at least 10 ft.

7. Construction details and sizes of all sewerage units, including structural details and specifications.

8. The highest groundwater levels, time of the year tests made, and how determined.

9. Location and yield or adequacy of the source, storage, and distribution of water. Actual or estimated daily water consumption and water pressures should also be given. Where the source of water is one or more wells or springs, the type of well or spring, strata penetrated, depth, and size, in addition to capacity and type of pump and motive power, should be included.

Design Details

The provision of bar screens or comminutors and grit chambers ahead of pumping equipment or settling tanks is strongly recommended.

Pumping equipment is readily accessible for inspection and servicing when located in a dry pump pit. For lift stations, the pneumatic ejectors and "bladeless" sewage pumps seem to be the most suitable for small plants. See pages 318 and 340.

If secondary treatment will be needed, primary treatment units should be designed with the water level at a sufficient height to permit gravity flow to trickling filters or sand filters and to the receiving stream without additional pumping.

Trickling filter units for small installations may be built at ground level. The stone may be contained by heavy fencing, concrete block with reinforcing, or cypress or pressure-treated pine staves held with iron hoops or similar arrangements. Make provision for recirculation for filter fly control.

The secondary settling tank should be constructed with sloping bottom. A bottom shaped like an inverted four-sided truncated regular pyramid will permit gravity collection and concentration of sludge for removal. Provide for a 2-ft hydrostatic head and proper diameter sludge drawoff pipe to give a flushing velocity when the control valve is opened and pumping started.

Chlorination of the final effluent can be accomplished in a chlorine contact tank or in the secondary settling tank. In small installations, manual-control continuous-feed chlorinators are commonly used. Proportional-feed automatic chlorinators are found to be more economical in the larger installations. Other means for chlorination include dosing tank and pump-activated controls.

The location of the treatment plant should take into consideration the type of plant and available supervision, the location of the nearest dwelling, the receiving watercourse, availability of submarginal low land not subject to flooding, prevailing winds and natural barriers, and the cost of land. A distance of 400 ft from the nearest dwelling is frequently recommended, although distances of 250 to 300 ft should prove adequate with good plant supervision. Some equipment manufacturers and designing engineers feel that covered sand filters, aeration-type plants, and high-rate trickling filters can be located closer to habitation without danger of odors or filter flies. Oxidation ponds and lagoons should be located at least ¼ to ½ mi from habitation.

Some of the more common flow diagrams for small sewage treatment plants are illustrated in Figure 4–29 to suggest the different possibilities, dependent on local factors.

Chlorination

Chlorine is added to sewage for a variety of purposes. One of its major uses is for the partial or complete destruction of pathogenic organisms including some viruses in sewage. The effectiveness of disinfection depends on the degree of treatment the sewage has received, the amount of chlorine used and residual chlorine maintained, mixing and retention period, and condition of the sewage. Sometimes chlorine is also added to sewage to control odors, undesirable growths, sewage flies, septicity,

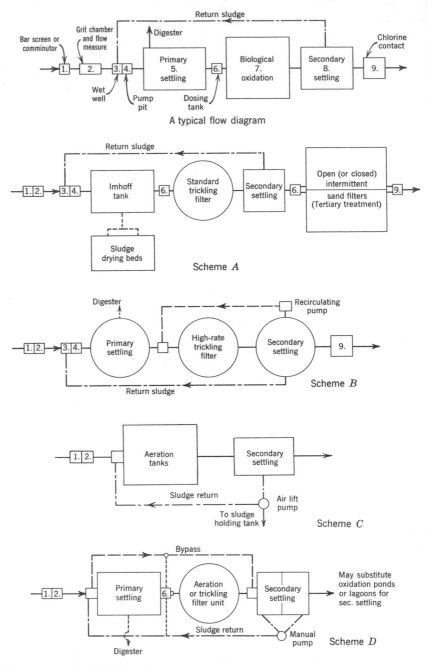

Figure 4–29 Typical flow diagrams.

and chemical or bacterial reactions unfavorable to the treatment process. Chlorination will also reduce the BOD of the treated effluent, roughly in the ratio of one part chlorine to two parts BOD.

Chlorine is available as a relatively pure liquid in steel cylinders under pressure having a net weight of 100 or 150 lb. Larger cylinders such as ton containers are also available. When the pressure is released, the liquid turns to a gas, in which form it is added (mixed with or without water) to sewage by means of a control and measuring device known as a chlorinator. Liquid chlorine is not ordinarily required or economical to use at very small sewage treatment plants. A separate gas-tight room above ground, a building with separate ventilation, an outside entrance, and special gas mask would be required. Calcium hypochlorite, which is a powder containing 70 percent or less available chlorine, or sodium hypochlorite, which is a solution containing 15 percent or less available chlorine, is more generally used. Both the powder and solution are mixed and diluted with water to make a 0.5 to 5.0 percent solution, in which form it is added to the sewage by means of a solution feeder known as a hypochlorinator. Positive feed hypochlorinators are preferred to other types because of their dependability.

Chlorination treatment of raw sewage is not reliable for the destruction of pathogenic organisms since solid penetration is limited. The required dosage of chlorine to produce a 0.5 mg/l residual after 15 min contact has been approximated in Table 4–21 for different kinds of sewage. Studies made using a domestic sewage show that less than 250 coliform organisms per 100 ml can be obtained in treated sewage 100 percent of the time if an orthotolidine chlorine residual of 2.1 to 4 mg/l is maintained in the effluent after 10 min contact.[26] Other experiments show that if the chlorine is first mixed with the sewage, and the treated sewage allowed to stand 10 min, a most probable number of 300 coliform organisms per 100 ml or less can be obtained 100 percent of the time if a chlorine residual of 1.1 mg/l is maintained in the effluent.[27] These tests also show that with no mixing at least twice the chlorine residual must be maintained in the treated sewage for 10 min to give results approximately equal to the results obtained with mixing. Another study shows that an MPN around 1000 per 100 ml can usually be expected in the effluent if the product of the combined chlorine residual (as measured by the OT test) and detention time in minutes (based on average

[26] H. Heukelekian and R. V. Day, "Disinfection of Sewage with Chlorine, III," *Sewage and Ind. Wastes,* **23,** No. 2, 155–163 (February 1951).
[27] Rolf Eliassen, Austin N. Heller, and Herman L. Krieger, "A Statistical Approach to Sewage Chlorination," *Sewage Works J.,* **20,** 1008–1024 (November 1948).

flow) is equal to or greater than 20.[28] Eliassen also reports a reduction in the MPN of aftergrowth values in a brackish water tidal basin, following discharge of a chlorinated combined sewage, of from 10 to 30 percent of those that would develop without chlorination.[29]

Chlorine should be added in a manhole, approximately 20 ft ahead of the chlorine contact tank, or in a mixing box, so as to provide good mixing of the chlorine and sewage. The chlorine contact tank should be designed to provide the required retention at peak hourly flows. This and the chlorine residual is usually specified by the regulatory agency. Chlorine contact tanks should be constructed with over-and-under or round-the-end baffles or equivalent obstruction to assure the required contact period, without short-circuiting. Provision should be made in the design for the collection of samples. Sometimes the required contact time can be obtained in a long outfall sewer or by adding chlorine in a final sedimentation tank when secondary treatment is provided.

Trickling Filter

A trickling filter may be used following a primary settling tank, a septic tank, or Imhoff tank to provide secondary treatment of the sewage. Habitation should not be closer than 400 ft. Some odors and filter flies can be expected. A receiving stream providing adequate dilution and supervision over operation are required. Seeding of the filter stone is necessary before good results are produced. High BOD reduction is obtained within 7 days of starting a trickling filter, but as long as 3 months may be required to obtain equilibrium, including high nitrification.[30]

Small standard-rate trickling filters are usually 6 ft deep and designed for a dosage of 200,000 to 300,000 gpd/acre-ft, or not more than 1,800,000 gal for a 6-ft deep filter. Filter loading is also expressed, with greater accuracy, in terms of 5-day biochemical oxygen demand in the sewage applied to the filter. It is usually assumed that 35 percent of the BOD in a raw sewage is removed by the primary settling unit. Standard-rate trickling filters are dosed at 200 to 600 lb of BOD/acre-ft/day. Average loadings are 400 lb in northern states and 600 lb in southern states. Since dosage must be controlled, dosing siphons or tipping trays

---

[28] Nicholas W. Classen, "Chlorination of Wastewater Effluents," *Public Works,* **100,** 63–66 (January 1969).

[29] Rolf Eliassen, "Coliform Aftergrowths in Chlorinated Storm Overflows," *J. Sanitary Engineering Division,* ASCE, Vol. 94, No. SA2, Proc. Paper 5913 (April 1968), pp. 371–380.

[30] G. R. Grantham and J. G. Seeger, Jr., "Progress of Purification During the Starting of a Trickling Filter," *Sewage and Ind. Wastes,* **23,** 1486–1492 (December 1951).

may be used for very small jobs and siphons or pumps with revolving distributors or stationary spray nozzles on the usual job. Continuous dosage at a higher rate, with recirculation of part of the effluent, may be suitable where good supervision is available and operation can be controlled to produce the intended results. Filter flies are reduced.

A trickling filter should be followed by a secondary settling or humus tank; this unit will require the removal of sludge at least twice a day. The sludge is removed by pumping or by gravity flow if possible to the septic tank, Imhoff tank, or sludge digester, depending on the plant design. The discharge of the raw sludge to a sand drying bed is not advisable, as sludge drying will be slow and odors will result.

Chlorination of the final effluent, for odor control or for disinfection of the sewage for bacterial reduction, is additional treatment that is often required. Trickling filter treatment can be supplemented by sand filtration or an oxidation pond where a higher quality effluent is necessary.

A typical design of an Imhoff tank standard-rate trickling filter plant is shown on pages 328 and 338.

Extended Aeration

Extended aeration plants, also referred to as aerobic digestion plants, have particular application for relatively small installations serving schools, trailer parks, motels, shopping centers, and the like. Some basic design data are given below and in Table 4–18.

Average Sewage Flow—400 gal/dwelling or 100 gpd/capita.
Screening and comminutor—recommended, bar screen minimum.
Aeration Tanks—at least two to treat flows greater than 40,000 gpd, 24-hr detention period at average daily flow, not including recirculation.
Air Requirements—3 cfm/ft of length of aeration tank, or 2000 ft³/lb of BOD entering the tank daily, whichever is larger. Additional air is required if air is needed for air-lift pumping of return sludge from settling tank.
Settling Tanks—at least two to treat flows greater than 40,000 gpd, 4-hr·detention period based on average daily sewage flow, not including recirculation. For tanks with hopper bottoms, upper third of depth of hopper may be considered as effective settling capacity.
Rate of Recirculation—at least 1:1 based on average daily flow.
Measurement of Sewage Flow—by V-notch weir or other appropriate device. Recording devices required for larger installations.
Sludge Holding Tanks—provide 8 ft³/capita. Such tanks are recommended for all plants but are not required when aeration is by diffused air from blowers. Sludge holding tanks are required when aeration tanks are equipped with mechanical aerators.

Daily operation control is essential. Air blowers must be operated continuously and sludge returned. Clogging of the air lift for return sludge is a common cause of difficulty. Grease that accumulates on the surface of settling tanks should be skimmed off and disposed of separately, not to the aeration tank. Aeration tubes or orifices require periodic cleaning. Odors should be minimal. See also page 354.

Waste Stabilization Pond

In areas where ample space is available, preferably 1000 ft or more from habitation, with consideration to the prevailing winds, a waste stabilization pond may be a relatively inexpensive and practical solution to a difficult problem. Construction cost (1966) has been estimated at $15 to $30/capita. It is reported that small systems at resorts or motels designed with a septic tank ahead of the oxidation pond never produced an odor problem.[31] Where indicated, primary treatment with grit chamber, comminutor and rack, and duplicate ponds arranged for series or parallel operation are recommended. A BOD removal of 85 to 90 percent is not unusual. Pond performance is affected by temperature, solar radiation, wind speed, loading, detention time, and other factors.[32] A summary of design criteria for waste stabilization ponds follows.

Pond loading	20 lb of BOD/acre/day, or 370 ft²/capita. The loading may be increased to 50 lb of BOD or more if approved.
Detention time	90 to 180 days; as little as 30 days in warm climates; accommodate entire winter flow.
Liquid depth	5 ft plus 2 ft freeboard. Minimum liquid depth 2 ft.
Embankment	top width 6 to 8 ft; inside and outside slope 3 horizontal to 1 vertical. Use dense impervious material; prepare bottom surface. Liner of clay soil, asphaltic coating, bentonite, plastic or rubber membrane, or other material required if seepage can be expected.
Pond bottom	level, impervious, no vegetation. Soil percolation should be less than $\frac{1}{4}$ in./hr after saturation.
Inlet	4-in. diameter minimum, at center of square or circular pond; at $\frac{1}{3}$ point if rectangular, with length not more than twice width. Submerged inlet 1 ft off bottom on a concrete pad or at least 1.5 ft above highest water level.
Outlet	4-in. minimum diameter; controlled liquid depth discharge using baffles, elbow or tee fittings; drawoff about 6 in. below water surface and so as to avoid short-circuiting; permit drainage of pond and discharge to concrete or paved gutter.
General	round pond corners; provide fencing, warning signs, and means for flow measurement. Locate in isolated area. Seed top and sides of embankment to grass from above waterline. Keep grass cut and prevent growth of weeds, trees, and shrubs. Odors may occur after ice break-up. Lower pond level in the autumn and in the spring. Fence in pond area.

[31] Harvey F. Ludwig, "Industry's Idea Clinic," *J. Water Pollution Control Fed.*, **36**, 937 (August 1964).
[32] L. W. Canter and A. J. Englande, Jr., "States' Design Criteria for Waste Stabilization Ponds," *J. Water Pollution Control Fed.*, **42**, 1840 (October 1970).

The Soil Conservation Service, local health department, or other regulatory agency may be of assistance in connection with needed soils studies, design, and effluent standards to be met.

The practicability of using waste stabilization ponds, lagoons, or spray irrigation should be investigated in light of local conditions. The effect of possible odor nuisance or hazard should be evaluated before a treatment process is adopted. These processes should not be dismissed too quickly as they may sometimes provide an acceptable answer when no other treatment is practical.

Low-Cost Sanitation

A study by Stander and Meiring in South Africa shows that waterborne sewage disposal systems including an oxidation pond can be provided at a cost that compares very favorably with a removal pail privy system.[33] Where aqua privies (with retention tanks) are acceptable, it is possible to collect the tank effluents and carry the wastewater to an oxidation pond for treatment. Operation is improved by connecting the shower and washbasin drains to the aqua privy to maintain the water seal and thus prevent odors and fly and mosquito breeding. Since solids are removed in the aqua privy tank, 4-in. sewers designed for a flow velocity of 1 fps may be used. The flat grade makes possible reduced excavation cost. A 4-in. sewer designed for 15 gpd/capita and for peaks three times mean flow can serve a population of up to 1000. Oxidation ponds can also be used for the disposal of night soil. A loading of 145 lb of BOD/acre/day appears reasonable. Oxidation pond design details are given on page 326. Other low-cost systems include irrigation systems and oxidation ditches.

Inspection During Construction

The best of design is no better than the construction. On installations for which engineering or architectural plans are prepared, it is advisable to make arrangements for retaining competent inspection and supervision during construction. This would include the checking of grades and elevations, inspection of quality of material and construction, infiltration tests, and full compliance with the treatment plant, pumping station, and sewerage plans and specifications. On small installations the regulatory

[33] G. J. Stander and P. G. Meiring, "Employing Oxidation Ponds For Low-Cost Sanitation," *J. Water Pollution Control Fed.*, **37**, No. 7, 1025–1033 (July 1965). Also L. J. Vincent, W. E. Algie, and G. van R. Marais, "A System of Sanitation for Low Cost, High-Density Housing," Document submitted to WHO by the African Housing Board, Ridgeway, Lusaka, Northern Rhodesia, 1961, 29 pp., 18 figures.

agency, if properly staffed, may be able to perform some of the inspection—but only in an advisory capacity for compliance with design.

Operation Control

A common weakness in both small and large sewage treatment plants is failure to properly operate and maintain a well-designed plant. It is important for the design engineer to impress upon the owner, be it an individual or a municipality, that sewage treatment plants require daily attention by a qualified person or persons. Repairs should be made promptly, and preventive maintenance should be the rule. If this is not done, the money spent is wasted; major damage requiring expensive repairs can result before the cause is detected, and the purpose of the treatment plant to prevent water pollution is nullified.

Wherever a sewage treatment plant is in use, an operation report should be kept and entries made daily. The regulatory agency usually provides forms for this purpose. This will help assure daily inspection and continuous operation of the plant as it was designed to function. An equipment maintenance schedule or checklist is also found to be very worthwhile in reducing expensive repairs and equipment replacement. A complete set of spare parts for pumps, special dosing devices, and chlorinators are necessary if extended interruptions are to be kept at a minimum.

TYPICAL DESIGNS OF SMALL PLANTS

Design for a Small Community

Three design analyses are given below for a plant to serve 150 persons at 100 gcd = 15,000 gpd.

1. *Standard-rate trickling filter plant with Imhoff tank.* (Design based on average daily flow of 15,000 gpd.)

  a. Flowing through channel provides 2.5 hr detention.

$$\frac{15,000}{24} \times 2.5 = 1560 \text{ gal} = 209 \text{ ft}^3$$

  b. Sludge storage at 5 ft³ per capita = 5 × 150 = 750 ft³.
  c. Sludge drying beds at 1.25 ft² per capita = 150 × 1.25 = 188 ft².
  d. Trickling filter loading at 400 lb of BOD/acre-ft = 0.25 lb per yd³. Loading based on 0.17 lb of BOD/capita with 35 percent removal in primary settling = 150 × 0.17 × 0.65 = 16.6 lb/day. Filter volume required = $\dfrac{400}{43,560} = \dfrac{16.6}{x}$; $x =$

1800 ft³. Hence the required filter diameter, assuming a 6-ft depth $= D =$
$\sqrt{\dfrac{180 \times 4}{\pi \times 6}} = 19.5$ ft, say 20 ft. The volumetric loading $= \dfrac{15,000}{\dfrac{\pi \times 20 \times 20}{4}} = \dfrac{x}{43,500}$,

which $= 2,080,000$ gpd/acre on a 6-ft deep filter.

e. Final settling provides 2 hr detention. $\dfrac{15,000}{24} \times 2 = 1250$ gal $= 167$ ft³.

With a surface settling rate $= 1000$ gpd/ft² $= \dfrac{180 \times \text{tank depth}}{2\text{-hr detention}}$; tank depth $=$ 11.1 ft.

f. If the BOD in the raw sewage is 200 mg/l, and the Imhoff tank removes 35 percent, the applied BOD $= 0.65 \times 200 = 130$ mg/l. According to the National Research Council Sanitary Engineering Committee formulae* a filter loaded at 400 lb of BOD per acre-ft will produce an average settled effluent containing 14 percent of that applied, or $0.14 \times 130 = 18$ mg/l.

2. High-rate trickling filter plant with Imhoff tank.

a. Flowing through channel same as with standard rate filter $= 209$ ft³.

b. Sludge storage at 8 ft³/capita $= 8 \times 150 = 1200$ ft³.

c. Sludge drying beds at 1.50 ft²/capita $= 150 \times 1.50 = 225$ ft².

d. Trickling filter loading at 3000 lb of BOD/acre-ft $= 1.86$ lb/yd³. Loading based on 0.17 lb of BOD per capita with 35 percent removal in primary settling $= 150 \times 0.17 \times 0.65 = 16.6$ lb/day. The BOD in the raw sewage is $150 \times 0.17 = 25.5$ lb. Filter volume required $= \dfrac{3000}{43,560} = \dfrac{25.5}{x}$; $x = 370$ ft³.

Hence the required filter diameter, assuming 3.25 ft depth $= D = \sqrt{\dfrac{370 \times 4}{\pi \times 3.25}} =$ 12.0 ft. The volumetric surface loading on a 12.0-ft diameter filter with influent $+$ recirculation ($I + R = 1 + 1 = 2$) of $2(15,000) = 30,000$ gal/day is

$$\frac{x}{43,560} = \frac{30,000}{\dfrac{\pi \times 12 \times 12}{4}}$$

$x = 11,500,000$ gpd/acre on a 3.25-ft deep filter.

e. Final settling provides 2-hr detention at flow $I + R$.

$$\frac{30,000}{24} \times 2 = 2500 \text{ gal} = 334 \text{ ft}^3$$

f. Without recirculation, an applied BOD of 130 mg/l (0.65×200), at a rate of 3000 lb/acre-ft will be reduced to $0.32 \times 130 = 42$ mg/l in the settled effluent.

* $E = \dfrac{100}{1 + 0.0085 \sqrt{u}}$; $E =$ percent BOD removed, standard filter and final clarifier. $u =$ Filter loading in pounds BOD per acre-ft.

With recirculation of $R/I = 1$ the efficiency of the high-rate filter and clarifier can be determined from the following formulas:[34]

$$F = \frac{1 + \dfrac{R}{I}}{\left(1 + 0.1\dfrac{R}{I}\right)^2}$$

where F = recirculation factor
R = volume of sewage recirculated = 1
I = volume of raw sewage = 1

$$F = \frac{1 + 1}{(1 + 0.1)^2} = 1.65$$

and from

$$u = \frac{w}{VF}$$

where u = unit loading on high-rate filter in lb of BOD/acre-ft
w = total BOD to filter in lb/day = 16.6
V = filter volume in acre-ft based on raw sewage strength = 0.0084
F = recirculation factor = 1.65

$$u = \frac{16.6}{0.0084 \times 1.65} = 1183 \text{ lb/acre-ft}$$

and from

$$E = \frac{100}{1 + 0.0085 \sqrt{u}}$$

where E = percent BOD removed by a high-rate filter and clarifier.

$$E = \frac{100}{1 + 0.0085 \sqrt{1183}} = 77 \text{ percent}$$

Hence, the BOD will be reduced to $(1 - 0.77)130 = 30$ mg/l.

3. Intermittent sand filter plant with Imhoff tank or septic tank.

a. Flowing through channel of Imhoff tank provides 2.5 hr detention = 209 ft³.
b. Sludge storage at 4 ft³ capita = 4 × 150 = 600 ft³.
c. Sludge drying bed provides 188 ft².
OR: Septic tank provides 24 hr detention = 15,000 gal = 2000 ft³.
d. Sand filter, covered, designed for loading of 50,000 gpd/acre.

[34] W. A. Hardenbergh, *Sewerage and Sewage Treatment,* 3rd ed., International Textbook Co., Scranton, Pa., 1950, p. 328.

Filter area is

$$\frac{50,000}{43,560} = \frac{15,000}{x}; x = 13,000 \text{ ft}^2.$$

If filter is open, the required area = 6550 ft². Make filters in two sections. Provide dosing tank to dose each covered filter section at volume equal to 75 percent of the capacity of the distributor laterals or to dose each open filter section to depth of 2 to 4 in.

If the efficiency of BOD removal of a sand filter is 90 percent, the BOD of the effluent would be 130 × 0.10 = 13 mg/l.

Children's Camp Design

A typical sewerage layout to serve a camp dining room and central bathhouse is shown in Figure 4–30. This boys' camp is assumed to be located on the watershed of a public water supply. The soil is a clay loam. There are no unusual plumbing fixtures at the camp. A basis for the design is given below.

1. Camp capacity is 96 campers and 27 staff = 123 total   123 total
2. Design flow based on 40 gal/capita,

$$123 \times 40 = 4920 \text{ gpd}$$ 4920 gpd

3. Septic-tank capacity designed for a 12-hr detention period. Total flow takes place in 16 hr. Liquid volume = x.

$$\frac{4920 \text{ gal}}{16 \text{ hr}} = \frac{x \text{ gal}}{12 \text{ hr}}; \quad x = 3690 \text{ gal capacity}$$ 3690 gal

Make septic tank 6′ × 17′ × 5′ liquid depth.   See Table 4–7 and Figure 4–9.

4. Grease trap designed for 2½ gal/person served at mealtime.

$$\text{Capacity} = 2\tfrac{1}{2} \times 123 = 308 \text{ gal}$$ 308 gal

Make grease trap 2½′ × 7′ × 2½′ liquid depth. See Figure 4–7.

5. Sand filter designed for filtration rate of 1.15 gpd/ft²

$$\text{The total required area} = \frac{4920}{1.15} = 4278 \text{ ft}^2$$ 4278 ft²

Make in two sections, each 2160 ft² = 2(30 × 72). Each section will consist of 5 distributors 69 ft long, 6 ft apart. See Figures 4–22 and 4–30.

6. Dosing tank to be provided with alternating siphons. Volume of dose is to be 60 percent of volume of 4-in. distributors in each section = (5 × 69) × (0.653 × 0.60) = 135 gal   135 gal

7. Size of siphon = twice maximum flow to the dosing tank. Probable maximum flow assumed as one-half daily flow in one

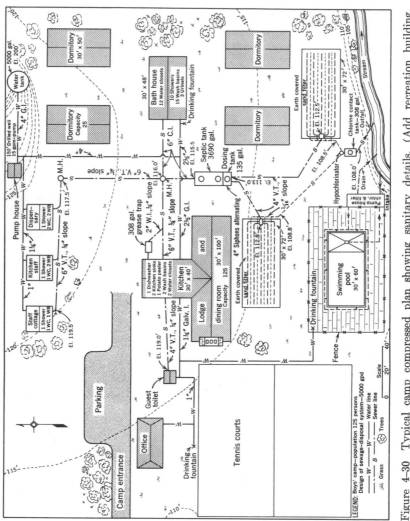

Figure 4-30 Typical camp compressed plan showing sanitary details. (Add recreation building, craft shop, nature lodge, garage, recreation area, council ring, etc.)

hour. $\dfrac{4920 \times \frac{1}{2}}{60} = 41$ gpm. (Check with fixture unit basis.) A
4-in siphon has a minimum discharge of 102 gpm. Use 4-in. siphon;
it has a drawdown of 17 in. If the dosing tank is made 6 ft wide,
and the discharge liquid depth is 17 in. or 1.41 ft, the dosing tank
length is

$$\frac{135}{7.48} \times \frac{1}{6} \times \frac{1}{1.41} = 2.14 \text{ ft} \qquad\qquad 2.14 \text{ ft}$$

See Figure 4–27.

8. Provide a chlorine contact tank giving a minimum detention
of 15 min. Say flow conditions given in paragraphs 6 and 7 are
leveled off through filter and reach contact tank at average rate
of 20.5 gpm under peak flow conditions; make the contact tank
capacity equal to

$$20.5 \times 15 = 308 \text{ gal} \qquad\qquad 308 \text{ gal}$$

9. Chlorination treatment will be required to fully protect the
receiving stream, which serves as a source of domestic water supply. Provide a hypochlorinator for the flow to be treated, having
positive feed and operating continuously, with point of applica-
tion in inspection manhole receiving filter drainage, 20 ft ahead   Use a positive feed
of chlorine contact tank.   hypochlorinator

10. Increase the volume of the septic tank by 50 percent and
the area of the sand bed by 20 percent if a kitchen garbage grinder
is installed.

Referring to the above design, an alternate design based on soil percolation tests in clay loam is also investigated. Tests made in the proposed
area for sewage disposal gave percolations of 1 in. in 40 min, 1 in. in 50
min, 1 in. in 50 min, 1 in. in 10 min, and 1 in. in 60 min. Discard the 10-
min test as not being representative. Use 1 in. in 50 min or 0.6 gpd/ft²
in design. A design based on a subsurface tile field absorption system
for the camp mentioned above follows:

1. Camp capacity   123 total
2. Design flow   4920 gpd
3. Septic-tank capacity   3690 gal
4. Grease trap capacity   308 gal
5. Tile field area required:

$$\frac{4920}{0.6} = 8200 \text{ ft}^2 \qquad\qquad 8200 \text{ ft}^2$$

Make in *two* sections *each* having an area of 4100 ft². This can be obtained
if each section consists of 2050 lineal ft of tile in trenches 2 ft wide, 1640
lineal ft of tile in trenches $2\frac{1}{2}$ ft wide, or 1367 lineal ft of tile in trenches

3 ft wide (14 laterals each 98 ft long). Space trenches respectively 6, 7½, or 9 ft apart.

6. Dosing tank to be provided with alternating siphons. Volume of dose is to be 60 percent of volume of 4-in. distributors in each section.

$$= 2050 \times 0.653 \times 0.6 = 803 \text{ gal} \qquad\qquad 803 \text{ gal}$$
$$\text{(for 2-ft wide trench)}$$

$$\text{or } 1640 \times 0.653 \times 0.6 = 643 \text{ gal} \qquad\qquad 643 \text{ gal}$$
$$\text{(for 2½-ft wide trench)}$$

$$\text{or } 1367 \times 0.653 \times 0.6 = 535 \text{ gal} \qquad\qquad 535 \text{ gal}$$
$$\text{(for 3-ft wide trench)}$$

With a 6-ft wide dosing tank and a 4-in. siphon having a drawdown of 1.41 ft, for an 803-gal discharge, the tank length is

$$\frac{803}{7.48} \times \frac{1}{6} \times \frac{1}{1.41} = 12.7 \text{ ft,}$$

for a 643 gal discharge the tank length is

$$\frac{643}{7.48} \times \frac{1}{6} \times \frac{1}{1.41} = 10.2 \text{ ft,}$$

and for a 535-gal discharge the tank length is

$$\frac{535}{7.48} \times \frac{1}{6} \times \frac{1}{1.41} = 8.5 \text{ ft.}$$

7. Size of siphon should give twice the maximum flow to the dosing tank. Use a 4-in. siphon.

8. Increase the volume of the septic tank by 50 percent and the area of the tile field by 20 percent if a kitchen garbage grinder is installed.

Compare the cost of installation, operation, and maintenance of the sand filter system including chlorination and the subsurface tile field system to determine which system to install. Other alternatives such as a trickling filter or aeration unit may also be studied and compared before a decision is made.

Specific construction details of grease traps, manholes, septic tanks, dosing tanks, sand filters, distribution boxes, and chlorine contact tanks are also given in Chapter 4.

Town Highway Building Design

1. Population:　70 persons
2. Fixtures:　　5 showers
　　　　　　　　5 water closets (flush valve)
　　　　　　　　6 washbasins
　　　　　　　　2 urinals (flush valve)

3. Soil tests: Percolation zero, tight clay soil
4. Design flow—Different bases:
    a. Fixture unit basis (see Table 4–4)

       5 showers @ 3   = 15
       5 water closets @ 8 = 40 (flush valve operated)
       6 washbasins @ 2 = 12
       2 urinals @ 4   = 8
          Total   = 75 fixture units

For 8-hr operation, say one fixture unit is 35 gal, hence 75 × 35 = 2625 gpd.
    b. Usage per capita basis
       70 persons @ 30 gcd = 2100 gpd
    c. Fixture hourly flow basis (from Table 4–3):

       5 showers @ 150 gal/hr   = 750
       5 water closets @ 36 gal/hr = 180
       6 washbasins @ 15 gal/hr   =   90
       2 urinals @ 10 gal/hr   =   20
          Total   1040 gal/hr

With an hourly peak in morning and afternoon, and usage during day, say practically all flow takes place in $2\frac{1}{2}$ hr.

$$\text{Daily flow} = 1040 \times 2\frac{1}{2} = 2600 \text{ gpd}$$

    d. The average estimated flow $= \dfrac{2625 + 2100\ 2600}{3} = 2442$ gpd average, say

2500 gpd.

5. Septic tank: Since all flow takes place in 8 hr, provide a minimum 2500-gal septic tank. (See Table 4–7 and Figure 4–9.)
    Make septic tank 6 ft wide, $14\frac{1}{2}$ ft long, 4 ft liquid depth. Dimensions are inside.
6. Sand filter:

$$\text{Required area} = \frac{2500}{1.15} = 2170 \text{ ft}^2$$

    Make bed: 36′ × 60′. (See Figure 4–22 and 4–23.)
7. Dosing tank: Dose field to 70 percent of interior capacity of distributors, which extend to within 18 in. of end of bed, 6 ft o.c., 3 ft from edge of bed.

6 laterals × 57′ long × 0.653 (vol. 1′ of 4″ pipe) × 70 percent
                                   = 6 × 57 × 0.653 × 0.7 = 156 gal

8. Size of siphon: Siphon should preferably have minimum capacity $1\frac{1}{2}$ to 2 times peak flow. Using National Bureau of Standards probability curves in Figure 3–20, with 75 fixture units; demand = ±60 gpm. Use a 4-in. Miller siphon, which has a drawdown of 17 in. and a total loss head from high water to invert of discharge pipe of $25\frac{1}{2}$ in. (See Figure 4–27.)
    Make dosing tank 6 ft wide and 2.45 ft long, attached to septic tank, with 17-in. liquid drawdown.
9. Size of chlorine contact tank, if required: Provide for at least 15 min detention. Make tank equal to dosing tank dose = 156 gal.

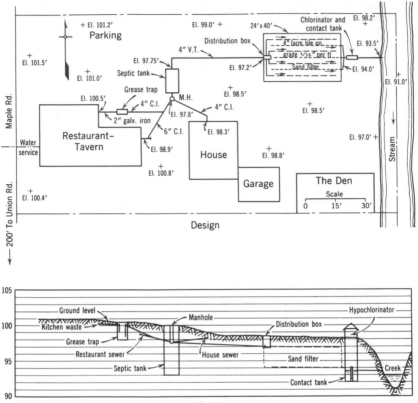

Design

Profile

Figure 4–31 Typical plan and profile of restaurant treatment plant.

House, two-bedroom, at 150 gal = 300 gpd

Restaurant—fixtures:

  2 flush toilets (tank) = 10 fixture units,
  2 wash basins   = 4 fixture units,
  2 kitchen sinks   = 8 fixture units.

          Total   22 fixture units.

One fixture unit = 35 gal per 8-hr operation.

Flow from 22 fixture units = 22 × 35 = 770 gpd.

        Total flow = 770 + 300 = 1070 gpd.

Soil tests show no percolation.

(If period of operation is 16 hr increase fixture unit
  flow from 35 to 70 gal.)

Septic tank capacity = 1,100 gal.

Grease trap = seat. cap. × 3 gal.

         = 40 × 3 = 120 gal.

$$\text{Sand filter} = \frac{1070 \text{ gal}}{1.15 \text{ gpd/ft}^2}$$

         = 930 ft²

Chlorine contact tank = 60 gal,
  15 min detention at peak flow.

NOTE: See sheet 2 for details of
  treatment units. See table
  for fixture unit values.

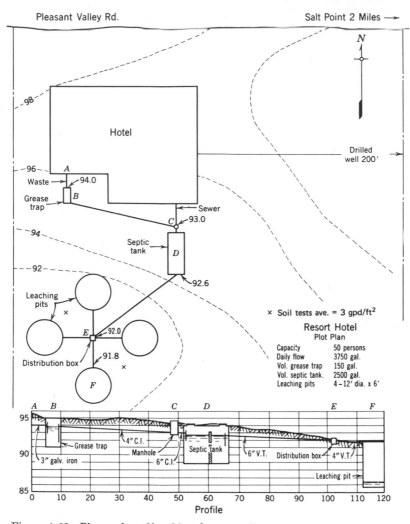

Figure 4–32   Plan and profile of hotel sewage disposal system.

Figure 4–31 shows a similar design for a sand filter to serve a small resturant and Figure 4–32 shows a leaching pit system.

Elementary School Design

1. Design basis: No. persons—600.
Soil tests—rock at 12 to 30 in.; soil gravelly loam with clay; shallow soil tests = 1 in. in 20 min. Suitable porous fill containing topsoil and gravelly loam is required to bring bottom of leaching system at least 2 ft above rock. Use assumed percolation rate of 0.5 gpd/ft². Pumping is required!

2. Design flow: 600 persons at 20 gcd = 12,000 gpd.

3. Septic tank: Minimum size to provide 12-hr detention period, assuming all flow takes place between 8:30 A.M. and 5:30 P.M., a period of 9 hr = x. Then $\dfrac{12,000}{9} = \dfrac{x}{12}$; $x = 16,000$ gal.

4. Tile field:

$$\frac{12,000 \text{ gpd}}{0.5 \text{ gpd/ft}^2} = 24,000 \text{ sq ft} = 12,000 \text{ lineal ft in 24-in. trench}$$

Divide into three sections, 4000 lineal ft each

5. Tile field dose:

4000 (lineal ft) $\times$ 0.653 (gal/1-ft 4-in. tile) $\times$ 0.75 (75 percent dose)
$$= 1960 \text{ gal (make pump sump ample size)}$$

6. Pump capacity: Probable flow on fixture basis using Table 4–4.

5 urinals @ 4	=	20
28 water closets @ 8	=	224
38 washbasins @ 2	=	76
27 service sinks @ 3	=	81
12 showers @ 3	=	36
4 drinking fountains @ ½	=	2
Total		439 fixture units

Probable maximum flow = 135 gpm. See Figure 3–20. Make pump capacity 150 gpm.

7. Controls: Provide three 150 gpm pumps, each discharging to a separate tile field section, and one standby hooked up and valved for series operation to each tile field. Provide high-water alarm and timer in series with motor to shut off and throw in the next pump after 13 min of continuous operation.

NOTE: 1. No receiving stream or sewer in vicinity.

2. If a kitchen garbage grinder is installed increase the volume of the septic tank by 50 percent and the area of the tile field by 20 percent, and provide grease trap on kitchen waste line ahead of septic tank.

Design for a Subdivision (See Figures 4–33, 4–34, and 4–35)

(Imhoff tank, standard-rate filter with secondary settling.)

1. Design population: 100 one-family homes = 350 persons.

2. Sewage flow: 100 gpd/capita = 35,000 gpd.

3. Strength of sewage: 200 mg/l 5-day BOD or 0.17 lb 5-day BOD/capita.

4. Design flow: Assume total flow reaches plant in 16 hr.

5. Imhoff tank design: Required volume of settling compartment to provide 2½ hr detention:

$$\frac{35,000}{16} \times 2.5 \times \frac{1}{7.48} = 732 \text{ ft}^3$$

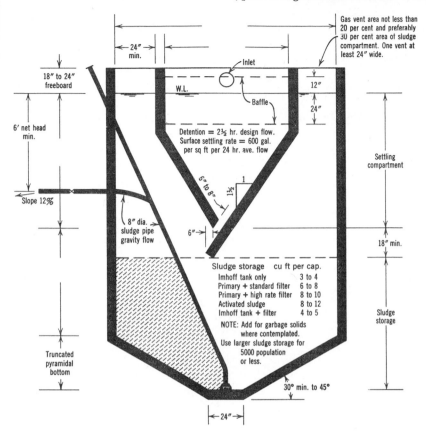

Gas vent area not less than
20 per cent and preferably
30 per cent area of sludge
compartment. One vent at
least 24" wide.

24"
min.

Inlet

18" to 24"
freeboard

W.L.

12"

Baffle

24"

6' net head
min.

Detention = 2½ hr. design flow.
Surface settling rate = 600 gal.
per sq ft per 24 hr. ave. flow

Settling
compartment

6" to 8"

1

1½

Slope 12%

8" dia.
sludge pipe
gravity flow

6"

18" min.

Sludge storage   cu ft per cap.

Imhoff tank only   3 to 4
Primary + standard filter   6 to 8
Primary + high rate filter 8 to 10
Activated sludge   8 to 12
Imhoff tank + filter   4 to 5

NOTE: Add for garbage solids
where contemplated.
Use larger sludge storage for
5000 population
or less.

Sludge
storage

Truncated
pyramidal
bottom

30° min. to 45°

24"

Figure 4–33   Section through Imhoff tank, with design details.

Required sludge storage to provide 8 ft³ per capita:

$$350 \times 8 = 2800 \text{ ft}^3$$

6. Trickling filter design: Volume of filter stone required to provide a loading of 400 lb of BOD/acre-ft/day and not more than 300,000 gpd/acre-ft. Organic loading basis, assuming 35 percent BOD removal in the Imhoff tank = x.

$$\frac{0.65 \times 200}{1,000,000} = \frac{x}{35,000 \times 8.34}; \quad x = 38.0 \text{ lb of BOD/day}$$

Volume of filter stone required based on organic load = Y_1:

$$\frac{400}{43,560} = \frac{38.0}{Y_1}; \quad Y_1 = 4130 \text{ ft}^3$$

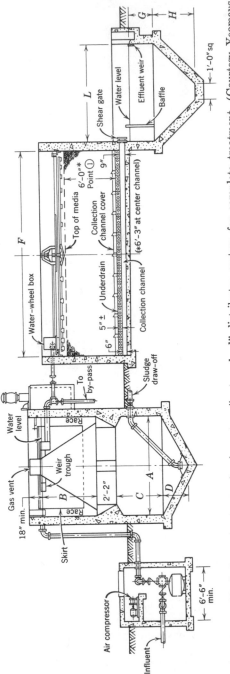

Figure 4-34 Typical arrangement of spiragester—"water-wheel" distributor system for complete treatment. (Courtesy Yeomans Brothers Co. Melrose Park, Ill.)

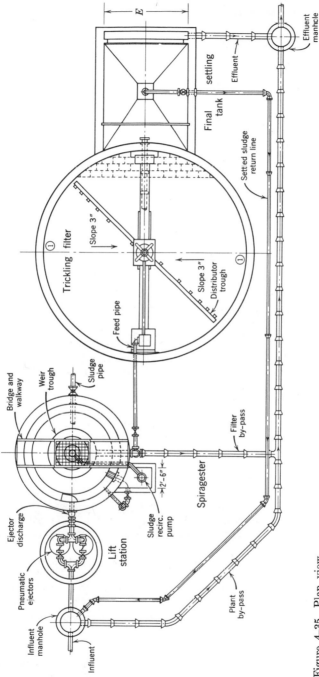

Figure 4-35 Plan view.

341

Volume of filter stone required based on volumetric load = Y_2:

$$\frac{300,000}{43,560} = \frac{35,000}{Y_2}; \qquad Y_2 = 5070 \text{ ft}^3$$

Diameter of 6-ft deep filter to provide, say, 5000 ft³ = D:

$$5000 = \frac{3.14 \times D^2}{4} \times 6; \qquad D = \sqrt{\frac{5000 \times 4}{3.14 \times 6}} = 32.6 \text{ ft}$$

If two filters proposed, each filter would have a diameter equal to

$$\sqrt{\frac{2500 \times 4}{3.14 \times 6}} = 23 \text{ ft}$$

7. Secondary settling: Required volume above sludge hopper to provide 2-hr detention:

$$\frac{35,000}{16} \times 2 \times \frac{1}{7.48} = 585 \text{ ft}^3$$

With a surface settling rate = 1,000 gpd/ft² = $\dfrac{180 \times \text{tank depth}}{\text{2-hr detention}}$; depth = 11.1 ft. Diameter = 8.2 ft. Make hopper bottom for temporary sludge storage and slope walls one on one. Pump sludge to Imhoff tank inlet.

8. Sludge drying bed: Provide 1.25 ft² of open sand bed per capita. The required area = 350 × 1.25 = 438 ft². Provide two beds each 11′ × 20′.

Toll Road Service-Area Design

1. Wastes containing high concentrations of grease and detergents, typical of superhighway service areas, cannot be successfully treated on sand filters unless first given adequate preliminary treatment. High concentrations of detergents cause the emulsification of grease and an increase in the amount of material in colloidal suspension, thereby resulting in a carry-over of grease and solids with consequent clogging. On the other hand, if the design in a small installation provides for proper grease separation and primary septic tank treatment of 2 to 3 days, the carry-over of solids is reduced. Careful maintenance is of course essential. An Imhoff tank plant preceded by adequate grease separation, such as a septic-tank-type grease trap, has merit in the larger installation.

2. Dixon and Kaufman report satisfactory results if the sewage is passed through an anaerobic digestion tank before being given high-rate trickling filter treatment with recirculation at a high ratio of final effluent to the primary settling tank following the digester.[35] A 24-hr displacement period based on average daily sewage flow is provided in the digester maintained at a temperature of 70°F. A 40 percent BOD reduction is reported in the digester.

[35] Gale G. Dixon and Herbert L. Kaufman, "Turnpike Sewage Treatment Plants," *Sewage & Ind. Wastes*, **28**, No. 3, 245–254 (March 1956).

3. Design of service areas in connection with toll roads was based on a BOD of 500 mg/l, suspended solids of 300 mg/l, grease of 130 mg/l, and commercial detergent concentration of 350 mg/l. The following design flows were used:

Flows	Per Counter Seat	Per Table Seat
Daily average	350 gpd	150 gpd
Peak day	630	270
Extreme peak	1890	810
Minimum flow	0	0

4. Experience with superhighway service and restaurant-area sewage treatment plants indicates that the design flows and sewage strengths given above are somewhat high. It also appears that because of the extreme variations in flows there is considerable advantage in the recirculation of primary and secondary settling tank effluents to the primary inlet. This would prevent the development of anaerobic conditions in these tanks and also add to the degree of treatment.

5. Kansas turnpike sewage treatment plant design used the following criteria:[36]
Flow = 350 gpd/counter seat plus 150 gpd/table seat. Ten percent of the cars passing a service area will enter and will contribute 15 to 20 gal/person or customer. The design flow is assumed to equal the water consumption. The flow is 200 percent of the daily average at noon and 160 percent of the daily average at 6 P.M.

Sewage characteristics: BOD = 600 mg/l, suspended solids = 300 to 450 mg/l with 90 percent volatile matter, pH = 9.5, grease = 100 mg/l, active detergent = 100 mg/l, temperature = 70 to 85°F.

Treatment: To consist of comminutor, Imhoff tank with overflow rate 1000 gpd/ft², trickling filter loaded at not greater than 650 lb BOD/acre-ft, final settling overflow rate not greater than 1000 gpd/ft² based on maximum rate of raw sewage flow, continuous sludge removal, and recirculation not less than 1 to 1. Sand filter dosage not greater than 80,000 gpd/acre when open, and 40,000 gpd/acre when covered. Sand ES 0.5 to 0.75 mm. Grease trap on kitchen waste line, 3 compartments and 500-gal capacity.

6. A study of New York State Thruway service areas showed the following:[37]

Average water use—9.16 gal/vehicle stopping with a range of 7.38 to 11.58.

Average stay in parking area—30 min/vehicle, 25 min for coffee shop and 20 min for snack bar.

Peak hour factor—8 percent of peak-day traffic.

Peak capacity of women's facilities—52 persons/hr for water closet.

Rest room usage—53 percent women at peak periods; 42.5 percent at off-peak periods.

[36] Clifford Sharp, "Kansas Turnpike Sewage Treatment Plants," *Public Works,* 88, 142 (August 1957).
[37] Clayton H. Billings and Irwin P. Sanders, "Determining the Future Needs of Highway Rest Areas," *Public Works,* 97, 108–111 (October 1966).

7. Another report gives the following information:[38]
Average flow from eight service areas—100 gpm.
Recirculation flow—100 gpm.
5-day BOD—500 mg/l.
Suspended solids—200 mg/l.
2-stage trickling filter—recirculation through primary settling tank and directly to secondary filter for continuous dosage; raw sewage enters digester for anaerobic digestion of grease and detergents.
Removal—BOD 95 to 99 percent, suspended solids 90 to 99 percent.
Final settling tank—detention 127 min, surface settling rate 841 gpd/ft².
Chlorine contact time—21 min.
Sludge drying bed—1.1 ft²/capita.
Combined settling digestion tank—4.7 ft³/capita, surface settling rate 568 gpd/ft², detention 400 min.
Primary trickling filter—159 gpd/ft², BOD 1.5 lb/yd³.
Secondary trickling filter—159 gpd/ft², BOD 0.39 lb/yd³.

SEWAGE WORKS DESIGN—LARGE SYSTEMS

General

The need for sewerage studies that take into consideration the broad principles of comprehensive community planning is discussed in Chapter 2. Also discussed is the importance of regional and area-wide sewerage planning (preliminary) that recognizes the extent of present and future service areas, the established water quality and effluent standards, and alternative solutions with their first costs and total annual costs. This information is needed to assist local officials in making a decision to proceed with the design and construction of a specific sewerage system, including treatment plant. These are essential first steps to ensure that the proposed construction will meet community, state, and national goals and objectives.

The degree and type of treatment to be provided is dependent on many factors (See Figure 4–1), a major one being the water quality standards established for the receiving water. A classification system for freshwaters and saltwaters is given in Tables 4–14 and 4–15. Also important, however, is the future as well as the existing upstream and downstream water usage, the minimum flows,[39] the types of sewers and wastewater characteristics, the assimilative capacity of the receiving waters, and the capability of the community to finance, operate, and maintain the facility as intended.

---

[38] Henry W. Haunstein, "Sewage and Treatment for Turnpike Service Areas," *Water and Sewage Works,* **107,** No. 3, 89–90 (March 1960).
[39] Minimum average seven-consecutive-day flow of the receiving stream once in ten years.

The design details of large sewage treatment plants and sewer systems is beyond the scope of this text. Some of the major design elements, however, are given here for general information. Federal and state regulatory agencies have recommended standards and guidelines.[40]

A preliminary basis of design, which has been used for many years where drinking water supplies are not directly involved and other local factors are not adversely affected, is given below. This dilution principle *by itself* is no longer acceptable in the United States. See also pages 261 and below.

Dilution Water Available*	Required Degree of Treatment	Dilution Factor†
3.5 to 5.0 ft²/sec or more	Effective sedimentation	22.6 –32.3
2.0 to 3.0	Chemical precipitation	12.9 –19.4
1.0 to 2.0	High-rate trickling filters	6.5 –12.9
1.0 to 1.5	Conventional trickling filters	6.5 – 8.7
0.5 to 1.0	Activated sludge	3.2 – 6.5
0.1 to 0.5	Intermittent sand filter	0.65– 3.2

* Per 1000 equivalent population.
† Based on 100 gpd/capita.

The dilution water available is in terms of $ft^3/sec/1000$ equivalent population, with the water 100 percent saturated with oxygen. Under special conditions a lesser volume of dilution water may be sufficient to prevent the development of unsatisfactory conditions, such as when the stream has a turbulent flow or joins a larger watercourse after only a few hours' flow. On the other hand 5 times the given dilution may be required if flow is through a densely populated area. In the final analysis the stream usage or classification will determine the degree of treatment required. Present opinion in the United States is that all sewage should receive a minimum of secondary treatment.

In Britain, the Ministry of Housing and Local Government reaffirmed in 1966 the Royal Commission's "general standard" as a "norm" for sewage effluents: 5-day BOD 20 mg/l and suspended solids 30 mg/l with a dilution factor of 9 to 150 volumes in the receiving watercourse

[40] *Recommended Standards for Sewage Works,* Great Lakes–Upper Mississippi River Board of State Sanitary Engineers, 1968 edition; *Federal Guidelines—Design, Operation and Maintenance of Waste Water Treatment Facilities,* Federal Water Quality Administration, U.S. Dept. of the Interior, September 1970.

TABLE 4-14   NEW ENGLAND INTERSTATE WATER POLLUTION
QUALITY FOR
(As Revised and Adopted

Water Use Classes	Description	STANDARDS OF WATER QUALITY	
		Dissolved Oxygen	Sludge Deposits, Solid Refuse, Floating Solids, Oils, Grease and Scum
Class A	Suitable for water supply with treatment by disinfection only, and all other water uses; character uniformly excellent. (See Notes 1 and 3.)	As naturally occurs.	None other than of natural origin.
Class B	Suitable for bathing and other primary contact recreation. Acceptable for public water supply with appropriate treatment. (See Note 3.) Suitable for agricultural and certain industrial process cooling uses. Suitable as an excellent fish and wildlife habitat. Excellent aesthetic value.	Minimum 5 mg/l at any time. Normal seasonal and diurnal variations above 5 mg/l will be maintained.	None Allowable.
Class C	Suitable for fish and wildlife habitat, boating, fishing, and certain industrial process and cooling uses.	Minimum 5 mg/l any time. Normal seasonal and diurnal variations above 5 mg/l will be maintained. For sluggish, eutrophic waters, not less than 3 mg/l at any time. Normal seasonal and diurnal variations above 3 mg/l will be maintained.	None (See Note 2.)
Class D	Suitable for navigation, power, certain industrial processes and cooling uses, and migration of fish. Good aesthetic value.	2 mg/l at any time. Normal seasonal and diurnal variations above 2 mg/l will be maintained.	None (See Note 2.)

Notes: 1. Class A waters reserved for water supply may be subject to restricted use by state and local regulation.

2. Sludge deposits, floating solids, oils, grease, and scum shall not be allowed except in that amount that may result from the discharge of appropriately designed and operated sewage and/or industrial waste treatment plants.

3. Waters shall be free from chemical and radiological constituents in concentrations or combinations that would be harmful to human, animal, or aquatic life for the most sensitive and governing water class use. In areas where fisheries are the governing considerations and approved limits have not been established, bioassays shall be performed as required by the appropriate agencies. For public drinking water supplies the raw water sources must be of such quality that the U.S. Public Health Service or higher appropriate state agency limits for finished water can be met after conventional water treatment.

4. Water quality analyses on interstate waters should include tests for both total and fecal coliform supported by sanitary surveys for the development of background data and

CONTROL COMMISSION CLASSIFICATION AND STANDARDS OF
INTERSTATE WATERS
December 12, 1969)

–INLAND WATERS

Color and Turbidity	Coliform Bacteria	Taste and Odor	pH	Allowable Temperature Increase
None other than of natural origin.	Total coliforms: not to exceed a monthly arithmetic mean of 100 per 100 ml. (See Note 4.)	None other than of natural origin.	As naturally occurs	None other than of natural origin.
None in such concentrations that would impair any usages specifically assigned to this Class.	Total coliforms: not to exceed a monthly median of 1,000 per ml nor more than 2,400 per 100 ml in more than 20% of samples collected. (See Note 4.)	None in such concentrations that would impair any usages specifically assigned to this Class nor cause taste and odor in edible fish.	6.5 – 8.0	Only such increases that will not impair any usages specifically assigned to this Class. (See Note 5.)
None in such concentrations that would impair any usages specifically assigned to this Class.	(See Note 4.)	None in such concentrations that would impair any usages specifically assigned to this Class nor cause taste and odor in edible fish.	6.0 – 8.5	Only such increases that will not impair any usages specifically assigned to this Class. (See Note 5.)
None in such concentrations that would impair any usages specifically assigned to this Class.	None in such concentrations that would impair any usages specifically assigned to this Class.	None in such concentrations that would impair any usages specifically assigned to this Class.	6.0 – 9.0	Only such increases that will not impair any usages specifically assigned to this Class. (See Note 5.)

the establishment of base lines for these two parameters for the specific waters involved.

5. The resultant temperature of a receiving water shall not exceed that for the most sensitive aspect of the most sensitive specie of aquatic life in that water. Normal seasonal and diurnal temperature variations that existed before the addition of heat of artificial origin shall be maintained. In addition: (a) For streams, the allowable total temperature rise shall not exceed 5° F above ambient unless it can be demonstrated to the satisfaction of the state regulatory agencies that greater rises at various times will not be harmful to fish, other aquatic life, or other uses. (b) For lakes, there shall be no discharge or withdrawal of cooling waters from the hypolimnion unless it can be demonstrated to the satisfaction of the state regulatory agencies that such discharges or withdrawals will not be harmful to fish, other aquatic life, or other uses. A heated discharge to a lake shall not raise the temperature more than 3° F at the surface immediately outside a designated mixing zone. (More restrictive requirements may be adopted by a state if deemed necessary to meet the Class use of the receiving waters.)

TABLE 4-15   NEW ENGLAND INTERSTATE WATER POLLUTION
QUALITY FOR COASTAL
(As Revised and Adopted

		STANDARDS OF WATER QUALITY	
Water Use Classes	Description	Dissolved Oxygen	Sludge Deposits, Solid Refuse, Floating Solids, Oils, Grease, and Scum
Class SA	Suitable for all seawater uses including shellfish harvesting for direct human consumption (approved shellfish areas), bathing, and other water contact sports; excellent aesthetic value.	Not less than 5.0 mg/l at any time. Normal seasonal and diurnal variations above 5 mg/l will be maintained.	None allowable
Class SB	Suitable for bathing, other recreational purposes, industrial cooling, and shellfish harvesting for human consumption after depuration; excellent fish and wildlife habitat; good aesthetic value.	Not less than 5.0 mg/l at any time. Normal seasonal and diurnal variations above 5 mg/l will be maintained.	None allowable
Class SC	Suitable fish, shellfish, and wildlife habitat; suitable for recreational boating and industrial cooling; good aesthetic value.	Not less than 5 mg/l during daylight hours nor less than 4 mg/l at any time. Normal seasonal and diurnal variations above 4 mg/l will be maintained.	None except that amount that may result from the discharge from a waste treatment facility providing appropriate treatment.
Class SD	Suitable for navigation, power, certain industrial processes and cooling uses, and migration of fish; good aesthetic value.	A minimum of 2 mg/l at any time. Normal seasonal and diurnal variations above 2 mg/l will be maintained.	None except for such small amounts that may result from the discharge of appropriately treated sewage and/or industrial waste effluents.

*Coastal and marine waters are those generally subject to the rise and fall of the tide.
Notes: S.1 Surveys to determine coliform concentrations shall include those areas most probably exposed to fecal contamination during the most unfavorable hydrographic and pollution conditions. Water quality analyses on interstate marine and coastal waters should include tests for both total and fecal coliform supported by sanitary surveys for the development of background data and the establishment of a base line for these two parameters for the specific waters involved.

S.2 Waters shall be free from chemical and radiological concentrations or combinations that would be harmful to human, animal, or aquatic life or that would make the waters unsafe or unsuitable for fish or shellfish or their propagation, impair the palatability of same, or impair the water for any other usage. In areas where fisheries are the governing considerations and approved limits have not been established, bioassays shall be performed as required by the appropriate agencies.

S.3 The temperature of a receiving water shall not exceed that shown here immediately outside a designated mixing zone: (a) Coastal Waters. The water temperature at the

CONTROL COMMISSION CLASSIFICATION AND STANDARDS OF
AND MARINE WATERS
December 12, 1969)

—MARINE WATERS*

Color and Turbidity	Coliform Bacteria	Taste and Odor	pH	
None in such concentrations that would impair any usages specifically assigned to this Class.	Not to exceed a median MPN of 70 and not more than 10% of the samples shall ordinarily exceed an MPN of 230 per 100 ml for a 5-tube decimal dilution or 330 per 100 ml for a 3-tube decimal dilution. (See Note S.1.)	None allowable	6.8 — 8.5	
None in such concentrations that would impair any usages specifically assigned to this Class.	Not to exceed a median value of 700 per 100 ml and not more than 2,300 per 100 ml in more than 10% of the samples collected in a 30-day period. (See Note S.1.)	None in such concentrations that would impair any usages specifically assigned to this Class and none that would cause taste and odor in edible fish or shellfish.	6.8 — 8.5	Allowable temperature increase. (See Note S.3.) Chemical constituents. (See Note S.2.)
None in such concentrations that would impair any usages specifically assigned to this Class.	None in such concentrations that would impair any usages specifically assigned to this Class.	None in such concentrations that would impair any usages specifically assigned to this Class and none that would cause taste and odor in edible fish or shellfish.	6.8 — 8.5	Radiological constituents. (See Note S.2.)
None in such concentrations that would impair any usages specifically assigned to this Class.	(See Note S.1.)	None in such concentrations that would impair any usages specifically assigned to this Class and none that would cause taste and odor in edible fish or shellfish.	6.0 — 9.0	

surface of coastal waters shall not be raised more than 4°F over the monthly means of maximum daily temperatures from October through June nor more than 1.5°F from July through September. (b) Estuaries or Portions of Estuaries. The water temperature at the surface of an estuary shall not be raised to more than 90°F at any point provided further, at least 50 percent of the cross-sectional area and/or volume of the flow of the estuary including a minimum of 1/3 of the surface as measured from water edge to water edge at any stage of tide, shall not be raised to more than 4°F over the temperature that existed before the addition of heat of artificial origin or a maximum of 83°F, whichever is less. However, during July through September, if the water temperature at the surface of an estuary before the addition of heat of artificial origin is more than 83°F, an increase in temperature not to exceed 1.5°F, at any point of the estuarine passageway as delineated above, may be permitted. (More restrictive requirements may be adopted by a state if deemed necessary for the protection of the most sensitive aspect of the most sensitive specie of aquatic life in that water and/or to meet the Class use of the receiving water.)

having not more than 4.0 mg/l BOD.[41] A higher effluent standard of 10 mg/l BOD and suspended solids may be required if indicated. In any case sewage effluents should not contain any matter likely to render the receiving stream poisonous or injurious to fish.

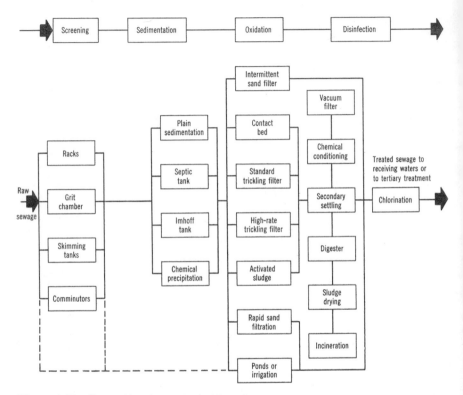

Figure 4–36   Conventional sewage treatment processes.

Sewage treatment processes, and bases of design, are summarized in simplified form in Figure 4–36 and Tables 4–16, 4–17, and 4–18. They are not meant to be complete. Recommended standards for the design and preparation of plans and specifications for sewage works are given in governmental standards and various texts.[42] Plant efficiencies are given below. A secondary sewage treatment process is shown in Figure 4–37. Typical flow diagrams are shown in Figure 4–29.

[41] W. R. Saunders, "Water and Wastes, The Public Health Viewpoint," *Royal Society of Health J.* 247–250 (September/October 1968).
[42] *Recommended Standards for Sewage Works and Federal Guidelines,* op. cit.

TABLE 4-16 PROBABLE CHLORINE DOSAGES TO GIVE A RESIDUAL OF AT LEAST 0.5 mg/l AFTER 15 MIN RETENTION IN AVERAGE SANITARY SEWAGE OR SEWAGE EFFLUENT

Type of Sewage Effluent	Suggested Chlorine Dosages in Milligrams per Liter*					
	N. Y. State (1)	Dunham (2)	Ehlers and Steel (3)	Griffin (4) (5)		Imhoff and Fair (6)
Raw sewage	24			6–12 fresh to stale 12–25 septic	6–25	6–25 fresh to stale and strength
Septic tank		10–25		12–24		—
Imhoff tank or settled sewage	18	5–20	10 or more	5–10 fresh to stale 12–40 septic	5–20	
Trickling filter	12	3–15	3–7	3–5 normal 5–10 poor	3–20	
Activated sludge			5	2–4 normal 5–8 poor	2–20	
Intermittent sand	6	2	2	1–3 normal 3–5 poor	1–10	
Chemical precipitation				3–6	3–20	

(1) New York State Health Dept. Bull. No. 1, Albany, N.Y., 1947.
(2) *Military Preventive Medicine*, Military Publishing Co., Harrisburg, Pa., 1940.
(3) *Municipal and Rural Sanitation*, McGraw-Hill Book Co., Inc., New York, 1965.
(4) *Public Works Magazine*, Ridgewood, N.J. (October 1949), p. 35.
(5) *Operation of Wastewater Treatment Plants*, Water Pollution Control Federation, Washington, D.C., 1970, p. 144.
(6) *Sewage Treatment*, John Wiley & Sons, Inc., New York, 1956.
*12 mg/l = 1 lb per 10,000 gal. Each mg/l chlorine in sewage effluent reduces the BOD about 2 mg/l.

351

TABLE 4–17   SEWAGE TREATMENT

	DE-
Preliminary	**Settling**

Racks:

Area: 200% plus sanitary sewer; 300% plus combined sewer. Bar space: 1″ to 1½″, 2″, or 3″.

Screens:

Net submerged area: 2 ft² per mgd for sanitary sewer; 3 ft² per mgd for combined sewer. Slot opening: $\frac{1}{8}$″ min.

Grit chamber:

Sewage velocity: 1 fps mean, ½ fps min. Detention: 45 to 60 sec, floor 1 ft below outlet. Min. of 2 channels.

Skimming tank:

Air or mechanical agitation with or without chemicals. Detention: 20 min for grease removal, 5 to 15 min for aeration, 30 min for flocculation.

Comminutors:

Duplicate or bypass.

Flow basis:

100 gal per capita plus industrial wastes. Usual to assume total flow reaches small plants in 16 hr.

Plain sedimentation:

Detention period: 2 to 2.5 hr, 1.5 hr with activated sludge. Surface settling rate: 600 gpd/ft². Use rate of 900 with secondary treatment. Hopper slope: 1 on 1. Mech. cleaned sump: 60° slope. Sludge pipe: 6 in. min.

Imhoff tank:

Detention period: 2 to 2.5 hr. Gas vent: 20% total area of tank min. Bottom slope: 1½ vert. to 1 horizontal. Sludge compartment: 3 to 4 ft³ per capita 18 in. below slot; 6 to 10 ft³ per cap. second treatment. Bottom slope: 1 on 1 or 2. Slot & overlap. 8 in. Sludge pipe: 8 in. min. under 6 ft head. Velocity: 1 fpm. Surface settling rate: 600 gpd/ft².

Chemical precipitation:

Ferric chloride. Ferrous sulfate. Ferric sulfate. Lime. Chemical mix.

Final settling tank:

Similar to primary tank: but use surface settling rate of 800 after trickling filter and 600 to 800 after activated sludge treatment. Use a weir loading of not more than 10,000 gpd per lin ft of weir; for flows greater than 1.0 mgd use up to 15,000 gal per day.

Note: Surface settling rate = gpd/ft² = $\dfrac{180 \times \text{tank depth in ft}}{\text{detention in hr}}$

*Sludge digestion will require 68 days at 60°F, 37 days at 71°, 33 days at 86°, 24

PLANT DESIGN FACTORS

SIGN

Biological	Sludge

Intermittent sand:
Filter rate: 50,000 to 100,000 gpad with plain settling and 400,000 gpad with trickling filter or activated sludge. Sand: 24 in. all passing ¼-in. sieve, eff. size 0.35−0.6 mm. Unif. coef. less than 3.5.

Contact bed:
Filter rate: 75,000 to 100,000 gpad per ft.

Trickling filter:
Standard rate: 400 to 600 lb BOD per acre-ft per day; or 2 to 4 mgad, 6 ft depth. High rate: 3000 + lb BOD per acre-ft per day, or 10 to 30 mgad for 3 to 6-ft depth.

Activated sludge:
See Table 4–18.

Rapid filtration:
1 to 2 gpm per ft².

Surface irrigation:
Raw sewage 2,000 to 4,000 gpad. Settled sewage 5,000 to 10,000 gpad.

Stabilization pond:
20 to 50 lb BOD per acre per day, 3- to 5-ft liquid depth, center inlet; variable withdrawal depth, 3-ft freeboard; detention 30 to 180 days; multiple units; winter flow retention. Use up to 50 lb BOD loading in mild climate and 15 to 20 in cold areas.

Digester: *
Capacity: with plain sedimentation 2 to 3 ft³ per cap. heated, or 4 to 6 unheated. With standard trickling filter 3 to 4 ft³ heated and 6 to 8 unheated; 4 to 5 ft³ heated and 8 to 10 ft³ unheated with a high rate filter. With activated sludge 4 to 6 ft³ per cap. heated and 8 to 12 ft³ unheated. Bottom slope: 1 on 2.

Sludge drying bed:
Open: 1 ft² per capita with plain sedimentation, 1.5 ft² with trickling filter. 1.75 ft² with activated sludge, and 2 ft² with chemical coagulation. Glass covered: reduce area by 25%.

Vacuum filtration:
Lbs per ft² per hr, dry solids. Prim. 4−7, trickling filter 3−6, activated sludge 2−5, fresh prim. 4−8; using cloth or coil, respectively.

Centrifuging:

Wet combustion:

Incineration:
Tons per hr depending on moisture and solids content. Temperature 1250° to 1400°F.

Gas:
A properly operated heated digester should produce about 1 ft³ of gas per capita per day from a secondary treatment plant and about 0.8 ft³ from a primary plant. The fuel value of the gas (methane) is about 640 Btu per ft³.

days at 95°, 14 days at 113°. Temperature of 140°F causes caking on pipes.

TABLE 4-18 ACTIVATED SLUDGE PLANT DESIGN FACTORS

Process	Plant Design Flow (mgd)	Aeration Retention Period (hr)	Min Air ft³/lb BOD Aer. Tank Load*	Plant Design (lb BOD/day)	Aerator Loading, lb (BOD/1,000 ft³)	MLSS/lb BOD†	Detention Time (hr)‡	Surface Set. (rate gpd/ft²)
Conventional	To 0.5	7.5	All	To 1,000	30	2/1 to	3.0	600
	0.5 to 1.5	7.5 to 6.0	1,500	1,000 to 3,000	30 to 40	4/1	2.5	700
	1.5 up	6.0		3,000 up	40		2.0	800
Modified or "high rate"	To 0.5	All	400 to	All	All	1/1 or	3.0	600
	0.5 to 1.5	2.5 up	1,500§	2,000 up	100	less	2.5	700
	1.5 up						2.0	800
Step aeration	0.5 to 1.5	7.5 to 5.0	All	1,000 to 3,000	30 to 50	2/1 to	2.5	700
	1.5 up	5.0	1,500	3,000 up	50	5/1	2.0	800
Contact stabilization	To 0.5	3.0‖	All	To 1,000	30	2/1 to	3.6	500
	0.5 to 1.5	3.0 to 2.0‖	1,500	1,000 to 3,000	30 to 50	5/1	3.0	600
	1.5 up	1.5 to 2.0‖		3,000 up	50		2.5	700
Extended aeration#	To 0.05	All	All	All	All	10/1	4.0	300
	0.05 to 0.15	24	2,000	20 per 1,000 ft³	12.5	to	3.6	300
	0.15 up					20/1	3.0	600

Source: Adapted from *Recommended Standards for Sewage Works*, Great Lakes–Upper Mississippi River Board of Sanitary Engineers, 1968 ed.

* Diffused air systems. Aeration equipment shall be capable of maintaining 2 mg/l dissolved oxygen in the mixed liquor for all systems.

† MLSS/lb BOD Normally recommended values at ratio of mixed liquor suspended solids under aeration to BOD loading. All BOD is 5-day. Rate of return sludge shall be variable as needed.

‡ For final settling tank design.

§ Depending on BOD removal expected; may range from 50–90 percent of that applied.

‖ Contact zone — 30–35 percent of total aeration capacity. Reaeration zone comprises the balance of the aeration capacity.

For schools, trailer parks, motels, nursing homes. Efficiency about 70 percent BOD removal.

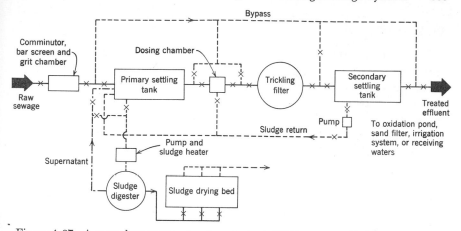

Figure 4-37   A secondary sewage treatment plant. (Units are usually in duplicate.)

Sewage Treatment Plant Unit Combinations and Efficiencies

	Total percent Reduction —Approximation	
Treatment Plant	Suspended Solids	Biochemical Oxygen Demand
Sedimentation plus sand filter	90 to 98	85 to 95
Sedimentation plus standard trickling filter, 600 lb BOD per acre-foot maximum loading	75 to 90	80 to 95
Sedimentation plus single stage high rate trickling filter	50 to 80	35 to 65*
Sedimentation plus two stage high rate trickling filter	70 to 90	80 to 95*
Activated sludge	85 to 95	85 to 95
Chemical treatment	65 to 90	45 to 80
Pre-aeration (1 hr) plus sedimentation	60 to 80	40 to 60
Plain sedimentation	40 to 70	25 to 40
Fine screening	2 to 20	5 to 10
Stabilization (aerobic) pond	—	70 to 90
Anaerobic lagoon	70	40 to 70

* No recirculation. Efficiencies can be increased within limits by controlling organic loading, efficiencies of settling tanks, volume of recirculation, and the number of stages; however, effluent will be less nitrified than from standard rate filter, but will usually contain dissolved oxygen. Filter flies and odors are reduced. Study first cost plus operation and maintenance.

Plans and Report

Plans of the area to be sewered should be complete and should include specifications and an engineering report giving the problem, objectives, and design details. The plans must be prepared by a licensed professional engineer, drawn to a scale of 1 in. equal to not less than 100 ft or more than 300, with contours at 2- to 10-ft intervals. The discussion that immediately follows deals with sanitary sewers. Combined sewers should be redesigned as storm and sanitary sewers insofar as possible; extensions should be separate sewers.

Sewers

The location of the sewers, with surface elevations at street intersections and changes in grade, are indicated on the general plan. The size of sewers, outfalls, slope, length between manholes, and invert elevations of sewers and manholes to the nearest 0.01 ft are also shown on the general plans. For all sewer laterals 12 in. or larger and all main, intercepting, and outfall sewers, profiles including manholes and siphons, stream crossings, and outlets must be included. The horizontal scale must be at least equal to that on the general plan and the vertical scale not smaller than 1 in. equal to 10 ft. Detail plans to a suitable scale are required of all appurtenances, manholes, flushing manholes, inspection chambers, inverted siphons, regulators, pumping stations, and any other devices to permit thorough examination of the plans and their proper construction. The total drainage area and the area to be served by sewers should be shown on a topographic plan.

Diameter of Sewer, inches	Grade, percent			Capacity Full, in mgd	Population at 250 gpd/capita	Served* at 400 gpd/capita
	$n = 0.013$	$n = 0.012$	$n = 0.011$			
4	0.65	0.625	0.60	0.12	—	—
6	0.60	0.51	0.42	0.27	—	—
8	0.40	0.32	0.25	0.46	1840	1150
10	0.28	0.23	0.18	0.72	2880	1800
12	0.22	0.18	0.14	1.00	4000	2500
15	0.15	0.125	0.10	1.60	6400	—
18	0.12	0.10	0.08	2.30	9200	—
21	0.10	0.08	0.063	3.20	12,800	—
24	0.08	0.066	0.053	4.00	16,800	—

At minimum slope, $n = 0.013$, velocity 2 fps. * Courtesy Joseph A. Kestner, Jr., Consulting Engineers, Troy, N.Y.

Sewers are usually designed for a future population 30 to 50 yr hence and for a per capita flow of not less than 400 gpd for submains and laterals and 250 gpd for main, trunk, and outfall sewers, plus allowance for industrial wastes. Intercepting sewers are designed for not less than 350 percent of the average dry weather flow.

Street sewers should not be less than 8 in. in diameter and at a depth sufficient to drain cellars, usually 6 to 8 ft. Vitrified clay or concrete sewers are designed for a mean velocity of 2 fps when flowing full or half full, based on Kutter's formula, with $n = 0.013$. When cement asbestos pipe or cement- or enamel-lined cast-iron pipe is used, $n = 0.011$ or $n = 0.012$ may be used in design. In general, sewers should be laid on grades not less than those in the chart shown on page 356.

In some special situations there may be justification for the use of 6-in. sanitary sewers with full knowledge of its limitations.[43] Compression-type joints for pipe 4 to 12 in. in diameter are highly recommended. Vitrified clay and cast-iron pipe have a life of more than 50 yr.

Manholes should be not more than 300 ft apart on 12-in. pipe or smaller and not more than 400 ft apart for 15-in. or larger pipe. Some standards permit a manhole spacing of 500 ft for sewers 18 to 30 in., and 400-ft spacing for sewers 8 to 12 in.

Wye connections should be installed for each existing and future service connection as the sewer is being laid. A tight connection must be assured.

Inverted Siphons

Where required, inverted siphons should be not less than 6 in. in diameter nor less than 2 in number. Sufficient head should be available to provide velocities of not less than 3 fps in sanitary sewer or 5 fps in combined sewers. Accessibility to each siphon and diversion of flow to either one would permit inspection and cleaning.

Pumping Stations

Pumping plants should contain at least three pumping units of such capacity to handle the maximum sewage flow with the largest unit out of service. The pumps should be selected so as to provide as uniform a flow as possible to the treatment plant. Two sources of motive power

[43] *Small-Size Pipe for Sanitary Lateral Sewers,* Building Research Advisory Board, Division of Engineering and Industrial Research, National Academy of Science, National Research Council, Washington, D.C., February 28, 1957.

should be available. Small lift stations should have duplicate pumping equipment or pneumatic ejectors with auxiliary power.

In all cases raw sewage pumps should be protected by screens or racks unless special devices are approved. Proper housing for electric motors above ground and in dry wells should provide good ventilation, preferably forced air, and accessibility for repairs and replacements. Wet wells or sumps should have sloping bottoms and provide for convenient cleaning. See Figure 4–28. Select water-level pump controls with care because they are the most frequent cause of pump failure.[44]

Sewage Treatment

Sewage treatment works should be designed for a population at least 10 years in the future, although 15 to 25 years is preferred, and a per capita flow of not less than 100 gpd plus acceptable industrial wastes. Actual flow studies should govern. Plants should be accessible from highways but as far as practical from habitation and wells or sources of water supply. The required degree of treatment should be based on the water quality standards and objectives established for the receiving waters and other factors, as pointed out earlier.

The two major sewage treatment design parameters are the 5-day biochemical oxygen demand and suspended solids of the wastewater to be treated and the removal expected of the treatment process. The processes, efficiencies, and design factors are shown in Figure 4–36, and Tables 4–16, 4–17, and 4–18. The 5-day BOD is usually assumed to be 0.17 lb/capita/day and the suspended solids 0.20 lb with the daily flow 100 gpd/capita. Studies at 78 cities suggest 0.20 lb BOD, 0.23 lb suspended solids, and 135 gpd/capita as being more representative. Design should be based on actual wastewater strength and flow.[45]

Scale drawings of the units comprising the sewage treatment works, together with such other details as may be required to permit review of the design, examination of the plans, and construction of the system in accordance with the design, must be prepared and submitted to the regulatory agency having jurisdiction. Provision for measuring the flow and sampling the sewage and an equipped laboratory for examinations to control operation should be included in the original plans. The design must include disinfection of the plant effluent if the receiving stream or body of water in the vicinity of the outfall is used for water supply,

[44] George M. Ely, Jr., "Wastewater Pumping Station Design Criteria," *J. Water Pollution Control Fed.*, **37**, 1437 (October 1965).
[45] Raymond C. Loehr, "Variation of Wastewater Quality Parameters," *Public Works*, **99**, 81–83 (May 1968).

shellfish propagation, recreation, or other purposes that may be detrimentally affected by the sewage disposal.

Hydraulic Overloading of Sewers and Treatment Plant

McKee, Palmer and Johnson have shown that the average dry weather flow of sanitary sewers is approximately equal to the discharge rate of runoff from a rainfall having an intensity of 0.01 in./hr.[46] But 0.02 to 0.03 in. of rainfall is needed to wet the ground before there is a runoff. Camp concludes that the rate of flow in a combined sewer is approximately equal to 100 times the rainfall intensity in in./hr times the dry weather flow, up to the capacity of the sewer.[47] Therefore if the average dry weather flow is 1 mgd and the rainfall intensity is 2 in./hr for a sufficient time to cause runoff from the area under consideration, the rate of flow would be 200 mgd. The economic futility of trying to design a treatment plant to treat combined sewer flows is obvious. On the other hand, combined sewers contribute significant amounts of sediment, oil, salts, and organic matter. Discharge, or overflow from combined sewers, immediately after a heavy rain may have a BOD of several hundred milligrams per liter and MPN of hundreds of thousands. Even after the initial flushing the pollution discharged is still substantial. Hence complete stream pollution control cannot be realized unless combined sewer overflows are retained and/or adequately treated before discharge.

The importance of sewer inspection before backfilling has not been given sufficient emphasis. Too many properly designed plants have created problems almost as serious as those they were intended to correct. Some plant flows equal or exceed the design flow before the system is completed because of improper sewer and house connection construction; roof, footing, and area drainage connection; street drainage connection; and similar practices. Inspection during construction should assure tight service connections and sewer construction by requiring full compliance with pipe, joint, bed, and backfill specifications.

Infiltration flow tests in wet ground of each line before acceptance, or exfiltration flow tests in dry ground, will determine compliance with

[46] M. E. McKee, "Loss of Sanitary Sewage through Storm Water Overflows," *J. Boston Society Civil Engineers,* 55 (April 1947); Clyde L. Palmer, "The Pollutional Effects of Storm Water Overflows from Combined Sewers," *Sewage and Industrial Wastes,* 22, 154 (February 1950); Frank C. Johnson, "Nation's Capital Enlarges its Sewerage System," *Civil Engineering,* 56 (June 1958).

[47] Thomas R. Camp, "Chlorination of Mixed Sewage and Storm Water," Annual Convention, ASCE, Boston, October 11, 1960.

specifications. The exfiltration test is a more severe test; large volumes of water are needed and must be disposed of.[48] Under a head of 2 ft of water the exfiltration was found to be 4.8 percent greater than shown by the infiltration test; under a 4-ft head, 19.8 percent greater; and 27.3 percent greater under a 6-ft head. Low pressure air test at 3.0 psi with air loss not greater than 0.0030 cfm/ft^2 of internal pipe surface has certain advantages and can also be used.[49]

The total leakage should not exceed 150 gpd/mi/in. of internal diameter of pipe over a 24-hr test period, in a well-laid line. *Recommended Standards for Sewage Works* states that leakage shall not exceed 500 gpd/mi/in. of pipe diameter.[50] A figure of 200 to 400 gal/acre of sewered area/day is also used. Rubber ring joints or the equivalent on vitrified clay tile, concrete, and asbestos cement pipe, and hot-poured joints in dry trenches can give good results.

A community, camp, institution, factory, school, or other establishment that constructs sanitary sewers and treatment plants should immediately set up rules and regulations prohibiting sewer connections by other than responsible individuals. This must be supplemented by effective enforcement to guarantee exclusion of groundwater, surface water, and rainwater from the sanitary sewer system in order to protect the investment made and accomplish the objective of the treatment plant. Sanitary sewers that, in effect, become combined sewers almost always cause backing up in basements after storms and sometimes overflow through manholes onto the street and into stores, basements, and so on, with resultant damage to oil burners, electric motors, and personal property. Cellars remain damp and become contaminated with sewage. Treatment plants become overloaded, requiring the bypassing of diluted sewage to the receiving stream or body of water, resulting in danger to water supplies, recreational area, fish life, and property values. This, of course, nullifies in part the purpose of sewage treatment. In recent years greater attention has been given to the proper construction of sanitary sewers and to the elimination of roof leader connections to the sanitary sewer, a major cause of hydraulic overloading.[51]

The effect of roof leader and submerged manhole flows is more fully

[48] Sherwood Borland, "New Data on Sewer Infiltration-Exfiltration Ratios," *Public Works,* **87**, 7–8 (September 1956).

[49] Roy Edwin Ramseier and George Charles Riek, "Experience in Using the Low-Pressure Air Test for Sanitary Sewers," *J. Water Pollution Control Fed.,* **38**, 1625–1633 (October 1966).

[50] op. cit.

[51] Gerald L. Peters and A. Paul Troemper, "Reduction of Hydraulic Sewer Loading by Downspout Removal," *J. Water Pollution Control Fed.,* **14**, 63–81 (January 1969).

appreciated by some comparisons. A 4-in. rainfall in 24 hr can be expected in the United States roughly south of the Ohio River, once in 5 yr. If 180 dwellings (1000 ft² roof area) connected their roof leaders to the sanitary sewer, the resultant flow would equal the capacity of an 8-in. line; that is, 460,000 gpd with a slope of 0.4 ft/100 ft. Five manhole covers with six vent and pick holes, under 6 in. of water, will also admit enough water to fill an 8-in. sewer.

Cost of Sewerage and Treatment

Cost estimates can vary widely because of location, labor and material costs, volume of construction, season of the year, local characteristics, and degree of treatment required. In general the cost can be divided into two parts: treatment and collection; operation and maintenance. Financing methods are discussed in Chapter 2.

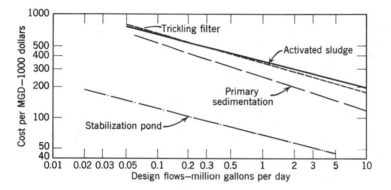

Figure 4–38 Sewage treatment plant construction costs for plants of various size, according to treatment method. Costs represented above are based on the Public Health Service Index, 1962, on data from "Sewage Treatment Construction Costs" by P. B. Rowan, K. H. Jenkins, and D. W. Butler, *J. Water Pollution Control Fed.*, **32**, 594 (1960), and on similar curves, published by the Office for Local Government, State of New York. The curves represent construction costs only and do not include land, engineering, legal, or other administrative costs; nor does it include interceptors, lift stations, or outfall sewer.

The cost of a particular type of sewage treatment should take into consideration the total annual cost. The sum of the principal and interest to pay off a bond issue for construction, plus the cost of operation and maintenance (O and M) should be determined to make a proper cost comparison. An analysis of O and M costs is given in Table 4–19. Sewage treatment plant construction costs are given in Figure 4–38. The economy of scale is also very apparent.

TABLE 4-19 ESTIMATED SEWAGE TREATMENT OPERATION AND MAINTENANCE COST BASED ON PE*

Type of Plant	No.	Total Annual Cost for Population Equivalent* (1965–1967 Dollars)									Correlation Coefficient
		1,000	2,500	5,000	10,000	25,000	50,000	125,000	250,000	500,000	
Conventional primary	163	3,770	7,410	12,300	20,600	40,400	67,300	132,000	220,000	367,000	0.83
Imhoff	34	3,210	5,940	9,490	15,100	–	–	–	–	–	–
Primary with oxidation lagoon	79	1,780	2,960	4,360	6,440	–	–	–	–	–	0.67
Waste stabilization pond	624	907	1,550	2,320	3,470	5,920	–	–	–	–	0.54
Standard-rate trickling filter	104	3,190	6,030	9,740	15,800	29,700	–	–	–	–	0.74
High-rate trickling filter	289	4,350	7,840	12,300	19,200	34,500	53,900	97,100	–	–	0.76
HRTF with oxidation lagoon	36	2,490	4,570	7,250	11,500	21,000	–	–	–	–	0.72
Activated sludge	118	5,150	10,300	17,300	28,500	58,200	98,200	196,000	330,000	557,000	0.89
Extended aeration	97	4,680	9,490	16,200	27,700	56,100	–	–	–	–	0.80
Contact aeration	54	5,320	10,100	16,500	27,700	51,400	–	–	–	–	0.85

Source: Robert L. Michel, "Costs and Manpower for Municipal Wastewater Treatment Plant Operation and Maintenance 1965–1968" *J. Water Pollution Control Fed.* 42, 1883 (November 1970).

*PE is based on 0.17 lb BOD being equal to one population equivalent. Cost includes sum of salaries and wages, electricity, chemicals, other supplies, and miscellaneous items required to operate a plant for the given year. The economy of scale is apparent. Vacuum filters were found to be considerably more expensive to operate and maintain than sand beds or lagoons for sludge dewatering.

Sewage Treatment

Figure 4–38 gives the approximate cost of sewage treatment plants by types and size. To this cost must be added the cost of site development, engineering and legal fees, interest, lateral sewers, collecting mains, manholes, lift stations, and house connections.

Sewerage

The approximate sewer cost including manholes, labor, and material is shown in Table 4–20. Some additional unit costs (1970) are given below.

TABLE 4-20   SEWER COST—VITRIFIED CLAY PIPE, MARCH 1962

Pipe Size	Estimated Cost (in dollars) at Depth Stated							
(in.)	8 ft	9 ft	10 ft	12 ft	14 ft	16 ft	18 ft	20 ft
8	11.50	12.00	13.00	14.50	16.75	19.00	22.00	25.00
10	12.50	13.00	14.00	15.75	18.00	20.00	23.00	26.00
12	13.25	14.00	14.75	16.50	18.75	21.00	24.00	27.00
15	14.00	15.00	15.75	18.00	20.25	23.00	26.00	29.50
18	15.50	16.00	17.25	19.50	22.00	25.00	28.50	32.00
21	17.00	18.00	19.00	21.50	24.50	27.50	31.00	35.00
24	19.00	20.00	21.50	24.00	27.00	31.00	35.00	39.00
30	24.50	26.00	27.00	30.00	34.00	38.00	42.00	47.00

Source: *Study of Needs for Sewage Works,* Constructing Economical Sewage Works, Guide for Municipal Officials, Rep. No. 3, Office for Local Government, State of New York, Albany, October 1962.

Note: Prices include work in urban areas, excavation and backfill of dry stable material, pipe and pipe laying, manholes, and an average surface restoration. Cost to replace macadam estimated at $4.70 and concrete $7.85/lineal ft for a 12 in. or smaller sewer, $7.20 and $12.05 for a 24-in. sewer, and $8.65 and $14.50 for a 36-in. sewer. Curb and gutter replacement is $4.60/lineal ft additional. ENR construction cost index = 1000.

Manhole, standard:	0 to 8 ft depth, each	$350 to 500
	8 to 10 ft	450 to 600
	10 to 12 ft	550 to 700
	12 to 14 ft	600 to 800
House connection: 6″ clay pipe, per ft		$5.00 to 6.00
Average connection		300 to 400
8″ sewer, clay pipe:	0 to 8 ft cut, per ft	$ 7.00
	8 to 10 ft	8.00
	10 to 12 ft	10.00
	12 to 14 ft	12.00
10″ sewer, clay pipe:	0 to 8 ft cut, per ft	$ 8.00
	8 to 10 ft	9.00
	10 to 12 ft	11.00

12″ sewer, clay pipe:	0 to 8 ft cut, per ft	$11.00
	8 to 10 ft	12.00
	10 to 12 ft	13.00
15″ sewer, clay pipe:	0 to 8 ft cut, per ft	$12.00
	8 to 10 ft	13.00
	10 to 12 ft	14.00

8- × 6-in. clay pipe Y branches, each .... $ 8.00 to 11.00
10- × 6-in. clay pipe Y branches, each .... 11.00 to 12.00
Rock excavation, per yd^3: blasting .... $31.00 to 35.00
          soft rock .... 11.00 to 15.00
          air hammer 160
Restoring gravel and shoulder pavement, per lineal ft $2.00
Restoring asphalt concrete pavement, per lineal ft 3.00
Pumping station, below ground, including standby generator,
  dehumidifier, sump pump, and all necessary controls:[52]
    1.0 to 3.0 mgd .... $50,000 to 75,000
    0.25 to 1.0 .... 35,000 to 50,000
    0.05 to 0.25 .... 30,000 to 35,000
Pumping station, above ground, including building and all
  necessary controls and equipment:[52] .... $125,000 to 150,000

Septic tank systems

The cost of an individual septic-tank leaching system will vary with the size of the dwelling, local conditions, and the type and size of leaching or treatment system. Approximate cost estimates (1970), including labor and material, are given for rough comparisons.

Septic tank:	500 gal	$140.00
(precast)	600	150.00
	750	155.00
	900	160.00
	1200	200.00
	1500	355.00
Grease trap:	600 gal	$155.00
(precast)	1000	180.00
	1500	400.00
Excavation for septic tank or grease trap		75.00

Leaching pit or dry well, with 8-in. washed stone, precast units,
  including cover; inside diameter:
    4 ft diameter and 5 ft deep $200.00
    6 ft diameter and 8 ft deep 325.00
    8 ft diameter and 4 ft deep 285.00
Distribution box: 6 outlets (15 × 29 in.) .... $ 20.00
  (precast)
          12 outlets (26 × 48 in.) 60.00

[52] Courtesy Joseph A. Kestner, Jr., Consulting Engineer, Troy, N.Y., 1970.

Curtain drain, 5 ft deep, including washed stone, per ft:	$ 3.00
Tile field, including washed stone, per ft:	
12-in. trench	$ 1.20
18	1.40
24	1.60
30	1.70
36	1.80
Subsurface sand filter system, 390 to 520 ft², including septic tank, contact tank, and chlorinator	$1600.00 to 2000.00
Evapotranspiration system, 10,000 ft², including septic tank	$1500.00 to 2200.00
Pipe, per ft:	
4-in. cast iron	$ 2.00
4-in. clay tile	1.00
4-in. composition	1.85

EXCRETA DISPOSAL

Privies and Latrines

Many types of privies and latrines have been in use all over the world for many years. If properly located, constructed, and maintained they can be adequate, economical, and sanitary devices for the disposal of human excreta where water-borne sewage disposal systems are not provided or not practical. The suitability, location, construction, and maintenance of various types of privies and latrines are summarized in Table 4–21. Figures 4–39 and 4–40 illustrate essential design and construction features.

INDUSTRIAL WASTES

Industrial Wastes Surveys

The design of new plants should incorporate separate systems for sewage, cooling water, waste, and storm- or surface-water drainage. In some cases combination of sewage and liquid waste is possible, but this is dependent on the type and volume of waste. At an existing plant and at a new plant, a flow diagram should be made showing every step in a process, every drain, sewer, and waste line. Dye or temperature tests will help confirm the location of flows. Radioactive isotope tracers might also be used under controlled conditions. Where pipelines are exposed, they can be painted definite colors to avoid confusion. In general, clean water from coolers, roof leaders, footing and area drains, and ice machines can be disposed of without treatment. Water conservation such as use of cooling towers and recirculation of the water, use of air-cooled

TABLE 4-21   SANITARY EXCRET

Facility	Suitability	Location
Sanitary earth pit privy	Where soil available and ground-water not encountered. Earth can be mounded up if necessary to bring bottom of pit 2 ft above groundwater or rock.	Downgrade, 100 ft or more fron sources of water supply; 100 ft from kitchens; 50 to 150 ft of users; at least 2 ft above ground water; 50 ft from lake, stream.
Masonry vault privy	To protect underground and surface-water supplies.	Downgrade and 50 ft or more from sources of water supply; 100 ft from kitchens; 50 to 150 ft of users.
Septic privy (Lumsden, Roberts, and Stiles: LRS privy.)	Where cleaning of pit is a problem and odors unimportant.	Same as pit privy.
Excreta disposal pit	For disposal of pail privy and chemical toilet contents.	Downgrade and 200 ft from sources of water supply; 100 ft from kitchen.
Chemical toilet (cabinet and tank-type)	A temporary facility. To protect water supply, where other method impractical. Temporary camp, vehicle, boat.	May adjoin main dwelling. Tank type same as masonry vault privy
Incinerator toilet	Where electricity or gas avail.	Within the facility.
Recirculating toilet	Airplanes, boats, fairgrounds, camps.	Within the facility.
Removable pail privy (bucket latrine)	A temporary facility; to protect water supply, where pit privy impractical.	Same as masonry vault privy.
Portable box, earth pit, latrine	At temporary camps.	Same as pit privy. Army recommends latrine 100 yd from kitchens.
Bored-hole latrine	In isolated place or when primitive, inexpensive, sanitary facility is needed.	Same as earth pit privy.
Straddle trench latrine	At temporary camp for less than one day.	Same as earth pit privy.
Cat hole	On hikes or in field.	Same as earth pit privy.
Squatting latrine	Where local conditions and customs permit.	Same as pit privy.

Criteria: Confines excreta; excludes insects, rodents, and animals; prevents contamina-
Note: If privy seat is removable and an extra seat is provided, it is easier to scrub seats place of improvised crudely made wooden seats. Deodorants that can be used if needed Keep privy pits dry. Solutions for chemical toilets include lye (potassium hydroxide), sulfate (1 lb in 2½ gal water), and a chlorinated benzene.

POSAL METHODS

Construction	Maintenance
p pit; insects, rodents, and animals ex-ded; surface water drained away; clean-e material; attractive; ventilated pit and ding. Pit 3' × 4' × 6' deep serves average ily 3 to 5 yr.	Keep clean and flytight; supply toilet paper. Apply residual fly spray to structure and borax, fuel oil, or kerosene to pit. Natural decay and desiccation of feces reduce odors. Keep waste water out. Scrub seat with hot water and detergent.
tertight concrete vault; flytight building; anable material; ventilated vault and lding. Capacity of 6 ft^3 per person ade-ate for 1 yr.	Keep clean, flytight, and attractive. Supply toilet paper. Apply residual spray. Clean pit when contents approach 18 in. of floor. Scavengers can be used.
tertight vault with tee outlet to leaching , gravel trench, filter, vault, etc. Provide acity of 250 gal plus 20 gal for each per-over 8 yrs.	Add 2 gal water per seat. Keep clean and flytight. Supply toilet paper. Agitate after use. Clean vault when depth of sludge and scum = 12 to 18 in.
ored pit with open-joint material. Tight and access door.	Keep flytight and clean. Drain surface water away.
me as masonry vault privy. Tank may be avy gauge metal with protective coating. vide capacity of 125 to 250 gal per seat.	Use ¼ lb lye for each ft^3 of vault capacity made up to 6 in. liquid depth in vault, or 25 lb caustic per seat in 15 gal water. Keep clean. Clean vault when 2/3 to 3/4 full. Odor control. Empty and recharge as direct-ed.
closed compartment.	Keep clean and supply toilet paper.
closed prefabricated unit.	Keep clean. Empty contents in approved facility and recharge with chemical.
me as masonry vault privy. Provide easily aned pails.	Provide collection service, excreta disposal pit, and cleaning facilities, including hot water (backflow preventer), long-handled brushes, detergent, drained concrete floor.
rth pit with portable prefabricated box.	Same as earth pit privy. Provide can cover to keep toilet tissue dry.
ored hole 14 to 18 in. diameter and 15 to ft deep with bracing if necessary. Seat ructure may be oil drum, box, cement or y tile riser with seat, or use squatting ate. Platform around hole.	Same as earth pit privy. Line upper 2 ft of hole; in a caving formation line hole to sup-port earth walls.
rench 1 ft wide, 2½ ft deep, and 4 ft long r 25 men.	Frequent inspection. Keep excreta covered. Provide toilet paper with waterproof cover.
ole about 1 ft deep.	Carefully cover hole with earth.
milar to privy or bored-hole latrine. See etches.	Same as privies and latrines.

on of water supply; provides convenience and privacy; clean and odor-free. d set aside to dry. A commercial plastic or composition-type seat is recommended in clude chlorinated lime, cloroben, iron sulphate, copperas, activated carbon, and pine oil. ustic soda or potash (sodium hydroxide), chlorinated lime (1 lb in 2½ gal water), copper

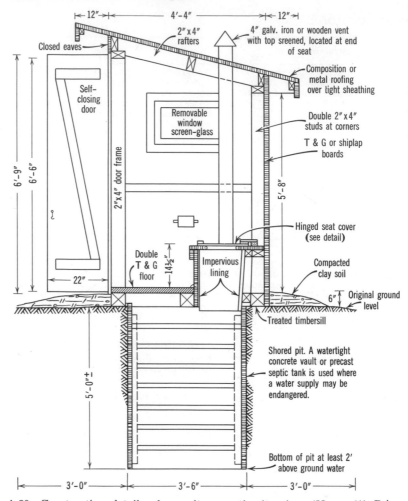

Figure 4–39 Construction details of a sanitary earth pit privy. (NOTE: (1) Privy construction may also be concrete or cinder block, brick, stone, or other masonry, with reinforced concrete floor and riser. (2) Make privy 3 ft wide for one seat or 4 ft wide for two seats. (3) Locate privy within 150 ft of users, 100 ft or more from kitchens, 50 ft from any lake or watercourse, and not in the direct line of drainage to or closer than 200 ft of any water supply.)

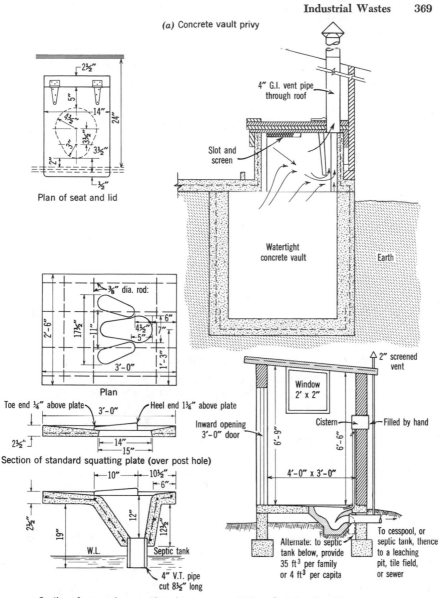

(a) Concrete vault privy

4" G.I. vent pipe
through roof

Slot and
screen

Watertight
concrete vault

Earth

Plan of seat and lid

⅜" dia. rod:

Plan

Toe end ¼" above plate   Heel end 1¼" above plate

Section of standard squatting plate (over post hole)

Inward opening
3'-0" door

Section of aqua privy squatting plate

W.L.   Septic tank

4" V.T. pipe
cut 8½" long

Window
2' x 2'

Cistern   Filled by hand

4'-0" x 3'-0"

2" screened
vent

To cesspool, or
septic tank, thence
to a leaching
pit, tile field,
or sewer

Alternate: to septic
tank below, provide
35 ft³ per family
or 4 ft³ per capita

(b) Pour flush (water seal) latrine

Figure 4–40 (a) Concrete vault and (b) squatting type latrines. (Squatting latrine types adapted from O. J. S. Macdonald, *Small Sewage Disposal Systems*, H. K. Lewis & Co., Ltd., London, England, 1951.)

exchangers, and wastewater pretreatment and reuse will reduce the wastewater problem.

A knowledge of the industrial process is fundamental in the study of a waste problem. The volume of flows from each step in a process, the strength, chemical characteristics, temperature, source, and variations in flow are some of the details to be obtained. The existing or proposed methods of disposal, opportunities for wastage, drippage, and spillage should be included. Sometimes revision of a chemical process can eliminate or reduce a waste problem. Possible waste-control measures are salvage, in-plant waste reduction, reclamation, concentration of wastes, flow equalization, and new methods.

A method of simplifying a waste problem is to spread its disposal over 24 hr rather than over a 4- or 6-hr period. This is accomplished by the use of a holding tank to equalize flows and strength of waste, accompanied by a constant uniform discharge over 24 hr. Where needed, aeration will prevent septic conditions and odors. If necessary, chemical mixing can be incorporated, followed by settling and uniform drawoff of supernatant by means of a flexible or swing joint pipe. The pipe may be lowered uniformly by mechanical means such as a clock mechanism or motor, or a float with a submerged orifice, to give the desired rate of discharge. The mass diagram approach may be used to determine the required storage to give a constant flow over a known length of time. Sludge is drawn from the hopper bottom of the settling tank to a drying bed or to a treatment device before disposal.

After investigating all possibilities of reducing the wastewater problem at the source, the next step would be provision of the required degree of treatment, based on existing laws and standards and competent engineering advice. Receiving waters' classification and minimum average seven-consecutive-day flow once in ten years of the receiving stream will largely determine the type and degree of treatment required. Black proposed that the approving agency consider the following items in its review of engineering reports, plans, and specifications for industrial waste treatment systems:[53]

Engineering Report (Part I).
Project Delineation
　　1. Type of Industry
　　2. Waste-Pollution Load
　　3. Receiving Waters
　　4. Waste Treatment Requirements

[53] Hayse H. Black, *Items Considered in Review of Engineering Reports, Plans, Specifications for Industrial Waste Treatment Systems,* New York State Dept. of Health, Albany, January 1969.

   5. Waste Abatement Plans
   6. Map of Environment
   7. Plot Plan and Hydraulic Profile
Plant and Process Description
   8. General Description of Factory
   9. Description of Principal Wet Processes
  10. Process Flow Diagram Showing Sources of Liquid Waste
  11. Sewer Map and Process Connections
  12. Liquid Waste Control Measures
  13. Experimental and New Processes
  14. By-Product Recovery Systems
Factory Operations
  15. Finished Products and Processes
  16. Production in 24 Hours—Rated and Actual
  17. Principal Raw Materials
  18. Sources of Water Supply
  19. Water Quantity and Quality Requirements
  20. Process Water Reuse
  21. Employees, Shifts, Days Per Week
  22. Expansion—Planned and Potential
Development of Design Criteria
  23. Comprehensive Industrial Waste Surveys
  24. Parameters Specific for the Industry
  25. Segregation of Cooling Water
  26. Diversion of Storm Water
  27. Inplant Improvements and Waste Reduction
  28. Waste Characterization and Treatability Studies
  29. Pilot Plant Investigations
  30. Avoidance of Nuisance to the Environment
  31. Design Liberal and Flexible
Combined Treatment With Domestic Waste
  32. Standards and Regulations for Controlling the Use of Municipal Sewers
  33. Pretreatment at the Factory
  34. Surcharge Agreements
  35. Industrial Effluent Monitoring Systems
  36. Influence on Sewage Treatment Efficiency
Engineering Report (Part II).
  37. Solids Removal and Disposal
  38. Chemical Precipitation
  39. Chemical Treatment
  40. Aerobic Biological Treatment
  41. Anaerobic Biological Treatment
  42. Role of Incineration
  43. Alternate Proposals
  44. Chronological Steps for Submission

The types of industrial wastes are of course numerous. Treatment processes vary from solids removal and disposal to involved chemical and biological processes including incineration, chemical neutralization and ion exchange. In view of the complexity of the problem only those industrial wastes more commonly encountered in connection with small plant operations will be discussed. More detailed information concerning the treatment of specific wastes can be obtained from standard texts, periodicals, and other publications devoted to this subject.

A common method of expressing the strength of wastes is in milligrams per liter or parts per million of 5-day BOD, suspended solids, and chlorine demand. Another parameter is the chemical oxygen demand or COD. These are by no means the only measures of waste strength, as the particular waste and its characteristics must be considered.

The BOD of an organic waste is frequently converted to its population equivalent. Since the waste from one person is said to equal 0.17 lb of 5-day BOD, the population equivalent X of say 1000 gal of a waste having a BOD of 600 mg/l is expressed by the proportion

$$\frac{600}{1,000,000} = \frac{X(0.17)}{1000 \times 8.34}, \qquad \text{or } X = 29 \text{ persons}$$

To determine the suspended-solids population equivalent, substitute 0.20 for 0.17.

The actual volume and strength of a waste should be individually determined; some approximations are given in Table 4-22 to show how wastes from different processes can vary. Many municipalities levy a charge for the handling and treatment of industrial wastes to help pay the cost of operating and maintaining the municipal treatment works. The basis for rental or assessment of cost should be volume and strength of the waste as determined by periodic analyses. Measures used are COD, BOD, chlorine demand, and other parameters. It is also common to require pretreatment in case the characteristics of the waste exceed certain predetermined values, the waste as released is not amenable to treatment in a municipal treatment plant, or the waste would cause a hazardous condition in the sewers.

The Water Pollution Control Federation's Manual of Practice No. 3, entitled "Regulation of Sewer Use," suggests standards and regulations for local sewer ordinances.[54] This manual provides sound regulations to exclude hazardous materials, protect sewers, and control the discharge of wastes to the sewer that may upset the municipal treatment plant operation. WPCF Manual of Practice No. 11 states that these wastes are not normally acceptable: pulp and paper, beet or cane sugar, rayon, oil refining, coke and gas manufacturing, metal plating, synthetic and

[54] Washington, D.C., 1963.

reclaimed rubber, sawmills, veneering, mines, gravel operations, and wool scouring.[55]

Milk Wastes

Solution of a dairy waste problem starts with waste-saving and waste prevention. If judiciously carried out the waste can be drastically reduced, thereby making treatment of the waste remaining simpler and more practical of solution. Some of the saving and prevention methods found effective follow.

1. Reduce the volume and strength of floor wastes. Obtain management and employee cooperation to prevent overflow of cans and vats, spillage, leaks, and drips. Collect can, pipeline, and tank drips, and equipment concentrated prerinse for salvage as animal feed or similar use. Make use of automatic shutoff nozzles on hoses.

2. Encourage bulk tank pickup rather than 10-gal cans, and in-place cleaning of dairy equipment and sanitary pipelines to reduce volume of waste.

3. Eliminate steam-water mixing tees for making hot water. Use a water-heating system and distribute a tempered water for utility purposes.

4. Separate and dispose separately strong wastes, wash water, cooling water, and sanitary sewage. Discharge sulphuric acid used in butterfat testing into ash or cinder pit. Salvage caustic solution from bottle washer and rinse, after removing suspended matter.

5. Skim milk, buttermilk, cheese whey, whole milk, and milk products have very high BOD and hence are expensive to treat. If not used for by-product, investigate possibilities of disposal in a lagoon or oxidation pond located in an isolated area or, if permissible, in sewage treatment plant digester provided the volume can be absorbed.

6. Use evaporator and drying equipment, or entrainment separator, where by-product production is possible.

All dairy plants should be equipped with a combination sand trap grease trap and a screen with ⅜-in. openings. It is good practice to incorporate a weir or other device for estimating flows. A trap 2½′ × 2½′ × 8′ long is adequate for a 1000 gal/hr flow.[56]

A convenient method of dairy-waste disposal is piping to a municipal sewer system if this is practical, as biological treatment plants can handle moderate amounts of dairy wastes. It is not advisable to dump large quantities of wastes into a sewer unless the sewage treatment plant operator is first notified and signifies he is prepared to accept the waste. Large quantities of whey are detrimental. In any case, by keeping the

[55] *Operation of Wastewater Treatment Plants,* Manual of Practice No. 11, Water Pollution Control Federation, Washington, D.C., 1970.
[56] H. A. Trebler and H. G. Harding, "Fundamentals of the Control and Treatment of Dairy Wastes," *Sewage & Ind. Wastes* 27, No. 12 (December 1955), pp. 1369–1382.

TABLE 4–22 TYPICAL WASTES FROM INDUSTRIAL OPERATIONS

Industry	Production Unit	Flow (gal/production unit)	BOD (mg/l)	SS (mg/l)	Population Equivalent BOD (per production unit)	SS (per production unit)
Meat packing						
Stockyard	1 acre	25,000	65	175	80	180
Slaughterhouse, hog	One hog	143	1,048	717	8	4
Slaughterhouse, cattle	One hog	157	998	820	8	5
Slaughterhouse mixed	One hog	160	2,200	930	18	6
Packing plant, mixed	One hog	550	900	650	24	14
Poultry	1,000 lb	2,200	–	1,460	300	130
Milk products						
General dairy	1,000 lb raw milk	340	570	540	10	7
Receiving station	1,000 lb raw milk	175	500	280	4	2
Bottling plant	1,000 lb raw milk	250	500	290	6	3
Creamery	1,000 lb raw milk	110	1,250	660	6	3
Condensery	1,000 lb raw milk	150	1,300	750	7	4
Condensery vacuum pan	1,000 lb raw milk	1,500	25	–	2	–
Dry milk powder	1,000 lb raw milk	300	480	500	6	6
Cheese factory	1,000 lb raw milk	200	1,600	750	16	6
Canning						
Fruits						
Apricots	100 cases No. 2 cans	8,000	1,020	260	416	83
Cherries, sweet	100 cases No. 2½ cans	16,500	470	20	380	13
Cherries, sour, pitted	100 cases No. 2½ cans	16,500	750	–	600	–
Grapefruit	100 cases No. 2 cans	5,600	1,850	270	500	60
Grapefruit juice	100 cases No. 2 cans	500	310	170	8	3
Pears	100 cases No. 2½ cans	6,500	1,340	400	440	104
Peaches	100 cases No. 2½ cans	6,500	1,340	400	440	104

TABLE 4–22 TYPICAL WASTES FROM INDUSTRIAL OPERATIONS (cont'd)

Industry	Production Unit	Flow (gal/production unit)	BOD (mg/l)	SS (mg/l)	Population Equivalent BOD (per production unit)	Population Equivalent SS (per production unit)
Canning (cont'd)						
Fruits (cont'd)						
Prunes and plums	100 cases No. 2½ cans	8,000	1,020	260	410	83
Apples	100 cases No. 10 cans	11,600	1,110	—	900	—
Grapes	Ton of raw grapes	100	720	—	3.6	—
Tomatoes, whole	100 cases No. 2 cans	800	4,000	2,000	150	640
Tomatoes, products	100 cases No. 2 cans	7,000	1,000	560	350	155
Tomatoes, puree	100 cases No. 2 cans	800	2,900	950	110	32
Vegetables						
Asparagus	100 cases No. 2 cans	7,000	100	30	35	9
Beans, green	100 cases No. 2 cans	3,500	200	60	35	9
Beans, lima	100 cases No. 2 cans	25,000	190	420	240	420
Beans, wax	100 cases No. 2 cans	2,600	240	60	20	6
Beans, pork and	100 cases No. 2 cans	3,500	920	225	160	31
Beets	100 cases No. 2 cans	3,700	2,600	1,530	480	226
Carrots	100 cases No. 2 cans	3,700	1,630	1,830	300	271
Corn, cream	100 cases No. 2 cans	2,500	620	300	75	30
Corn, kernel	100 cases No. 2 cans	2,500	2,000	1,250	250	125
Peas	100 cases No. 2 cans	2,500	1,700	400	210	40
Sauerkraut	100 cases No. 2 cans	300	6,300	630	100	1
Spinach	100 cases No. 2 cans	16,000	620	580	480	370
Pumpkin	100 cases No. 2½ cans	2,500	6,400	1,850	800	185
Dried and dehydrated vegetables						
Potatoes, abrasive peeled	Ton of raw potatoes	500	4,000	—	120	—

TABLE 4–22 TYPICAL WASTES FROM INDUSTRIAL OPERATIONS (cont'd)

Industry	Production Unit	Flow (gal/production unit)	BOD (mg/l)	SS (mg/l)	Population Equivalent BOD (per production unit)	Population Equivalent SS (per production unit)
Vegetables (cont'd)						
Dried and dehydrated vegetables (cont'd)						
Potatoes, spray washed	Ton of raw potatoes	2,500	1,000	—	120	—
Beet sugar						
Flume water	Ton of raw beets	2,200	200	800	22	74
Process of water	Ton of raw beets	660	1,230	1,100	40	29
Lime cake drainage	Ton of raw beets	75	1,420	450	5	1
Steffens waste	Ton of raw beets	120	10,000	7,700	60	37
Brewery						
Grain dewatered	Bbl. of beer	470	1,200	650	19	13
Grain sold wet	Bbl. of beer	470	800	450	12	9
Distillery						
Combined	Bu of grain	600	1,230	350	35	13
Molasses	Gal 100 proof	8	33,000	3,270	12	1
Paper mill						
Unbleached	Ton of paper	39,000	19	450	26	704
Bleached	Ton of paper	47,000	24	156	40	282
Paperboard	Ton of paper	14,000	121	560	97	447
Strawboard	Ton of paper	26,000	965	1,790	1,230	1,871
Deinking	Ton of paper	83,000	300	—	1,250	—

TABLE 4-22 TYPICAL WASTES FROM INDUSTRIAL OPERATIONS (cont'd)

Industry	Production Unit	Flow (gal/production unit)	BOD (mg/l)	SS (mg/l)	Population Equivalent	
					BOD (per production unit)	SS (per production unit)
Pulp mill						
Groundwood	Ton of pulp	5,000	645	–	16	–
Soda	Ton of pulp	85,000	110	1,720	460	5,849
Kraft	Ton of pulp	64,000	123	–	390	–
Sulfite	Ton of pulp	60,000	443	–	1,330	–
Tanning						
Vegetable	100 lb of hides	800	1,200	2,400	48	77
Chrome	100 lb of hides	3,470	460	3,760	80	520
Laundry	100 lb of clothes	1,485	320	–	24	8
Oil refining	Bbl crude	770	20	50	0.6	1

Note: The above data in this table have been taken largely from Supplement "D," Ohio River Pollution Survey Report, U. S. Public Health Service (1942), 2.5 hog units = 1 head of cattle; 1 hog unit = 1 lamb = 1 calf; all in terms of waste; lb × 0.454 = kg; gal × 0.003785 = cu m.

Source: *Operation of Wastewater Treatment Plants*, Manual of Practice No. 11, Water Pollution Control Federation, Washington, D.C., 1970).

volume and strength of the waste low, it will be possible to realize a saving should a charge be made by the municipality.

If it is not possible to discharge dairy wastes directly to a municipal sewer, either partial or complete treatment may be required. Some of the treatment and disposal methods available are outlined below. A coarse screen and grit chamber should generally precede the treatment and disposal process.

Where a *dilution* of at least 50 to 1 is available it may be possible to discharge the waste to a stream through a holding tank with a controlled discharge. Aeration or chlorination and recirculation during the holding period will prevent souring of the waste. Aeration can be accomplished by means of splash plates, aspirator, cascading, or other simple method.

A shallow oxidation pond, 2–5 ft liquid depth, is an inexpensive and effective way of treating dairy wastes. A design based on a loading of 25 lb BOD per acre per day, preceded by a settling tank providing 72 hr detention or an anaerobic digestion chamber, should not cause a nuisance. See pages 326 and 327.

Ridge and furrow irrigation is another effective disposal method. Careful initial grading and maintenance of the furrows (1 to 3 ft deep, 1 to 3 ft wide, spaced 3 to 15 ft apart), with resting of sections is necessary. A loading of 3600 to 30,000 gpd/acre is used, depending on the prevailing temperature and soil percolation. Odor nuisance is likely in warm weather. Oxidation ditches with brush or paddle rotors to provide mixing and aeration can also be used, but cost may be a limiting factor.

Spray irrigation with No. 80S Rainbird or ¾-in. Skinner sprinkler to pasture, field crops, or wooded areas at the rate of about 2000 to 6000 or more gpd/acre can also be used.[57] In favorable soil 10,000 to 15,000 gpd may be applied. The irrigation laterals and spray nozzles should be designed to give a hydraulic loading that can be safely absorbed. Sprinkler nozzle pressure may vary from 30 to 60 psi and discharge from 10 to 75 gpm. Intermittent dosing with extended rest periods and grading of the land to prevent ponding will prevent fly and odor nuisances. Periodic cultivation of the land is usually necessary.

Conventional open sand filters can give satisfactory results for a time if preceded by a settling tank providing at least three days' detention. A loading of 50 to 100 lb of BOD/acre/day and 50,000 to 120,000 gal/acre/day are used in design. Odors can be expected.

---

[57] Frank J. McKee, "Dairy Waste Disposal by Spray Irrigation," *Sewage and Industrial Wastes,* **29,** No. 2, 157–164 (February 1957).

High-rate trickling filters are one of the preferred treatment devices. Two complete stages in series are needed if a high-quality effluent is required. A low rate trickling filter (6 ft deep), loaded at the rate of one mgd/acre, will also give good results but will be accompanied by some odors and filter flies. Small rock or lath filters can be used for small flows. High-rate filters may be loaded at six times standard rate filters (3.0 lb BOD/yd^3).

Aeration. For complete oxidation, aeration of the wastes for 24 hr at a temperature of 80 to 90°F accompanied by good agitation and 3 to 4 hr settling provides good treatment. For activated sludge treatment provide 3 to 6 hr detention. If batch treatment is proposed, provide ½ to 2 days' capacity, depending on the required degree of treatment, and air agitation to prevent septic action. Aerobic treatment causes a breakdown of milk wastes to CO_2 and H_2O with very little sludge production. Complete oxidation requires 1.25 lb of oxygen/pound of milk solids; initial assimilation requires 0.50 lb.

A settling tank and shallow tile field can give satisfactory results at small plants if properly designed and maintained. Seeding of a new tank with lime and septic sludge, maintenance of the wastes at a temperature of 90°F, provision of *3 to 6 days' retention,* and design of a tile field based on percolation tests are important considerations. If the settling tank is not cleaned regularly the tile field will soon fail.

Chemical precipitation is a suitable treatment method, but it is not advised for small plants unless intelligent operation can be assured; it is expensive. The Mallory and Guggenheim processes give satisfactory results. In the Mallory process, activated sludge plus chemical precipitation with lime and alum is used. In the Guggenheim process, storage with aeration, lime, ferric salt, return sludge, 4 hr aeration, and settling are used.

Combinations of processes are possible depending on the local conditions, volume and type of waste, personnel available, and other factors. Characteristics of various types of dairy wastes are given in Table 4–22.

Poultry Wastes

As in all waste treatment operations, the most important step is good housekeeping. The major source of waste has been found to be the battery and feeding rooms. This is followed by the wastes from killing, scalding and picking operations.

The waste reduction measures that are possible include:

1. Blood collection from the killing bench or trough, as blood has a very high BOD. An electric stunning plate is practical in reducing the spattering of blood. Running water to continuously carry away the blood and offal will greatly increase

the volume of the waste and hence the cost of treatment. Squeegee the killing bench or trough and floor to concentrate the blood, feathers, and offal. Collect the waste in a container for separate disposal, then wash all surfaces with water.

2. Use of securely fastened but removable screens or plates with ⅛-in. opening at all floor drains. This will keep feathers, wax, and large particles from flowing into the sewer, which would add to the nuisance problem.

3. Study of the scalding and chilling operations to reduce the volume of water caused by carelessness. Opportunities for improved or revised operations should not be overlooked.

4. Collection of wastes from the evisceration process in separate cans. The entrails should not be permitted to enter the sewer.

5. Elimination of running hoses. Although desirable for sanitary reasons, the increase in waste volume will make the cost of treatment excessive. Substitute routine clean-up several times during the day.

6. Blood from poultry slaughtering may also be conveniently collected in floor catch-basins. But a routine procedure of skimming the coagulated blood for collection in containers and separate disposal must be in effect if this method is to be effective.

When arrangements cannot be made for discharge into a sewerage system, separate treatment should be considered. This should include holding tanks to equalize peak flows and screens. Holding tanks may require aeration or chlorination to prevent septic action. Grit chambers, if provided, should include aeration to separate out entrained organic matter. Pretreatment should include stationary or mechanical screens depending on the size of the plant and grease removal.

Treatment and disposal processes include trickling filters, activated sludge, extended aeration, chemical precipitation, spray irrigation, ridge and furrow, lagooning, and oxidation ponds (50 lb BOD/acre/day). The activated sludge process requires careful supervision and precautions against shock loads. The chemical precipitation process is expensive and also requires intelligent supervision.[58] Sludge digesters should be heated. Lagoons cause odors.

A typical spray irrigation plant may consist of a 75-gal screening tank equipped with a stationary or moving 10 to 20 mesh screen, followed by a storage tank providing 2 hours' detention and a pump to carry the wastes to the irrigation area or to a bypass to a lagoon.[59] The spray irrigation system would consist of a 125 gpm pump against a 90-foot head, and 3- and 4-inch aluminum laterals provided with revolving $7/32$ in. "Rainbird" spray heads spaced 60 ft apart. Application rates

[58] Ralph Porges and Edmund J. Struzeski, Jr., *Wastes From the Poultry Processing Industry*, U.S. Public Health Service, Dept. of HEW, SEC TR W62-3, Cincinnati, Ohio, 1962.

[59] James W. Bell, "Spray Irrigation for Poultry and Canning Wastes," *Public Works*, **86**, 111–112 (September 1955).

vary from 0.125 in. per day per acre with tight clay loam to 0.6 in. per day per acre with a sandy loam having a soil percolation rate of 1 in. in 5 min.

Canning Wastes

Most canneries operate on a seasonal basis, usually during the dry months of the year when stream dilution is least. The plants are commonly in rural areas near the supplying farms. Products packed may be limited to a few or to a full line of diversified products. The wastes consist largely of wash water plus soil, leaves, and parts of the products. The objectionable parts are the decomposing organic solids in suspension. Treatment should include screening and one of the following methods, depending on the stream requirements of the water pollution control agency.

Screening. A 40-mesh mechanically cleaned (screw or bucket type) screen, or perforated plate with $\frac{1}{16}$-in. diameter holes, and water jets at not less than 40 psi are commonly specified. The area of screen is based on the product; 10 ft²/1000 cases of peas or 38 ft²/1000 bushels of tomatoes per 12-hr operation. A 20-mesh screen is suggested for tomato wastes. The manufacturers of screening equipment should be consulted for their recommendations in connection with specific products. Provide for a conveyor and loading hopper. Larger screen openings may be considered, depending on the treatment and disposal proposed.

Dilution, not less than 1 to 100.

Biological filtration. A trickling filter dosed at 0.5 mgd/acre removes 80–95 percent 5-day BOD. The effluent will not cause a nuisance with a dilution of 1 to 5, provided the stream is not seriously polluted. At a filtration rate of 2 mgd/acre 50 to 70 percent BOD can be removed. This effluent can be absorbed satisfactorily by a stream providing a dilution of 1 to 20. Treatment with lime every 2 days at the rate of 100 lb per 100,000 gal of wastes will help prevent fungi clogging. Plants should be at least 200 feet from dwellings.

A *high-rate filter* with one settling tank providing 1-hr detention, based on the pump capacity, is also used. The loading could be about 1500 lb of BOD per acre-ft with the pump capacity at least six times the maximum rate of waste flow at the plant. This has been recommended by the National Canners' Association. A removal of 70 percent for beet waste and up to 88 percent for tomato waste is reported.

Chemical precipitation. Chemical precipitation is particularly adaptable to canneries operating for short seasons. The fill and draw procedure is common. At least two hopper-bottom tanks are provided, each having a capacity to hold the maximum 4-hr waste flow. In this method, 30 min. is allowed for chemical feed, mix, and flocculation; 2 hr are allowed for settling, and $1\frac{1}{2}$ hr are allowed for the drawing of sludge on a sand-drying bed and for the discharge of the treated waste. The chemical feed usually consists of 5 to 10 lb of lime in combination with ferric sulfate, ferric chloride, alum, zinc chloride, or, in the case of tomatoes or carrots, lime alone. The expected BOD reduction is 33 to 75 percent, depending on the chemicals used. Control tests to determine the best chemical dosage should be made daily.

Spray irrigation. A holding tank providing about 2-hr storage is recommended following a stationary or moving 10- to 20-mesh screen. The waste is then pumped to the header and spray nozzles on lateral distributors at a rate that can be satisfactorily absorbed by the irrigation area. Portable 3- and 4-in. aluminum pipe and revolving Rainbird $\frac{7}{32}$-in. spray nozzles can be used. With a flow of 125 gpm and a head of 90 ft, a 60-ft diameter spray can be obtained. An application rate of 0.125 in. per acre per day, 0.079 gal/ft², has been used where the soil is a tight clay to as much as 0.6 in., and 0.374 gal/ft² where the soil is a sandy loam having a soil percolation of 1 in. in 5 min. Design should include bypass to a lagoon.

Irrigation. Fields can also be used where the soil is porous. The field is plowed and trapezoidal channels, ditches, or furrows are formed having a depth of about 12 to 18 in. Careful grading and intermittent dosing will help assure satisfactory results. The dosage and rest periods should be based on the permeability of the soil. Ponding must be avoided.

Lagoons. Lagoons cause odors that carry $\frac{1}{2}$ to 1 mi. On the other hand, stream pollution can be completely eliminated. The most satisfactory treatment of wastes to be lagooned is the application of sodium nitrate. Agricultural sodium nitrate contains about 55 percent available oxygen, which helps maintain aerobic conditions during the decomposition of the wastes. Satisfaction of 20 percent of the BOD of the daily waste by the application of sodium nitrate should prevent a nuisance. The lagoon should have a capacity 25 percent greater than the total waste for the year, and the depth should be between 3 and 5 ft. About 25 percent of the waste should be left in the lagoon to provide seeding organisms for the next year.

The treatment methods mentioned can be expanded to include sand filtration, sedimentation, oxidation ponds, patented processes, and combinations of the methods given to produce the desired degree of treatment.

Packing-House Wastes

The treatment of slaughterhouse and meat-packing plant wastes can be a very difficult problem, particularly if good housekeeping practices are not followed. The reduction of losses requires proper equipment, a properly designed system of sewers, and constant vigilance. Some of the more common waste-prevention or reduction procedures are outlined below.

By-product production. Edible fats, inedible fats, hides, skins, pelts, tankage, meat scraps, and the condensate from the steam rendering of fats (called "stick"), dried blood, bone and bone products, hog hair and bristles, horns and hoofs, intestines, and glands are among the principal by-products. The salvaging of this material can reduce the amount of wastes, but can also add wastes incidental to the production of the by-products.

Waste reduction. In addition to the salvaging of as much material as possible as explained above, other steps can be taken to reduce the volume and strength of the wastes. The methods used include the following:

Grease salvage. Screens, vacuators, baffled skimming basins, or traps should be used on all waste lines in rooms or spaces where greasy waste water results. Grease basins or separators are designed to provide ¾ to 1 hr detention and be easily cleaned. Depths of 4 to 6 ft and a flow velocity of 1 to 2 ft/min are suggested. It is sometimes practical to install one large grease separator on a main drain. In any case, grease separators or traps, skimming basins, and screens require cleaning at least once a day. Other wastes that do not contain grease, including condenser water, paunch manure, boiler plant wastes, and wash water should be bypassed around grease separators or basins.

Wastes in yards and pens. Areas for the temporary holding of animals should be first cleaned by dry methods and then washed down.

Blood. All kill blood should be collected for separate disposal or use. Squeegees can be used to collect blood on floors and tables for transferral to the blood storage cans or tanks.

Paunch manure. A fine screen is recommended to collect paunch manure, sludge, hair, and trimmings. Special screens suitable for the purpose, either manually or mechanically cleaned, should be investigated. Fleshings and cuttings, grease, and solid particles are best collected by dry cleaning of floors.

Waste treatment. After all practical steps have been taken to reduce the volume and strength of the waste, consideration should be given to the degree and type of treatment required. This is usually dependent on the strength of the wastes, availability of municipal sewers and type of treatment, dilution available, the stream usage and the existent legal regulations. A special study made by Mohlman showed that the water consumption from packing houses varied from 1900–8700 gal per ton of kill and averaged 4130.[60] The BOD averaged 28.9 lb per ton of kill,

[60] F. W. Mohlman, "Packing House Industry," *Industrial and Engineering Chemistry,* **39,** No. 5, 637–641 (May 1947).

suspended solids 22.7 lb, nitrogen 3.49 lb, and grease 2.64 lb. The treatment and disposal methods possible are:

1. *Disposal to a municipal sewerage system* if a satisfactory agreement can be made with the responsible officials. Screening, air flotation, and sedimentation may be required before discharge to the sewers. Pretreatment in an anaerobic pond is also feasible.[61]

2. *Coarse screens or fine screens* to take out hair, meat or fat particles, manure, and floating solids. Suspended solids removal of 80 percent is possible if followed by a grit chamber, flocculation and sedimentation.

3. *Imhoff tank.* With 2 hours' detention, a suspended solids removal of approximately 65 percent and BOD removal of 35 percent can be expected.

4. *Trickling filters.*[62] Filtration at a rate of 0.5 to 1 million gal per acre per day can give an overall removal of about 85 percent. The effluent can be expected to be well nitrified and the operation relatively trouble-free.

5. *Double filtration.*[63] Two filters in series, with the first a high-rate 3 million gal per acre per day washable filter, and the second operated at 1.5 million gal per acre per day can yield a 95 percent BOD reduction. Such a plant would include fine screens, grit removal, 15-min grease flotation, 40-min flocculation, 2-hr primary settling, 2-hr secondary settling, and final settling. Single filtration at the high rate would give between 65 and 75 percent overall removal.

6. *Activated sludge.*[64] Satisfactory results are obtained with 9 hours' aeration and 3.5 ft³ of air per gal of waste. In cold months air is increased to 4 or 5 ft³. Suspended solids and 10-day BOD removal averaged 95 percent.

7. *Anaerobic and aerobic stabilization ponds.*[65] Complete treatment can be obtained with anaerobic ponds 8 to 17 ft deep, loaded at 0.011 to 0.015 lb BOD per day per ft³ followed by conventional aerobic stabilization ponds loaded at 50 to 280 lb BOD per day per acre. Complete treatment can also be obtained in an aerobic stabilization pond loaded at 50 lb BOD per day per acre.

8. *Chemical precipitation.* Ferric chloride, zinc chloride, chlorine, ferric sulfate, ferric salts plus lime, and ferric chloride plus chlorine have been used for the chemical treatment of packing house wastes. The cost of chemicals and operation are significant. Secondary treatment on trickling filters or in the activated sludge process produces a high-quality effluent. This method is used by small plants.[66]

9. *Anaerobic digestion.* Schroepfer et al. report a 95 percent BOD and 90 percent suspended-solids removal with anaerobic digestion at a temperature of 95°F, 60 turnovers per day, a loading of 0.20 lb of BOD per ft³ of digester volume per day, and a 12-hr total detention in the digester.[67] The suspended solids of the

[61] A. J. Steffen, "Waste disposal in the meat industry/2," *Water and Wastes Engineering/Industrial,* C-1-4 (May 1970).

[62] *An Industrial Waste Guide to the Meat Industry,* U.S. Public Health Service Publication No. 386, Dept. of HEW, Washington, D.C., Revised 1965.

[63] Ibid.

[64] Ibid.

[65] Ibid.

[66] A. J. Steffen, "Waste disposal in the meat industry/1," *Water and Wastes Engineering/Industrial,* B-20-2 (March 1970).

[67] George J. Schroepfer, W. J. Fullen, A. S. Johnson, N. R. Ziemke, and J. J. Anderson, "The Anaerobic Contact Process as Applied to Packinghouse Wastes," *Sewage and Industrial Wastes,* **27,** 460–486 (April 1955).

mixed liquor was 15,000 mg/l. An 18-in. vacuum was maintained on the digester to degassify the sludge. Other investigators suggest an equalizing tank, a digester loading of 0.15 lb of BOD and supplementary treatment consisting of a high-rate trickling filter loaded at 2600 lb of BOD per acre-ft, and final settling providing a surface settling rate of 800 gal/ft^2 per day. Excess sludge is lagooned.

10. *Various combinations* of the treatment processes mentioned may be used, including anaerobic ponds, aerobic ponds, and anaerobic-aerobic systems. Final effluent can also be disposed of to an oxidation pond or by irrigation in isolated areas. Aeration and the addition of air to the waste under pressure can also be used.

Laundry Wastes

The treatment of laundry wastes where public sewers are not available is dependent on location, availability of suitable stream, soil percolation characteristics, volume of the waste, and similar factors.

1. Small operations can use a plant consisting of a grease trap to remove floatable grease, a dosing tank, trickling filter with recirculation, and settling. Dosing can be provided by means of a siphon, pump, or tipping bucket.

2. Where permeable soil is available and groundwater supplies are not endangered, a plant consisting of a large grease trap, trickling filter, grease trap, and leaching pits can give satisfactory results for a period of time. The first grease trap will remove floating grease, and the second trap will remove emulsified grease that has been agglomerated in passing through the trickling filter. The filter stone will have to be cleaned or replaced after extended use; the traps will require frequent cleaning.

3. A modification of (2) is the substitution of chemical flocculation for the trickling filter.

4. Another treatment system consists of chemical precipitation using sulfuric acid to reduce the pH to about 6.0, alum to coagulate, followed by sedimentation, high-rate trickling filter, and secondary settling.

BIBLIOGRAPHY

An Industrial Waste Guide to the Meat Industry, U.S. Public Health Service Pub. No. 386, Dept. of HEW, Washington, D.C., 1965, 14 pp.

An Industrial Waste Guide to the Milk Processing Industry, U.S. Public Health Service Pub. No. 298, Dept. of HEW, Washington, D.C., 1959, 22 pp.

Fair, Gordon M., and Geyer, John C., *Elements of Water Supply and Waste-Water Disposal*, John Wiley & Sons, Inc., New York, 1958, 615 pp.

Fair, Gordon M., Geyer, John C., Okun, Daniel A., *Water and Wastewater Engineering*, John Wiley & Sons, Inc., New York, 1966.

Glossary Water and Wastewater Control Engineering, prepared by Joint Editorial Board representing APHA, ASCE, AWWA, WPCF, American Public Health Association, Washington, D.C., 1969, 387 pp.

Hardenbergh, William A., *Sewerage and Sewage Treatment*, 3rd ed., International Textbook Co., Scranton, Pa., 1950, 467 pp.

Imhoff, Karl, and Fair, Gordon Maskew, *Sewage Treatment*, John Wiley & Sons, Inc., New York, 1956, 338 pp.

Individual Sewage Disposal Systems, Recommendations of Joint Committee on Rural Sanitation, U.S. Public Health Service Reprint No. 2461, Dept. of HEW, Washington, D.C., 1947, 33 pp.

Industrial and Engineering Chemistry, **39,** No. 5 (May 1947), American Chemical Society, Easton, Pa.

Kiker, John E., Jr., "Subsurface Sewage Disposal," Bull. No. 23, *Florida Eng. & Ind. Exp. Sta.,* University of Florida, Gainesville (December 1948), 72 pp.

Manual of Septic-Tank Practice U.S. Public Health Service Pub. No. 526, Dept. of HEW, Washington, D.C., 1967, 92 pp.

Manuals of Practice, *Regulation of Sewer Use*—No. 3, 1963; *Sewer Maintenance*—No. 7, 1966; *Sewage Treatment Plant Design*—No. 8, 1959; *Design and Construction of Sanitary and Storm Sewers*—No. 9, 1960; and *Operation of Wastewater Treatment Plants*—No. 11, 1970, Water Pollution Control Federation, Washington, D.C.

National Specialty Conference on Disinfection, American Society of Civil Engineers, 345 East 47th Street, New York, N.Y., 1970, 705 pp.

Olson, Gerald W., *Application of Soil Survey to Problems of Health, Sanitation, and Engineering,* Cornell University, New York State College of Agriculture, Ithaca, N.Y., March 1964, 77 pp.

Recommended Standards for Sewage Works, Great Lakes–Upper Mississippi River Board of State Sanitary Engineers, Health Education Service, Albany, N.Y., 1968 ed., 97 pp.

Wagner, E. G., and Lanoix, J. N., "Excreta Disposal for Rural Areas and Small Communities" *WHO Mon. Ser.* No. 39, Geneva, 1958, 187 pp.

Water Quality Criteria, Federal Water Pollution Control Administration, U.S. Dept. of the Interior, Washington, D.C., April 1, 1968, 234 pp.

5

SOLID WASTE MANAGEMENT

COMPOSITION, STORAGE, AND COLLECTION

General

Aesthetic, land use, health, water pollution, air pollution, and economic considerations make proper solid waste storage, collection, and disposal municipal, corporate, and individual functions that must be taken seriously by all. Indiscriminate dumping of solid waste and failure of the collection system in a populated community for 2 or 3 weeks would soon cause many problems. Odors, flies, rats, roaches, crickets, wandering dogs and cats, and fires would dispel any remaining doubts of the importance of proper solid waste storage, collection, and disposal.

The complexities of solid waste management are not readily appreciated. Reference to Figure 5–1, however, will give a helpful perspective. As can be seen, there are numerous sources and types of solid wastes ranging from the home to the farm and from garbage to radioactive wastes, junked cars, and industrial wastes. Handling involves storage, collection, transfer, and transport. Processing includes incineration, densification, composting, separation, treatment, and energy conversion. Disposal methods show the environmental interrelation of air, land, and water, and the place of salvage and recycling. All these steps introduce constraints—social, political, economic, technological, ecological, legal, information, and communications—that must be considered in the analysis of the problem and in coming up with acceptable solutions. The discussion that follows will treat most of these factors in some depth. Chapter 2 contributes additional information on the planning and implementation aspects.

Composition, Weight, and Volume

The solid waste of a municipality may consist of garbage, rubbish, incinerator residue, as well as dead animals, business wastes, demolition materials, street sweepings, and abandoned automobiles. The general classifications are shown in Table 5–1.

Various estimates have been made of the quantity of solid waste generated and collected per person per day. The amount of municipal solid waste *generated* in 1968 was estimated to be 7 lb/capita/day and was expected to increase at about 4 percent/yr. The amount collected was found to average 5.32 lb (Table 5–2). Averages are subject to adjustment dependent on many local factors, including time of the year; habits, education, and economic status of the people; whether urban or rural area; and location. The estimates should not be used for design purposes. Each community should be studied and actual weighings made to obtain representative information.

The volume occupied by solid waste under certain conditions determines the number and size or type of refuse containers, collection vehicles, and transfer stations. Transportation systems and land requirements for disposal are also affected. For example, Kaiser found that loose refuse weighed 108 to 135 lb/yd^3.[1] Under pressure equivalent to 30 ft of refuse, the weight was 349 lb when the moisture content was 27 percent and 480 lb when the moisture was 42 percent. Settling caused a 7-percent increase in moisture. The composition of ordinary refuse is shown in Table 5–4. Solid waste characteristics and production rates from various sources are given in Tables 5–1 to 5–9.

Storage

Where refuse is temporarily stored on the premises, an adequate number of suitable containers should be provided to store the refuse accumulated between collections. Watertight rust-resistant containers with tight-fitting covers are needed. Containers for incinerator residue and ashes must, in addition, be fire resistant. Cans for ordinary refuse and ash cans up to 20- or 32-gal capacity and rubbish containers up to 50-gal capacity are practical sizes. The 10-gal can for kitchen wastes alone is a good household size for the average family when there is twice-a-week collection. In any case the weight and shape of containers must be kept within the limits that can be easily and conveniently handled by the collection crew. The weight should preferably not exceed 70 lb.

[1] Elmer Kaiser, "Refuse is the sweetest fuel," *American City*, 116–118 (May 1967).

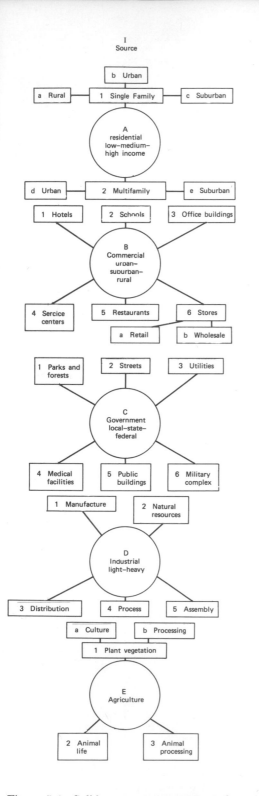

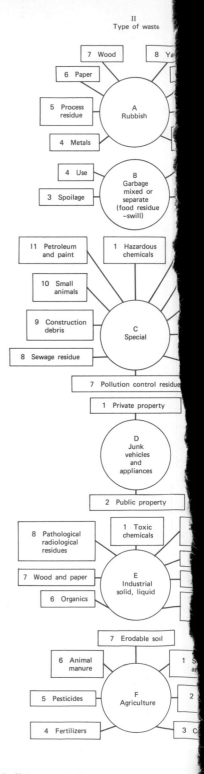

Figure 5-1   Solid waste management study matrix. (*Policies for Solid Waste Management*

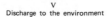

V
Discharge to the environment

VI
Constraints

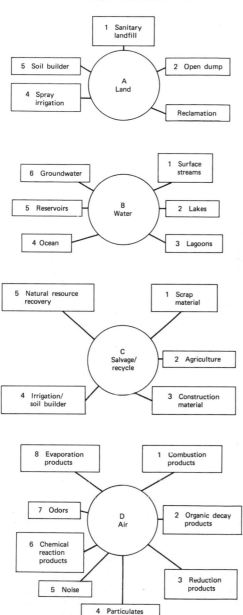

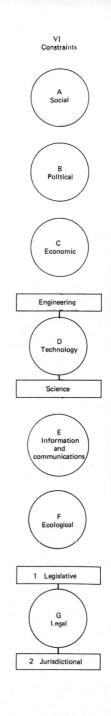

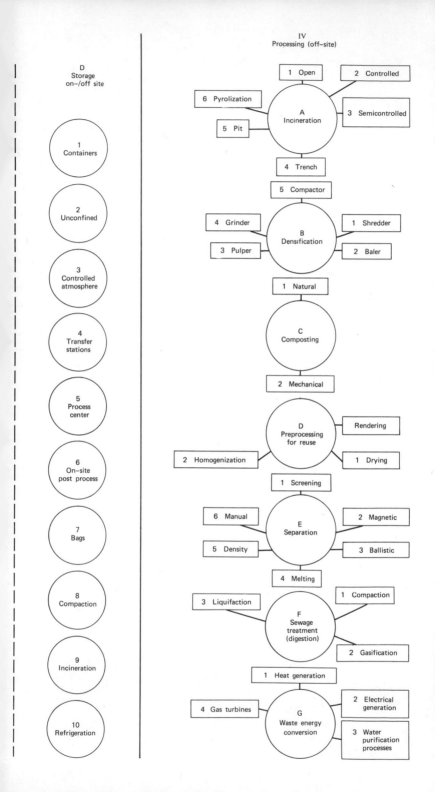

Pub. No. 2018, Dept. of HEW, Washington, D.C., 1970.)

III
Handling

d trimmings

1 Glass

2 Rags and clothes

3 Plastics and
   rubber

1 Preparation

2 Storage

2 Street
   sweepings

3 Bulky
   materials

4 Pathological
   residues

5 Radiosotopes

6 Demolition
   debris

2 Hazardous
   materials

3 Chemicals

4 Minerals

5 Inorganic
   residue

mall – large
imals

Process
waste

rop residue

A–B
On–site

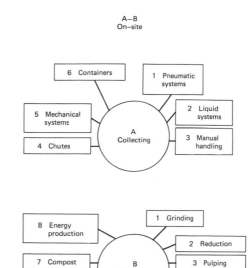

6 Containers

1 Pneumatic
   systems

5 Mechanical
   systems

2 Liquid
   systems

4 Chutes

A
Collecting

3 Manual
   handling

8 Energy
   production

1 Grinding

7 Compost

2 Reduction

B
Processing

3 Pulping

6 Separation

4 Incineration

5 Densification

Refrigeration

C
Transpo
off–s

1
Mech
deliv

2
Liq
trans

3
Pneu
trans

4
Airbo
vehic

5
Auton
transp

6
Railw
transp

7
Water
trans

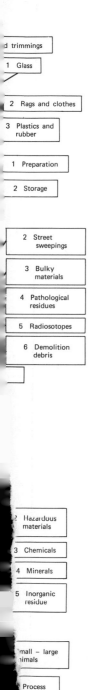

National Academy of Engineering–National Academy of Sciences, U.S. Public Healt

TABLE 5-1 REFUSE MATERIALS BY KIND, COMPOSITION, AND SOURCES

Kind	Composition	Sources
Garbage	Wastes from preparation, cooking, and serving of food; market wastes; wastes from handling, storage, and sale of produce.	Households, restaurants, institutions, stores, markets.
Rubbish	Combustible: paper, cartons, boxes, barrels, wood, excelsior, tree branches, yard trimmings, wood furniture, bedding, dunnage. Noncombustible: metals, tin cans, metal furniture, dirt, glass, crockery, minerals.	Same as garbage.
Ashes	Residue from fires used for cooking and heating and from on-site incineration.	Same as garbage.
Street refuse	Sweepings, dirt, leaves, catch-basin dirt, contents of litter receptacles.	Streets, sidewalks, alleys, vacant lots.
Dead animals	Cats, dogs, horses, cows.	Same as street refuse.
Abandoned vehicles	Unwanted cars and trucks left on public property.	Same as street refuse.
Industrial wastes	Food-processing wastes, boiler-house cinders, lumber scraps, metal scraps, shavings.	Factories, power plants.
Demolition wastes	Lumber, pipes, brick, masonry, and other construction materials from razed buildings and other structures.	Demolition sites to be used for new buildings, renewal projects, expressways.
Construction wastes	Scrap lumber, pipe, other construction materials.	New construction, remodeling.
Special wastes	Hazardous solids and liquids: explosives, pathological wastes, radioactive materials.	Households, hotels, hospitals, institutions, stores, industry.
Sewage treatment residue	Solids from coarse screening and from grit chambers; septic-tank sludge.	Sewage treatment plants, septic tanks.

From *Municipal Refuse Disposal,* Institute for Solid Wastes of American Public Works Association and Bureau of Solid Waste Management, Public Administration Service, U.S. Dept. of HEW, Chicago, Illinois, 1970, p. 13.

TABLE 5-2 AVERAGE SOLID WASTE COLLECTED, 1968
(pounds per person per day)

Solid-waste type	Urban	Rural	National
Household	1.26	0.72	1.14
Commercial	0.46	0.11	0.38
Combined	2.63	2.60	2.63
Industrial	0.65	0.37	0.59
Demolition, construction	0.23	0.02	0.18
Street and alley	0.11	0.03	0.09
Miscellaneous	0.38	0.08	0.31
Totals	5.72	3.93	5.32

Source: Anton J. Muhich, "Sample Representativeness and Community Data," *Proceedings, Institute for Solid Wastes,* American Public Works Association, Chicago, Ill., 1968.

TABLE 5-3 SOLID WASTE DISPOSAL IN NEW YORK STATE,* 1968
(pounds per capita per day)

Type of Waste	Range	Median
Residential	2.1 to 2.6	2.3
Commercial	0.4 to 2.4	0.9
Demolition and Construction	0.1 to 1.3	0.5
Industrial	0.2 to 1.4	0.3
Street Sweepings	0.1 to 0.8	0.4
Institutional	0.1 to 0.2	0.1
Trees and Brush	0.2 to 0.4	0.3
Other	0.3 to 0.9	0.6
Total Disposal	3.5 to 6.6	5.7

*Weighings from 8 counties, population 104,000 to 983,000. *Sanitary Landfill—Planning, Design, Operation, Maintenance,* New York State Dept. of Health, Albany, 1969.

TABLE 5-4 APPROXIMATE COMPOSITION OF RESIDENTIAL SOLID WASTES*

Component	Percent by weight
Food waste	15
Paper products	50
Plastics, rubber, leather	3
Rags	2
Metals	8
Glass and ceramics	8
Wood	2
Garden wastes	5
Rocks, dirt, misc.	7

*Moisture content approximately 30 percent.

TABLE 5-5   APPROXIMATE SOLID WASTE PRODUCTION RATES

Source of Waste	lb per Day
Municipal	5.5 per capita
Household	2.3 per capita, or
	5.0 per capita plus 1 lb per bedroom†
Apartment building‡	4.0 per capita per sleeping room†
Seasonal home	2.5 per capita*
Resort	3.5 per capita*
Camp	1.5 per capita*
School, general	2.1 per capita
Grade school	0.25 per capita plus 10 lb per room†
High school	0.25 per capita plus 8 lb per room†
Institution, general	2.5 per bed
Hospital	8.0 per bed†
Nurses' or interns' home	3.0 per bed†
Home for aged	3.0 per bed†
Rest home	3.0 per bed†
Hotel, first class	3.0 per room†
Medium class	1.5 per room†
Motels	2.0 per room†
Day use facility, resort	0.5 per capita*
Trailer camp	6 to 10 per trailer†
Commercial building, office	1.0 per 100 ft² †
Department store	40 per 100 ft² †
Shopping center	Survey required†
Supermarket	9.0 per 100 ft² †
Restaurant	2.0 per meal†
Drugstore	5.0 per 100 ft² †
Retail and service facility	13.0 per 1000 ft² †
Wholesale and retail facility	1.2 per 1000 ft² †
Industrial building, factory	Survey required
Warehouse	2.0 per 100 ft² †
National Forest recreation area§	
Campground	1.26 ± 0.08 per camper
Family picnic area	0.93 ± 0.16 per picnicker
Organized camps	1.81 ± 0.39 per occupant
Rented cabin, with kitchen	1.46 ± 0.31 per occupant
Lodge, without kitchen	0.59 ± 0.64 per occupant
Restaurant	0.71 ± 0.40 per meal served
Overnight lodge, winter sports area	1.87 ± 0.26 per visitor
Day lodge, winter sports area	2.92 ± 0.61 per visitor
Swimming beach	0.04 ± 0.01 per swimmer
Concession stand	0.14 per patron
Job Corps, Civilian Conservation Corps Camp,	
Kitchen Waste	2.44 ± 0.63 per corpsman
Administrative and Dormitory	0.70 ± 0.66 per corpsman

*Environmental Health Practice in Recreational Areas, U.S. Public Health Service Pub. No. 1195, Dept. of HEW, Washington, D.C., 1965.

†I.I.A. Incinerator Standards, Incinerator Institute of America, New York, May 1966. Up to 20–30 lb per bed per day at teaching hospitals, total. Total solid waste production at general hospitals is estimated at 12 to 15 pounds per bed with an average patient occupancy of 80 percent.

‡A study at a low-income high-rise multifamily housing project shows the average daily refuse disposal via trash chutes to be 1.48 lb per capita, or 6.23 lb per dwelling unit, and 5.6 lb/ft³, not including yard or bulky wastes. R. Barry Ashby, "Experiment in New Haven," Waste Age (November–December 1970), p. 22.

§Source: Charles S. Spooner, Solid Waste Management in Recreational Forest Areas, U.S. Environmental Protection Agency, Washington, D.C., 1971. Average rate of waste generation 90 percent confidence interval.

TABLE 5-6   MISCELLANEOUS SOLID WASTE GENERATION

Type	Amount
Tires	0.6 to 1.0 tires discarded per capita per year
Waste oil	2.5 gal per vehicle per year
Sewage sludge, raw	0.4 tons per day per 1,000 people, dewatered to 25% solids, with no garbage grinders 0.8 tons per day per 1,000 people, dewatered to 25% solids, with 100% garbage grinders
Sewage sludge, digested	0.25 tons per day per 1,000 people, dewatered to 25% solids, or 3 lb per capita per day dry solids
Water supply sludge	200 pounds per million gal of raw water, on a dry-weight basis, with conventional rapid-sand filtration using alum; raw water with 10 JU
Scavenger wastes	0.3 gal per capita per day
Pathological wastes	0.6 to 0.8 lb per bed per day—hospital 0.4 to 0.6 lb per bed per day—nursing home

Source: *Comprehensive Solid Waste Planning Study—1969*, Herkimer-Oneida Counties, Malcolm Pirnie Engineers, State of New York Dept. of Health, Albany.

Plastic liners for cans reduce the need for cleaning and keep down odors and fly breeding. Paper and heavy plastic sacks, either alone or mounted on special frames, are also being used. Can liners and sacks simplify and expedite collection.

Satisfactory storage and can-washing platforms for commercial and institutional operations are shown in Figures 5–2 and 5–3. Where outside platforms are impractical or undesirable, a refuse storage room may be used. Such a room should have a drained concrete floor, coved base, and waterproof wall at least to splash height. A hot-water hose bib outlet and ample natural or artificial light and ventilation are essential. The temperature of the storage room should be kept below 50°F for odor control.

Garbage cans used for the storage of wet garbage require daily cleaning; but where garbage is drained and wrapped before being deposited, less frequent cleaning is needed. Odor and fly nuisances and difficulty due to freezing are also reduced.

Garbage stands should not be screened nor should cans be whitewashed or painted. Stands should be convenient to the kitchen, in an airy shaded location away from children and stray cats and dogs. If the cans, platform, and surrounding ground are kept clean and the cans are kept

TABLE 5-7 AGRICULTURAL SOLID WASTES—WASTE PRODUCTION RATES

Category	Annual Waste Production Rate
Wet manures	
Chickens (fryers)	6.4 tons/1,000 birds*
Hens (layers)	67 tons/1,000 birds*
Hogs	3.2 tons/head*
Horses	12 tons/head†
Beef cattle (feedlot)	10.9 tons/head*
Dairy cattle	14.6 tons/head*
Fruit and nut crops	
Class 1 (grapes, peaches, nectarines)	2.4 tons/acre‡
Class 2 (apples, pears)	2.25 tons/acre‡
Class 4 (plums, prunes, misc.)	1.5 tons/acre‡
Class 5 (walnuts, cherries)	1.0 tons/acre‡
Field and row crops	
Class 1 (field and sweet corn)	4.5 tons/acre‡
Class 2 (cauliflower, lettuce, broccoli)	4.0 tons/acre‡
Class 3 (sorghum, tomatoes, beets, cabbage, squash, brussel sprouts)	3.0 tons/acre‡
Class 4 (beans, onions, cucumbers, carrots, peas, peppers, potatoes, garlic, celery, misc.)	2.0 tons/acre‡
Class 5 (barley, oats, wheat, milo, asparagus)	1.5 tons/acre‡

Source: Roy F. Weston, *A Statewide Comprehensive Solid Waste Management Study,* New York State Dept. of Health, Albany, February 1970.

*Raymond C. Loehr, "Animal Wastes—A National Problem," *J. Sanitary Engineering Division,* Proceedings, American Society of Civil Engineers (April 1969).

†Estimated.

‡"Status of Solid Waste Management in California," California Dept. of Public Health, Sacramento, September 1968.

tightly covered, there should be no cause for complaint. Built-in garbage or trash boxes, bins or sheds are not recommended; the storage structure becomes soiled, is difficult to keep clean, and becomes an odor nuisance and a rat and fly feeding and breeding ground. Removable bulk containers and compactors are recommended for institutional type operations.

Can Washing

If a properly drained can-washing platform is provided with a hot-water hose bib outlet, cleaning is simplified. The hose bib outlet should be provided with a backflow preventer. Hot water above a temperature of 120°F is necessary to dissolve grease. Grease and adhering food particles can be removed by scrubbing the can with a long-handled stiff-bris-

TABLE 5-8   INDUSTRIAL SOLID WASTE PRODUCTION RATES*

SIC Code	Industry	Waste Production Rate (tons/employee/year)
201	Meat processing	6.2
2033	Cannery	55.6
2037	Fozen foods	18.3
Other 203	Preserved foods	12.9
Other 20	Food processing	5.8
22	Textile mill products	0.26
23	Apparel	0.31
2421	Sawmills and planing mills	162.0
Other 24	Wood products	10.3
25	Furniture	0.52
26	Paper and allied products	2.00
27	Printing and publishing	0.49
281	Basic chemicals	10.00
Other 28	Chemical and allied products	0.63
29	Petroleum	14.8
30	Rubber and plastic	2.6
31	Leather	0.17
32	Stone, clay	2.4
33	Primary metals	24
34	Fabricated metals	1.7
35	Nonelectrical machinery	2.6
36	Electrical machinery	1.7
37	Transportation equipment	1.3
38	Professional and scientific inst.	0.12
39	Miscellaneous manufacturing	0.14

*Consultant's Analysis.

Source: Roy F. Weston, *A Statewide Comprehensive Solid Waste Management Study,* New York State Dept. of Health, Albany, February 1970.

TABLE 5-9   RELATION OF DOMESTIC SOLID WASTE PRODUCTION TO POPULATION DENSITY

Persons/mi^2	lb per capita per day
10 to 200	2.4 to 3.1
200 to 2,000	2.8 to 3.8
2,000 to 7,000	3.2 to 4.5
7,000 to 10,000	5.0 to 5.5

Source: Adapted from Garret P. Westerhoff and Robert M. Gruninger, "Population Density vs. Per Capita Waste Production," *Public Works* (February 1970), p. 87.

Note: Increase in pounds per capita with population density is related to increase in commercial and industrial wastes and street sweepings.

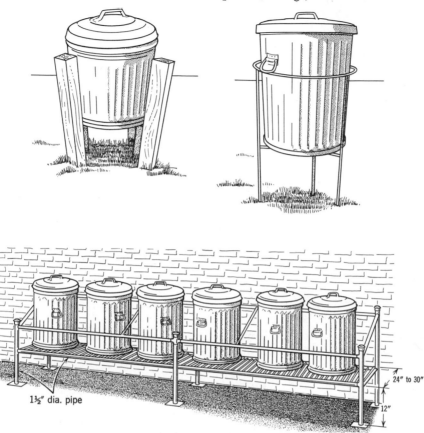

Figure 5–2 Good storage practices. (From *Refuse Collection and Disposal for the Small Community*, U.S. Public Health Service and APWA, Washington, D.C., November 1953.)

tled brush and hot water. Then the container should be rinsed with a hot-water spray and allowed to dry. The platform drainage should pass through a grease trap and enter a sewerage system.

If a washing platform is not available, a detergent can be added to a small quantity of warm water in a pail to be used to scrub the inside and outside of the can. The waste water should be collected in one garbage can and either disposed of with the garbage or discharged into a sewerage system ahead of a grease trap.

Waste water and particles of garbage should not be allowed to run on the surface of the ground, as an odor nuisance and fly-breeding problem will result.

Where steam is available, emptied garbage cans can be inverted over

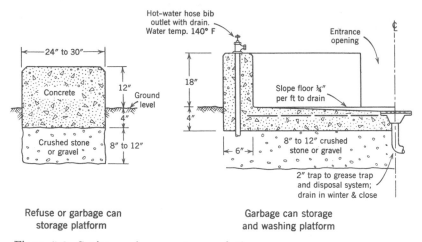

Figure 5–3   Sanitary refuse can storage platforms.

steam jets to loosen adhering garbage. The steam jet can also be used for final "sterilization" somewhat similar to the handling of milk cans, after the cans have been cleaned.

As previously stated, a can liner will reduce the frequency for can cleaning, simplify collection, and reduce fly breeding and odor potential.

Apartment House and Institution Compactors, Macerators, and Pneumatic Tubes

The disposal and removal of solid wastes from apartment houses and institutions have challenged the designer for some time. There is a real need to reduce air pollution from on-site incinerators. Compaction of wastes reduces the storage and handling volume. Maceration of wastes to a slurry form and transport to a central dewatering point, and pneumatic transport, make possible central collection for processing, incineration, or disposal. Each of these systems, and combinations, offer potential economies and conveniences as well as environmental sanitation improvements.

The limitations of on-site incinerators are well known and are discussed separately. Apartment-house and individual-home compactors are now commercially available. Maceration of solid waste to a slurry at a central processing station followed by dewatering of the resulting pulp is being refined. Although there is a volume reduction, the added moisture weight increases handling and haul costs. Odors can also be a problem. The pumping of crushed or shredded solid waste as a slurry in a pipeline, its hydraulic characteristics and feasibility, are in the experimental stage. The pneumatic transport of solid wastes in large pipelines (20 to 24 in.) from an apartment-house complex to an incinerator also providing heat for space heating is a reality in Sweden.[2]

Collection

Collection cost has been estimated to represent about 80 percent of the total cost of collection and disposal by sanitary landfill and 60 percent when incineration is used. In 1968 collection cost alone was found to average $10 to $13/ton.

The frequency of collection will depend on the quantity of solid waste, time of the year, socioeconomic status of area served, and municipal or contractor responsibility. In business districts refuse, including garbage from hotels and restaurants, should be collected daily except Sundays. In residential areas, twice-a-week refuse collection during warm months of the year and once a week at other times should be the maximum permissible interval. Slum and ghetto areas usually require at least twice-a-week collection. The receptacle should be either emptied directly into the garbage truck or carted away and replaced with a clean container. Refuse transferred from can to can will invariably cause spilling, with resultant pollution of the ground and attraction of flies. If other than curb pickup is provided, the cost of collection will be high. Some property owners are willing to pay for this extra service.

Collection equipment that simplifies the collection of refuse and practically eliminates cause for legitimate complaint is available. The tight-body open truck with a canvas or metal cover is being replaced by the automatic loading truck with packer to compact refuse dumped in the truck during collection except for the collection of bulky items. Compaction-type bodies have twice the capacity of open trucks and a convenient loading height. Low-level closed-body trailers to eliminate the strain of lifting cans are also available. The density of refuse under various conditions is shown in Table 5–10.

In some places closed-body metal bins are centrally located at a housing project, trailer camp, large hotel or restaurant, commercial establishment, and so forth. Refuse is deposited in the bin by the individual. The bin is picked up on a previously established schedule by the collector. The number and size of the collection vehicles and the number of pickups in residential and business areas for communities of different population will vary with location, affluence, and other factors.

The people must understand that a good refuse-collection service also requires citizen cooperation in the provision and use of proper receptacles

[2] Iraj Zandi and John A. Hayden, "Collection of Municipal Solid Wastes in Pipelines," ASCE National Meeting on Transportation Engineers, San Diego, Calif., February 19–23, 1968.

TABLE 5–10   WEIGHT OF REFUSE UNDER CERTAIN CONDITIONS

Condition of Refuse	Weight (lb/yd^3)
Loose refuse at curb	125 to 240
Normal compacted refuse in a sanitary landfill	750 to 850
Well-compacted refuse in a sanitary landfill	1,000 to 1,250
In compactor truck	300 to 600
Shredded refuse, uncompacted	600
Shredded refuse, compacted	1,600
Compacted and baled	1,600 to 3,200
Apartment-house compactor	700
In incinerator pit	300 to 550

in order to keep the community clean and essentially free of rats, flies, and other vermin. Collection cost in suburban areas has been estimated (1968) at about $2.50 per family per month for once-a-week service. Haul distance to the disposal facility must be taken into consideration in making a cost analysis. In some highly urbanized areas it is economical to reduce haul distance by providing large, specially designed trailers at transfer stations. In rural areas, container stations can be established at central locations. Tables 5–11 and 5–13 show some cost comparisons.

Transfer Station

The urban areas around cities have been spreading, leaving fewer nearby acceptable solid waste disposal sites. This has generally made necessary the construction of incinerators in cities or in the outskirts, or the transportation of wastes longer distances to new sites. However, as the distance from the centers of solid waste generation increases, the cost of direct haul to a site increases. A "distance" is reached (in terms of time) when it becomes less expensive to construct an incinerator, or a transfer station, near the center of solid waste generation where wastes from collection vehicles can be transferred to large tractor-trailers for haul to more distant disposal sites.

A comparison of direct haul versus use of a transfer station, for various distances, is shown in Table 5–11. This type of information is also useful in making an economic analysis of potential landfill sites, which should properly be considered with the physical development and social factors involved in site selection.

A transfer station should be located and designed with the same care as described for an incinerator. Drainage of paved areas and adequate water hydrants for maintenance of cleanliness and for fire control are equally important. See Figure 5–4.

If the cost of disposal by sanitary landfill is added to the figures

TABLE 5-11   COST (PER TON) COMPARISON OF DIRECT HAUL
AND USE OF TRANSRER STATION (1970)

| Method | Round-Trip Distance, in mi | | | | |
	10	20	30	40	60
Direct haul*	$1.80	3.60	5.40	7.20	10.80
Transfer station and haul†	$2.60	3.20	3.80	4.40	5.60

*Assumes a direct-haul cost of $.18 per ton-mile in a 20-yd³ compactor truck (approximately 5-ton capacity) with a 2-man crew.·

†Assumes a transfer station and operation cost of $2.00 per ton and a haul cost of $.06 per ton-mile in a 65-yd³ tractor-trailer (approximately 17-ton capacity) with a 1-man operator. Refuse density in compactor truck and tractor-trailer is 450 to 500 lb/yd³. Maximum allowable highway gross vehicle weight is 72,000 lb.

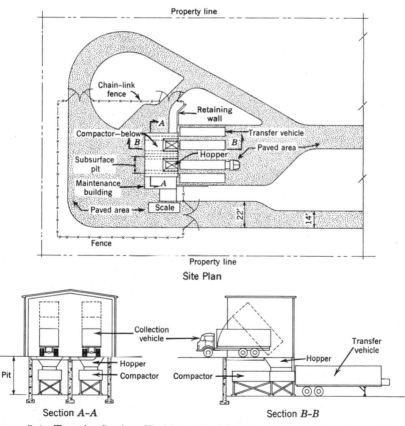

Figure 5-4   Transfer Station. Herkimer–Oneida Counties comprehensive solid waste study. Malcolm Pirnie Engrs., State of New York Dept. of Health, Albany, July 1969.

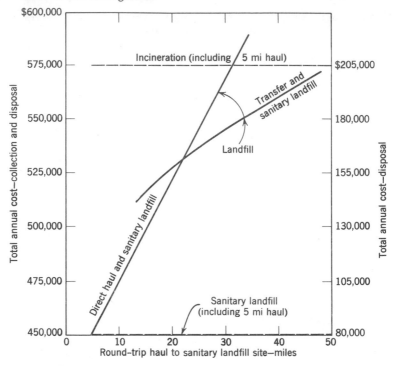

Figure 5–5 Effect of haul distances to site on cost of disposal by sanitary landfill compared to cost of disposal by incineration. 1964. Population 50,000—Density 12 people per acre. (From *Municipal Refuse Collection and Disposal,* Office for Local Government, Albany, N.Y., 1964.)

Example: Annual Cost (Dollars) Collection and Disposal

Incineration	Sanitary Landfill			Cost Difference
(5-mile Haul)	Haul	Direct	Transfer	Additional Cost of Incineration
$575,000	5	$450,000	NA	$125,000
	20	520,000	NA	55,000
	22	535,000	$535,000	40,000
	35	NA	550,000	25,000
	48	NA	575,000	equal

Cost Basis—
  Collection (5-mile Haul)—$370,000
  Incineration—$205,000
  Landfill—$80,000

Note: Each cost figure includes $370,000 for collection.
N.A.—Not applicable

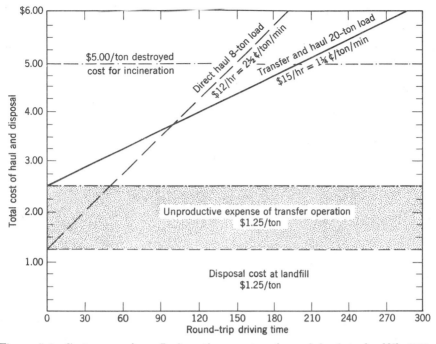

Figure 5-6 Cost comparison. Incineration vs. transfer and haul to landfill, 1968. (Courtesy Sanitation Districts of Los Angeles County, John D. Parkhurst, Chief Engineer and General Manager.)

in Table 5-11, one can compare the total cost of refuse transfer, transportation, and disposal by sanitary landfill with incineration. For example, if the total cost of modern incineration is $8.00/ton and sanitary landfill cost is $2.50/ton, the cost of incineration will become less expensive when the round-trip distance from a centrally located transfer station to a sanitary landfill exceeds 60 mi under the conditions stated. This is also illustrated in the hypothetical example shown in Figure 5-5 for a population of 50,000. Figure 5-6 shows a similar comparison for the sanitation districts of Los Angeles County, in which distance is shown in terms of times of travel to the disposal site.

Rail haul and barging to sea also involve the use of transfer stations. They may include one or a combination of grinding, compaction to various densities, and baling. Table 5-13 gives some cost estimates.

TREATMENT AND DISPOSAL OF SOLID WASTES

Solid waste disposal methods include the open dump, hog feeding, incineration, grinding and discharge to a sewer, milling, compaction, sanitary landfill, dumping and burial at sea, reduction, composting, pyrolization, wet oxidation, and anaerobic digestion. The common acceptable refuse disposal and treatment methods are incineration, sanitary landfill, and, in some parts of the world, composting. Reclamation and recycling are receiving greater attention.

Open Dump

The open dump is all too common and needs no explanation. It is never satisfactory, as usually maintained. Refuse is generally spread over a large area, providing a source of food and harborage for rats, flies, and other vermin. It is unsightly, an odor and smoke nuisance, a fire hazard, and often a cause of water pollution. It should be eliminated or its operation changed to a sanitary landfill.

Hog Feeding

Where garbage is fed to hogs, careful supervision is necessary. The spread of trichinosis to man, hog cholera, the virus of foot-and-mouth disease, and vesicular exanthema in swine is encouraged when uncooked garbage is fed to hogs. In some instances tuberculosis, swine erysipelas, and stomatitis may also be spread by raw garbage. The boiling for 30 min of all garbage fed to hogs will prevent transmission of trichinosis and economic loss to the swine industry due to hog illness and death.

The federal interstate quarantine regulations require the heat treatment of garbage transported across state boundaries for hog feed. Methods for heating garbage are given in *Equipment for the Heat-Treatment of Garbage to be Used for Hog Feed*, issued jointly by the U.S. Department of Agriculture and the Federal Security Agency in 1952. Most states have the same requirements, but enforcement is weak and in some instances ineffective. For hog feeding to be satisfactory (in addition to the cooking of garbage) it is necessary to rat-proof concrete feeding platforms and structures; remove manure and leftover waste daily; dispose of the waste by sanitary landfill or incinerate or compost the waste; clean the hog pens and flush the feeding platforms and troughs frequently. Wastewater should discharge to a disposal system that will not pollute receiving waters or become a nuisance. More often than not, these precautions are neglected. As a result fly and rat breeding is supported and bad odors are common.

Grinding

The grinding of garbage is fast becoming a common method of garbage "disposal". It is highly recommended from a convenience and public health standpoint, but the disposal of other refuse remains to be handled. The putrescible matter is promptly removed, thereby eliminating this as a source of odors and food for rats, flies, and other vermin. In one system, the home grinder is connected to the kitchen-sink drain. Garbage is shredded into small particles while being mixed with water and is discharged to the house sewer. In another system garbage is collected as before but dumped into large, centrally located garbage-grinding stations that discharge garbage to the municipal sewerage system. In small communities, the garbage-grinding station may be located at the sewage treatment plant. The strength of the sewage is increased and additional sludge digestion and drying facilities will be required when a large amount of garbage is handled.

Disposal at Sea

Where dumping at sea is practiced, all garbage and other refuse is dumped into large garbage scows or barges. The scows are towed by tugs and the garbage is taken out to sea and dumped a sufficient distance out to prevent the refuse being carried back to shore and causing a nuisance. Bad weather conditions hamper this operation and unless this method is kept under very careful surveillance, abuses and failures will result. Because of the cost of maintaining a small navy and difficulties in satisfactorily carrying out this operation, coastal cities have reverted to sanitary landfill and incineration. In recent years, consideration has been given to the compaction of refuse to a density greater than 66.5 lb/ft^3 prior to transport and then disposal by burial in the ocean at depths greater than 100 ft, based on oceanographic conditions, to ensure there will be no mixing with surface water. More research is needed to determine stability of the compacted refuse and effect on marine life. See "high-density compaction."

Garbage Reduction

In the reduction method of garbage disposal, the garbage is cooked under pressure. Fats melt out and are separated from the remaining material. The fat is used in the manufacture of soaps or glycerines and the residue is dried, ground, and sold for fertilizer or cattle feed. Odor complaints are associated with this process and, where a solvent such as naphtha is used to increase the extraction of fat, a greater fire hazard exists. The use of synthetic detergents and chemical fertilizers and high operating costs have led to the abandonment of this process.

Composting

Composting is the controlled decay of organic matter in a warm, moist environment by the action of bacteria, fungi, molds, and other organisms. This may be an aerobic and/or anaerobic operation. Moisture is maintained at 40 to 65 percent; 50 to 60 percent is best. Composition of the refuse, disposal of refuse not composted, demand for compost and salvaged material, odor production and control, public acceptance, and total cost are factors to be carefully weighed. Compost is a good soil conditioner but a poor fertilizer. Compost characteristics are given in Table 5–12. The process is very attractive, but because of the factors mentioned has not met with success in the United States.

The composting operation involves a combination of steps. These may include (1) weighing, (2) separation of noncompostables and salvage by hand and by a magnetic separator, (3) size reduction to 2 in. or less by means of a shredder, grinder, chipper, rasp mill, hammermill, (4) ballistic and magnetic separation, (5) biological digestion by any one of a number of composting methods described below, (6) screening and possible standardization of fertilizer value, and (7) disposal by bagging for sale or transporting to a sanitary landfill.

The Beccari method of garbage treatment has been used in Europe for many years. The garbage is placed in tightly sealed tanks and allowed to digest 10 days without air and then 10 to 20 days in the presence of air. Drainage from the garbage is collected at the bottom of the tank and is recirculated back over the garbage if necessary to keep it moist. The digested residue is relatively stable.

Naturizer composting uses sorting, grinding and mixing, primary and secondary composting including three grinding operations, aeration, and screening. Digested sewage sludge, raw sewage sludge, water, or segregated wet garbage is added at the first grinding for dust and moisture control. The total operation takes place in one building in about 6 days.[3]

The Dano composting plant consists of sorting, crushing, biostabilization in a revolving drum to which air and moisture are added, grinding, air separation of nonorganics, and final composting in open windrows. Temperatures of 140°F are reached in the drum. Composting can be completed in 14 days by turning the windrows after the fourth, eighth, and twelfth days.[4] Longer periods are required if the windrows are not kept small, turned, and mixed frequently and if grinding is not thorough.

Studies reported at the University of California give time and temperature of composting separated municipal refuse alone and with raw and digested sludge.[5] The process involves sorting, grinding, stacking, aera-

[3] P. H. McGauhey, "Refuse Composting Plant at Norman, Oklahoma," *Compost Science,* **1,** No. 3, 508 (Autumn 1960).

[4] Clarence G. Golueke, "Composting Refuse at Sacramento, California," *Compost Science,* **1,** No. 3, 12–15 (Autumn 1960).

[5] C. G. Golueke and H. B. Gotaas, "Public Health Aspects of Waste Disposal by Composting," *American J. Public Health,* **44,** No. 3 339–348 (March 1954).

TABLE 5-12 COMPOSTING CHARACTERISTICS

Condition	Composting Time in Days					
	0	10	20	30	40	50
Temperature, deg. F	30 to 100	140 to 147	142 to 152	142 to 151	138 to 145	130 to 135
Carbon—% of dry weight	45	43	42	41	40	38
Nitrogen—% of dry weight	0.6	0.7	0.72	0.73	0.74	0.75
C/N ratio %	80	62	58	57	56	55
pH*	5.8	6.3	8.3	8.3	8.3	8.3
Compost moisture—% of wet weight	50 to 55	48 to 53	47 to 52	46 to 50	44 to 47	41 to 43

Optimum composting period somewhat longer than seven weeks.

Source: extracted from O. W. Kochtitsky, W. K. Seman, and John S. Wiley, "Municipal Composting Research at Johnson City, Tennessee," *Compost Science*, 9, No. 4 (Winter 1969).

*At 18-in. depth, Windrow Composting.

tion by turning or adding air, and regrinding. About 28 percent of the refuse is rejected and 72 percent is used in composting. A temperature of 158 to 178°F is reached after about 3 days. This temperature is maintained for about 3 to 5 days, which is believed to be adequate to kill pathogens and parasites exposed to these temperatures. After 10 or 12 to 21 days, when the temperature of the compost drops to 122 to 131°F, the process is completed. Aerobic conditions prevent odors and accelerate the action.

Experimental studies show that the degree of decomposition of compost can be estimated by the maximum temperature reached by the refuse.[6] For example, a temperature of about 150°F can be reached in 2 to 7 days of operation; a temperature of 140 to 158°F in 7 to 21 days; a temperature of 112 to 140°F in 21 to 28 days; a temperature of 86 to 113°F in 28 to 180 or more days. Refuse going through these temperatures can be classified in terms of decomposition as slight, moderate, medium, good. As the paper content increases, time for decomposition increases.

Pathogenic organisms exposed to the higher temperatures for the times indicated will be destroyed. However, because of the nature of refuse and the range in temperature between the outside and inside of a mass of compost, this cannot be guaranteed.

The cost of composting in 1965 was estimated to be $4.00 to $8.00/ton, not including cost of disposing noncompostable wastes or value of compost salvage. The cost in 1969 was estimated to be $7 to $10/ton for a 100- to 200-ton/8 hr plant and $5 to $8 for a 200- to 500-ton/8 hr plant.[7]

Incineration

Incineration is a controlled combustion process for burning solid, liquid, or gaseous combustible waste to gases and a residue containing little or no combustible material *when properly carried out*. It is a volume reduction process suitable for about 70 percent of the municipal solid wastes. This subject is discussed at greater length in pages 411 to 422. Cost estimates are given in Table 5–13 and Figure 5–7.

---

[6] G. Niese, "Experiments to Determine Degree of Decomposition of Refuse Compost by Its Self-Heating Capability," Agricultural Microbiology Institute, Justus-Liebig-Universitat, Giessen, *Information Bull.* No. 17, 18 (May–August 1963), PHS.
[7] Garret P. Westerhoff, "A Current Review of Composting," *Public Works*, **100**, 87–90 (November 1969).

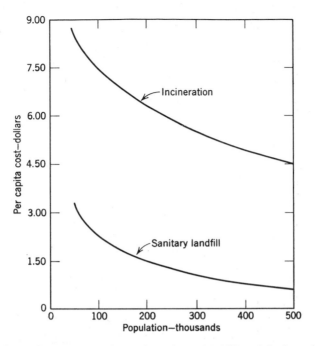

Figure 5-7   General cost comparison of sanitary landfill and incineration. (Costs include operation and fixed charges ±30%.)

Sanitary Landfill

Sanitary landfilling is an engineered method of disposing of solid wastes on land by spreading them in thin layers, compacting them to the smallest practical volume, and covering them with soil each working day in a manner that protects the environment.[8] The approximate cost of sanitary landfill versus incineration for various populations is shown in Figure 5–7. This subject is discussed further in pages 423 to 448.

Pyrolization

Pyrolization as applied to solid wastes (metal and glass removed) is an experimental thermochemical process for conversion of complex organic solids, in the absence of added oxygen, to water, combustible gases, tarry liquids, and a stable residue. Intermediate products may

[8] Dirk R. Brunner and Daniel J. Keller, *Sanitary Landfill Design and Operation,* U.S. Environmental Protection Agency, Washington, D.C., 1971.

TABLE 5-13   SOLID WASTE HANDLING AND DISPOSAL COSTS
(1968–1970)

Method	Cost in $ per ton	Capital Cost $ per ton
Collection	10.00 to 13.00	—
Sanitary landfill*	0.75 to 2.50	8.00 to 14.00
Incineration†	4.00 to 8.00	8,000 to 12,500
Composting	5.00 to 10.00	8,000 to 12,000
Pyrolizer‡	4.00 to 8.00	5,000 to 10,000
Transfer station	1.00 to 2.00	1,000 to 1,500
Haul by trailer and sanitary landfill	3.50 to 6.00	—
Compact, bale, rail haul to sanitary landfill	4.00 to 6.00	—
Rural container station§	7.00 to 10.00	—
Haul cost, 20-yd^3 compactor‖	0.20 to 0.25 per mi.	—
Compaction, rail haul, disposal to sanitary landfill¶	7.15 to 7.60	—
Compaction	1.00 to 1.35	9,000 to 12,000
Shredding	2.00 to 4.00	—
Shred, compact, bale	5.40 to 6.50	—
Barging to sea	4.00 to 6.00	—

*Capital cost includes site, fencing, roadway, scale, hammermill or shredder, tractors, trucks, engineering based on annual capacity. Can reduce by one-half if fencing, hammermill or shredder, and miscellaneous are omitted.

†The cost is estimated to vary from $9,000 to 15,000 for construction cost per ton per day rated capacity including precipitator, shredder, odor and dust control, improved furnace and grate design. For a 1,000-ton-per-day plant, the construction is estimated at $12,500 per ton per day capacity; the total annual cost including amortization at 4% interest, 40-year life of building and 15-year life of building is $7.80 per ton processed. [Casimir A. Rogus, *APWA Reporter,* 3–5 (March 1969).]

‡Experimental with solid wastes.

§Includes cost to build and operate.

‖Will vary with compactor or trailer size, men per crew, and actual travel time to disposal site.

¶Cost per ton estimated at $2.00 per ton for rail haul, $.25 for car leasing, $.75 for sanitary landfill disposal, and $4.00 for compaction and baling.

be collected or may be used to contribute heat to support the process. The end products would be carbon, water and carbon dioxide if carried to completion. If the raw material contains sulfur and nitrogen, these oxides will be formed with resultant air pollution unless provision is made for their removal. Temperatures of 900 to 1700°F have been used. In a variation, some oxygen and a temperature up to 2100°F is used. It is a process of destructive distillation similar to that used for making charcoal and for the recovery of organic by-products such as turpentine, acetic acid, and methanol from wood.[9] The Lantz converter and the Urban Research and Development Corporation unit are variations of the pyrolytic process. The Lantz system uses a temperature of 1200 to 1400°F with ground refuse.

High-Temperature Incineration

High-temperature incineration is carried out at 3400°F (Melt-zit), 2500 to 2600°F (FLK Slagging Incinerator, Germany), 2600 to 3000°F (Torrax system), 3000°F (American Design and Development Corporation). These units are in the developmental and pilot stage. Combustibles are destroyed and noncombustibles are reduced to slag or sandlike grit. A 97 percent volume reduction is reported. Special provision must be made for air pollution control.

Wet Oxidation

Wet oxidation and anaerobic digestion of refuse are in the experimental stage. In wet oxidation, the refuse can be processed under high pressures and temperatures or the refuse can be ground and aerated while in suspension in a liquid medium. In anaerobic digestion, decomposable material is separated from refuse, ground, and then digested at controlled temperature in the absence of air. The resultant gases are mostly methane and carbon dioxide, and the residue can serve as a soil conditioner.

Size Reduction

Shredders, hoggers, and chippers are devices that reduce refuse, bulky items, and other solid wastes to a manageable size for disposal by landfill or incinerator. Shredding has aesthetic and density benefits at higher cost.

High-Density Compaction

High-density compaction of solid wastes is accomplished by compression to a density of more than 66.5 lb/ft^3. The resulting bales may be enclosed in chicken wire, hot asphalt, vinyl plastic, plain or reinforced cement, or welded sheet steel depending on the method of disposal or intended use of the bales or blocks. Rogus, reporting on the Tezuka-Kosan process in Tokyo, stated that liquid release during compression ranged from 2 to 5 percent by weight and that it will require treatment. The bales had a density of 70 to 109 lb/ft^3; they sank in sea water; had good structural cohesiveness; resisted corrosion; were reasonably

[9] Donald A. Hoffman and Richard A. Fitz, "Batch Retort Pyrolysis of Solid Municipal Wastes," *Environmental Science and Technology*, 2, 1023–1026 (November 1968).

free of odors and insect and rodent hazard; and showed no evidence of aerobic or anaerobic decomposition.[10] This system has not yet been adopted in the United States but offers attractive possibilities for solid waste reuse, economic hauling, and more acceptable disposal.

Disposal of Animal Wastes

Runoff from beef cattle feedlots and from land disposal of liquid cow manure, swine wastes, poultry manure, and wastes from other concentrated livestock production can contaminate surface water and groundwater supplies, kill fish, destroy aquatic biota, and in general degrade water quality. The odors associated with the handling and disposal of these wastes contribute to the problem and may determine the disposal method and its location. The primary control measure should be reduction or prevention of pollution at the source, followed by reclamation and reuse of wastes, composting and reuse, separation of solids and liquids, or treatment and disposal to prevent water pollution or other environmental degradation.

Systems for treatment and disposal include field spreading, plow-furrow-cover, irrigation, aerobic digestion, anaerobic digestion, lagoon, aerated lagoon, oxidation ditch, lagoon and oxidation ditch, oxidation ditch and lagoon, drying, incineration, and wet oxidation. Each method has limitations and special application. The disposal of large quantities of animal wastes introduces special problems requiring individual study. Loehr suggests some alternatives.[11] More research is needed.

Where a small number of animals are kept, the manure should be collected each morning (not less than twice weekly) and spread on the fields, composted in a properly maintained compost pile, or stored in a flytight bin. Well-fed saddle horses pass 15 to 30 lb of manure and 4 to 7 qt of urine per day.[12] Cows will produce about twice as much. See also Table 5–7.

Composting of manure takes advantage of the inclination that larvae have to move out of and away from moist manure in search of a dry place to pupate. Therefore, if manure is stacked on a slatted platform over about 12 in. of water on which kerosene has been sprayed, the crawling larvae will fall out of the compost heap into the water and

---

[10] Casimir A. Rogus, "High Compression Baling of Solid Wastes," *Public Works*, **100**, 85–89 (June 1969).

[11] Raymond C. Loehr, "Alternatives for the Treatment and Disposal of Animal Wastes," *J. Water Pollution Control Fed.*, **43**, 668 (April 1971).

[12] Seymour Barfield, "Practical Fly Control for Horse Keeping on Hillside Residential Lots," *J. Environmental Health*, **28**, No. 4 (January–February 1966).

drown. Waste oil may be used instead of water, and instead of a slatted platform a well-compacted earth or concrete base surrounded by a tight earth or concrete channel containing 12 in. or more of water or oil may be used.

Manure is commonly spread on the field as a fertilizer in thin scattered layers to dry. If wet manure is plowed under, fly eggs almost always present will hatch and cause a nuisance. If weather conditions do not permit spreading of manure, it must be stored in an acceptable manner until it can be disposed of.

INCINERATION

General

A properly designed and controlled incinerator is satisfactory for burning combustible refuse provided air pollution standards can be met. Continuous operation six or seven days a week and a high controlled temperature are needed for efficiency, prevention of excessive air pollution, and odor control.

An operating design temperature range of 1500 to 1800°F is generally recommended. Excessively high temperatures and extreme variations cause cracking and spalling, with rapid deterioration of fire tile and brick linings (refractories). Batch feed or one-shift operation promote spalling and loosening of tile linings. Other types of lining and design may permit higher operating temperatures. Actually, temperature in the furnace will range from 2100 to 2500°F. When the gases leave the combustion chamber the temperature should be between 1400°F and 1800°F and the gas entering the stack 1000°F or less. The temperature will have to be lowered to 450 to 500°F before the gas is filtered or to 600°F or less if electrical precipitators are used.

Refuse storage bins or pits providing at least 3 days' storage are necessary to provide sufficient refuse for a continuous period of operation. Incineration is not generally recommended for small towns, villages, apartment buildings, schools, institutions, camps, and hotels unless good design and supervision can be assured and cost is not a factor. Past experience indicates that incinerators generally are not economically feasible for communities of less than 50,000 to 100,000 population.

Supplemental air pollution control equipment is now required on all incinerators. Additional fuel is needed when 30 percent or less of the refuse is rubbish or when the refuse contains more than 50 percent moisture; however, this is not a problem with ordinary refuse in the more developed areas of the world.

Nonincinerable Waste and Residue

Nonincinerable and bulky refuse may amount to 20 to 30 percent of the total weight of refuse collected (15 to 30 percent by volume). The residue after burning is 15 to 25 percent of the original weight (10 to 20 percent by volume). Table 5–14 shows the physical and chemical characteristics of incinerator solid waste. Food for rats and flies all too frequently is still in the residue, although proper design and operation can greatly minimize this problem. In addition, there is always the problem of what to do with nonburnable refuse, such as street cleanings, construction and demolition wastes, abandoned automobiles, industrial wastes, junk, trees, and trimmings, and so on. Burial of the incinerator residue and noncombustible refuse in a sanitary landfill can eliminate the objections mentioned. Where incinerators are carefully operated, the residue can be used as cover material. A land area (volume) of about one-quarter to one-third that required for sanitary landfill should be set aside for the burial of bulky solid wastes and incinerator residue, adequate for about 20 or more years in the future, based on population projections and industrial development. Salvage operations carried on

TABLE 5–14   PHYSICAL AND CHEMICAL CHARACTERISTICS OF INCINERATOR SOLID WASTE*

Constituents	Percent by Weight (as Received)
Proximate analysis	
Moisture	15 to 35
Volatile matter	50 to 65
Fixed carbon	3 to 9
Noncombustibles	15 to 25
Ultimate analysis	
Moisture	15 to 35
Carbon	15 to 30
Oxygen	12 to 24
Hydrogen	2 to 5
Nitrogen	0.2 to 1.0
Sulfur	0.02 to 0.1
Noncombustibles	15 to 25
Higher heating value	*Btu/lb (as received)* 3,000 to 6,000

Source: J. DeMarco, D. J. Keller, J. Leckman, and J. L. Newton, *Incinerator Guidelines 1969*, Public Health Service Pub. No. 2012, U.S. Dept. of HEW, Washington, D. C., 1969, p. 6.
   *Principally residential-commercial waste excluding bulky waste.

at the incinerator are often economical and make possible elimination of the annoying on-site separation of refuse when this practice is still required.

The cost of incineration can vary greatly dependent on the factors included in the cost and the method of reporting—the operation, maintenance, air pollution control requirements, amortization. Some comparative costs are given in Table 5–13.

Site Selection, Plant Layout, and Building Design[13]

It is extremely important that a careful investigation be made of the social, physical, and economic factors involved when incineration is proposed. Some of the major factors are:

1. Public acceptance in relation to the surrounding land use and precautions to be taken in location and design to offset public objections should be considerations. A location near the sewage treatment plant, for example, may not meet with much objection. Heat utilization for sludge drying or burning and use of treated wastewater for cooling are possibilities.

2. Site suitability in reference to foundation requirements, prevailing winds, topography, surface water and groundwater, floods, adjacent land uses, and availability of utilities should be considered. A central location to the source of wastes for minimum haul distance, smooth movement of traffic in and out of the site, and location readily accessible from major highways without interrupting traffic are important considerations.

3. Plant layout should be arranged to facilitate tasks to be performed and provide for adequate space, one-way traffic, parking area, paving, drainage, equipment maintenance and storage.

4. Building design should be attractive and provide adequate toilets, showers, locker room, and lunchroom. A control room, administrative offices, weighmaster office, maintenance and repair shops, and laboratory should be included. Adequate lighting contributes to attractiveness, cleanliness, and operating efficiency. Good landscaping will promote public acceptance.

5. Also to be evaluated are the availability and cost of providing electric power, water supply, sanitary sewers, and pretreatment required before plant wastewater can be discharged to the sewer, and availability of storm sewers, telephone, fuels.

6. The proposed method and cost of handling bulky and nonincinerable wastes should be taken into consideration when incineration is proposed. Also to be determined are the location and size of the sanitary landfill and its ability to receive incinerator residue as well as the bulky and nonincinerable solid wastes.

Incinerator Design

The basic components of an incinerator are shown in Figure 5–8. Incinerator design parameters are summarized in Table 5–15.

[13] Jack Demarco, Daniel J. Keller, Jerold Leckman, and James L. Newton, *Incinerator Guidelines—1969*, U.S. Public Health Service Pub. No. 2012, Dept. of HEW, Washington, D.C., 1969, 98 pp.

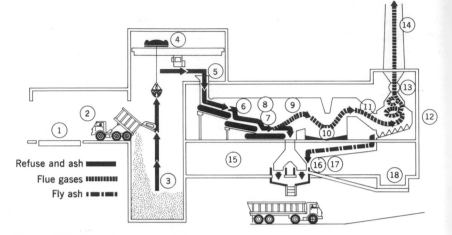

Figure 5–8   Basic incinerator design.

1. Scales
2. Tipping floor
3. Storage bin (Pit)
4. Bridge crane
5. Charging hopper
6. Drying grates
7. Burning grates

8. Primary Combustion chamber
9. Secondary combustion chamber
10. Spray chamber
11. Breeching
12. Cyclone dust collector

13. Induced draft fan
14. Stack
15. Garage—storage
16. Ash conveyors
17. Forced draft fan
18. Fly ash settling chamber

(From *Solid Waste Management, 5 design and operation,* prepared by National Association of Counties Research Foundation for U.S. Public Health Service, Dept. of HEW, Washington, D.C.)

Design Capacity. Incinerators are rated in terms of tons of burnable or incinerable waste per day. For example, an incinerator having a furnace capacity of 600 tons/day can theoretically handle 600 tons in 24 hr with three-shift operation; 400 tons in 16 hr with two-shift operation; and 200 tons in 8 hr with one-shift operation. Hence if 400 tons of incinerable wastes collected per day are to be incinerated in 8 hr, an incinerator with a rated capacity of 1200 tons/day will be required plus a 15-percent downtime allowance for repairs. Consideration must also be given, in determining design capacity, to daily and seasonal variations, which will range from 85 to 115 percent of the median.[14]

The least expensive operation for a particular community would be

[14] "Municipal Incineration of Refuse: Foreword and Introduction," Progress Report of the Committee on Municipal Refuse Practices, *J. Sanitary Engineering Division,* ASCE, 13–26 (June 1964).

TABLE 5-15   INCINERATOR DESIGN PARAMETERS*

Item	Design Factor
Storage pit	Volume of 1 to 1½ times the daily rated capacity; not less than 3 days' refuse collection recommended.
Grate loading	15 to 25 lb/ft²/hr burning rate (commercial type); 50 to 70 lb/ft²/hr burning rate, 55 lb average (other).
Grate area	Lb/hr solid waste to be burned ÷ lb/ft²/hr solid waste the grates are capable of burning.
Combustion chamber, total	35 to 40 ft³ (total)/ton rated capacity.
Air-pollution control	Depending on local air pollution control regulations, 0.5 lb of dust/1000 lb of flue gas adjusted to 50% excess air or less. Dust emission can be reduced to less than 0.2 lb/1000 lb of flue gas.
Secondary, or combustion, chamber	Gas velocity of 10 to 40 fps.
Subsidence chamber	Gas velocity 5 to 10 fps.
Breeching	Gas velocity 20 to 40 fps.
Stack	Draft 2 to 4 in. water; gas velocity 25 to 50 fps.
Fuel value of refuse	9,000 to 10,000 Btu/lb of dry combustible solids or 3,000 to 6,000 Btu/lb of ordinary refuse (± 30% moisture). Determine Btu value for each community.
Noncombustible refuse (Nonincinerable, bulky)	20 to 30% of refuse collected, by weight. 15 to 30% of refuse collected, by volume.
Residue	15 to 25% of original weight and 10–20% of original volume.
Operating temperature	1400 to 1800°F normal, 1700 to 1800°F optimum. Higher temperature recommended by some designers using special firewalls.
Supplemental fuel	When rubbish is less than 30%, or when refuse contains more than 50% moisture.
Design life	30 to 40 yr for plant; 15 yr for equipment.
Downtime for repairs	15%.
Process water	1,000 to 2,000 gal/ton solid waste processed. For residue quenching, ash conveying, wetted baffle dust collection. Can be reduced 50 to 80 percent with water treatment and recirculation. Residue is very abrasive and quench water is very corrosive.

*For specific design details see *I.I.A. Incinerator Standards,* Incinerator Institute of America, New York, May 1966.

determined by comparing the total annual cost, including operating costs and fixed charges on the capital outlay, for each method. It will generally be found that, for large cities, three-shift operation will be the least expensive. The two- or one-shift operation will be somewhat cheaper for the smaller community. The relative cost of maintenance, however, will be higher and the efficiency poor because of starting and shutting down of the furnace, with accompanying refractory brick spalling due to differential expansion and air pollution from fly ash.

Chimney. High chimneys, 150 to 200 ft above ground level, are usually constructed to provide natural draft and air supply for combustion. Stack heights of 300 to 600 ft are not uncommon. Discharge of gases at these heights also facilitates dilution and dispersal of the gases. In some designs short stacks are used for aesthetic reasons, and the equivalent effective stack height is obtained by induced draft. Meteorological conditions, topography, adjacent land use, air pollution standards, and effective stack height should govern. See chapter 6.

Types of Furnaces

The common types of furnaces are the rectangular furnace (Figure 5–10), the vertical circular furnace (Figure 5–9), the rotary kiln furnace (Figure 5–11), and the rectangular furnace with water walls (Figure 5–12). In the rectangular furnace two or more grates are arranged in tiers. The vertical circular furnace is fed from the top directly onto a circular grate. The rotary kiln furnace incorporates a drying grate ahead of a rotary drum or kiln where burning is completed. Water wall furnaces substitute water-cooled tubes for the exposed furnace walls and arches. Other types of furnaces are available. All furnaces should be designed for continuous feed.

Combustion essentials. The three essentials for combustion are time, temperature, and turbulence. There must be sufficient time to drive out the moisture, the temperature must be raised to the ignition point, and there must be sufficient turbulence to ensure mixing of the gases formed with enough air to completely burn the volatile combustible matter and suspended particulates. The combustion process involves first, drying, volatilization, and ignition of the solid waste; second, combustion of unburned furnace gases, elimination of odors, and combustion of carbon suspended in the gases. This second step requires a high temperature, at least 1400 to 1800°F, sufficient air, and mixing of the gas stream to give it turbulence until burning is completed.

Incinerators are of the batch or intermittent feed type and the continuous feed type. Air pollution and odors are associated with the first type

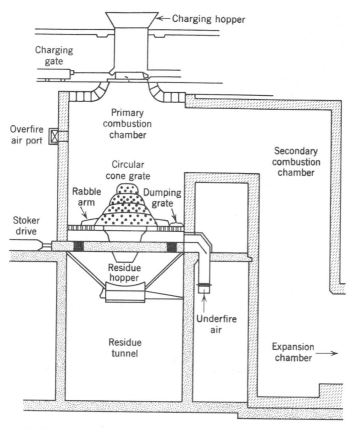

Figure 5-9   Vertical circular furnace. (From J. DeMarco, D. J. Keller, J. Leckman, and J. L. Newton, *Incinerator Guidelines 1969*, U.S. Public Health Service Pub. 2012, Dept. of HEW, Washington, D.C., p. 27.)

due to intermittent charging and changes in temperature and air supply. A continuous feed furnace permits addition of well-mixed refuse at a uniform rate with controlled temperature and air supply.

Furnace walls. Modern furnace walls are usually lined with tile or have water walls. With tile refractories, repairs can be readily made without the need for expensive and time-consuming rebuilding of entire solid brick walls found in old plants. Special plastic or precast refractories can be used for major or minor repairs. Water walls in a furnace actually consist of water-cooled tubes that serve as heat exchangers, thereby reducing the outlet gas temperature and simplifying dust collec-

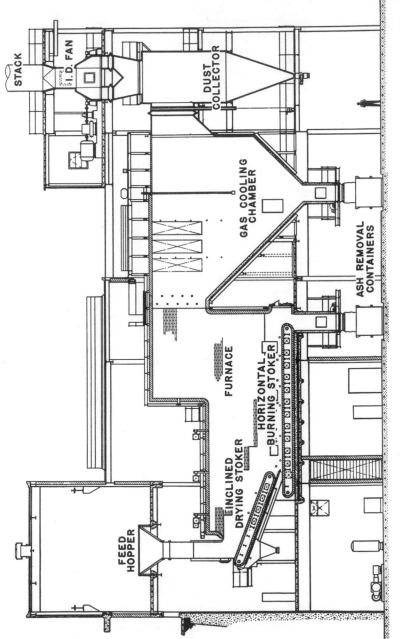

Figure 5-10 Modern continuous-feed, refractory-lined incinerator with traveling-grate stokers (rectangular type). (From Richard C. Corey, *Principles and Practices of Incineration*, Wiley-Interscience, New York, 1969, p. 183.)

418

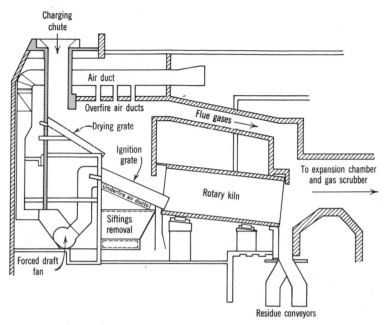

Figure 5–11   Rotary kiln furnace. (From J. DeMarco, D. J. Keller, J. Leckman, and J. L. Newton, *Incinerator Guidelines 1969*, U.S. Public Health Service Pub. 2012, Dept. of HEW, Washington, D.C., p. 30.

tion. The tubes also cover and protect exposed furnace walls and arches. Less air is required: 100 to 200 percent excess air for refractory walls compared to less than 80 percent for water walls. External pitting of the water-cooled tubes may occur if the water temperature drops below 300°F due to condensation of the corrosive gases. Internal tube corrosion must also be prevented by recirculation of conditioned water.

Control of Incineration

The poor image that incineration has in the eyes of many people is due largely to the failure to control operation, with resultant destruction of the equipment and air pollution. A properly designed and operated incinerator requires control instrumentation for temperature, draft pressures, smoke emission, weights of solid wastes coming in and leaving the plant, and air pollution control equipment.

Temperature. It is necessary for control purposes to monitor the incoming air and gases leaving the combustion chamber at the settling chamber outlet, at the cooling chamber outlet, at the dust collector inlet and outlet, and the stack temperature. Furnace temperature can be controlled

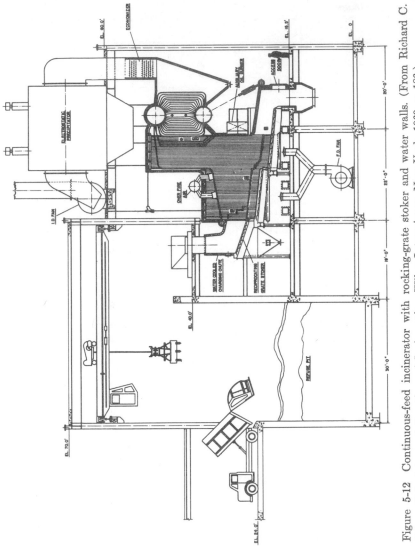

Figure 5-12 Continuous-feed incinerator with rocking-grate stoker and water walls. (From Richard C. Corey, *Principles and Practices of Incineration*, Wiley-Interscience, New York, 1969, p. 186.)

by adjusting the amount of overfire or underfire air. The temperature of the gases leaving the furnace are cooled by spraying with water (causes a white stack plume unless the flue gas is reheated before discharge), by dilution with cool air (high equipment cost to handle large volumes of diluted gases), or by passing through heat exchangers (ready market for heat, steam or high-temperature water needed). Gas scrubbers using water sprays can be used to cool the effluent gas so that an induced-draft fan can be used to reduce the chimney height; large particulates can also be removed.

Draft pressure to control the induced-draft fan and the chimney draft. Measurements should be made at the underfire air duct, overfire air duct, stoker compartment, sidewall air duct, sidewall low furnace outlet, dust collector inlet and outlet, and induced fan inlet. Control of underfire air can provide more complete combustion with less fly-ash carryover up the stack.

Smoke density. The smoke emission can be controlled by continuous measurement of the particulate density in the exhaust gas. A photoelectric pickup of light across the gas duct is used, preferably located between the particulate collector and the induced fan duct.

Weigh station. Platform scales to weigh and record the incoming solid waste and outgoing incinerator residue, fly ash, siftings, and other materials are generally required.

Instrumentation should include devices to keep record of overfire and underfire air flow rates; temperatures and pressures in the furnace, along gas passages, in the particulate collectors, and in the stack; electrical power and water use; and grate speed.

Odor control requires complete combustion of hydrocarbons, that is, excess air and a retention time of 0.5 seconds at 1500°F (above 1400°F at the exit of the furnace). Adequate dilution of gases leaving the stack by an effective stack height (actual stack height plus plume rise) is another possible method for odor control, but its effectiveness is related to meteorological conditions and persistence of the odors. Wet scrubbers can also be used to absorb odors while removing particulates.

Other gaseous emissions in addition to carbon dioxide, water vapor, and oxygen emissions include sulfur oxides, nitrogen oxides, carbon monoxide, and hydrogen chloride. Hydrogen chloride is released when plastic polyvinyl chloride is burned. The hydrogen chloride can cause corrosion of air pollution control equipment. Wet scrubbers are believed to be effective in removing the hydrogen chloride gas as well as fly ash.

Particulate emissions can be controlled by settling chambers, wetted baffle spray system, cyclones, wet scrubbers, electrostatic precipitators, and fabric filters. Their efficiencies and other details are discussed in

Chapter 6. Apparently only wet scrubbers, electrostatic precipitators and bag filters can meet an air pollution code requirement of less than 0.5 lb of particulates per 1000 lb of flue gas at 50 percent excess air. Cyclones in combination with other devices might approach the standard.

On-Site Incineration[15]

When possible a large municipal incinerator should be used in preference to a small on-site incinerator. Better operation at lower cost with less air pollution can usually be expected. However, on-site incinerators are used in homes, apartment houses, hospitals, schools, and commercial and industrial establishments. In recent years their use is being severely limited by air pollution control requirements. In Los Angeles, on-site incinerators are prohibited. In the City of New York, their use was required in buildings with more than four dwelling units, but strict air pollution standards make incinerators feasible only in large structures. Chutes and compaction units are finding application.

Residential incinerators need supplementary fuel to burn domestic garbage satisfactorily. After-burners are used to obtain smokeless and odorless operation. They are not entirely reliable in view of variable design, operation, and maintenance. Convenience, cleanliness, and reduced storage facilities make on-site units attractive.

Multiple-dwelling incinerators. In the single-flue type the refuse drops directly into the furnace. If a double flue is provided, the flue gases are separated from the charging duct. In a chute-fed incinerator, the refuse is dumped in a chute through hopper doors on each floor and are collected in a basement bin. The refuse is then fed into the incinerator mechanically or by hand. Auxiliary burners are needed in all types to maintain adequate temperature and attain complete combustion. Existing incinerators, unless designed to meet present-day standards, require major alterations.

Commercial and industrial incinerators are of the retort and in-line multiple-chamber types. The single-chamber incinerator is unsatisfactory. Auxiliary heat is usually needed to compensate for marginal or poor operation and for the burning of certain wastes. Some units need to be redesigned in order to meet modern air pollution control standards. The incineration of hazardous or toxic solid and liquid industrial wastes calls for special design and carefully controlled operation, with attention to permissible stack emissions, residue handling, and protection of the operator.

[15] Richard C. Corey, *Principles and Practices of Incineration,* Wiley-Interscience, New York, 1969.

SANITARY LANDFILL

Introduction

Sanitary landfill is a controlled method of refuse disposal in which refuse is dumped in accordance with a preconceived plan, compacted, and covered during and at the end of each day. *It is not an open dump.* The nuisance conditions associated with an open dump such as smoke, odor, unsightliness, and insect and rodent problems are not present in a properly designed and operated sanitary landfill. A sanitary landfill is as much an engineering project as is construction of a building. It is a well organized, well planned, and well executed construction project to dispose of solid wastes without causing health hazards or nuisance conditions.

Municipalities have an obligation to provide or make available, by whatever means feasible, facilities and services to accomplish this objective. The purpose of this section is to set forth the proper methods and procedures necessary to ensure that a solid waste disposal project is a community asset rather than liability.

Sanitary Landfill Planning and Design[16]

Legal Requirements

State and local sanitary codes or laws usually require that a new refuse disposal area not be established until the site and method of proposed operation have been approved in writing by the department having jurisdiction. The department should be authorized to approve a new refuse disposal area and require such plans, reports, specifications, and other data as are necessary to determine whether the site is suitable and the proposed method of operation feasible. Intermunicipal planning and operation on a county or regional basis should be given very serious consideration before a new refuse disposal site is acquired because larger operations result in more efficient and lower unit costs of operations.

Intermunicipal Cooperation—Advantages

County or regional area-wide planning and administration for solid waste collection, treatment, and disposal can help overcome some of the seemingly insurmountable obstacles to satisfactory solution of the

[16] Joseph A. Salvato and William G. Wilkie, "New York *Plans* for Solid Waste," *J. Environmental Health*, **32**, No. 2, 202–208 (September/October 1969).

problem. Some of the advantages of county or regional area-wide solid wastes management are the following:

1. It makes possible comprehensive study of the total area generating the solid wastes and consideration of area-wide solution of common problems on short-term and long-term bases. This can also help overcome the mutual distrust that often hampers joint operations among adjoining municipalities.

2. There is usually no more objection to one large site operation than to a single town, village, or city operation. Coordinated effort can therefore be directed to overcoming the objections to one site and operation, rather than to each of several town, village, and city sites.

3. The unit cost for the disposal of a large volume of solid waste is less. Duplication of engineering, overhead, equipment, labor, and supervision is eliminated. Figure 5-7 gives some cost comparisons.

4. Better operation is possible in an area-wide service, as adequate funds for proper supervision, equipment, and maintenance can be more easily provided.

5. More sites can be considered. Some municipalities would have to resort to the more costly method of incineration because suitable landfill sites may not be available within the municipality.

6. County or regional financing for solid waste disposal often costs less, as a lower interest rate can usually be obtained on bonds because of the broader tax base.

7. A county agency or a joint municipal survey committee, followed by a county or regional planning agency, and then an operating department, district or private contractor, is a good overall approach because it makes possible careful study of the problem and helps overcome interjurisdictional resistance.

Social and Political Factors

An important aspect of refuse disposal site selection, in addition to engineering factors, is the evaluation of public reaction and education of the public so that understanding and acceptance are developed. A program of public information is also needed. Equally important are the climate for political cooperation, cost comparison of alternative solutions, available revenue, aesthetic expectations of the people, organized community support, and similar factors.

Films and slides that explain proper sanitary landfill operations are available from state and federal agencies and equipment manufacturers. Sites having good operations can be visited to obtain firsthand information.

Experience shows that where open dumps have been operated, there will be opposition to almost any site proposed for sanitary landfill or an incinerator for that matter. However, local officials will have to study all of the facts and make a decision to fulfill their responsibility and exercise their authority for the public good in spite of any expressed opposition. Usually the critical factor is convincing the public that a nuisance-free operation will in fact be conducted.

Planning

Local officials can make their task easier by planning ahead together on a county or regional basis for 20 to 40 yr in the future and by acquiring adequate sites at least 5 yr prior to anticipated needs and use. The availability of federal and state funds for planning for collection, treatment, and disposal of refuse on an area-wide basis such as a county should be explored. The planning will require an engineering analysis of alternative sites including population projections, volume, and characteristics of all types of solid wastes to be handled; cost of land and site preparation; expected life of the site; haul distances from the sources of refuse to the site; cost of operation; and possible value of the finished sites. Consideration would be given to the climate of the region, prevailing winds, zoning ordinances, geology, and topography. Location and drainage to prevent surface water or groundwater pollution, access roads to major highways, and availability of suitable cover material are other considerations.

Once a decision is made it should be made common knowledge and plans developed to show how it is proposed to reclaim or improve and reuse the site upon completion. This should include talks, slides, news releases, question-and-answer presentations, and inspection of good operations. Artist's renderings are very helpful in explaining construction methods and final use of the land. See Figures 5–13 to 5–17.

The general planning is followed by specific site selection and preparation. Site preparation requires that an engineering survey be made and a map drawn at a scale of not less than 200 ft to the in. with contours at 2-ft intervals, showing the boundaries of the property, location of structures within 1000 ft, adjoining ownership, topography, soil borings, groundwater levels, prevailing winds, and drainage. Proposed year-round access roads, direction of operation, borrow areas, finished grade and drainage with allowance for settlement, proposed windrows and fences, housing for men and equipment (trailer or building including locker room, showers, water supply, sewage disposal, or privies), office and telephone are also shown. See Figure 5–15.

It is essential that the refuse site be planned as an engineering operation and that it be under close engineering direction to assure that operation and maintenance follow the proposed plan and that the site does not degenerate into an open dump. The sanitary landfill supervisor should keep a daily record of the type and quantity of solid wastes received, sources of wastes, equipment usage and repairs, personnel mandays, and other pertinent data. He should be thoroughly familiar with the plan and be supplied the personnel and equipment to carry out the work properly. Consideration should be given to an attractive entrance and approach road.

Location

The site location directly affects the total refuse collection and disposal cost. If the site is remotely situated, the cost of hauling to the site may become high and the total cost uneconomical. It has been established that the normal economical hauling distance to a refuse disposal site is 10 to 15 mi, although this will vary depending on the volume of refuse and other factors. Actually, the hauling time is more important than the hauling distance. The disposal site may be as far as 30 to 40 mi away if a transfer station is used. Rail haul and barging introduce other possibilities. The cost of transferring the refuse per ton is used to compare refuse collection and disposal costs and to make an economic analysis. See pages 399 to 401.

Accessibility

Another important consideration in site selection is accessibility. A disposal area should be located near major highways in order to facilitate use of existing arterial roads and lessen the hauling time to the site. Highway wheel load and bridge capacity and underpass and bridge clearances must also be investigated. It is not good practice to locate a landfill in an area where collection vehicles must constantly travel through residential streets in order to reach the site. The disposal area itself should normally be located at least 500 ft from habitation, although lesser distances have been successfully used to fill in low areas and improve land adjacent to residential areas for parks, playgrounds, or other desirable uses. Where possible, a temporary attractive screen should be erected to conceal the operation.

In order that vehicular traffic may utilize the site throughout the year, it is necessary to provide good access roads to the site so that trucks can move freely into and out of the site during all weather conditions and at all seasons of the year. Poorly constructed and maintained roads to a site can create conditions that cause traffic tie-ups and time loss for the collection vehicles.

Land Area (Volume) Required

The amount of refuse that will be produced by the communities served by the disposal area must be estimated in order to determine the amount of land that is needed. Land area for sanitary landfill should provide for a 20- to 40-yr period. Since the population in an area will not usually remain constant, it is essential that population projections be taken into account. The total average amount of refuse collected by a community was estimated in 1968 to be 5.32 lb per capita per day. However, it is estimated that this average figure will increase at the rate of 4 percent/yr. These factors plus the probable solid waste contributions by industry and agriculture must be taken into account in planning for needed land. Disposal of industry wastes may double land requirements. *Refuse production will vary, hence each community should make its own estimates.*

The space needed for refuse disposal is a function of population served, per capita refuse contribution, density of the refuse in place, total amount of earth cover used, and time in use. This may be expressed as

$$Q = \frac{peck}{d}, \text{ in which}$$

Q = space or volume needed in acre-ft per year.

p = population served.

e = ratio of earth to compacted fill. Use 1.25 if one part earth is used to four parts fill; use 1.20 if one part earth is used to five parts of fill; use 1.0 if no earth is used.

c = pounds collected per capita per day.

$k = 0.226 = \dfrac{365 \text{ days/yr} \times 27 \text{ ft}^3/\text{yd}^3}{43,560 \text{ ft}^3/\text{acre-ft}}$.

d = density of compacted fill. A density of 800 to 1000 lb/yd^3 is readily achieved with proper operation; 600 or less is poor; 1200 or more is very good.

Depth to Rock and Groundwater

In order to determine the depth at which a sanitary landfill can be operated, the location of bedrock, the groundwater table, and finished grade must be determined. This will require borings or test holes over the area under consideration. The location of bedrock and the highest groundwater table is of utmost importance in planning for a refuse disposal area. The bottom of sanitary landfills must be well above the high groundwater level and bedrock.

Prevention of Groundwater and Surface Water Pollution[17]

Experience indicates that the usual mixed refuse that is placed no deeper than 3 to 5 ft above the groundwater table and bedrock will

---

[17] Joseph A. Salvato, William G. Wilkie, and Berton E. Mead, "Sanitary Landfill Leachate Prevention and Control," *J. Water Pollution Control Fed.*, **43**, (October 1971), pp. 2084–2100.

not present any serious hazard of groundwater pollution provided surface water is drained off the site.

If refuse is placed below the groundwater table or directly on bedrock, serious pollution of the groundwater can result. This pollution will mainly be the result of leachings from the refuse in contact with water and the transfer of gases such as carbon dioxide, methane, hydrogen sulfide, nitrogen, and ammonia (by diffusion and convection), which are produced during refuse decomposition. The gases entering the groundwater accelerate the dissolving of calcium, magnesium, iron, and other substances that are undesirable at high concentrations in water.

The runoff from the drainage area tributary to the refuse disposal site must be determined to ensure that the surface water drainage is properly diverted to prevent flooding, erosion, infiltration, and groundwater pollution. The topography and soil cover should be carefully examined to be sure that there will be no obstruction of natural drainage channels. Obstructions could create flooding conditions and excessive infiltration during heavy rains and snow melt. Flooding conditions can erode the cover material, expose the refuse, and cause the rapid travel of dissolved organic and chemical pollutants through the refuse to the groundwater table. The velocity of groundwater flow will be increased as a result of the added flood water. In addition to water pollution problems, the erosion can create a condition in which insects and rodents can breed and find harborage and it permits the escape of odors.

Sites should be at least 200 ft from streams, lakes, or other bodies of water and well above groundwater. If a geological and hydrological study is made and special arrangements are made to contain and prevent the travel of pollution, a lesser distance may be used if acceptable to the agencies having jurisdiction. In any case special care must be exercised to provide for surface water and groundwater drainage and natural runoff.

Cover Material

The site should provide suitable cover material. Since the advancing face of a sanitary landfill should be covered with at least 6 in. of earth daily, and 2 ft of earth is required on the final surface, a sufficient quantity of earth should be available for the entire operation for the life of the site. See Figure 5–13 and Table 5–16. A rough estimate for cover material requirements is 1 yd³ per capita per year. Another estimate is that the volume of required earth cover is $\frac{1}{5}$ to $\frac{1}{4}$ the volume of compacted refuse. Actually, the amount of cover material required is a function of the depth of the compacted fill. From experience, the most suitable soil for cover material is one that is easily worked and

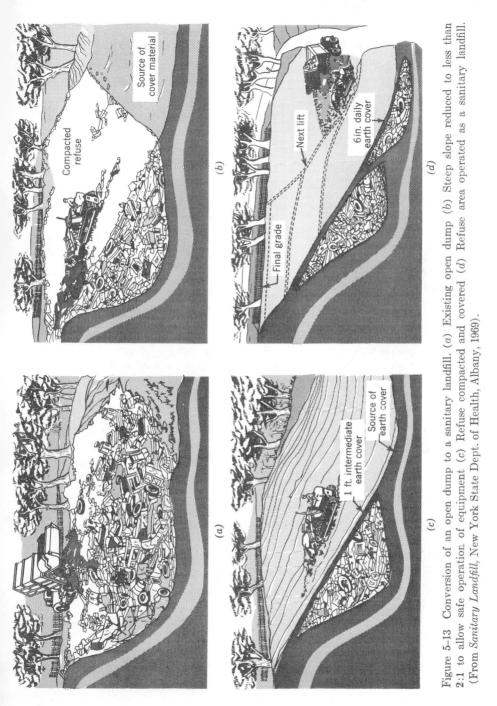

(a)   *(b)*

(c)   *(d)*

Source of cover material

Compacted refuse

Final grade

Next lift

6 in. daily earth cover

1 ft. intermediate earth cover

Source of earth cover

Figure 5-13   Conversion of an open dump to a sanitary landfill. (*a*) Existing open dump (*b*) Steep slope reduced to less than 2:1 to allow safe operation of equipment (*c*) Refuse compacted and covered (*d*) Refuse area operated as a sanitary landfill. (From *Sanitary Landfill*, New York State Dept. of Health, Albany, 1969).

TABLE 5-16   TYPE AND DEPTH OF EARTH COVER

Type of Cover	Depth	Period Left Open
Daily	6 in.	Less than a week
Intermediate	12 in.	Week to a year
Final	24 in. or more	More than a year

yet minimizes infiltration; however, this is not always available. In any case the soil generally available can be made acceptable. It is good practice to stockpile material for cold weather operation and access road maintenance. Residue from a properly operated incinerator might be used for cover if it is free from organic material. Cover material between lifts, usually 8 to 10 ft deep, should be at least 12 in. of compacted earth.

Fire Protection

The availability of fire protection facilities at a site should also be considered since fire may break out at the site without warning. Protective measures may be a fire hydrant near the site with portable pipe or fire hose, a watercourse from which water can be readily pumped, a tank truck, or an earth stockpile. The best way to control deep fires is to separate the burning refuse and dig a fire break around the burning refuse using a bulldozer. The refuse is then spread out so it can be thoroughly wetted down or smothered with earth. Limiting the refuse cells to about 200 tons, with a depth of 8 ft and 2 ft of compacted earth between cells (cells 15×112.5 ft or 20×85 ft assuming 1 yd^3 of compacted refuse weighs 800 lb), will prevent the spread of underground fires. The daily 6-in. cover will also minimize the start and spread of underground fires. Fires are a rare occurrence at a properly compacted and operated sanitary landfill.

Weigh Station

It is desirable to construct a weigh station at the entrance to the site. Vehicles can be weighed upon entering and, if necessary, billed for use of the site. Scales are required to determine tonnage received, unit operation costs, relation of weight of refuse to volume of in-place refuse, area work loads, personnel, collection rates, organization of collection crews, and need for redirection of collection practices. However, the cost involved in construction of a weigh station cannot be always justified for a small sanitary landfill handling less than 50 to 100 tons/

day. Nevertheless, estimates of volume and/or weights received should be made and records kept on a daily or weekly basis to help evaluate collection schedules, site capacity, usage, and so forth. At the very least, an annual evaluation is essential.

Conversion of a Dump to a Sanitary Landfill

The typical open dump may be in a relatively flat low area or on a steep embankment. If the area (volume) available is limited, the dump should be graded and covered. A rat poisoning program should be instituted 2 weeks before the dump is covered and abandoned. If the site is adjacent to a stream, the refuse should be removed for an appropriate distance back from the high-water level to allow for the construction of an adequate and substantial, protected earth dike between the dump area and stream high-water level. The site is then finished off by covering with at least 2 ft of compacted earth.

Where adequate land is available and the site is suitable, steps should be taken to convert the dump into a properly operated and maintained sanitary landfill as described under "Sanitary Landfill Methods." A rat-poisoning program should be carried on for 2 weeks before the conversion of the dump into a sanitary landfill. The conversion should be preceded by a plot layout drawn to scale showing the available land and its sequence of use. A plan of operation, supervision, and maintenance should accompany the layout. Conversion of a suitable existing site overcomes the problems associated with the selection of a new site and provides an opportunity to demonstrate dramatically the difference between an open dump and a sanitary landfill. The sequence for conversion of a dump to a sanitary landfill is shown in Figure 5–13.[18]

Sanitary Landfill Methods

General

There are many methods of operating a sanitary landfill. The most common are the trench, area, ramp, valley, and low-area fill methods. The trench method has the advantage of providing a more direct dumping control, which is not always possible with the area method. Since a definite place is designated for dumping, the scattering of refuse by wind is minimized and trucks can be more readily directed to the trench. The area method is more suitable for level ground. Here it is necessary

[18] See also *Closing Open Dumps* and *Open Dump Closing—Alternative Procedures,* Environmental Protection Agency, Solid Waste Management Office, Cincinnati, Ohio, 1971.

to strip and stockpile sufficient cover material to meet the total need for earth cover; if this is not possible, earth must be hauled in. There are many variables. In all cases refuse should be spread and compacted in 12- to 18-in. layers as dumped and promptly covered, as explained previously under "Cover Material." The spreading and compaction should be on a 30-percent slope.

Trench Method

The trench method is used primarily on level ground, although it is also suitable for moderately sloping ground. In this method, a trench is constructed by making a shallow excavation and using this excavated material to form a ramp above the original ground. Refuse is then methodically placed within the excavated area, compacted, and covered with suitable material at the end of the day's operation. Earth for cover material may be obtained from the area where the next day's refuse will be placed; hence a trench for the next day's refuse is completed as cover material is being excavated. Trenches are made 20 to 25 ft wide. The depth of fill is determined by the established finished grade and depth to groundwater or rock. If trenches can be made deeper, more efficient use is made of the available land area. Figure 5–14 shows the trench method.

Area or Ramp Method

On fairly flat and rolling terrain, the area method can be utilized by using the existing natural slope of the land. The width and length of the fill slope are dependent on the nature of the terrain, the volume of refuse delivered daily to the site, and the approximate number of trucks that must be unloading at the site at one time. Side slopes are 30 percent; width of fill strips and surface grades are controlled during operation by means of line poles and grade stakes. The working face should be kept as small as practical to take advantage of truck compaction, restrict dumping to a limited area, and avoid scattering of debris. In the ramp method, earth cover is scraped from the base of the ramp. In the area method, cover material is hauled in from a nearby stockpile or from some other source. See Figure 5–15.

Low-Area Method

The sanitary landfill can also be used to improve marginal and hazardous areas, such as lowlands, depressions, swamps, and pits provided the spread of surface water and groundwater pollution is prevented and filling is not prohibited. The same basic method is used in this operation as in the area method except that cover material may have

Figure 5-14 Sanitary landfill, trench method.

to be brought in. In addition, some means must be provided to prevent the scattering of refuse throughout the area. Depending on the condition of the area, this can be done by diking or "fencing."

In wet areas, cover material may be obtained by the use of a dragline working ahead of the active face. The dragline can stockpile wet material in advance in order to let the material drain. This material will then be more stable and more easily handled by equipment for refuse cover.

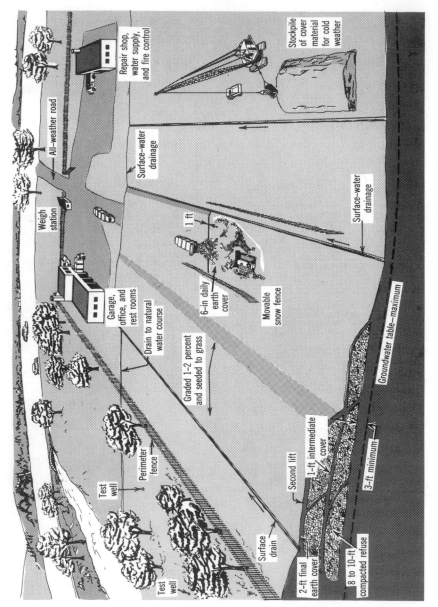

Figure 5-15 A sanitary landfill in a flat area. (Adapted from *Sanitary Landfill*, New York State Dept. of Health, Albany, 1969.)

434

In this method the slope on the active face of the refuse must be gentle enough to allow a tractor to operate and compact refuse without becoming bogged down in mud. See Figure 5–16.

If it is proposed to use a low or wet area, arrangements must first be made to divert and drain surface water to lower the groundwater table. If the design engineer can justify that there will be no contravention of surface water or groundwater standards, a low-area fill may be approved. In such instances, it is considered good practice to dump refuse in the cool months only when the temperature is less than 50°F. Prompt covering with earth will prevent odor nuisances that may arise from the decomposition of dumped refuse during hot weather. Another method is to construct a watertight dike to isolate the fill area from the adjoining body of water and pump out the trapped water to facilitate filling with demolition and other inert material to above the surrounding water level. Park areas, marinas, and open spaces can be reclaimed along water courses and coastal areas by this method, but a solid protected earth dam must be constructed along the shore line to prevent leachate seepage and erosion.

Valley or Ravine Area Method

In valleys and ravines, the area method is usually the best method of operation. In those areas where the ravine is deep, the refuse should be placed in "lifts" from the bottom up with a depth of 8 to 10 ft. Greater depths are also used. Cover material is obtained from the sides of the ravine. It is not always desirable to extend the first lift the entire length of the ravine. It may be desirable to construct the first layer for a relatively short distance from the head of the ravine across its width. The length of this initial lift should be determined so that about one year's settlement can take place before the next lift is placed, although this is not essential if operation can be carefully controlled. Succeeding lifts are constructed by trucking refuse over the first lift to the head of the ravine. When the final grade has been reached (with allowance for settlement), the lower lift can be extended and the process repeated.

Care must be taken to avoid pollution of both surface waters and groundwaters. This can be done by intercepting and diverting groundwater away from the fill area or by directing its flow through pipes of suitable size. See Figure 5–17.

Special provision should be made for temporary surface water drainage and runoff to prevent erosion, wash out of refuse, and contact of surface water with the filled area. In some cases, it may be practical to provide a diversion ditch to carry most of the runoff away from the fill area.

Figure 5-16 Lowland sanitary landfill. (From *Sanitary Landfill*, New York State Dept. of Health, Albany, 1969.)

Prevailing winds

Compactor truck unloading

Cover material stockpiled for draining

Dragline excavating cover material

Tractor for earth movement and compacting

Lowland

Demolition and other inert material filled to above high-water level and compacted before dumping ordinary refuse

6-in. daily cover

2-ft earth cover

8-ft

Completed refuse cells, compacted and covered

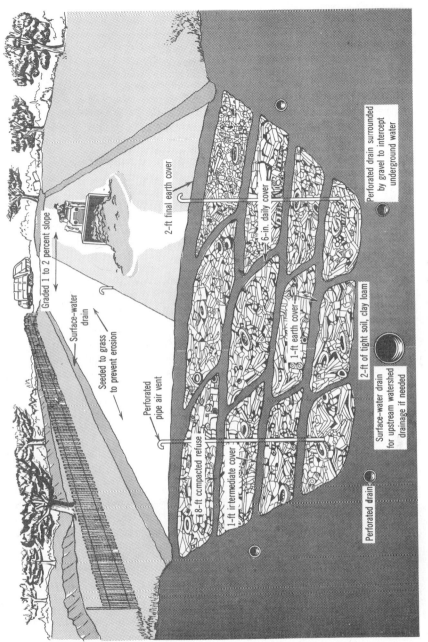

Graded 1 to 2 percent slope

Surface–water drain

Seeded to grass to prevent erosion

Perforated pipe air vent

8-ft compacted refuse

1-ft intermediate cover

2-ft final earth cover

6-in. daily cover

1-ft earth cover

2-ft of tight soil, clay bam

Perforated drain

Surface–water drain for upstream watershed drainage if needed

Perforated drain surrounded by gravel to intercept underground water

Figure 5-17 Sanitary landfill in ravine—fill starts at upper end of ravine bottom.

437

In other cases, it may be necessary to "line" the sides and bottom of the area perimeter with a layer of coarse gravel or other pervious material to allow groundwater and runoff to drain without contact with the refuse. As with other fill methods, it is important to grade during operations to avoid ponding on the surface of, or seepage into, the completed lifts.

Equipment for Disposal by Sanitary Landfill

Equipment—General

In order to attain proper site development and ensure proper utilization of the land area, it is necessary to have sufficient proper equipment available at all times at the site. One piece of refuse-compaction and earth-moving equipment is needed for approximately each 80 loads per day received at the refuse site. The type of equipment should be suitable for the method of operation and the prevailing soil conditions. Additional standby equipment should be available for emergencies, breakdowns, and equipment maintenance. See Figure 5–18 and Table 5–17.

Tractors

Tractor types include the crawler, rubber-tired and steel-wheeled types equipped with bulldozer blade, bullclam or front-end loader. The crawler tractor with a front-end bucket attachment is an all-purpose piece of equipment. It may be used to excavate trenches, place and compact refuse, transport cover material, and level and compact the completed portion of the landfill. Some types can also be used to load cover material into trucks for transportation and deposition near the open face.

A bulldozer blade on a crawler tractor is good for landfills where hauling of cover material is not necessary. It is well suited for the area method landfill where cover material is taken from nearby hillsides. It can also be used for trench method operation where the trench has been dug with some other type of equipment. A bulldozer is normally used in conjunction with some other type of earth-moving equipment, such as a scraper, where earth is hauled in from a nearby source.

The life of a tractor is figured at about 10,000 hr. Contractors usually depreciate their equipment over a 5-year period. On a landfill, if it is assumed that the equipment would be used 1000 hr a year, the life of the equipment could be 10 yr. After 10 yr, operation and maintenance costs can be expected to approach or exceed the annual cost of new equipment. Lesser life is also reported. Equipment maintenance and operator competence will largely determine equipment life.

The size of the tractor needed at the sanitary landfill is dependent

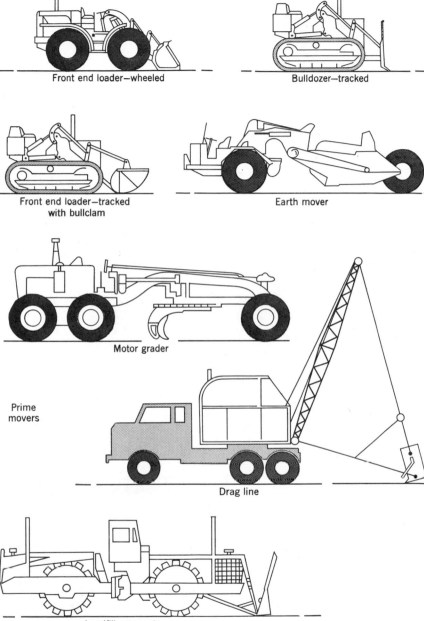

Front end loader—wheeled

Bulldozer—tracked

Front end loader—tracked
with bullclam

Earth mover

Motor grader

Prime
movers

Drag line

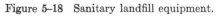

Landfill compactor

Figure 5–18 Sanitary landfill equipment.

TABLE 5-17 AVERAGE EQUIPMENT REQUIREMENTS

Population	Daily tonnage	No.	Type	Equipment Size in lb	Accessory*
0 to 15,000	0 to 46	1	tractor crawler or rubber-tired	10,000 to 30,000	dozer blade landfill blade front-end loader (1- to 2-yd)
15,000 to 50,000	46 to 155	1	tractor crawler or rubber-tired	30,000 to 60,000	dozer blade landfill blade front-end loader (2- to 4-yd) multipurpose bucket
		*	scraper dragline water truck		
50,000 to 100,000	155 to 310	1 to 2	tractor crawler or rubber-tired	30,000 or more	dozer blade landfill blade front-end loader (2- to 5-yd) multipurpose bucket
		*	scraper dragline water truck		
100,000 or more	310 or more	2 or more	tractor crawler or rubber-tired	45,000 or more	dozer blade landfill blade front-end loader multipurpose bucket
		*	scraper dragline steel-wheel compactor road grader water truck		

Source: *Sanitary Landfill Facts*, U.S. Public Health Service Pub. No. 1792, Dept. of HEW, Washington, D.C., 1970, p. 21.

on population. A rule that has been used is that a community with a population of less than 10,000 requires a $1\frac{1}{8}$ yd³ bucket on a suitable tractor. Communities with a population between 10,000 and 30,000 should have a $2\frac{1}{4}$ yd³ bucket, and populations of 30,000 to 50,000 should have at least a 3 yd³ bucket. Larger populations will require a combination of earth-moving and compaction equipment depending on the site and method of operation. A heavy tractor (D-8) can handle up to 200 tons of refuse per day, although 100 to 200 tons per day per piece of equipment is a better average operating capacity.

Many small rural towns have earth-moving equipment that they use on highway maintenance and construction. For example, a rubber-tired loader with special tires can be used on a landfill that is open 2 days a week. On the other 3 days the landfill can be closed (with fencing and locked gate), and the earth-moving equipment can be used on regular road construction work and maintenance. The people and contract users of the site should be informed of the part-time nature of the operation so as to receive their cooperation. The public officials responsible for the operation should establish a definite schedule for the assignment of the equipment to the landfill site to make sure the operation is always under control and maintained as a sanitary landfill.

Other Equipment

The dragline is well adapted for work in low areas where soil is difficult to work. This piece of equipment is excellent for digging trenches, stockpiling cover material from swampy areas, and placing cover material over compacted refuse. An additional piece of equipment is necessary to spread and compact the refuse and cover material.

Although not commonly used, the backhoe is suitable for digging trenches on fairly level ground and the power shovel is suitable for loading trucks with cover material.

In large operations scrapers can be used for the short haul of cover material to the site when adequate cover is not readily available nearby. Dump trucks may also be needed where cover material must be hauled in from some distance. Other useful equipment is a grader, sheepsfoot roller, and a water tank truck equipped with a sprinkler to keep down dust or a power sprayer to wet down the refuse to obtain better compaction.

Equipment Shelter

In cold areas it is necessary to construct an equipment shelter at the site. This will protect equipment from the weather and possible vandalism. The shelter can also be used to store fire-protection equipment and other needed materials. Operators of sanitary landfills have found a shelter to be of great value during the cold winter months since there is much less difficulty in starting motorized equipment.

Operation and Supervision

Operation Control

The direction of operation of a sanitary landfill should be with the prevailing wind. This will prevent the wind from blowing refuse back toward the collection vehicle and over the completed portion of the landfill. In order to prevent excessive wind scattering of refuse throughout the area, snow fencing or some other means of containing papers should be provided. The fencing can be utilized in the active area and then moved as the operation progresses. In some instances the entire area is fenced. Other sites have natural barriers around the landfill, such as is the case in heavily wooded areas. It is desirable to design the operation so that the work area is screened from the public line of sight. By taking this into consideration, scavenger wastes can be disposed of with the dry refuse without causing any particular problem.

The disposal of certain industrial and scavenger wastes at a sanitary landfill must take into consideration possible groundwater and surface-water pollution, insect and rodent breeding. Industrial wastes that introduce hazards, such as toxicity, explosion, and flammability must be evaluated and adequate protection provided. If there is doubt as to the method of handling such material, the industry involved and the control agency should be contacted for advice before the wastes are accepted. Pretreatment or preparation of industrial wastes for disposal may be required, and this is a proper responsibility of industry.

Large items such as refrigerators, ranges, doors, and so forth, can be placed directly into a sanitary landfill. However, in small operations this may be undesirable since these items are bulky and require considerable landfill volume and cover material. In these cases it is not objectionable to place such items in a separate area of the landfill and cover them periodically rather than daily. Compression of bulky objects will improve compaction of the fill, reduce land volume requirements, and allow more uniform settlement with less compaction. The cost of disposal, however, will be increased.

Drivers of small trucks and private vehicles carrying rubbish and other solid wastes interfere with the operation of a sanitary landfill. To accommodate these individuals on weekends and avoid traffic and unloading problems during the week, it is good practice to provide a special unloading area adjacent to the sanitary landfill entrance. Figure 5–19 shows a possible arrangement.

Personnel and Operation

Proper full-time supervision is necessary in order to control dumping, compaction, and covering. The supervisor should erect signs for direction

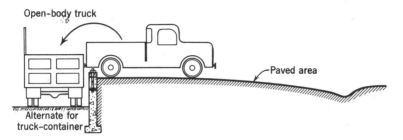

Figure 5–19 Small-load individual unloading area at side of entrance area to sanitary landfill.

of traffic to the proper area for disposal. It is essential that he be present at all hours of operation to ensure that the landfill is progressing according to plan. Days and hours of operation should be posted at the entrance to the landfill. A locked gate should be provided at the entrance to keep people out when closed. It is also advisable to inform the public of the days and hours of operation.

In supervising an operation, the length of the open face should be controlled since too large an open face will require considerably more cover material at the end of day's operation. Too small an open face will not permit sufficient area for the unloading of the expected number of collection vehicles that will be present at one time. After vehicles have deposited the refuse at the top or preferably at the base of the ramp as directed, the refuse should be spread and compacted from the bottom up into a 12 to 18-in. layer with a tractor. *This should be done continually during the day's operation* to ensure good compaction, vermin and fire control. If refuse is allowed to pile up without spreading and compaction for most of the day, proper compaction will not be achieved, resulting in uneven and excessive settling of the area and extra maintenance of the site after the fill is completed. At the end of each day, the refuse should be covered with at least 6 in. of earth. For final cover of refuse, at least 2 ft of earth is required.

Adequate personnel are needed for proper operation. Depending on the size of the community, there should be a minimum of one man at a site and six men per 1000 yd^3 dumped per day that the site is open.

No Burning or Salvaging

Burning at a landfill site must not be permitted since air pollution problems due to smoke, fly ash, odors, and so on, would result. A nuisance condition would be created and public acceptance of the operation endangered. Air pollution standards and sanitary codes generally prohibit open burning. Fire regulations also prohibit open uncontrolled burning.

Limited controlled burning may be permitted in some emergency cases, but special permission would be required from the air pollution control agency, health department, and local fire chief. Arrangements for fire control, complete burning in one day, control over material to be burned (no rubber tires or the like), and restrictions for air pollution control would also be required.

Salvaging at sanitary landfills is not recommended since, as usually practiced, it interferes with the operation. It will slow down the entire operation and thus result in time loss. Salvaging can also result in fires and unsightly stockpiles of the salvage material in the area.

Area Policing

Since wind will blow papers and other refuse around the area as the trucks are unloading, it will be necessary to clean up the area and access road at the end of each day. One of the advantages of portable snow fencing is that it will usually confine the papers near the open face and thereby make the policing job easier and less time consuming.

At many sanitary landfills, dust will be a problem during dry periods of the year. A truck-mounted water sprinkler can keep down the dust and can also be used to wet down dry refuse to improve compaction. The bulldozer operator should be protected by a dust mask or similar device.

Insect and Rodent Control

An insect and rodent control program is not usually required at a properly designed and operated landfill. However, from time to time certain unforeseen conditions may develop that will make control necessary. For this reason, prior arrangement should be made to take care of such emergencies until the proper operating corrections can be made. Prompt covering of refuse is necessary.

Site Completion and Abandonment Precautions

If a disposal site is to be abandoned or closed, the users, including contractors, should be notified and an alternate site designated. A rat-poisoning program should be started 2 weeks before the proposed closing of a dump site that has not been operated as a sanitary landfill and continued until the dump has been completely closed. The site should be closed off so as to be inaccessible and covered with at least 2 ft of compacted earth, on top and on exposed sides, and then posted to prohibit further dumping. See Chapter 10.

Maintenance

Once a sanitary landfill, or a lift of a landfill, is completed or partially completed it will be necessary to maintain the surface in order to take care of differential settlement. Settlement will vary, ranging up to approximately 20 to 30 percent, depending on the compaction, depth, and character of refuse. Maintenance of the cover is necessary to prevent ponding and excessive cracking, which will permit gases to escape and insects and rodents to enter the fill and multiply.

It is also necessary to maintain proper surface water drainage to prevent the seepage of contaminated leachate water through the fill to the groundwater table or to the surface. A 1 to 2-percent grade with culverts and lined ditches as needed are essential. The formation of water pockets is objectionable since vehicular traffic over these puddles will wash away the final cover from the refuse and cause trucks to bog down. The maintenance of access roads to the site is also necessary to prevent the formation of potholes, which will slow down vehicles using the site. It is good practice to finish off the landfill site with 2 ft of suitable earth cover graded to gently drain surface water and with at least 6 in. of topsoil seeded to grass to prevent wind and water erosion. Four feet or more earth cover is needed if the area is to be landscaped, but the amount of cover depends on the plants to be grown. The carbon dioxide and methane gases generated in a landfill may interfere with vegetation root growth if not adequately diffused or collected and disposed of through specially provided sand, gravel, or porous pipe vents, because oxygen penetration to the roots is necessary.

Use of Completed Sanitary Landfill

A sanitary landfill plan should provide for a specific use of the area after completion. It is expensive to excavate and regrade such an area. Final grades for a sanitary landfill should be established in advance to meet the needs of the proposed future use. For example, the use of the site as a golf course can tolerate rolling terrain while a park, playground, or storage lot would be best with a flat graded surface. In planning for the use of such an area, permanent buildings or habitable dwellings should not be constructed over the fill since gas production beneath the ground may escape into sewers and into the basements of such dwellings and may reach explosive levels. Some structures that would not require excavation, such as grandstands, equipment shelters, and so on, can be built on a sanitary landfill with little resulting hazard. Buildings constructed on sanitary landfills can be expected to settle unevenly unless special foundation structures such as pilings are provided; but special provisions must be made to take care of gas production. When the final use is known beforehand, selected undisturbed ground islands or earth-fill building sites are usually provided to avoid these problems. Lateral movement of gases, particularly carbon dioxide and methane, must also be taken into consideration.

Summary of Recommended Operating Practices

1. The sanitary landfill should be planned as an engineering project, operated and maintained by qualified personnel under technical direction, without causing air or water pollution, health hazards, or nuisance conditions.

2. The face of the working fill should be kept as narrow as is consistent with the proper operation of trucks and equipment in order that the area of exposed waste material be kept to a minimum.

3. All refuse should be spread as dumped and compacted into 12 to 18-in. thick layers as it is hauled in. Operate tractor up and down slope (3-1) of fill to get good compaction—four to five passes.

4. All exposed refuse should be covered with 6 in. of earth at the end of each day's operation.

5. The final earth covering for the surface and side slopes should minimize infiltration, be compacted, and be maintained at a depth of at least 24 in.

6. The final level of the fill should provide a 1 to 2-percent slope to allow for adequate drainage. Side slopes should be as gentle as possible to prevent erosion. The top of the fill and slopes should be promptly seeded. Drainage ditches and culverts are usually necessary to carry away surface water without causing erosion.

7. The depth of refuse should usually not exceed an average depth of 8 to 10 ft after compaction. In a landfill where successive lifts are placed on top of the preceding one, special attention should be given to obtain good compaction and proper surface water drainage. A settlement period of preferably 1 yr should be allowed before the next lift is placed.

8. Control of dust, wind-blown paper, and access roads should be maintained. Portable fencing and prompt policing of the area each day after refuse is dumped are necessary. Design the operation if possible so that it is not visible from nearby highways or residential areas.

9. Salvaging, if permitted by the operator of the refuse disposal area, should be conducted in such a manner as not to create a nuisance or interfere with operation. Salvaging is not recommended at the site.

10. A separate area or trench may be desirable for the disposal of such objects as tree stumps, large limbs, refrigerators, water tanks, and so on.

11. Where necessary, provision should be made for the disposal, under controlled conditions, of small dead animals and septic tank wastes. These should be covered immediately. The disposal of large quantities of sewage sludge, industrial or agricultural wastes, toxic or flammable materials should not be permitted unless study and investigation show that the inclusion of these wastes will not cause a hazard or nuisance.

12. An annual or more frequent inspection maintenance program should be established for completed portions of the landfill to ensure prompt repair of cracks, erosion, and depressions.

13. Sufficient equipment and personnel should be provided for the digging, compacting, and covering of refuse. Daily records should be kept, including type and amount of solid wastes received. *At least annually,* an evaluation should be made of the weight of refuse received and volume of refuse in place as a check on compaction and rate at which the site is being used.

14. Sufficient standby equipment should be readily available in case there is a breakdown of the equipment in use.

15. The breeding of rats, flies, and other vermin; release of smoke and odors; pollution of surface waters and groundwaters; and causes of fire hazards are prevented by proper operation, thorough compaction of refuse in 12 to 18-in. layers, daily covering with earth, proper surface water and groundwater drainage, and good supervision.

Modified Sanitary Landfill

A variation of the sanitary fill method of refuse disposal can be used at rural communities, camps, hotels, and other places where the quantity of refuse is small. This method consists of disposing of garbage and other refuse at one end of a trench. A bulldozer can be rented or borrowed from a contractor or the highway or street department to excavate the trench in advance. The trench may be 4 to 6 ft deep, 8 to 12 ft wide, and as long as needed, depending on the desired finished grade and quantity of refuse and equipment available. A trench 6 ft deep, 10 ft wide, and 20 ft long should serve a camp of 100 for at least one season. The earth dug out of the trench should be piled alongside the trench so the refuse can be covered easily at the end of each day. The refuse is pushed to one end of the trench and compressed with a small bulldozer or tractor equipped with a blade or bucket. A camp truck or jeep provided with a homemade blade is also suitable. As the trench is filled, the refuse is completely covered with 24 in. of earth, which is packed down by driving over it.

Where no collection service or community-operated sanitary refuse disposal area is provided, such as at a family summer camp, garbage, bottles, and tin cans from private dwellings can be satisfactorily disposed of on one's own property for a time. A trench should be dug about 4 ft deep and 2 ft wide. The garbage is dumped at one end of the trench; cans, cartons, and bottles are smashed with a post and then covered with about 6 in. of earth. When the garbage approaches within about 12 in. of the ground surface, the pit should be filled in with earth tamped in layers to provide an earth cover of 24 in. over the garbage.

BIBLIOGRAPHY

Corey, Richard C., *Principles and Practices of Incineration*, Wiley-Interscience, New York, 1969, 297 pp.

DeMarco, Jack, Keller, Daniel J., Leckman, Jerold, and Newton, James L., *Incinerator Guidelines—1969*, U.S. Public Health Service Pub. No. 2012, Dept. of HEW, Washington, D.C., 1969, 98 pp.

Ellis, H. M., Gilbertson, W. E., Jaag, O., Okun, D. A., Shuval, H. I., and Sumner J., *Problems in Community Wastes Management, WHO Pub. Health Pap.* No. 38, Geneva, 1969, 89 pp.

Golueke, C. G., *Comprehensive Studies of Solid Wastes Management*, Abstracts and Excerpts from the Literature, University of California, Berkeley, July 1969, 147 pp.

Golueke, C. G., and McGauhey, P. H., *Comprehensive Studies of Solid Wastes Management*, University of California, Berkeley, January 1969, 245 pp.

Municipal Refuse Collection and Disposal, Office for Local Government, Albany, N.Y., 1964, 69 pp.

Municipal Refuse Disposal, American Public Works Association, Public Administration Service, Chicago, 1970, 535 pp.

Proceedings: National Conference on Solid Waste Research, University of Chicago Center for Continuing Education, Chicago, December 1963, 228 pp.

Proceedings: The Surgeon General's Conference on Solid Waste Management, U.S. Public Health Service Pub. No. 1729, Dept. of HEW, Cincinnati, Ohio, 1967, 194 pp.

Proceedings: The Third National Conference on Air Pollution, U.S. Dept. of HEW, Washington, D.C., December 12–14, 1966, 667 pp.

Refuse Collection Practice, American Public Works Association, Public Administration Service, Chicago, 1966, 526 pp.

Sanitary Landfill, New York State Dept. of Health, Albany, 1969, 40 pp.

Sorg, Thomas J., and Hickman, Lanier, H., *Sanitary Landfill Facts*, U.S. Public Health Service Pub. No. 1792, Dept. of HEW, Washington, D.C., 1970, 30 pp.

Solid Waste Management, Office of Science and Technology, Washington, D.C., May 1969, 111 pp.

Solid Waste Management, National Association of Counties Research Foundation, Washington, D.C., 1970. Ten Guides.

Solid Wastes Management, Proceedings of the National Conference, University of California, Davis, April 4 and 5, 1966, 214 pp.

Waste Management and Control, Committee on Pollution, National Academy of Sciences–National Research Council, Washington, D.C., 1969, 257 pp.

6

AIR POLLUTION CONTROL

THE PROBLEM AND ITS EFFECTS

Air pollution is the presence of solids, liquids, or gases in amounts that
are injurious or detrimental to man, animals, plants, or property; or
that unreasonably interfere with the comfortable enjoyment of life and
property. The composition of clean air is shown in Table 6–1. The effects
of air pollution are influenced by the types and quantity of pollutants
and possible synergism, wind speed and direction, topography, sunlight,
precipitation, vertical change in air temperature, photochemical reactions, and susceptibility of the individual to specific contaminants—
singularly and in combination.

Health Effects

Man is dependent on air. He breathes about 35 lb of air/day as compared to the consumption of 3–4 lb of water and 1½ lb (dry) of food.
Pollution in the air places an undue burden on the respiratory system,
which contributes to increased morbidity and mortality, especially among
susceptible individuals in the general population. Health effects have
been discussed in Chapter 1 as well.

Some well-known air pollution episodes are given in Table 6–2. The
illnesses were characterized by cough and sore throat, irritation of the
eyes, nose, throat, and respiratory tract, plus stress on the heart. The
weather conditions were typically fog, temperature inversion, and nondis-
persing wind.

It should be noted that whereas smoking is a major contributor to
respiratory disease in the smoker, air pollution, climate, age, sex, and
socioeconomic conditions affect the incidence of respiratory disease in
the general population.

Economic Effects

Pollutants in the air cause damage to property, equipment, and facilities. Sulfur pollution attacks copper roofs and zinc coatings; steel corrodes two to four times faster in urban and industrial areas; the usual electrical equipment contacts become unreliable unless serviced frequently; clothing fabric is weakened; paint pigments are destroyed; and building surfaces, materials, and works of art are corroded. In addition particulates (including smoke) in polluted air cause erosion, accelerate corrosion, and soil clothes, buildings, cars, and other property, making more frequent cleaning and use of air-filtering equipment necessary. Ozone reduces the useful life of rubber, discolors dyes, and damages textiles.

Air pollutants intercept part of the light from the sun, thereby increasing use of electricity in daytime. Unburned fuel coming out of a chimney

TABLE 6-1   COMPOSITION OF CLEAN, DRY AIR NEAR SEA LEVEL

Component	% by Volume	Content ppm
Nitrogen	78.09%	780,900 ppm
Oxygen	20.94	209,400
Argon	.93	9,300
Carbon dioxide	.0318	318
Neon	.0018	18
Helium	.00052	5.2
Krypton	.0001	1
Xenon	.000008	0.08
Nitrous oxide	.000025	0.25
Hydrogen	.00005	0.5
Methane	.00015	1.5
Nitrogen dioxide	.0000001	0.001
Ozone	.000002	0.02
Sulfur dioxide	.00000002	.0002
Carbon monoxide	.00001	0.1
Ammonia	.000001	.01

From: *Cleaning Our Environment—The Chemical Basis for Action,* American Chemical Society, 1969.

Sources: C. E. Junge, "Air Chemistry and Radioactivity" Academic Press, New York, 1963, p. 3; A. C. Stern, ed., "Air Pollution," Vol. 1, 2nd ed., Academic Press, New York, 1968, p. 27; E. Robinson and R. C. Robbins, "Sources, Abundance, and Fate of Gaseous Atmospheric Pollutants," prepared for American Petroleum Institute by Stanford Research Institute, Menlo Park, Calif., 1968.

Note: The concentrations of some of these gases may differ with time and place, and the data for some are open to question. Single values for concentrations, instead of ranges of concentrations, are given above to indicate order of magnitude, not specific and universally accepted concentrations.

TABLE 6-2   SOME MAJOR AIR POLLUTION EPISODES

Location	Excess Deaths	Illnesses	Causative Agents
Meuse Valley, Belgium			
December 1930	63	6,000	Probably SO_2 and oxidation products with particulates from industry—steel and zinc.
Donora, Pa.			
October 1948	20	7,000	Not proven. Particulates and oxides of sulfur high; probably from industry—steel and zinc.
Poza Rica, Mexico			
1950	22	320	H_2S escape from a pipeline.
London, England			
December 1952	4,000	Increased	Not proven. Particulates and oxides of sulfur high; probably from household coal-burning.
January 1956	1,000	—	—
December 1957	750	—	—
January 1959	200–250	—	—
December 1962	700	—	—
January 1963	700	—	—
December 1967	800–1,000	—	—
New York, N.Y.			
November 1953	165	—	Increased pollution.
October 1957	130	—	Increased pollution.
January-February 1963	200–400	—	SO_2 unusually high (1.5 ppm maximum).
November 1966	152–168	—	Increased pollution.
New Orleans, La.			
October 1955	2	350	Unknown.
1958	—	150	Believed related to smouldering city dump.

or auto exhaust pipe as black smoke is wasted energy. Damage by air pollutants to plants, trees, livestock, and food crops is estimated at $500 million a year. Add to this other losses, waste, and added costs and the total bill in the United States is estimated at over $65 per capita per year. However, some feel that accurate estimates of cost are difficult to identify and compile and until more is known should not be used.[1]

[1] Allen V. Kneese, "How Much Is Air Pollution Costing Us in the United States," *Proceedings: The Third National Conference On Air Pollution*, U.S. Dept. of HEW, Washington, D.C. (December 12–14, 1966). See also "Air Pollution," *WHO Mon. Ser.* No. 46, Columbia University Press, New York, 1961.

Effects on Plants

It has been suggested that plants be used as indicators of harmful contaminants because of their greater sensitivity to certain specific contaminants. Hydrogen fluoride, sulfur dioxide, smog, ozone, and ethylene are among the compounds that can harm plants. Assessment of damage shows that the loss can be significant, although other factors such as soil fertility, temperature, light, and humidity also affect production. Among the plants that have been affected are truck garden crops (New Jersey), orange trees (Florida), orchids (California), and various ornamental flowers, shade trees, evergreen forests, alfalfa, grains, tobacco, citrus, and lettuce (Los Angeles), and many others. In Czechoslovakia more than 300 square miles of evergreen forests are reported severely damaged by sulfur dioxide fumes.[2] Sulfur dioxide, hydrogen fluoride, and the Los Angeles and London type smogs seem to be the pollutants causing the major effects on plants.[3]

Effects on Animals

Fluorides have caused crippling skeletal damage to cattle in areas where fluorides fall out or are absorbed by the vegetation. Animal laboratory studies show deleterious effects from exposure to low levels of ozone, photochemical oxidants, and peroxyacyl nitrates[4] (PAN). Lead and arsenic have also been implicated in the poisoning of sheep, horses, and cattle. About 50 percent of the animals exposed to hydrogen sulfide in the Poza Rica, Mexico, incident died (see Table 6–2).

Aesthetic, Climatic, and Related Effects

Insofar as the general public is concerned smoke, dust, and haze, which are easily seen, cause the greatest concern. Reduced visibility not only obscures the view but is also an accident hazard to air, land, and water transportation. Soiling of statuary, clothing, buildings, and other property aggravates the public to the point of demanding action on the part of public officials and industry. Correction of the air pollution usually results in increased product cost to the consumer.

Haze, dust, smoke, and soot reduce the amount of solar radiation (ultraviolet light) reaching the surface of the earth. Aerosol emissions from jet planes also intercept some of the sun's rays. However, the solar radiation as light energy that does reach the earth is absorbed

---

[2] Associated Press, Prague, Czechoslovakia, May 30, 1970.
[3] Los Angeles smog is photochemical and oxidizing; London smog is a mixture of smoke, fog, and sulfur oxides.
[4] Also cause eye irritation.

and reradiated back to the atmosphere as heat energy. But the carbon dioxide in the lower atmosphere tends to trap the heat, causing a warming of the atmosphere (greenhouse effect). Hence the reduction in solar radiation or cooling effect of the haze, dust, smoke, and soot is offset by the warming effect of the carbon dioxide. Which effect will prevail is not known.

Air pollution, both natural and man-made, affects the climate in other ways. Dust and other particulate matter in the air provide nuclei around which condensation takes place, forming droplets and thereby playing a role in snowfall and rainfall patterns.

Certain malodorous gases interfere with the enjoyment of life and property. In some instances individuals are seriously affected. The gases involved include hydrogen sulfide, aldehydes, phenols, polysulfides, and some olefins. Equipment is available to reduce these gases to less objectionable compounds.

SOURCES AND TYPES OF AIR POLLUTION

The sources of air pollution may be man-made, such as the internal combustion engine, or natural, such as plants (pollens). The pollutants may be in the form of particulates,[5] aerosols,[6] and gases,[7] or as microorganisms. Included are pesticides, odors, and radioactive particles carried in the air.

Man-made Sources

Air pollution in the United States is the result of industrialization and mechanization. The major sources and pollutants are shown in Tables 6–3 and 6–4.[8] It can be seen that carbon monoxide is the principal pollutant by weight and that the motor vehicle is the major contributor, followed by industry and power plants. However, in terms of hazard it

---

[5] From less than 0.01 to 1000 μ in size; generally smaller than 50 μ. Smoke is generally less than 0.1 μ size soot or carbon particles. Those below 10 μ can penetrate respiratory tract; particles less than 3 μ reach deep parts of the lung.

A micron is $\frac{1}{1000}$ of a millimeter or $\frac{1}{25,000}$ of an inch. Particles of 10 μ and larger in size can be seen with naked eye.

[6] Usually particles 50 μ to less than .01 μ in size; generally less than 1 μ in diameter.

[7] Organic gases such as hydrocarbons, aldehydes, ketones, and inorganic gases such as oxides of nitrogen and sulfur, carbon monoxide, hydrogen sulfide, ammonia, and chlorine.

[8] *The Sources of Air Pollution and Their Control*, U.S. Public Health Service Pub. No. 1548, Dept. of HEW, Washington, D.C., 1966, and *Nationwide Inventory of Air Pollutant Emissions* 1968, U.S. Public Health Service, NAPCA Pub. No. AP-73, Dept. of HEW, Washington, D.C., August 1970.

TABLE 6-3   SOURCES AND EMISSIONS OF AIR POLLUTANTS, MILLIONS OF TONS PER YEAR — 1966

Source	Total*	Carbon monoxide	Sulfur oxides	Nitrogen oxides	Hydro-carbons	Partic-ulates
Motor vehicles†	86 (61%)	66	1	6	12	1
Industry‡	23 (16%)	2	9	2	4	6
Power plants§	20 (14%)	1	12	3	<1	3
Space heating‖	8 (6%)	2	3	1	1	1
Refuse disposal¶	5 (3%)	1	<1	<1	1	1
Totals	142	72	26⁻	13⁻	19⁻	12

Source: *The Sources of Air Pollution and Their Control,* U.S. Public Health Service Pub. No. 1548, Dept. of HEW, Washington, D.C., 1966.

* Totals show relative weights and do not indicate hazard or severity of problem in a particular area.

† Automobile tail-pipe exhaust, crankcase vent, evaporation from gas tank and carburetor; also tetra-ethyl lead and acids.

‡ Pulp and paper mills, iron and steel mills, petroleum refineries, smelters, inorganic chemical manufactures such as fertilizer, and organic chemical manufactures such as synthetic rubber. Also grinding, crushing, pulverizing, sanding, and cutting operations.

§ Burning coal and oil containing sulfur as an impurity.

‖ To heat homes, apartments, and offices.

¶ By inefficient incineration and burning dumps; also burning of agricultural wastes.

TABLE 6-4   SOURCES AND EMISSIONS OF AIR POLLUTANTS, MILLIONS OF TONS PER YEAR — 1968

Source	Total*	Carbon monoxide	Sulfur oxides	Nitrogen oxides	Hydro-carbons	Partic-ulates
Transportation	90.5 (42%)	63.8	0.8	8.1	16.6	1.2
Fuel combustion in stationary sources	45.9 (21%)	1.9	24.4	10.0	0.7	8.9
Industrial processes	29.3 (14%)	9.7	7.3	0.2	4.6	7.5
Solid waste disposal	11.2 (5%)	7.8	0.1	0.6	1.6	1.1
Miscellaneous	37.3 (17%)	16.9	0.6	1.7	8.5	9.6
Totals	214.2	100.1	33.2	20.6	32.0	28.3

*Totals show relative weights and do not indicate hazard or severity of problem in a particular area.

Source: *Nationwide Inventory of Air Pollutant Emissions 1968,* U.S. Public Health Service, NAPCA Pub. No. AP-73, Dept. of HEW, Washington, D.C., August 1970. Totals are for the most part higher than 1966 estimates primarily because of the inclusion of sources not previously considered, that is, forest fires, burning of coal refuse banks, and an increased number of industrial process sources. Therefore the data in Table 6-4 is more representative of sources and emissions of air pollutants than Table 6-3.

is not the tons of pollutant that is important but the toxicity or harm that can be done by the particular pollutant released.

Agricultural spraying of pesticides, orchard heating devices, exhausts from various commercial processes, rubber from tires, mists from spray-type cooling towers, and use of cleaning solvents and household chemicals add to the pollution load. Other potential sources are wastewater treatment plants, stack pollution from processing of nuclear fuels, nuclear reactors, and nuclear explosions.

Natural Sources

Discussions of air pollution frequently overlook the natural sources. These include dust; plant and tree pollens; arboreal emissions; bacteria and spores; gases and dusts from forest and grass fires; ocean sprays and fog; ozone and nitrogen dioxide from lightning; ash and gases (SO_2, HCl, HF, H_2S) from volcanoes; natural radioactivity; and micro-organisms such as bacteria, molds, or fungi from plant decay. Most of these are beyond control or of limited significance. Ozone is formed by the action of sunlight on automobile exhaust, but it is also generated during electrical storms and by ionizing radiation.

Types of Air Pollutants

The types of air pollutants are related to the original material used for combustion or processing and the impurities it contains.

Most combustible materials are composed of hydrocarbons. In the ideal combustion, carbon dioxide and water are given off and may be expressed as $C_xH_y + O_2 = CO_2 + H_2O$. If the combustion is inefficient, unburned hydrocarbons, smoke, carbon monoxide, and to a lesser degree aldehydes and organic acids are released. At elevated temperatures, more oxides of nitrogen resulting from the oxidation of atmospheric nitrogen are formed and given off.

Impurities in combustible hydrocarbons (coal and oil), such as sulfur, combine with oxygen to produce SO_2 and SO_3 when burned. Oxides of nitrogen, from high-temperature combustion, are released mostly as NO_2 and NO. The source of nitrogen is principally the atmosphere rather than an impurity in fuel. Fluorides and other fuel impurities may also be carried out with the hot stack gases. Additional knowledge is needed to determine the effects of air pollutants singly and in combination. Those that need more systematic investigation are the following:

"Highest priority: Sulfur dioxide, carbon monoxide, carbon dioxide, fluoride, ozone, sulfuric acid droplets, oxides of nitrogen, carcinogens (various types), peroxyacyl nitrates, gasoline additives including lead, and asbestos particles.

"High priority: benzene and homologues, alkyl nitrates, alkyl nitrites, aldehydes, ethylene, pesticides, auto exhaust (raw), amines, mercaptans, hydrogen sulfides, and beryllium particles."[9]

SAMPLING AND MEASUREMENT

Air-sampling devices are used to detect and measure smoke, particulates, and gases. The equipment selected and used is determined by the problem being studied and the purpose to be served. Representative samples must be collected and readings or analyses standardized to obtain valid data. Supporting meteorological and other environmental information is needed to properly interpret the data collected. Continuous sampling equipment should be selected with great care and its reliability demonstrated to ensure that it will perform the assigned task with a minimum of calibration and maintenance. Reliable instruments are reported to be available for the continuous monitoring of carbon dioxide, carbon monoxide, and oxygen in stacks. Other instruments such as for smoke, hydrocarbons, and sulfur are also available.

Some instruments for atmospheric measurement of pollution are quite elaborate. The simplest readily available instrument should be selected that meets the required sensitivity and specificity. Service, maintenance, the frequency calibration is needed, and time required to collect information are important. A continuous sampling instrument surrounded by simple manually operated stations strategically located may often yield adequate information.

Measurement of Materials' Degradation

The direct effects of air pollution can be observed by exposing various materials to the air at selected sampling stations, such as at dustfall jar, lead peroxide candle, high-volume sampler, and sticky paper stations. Degradation of materials is measured for a selected period on a scale of 1 to 10 with 1.0 representing the least degradation and 10.0 the worst, as related to the sample showing the least degradation. Materials exposed and conditions measured include steel corrosion, dyed fabric (nonspecific) for color fading, dyed fabric (NO_x-sensitive), dyed fabric (ozone-sensitive), dyed fabric (fabric soiling), dyed fabric (SO_2-sensitive), silver tarnishing, nylon deterioration, rubber cracking (crack depth), leather deterioration, copper pitting, and others. The samples are exposed for a selected period, such as rubber, 7 days at a time; silver, 30 days

[9] *Restoring the Quality of Our Environment,* Report of Environmental Pollution Panel, President's Science Advisory Committee, Washington, D.C., November 1965.

at a time; nylon, 30 or 90 days at a time; cotton, 90 days at a time; steel, 90 days and one year at a time; zinc one year at a time. Shrubs, trees, and other plants sensitive to certain contaminants or pollutants can also be used to monitor the effects of air pollution.

Smoke Measurement

The Ringelmann Smoke Chart[10] consists of five[11] rectangular grids produced by black lines of definite width and spacing on a white background. When held at a distance, about 50 ft from the observer, the grids appear to give shades of gray between white and black. The grid shadings are compared with the pollution source, and the grid number closest to the shade of the pollution source is recorded. About 30 observations are made in 15 min, and a weighted average is computed of the recorded Ringelmann numbers. The chart is used to determine whether smoke emissions are within the standards established by law; the applicable law is referenced to the chart. The system cannot be applied to dusts, mists, and fumes. Inspectors need training in making smoke readings.

The Public Health Service Guide for Smoke Inspectors uses a piece of photographic film, corresponding to Ringelmann numbers, with four adjacent rectangular degrees of darkness plus a fifth black area.

The automatic smoke sampler collects suspended material on a filter tape that is automatically exposed for predetermined intervals over an extended period of time. The opacity of the deposits or spots on the tape to the transmission or reflectance of light from a standard source is a measure of the air pollution. The equipment is used primarily to indicate the dirtiness of the atmosphere.

Particulate Sampling

A comparative summary of particulate sampling devices is given in Table 6–5. Some are briefly described below.

Sedimentation and settling devices include fallout or dustfall jars, settling chambers or boxes, Petri dishes, coated metal sheets or trays, and gum-paper stands for the collection of particulates that settle out. Vertically mounted adhesive papers or cylinders coated with petroleum

---

[10] Bureau of Mines Information Circular No. 7718, Publications Distribution Section, 4800 Forbes Street, Pittsburgh, Pa., August 1955.

[11] Reduced to four grids or charts in the United States. The width or thickness of lines and their spacing in each grid or chart varies. A handy reduction of the Ringelmann Smoke Chart is the Power's Microringelmann available from Power, 33 West 42nd Street, New York, N.Y. 10036. Price is 35 cents.

TABLE 6–5 COMPARATIVE SUMMARY OF PARTICULATE
SAMPLING DEVICES

Instruments	Method of Quantitation	Skill in Operation	Skill in Quantitation	Remarks
Sedimentation				
Dust jars, screens, glass plates, boxes	Count, weight, chemical analysis	Considerable	Considerable	Simple equipment readily available. Useful for qualitative examination and collection for chemical identification and size analysis. Long settling times needed for submicron particles. Accepted method for pollen collection.
Impingement				
Cascade impactors	Count, chemical, micrurgic,* weight	Some	Considerable	Provides rapid means of fractioning aerosols. Not well adapted to high concentrations if particles are to be counted. Glass or machined parts may be used.
Single-jet impactors	Count, chemical, micrurgic,* weight	Some	Considerable	Provides efficient collection for particles down to $1\ \mu$. Collects smaller particles less efficiently. Useful for distinguishing between solid and liquid aerosols. Glass or machined parts may be used.
Greenburg-Smith impinger	Count, chemical analysis	Some	Considerable	Can be used for soluble gases and particulate material. Commercially available in compact portable form.
Fine-fiber impinger	Count, chemical, micrurgic*	Considerable	Considerable	Simple equipment, light-weight and compact. Highly useful for distinguishing between liquid and solid aerosols.
Filtration				
Sugar or salicylic acid, or naphthalene, packed containers	Count, chemical analysis	Some	Some	Provides means of separating filter body from collected material. Not suitable for wet or hot atmospheres. Sugar and salicylic acid removed by dissolving, naphthalene by evaporation.
Paper thimbles or discs	Weight, chemical analysis	Some	Considerable for trace quantities. Less for gross amounts and weight.	Can measure quantitatively by discoloration. Inexpensive, readily available. Can be impregnated with indicators to show acid mists.

*From micrurgy, broadly: the practice of using minute tools in a magnified field. (Webster's Third New International Dictionary, 1968)

TABLE 6-5   COMPARATIVE SUMMARY OF PARTICULATE
SAMPLING DEVICES (cont'd)

Instruments	Method of Quantitation	Skill in Operation	Skill in Quantitation	Remarks
			Filtration (cont'd)	
Cotton, glass, asbestos, or wool mats	Weight, chemical analysis	Some	Considerable for trace quantities. Less for gross amounts and weight.	Adaptable to wide range of aerosol separation from gases, for weight analysis, and under some conditions, chemical analyses. Difficulty frequently encountered in separation of aerosol from filter body for detailed analysis. May be built for large or small sampling rates. Inexpensive, readily available.
			Centrifugation	
Cyclones	Weight, chemical analysis, count, screen analysis	Some	Considerable	May be built in wide range of sizes and of temperature- and corrosion-resistant materials. Collected material readily removed. Inexpensive, readily available.

Source: *Guide to the Appraisal and Control of Air Pollution,* American Public Health Association, Inc., 1740 Broadway, New York, 1969.

jelly can indicate the directional origin of contaminants. Dustfall is usually reported as milligrams per square centimeter per month. Particulates can also be measured for radioactivity.

High-volume samplers pass a measured high rate of air through a special filter paper, usually for a 24-hr period. The filter paper is weighed before and after exposure, and the change in weight is a measure of the suspended particulate matter in $\mu g/m^3$ of air filtered. The particulates can be analyzed for weight, particle size, and composition (such as benzene solubles, nitrates, and sulfates), and for radioactivity.

Inertial or centrifugal collection equipment operates on the cyclone collection principle. Large particles above 1 μ in diameter are collected, although the equipment is most efficient for the collection of particles larger than 10 μ.

Impingers separate particles by causing the gas stream to make sudden changes in direction in passing through the equipment. The wet impinger is used for the collection of small particles, the dry impinger for the larger particles. In the dry impinger a special surface is provided on which the particles collide and adhere.

In the *cascade impactor* the velocities of the gas stream vary, making possible the sorting and collection of different size particles on special microscopic slides. Particulates in the range of 0.7 to 50 μ are collected.

Electrostatic precipitator-type sampling devices operate on the ionization principle using a platinum electrode. Particles less than 1 μ in size collect on an electrode of opposite charge and then are removed for examination.

Nuclei counters measure the number of condensation nuclei in the atmosphere. They are a useful reference for weather commentators. A sample of air is drawn through the instrument, raised to 100 percent relative humidity, and expanded adiabatically, with resultant condensation on the nuclei present. The droplets formed scatter light in proportion to the number of water droplets, which are counted by a photomultiplier tube. Concentrations of condensation nuclei may range from 10 to 10,000,000 particles per cubic centimeter.[12] Condensation nuclei are believed to result from a combination of natural and man-made causes, including air pollution. A particle count above 50,000 is said to be characteristic of an urban area.

Pollen samplers generally use petroleum-jelly-coated slides placed on a covered stand in a suitable area. The slides are usually exposed for 24 hr, and the pollen grains are counted with the aid of a microscope. The counts are reported as grains per square centimeter. See Chapter 10 for ragweed control and sampling.

Gas Sampling

Gas sampling requires separation of the gas or gases being sampled from other gases present. The temperature and pressure conditions under which a sample is collected must be accurately noted. The pressure of a gas mixture is the sum of the individual gas pressures, as each gas has its own pressure. The volumes of individual gases at the same pressure in a mixture are also additive. Concentrations of gases when reported in terms of parts per million are by volume rather than by weight. Proper sampling and interpretation of results requires competency and experience, knowledge concerning the conditions under which the samples are collected, and an understanding of the limitations of the testing procedures. Instruments and equipment for gas sampling include:

Lead peroxide candle for the sampling of sulfur compounds. It is actually a glass tube that has wrapped around it a 4-in. wide gauze coated with a paste containing PbO_2. Sulfur compounds in the atmosphere (SO_2) react with PbO_2 to form $PbSO_4$. Measurement is in sulfate ions as milli-

[12] Alexander Rihm, "Air Pollution," in *Dangerous Properties of Industrial Materials*, N. Irving Sax, ed., Van Nostrand Reinhold Co., New York, 1969.

grams per square centimeter of peroxide surface per 30 days. This is usually an indication of the sulfur dioxide in the atmosphere over the period sampled; it does not give the maximum concentration at any time. A special coated petri dish is also used to measure sulfur compounds.

Detector tube containing a chemical that reacts colorimetrically with a specific contaminant drawn through the tube is used to indicate the presence and concentration of the contaminant.

Impregnated tape samplers, similar to the automatic smoke sampler, are used to measure contaminants to which the filter tape has been sensitized. When impregnated with lead acetate, hydrogen sulfide is measured. The air to the filter tape is prefiltered.

Other samplers include the hydrogen peroxide analyzer (sulfur dioxide), electroconductivity analyzer (sulfur dioxide), potentiometric analyzer (sulfur dioxide), continuous wet chemistry instruments (also measure sulfur dioxide, nitrogen oxide, nitrogen dioxide, aldehydes, and oxidants), infrared analyzer (carbon monoxide, hydrocarbons, and other contaminants), hydrogen flame ionization sampler (hydrocarbons), and the coulometric ozone sampler.

The collection of stack samples, such as fly ash and dust emissions, requires special filters or thimbles of known weight and a measure of the volume of gases sampled. The sample must be collected at the same velocity at which the gases normally pass through the stack. The gain in weight divided by the volume of gases sampled corrected to 0°C (or 21°C) temperature and 760 mm mercury gives a measure of the dust and fly ash going out of the stack, usually as grains per cubic foot. When a series of samples is to be collected or measurements made, a "sampling train" is put together. It may consist of a sampling nozzle, several impingers, a freeze-out train, a weighed paper thimble, dry gas meter, thermometers, and pump.

A common piece of equipment for boiler and incinerator stack sampling is the Orsat apparatus. By passing a sample of the stack gas through each of three different solutions the percent carbon dioxide, carbon monoxide, and oxygen constituents in the flue gas is measured. The remainder of the gas in the mixture is usually assumed to be nitrogen.

Tracer materials may be placed in a stack to indicate the effect of a pollution source on the surrounding area. The tracer may be a fluorescent material, a dye, a compound that can be made radioactive, a special substance or chemical, or a characteristic odor-producing material. The tracer technique can be used in reverse, that is, to detect the source of a particular pollutant provided there are no interfering sources.

ENVIRONMENTAL FACTORS

The behavior of pollutants released to the atmosphere is subject to diverse and complex environmental factors associated with meteorology and topography. Meteorology involves the physics, chemistry, and dynamics of the atmosphere and includes many direct effects of the atmosphere on the earth's surface, ocean, and life. Topography refers to both the natural and man-made features of the earth's surface. The pollutants can be either accumulated or diluted, depending on the nature and degree of the physical processes of transport, dispersion, and removal and the chemical changes taking place.

Within the scope of this text, it is not intended to provide a complete technical understanding of all the meteorological and topographical factors involved, but to provide only an insight into the relationships to air pollution of the more important processes.

Meteorology

The meteorological elements that have the most direct and significant effects on the distribution of the air pollutants are wind, stability, and precipitation. It is therefore important to have a continuing base line of meteorological data, including these elements, to interpret and anticipate probable effects of air pollution emissions. The National Weather Service (formerly U.S. Weather Bureau) is a major source of information.

Wind

Wind is the motion of the air relative to the earth's surface. Although it is three-dimensional in its movement, generally only the horizontal components are denoted when used because the vertical component is very much smaller than the horizontal. This motion derives from the unequal heating of the earth's surface and the adjacent air, which in turn gives rise to a horizontal variation in temperature and pressure. The variation in pressure (pressure gradient) constitutes an imbalance in forces so that air motion from high toward low pressure is generated.

The uneven heating of the surface occurs over various magnitudes of space resulting in different magnitudes of organized air motions (circulations) in the atmosphere. Briefly, in descending order of importance, these are:

1. The primary or general (global) circulation associated with the large-scale, hemispheric motions between the tropical and polar regions.

2. The secondary circulation associated with the relatively large-scale motions of migrating pressure systems (highs and lows) developed by the unequal distribution of large land and water masses.

3. The tertiary circulation (local) associated with small-scale variations in heating, such as valley winds and land and sea breezes.

For a particular area, the total effect of these various circulations establishes the hourly, daily, and seasonal variation in wind speed and direction. With respect to a known source or distribution of sources of pollutants, the frequency distribution of wind direction will indicate toward which areas the pollutants will be most frequently transported. It is customary to present long-term wind data at a given location graphically in the form of a "wind rose," an example of which is shown in Figure 6–1.

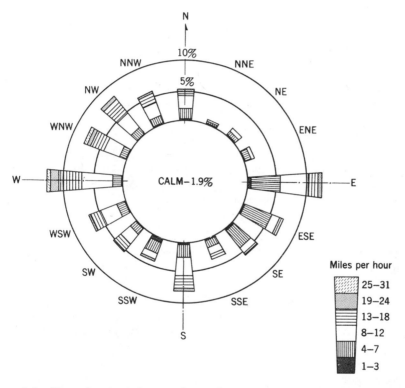

Figure 6–1 Example of wind rose, for a designated period of time, by month, season, or year.)

NOTE: 1. The positions of the spokes show the direction from which the wind was blowing.
   2. The total length of the spoke is the percent of the time, for the reporting period, that the wind was blowing from that direction.
   3. The length of the segments into which each spoke is divided is the percent time the wind was blowing from that direction at the indicated speed in mph.
   4. Horizontal wind speed and direction can vary with height.

The concentration resulting from a continuous emission of a pollutant is inversely proportional to wind speed. The higher the wind speed the greater the separation of the particles or molecules of the pollutant as they are emitted, and vice versa. This is shown graphically in Figure 6–2a. The wind speed therefore is an indicator of the degree of dispersion

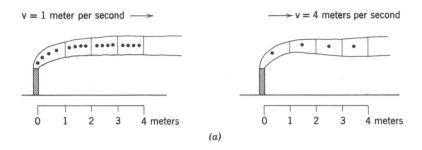

Figure 6–2a   Effect of wind speed on pollutant concentration from constant source. (Continuous emission of 4 units per second.)

of the pollutant and contributes to the determination of the area most adversely affected by an emission. Although an area may be located in the most frequently occurring downwind direction from a source, the wind speeds associated with this direction may be quite high so that resulting pollutant concentrations will be low as compared to another direction occurring less frequently but with lower wind speeds.

Smaller in scale than the tertiary circulation mentioned above there is a scale of air motion that is extremely significant in the dispersion of pollutants. This is referred to as the micrometeorological scale and consists of the very short-term, on the order of seconds and minutes, fluctuations in speed and direction. As opposed to the "organized" circulations discussed above, these air motions are rapid and random and constitute the wind characteristic called "turbulence." The turbulent nature of the wind is readily evident upon watching the rapid movements of a wind vane. These air motions provide the most effective mechanism for the dispersion or dilution of a cloud or plume of pollutants. The turbulent fluctuations occur in both the horizontal and vertical directions. The dispersive effect of fluctuations in horizontal wind direction is shown graphically in Figure 6–2b.

Turbulent motions are induced in the air flow in two ways: (1) by thermal convective currents resulting from heating from below (thermal turbulence) and (2) by disturbances or eddies resulting from the passage of air over irregular, rough ground surfaces (mechanical turbulence).

It may be generally expected that turbulent motion and, in turn, the

·dispersive ability of the atmosphere would be greatly enhanced during a period of good solar heating and over relatively rough terrain.

Another characteristic of the wind that should be noted is that wind speed generally increases with height in the lower levels. This is due

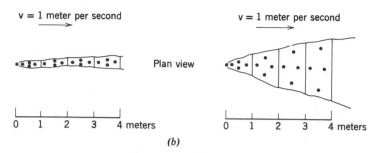

(b)

Figure 6–2b   Effect of variability of wind direction on pollutant concentration from constant source. (Continuous emission of 4 units per second.)

to the decrease with height of the "frictional drag" effect of the underlying ground surface.

Stability and Instability

The stability of the atmosphere is its ability to enhance or suppress vertical air motions. Under unstable conditions the air motion is enhanced, and under stable conditions the air motion is suppressed. The conditions are determined by the vertical distribution of temperature.

In vertical motion, parcels of air are displaced. Due to the decrease of pressure with height, a parcel displaced upward will encounter decreased pressure and expand. If this expansion process is relatively rapid or over a large area so that there is little or no exchange of heat with the surrounding air or by a change of state of water vapor, the process is dry adiabatic and the parcel of air will be cooled. Likewise, if the displacement is downward so that an increase in pressure and compression is experienced, the parcel of air will be heated.

This rate of cooling with height is the *dry adiabatic process lapse rate* and is approximately $-1°C/100$ m $(-5.4°F/1000$ ft$)$.

The *prevailing* or *environmental lapse rate* is the decrease of temperature with height that may exist at any particular time and place. It can be shown that if the decrease of temperature with height is greater than $-5.4°F/1000$ ft, parcels displaced upward will attain temperatures higher than their surroundings. Air parcels displaced downward will attain lower temperatures than their surroundings. The displaced parcels will tend to continue in the direction of displacement. Under these conditions, the vertical motions are enhanced and the layer of air is defined as "unstable."

Furthermore, if the decrease of temperature with height is less than —5.4°F/1000 ft, it can be shown that air parcels displaced upward attain temperatures lower than their surroundings and will tend to return to their original positions. Air parcels displaced downward attain higher temperatures than their surroundings and also tend to return to their original position. Under these conditions, vertical motions are suppressed and the layer of air is defined as "stable."

Finally, if the decrease of temperature with height is equal to —5.4°F/1000 ft, displaced air parcels attain temperatures equal to their surroundings and tend to remain at their position of displacement. This is called "neutral stability."

Inversions

Up to this point, the prevailing temperature distribution in the vertical has been referred to as a "lapse rate," which indicates a decrease of temperature with height. However, under certain meteorological conditions, the distribution can be such that the temperature increases with height within a layer of air. This is called an "inversion" and constitutes an extremely stable condition.

There are three types of inversions that develop in the atmosphere: radiational (surface), subsidence (aloft), and frontal (aloft).

Radiational inversion is a phenomenon that develops at night under conditions of relatively clear skies and very light winds. The earth's surface cools by reradiating the heat absorbed during the day. In turn the adjacent air is also cooled from below so that within the surface layer of air there is an increase of temperature with height.

Subsidence inversion develops in high-pressure systems (generally associated with fair weather) within a layer of air aloft when the air layer sinks to replace air that has spread out at the surface. Upon descent the air heats adiabatically, attaining temperatures greater than the air below.

A condition of particular significance is the subsidence inversion that develops with a stagnating high-pressure system. Under these conditions the pressure gradient becomes progressively weaker so that the winds become very light, resulting in a great reduction in the horizontal transport and dispersion of pollutants. At the same time the subsidence inversion aloft continuously descends, acting as a barrier (lid) to the vertical dispersion of the pollutants. These conditions can persist for several days so that the resulting accumulation of pollutants can cause a serious health hazard.

Frontal inversion forms when air masses of different temperature characteristics meet and interact so that warm air overruns cold air.

There are many and varied effects of stability conditions and inversions on the transport and dispersion of pollutants in the atmosphere. In general enhanced vertical motions under unstable conditions increase the turbulent motions, thereby enhancing the dispersion of the pollutants. Obviously the stable conditions have the opposite effect.

Stack emissions in inversions—depending on the elevation of emission with respect to the distribution of stability in the lower layers of air, the behavior of the plumes can be affected in many different ways. Pollutants emitted within the layer of a surface-based (radiational) inversion can develop very high and hazardous concentrations at the surface level. On the other hand, when pollutants are emitted from stacks at a level aloft within the surface inversion, the stability of the air tends to maintain the pollutant at this level, preventing it from reaching the surface. However, after sunrise and continued radiation from the sun resulting in heating of the earth's surface and adjacent air, the inversion is "burned off." Once this condition is reached, the lower layer of air has become unstable and all of the pollutant that has accumulated at the level aloft is rapidly dispersed downward to the surface. This behavior is called "fumigation" and can result in very high concentrations during the period.

Precipitation

Precipitation constitutes an effective cleansing process of pollutants in the atmosphere in three ways: washing out or scavenging of large particles by falling raindrops or snowflakes (washout); accumulation of small particles in the formation of raindrops or snowflakes in clouds (rainout); and removal of gaseous pollutants by dissolution and absorption.

The most effective and prevalent process is the washout of large particles, particularly in the lower layer of the atmosphere where most of the pollutants are released. The efficiencies of the various processes depend on complex relationships between the properties of the pollutants and the characteristics of the precipitation.

Topography

The topographic features of a region include both the natural (hills, valleys, oceans, rivers, lakes, foliages, and so on) and man-made (cities, bridges, roads, canals, and so on) elements distributed within the region. These elements, per se, have little direct effect on pollutants in the atmosphere. The prime significance of topography is its effects on the meteorological elements. As stated previously, the variation in the distribution of land and water masses gives rise to various types of circulations. Of particular significance are the local or small-scale circulations that develop. These circulations can contribute either favorably or unfavorably to the transport and dispersion of the pollutants.

Along a coastline during periods of weak pressure gradient, intense heating of the land surface as opposed to the lesser heating of the contiguous water surface develops a temperature and pressure differential that generates an onshore air circulation. This circulation can extend to a considerable distance inland. At times during stagnating high-pressure systems, when the transport and dispersion of pollutants have been greatly reduced, this short-period afternoon increase in airflow may well prevent the critical accumulation of pollutants.

In valley regions, particularly in the winter, intense *surface inversions* are developed by the drainage down the slopes of air cooled by the radiationally cooled valley wall surfaces. Bottom valley areas that are significantly populated and industrialized can be subject to critical accumulation of pollutants during these periods.

The increased roughness of the surface created by the widespread distribution of buildings throughout a city can significantly enhance the turbulence of the airflow over the city, thereby improving the dispersion of the pollutants emitted. But at the same time the concrete, stone, and brick buildings and asphalt streets of the city act as a heat reservoir for the radiation received from the sun during the day. This, plus the added heat from nighttime space heating during the cool months of the year, creates a temperature and pressure differential between the city and the surrounding rural area so that a local circulation inward to the city is developed. The circulation tends to concentrate the pollutants in the city. This phenomena is called the "urban heat island effect."

Areas on the windward side of mountain ranges can expect added precipitation due to the forced rising, expansion, and cooling of the moving air mass with resultant release of available moisture. This increased precipitation serves to increase the removal of the pollutants.

It is apparent, then, that topographical features can have many and diverse effects in the meteorological elements and the behavior of pollutants in the atmosphere.

AIR POLLUTION SURVEYS

Air pollution survey of a region having common topographical and meteorological characteristics is a necessary first step before a meaningful air resources management plan and program can be established. The survey includes an inventory of source emissions and a contaminant and meteorological sampling network, supplemented by study of basic demographic, economic, land use, and social factors.

Inventory

The inventory includes the location of emission sources and identification of the processes involved; the air pollution control devices installed and their effectiveness; the pounds or tons of specific air pollutants emitted per day, week, month, and year. An estimate can then be made of the total pollution burden on the atmospheric resources of any given air basin.[13] Tables have been developed to assist in the calculation of the amounts and types of contaminants released; they can also be used to check on information received through personal visits, questionnaires, telephone calls, governmental reports, and the technical and scientific literature.[14] Additional sources of information are the complaint files of the health department and other agencies, university studies, state and local chamber of commerce reports and files, and results of traffic surveys.

Air Sampling

Air and meteorological sampling equipment located in the survey area will vary, depending on a number of factors such as land area, topography, population densities, industrial complexes, and the manpower, and budget available. A minimum number of stations are necessary to obtain meaningful data.

Specific sampling sites for a comprehensive survey are selected on the basis of (1) objective, scope, and budgetary limitations; (2) accessibility for year-round operation, availability of reliable electrical power, amount and type of equipment available, program duration, and personnel available to operate stations; (3) meteorology of the area, topography, adjacent obstructions, and vertical and horizontal distribution of equipment; (4) sampler operator problems, space requirements, protection of equipment and site, possible hazards, and public attitude toward the program.[15]

Basic Studies and Analyses

Basic studies include population densities and projections; land use analysis; mapping; economic studies and proposals including industrial-

---

[13] W. E. Jackson and H. C. Wohlers, "You need an air-pollution inventory," *American City* (October 1967).

[14] Harry H. Hovey, Arnold Risman, and John F. Cunnan, "The Development of Air Contaminant Emission Tables for Nonprocess Emissions," *J. Air Pollution Control Assoc.*, **16**, 362–366 (July 1966).

[15] "Introduction to Air Pollution Control," New York State Dept. of Health, Albany, March 1964.

ization, transportation systems, community institutions, environmental health and engineering considerations, relationship to federal, state, and local planning, and related factors. Liaison with other planning agencies can be helpful in obtaining needed information that may already be available. See Chapter 2.

When all the data from the emission inventory, air sampling, and basic studies are collected, analyzed, and evaluated, a report is usually prepared. The analysis step should include calculations to show how the pollutants released to the atmosphere are dispersed and their possible effects under existing conditions and with future development.

Mathematical models could be developed, based on the data collected, to estimate the pollution levels that might result under various emission, topographical, and meteorological conditions. A data bank and a system for the collection and retrieval of information would generally be indicated. The approximate cost to achieve selected levels of air quality (for health, aesthetic, plant, and animal considerations) and the possible effect on industrial expansion, transportation modes and systems, availability and cost of fuel, community goals and objectives should be determined.

The report would recommend air-quality standards or objectives to areas in the region studied based on existing and proposed land uses. This will require consultation and coordination with state and local planning agencies.[16]

Short- and long-term objectives and priorities should be established to achieve the desired air quality. Recommendations to reduce air pollution might include control of pollutant emissions and limits in designated areas and under hazardous weather conditions and predictions; time schedules for starting control actions; control of fuel composition; requirement of plans for new or altered emission sources and approval of construction for compliance with emission standards; denial of certain plan approvals and prohibition of activities, or requirement of certain types of control devices; and performance standards to be met by existing and new structures and facilities.

The report is then formally submitted to the regulatory agency, board, or commission for further action. It would generally include recommendations for needed laws, rules, and regulations and administrative organization and staffing for the control of existing and new sources of air pollution.

---

[16] Vernon G. MacKenzie, "Management of Our Air Resources," paper presented at the Growth Conference on Air, Land, and Water Resources, University of California at Riverside, October 7, 1963.

AMBIENT AIR QUALITY STANDARDS[17]

Topographic, meteorological, and land use characteristics of areas within an air region will vary. The social and economic development of an area will result in different degrees of air pollution and demands for air quality. Because of this it is practical and reasonable to establish different levels of air purity for certain areas within a region. However, any standards adopted must assure, at a very minimum, no adverse effects on human health.[18]

Federal Standards

On December 31, 1970, the Federal Air Quality Act of 1967 (Public Law 90-148) was amended by Public Law 91-604. The act requires that the administrator of the Environmental Protection Agency develop and issue to the states criteria of air quality for the protection of public health and welfare and further specifies that such criteria shall reflect the latest scientific knowledge useful in indicating the kind and extent of all identifiable effects on health and welfare that may be expected from the presence of an air contaminant, or combination of contaminants, in varying quantities.

The act requires the administrator to designate interstate or intrastate air quality control regions throughout the country as considered necessary to ensure adequate implementation of air quality standards. These regions are to be designated on the basis of meteorological, social, and

[17] Ambient air means that portion of the atmosphere, external to buildings, to which the general public has access.

[18] In the United States national primary and secondary ambient air quality standards have been promulgated effective April 30, 1971. Primary ambient air quality standards are those that, in the judgment of the Environmental Protection Agency Administrator, based on the air quality criteria and allowing an adequate margin of safety, are requisite to protect the public health. Secondary ambient air quality standards are those that, in the judgment of the administrator, based on the air quality criteria, are requisite to protect the public welfare from any known or anticipated adverse effects associated with the presence of air pollutants in the ambient air (on soil, water, vegetation, materials, animals, weather, visibility, personal comfort and well-being).

In England (Ministry of Housing and Local Government 1966B) the standard states, in part: "No emission discharged in such amount or manner as to constitute a demonstrable health hazard in either the short or long term can be tolerated. Emissions, in terms of both concentration and mass rate of emission, must be reduced to the lowest practicable amount."

In the Soviet Union the goal is protection from any agent in the atmosphere that can be demonstrated to produce physiological effect, even if the effect cannot be shown to be harmful.

political factors, which suggests that a group of communities should be treated as a unit.

The Federal Clean Air Act, as amended December 31, 1970 (P.L. 91-604) requires that the administrator of the Environmental Protection Agency promulgate national ambient air quality standards for sulfur oxides, particulate matter, carbon monoxide, photochemical oxidants, hydrocarbons, and nitrogen oxides. National primary and secondary ambient air quality standards were published in the Federal Register (Vol. 36, No. 4) on April 30, 1971. These standards are included in Table 6–6.

TABLE 6-6   FEDERAL AIR-QUALITY STANDARDS

Pollutant	Primary†	Secondary‡
Sulfur oxides (sulfur dioxide)		
Annual arithmetic mean	80 μg/m^3 (0.03 ppm)	60 μg/m^3 (0.02 ppm)
Maximum 24-hr concentration*	365 μg/m^3 (0.14 ppm)	260 μg/m^3 (0.1 ppm)
Maximum 3-hr concentration*	—	1,300 μg/m^3 (0.5 ppm)
Particulate matter		
Annual geometric mean	75 μg/m^3	60 μg/m^3
Maximum 24-hr concentration*	260 μg/m^3	150 μg/m^3
Carbon monoxide		
Maximum 8-hr concentration*	10 mg/m^3 (9 ppm)	10 mg/m^3
Maximum 1-hr concentration*	40 mg/m^3 (35 ppm)	40 mg/m^3
Photochemical oxidants		
Maximum 1-hr concentration*	160 μg/m^3 (0.08 ppm)	160 μg/m^3
Hydrocarbons		
Maximum 3-hr concen. (6–9 a.m.)*	160 μg/m^3 (0.24 ppm)	160 μg/m^3
Nitrogen dioxide		
Annual arithmetic mean	100 μg/m^3 (0.05 ppm)	100 μg/m^3

Sampling and analyzing shall be made in accordance with the "reference method" for the air pollutant as described in an appendix to Part 410—*National Primary and Secondary Ambient Air Quality Standards, Federal Register,* 36, No. 84 (April 30, 1971) or by an equivalent method that can be demonstrated to the administrator's satisfaction to have a consistent relationship to the reference method.

*Not to be exceeded more than once per year.
†To protect the public health.
‡To protect the public welfare.

The act requires each state to adopt and submit to the administrator within 9 months "a plan which provides for the implementation, maintenance, and enforcement of such national ambient air quality standards within each air quality control region (or portion thereof) within the

State. An additional period of no longer than 18 months may be allowed for adoption and submittal of that portion of a plan relating to implementation of secondary ambient air quality standards." States are expected to attain the national primary ambient air quality standards within 3 years after approval by the administrator of the state plan. An extension of 2 years may be granted upon application by a governor. State plans are also expected to provide for the attainment of the national secondary ambient air quality standards within a reasonable time. Both primary and secondary federal standards apply nationwide; however, state standards may be more stringent, except for motor vehicle emission standards, which are prescribed by law (California is exempt).

A State Classification System

New York State, for example, has established ambient-air-quality standards and a classification system that recognizes local variations and air-quality standards to meet local needs.[19] The land uses associated with the classification levels assigned to geographical areas of the state are broadly outlined as follows:

Level I—predominantly used for timber, agricultural crops, dairy farming, or recreation. Habitation and industry sparse.

Level II—predominantly single- and two-family residences, small farms, and limited commercial services and industrial development.

Level III—densely populated, primarily commercial office buildings, department stores, and light industries in small and medium metropolitan complexes, or suburban areas of limited commercial and industrial development near large metropolitan complexes.

Level IV—densely populated, primarily commercial office buildings, department stores, and industries in large metropolitan complexes, or areas of heavy industry.

CONTROLS

Air pollution involves a source such as a power-generating plant burning heavy fuel oil; a production by-product or waste such as particulates, vapors, or gases; release of pollutants into the atmosphere, such as smoke or sulfur dioxide; transmission by airflows; and receptors who are affected, such as people, animals, plants, structures, and clothing. Controls can be applied at one or more points between the source and the

[19] Classification Levels, Part 501, Air Quality Classification System, Environmental Conservation Law, Art. 2, Sec. 15. Effective November 1, 1971. This includes standards for sulfur dioxide, particulates, carbon monoxide, photochemical oxidants, hydrocarbons (non-methane), fluorides, and beryllium.

receptor, starting preferably at the source. The application of control procedures and devices is more effective when supported by public information and motivation, production and process revision, installation of proper air-cleaning equipment, regulatory persuasion, and, if necessary, legal action.

Source Control

Processes that are sources of air pollution include chemical reaction, evaporation, crushing and grinding, drying and baking, and combinations of these operations.

If the source of pollution is fuel for heating, the amounts and types of pollutants can be controlled by conversion to a fuel with less air pollution potential. Some examples of the types and amounts of contaminants from different types of fuels are given in Table 6–7.

TABLE 6–7   CONTAMINANT EMISSIONS (LB/1,000,000 BTU OF FUEL)

Contaminant	Bituminous Coal	Anthracite Coal	Residual Fuel Oil	Distillate Fuel Oil	Natural Gas
Solids	.630(A)*	.800	.112	0.085	.018
SO_2	1.407(S)†	1.52(S)	1.046(S)	1.120(S)	<.001
NO_2	.769	.160	.439	.365	.200
Organics	.037	.039	.020	.021	.020
Organic Acids	1.150	.595	.714	.765	.003
Aldehydes	<.001	<.001	.007	.014	.005
NH_3	.078	.040	.047	.050	.020
CO	.074	2.00	.001	.014	.004

Source: Personal communication from E. W. Davis, Division of Air Resources, New York State Dept. of Environmental Conservation, Albany, 1972.

*Contaminant emission in pounds =   .630 × ash content in percent.

†Contaminant emission in pounds = 1.407 × sulfur content in percent.

Production processes can be reviewed and revised to reduce waste and the pollutants produced at the source. This has been a fundamental step in the reduction of industrial wastewater pollution and can certainly be applied to industrial air pollution control.

The internal combustion engine is a major producer of air pollutants. A change from gasoline to another fuel or major improvement in the efficiency of the gasoline engine would attack that problem at the source. Control of lead in gasoline and gasoline evaporation during handling, from auto tanks and from carburetors, are other means of control at the source.

Proper design of basic equipment, provision of adequate solid waste collection service, elimination of open burning, and the upgrading or abandonment of inefficient apartment house, municipal, and commercial incinerators also attack the problem at the source.

Proper operation and maintenance of production facilities and equipment will not only often reduce air pollution but will save money. For example, air-fuel ratios can determine the amount of unburned fuel going up the stack, combustion temperature the strain placed on equipment when operated beyond rated capacity, and the competency of supervision can determine the quantity and type of pollutants released and the quality of the product.

Control Equipment

Control equipment is designed to remove or reduce particulates, aerosols (solid and liquid forms), and gaseous by-products resulting from inefficient design and operation.

The principles controlling design of aerosol collection equipment include the following:

1. Inertial entrapment by altering the direction and velocity of the effluent.
2. Increasing the size of the particles through conglomeration or liquid mist entrainment so as to subject the particles to inertial and gravitational forces within the operational range of the control device.
3. Impingement of particles on impact surfaces, baffles, or filters.
4. Precipitation of contaminants in electrical fields or by thermal convection.[20]

The collection of gases and vapors is based on the particular physical and chemical properties of the gases to be controlled.

Collectors and Separators

Some of the more common collectors and separators are identified below. These have application in mechanical operations for dust control such as in pulverizing, grinding, blending, woodworking, and in handling flour.

Settling chambers cause velocity reduction, usually to slower than 10 fps, and the settling of particles larger than 40 μ in diameter in trays that can be removed for cleaning. Special designs can intercept particles as small as 10 μ.

Cyclones impose a spiraling downward movement on the tangentially directed incoming dust-laden gas, causing separation of particles by centrifugal force and collection on the sides and bottom of the cylinder. Particle sizes collected range from 5 to 200 μ at gas flows of 30 to

[20] *Air Pollution Control Field Operations Manual*, U.S. Public Health Service, Pub. No. 937, Dept. of HEW, Washington, D.C. 20201, 1962.

25,000 ft³/min. Removal efficiency below 10 µ particle size is low. Cyclones can be placed in series or combined with other devices to increase removal efficiency. See Figures 6–3 and 6–4.

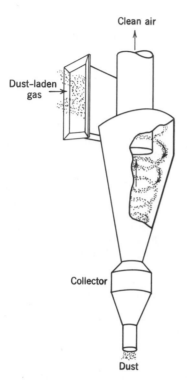

Figure 6–3   Flow of dust through cyclone. (From *Air Pollution Control Field Operations Manual,* U.S. Public Health Service Pub. No. 937, Dept. of HEW, Washington, D.C., 1962.)

Sonic collectors can be used to facilitate separation of liquid or solid particles in settling chambers or cyclones. High-frequency sound-pressure waves cause particles to vigorously vibrate, collide, and coalesce. Collectors can be designed to remove particles smaller than 10 µ.

Filters

Filters are of two general types—the baghouse and cloth-screen. The filter medium governs the temperature of the gas to be filtered, the particle size removed, capacity and loading, and durability of the filter. Filter operating temperatures vary from about 200°F for wool or cotton to 450 to 500°F for glass fiber.

A *baghouse filter* is shown in Figure 6–5. The tubular bags are 5 to 18 in. in diameter and from 2 to 30 ft in length. The air to be filtered is forced through a cooling system to an expansion chamber to remove large particles, and then through the bags where the smaller particles build up on the inside and, in so doing, increase the filtering efficiency. Periodic shaking of the bags (tubes) causes the material collected to fall off and restore the filtering capacity. The baghouse filter

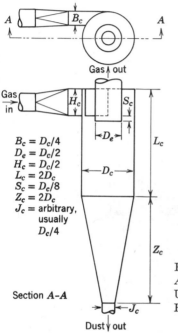

$B_c = D_c/4$
$D_e = D_c/2$
$H_c = D_c/2$
$L_c = 2D_c$
$S_c = D_c/8$
$Z_c = 2D_c$
$J_c = $ arbitrary, usually $D_c/4$

Section A–A

Figure 6–4   Diagram of cyclone separator. (From *Air Pollution Control Field Operations Manual,* U.S. Public Health Service Pub. No. 937, Dept. of HEW, Washington, D.C., 1962.)

has particular application in cement plants, heavy metallurgical operations, and other dusty operations. Efficiencies exceeding 95 percent and particle removal below 10 μ in size are reported, depending on the matt form and buildup.

Cloth-screen filters are used in the smaller grinding, tumbling, and abrasive cleaning operations. Dust-laden air is forced through one or more cloth screens in series. The screens are replaced as needed. Other types of filters use packed fibers, filter beds, granules, and oil baths.

Electrical Precipitators

Electrical precipitators have application in power plants, in cement plants, and in metallurgical, refining, and heavy chemical industries for the collection of fumes, dusts, and acid mists. Particles, in passing through a high-voltage electrical field, are charged and then attracted to a plate of the opposite charge, where they collect. The accumulated material falls into a hopper when vibrated. See Figure 6–6.

The gases treated may be cold or at a temperature as high as 1100°F; but 600°F or less is more common. Electrical precipitators are efficient

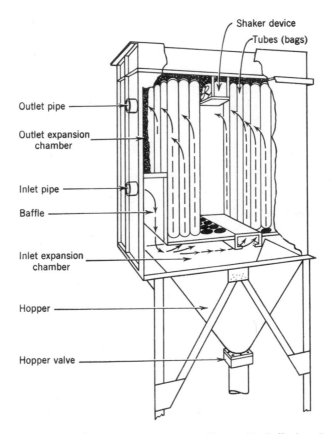

Figure 6–5 Simplified diagram of a baghouse. (From *Air Pollution Control Field Operations Manual*, U.S. Public Health Service Pub. No. 937, Dept. of HEW, Washington, D.C., 1962.)

for the collection of particles less than 0.5 μ in size; hence cyclones and settling chambers, which are better for the removal of larger particles, are often used ahead of precipitators. Single-stage units operate at voltages of 25,000 or higher; two-stage units (used in air conditioning) operate at 12,000 V (volts) in the first or ionizing unit and at 6000 V in the second collection unit.

Electrostatic precipitators are commonly used at large power stations to remove dust from flue gases. They are considered one of the most effective devices for this purpose.

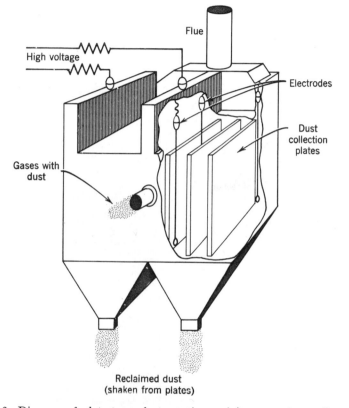

Figure 6-6 Diagram of plate-type electrostatic precipitator used to collect catalyst dust. (From *Air Pollution Control Field Operations Manual,* U.S. Public Health Service Pub. No. 937, Dept. of HEW, Washington, D.C., 1962.)

Scrubbers

Scrubbers are wet collectors generally used to remove particles that form as a dust, fog, or mist. A high-pressure liquid spray is applied to the gas passing through the washer, filter, venturi, or other device. In so doing the gas is cooled and cleaned. Although water is usually used as the spray, a caustic may be added if the gas stream is acidic.

Where the spray water is recirculated, corrosion of the scrubber, fan, and pump impeller can be a serious problem. Particle size collected may range from 40 μ to as low as 1 μ, with efficiency as high as 99 percent depending on the collector design. See Figures 6–7 and 6–8.

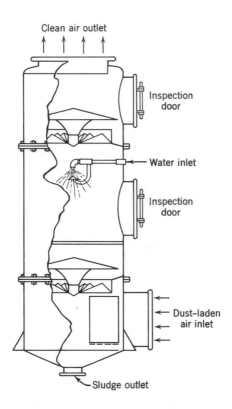

Figure 6–7 Centrifugal wash collector. (From *Air Pollution Control Field Operations Manual,* U.S. Public Health Service Pub. No. 937, Dept. of HEW, Washington, D.C., 1962.)

Afterburners

Afterburners are used to complete the combustion of unburned fuel, such as smoke and particulate matter, and to burn odorous gases. Apartment-house and commercial incinerators and meat-packing plant smokehouses are examples of smoke and particulate emitters. Rendering, packing house, refinery, paint and varnish, fish processing, and coffee roasting are examples of odor-producing operations. Afterburners usually operate at around 1200°F, but may range from 900°F to 1600°F depending on the ignition temperature of the contaminant to be burned. Catalytic afterburners may be used for the burning of lean mixtures of contaminated air and gas.

Other Collectors

Vapor conservation equipment is used to collect vapors escaping from the storage of volatile products such as gasoline. A sealed floating cover or a vapor recovery system is used. Vapors that can be condensed may be compressed back into a liquid; the noncondensable vapors are used as a fuel or are burned.

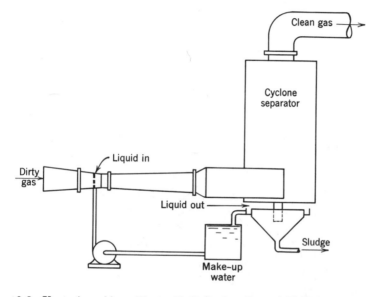

Figure 6–8   Venturi scrubber. (From *Air Pollution Control Field Operations Manual*, U.S. Public Health Service, Pub. No. 937, Dept. of HEW, Washington, D.C., 1962.)

Gases and liquids can also be collected by absorption and adsorption. In absorption, the gas or liquid to be treated is passed through a tower, where it comes in contact with a specific absorbing medium or spray that selectively removes the desired material. For example, oxides of nitrogen can be absorbed by water; hydrogen fluoride, by water and alkaline water solution. In adsorption the molecules of a gas, liquid, or dissolved substance to be treated are brought into contact with the adsorbent, such as activated carbon, aluminas, silicates, chars, or gels, which collect the contaminant in the pores or capillaries. The adsorbent is regenerated by steam; the contaminant is condensed and collected for proper disposal.

Dilution by Stack Height

Since wind speed increases with height in the lower layer of the atmosphere, the release of pollutants through a tall stack enhances the transport and dispersion of the material. The elevated plume is rapidly transported and dispersed downwind. This generally occurs at a rate faster than that of the dispersion toward the ground. The resulting downwind distribution of pollutant concentrations at the ground level is such that concentrations are virtually zero at the base of the stack, increase to a maximum at some downwind distance, and then decrease to negligible concentrations thereafter. This distribution and the difference due to stack height are shown schematically in Figure 6–9.

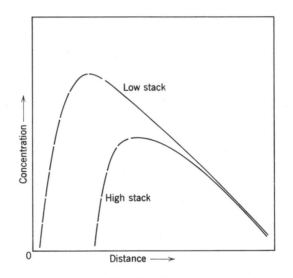

Figure 6–9 Variation of ground-level pollutant concentration with downwind distance.

It should be noted that the higher the stack, the lower the pollutant ground-level concentration, assuming that the emissions are equal. This applies to uncomplicated level terrain. Obviously, if the plume is transported to hill areas, the surfaces will be closer to the center of the plume and hence will experience higher concentrations.

It is general practice to use high stacks for the emission of large quantities of pollutants, such as in fossil-fueled power production, to reduce the ground-level effects of the pollutants. Stacks of 250 to 350 ft in height are not unusual, and some are as high as 800 ft. It should be recognized, however, that there is a practical limit to height beyond which cost becomes excessive and the additional dilution obtained is not significant.

Planning and Zoning

The implementation of planning and zoning controls requires professional stimulation and the cooperation of the state and regional planning agencies and the local county, city, village, and town units of government.

Although air pollution control should start at the source, the devices available and the economic, social, and political factors involved will limit the amount of control that can be realistically achieved in many instances. Hence a combination of planning and zoning means should be considered when adequate pollution elimination at the source is not feasible. These means could include plant siting downwind from residential, work, and recreation areas, with consideration given to climate and meteorological factors, frequency of inversions, topography, air movement, stack height, and adjacent land uses. Additional factors are distance separation, open-space buffers, designation of industrial areas, traffic and transportation control, and possible regulation of plant raw materials and processes. All these controls must recognize the present and future land use, especially the air quality needed for health and comfort, regardless of the land ownership.

It is possible to calculate and predict, within limits, the approximate effects of existing and proposed air pollution sources on the ambient air quality in a community.[21]

For example, a simple rough approximation can be made by use of the formula $P = Q/V$, in which:

P = the concentration of the pollutant in the air;
Q = the quantity or pounds of wastes emitted;
V = the volume of air into which the wastes are dispersed.

Obviously it is not readily possible to determine the amounts and types of all pollutants emitted, the volume of dilution air, and the influencing meteorological factors. The problem is made more complex when more than one pollution source is involved. But some reasonable assumptions can be made and conclusions drawn to guide one's judgment. These can be checked by the use of monitors to sample pollutants. The effects caused and the need for adjustments in permissible emissions can thus

[21] Maynard Smith, *Recommended Guide for the Prediction of the Dispersion of Airborne Effluents,* American Society of Mechanical Engineers, New York, 1968.

be determined. In a more sophisticated approach a diffusion model could be developed and applied.

The maintenance of air quality that meets established criteria requires regulation of the location, density, and/or type of plants and plant emissions that could cause contravention of air quality standards. This calls for local and regional land-use control and cooperation so that the construction of plants permitted would incorporate practices and control equipment that would not emit pollution that could adversely influence the air quality of the community in the airshed. See Tables 6–3, 6–4, and 6–7 and "Ambient Air Quality Standards," earlier in this chapter.

Monitoring of the air at carefully selected locations would continually inform and alert the regulatory agency of the need for additional source control and enforcement of emission standards. Conceivably, under certain unusually adverse weather conditions, a plant may have to take previously planned emergency actions to reduce or practically eliminate emissions for a period of time.

Air zoning establishes different air quality standards for different areas based on the most desirable and feasible use of land. As discussed earlier in this chapter, the New York State Air Quality Classification Standards, for example, recognizes four general classification levels for assignment to geographic areas of the state; they are based on the social and economic development and pollution potentials of the areas. The land uses and classification levels vary from very rural, Level I, to heavy industry, Level IV. Specific standards are established for each classification level. In all levels, however, protection of the health of people is paramount. Insofar as air zoning is concerned, an industry should be able to choose its location and types of emission controls so long as the air quality standards are not violated.

Although air zoning provides a system or basis for land use and development, sound planning can assist in greatly minimizing the effects of air pollution. A WHO Expert Committee suggests the following:[22]

1. The siting of new towns should be undertaken only after a thorough study of local topography and meteorology.

2. New industries using materials or processes likely to produce air contaminants should be so located as to minimize the effects of air pollution.

3. Satellite (dormitory) towns should restrict the use of pollution-producing fuels.

4. Provision should be made for green belts and open spaces to facilitate the dilution and dispersion of unavoidable pollution.

5. Greater use should be made of hydroelectric and atomic power and of natural

[22] *Environmental Health Aspects of Metropolitan Planning and Development, WHO Tech. Rep. Ser.* No. 297, Geneva, 1965.

gas for industrial processes and domestic purposes, thereby reducing the pollution resulting from the use of conventional fossil fuels.

6. Greater use should be made of central plants for the provision of both heat and hot water for entire (commercial or industrial) districts.

7. As motor transport is a major source of pollution, traffic planning can materially affect the level of pollution in residential areas.

It is apparent that more needs to be learned and applied concerning open spaces, bodies of water, and trees and other vegetation to assist in air pollution control. For example parks and green belts appear to be desirable locations for expressways because vegetation, in the presence of light, will utilize the carbon dioxide given off by automobiles and release oxygen. In addition highway designers must give consideration to such factors as road grades, speeds and elevations, interchange locations, and adjacent land uses as means of reducing the amounts and effects of automobile noise and emissions.

PROGRAM AND ENFORCEMENT

General

A program for air resources management should be based on a comprehensive area-wide air pollution survey including air sampling, basic studies and analyses, and recommendations for ambient air quality standards. The study should be followed by an immediate and long-term plan to achieve the community air quality goals and objectives, coupled with a surveillance and monitoring system and regulation of emissions. See Chapter 2 and pages 471 to 473 and 483 to 485.

MacKenzie proposes the following conclusions and decisions for the implementation of a study:[23]

1. Select air quality standards—possibly with variations in various parts of the area.

2. Cooperate with other community planners in allocating land uses.

3. Design remedial measures calculated to bring about the air quality desired. Such measures might include several or all of the following: limitations on pollutant emissions; variable emission limits for certain weather conditions and predictions; special emission limits for certain areas; time schedules for commencing certain control actions; control of fuel composition; control of future sources by requiring plan approvals; prohibition of certain plan approvals; prohibition of certain activities or requirements for certain types of control equipment; and performance standards for new land uses.

4. Outline needs for future studies pertaining to air quality and pollutant emissions and design systems for collection, storage, and retrieval of the resultant data.

[23] Op. cit.

5. Establish priorities among program elements and set dates for implementation.

6. Prepare specific recommendations as to administrative organization needed to implement the program; desirable legislative changes; relationships with other agencies and programs in the area and adjoining areas and at higher governmental levels; and funds, facilities, and staff required.

As in most studies, a continual program of education and public information supplemented by periodic updating is necessary.[24] People must learn that air pollution can be a serious hazard and be motivated to support the need for its control. In addition, surveys and studies must be kept current; otherwise the air resources management activities may be based on false or outdated premises.

International treaties, interstate compacts or agreements, and regional organizations are sometimes also needed to resolve air pollution problems that cross jurisdictional boundaries. This becomes more important as industrialization increases and as people become more concerned about the quality of their environment.

It becomes apparent that the various levels of government each have important complementary and cooperative roles to play in air pollution control.

The federal government role includes research into the causes and effects of air pollution as well as the control of international and interstate air pollution on request of the affected parties. It should also have responsibility for a national air-sampling network, for training, for preparation of manuals and dissemination of information, and for assisting the states and local governments. In the United States this has been done primarily through the U.S. Public Health Service in the Department of Health, Education, and Welfare, and now in the Environmental Protection Agency. Other federal agencies making major contributions are the U.S Weather Service; the Atomic Energy Commission, in relation to the effects of radioactivity; the Department of Agriculture, in relation to the effects of air pollution on livestock and crops; the Department of Interior; the Department of Commerce, including the National Bureau of Standards; and the Civil Aeronautics Administration.

The state role is similar to the federal role. It would include, in addition, the setting of statewide standards and establishment of a sampling network; the authority to declare emergencies and possession of appropriate powers during emergencies; the delegation of powers to local agencies for control programs; and the conducting of surveys, demonstration projects, public hearings, and special investigations.

[24] Alexander Rihm, Jr., "Public Relations in Air Pollution Control," paper presented at meeting of Public Relations Society of America, Albany, N.Y., November 16, 1967.

The role of local government is that delegated to it by the state and could include complete program implementation and enforcement.

Organization and Staffing

Organization and staffing will vary with the level of government, the legislated responsibility, funds provided, governmental commitment, extent of the air pollution, and other factors. Generally air pollution programs are organized and staffed on the state, county, and large-city, in addition to federal, levels. In some instances limited programs of smoke and nuisance abatement are carried out in small cities, towns, and villages as part of a health, building, or fire department program. Because of the complexities involved, competent direction, staff, and laboratory support is needed to carry out an effective and comprehensive program. A small community usually cannot afford and in fact might not have need for a full staff; but it could play a needed supporting

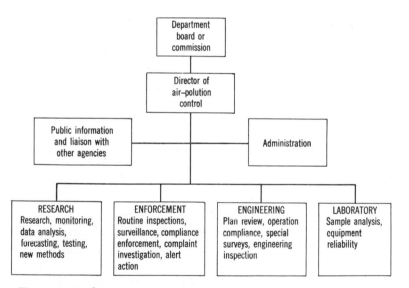

Figure 6–10 Air resources management functional organization chart.

role to the county and state programs. In this way uniform policy guidance and technical support could be provided and local on-the-spot assistance utilized. The local government should be assigned all those responsibilities it is capable of handling effectively.

An organization chart for an Air Resources Management Agency is shown in Figure 6–10. There are many variations.

Regulation

A combination of methods and techniques are generally used to prevent and control air pollution, after a program is developed, air quality objectives established, and problem areas defined. These include:

1. Public information and education.
2. Source registration.
3. Plan review and construction-operation approval.
4. Emission standards.
5. Monitoring and surveillance.
6. Technical assistance and training.
7. Inspection and compliance follow-up.
8. Conference, persuasion, and administrative hearing.
9. Rescinding or suspension of operation permit.
10. Legal action—fine, imprisonment, misdemeanor, injunction.

Effective regulation requires the development and retention of competent staff and the assignment of responsibilities, such as shown in Figure 6–10. In a small community the responsibilities would probably be limited to source location and surveillance, data collection, smoke and other visible particulate detection, complaint investigation, and abatement as an arm of a county, regional, or state enforcement unit.

Regulatory agencies usually develop their own procedures, forms, and techniques to carry out the functions listed above. Staffing, in addition to the director of air pollution control, may include one or more of the following: engineers, scientists, sanitarians, chemists, toxicologists, epidemiologists, public information specialists, technicians, inspectors, attorneys, administrative assistants, statisticians, meteorologists, electronic data processing specialists, and personnel in supporting services.

Detailed information on inspection and enforcement is given in *Air Pollution Control Field Operations Manual*[25] and in *Community Action Guides for Public Officials—Air Pollution Control.*[26] Additional information is also given in Chapter 12.

Important in regulation is the development of working relationships and memoranda of agreements with various public and private agencies. For instance governmental construction, equipment, and vehicles could set examples of air pollution prevention. The building department would ensure that new incinerators and heating plants have the proper air pollution control equipment. The police would enforce vehicular air pol-

---

[25] Melvin I. Weisburd and S. Smith Griswold, U.S. Public Health Service Pub. No. 937, Dept. of HEW, Washington, D.C., 1962.
[26] National Association of Counties Research Foundation, Washington, D. C.

lution control requirements. The fire department would carry out fire prevention and perhaps boiler inspections. The planning and zoning boards would rely on the director of air pollution control and his staff for technical support, guidance, and testimony at hearings. Equipment manufacturers would agree to sell only machinery, equipment, and devices that complied with the emission standards. The education department would incorporate air pollution prevention and control in its environmental health curriculum. Industry, realty, and chain-store management would agree to abide by the rules and police itself. Cooperative training and education programs would be provided for personnel responsible for operating boilers, equipment, and other facilities that may contribute to air pollution. These are but a few examples. With ingenuity many more voluntary arrangements can be devised to make regulation more acceptable and effective.

BIBLIOGRAPHY

Air Pollution, WHO Mon. Ser. No. 46, Columbia University Press, New York, 1961.

Air Pollution Control Field Operations Manual, compiled and edited by Melvin I. Weisburd and S. Smith Griswold, U.S. Public Health Service Pub. No. 937, Dept. of HEW, Washington, D.C. 20201, 1962.

Air Pollution Engineering Manual, compiled and edited by John A. Danielson, U.S. Public Health Service Pub. No. 999-AP-40, Dept. of HEW, Cincinnati, Ohio, 1967.

Air Pollution Experiments for Junior and Senior High School Science Classes, edited by Donald C. Hunter and Henry C. Wohlers, Air Pollution Control Association, Pittsburgh, Pa., 1969.

Air Pollution Primer, National Tuberculosis and Respiratory Association, New York, 1969.

Air Quality Criteria, staff report, Subcommittee on Air and Water Pollution, Committee on Public Works, U.S. Senate, Government Printing Office, Washington, D.C., 1968.

Byers, H. R., *General Meteorology,* 3rd ed., McGraw-Hill Book Co., New York, 1959.

Cleaning Our Environment—The Chemical Basis For Action, a report by the Subcommittee on Environmental Improvement, Committee on Chemistry and Public Affairs, American Chemical Society, Washington, D.C., 1969.

Community Action Guide for Public Officials—Air Pollution Control, National Association of Counties Research Foundation, Washington, D.C., 1969.

The Effects of Air Pollution, U.S. Public Health Service, Dept. of HEW, Washington, D.C. 20201, 1967.

Guide to the Appraisal and Control of Air Pollution, Program Area Committee on Air Pollution, American Public Health Association, Inc., 1015 Eighteenth Street, N.W. Washington, D.C., 1969.

"Introduction to Air Pollution Control," New York State Dept. of Health, Albany, March 1964.

Meteorology and Atomic Energy, edited by D. H. Slade, U.S. Atomic Energy Commission, Tech. Information Div., TID-24190, July 1968.

Meteorological Aspects of Air Pollution, U.S. Dept. of HEW, CP & EHS, NAPCA, Box 12055, Research Triangle Park, N.C.

Recommended Guide for the Prediction of the Dispersion of Airborne Effluents, edited by M. E. Smith, American Society of Mechanical Engineers, United Engineering Center, 345 East 47th Street, New York, 10017, May 1968.

Research into Environmental Pollution, WHO Tech. Rep. Ser., No. 406, Geneva, 1968.

Rihm, Alexander, "Air Pollution," in *Dangerous Properties of Industrial Materials,* N. Irving Sax, ed., Van Nostrand Reinhold Co., New York, 1969.

Stern, Arthur C., *Air Pollution,* Vol. 1, Air Pollution and Its Effects; Vol. 2, Analysis, Monitoring and Surveying; Vol. 3, Sources of Air Pollution and Their Control; Academic Press, New York, 1968. Revised 1971.

Sutton, O. G., *Micrometeorology,* McGraw-Hill Book Co., New York, 1953.

7

RADIATION USES AND PROTECTION

RADIATION FUNDAMENTALS

Definitions

Radiation is the emission of energy from a point of origin. Any electro-magnetic or particulate radiation capable of producing ions, directly or indirectly, by interaction with matter is referred to as ionizing radiation. The medical uses generally involve X rays and gamma rays as well as high-energy electrons and beta particles. Corpuscular emission from radioactive substances of other sources, commonly alpha particles, beta particles, and neutrons, are also forms of radiation.

The more common types of radiation are classified according to wave-length. The relationship of radiation wavelength, velocity, and frequency is shown by:

$$\text{wavelength of radiation} = \frac{\text{velocity of radiation}}{\text{frequency of radiation}}$$

The velocity of radiation varies with the medium through which it travels. The frequency of radiation is dependent on its source. Frequency is the number of occurrences of a periodic quantity as waves, vibrations, or oscillations in a unit of time usually expressed as cycles per second. Particles are not included.

The shorter the wavelength, the higher the frequency and energy. The longer the wavelength, the lower the frequency and energy. In general the shorter the wavelength of a particular type of radiation, the

491

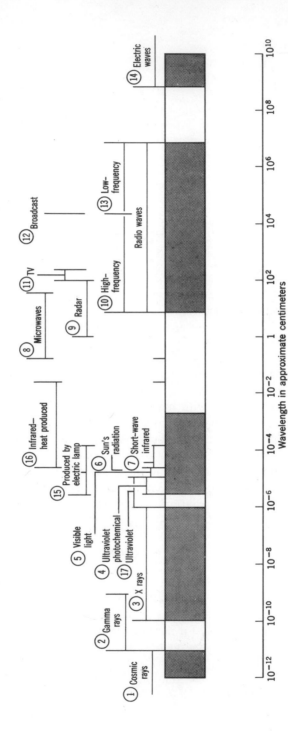

Figure 7–1 For legend see opposite page.

greater is its power of penetration. This is apparent by inspection of Figure 7–1, which also shows radiation sources in the electromagnetic spectrum.

The energy of ionizing radiation is measured as electron volts (eV), thousands of electron volts (keV), or millions of electron volts (MeV), whichever is applicable. An electron volt is the energy an electron gains in passing through a difference of potential of one volt and is equivalent to 1.6×10^{-12} ergs. The intensity of a beam of radiation is usually expressed in terms of ergs per square centimeter per second. It is the quantity of energy passing through a known area per unit of time. The quantity of radiation may be stated as ergs per square centimeter.

Roentgen (R)

The absorption of about 86 ergs of energy per gram of air represents one roentgen. It is equal to one electrostatic unit (esu) per 0.001293 grams of dry air. The roentgen is a measure of the ionization in air produced by exposure to X rays or gamma rays and is not applicable

---

Figure 7–1   Electromagnetic radiations.
1. Cosmic Rays. Reaching earth from sky.
2. Gamma Rays.
3. X rays. High-frequency oscillations produced by X-ray tubes.
4. Ultraviolet. Photochemical-photoelectric and fluorescent effects. Germicidal action and health maintenance by virtue of radiation absorbed by bacteria.
5. Visible. Seeing—discrimination of color and detail.
6. Sun's radiation reaching the earth.
7. Short-wave infrared. Heat therapy—drying.
8. Microwaves.
9. Radar.
10. High frequency.
11. Television.
12. Broadcast.
13. Low frequency.
14. Electric waves. Produced by electric generators.
15. Produced by electric lamps.
16. Infrared. Produced by heat.
17. Ultraviolet.

The electromagnetic spectrum. (Adapted from "Lamps and the Spectrum," General Electric, Nela Park, Cleveland, Ohio, and other sources.)

NOTE: The velocity or speed of radiation throughout the spectrum is that of light, or about 186,000 mi/sec.

The nature of all radiation is the same, and the difference lies only in the frequency and wavelength, i.e., frequency $\times$ wavelength = velocity. The range in wavelengths is from 1.2×10^{-7} meters for cosmic rays to a length of 3100 miles for 60-cycle electric current.

One micron = 10^{-4} cm = 10^4 angstroms = one micrometer. One cm = 10^4 microns = 10^8 Angstrom units (Å).

to particle radiations such as alphas, betas, and neutrons. Survey meters usually read in roentgens or milliroentgens per hour.

Rad

One rad represents the absorption of 100 ergs per gram of medium from any type of ionizing radiation. For water and soft tissue, with X rays and gamma rays (having energies between 100 keV and 3 MeV), an exposure to one roentgen is roughly equal to an absorbed dose of one rad. The rad replaced the rep as a unit of measure of radiation absorption.

Rem

The rem, short for *r*oentgen *e*quivalent *m*an, is intended to take into consideration the biological effect of different kinds of radiation from the same dose in rads. For medical radiation purposes, covering only X and gamma radiation, the number of rems may be considered equivalent to the number of rads or to the number of roentgens. A quality or damage factor is used to convert rads to rems. Some practical quality factors (QF) are:[1] 1.0 for X rays, gamma rays, electrons or positrons, energy >0.03 MeV; 1.0 for electrons or positrons, energy <0.03 MeV; 3 for neutrons, energy <10 keV; 10 for neutrons, energy >10 keV; 1 to 10 for protons; 1 to 20 for alpha particles; and 20 for fission fragments, recoil nuclei.

$$\text{rems} = \text{rads} \times QF.$$

Curie (Ci)[2]

The rate at which atoms of radioactive sources disintegrate is measured in curies. One curie is defined as 37,000,000,000 (3.7×10^{10}) disintegrations per second. The radioactivity of one gram of radium is approximately one curie.

The Atom

A brief elementary review of the characteristics of the atom is useful for understanding the effects of radiation. All atoms are composed of

[1] *Basic Radiation Protection Criteria,* Rep. No. 39, National Council on Radiation Protection and Measurement, 4201 Connecticut Avenue, N.W., Washington, D.C., January 15, 1971, p. 83.

[2] A picocurie (pCi) is a trillionth of a curie; a nanacurie (nCi) is a billionth of a curie; a microcurie (μCi) is a millionth of a curie; a millicurie (mCi) is a thousandth of a curie; a kilocurie (kCi) is a thousand curies; and a megacurie (MCi) is a million curies.

protons, electrons, and neutrons except the hydrogen atom, which contains a proton and an electron. The structure and components of an atom are shown in Figure 7-2.

When atoms combine to form compounds, the resulting molecule usually has the same number of electrons as it does protons and it is also

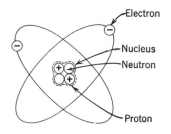

Helium, $_2He^4 = _zx^A$

Figure 7-2 Atomic structure. The electrons move in the electron cloud, orbit, or sheath and have a negative electrical charge. The protons have a positive electrical charge; the neutrons are neutral. The number of protons (Z) = atomic number of the element. The number of protons plus the number of neutrons is equal to the mass number (A). Isotopes of the same element have as many mass numbers as there are isotopes of the element, but the same atomic (Z) number. Atoms in their natural state are electrically neutral; the number of electrons outside the nucleus are the same as the number of protons within the nucleus.

electrically neutral. Since the electron is the lightest part of the atom, and each electron is not bound as tightly to the nucleus as the protons and neutrons are bound to each other in the nucleus, the electron is more mobile and can be removed from the atom or molecule without expending much energy. (The energy binding the electron to the nucleus is small.) When an electron is removed from an atom or a molecule, the resulting component has an electrical charge because the protons are then in excess of the electrons. The atom or molecule with the electron removed has an electrical charge and is, by definition, an ion. When the nucleus is split or fissioned, part of the energy holding the protons and neutrons together is released as heat.

Ions

Ions may be charged positively or negatively and may exist in crystals, liquid, and gases. Positive ions are produced by removing electrons from neutral atoms and molecules; negative ions are created when electrons attach themselves to neutral atoms or molecules. Any process by which a neutral atom or molecule loses or gains electrons, thereby resulting in a net charge, is known as ionization. Figure 7–3 illustrates the process.

The ionization resulting from radiation absorption has significant public health aspects. The production of ions within tissues can injure people,

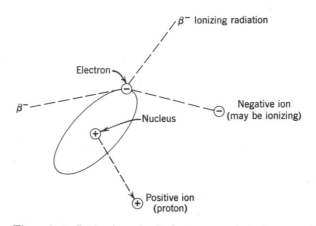

Figure 7–3   Ionization of a hydrogen atom by a beta particle β^-.

animals, and plants and cause genetic as well as somatic damage. The radiation is detected and measured through the ions created by the energy transfer. In practice, however, exposure to ionizing radiation is controlled by the use of shields of lead, concrete, and other materials.

Isotopes

There are also different members, or isotopes, of an element. The isotopes have the same atomic number (number of protons), or Z number, but different mass numbers (different number of neutrons in the nucleus), or A number. For example hydrogen ($_1H^1$), deuterium ($_1H^2$), and tritium ($_1H^3$) all have the same atomic, or Z number, of 1, but different A numbers. But they are all members of the hydrogen family. Uranium 238 contains 92 protons and 146 neutrons, while the isotope uranium 235 contains 92 protons and 143 neutrons. Oxygen has six known isotopes; tin has at least 23, yet the chemical properties of all isotopes of tin are for practical purposes the same.

Types of Radiation

The common types of radiation are X rays, gamma rays, alpha particles, beta particles, and neutrons. Each has its own characteristics. X rays and gamma rays are similar.

Alpha particles have large specific ionization values. Since they create many ions per unit of path length, they dissipate their energy rapidly and penetrate only 3 to 5 cm of air. A thin sheet of paper will stop the particle. Alpha particles have a positive electric charge, exactly twice that of an electron, and a mass of four, the same as a helium atom. The particles are normally a hazard to health only in the form of internal radiation.

Beta particles are light in weight and carry single charges. They are high-speed electrons that originate in the nucleus. Their specific ionization values are intermediate between those of alpha particles and gamma and X rays. They ionize slightly, dissipate their energies rather quickly, and are moderately penetrative but are stopped by a few millimeters of aluminum. Beta particles can be a health hazard either as internal or external radiation due to the ionization in the tissues.

X rays and gamma rays move with the speed of light. The only difference in this respect between gamma rays, X rays, and visible light is their frequency. See Figure 7–1. X rays and gamma rays ionize slightly in travel and are very penetrating compared to alpha and beta particles. They constitute the chief health hazard of external radiation, although gamma rays can be a hazard also as internal radiation. Whereas gamma rays come from the nucleus of an atom, X rays come from the electrons around the nucleus and are produced by electron bombardment. As is commonly known, when X rays pass through an object, they give a shadow picture of the denser portions on film or a special screen.

Neutrons are uncharged high-energy particles that may be given off under certain conditions and can have both physiological effects and the ability to make other substances radioactive. Neutrons present major problems in areas around nuclear reactors and particle accelerators.

A *radionuclide* is a radioactive isotope of an element. A radionuclide can be produced by placing material in a nuclear reactor or particle accelerator (cyclotron, betatron, Van de Graff generator) and bombarding it with neutrons. Radionuclides have particular application as tracers in many areas of medicine, industry, and research. Mixing a radionuclide with a stable substance makes possible study of the path it follows and the physical and chemical changes the substance goes through.

Half-life ($T_{1/2}$)

A characteristic common to all radionuclides is the decay or loss of activity as it ages. This loss is measured as the time it takes a radionuclide to lose half its activity and is referred to as its *radioactive half-life*. In general, seven half-lives will reduce the radioactivity of a radionuclide to about 1 percent of what it was when first measured. The loss of activity is usually measured in terms of disintegrations per second or minute, which can be readily converted to curies or millicuries. For example, if the radioactivity of a material with a half-life of 24 hr is a gamma emitter and emits radiation equivalent to 100 milliroentgens per hour (mR/h), it will be 50 mR/h after 24 hr, 25 mR/h after the second 24 hr, and 0.78 mR/h after the seventh 24-hr period. If the radioactivity of a material is not known, periodic measurements can be made and a curve constructed with radioactivity plotted on the vertical axis and time on the horizontal axis. The time for the material to lose half its radioactivity can then be read off the graph. The measure of half-life is useful to identify a particular radionuclide. Some common radioactive materials and characteristics are given in Table 7–1.

A distinction should be made between biological and effective half-life. The *biological half-life* is the time in which a living tissue, organ, or individual eliminates, through biologic processes, one-half of a given amount of substance that has been introduced into it. The *effective half-life* is the time in which a radioactive substance fixed in an animal body is reduced to half the original amount, resulting from the combined actions of natural radioactive decay and biological elimination. It may be derived from the formula:

$$\text{effective half-life} = \frac{\text{biological half-life} \times \text{radioactive half-life}}{\text{biological half-life} + \text{radioactive half-life}}$$

The maximum permissible body burdens and maximum permissible concentrations of radionuclides in air and in water for occupational exposure are given in Table 7–5.

Sources of Exposure

Sources of radiation are natural background, radioactive fallout from nuclear testing or use of nuclear devices, radiation from medical diagnosis and treatment, and radiation from industrial and other man-made sources.

Some radiations to which individuals are or might be exposed are given in Table 7–2 to help understand and better appreciate the significance of radiation dosages.

X-ray Machines and Equipment

One of the common types of ionizing radiation is the X ray. It is produced by fluoroscopic equipment, radiographic equipment (fixed and mobile), X-ray therapy equipment, dental X-ray machines, X-ray diffraction apparatus, industrial X-ray machines, the generally pro-

TABLE 7-1   RADIOACTIVE SOURCES OF EDUCATIONAL INTEREST

Radionuclide	Half-Life	Type of Radiation	Specific Gamma-Ray Constant, R/mCi-h at 1 cm*	"Generally Licensed," or "Exempt," Quantities Activity, microcuries	Exposure rate, mR/h at 1 m
Carbon 14	5730 yr	Beta		50	0.0003
Cesium 137	30 yr	Beta and gamma	3.0	1	0.0013
Cobalt 60	5.3 yr	Beta and gamma	12.9	1	0.0023
Gold 198	2.7 days	Beta and gamma	2.27	10	
Hydrogen 3 (Tritium)	12.3 yr	Beta		250	
Iodine 131	8.1 days	Beta and gamma	2.20	10	0.0022
Iridium 192	74.4 days	Beta and gamma	4.83	10	0.0048
Phosphorus 32	14.2 days	Beta		10	
Polonium 210	138 days	Alpha		0.1	
Silver 111	7.5 days	Beta and gamma	0.13	10	0.0002
Sodium 24	0.63 days	Beta and gamma	18.2	10	0.0187
Strontium 90	28 yr	Beta		0.1	
Sulfur 35	87 days	Beta		50	
Radium (in equilibrium with Radium A, B, C)	1620 yr	Alpha, beta, gamma	8.25 with 0.5 mm Pt filter	0.1	0.00008 with 0.5 mm Pt filter

*Roentgens per hour for a point source of 1 millicurie at a distance of 1 centimeter.

Source: *Radiation Protection in Educational Institutions*, Rep. No. 32, National Council on Radiation Protection and Measurements, Washington, D.C., 1966.

499

TABLE 7–2   SOME RADIATIONS TO WHICH WE ARE OR MIGHT BE EXPOSED

Peacetime occupational allowable exposure	300 mR/wk, avg. in 13 wk (see Table 7–4)
Cosmic rays	0.1 mR/day (gamma)
Naturally occurring radioactive substances in water, air, soil	0.1 to 0.3 mR/day*
Natural environment – Denver	1.4 mR/day, incl. cosmic
San Francisco	0.4 mR/day, incl. cosmic
Human red blood cells	0.1 mR/day
Sun glasses – cornea glasses tinted with uranium oxide	1 to 10 mR/hr (beta and alpha)
Chinaware	5 mR/hr (beta and gamma)
Radium dial watch	100 mR/day (beta and gamma at surface)
Common X rays.†	
Chest	10 to 250 mR/film or higher
Chest fluoroscopic	1,000 mR/film
Pregnancy examination‡	9,000 mR
Lumbar spine, lateral	5,700 mR
Lumbar spine, AP	1,500 mR
Abdomen	1,300 mR
Fluoroscopic	10,000 mR/min
Gastrointestinal examination	5,000 to 50,000 mR
Dental§	200 to 1,000 mR/film
Shoe-fitting fluoroscope machine‖	500 to 5,000 mR/sec
Nuclear fallout (through 1962)	100 mR/30 yr, (0.02 to 0.5R)
In wooden homes	0.25 to 0.50 mR/day
In masonry homes, brick, stone, concrete block¶	0.30 to 1.50 mR/day

Reference: Various sources. See definitions of roentgens, rems, and rads in this chapter.

  * Exposure increases slightly with altitude and varies with soils and rock natural radioactivity content. The main sources of background radiation are potassium 40, thorium, uranium, and their decay products, including radium and cosmic rays. Average 125 mrem (100 to 400) per year total body.

  † May be exceeded with poor practice or equipment, or if medically indicated.

  ‡ With special care to permissible period; should be strongly discouraged.

  § A whole-mouth X ray giving a dose of 50 R can be reduced to 5 R if an aluminum filter (2mm) is used in the X ray machine.

  ‖ Should be prohibited.

  ¶ Will vary with radiation content of the material.

hibited shoe-fitting fluoroscope, and other ionizing radiation-producing equipment. Gamma-beam therapy equipment, including a sealed source capsule in a protective housing, is another source of ionizing radiation (gamma rays); gamma rays differ from X rays in the manner in which they are produced, the source (cesium 137, cobalt 60, gold 198, iridium 192, radium 226), or from where in the atom they originate.

The X-ray machine is essentially a particle accelerator in which the electrons (negative charge) are first generated by a hot filament, then accelerated through a vacuum by means of a high voltage, and abruptly stopped by a "target" that usually consists of a material of high atomic number. Tungsten is usually used because of its high melting point and high atomic number. See Figure 7–4.

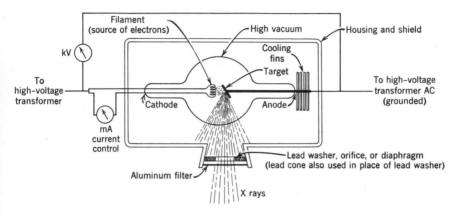

Figure 7–4   An X-ray machine.

As electrons enter the surface layers of the target, they are slowed down by collision with the nuclei and electrons in the target and are diverted from their original direction of motion. Each time the electron suffers a deceleration, energy in the form of X rays is produced. The energy of the X ray that results will depend on the deceleration that occurs. If brought to rest in a single collision, all its energy will appear in one burst of high energy and short wavelengths. If the electron encounters a less drastic collision, a longer wavelength results and lower-energy rays will be produced. Therefore energies of a wide range are produced from an X-ray tube.

The strength and quantity of X rays are usually expressed in terms of electrical current between the filament and target within the X-ray tube. The energy is measured in thousands of volts or kilovolts (keV). The quantity of X rays produced is directly proportional to the current through the tube and is measured as thousandths of an ampere or milli-ampers (mA). The dose or energy absorbed by an irradiated object is a function of the kV and mA of the machine as well as other factors.

Sources of X radiation that can operate at a voltage above 10 kV produce more penetrating radiation and therefore may be more hazardous than sources less than 10 kV, *if not adequately shielded.* These include high-voltage television projection systems, color television sets, electron microscopes and power sources, high-power amplifying tubes producing intense microwave fields, radio-transmitting tubes, high-voltage rectifier tubes, and other devices producing penetrating electromagnetic radiation.

Radioactive Materials

Other sources of ionizing radiation are radionuclides used in medicine, research, and industry as previously mentioned. Radium in equilibrium with its daughter products was widely used but is being replaced by other radionuclides. It is an alpha, beta, and gamma emitter and requires special shielding and care in handling. But sealed radium sources are used primarily as gamma emitters because the alpha and beta components of the radiation are stopped by the encapsulation material. Some of the other commonly used, naturally occuring radioactive materials are uranium, polonium, actinium, and thorium.

Artificially produced radionuclides generally include those made either through the fissioning process in a nuclear reactor or by bombardment of nonradioactive isotopes in high-energy accelerators or reactors. Many hundreds of isotopes have been produced.

The widespread use of radioactive materials and radionuclides in research, medical diagnosis, laboratories, industry, and educational institutions requires that precautions be taken in their storage, use, and disposal. Ingestion may occur through contaminated hands, cigarettes, or food. Radioactive vapor, dust, gas, or spray may be inhaled. Medical uses can present a hazard to technicians, patients, and others through the improper handling of radionuclides and contaminated wastes.

BIOLOGICAL EFFECTS OF RADIATION

Ionizing Radiations

Radiation which is capable of producing ionization and which is absorbed can cause injury or other damage. As the radiation penetrates the organism, it comes into contact with atoms and molecules in the living protoplasm altering the structure or electrical charges with formation of other molecules and substances which in turn may cause other effects.

Somatic and Genetic Effects

The biological effects of radiation on all living organisms, including human beings, are termed *somatic,* meaning effects occuring during the lifetime of the exposed organism, as opposed to *genetic,* which are the effects on generations yet unborn.[3] Somatic or genetic effects from radiation exposure may not be evident for months, years, or a lifetime. Even relatively large doses usually show no immediate apparent injury. But many cells, tissues, and organs of the body are interdependent, and the destruction of one may eventually result in poor functioning or death of the other. All cells, with the exception of germ cells, can apparently replace themselves or recover to some extent from radiation exposure, if the dose is not excessive and is administered in small increments over an extended period of time.

In assessing radiation hazard the sum total of all exposures should be considered, that is, natural background, medical, and occupational exposures, and radiation ingested through the air, water, and food. Exposure of the gonads (ovaries or testes), egg, or sperm is necessary to cause genetic effects from ionizing radiations. Everyone is subjected to natural background radiation, as well as heat and trace amounts of chemicals that cause a small number of so-called spontaneous mutations in genes. The genes determine all inherited characteristics. Hence any additional radiation of the reproductive cells should be avoided to keep down further mutations. The genetic harm done by radiation is cumulative up to the end of the reproductive period and depends on the total accumulated gonadal dose. Not all mutant genes are equally harmful. Some may cause serious harm, most cause very little or no apparent damage. Nevertheless, radiation exposure of large population groups should be kept as low as practicable since any mutation is generally considered harmful. It is estimated, for example, that a person receives on the average 125 millirems/yr from natural sources[4] and 50 to 70 millirems/yr (genetically significant dose—GSD)[5] from medical

---

[3] "Biological Effects of Radiation," U.S. Public Health Service, Dept. of HEW, Washington, D.C., 1963.
[4] Report of the United Nations Scientific Committee on the Effects of Atomic Radiation, 1962.
[5] The genetically significant dose which, if received by every member of the population, would be expected to produce the same total genetic injury to the population as do the actual doses received by the various individuals. (*Basic Radiation Protection Guide,* NCRP Report No. 39, National Council on Radiation Protection and Measurement, Washington, D.C., January 15, 1971, p. 13.)

exposures.[6] The actual amounts received in different parts of the world will vary with altitude, background, medical practices, and other factors. See also Table 7–2.

Knowledge of the long-term effects on human beings of both acute doses (large doses received in a short period of time) and chronic doses (those received frequently or continuously over a long period of time), from either external or internal sources of radiation, is far from complete. In order to give some appreciation of the relative health hazard associated with natural radiation, fallout radiation, and similar effects from all other causes in the United States, the Federal Radiation Council gave the following summary estimates for the next 70 years:[7]

	Leukemia	Bone Cancer
Total number of cases from all causes................	840,000	140,000
Estimated number caused by natural radiation........	0 to 84,000	0 to 14,000
Estimated number of additional cases from all nuclear tests through 1962...............................	0 to 2,000	0 to 700

Factors to Consider

Some of the factors that determine the effect of radiation on the body are the following:

1. Total amount and kind of radiation absorbed, external and internal.
2. Rate of absorption.
3. Amount of the body exposed.
4. Individual variability.
5. Relative sensitivity of cells and tissues; parts of the body exposed.
6. Nutrition, oxygen tension, metabolic state.

For example, a dose of 600 R whole body X radiation in one day would mean almost certain death. On the other hand, 600 R administered in weekly increments over a period of 30 yr may result in little or no detectable effect. This would amount to

$$\frac{600 \text{ R}}{30 \text{ yr} \times 52 \text{ wk}} = 0.385 \text{R/wk, or } 385 \text{ mR/wk.}$$

Large doses of radiation can be applied to local areas, as in therapy, with little danger. A person could expose a finger to 1000 R and experience a localized injury with subsequent healing and scar formation.

Effects of whole-body radiation are given in Table 7–3. It is known

---

[6] *Basic Radiation Protection Criteria,* op. cit.
[7] *Biological Effects of Radiation,* op. cit. Based on reports of the FRC.

that the blood-forming organs—spleen, lymph nodes, and bone mar-
row—are the most radiosensitive organs when the whole body is irradiated and that protection of these organs from exposure lessens the
whole-body effect.

TABLE 7-3   EFFECTS OF TOTAL-BODY RADIATION

Dose (roentgens)*			Probable Effect
0	to	50	No obvious effect except possible minor blood changes.
150	to	250	Vomiting and nausea for about 1 day, followed by other symptoms of radiation sickness in about 50% of personnel. No deaths expected.
200	to	400	Vomiting and nausea for about 1 day, followed by other symptoms of radiation sickness in about 50% of personnel. 10% deaths expected.
350	to	550	Vomiting and nausea in nearly all personnel on first day, followed by other symptoms of radiation sickness. About 25% deaths expected.
550	to	750	Vomiting and nausea in all personnel on first day, followed by other symptoms of radiation sickness. About 50% deaths within 1 month.
1,000			Vomiting and nausea in all personnel within 1 to 2 hr. Probably no survivors.
5,000			Almost immediate incapacitation. All personnel will be fatalities within 1 wk.

Source: *Fundamentals of Nuclear Warfare,* Special Text 3-154, U.S. Army Chemical
Corps Training Command, Fort McClellan, Ala., January 1961, p. 16.
*And/or rem in the case of neutrons.

Avoiding Unnecessary Radiation

Ideally, exposure to any ionizing radiation should be avoided. In real
life this is not possible or in some instances desirable. Background radiation cannot be eliminated, and there are times when medical uses of
radiation may be indicated. Research and industrial uses are also usually
indicated with proper controls.

To receive the many social and health benefits of radiation and keep
the risk minimal but within acceptable limits, maximum permissible
doses (MPD) have been proposed by the International Commission on
Radiological Protection (ICRP), the National Council on Radiation

Protection (NCRP), the Federal Radiation Council (FRC), and others. The functions of the FRC have been assumed by the Environmental Protection Agency under the President's Reorganization Plan Number

TABLE 7-4   RADIATION PROTECTION GUIDES

Type of Exposure	Condition	Dose (rem)
Radiation worker:		
Whole body, head and trunk, active blood-forming organs,	Accumulated dose	5 times number of years beyond age 18
gonads, or lens of eye	13 wk	3
Skin of whole body and	Yr	30
thyroid	13 wk	10
Hands and forearms, feet	Yr	75
and ankles	13 wk	25
Bone	Body burden	0.1 μCi of radium 226 or its biological equivalent
Other organs	Yr	15
	13 wk	5
Population:		
Individual	Yr	0.5 (whole body)
Average	30 yr	5 (gonads)

Source: *Radiological Health Handbook,* U.S. Public Health Service, Dept. of HEW, Washington, D.C., January 1970, p. 210.

Note: See Federal Radiation Council Report No. 1, May 1960, for details, also *Basic Radiation Protection Criteria,* NCRP Report No. 39, National Council on Radiation Protection and Measurements, Washington, D.C., January 15, 1971.

3 of 1970. Radiation protection guides are summarized in Table 7-4. Maximum permissible body burdens and maximum permissible concentrations of radionuclides in air and in water for occupational exposure are given in Table 7-5.

TYPES OF X-RAY UNITS AND COMMON HAZARDS

Medical and dental diagnostic examinations constitute a major source of radiation exposure to the individual. Therapeutic uses involve special considerations. Some of the common hazards associated with the use of medical and dental diagnostic and therapeutic machines are outlined below. There is far too much carelessness, familiarity, and ignorance involved in the use of radiation-producing equipment that should not be condoned. Medical diagnostic X-ray exposure is the most important source of population exposure. Education and motivation on the part of the medical profession can do more than anything else to reduce medical exposure.[8]

Dental X-ray Machines

These units are usually operated at 50 to 90 kVp at a tube current of 10 mA. Machine voltage should be properly calibrated. Fast, sensitive film and proper film-developing techniques are important. A 90-kVp unit will require 0.5 seconds exposure time compared to 1.5 seconds for a 50 to 75-kVp unit. A fast film will reduce exposure at skin surface by a factor of approximately 3 for a machine below 60 kVp and by approximately 9 for an 80-kVp or above machine as compared to slow film.[9] Common deficiencies are the following:

1. Diameter of X-ray beam exceeds 2.75 in. at the end of the cone. Requires use of proper diaphragm and collimators.
2. Exposure to the direct beam, if pointed toward occupied adjacent rooms, to the ceiling, or to the floor.
3. Scatter radiation, especially that from the patient. Plastic cones contribute to scatter; open-ended devices are recommended.
4. Filters missing; use 2.0 mm aluminum for 50 to 70-kVp unit, and 2.5 mm aluminum for unit above 70 kVp.
5. Multiple sources, particularly if several units are in the same room.
6. Operator in the same room as patient; no extension cord on timer button to make possible distance (at least 6 ft) protection.
7. Timer not accurate.
8. Use of slow emulsion speed film.

Fluoroscopy

Fluoroscopy units usually operate at voltages up to 100 kVp, usually about 80, at 3 to 5 mA. They should be used *only* where X-ray film will not provide the information. A film gives greater detail and a record

[8] Karl Z. Morgan, "Common Sources of Radiation Exposure," *American Engineer*, 38–42 (July 1968).
[9] K. L. Travis and J. E. Hickey, "A State Program for Reducing Radiation Exposure from Dental X-ray Machines," *Am. J. Public Health*, **60**, 1522–1527 (August 1970).

TABLE 7-5 MAXIMUM PERMISSIBLE BODY BURDENS AND MAXIMUM PERMISSIBLE CONCENTRATIONS OF RADIONUCLIDES IN AIR AND IN WATER FOR OCCUPATIONAL EXPOSURE

Radionuclide and Type of Decay	Organ of Reference (Critical Organ in Boldface)*	Maximum Permissible Burden in Total Body $q(\mu c)$	Maximum Permissible Concentrations			
			For 40-hr week		For 168-hr week†	
			$(MPC)_w$ $\mu c/cc$	$(MPC)_a$ $\mu c/cc$	$(MPC)_w$ $\mu c/cc$	$(MPC)_a$ $\mu c/cc$
$_1H^3$ (HTO or H_2O)(β^-) (Sol)	**Body tissue**	10^3	0.1	5×10^{-6}	0.03	2×10^{-6}
	Total body	2×10^3	0.2	8×10^{-6}	0.05	3×10^{-6}
(H_2^3) (Immersion)	**Skin**	—	—	2×10^{-3}	—	4×10^{-4}
$_6C^{14}(CO_2)(\beta^-)$ (Sol)	**Fat**	300	0.02	4×10^{-6}	8×10^{-3}	10^{-6}
	Total body	400	0.03	5×10^{-6}	0.01	2×10^{-6}
	Bone	400	0.04	6×10^{-6}	0.01	2×10^{-6}
(Immersion)	**Total body**	—	—	5×10^{-5}	—	10^{-5}
$_{15}P^{32}$ (β^-) (Sol)	**Bone**	6	5×10^{-4}	7×10^{-8}	2×10^{-4}	2×10^{-8}
	Total body	30	3×10^{-3}	4×10^{-7}	9×10^{-4}	10^{-7}
	GI (LLI)	—	3×10^{-3}	6×10^{-7}	9×10^{-4}	2×10^{-7}
	Liver	50	5×10^{-3}	6×10^{-7}	2×10^{-3}	2×10^{-7}
	Brain	300	0.02	3×10^{-6}	8×10^{-3}	10^{-6}
(Insol)	**Lung**	—	—	8×10^{-8}	—	3×10^{-8}
	GI (LLI)	—	7×10^{-4}	10^{-7}	2×10^{-4}	4×10^{-8}
$_{20}Ca^{45}$ (β^-) (Sol)	**Bone**	30	3×10^{-4}	3×10^{-8}	9×10^{-5}	10^{-8}
	Total body	200	2×10^{-3}	3×10^{-7}	7×10^{-4}	9×10^{-8}
	GI (LLI)	—	0.01	3×10^{-6}	4×10^{-3}	10^{-6}
(Insol)	**Lung**	—	—	10^{-7}	—	4×10^{-8}
	GI (LLI)	—	5×10^{-3}	9×10^{-7}	2×10^{-3}	3×10^{-7}
$_{24}Cr^{51}$ (ϵ, γ) (Sol)	**GI (LLI)**	—	0.05	10^{-5}	0.02	4×10^{-6}
	Total body	800	0.6	10^{-5}	0.2	4×10^{-6}
	Lung	10^3	1			

Isotope	Form	Organ					
$_{24}Cr^{51}$ (ϵ, γ) (cont'd)	(Sol)	Prostate	2×10^3	2	3×10^{-5}	0.5	10^{-5}
		Thyroid	4×10^3	3	6×10^{-5}	1	2×10^{-5}
		Kidney	8×10^3	6	10^{-4}	2	4×10^{-5}
	(Insol)	Lung	—	—	2×10^{-6}	—	8×10^{-7}
		GI (LLI)	—	0.05	8×10^{-6}	0.02	3×10^{-6}
$_{27}Co^{60}$ (β^-, γ)	(Sol)	GI (LLI)	—	10^{-3}	3×10^{-7}	5×10^{-4}	10^{-7}
		Total body	10	4×10^{-3}	4×10^{-7}	10^{-3}	10^{-7}
		Pancreas	70	0.02	2×10^{-6}	7×10^{-3}	6×10^{-7}
		Liver	90	0.03	10^{-6}	9×10^{-3}	5×10^{-7}
		Spleen	200	0.05	4×10^{-6}	0.02	2×10^{-6}
		Kidney	200	0.07	6×10^{-6}	0.03	2×10^{-6}
	(Insol)	Lung	—	—	9×10^{-9}	—	3×10^{-9}
		GI (LLI)	—	10^{-3}	2×10^{-7}	3×10^{-4}	6×10^{-8}
$_{30}Zn^{65}$ ($\beta^+, \epsilon, \gamma$)	(Sol)	Total body	60	3×10^{-3}	10^{-7}	10^{-3}	4×10^{-8}
		Prostate	70	4×10^{-3}	10^{-7}	10^{-3}	4×10^{-8}
		Liver	80	4×10^{-3}	10^{-7}	10^{-3}	5×10^{-8}
		Kidney	100	6×10^{-3}	2×10^{-7}	2×10^{-3}	7×10^{-8}
		GI (LLI)	—	6×10^{-3}	10^{-6}	2×10^{-3}	4×10^{-7}
		Pancreas	200	7×10^{-3}	3×10^{-7}	3×10^{-3}	9×10^{-8}
		Muscle	200	0.01	4×10^{-7}	4×10^{-3}	10^{-7}
		Ovary	300	0.01	5×10^{-7}	4×10^{-3}	2×10^{-7}
		Testis	400	0.02	6×10^{-7}	6×10^{-3}	2×10^{-7}
		Bone	700	0.04	10^{-6}	0.01	4×10^{-7}
	(Insol)	Lung	—	—	6×10^{-8}	—	2×10^{-8}
		GI (LLI)	—	5×10^{-3}	9×10^{-7}	2×10^{-3}	3×10^{-7}
$_{33}As^{76}$ (β^-, γ)	(Sol)	GI (LLI)	—	6×10^{-4}	10^{-7}	2×10^{-4}	4×10^{-8}
		Total body	20	0.4	5×10^{-6}	0.1	2×10^{-6}
		Kidney	20	0.6	8×10^{-6}	0.2	3×10^{-6}
		Liver	40	1	10^{-5}	0.4	5×10^{-6}
	(Insol)	GI (LLI)	—	6×10^{-4}	10^{-7}	2×10^{-4}	3×10^{-8}
		Lung	—	—	6×10^{-7}	—	2×10^{-7}

TABLE 7-5 MAXIMUM PERMISSIBLE BODY BURDENS AND MAXIMUM PERMISSIBLE CONCENTRATIONS OF RADIONUCLIDES IN AIR AND IN WATER FOR OCCUPATIONAL EXPOSURE (cont'd)

Radionuclide and Type of Decay		Organ of Reference (Critical Organ in Boldface)*	Maximum Permissible Burden in Total Body $q(\mu c)$	Maximum Permissible Concentrations			
				For 40-hr week		For 168-hr week†	
				$(MPC)_w$ $\mu c/cc$	$(MPC)_a$ $\mu c/cc$	$(MPC)_w$ $\mu c/cc$	$(MPC)_a$ $\mu c/cc$
$_{38}Sr^{89}$ (β^-)	(Sol)	**Bone**	4	3×10^{-4}	3×10^{-8}	10^{-4}	10^{-8}
		GI (LLI)	—	10^{-3}	3×10^{-7}	4×10^{-4}	9×10^{-8}
		Total body	40	2×10^{-3}	2×10^{-7}	7×10^{-4}	6×10^{-8}
	(Insol)	**Lung**	—	—	4×10^{-8}	—	10^{-8}
		GI (LLI)	—	8×10^{-4}	10^{-7}	3×10^{-4}	5×10^{-8}
$_{38}Sr^{90}$ (β^-)	(Sol)	**Bone**	2	4×10^{-6}	3×10^{-10}	10^{-6}	10^{-10}
		Total body	20	10^{-5}	9×10^{-10}	4×10^{-6}	3×10^{-10}
		GI (LLI)	—	10^{-3}	3×10^{-7}	5×10^{-4}	10^{-7}
	(Insol)	**Lung**	—	—	5×10^{-9}	—	2×10^{-9}
		GI (LLI)	—	10^{-3}	2×10^{-7}	4×10^{-4}	6×10^{-8}
$_{40}Zr^{95}$ (β^-, γ, e^-)	(Sol)	**GI (LLI)**	—	2×10^{-3}	4×10^{-7}	6×10^{-4}	10^{-7}
		Total body	20	3	10^{-7}	1	4×10^{-8}
		Bone	30	4	2×10^{-7}	2	6×10^{-8}
		Kidney	30	4	2×10^{-7}	2	6×10^{-8}
		Liver	40	6	3×10^{-7}	2	9×10^{-8}
		Spleen	40	7	3×10^{-7}	2	10^{-7}
	(Insol)	**Lung**	—	—	3×10^{-8}	—	10^{-8}
		GI (LLI)	—	2×10^{-3}	3×10^{-7}	6×10^{-4}	10^{-7}
$_{41}Nb^{95}$ (β^-, γ)	(Sol)	**GI (LLI)**	—	3×10^{-3}	6×10^{-7}	10^{-3}	2×10^{-7}
		Total body	40	10	5×10^{-7}	4	2×10^{-7}
		Liver	60	20	7×10^{-7}	6	3×10^{-7}
		Kidney	60	20	8×10^{-7}	6	3×10^{-7}
		Bone	80	20	9×10^{-7}	7	3×10^{-7}

Isotope		Organ					
$_{41}Nb^{95}$ (β^-, γ) (cont'd)	(Insol)	Lung	—	—	10^{-7}	—	3×10^{-3}
		GI (LLI)	—	3×10^{-3}	5×10^{-7}	10^{-3}	2×10^{-7}
$_{44}Ru^{106}$ (β^-, γ)	(Sol)	GI (LLI)	3	4×10^{-4}	8×10^{-8}	10^{-4}	3×10^{-8}
		Kidney	10	0.01	10^{-7}	4×10^{-3}	5×10^{-8}
		Bone	10	0.04	5×10^{-7}	0.01	2×10^{-7}
		Total body		0.06	7×10^{-7}	0.02	3×10^{-7}
	(Insol)	Lung	—	—	6×10^{-9}	—	2×10^{-9}
		GI (LLI)	—	3×10^{-4}	6×10^{-8}	10^{-4}	2×10^{-8}
$_{53}I^{131}$ (β^-, γ, e^-)	(Sol)	Thyroid	0.7	6×10^{-5}	9×10^{-9}	2×10^{-5}	3×10^{-5}
		Total body	50	5×10^{-3}	8×10^{-7}	2×10^{-3}	3×10^{-7}
		GI (LLI)		0.03	7×10^{-6}	0.01	2×10^{-6}
	(Insol)	GI (LLI)	—	2×10^{-3}	3×10^{-7}	6×10^{-4}	10^{-7}
		Lung	—	—	3×10^{-7}	—	10^{-7}
$_{55}Cs^{137}$ (β^-, γ, e^-)	(Sol)	Total body	30	4×10^{-4}	6×10^{-8}	2×10^{-4}	2×10^{-3}
		Liver	40	5×10^{-4}	8×10^{-8}	2×10^{-4}	3×10^{-8}
		Spleen	50	6×10^{-4}	9×10^{-8}	2×10^{-4}	3×10^{-8}
		Muscle	50	7×10^{-4}	10^{-7}	2×10^{-4}	4×10^{-8}
		Bone	100	10^{-3}	2×10^{-7}	5×10^{-4}	7×10^{-8}
		Kidney	100	10^{-3}	2×10^{-7}	5×10^{-4}	8×10^{-8}
		Lung	300	5×10^{-3}	6×10^{-7}	2×10^{-3}	2×10^{-7}
		GI (SI)		0.02	5×10^{-6}	8×10^{-3}	2×10^{-6}
	(Insol)	Lung	—	—	10^{-8}	—	5×10^{-9}
		GI (LLI)	—	10^{-3}	2×10^{-7}	4×10^{-4}	8×10^{-8}
$_{58}Ce^{144}$ (α, β^-, γ)	(Sol)	GI (LLI)	—	3×10^{-4}	8×10^{-8}	10^{-4}	3×10^{-8}
		Bone	5	0.2	10^{-8}	0.08	3×10^{-9}
		Liver	6	0.3	10^{-8}	0.1	4×10^{-9}
		Kidney	10	0.5	2×10^{-8}	0.2	7×10^{-9}
		Total body	20	0.7	3×10^{-8}	0.3	10^{-8}
	(Insol)	Lung	—	—	6×10^{-9}	—	2×10^{-9}
		GI (LLI)	—	3×10^{-4}	6×10^{-8}	10^{-4}	2×10^{-8}

TABLE 7-5 MAXIMUM PERMISSIBLE BODY BURDENS AND MAXIMUM PERMISSIBLE CONCENTRATIONS OF RADIONUCLIDES IN AIR AND IN WATER FOR OCCUPATIONAL EXPOSURE (cont'd)

Radionuclide and Type of Decay		Organ of Reference (Critical Organ in Boldface)*	Maximum Permissible Burden in Total Body $q(\mu c)$	Maximum Permissible Concentrations			
				For 40-hr week		For 168-hr wk†	
				$(MPC)_w$ $\mu c/cc$	$(MPC)_a$ $\mu c/cc$	$(MPC)_w$ $\mu c/cc$	$(MPC)_a$ $\mu c/cc$
$_{61}Pm^{147}$ (α, β^-)		**GI (LLI)**	—	6×10^{-3}	10^{-6}	2×10^{-3}	5×10^{-7}
		Bone	60	1	6×10^{-8}	0.5	2×10^{-8}
	(Sol)	Kidney	200	4	2×10^{-7}	2	7×10^{-8}
		Total body	300	7	3×10^{-7}	2	10^{-7}
		Liver	300	8	4×10^{-7}	3	10^{-7}
	(Insol)	**Lung**	—	—	10^{-7}	—	3×10^{-8}
		GI (LLI)	—	6×10^{-3}	10^{-6}	2×10^{-2}	4×10^{-7}
$_{73}Ta^{182}$ (β^-, γ)		**GI (LLI)**	—	10^{-3}	3×10^{-7}	4×10^{-4}	9×10^{-8}
		Liver	7	0.9	4×10^{-8}	0.3	10^{-8}
	(Sol)	Kidney	20	2	8×10^{-8}	0.7	3×10^{-8}
		Total body	20	2	9×10^{-8}	0.7	3×10^{-8}
		Spleen	30	4	10^{-7}	1	5×10^{-8}
		Bone	50	6	3×10^{-7}	2	9×10^{-8}
	(Insol)	**Lung**	—	—	2×10^{-8}	—	7×10^{-9}
		GI (LLI)	—	10^{-3}	2×10^{-7}	4×10^{-4}	7×10^{-8}
$_{77}Ir^{192}$ (β^-, γ)		**GI (LLI)**	—	10^{-3}	3×10^{-7}	4×10^{-4}	9×10^{-8}
		Kidney	6	4×10^{-3}	10^{-7}	10^{-3}	4×10^{-8}
	(Sol)	Spleen	7	4×10^{-3}	10^{-7}	10^{-3}	5×10^{-8}
		Liver	8	5×10^{-3}	2×10^{-7}	2×10^{-3}	6×10^{-8}
		Total body	20	0.01	4×10^{-7}	4×10^{-3}	10^{-7}
	(Insol)	**Lung**	—	—	3×10^{-8}	—	9×10^{-9}
		GI (LLI)	—	10^{-3}	2×10^{-7}	4×10^{-4}	6×10^{-8}

$_{79}\mathrm{Au}^{198}$ (β^-, γ)	(Sol)	GI (LLI)	20	2×10^{-3}	3×10^{-7}	5×10^{-4}	10^{-7}
		Kidney	30	0.07	3×10^{-6}	0.02	9×10^{-7}
		Total body	60	0.1	4×10^{-6}	0.04	2×10^{-6}
		Spleen	80	0.2	8×10^{-6}	0.07	3×10^{-6}
		Liver	—	0.3	10^{-5}	0.1	4×10^{-5}
	(Insol)	GI (LLI)	—	10^{-3}	2×10^{-7}	5×10^{-4}	8×10^{-3}
		Lung	—	—	6×10^{-7}	—	2×10^{-7}
$_{86}\mathrm{Rn}^{222}$ ‡ (α, β, γ)		Lung	—	—	3×10^{-8}	—	10^{-8}
$_{88}\mathrm{Ra}^{226}$ (α, β^-, γ)	(Sol)	Bone	0.1	4×10^{-7}	3×10^{-11}	10^{-7}	10^{-11}
		Total body	0.2	6×10^{-7}	5×10^{-11}	2×10^{-7}	2×10^{-11}
		GI (LLI)	—	10^{-3}	3×10^{-7}	5×10^{-4}	10^{-7}
	(Insol)	Lung	—	—	5×10^{-11}	—	2×10^{-11}
		GI (LLI)	—	9×10^{-4}	2×10^{-7}	3×10^{-4}	6×10^{-8}
$_{92}\mathrm{U}^{235}$ (α, β^-, γ)	(Sol)	GI (LLI)	0.03	8×10^{-4}	2×10^{-7}	3×10^{-4}	6×10^{-8}
		Kidney	0.06	0.01	5×10^{-10}	4×10^{-3}	2×10^{-10}
		Bone	0.4	0.01	6×10^{-10}	5×10^{-3}	2×10^{-10}
		Total body	—	0.04	2×10^{-9}	0.01	6×10^{-10}
	(Insol)	Lung	—	—	10^{-10}	—	4×10^{-11}
		GI (LLI)	—	8×10^{-4}	10^{-7}	3×10^{-4}	5×10^{-8}
$_{92}\mathrm{U}^{238}$ (α, γ, e^-)	(Sol)	GI (LLI)	5×10^{-3}	10^{-3}	2×10^{-7}	4×10^{-4}	8×10^{-8}
		Kidney	0.06	2×10^{-3}	7×10^{-11}	6×10^{-4}	3×10^{-11}
		Bone	0.5	0.01	6×10^{-10}	5×10^{-3}	2×10^{-10}
		Total body	—	0.04	2×10^{-9}	0.01	6×10^{-10}
	(Insol)	Lung	—	—	10^{-10}	—	5×10^{-11}
		GI (LLI)	—	10^{-3}	2×10^{-7}	4×10^{-4}	6×10^{-8}
$_{94}\mathrm{Pu}^{239}$ (α, γ)	(Sol)	Bone	0.04	10^{-4}	2×10^{-12}	5×10^{-5}	6×10^{-13}
		Liver	0.4	5×10^{-4}	7×10^{-12}	2×10^{-4}	2×10^{-12}
		Kidney	0.5	7×10^{-4}	9×10^{-12}	2×10^{-4}	3×10^{-12}
		GI (LLI)	—	8×10^{-4}	2×10^{-7}	3×10^{-4}	6×10^{-8}
		Total body	0.4	10^{-3}	10^{-11}	3×10^{-4}	5×10^{-12}
	(Insol)	Lung	—	—	4×10^{-11}	—	10^{-11}
		GI (LLI)	—	8×10^{-4}	2×10^{-7}	3×10^{-4}	5×10^{-8}

TABLE 7-5   MAXIMUM PERMISSIBLE CONCENTRATION OF UNIDENTIFIED RADIONUCLIDES IN WATER, $(MPCU)_w$ VALUES,* FOR CONTINUOUS OCCUPATIONAL EXPOSURE (cont'd)

Limitations	$\mu c/cm^3$ of water§
If no one of the radionuclides Sr^{90}, I^{126}, I^{129}, I^{131}, Pb^{210}, Po^{210}, At^{211}, Ra^{223}, Ra^{224}, Ra^{226}, Ra^{228}, Ac^{227}, Th^{230}, Pa^{231}, Th^{232}, and Th-nat is present, then the $(MPCU)_w$ is	3×10^{-5}
If no one of the radionuclides Sr^{90}, I^{129}, Pb^{210}, Po^{210}, Ra^{223}, Ra^{226}, Ra^{228}, Pa^{231}, and Th-nat is present, then the $(MPCU)_w$ is	2×10^{-5}
If no one of the radionuclides Sr^{90}, I^{129}, Pb^{210}, Ra^{226}, and Ra^{228} is present, then the $(MPCU)_w$ is	7×10^{-6}
If neither Ra^{226} nor Ra^{228} is present, then the $(MPCU)_w$ is	10^{-6}
If no analysis of the water is made, then the $(MPCU)_w$ is	10^{-7}

MAXIMUM PERMISSIBLE CONCENTRATION OF UNIDENTIFIED RADIONUCLIDES IN AIR, $(MPCU)_a$ VALUES,* FOR CONTINUOUS OCCUPATIONAL EXPOSURE

Limitations	$\mu c/cm^3$ of air‖
If there are no α-emitting radionuclides and if no one of the β-emitting radionuclides Sr^{90}, I^{129}, Pb^{210}, Ac^{227}, Ra^{228}, Pa^{230}, Pu^{241}, and Bk^{249} is present, then the $(MPCU)_a$ is	10^{-9}
If there are no α-emitting radionuclides and if no one of the β-emitting radionuclides Pb^{210}, Ac^{227}, Ra^{228}, and Pu^{241} is present, then the $(MPCU)_a$ is	10^{-10}
If there are no α-emitting radionuclides and if the β-emitting radionuclide Ac^{227} is not present, then the $(MPCU)_a$ is	10^{-11}
If no one of the radionuclides Ac^{227}, Th^{230}, Pa^{231}, Th^{232}, Th-nat, Pu^{238}, Pu^{239}, Pu^{240}, Pu^{242}, and Cf^{249} is present, then the $(MPCU)_a$ is	10^{-12}
If no one of the radionuclides Pa^{231}, Th-nat, Pu^{239}, Pu^{240}, Pu^{242}, and Cf^{249} is present, then the $(MPCU)_a$ is	7×10^{-13}
If no analysis of the air is made, then the $(MPCU)_a$ is	4×10^{-13}

TABLE 7-5 FOOTNOTES

Source: *Radiological Health Handbook*, U.S. Public Health Service, Dept. of HEW, January 1970.

* The abbreviations GI, S, SI, ULI, and LLI refer to gastrointestinal tract, stomach, small intestine, upper large intestine, and lower large intestine, respectively.

† It will be noted that the MPC values for the 168-hr week are not always precisely the same multiples of the MPC for the 40-hr week. Part of this is caused by rounding off the calculated values to one digit, but in some instances it is due to technical differences discussed in the ICRP report. Because of the uncertainties present in much of the biological data and because of individual variations, the differences are not considered significant. The MPC values for the 40-hr week are to be considered as basic for occupational exposure, and the values for the 168-hr week are basic for continuous exposure as in the case of the population at large.

‡ The daughter isotopes of Rn^{220} and Rn^{222} are assumed present to the extent they occur in unfiltered air. For all other isotopes the daughter elements are not considered as part of the intake and if present must be considered on the basis of the rules for mixtures.

§ Each $(MPCU)_w$ value is the smallest value of $(MPC)_w$ in table 1 for radionuclides other than those listed opposite the value. Thus these $(MPCU)_w$ values are permissible levels for continuous occupational exposure (168 hr/wk) for any radionuclide or mixture of radionuclides where the indicated isotopes are not present (i.e., where the concentration of the radionuclide in water is small compared with the $(MPC)_w$ value for this radionuclide). The $(MPCU)_w$ may be much smaller than the more exact maximum permissible concentration of the material, but the determination of this $(MPC)_w$ requires identification of the radionuclides present and the concentration of each.

‖ Use one-tenth of these values for interim application in the neighborhood of a controlled exposure area.

Note: These radionuclides were selected from National Bureau of Standards Handbook 69 (for sale by U.S. Government Printing Office, Washington, D.C.). This publication lists (for all radionuclides) the recommendations of the National Committee on Radiation Protection and Measurements for Maximum Permissible Body Burdens and Maximum Permissible Concentrations in Air and Water for Occupational Exposure. The handbook should be consulted for MPC and MPBB values of other nuclides or for information on derivation and limitations of these values.

515

for study, with only a small fraction (5 percent) of the patient exposure compared to fluoroscopy. Common deficiencies are the following:

1. The useful beam may extend beyond the fluoroscopic screen and its lead glass at maximum beam size and target screen distance. Shutter opening not controlled to minimum size needed for examination. If a full screen is needed, the shutter should be limited so there will be a visible dark margin of at least ¼ in. around the perimeter of the screen when the screen is about 15 in. from the table top.

2. Scatter radiation from the patient and the undersurface of the tabletop.

3. Timing device not functioning properly; maximum exposure to the patient is not kept below 10 R/min.

4. Absence of a cone between tube housing and tabletop.

5. Inadequate shielding of the tube enclosure against the direct beam.

6. Too short a target table distance.

7. Absence of leaded apron and gloves for physician and attendant.

8. Filter too thin or totally lacking.

9. Increased time of use due to failure to allow for eye accommodation (10 to 20 min.). Intensifying screens should be used. An image intensifier will permit use of a much lower (½ to 1/10) dose rate.

Radiography

Usual operating conditions are 40 to 135 kVp at currents up to 500 mA. Machine should be equipped with calibrated voltage meter and calibrated timing device. Fast, sensitive film and proper developing techniques should be used. Lead shielding may be necessary behind the cassette holder for chest and upright X rays, or at any other primary beam areas, if the "useful" beam penetrates into a waiting room or other occupied areas. Common deficiencies are the following:

1. Failure to confine the X-ray beam to the part of the patient's body being examined.

2. Scatter radiation around or over protective screens or into control booth where no door is provided.

3. Scattering under doors and at junction of walls with floor and ceiling if no lead baffle is provided.

4. Useful beam and scatter radiation passing through windows and into occupied regions nearby.

5. Holding the patient or film during exposure.

6. Filters missing.

Therapy Units Up to 10 MeV

Therapy units operate at energies up to 10 MeV and higher, depending on the type of equipment. Common deficiencies are the following:

1. Scattering around or into control booth, where no door is provided.

2. Scattering under doors and at junctions of walls with floor and ceiling if no lead baffle is provided.

3. Scattering from nearby buildings or from the floor of the room below if the treatment room floor is insufficiently shielded.

4. Leakage around doors and observation windows.

In addition to the hazards stated above, it is not unusual to find in X-ray installations loss of adequate shielding, removal or failure of safety devices, infrequent calibration, and electrical hazards. Lack of door interlocks or insufficient primary or secondary shielding may also present a hazard with therapy machines.

When developing a control program, many X-ray machines, although properly installed originally, will be found to be hazardous because of alterations to the building and removal of the original shielding. Filters become lost and lead diaphragms are missing. A concentration of X-ray machines in one building may also increase the scattered radiation to unsatisfactory levels. In some instances equipment may not have current and voltage meters.

Also used in medical therapy is the cobalt 60 teletherapy unit and particle accelerator. These units require specially designed rooms, facilities, shielding, and operating procedures.

Improper Use of X-ray Equipment

It is usually the opinion of health departments that the use of radiation on humans should be limited to usage by persons licensed to practice medicine, dentistry, podiatry, or osteopathy, and in some instances chiropody. This would eliminate, for example, the exposure of humans by unorthodox procedures such as fluoroscopy for the purpose of fitting shoes. This particular use of X rays, besides being considered completely unnecessary by orthopedic experts, is used almost exclusively on children, whose rapidly developing bodies are most susceptible to radiation damage. It is prohibited in most states. Fluoroscopic equipment has also been used as an advertising technique and on music students to study throat-muscle development. These and similar uses should not be permitted.

X-ray and Other Radiation Sources Used in Industrial and Commercial Establishments

Industrial and commercial X-ray devices include primarily the following:

1. Radiographic and fluoroscopic units used for the determination of defects in welded joints and in casting fabricated structures and molds.

2. Fluoroscopic units used for the detection of foreign material, as in packaged foods.

3. Ionizing radiation being considered for the pasteurization of foods, sterilization of medical supplies, and other purposes.

4. Antitheft and antisabotage fluoroscopic examinations. These devices should not be recommended or used.

5. Sealed sources for measuring the density or thickness of products and for determining liquid levels in closed tanks.

Radiation machines used in industry, training, and research include Van de Graff accelerators, particle accelerators, neutron generators, and other special X-ray units. These machines and the sources mentioned have the same hazards associated with them as do the medical and dental X-ray equipment. Radiographic units having sealed sources use cobalt 60, cesium 137, and iridium 192. The industrial radiographic X-ray machine has a poor operator and surrounding worker operating safety record.

Some industries have used antitheft, full-body fluoroscopic examinations of employees and similar examinations for antisabotage purposes. Unless such examinations are absolutely necessary and records kept as to exposure of personnel, it is possible for employees to receive excessive doses of radiation. Such devices should not be used by untrained personnel.

Other

Projection television tubes, diffraction analyzers, and, for that matter, any equipment using high voltages are capable of generating X rays. Even the home television tubes generate X rays, but have not become a hazard since the "soft" X rays generated are filtered out by the glass of the tube. However, if higher voltages are used than are allowed in order to get good images in color, these tubes may also become a hazard.

General Preventive Measures

Initiation of a routine inspection program will reveal a number of common defects. For instance, many dental installations may lack filters in the X-ray beam. Lead cones to define the area of the X-ray beam will not be found in some medical installations. In many cases X-ray beams on fluoroscopes will extend past the fluoroscopic screen and protective lead glass, thereby exposing the operator to the direct beam.

All these defects are very simple to find and need no instruments whatever. However, their importance should not be underestimated. For instance, the addition of filters to dental machines can reduce the surface dose to the patients by a factor of up to 10, as well as the resultant exposure to the operator. A filter in the tube port providing the equivalent filtration of approximately 2 mm of aluminum will absorb the soft, or less penetrating, radiations. This will reduce the radiation that is absorbed by the patient, reduce the stray radiation, and eliminate scatter haze on the film. The use of open-ended pointing devices on X-ray machines can eliminate much unnecessary irradiation of the patient by reducing the exposed area to the particular field of interest. These are very simple adjustments needing very little technical knowledge, but do a great amount of good.

Patient and individual exposure can be greatly reduced. The basic measures include: (1) proper collimation to limit the X-ray beam to the area needed (can reduce the average gonadal dose by 65%), (2) the use of fast film when film is used, (3) gonadal shielding, (4) better trained technicians,[10] and (5) continued education of those using or prescribing X ray. The more general use of machines with automatic collimation and other controls will also do much to reduce unnecessary exposure.

RADIUM AND OTHER RADIOACTIVE SOURCES

Radium occurs naturally. It is used extensively in medicine and in luminous compounds for markers and watch and instrument dials. Radium has a very long half-life (1622 yr), which on decay produces radon. Radon is an alpha emitter; in addition, the radon daughter decay products are alpha, beta, and gamma emitters. Radium as used in medicine is in the form of a loose powdered salt placed in a sealed capsule. There is a danger of the capsule rupturing or leaking due partly to the pressure exerted by the radon gas or by careless handling.

Because of the hazards associated with the use of radium, and loss of the capsule, an accountability procedure should be in effect at places where radium is stored. This should be supplemented by radiation protection equipment and techniques and periodic radiation safety surveys. If a capsule is lost or broken, the state and local health departments should be immediately notified.

Medical personnel involved in radium radiation therapy will be close to the source. They may therefore be unnecessarily exposed during the handling of the radium needle, during insertion or removal, and during the time the patient is being treated. It is important therefore that medical personnel receive special training, experience, and direct supervision if their duties include assisting in the therapeutic use of radium or radionuclides that are replacing radium.

The general safety precautions for radium apply equally as well to other sealed, encapsulated, or otherwise contained radioactive sources.

---

[10] Only New York, New Jersey, and California require education in radiation protection and certification of X-ray technologists as of 1971.

RADIATION PROTECTION

The basic tenet of radiation protection is to avoid all *unnecessary* exposure to ionizing radiations. When exposure is necessary because of medical reasons or work environment, then every known protection should be applied. The three principles for effecting external radiation protection are distance, time of exposure, and shielding. Also important are sanitation and the prevention of contamination. The safeguards will vary with the type and strength of the sources of radiation.

Distance

The further away a person is from a radiation source, the less will be the exposure he receives. This is particularly true for a point source of radiation as the subject exposure decreases inversely with the square of the distance from the source. For example, if at a distance of 5 ft one is exposed to A mR, at a distance of 10 ft the exposure would be $\frac{1}{4}A$ mR. This is known as the Inverse Square Law and is expressed as $I_1/I_2 = R_2{}^2/R_1{}^2$ in which R_1 and R_2 are any two distances, and I_1 and I_2 are the values of the intensity at the distances.

In practice the above principle also applies to other than point sources of radiation in view of the small sizes of the radiation sources and relatively large distances generally involved.

Time of Exposure

When exposure to radiation is necessary, the time of such exposure should be kept as low as practicable to accomplish a particular task. In the case of occupational or similar situations, the total exposure shall be kept below an individual's maximum permissible dose. This might mean transfer to another assignment where he will receive no radiation so as to stay within the limits shown in Tables 7–4 and 7–5.

Contamination Prevention

Materials that emit alpha and beta particles present a particularly dangerous hazard if ingested or breathed in because their specific ionization is very high and they irradiate the body continuously until they are eliminated. The seriousness of the hazard depends on the type of radioactive material, the type of radiation emitted, the energy of the radiation, its physical and biological half-life, and the radiosensitivity of the tissue and body organ where the isotope establishes itself. See Tables 7–1 and 7–5.

The objective must therefore be to keep the radioactive materials out of the body. This is accomplished by the use of proper procedures and good practices, such as using laboratory hoods, air filters, and exhaust systems; eliminating dry sweeping; making dry runs and wearing protective clothing; using respirators when indicated; using proper monitoring and survey instruments; and prohibiting eating and smoking where radioactive materials are handled or used.

Shielding

Shielding is the interposition of a dense attenuating matcrial between a source of radiation and the surroundings so as to adequately reduce or practically stop the travel of radiation. A shield may be used for

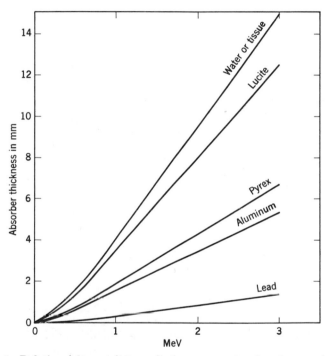

Figure 7–5 Relation between beta radiation energy level and range in aluminum and various other materials. [From Atomic Radiation, RCA Service Company, Government Services, Camden, N.J., 1959 (Contract Nr. AF 33(616)-3665).]

different purposes, such as to absorb X rays, gamma rays, neutrons, or intense heat. For neutrons, a hydrogenous material such as paraffin is used. The energy and type of the radiation to be attenuated is a basic factor in the selection of the shielding material or materials and their size.

The term "half-value layer" (HVL) is used to designate the thickness of a particular material that will reduce by one-half the intensity of radiation passing through the material. Lead and concrete are commonly used to shield X and gamma radiations. Glass or plastic are commonly used to eliminate beta radiation. See Figures 7–5 and 7–6.

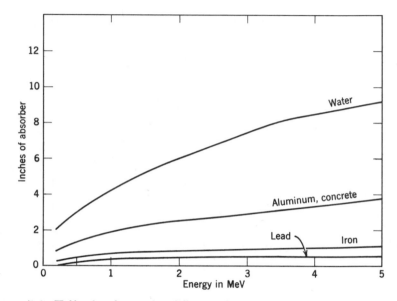

Figure 7–6 Half-value layers for different electromagnetic energies as related to thickness (inches) of iron and lead, aluminum, concrete, and water. [From *Atomic Radiation*, RCA Service Co., Government Services, Camden, N.J., 1959 (Contract Nr. AF 33(616)-3665).]

Radiation, however, will scatter and bounce off the floor and ceiling and from wall to wall. It can therefore be reflected around shields, around corners, over and under doors, and through ventilating transoms and windows. Shields must be placed so as to protect anyone who might have access to spaces above or below, or on any side of the source. It is generally better to place shields close to the source of radiation because this will reduce the required area and weight of the shield.

Some of the characteristics of alpha and beta particles, X rays and gamma rays, and neutrons are given earlier in this chapter. The ranges of alpha and beta particles in air are shown in Tables 7–6 and 7–7. Neutron shielding presents special problems around nuclear reactors and particle accelerators.

TABLE 7-6   RANGE OF BETA PARTICLES IN AIR

Energy (millions of electron volts)	Maximum Range of Beta Rays in Air (meters)
0.01	0.0022
0.02	0.0072
0.03	0.015
0.04	0.024
0.05	0.037
0.06	0.050
0.07	0.064
0.08	0.080
0.09	0.095
0.10	0.11
0.15	0.21
0.2	0.36
0.3	0.65
0.4	1.0
0.6	1.8
0.8	2.8
1.0	3.7
1.5	6.1
2.0	8.4
3.0	13.0
4.0	16.0
5.0	19.0

Source: *Atomic Radiation*, RCA Service Co., Government Services, Camden, N.J., 1959 (Contract Nr. AF 33(616)-3665).

TABLE 7-7   EMISSION ENERGY AND RANGES OF ALPHA PARTICLES IN AIR

Radioisotope	Alpha Emission Energy (in millions of electron volts)	Alpha Emission Range (in centimeters)
Thorium (Th^{232})	3.97	2.8
Radium (Ra^{226})	4.97	3.3
Radiothorium (Th^{228})	5.41	3.9
Radium A (Po^{218})	5.99	4.6
Thorium A (Po^{216})	6.77	5.6
Radium C′ (Po^{214})	7.68	6.9
Thorium C′ (Po^{212})	8.78	8.6

Source: *Atomic Radiation*, RCA Service Co., Government Services, Camden, N.J., 1959 (Contract Nr. AF 33(616)-3665).

Calculation of structural shielding requirements takes into consideration the work load, use factor, occupancy factor, and whether the area exposed is under control of the person responsible for radiation protection.

The work load, usually expressed in milliampere seconds per week, is the milliamperage of the current passing through the X-ray tube times the sum of the number of seconds the tube is in operation in one week.

The use factor is the part of the time a machine is in use that the useful beam may strike the wall, ceiling, or floor to be shielded. The occupancy factor is the time a person on the other side of a shielded wall, floor, or ceiling will be exposed to radiation when a unit is in operation. This factor will vary from 1 when there are people living or working in the adjoining space to $\frac{1}{16}$ when people may be exposed in a stairway or elevator.

Information for the calculation of shielding requirements is given in *Medical X-Ray and Gamma-Ray Protection for Energies Up to 10 MeV—Structural Shielding Design and Evaluation*[11] and in *Dental X-Ray Protection*.[12]

Radiation Protection Guides for Water, Food, and Air

The controlled use of nuclear energy will result in the release of small amounts of radioactivity to the environment. Nuclear weapons testing will add to this. Although there is governmental control of nuclear power plants and the industrial and medical uses of radioactive materials and devices, it is nevertheless essential to limit body intake and exposure from such sources to protect people who might be affected. The Federal Radiation Council, to meet this need, has provided guidelines for federal agencies carrying out activities that might affect individuals or groups of people. These guidelines have general application and are widely used.

The FRC guides consider three ranges of daily intake of radioactivity. The ranges and actions recommended to be taken with reference to specific radionuclides are given in Table 7–8. The council emphasizes that "there can be no single permissible or acceptable level of exposure without regard to the reason for permitting the exposure. It should be general practice to reduce exposure to radiation, and positive effort should be carried out to fulfill the sense of these recommendations. It is basic that exposure to radiation should result from a real determination of its necessity." (Federal Radiation Council, Radiation Protection Guid-

[11] National Council on Radiation Protection and Measurements, Rep. No. 34, 4201 Connecticut Ave., N.W., Washington, D.C. 20008 (March 2, 1970).
[12] Ibid., Rep. No. 35 (March 9, 1970).

TABLE 7-8   RANGES OF TRANSIENT RATES OF INTAKE
(PICOCURIES PER DAY) FOR USE IN GRADED SCALE OF
ACTION-RANGES I, II, III

Radionuclides	Range I*	Range II†	Range III‡
Radium 226	0 to 2	2 to 20	20 to 200
Iodine 131§	0 to 10	10 to 100	100 to 1,000
Strontium 89	0 to 200	200 to 2,000	2,000 to 20,000
Strontium 90	0 to 20	20 to 200	200 to 2,000

Source: Adapted from *Public Health Service Drinking Water Standards 1962*, U.S. Public Health Service Pub. No. 956, Dept. of HEW, Washington, D.C., 1963. The range for specific radionuclides is as recommended by the Federal Radiation Council. The radionuclide intake ranges recommended by the FRC are the sum of radioactivity from air, food, and water. Daily intakes were prescribed with the provision that dose rates be averaged over a period of one year.

* Periodic confirmatory surveillance as necessary in this range.

† Quantitative surveillance and routine control of useful applications of radiation and atomic energy such that expected average exposures of suitable samples of an exposed population group will not exceed the upper value of this range.

‡ Evaluation and application of additional control measures as necessary to reduce the levels to Range II or lower to provide stability at lower levels.

§ In the case of iodine 131, the suitable sample would include only small children. For adults, the Radiation Protection Guide for the thyroid would not be exceeded by rates of intake higher by a factor of 10 than those applicable to small children.

ance for Federal Agencies, Memorandum for the President, May 13, 1960. Reprint from *Federal Register* of May 18, 1960.)

Water

The FRC criteria have been observed in establishing the limits for radioactivity in the drinking water standards. These are given and explained in Table 7-9.

Nuclear reactors. The increasing use of nuclear reactors for power production requires cooling water. Although significant contamination of the water is extremely unlikely, the water used for this purpose will increase in temperature. It may first have to pass through a cooling tower before discharge to a stream or lake (if located inland) if the receiving body of water is unable to assimilate the added heat without causing ecological harm. It is also possible, however, that the heat energy going to waste can be put to productive use, such as for space heating; controlled fish propagation; irrigation to extend the agricultural growing season and increase crop production; water desalting; sewage evaporation; steam generation (boiling water reactor); and recreational pur-

TABLE 7-9   LIMITS OF RADIONUCLIDES IN DRINKING WATER

Radionuclide	Not to Exceed*	Routine Sampling†
Radium 226	3 p Ci/l‡	Yes
Strontium 90 §	10 p Ci/l‡	Yes
Gross beta activity	1,000 p Ci/l‖	Yes
Iodine 131	(Usually not found)	No
Strontium 89	(Not significant unless Sr 90 is also high)	No

Source: adapted from *Public Health Service Drinking Water Standards, 1962,* and *Manual for Evaluating Public Drinking Water Supplies,* Environmental Control Administration, U.S. Dept. of HEW, Cincinnati, Ohio, 1969.

  * p Ci/l = picocuries per liter.

  † Frequency based on likelihood of significant amounts being present.

  ‡ When these concentrations are exceeded, continued use may be permitted if surveillance indicates the total daily intakes from all sources (air, water, food) averaged over a period of one year are within limits recommended by the Federal Radiation Council for control action, i.e., 20 p Ci/day of radium 226 and 200 p Ci/day of strontium 90.

  § Principal source of strontium 90 in environment to date has been due to fallout from weapons tests.

  ‖ In the essential absence of alpha emitters and strontium 90. The limit for unidentified alpha emitters is taken as the listed limit for radium 226. When gross beta exceeds 1,000 p Ci/l the specific radionuclides present must be identified by complete analysis to establish the fact that the concentration of nuclides will not produce exposures greater than the recommended limits established in the Radiation Protection Guides, FRC, Rep. No. 1.

  Note: These limits are being reviewed (1972) by a technical committee of the Environmental Protection Agency and may be revised.

poses. Cooling water released to the environment should contain no radioactivity above that in natural background or that in Table 7-9.

Food

The amount of radioactivity in the diet should be kept at levels that will ensure protection of the general population in accordance with FRC reports, as stated earlier and in Table 7-8.

In Great Britain "working levels for assessment" are used in the same manner as "radiation protection guides" are used in the United States.[13] The working level for strontium is given as 130 pCi strontium 90/gram of calcium in the total diet of the general population. For individuals or small groups the working level may be three times that for the general

---

[13] *Radiation Levels: Air, Water and Food,* Royal Society of Health, London, S.W. 1, 1964.

population. In the case of iodine 131, the annual average exposure has been set at 130 pCi per liter of milk.

The main sources of radiation in the diet are probably from man-made fallout on agricultural land by iodine 131, strontium 89, strontium 90, and cesium 137. By comparison the amount contributed by inhalation and drinking water would be negligible except when there is a direct discharge into the water.

Milk would be the major source of iodine 131 and the isotopes of strontium and cesium. The radiation dose from the ingestion of strontium 90 is primarily determined by the ratio of strontium 90 to calcium in the total food intake. Although strontium 90 follows approximately the same route as calcium through the food chain, the deposition of strontium 90 in the bone depends on the amount of calcium in the diet. Both strontium and calcium will enter plants through the root system; but only one-seventh to one-tenth of the strontium as compared to calcium will be transferred by cattle to their milk, and that reaching the bone is about one-quarter of that in the diet.[14] Meat is also a significant source of cesium 137.

Radioiodine (iodine 131) may enter the body through consumption of fluid milk and fresh vegetables. It concentrates in the thyroid and increases the risk of thyroid cancer. Milk consumed within a few days after contaminated pasture grass is eaten by dairy cows presents the greater public health hazard. However, since iodine 131 has a relatively short half-life (8.14 days, physical), milk could be directed into manufacturing processes, such as in powdered milk, butter and cheese where it could be stored until safe. An alternative would be to put cattle on stored feed.

Strontium 90 is also hazardous because of its long biological half-life and because it deposits in the bone, increasing the risk of bone tumors and leukemia. Cesium 137 is not normally retained in the body for a long time, but it adds a small increment of exposure and increases the probability of genetic damage.[15]

Assessment of the significance of the dietary intake of radioactivity must take into consideration the total exposure including air, water, food (including fish), occupational, background, and medical exposures. In the United States, the Public Health Service and some state health departments routinely monitor the milk supply and the total dietary intake of radioactivity in controlled population groups up to age 18. The Food and Drug Administration also samples other foods.

---

[14] Ibid.

[15] Morgan, op. cit.; Herman E. Hilleboe and Granville W. Larimore, *Preventive Medicine*, W. B. Saunders Co., Philadelphia, 1965.

Air

Contamination of the air by radioactive materials can come from nuclear explosions (peaceful uses, testing, and war), nuclear reactors and fuels processing, accidental releases, and natural sources. Insofar as the general population is concerned, the maximum permissible exposure should be in the order of that due to the natural background level, approximately 100 to 150 mR/yr. Any activity in excess calls for action to reduce releases to achieve this goal. With this constraint in mind, the International Commission on Radiological Protection recommends, based on certain generally accepted assumptions, that occupational air exposure not exceed the following and that exposure of a small part of a population not exceed one-tenth of these concentrations (not including background, medical, or dental exposures) under the circumstances stated.[16]

10^{-9} μCi/ml—no alpha-emitting radionuclides and
(1000 pCi/m³)   beta-emitting radionuclides (Sr^{90}, I^{129}, Pb^{210}, Ac^{227}, Ra^{228}, Pa^{230}, Pu^{241}, Bk^{249}) present.

10^{-10} μCi/ml—no alpha-emitting radionuclides and
(100 pCi/m³)   beta-emitting radionuclides (Pb^{210}, Ac^{227}, Ra^{228}, Pu^{241}) present.

4×10^{-13} μCi/ml—no information as to composition
(0.4 pCi/m³)   of radioactivity.

Radioactive contaminants, with their half-lives, from the detonation of nuclear weapons include iodine 131, 8 days; strontium 90, 28 years; strontium 89, 53 days; cesium 137, 30 years; and cerium 144, 275 days. Atmospheric detonations should obviously be prohibited by international agreements.

Nuclear fuels processing plants also present special problems due to the reprocessing of irradiated nuclear reactor fuel in which uranium and plutonium are separated from the other fission products. In the process it is possible for radioactive liquid, gases, and particulates to be released. Reliance is placed on process control, filtration, dilution by high-stack emission, liquid retention, and dilution of low-level liquid wastes to keep releases to a minimum and within acceptable limits.

The production of nuclear fuels involves the washing and concentration of the ore with the danger of radon and thoron contamination of dust. In the separation of the isotopes of uranium and thorium, dusts of ura-

---

[16] *Radiation Levels: Air, Water and Food,* op. cit. (μCi/ml = microcurie per milliliter; pCi/m³ = picocurie per cubic meter.)

nium and thorium oxides or gases add to the potential hazard. Surveillance of waste discharges may be required.

In nuclear power plants air pollution is possible from the release of radioactive xenon and krypton produced by fission; the activity induced in gases in the air; and the release of fission products as a result of fuel-cladding ruptures. The significant releases are likely to be iodine 131, krypton 85, and tritium.

WASTE DISPOSAL

The disposal of radioactive waste should not endanger individuals or the environment. The guiding principle should be prevention and control of gaseous, liquid, and solid waste at the source followed by segregation and the indicated collection, treatment, and storage or disposal.

Collection and Treatment

Liquid and solid wastes may originate in radioisotope laboratories, chemical processing plants, water-cooled reactors, change rooms, and decontamination laundries. Solid wastes are collected in paper or plastic-lined containers, if of low activity, and then disposed of in an approved procedure such as by commercial waste disposal or incineration. High-activity solid wastes are placed in shielded containers and stored or disposed of in an approved manner.

Radioactive wastes may be treated by physical and chemical methods depending on the characteristics of the waste. Low-level liquid wastes are usually disposed of by dilution to reduce activity to below maximum permissible levels. High-level liquid or solid wastes are concentrated or stored. Liquid wastes are commonly concentrated by evaporation, by precipitation of a soluble material with an insoluble precipitate, and by ion exchange. Biological assimilation is also possible.

Storage and Disposal

After treatment and concentration high-level radioactive wastes are carefully packaged and, when indicated, fixed in an inert solid material in preparation for storage or disposal in a restricted area. Concentrated high-activity liquid wastes are stored in a specially constructed tank within a tank or a tank over a secondary container of equal volume that can be monitored for leakage. These require careful monitoring and a well-developed emergency plan as container failures have occurred. Liquid radioactive wastes may also be collected for storage and retention until sufficient decay has taken place to permit controlled discharge and disposal by dilution. Ground disposal of low- or intermediate-level wastes may be permitted under approved soil, rock, and groundwater conditions. Other disposal methods considered include dry natural caverns, deep mines, salt cavities, deep-well disposal, and ocean disposal; but each of these requires careful evaluation before being permitted.

Disposal of radioactive wastes is under the jurisdiction of the Atomic Energy Commission or state health department if the AEC has an agreement with the state.

ENVIRONMENTAL RADIATION SURVEILLANCE AND MONITORING

Monitoring Stations and Sampling

An environmental monitoring program in and around large nuclear facilities serves as a check on the various operations that might produce contamination.[17] It also permits evaluation of the exposure to the surrounding area and the need for preventive action. The program should be a cooperative one between the health authorities and the industry.

Industry has responsibility for internal housekeeping and for monitoring all its waste discharges in terms of types and quantities. Peripheral monitoring should be shared between the industry and the health authorities; distant monitoring, up to 40 mi, is done by the health agency. Maximum fallout from a stack will usually occur immediately after emission within a 1-mi radius.

Solid, liquid, and gaseous radioactive materials can be carried considerable distances by air and water and may adversely affect plant and animal life as well as people. It is essential therefore that a surveillance and monitoring program be maintained in and around installations from which radiation release is possible. These include nuclear power plants, fuel processing plants, uranium milling industries, university reactors, and certain industries and laboratories. The surveillance monitoring should be carefully planned with the advice and assistance of radiological health specialists and physicists, biologists, meteorologists, environmental engineers, geologists, and others familiar with the problem and the local area. A practical guide for the planning, operation, and evaluation of an effective program for the routine surveillance of radioactivity around nuclear facilities is available from the Public Health Service.[18] The sam-

[17] Dade W. Moeller and Abraham S. Goldin, "Environmental Protection for Nuclear Applications," *J. Sanitary Engineering Division*, ASCE, 373–385 (June 1969).
[18] *Routine Surveillance of Radioactivity Around Nuclear Facilities*, U.S. Public Health Service Pub. No. 999-RH-23, Dept. of HEW, Washington, D.C., December 1966.

pling for resulting contamination should include air, water, milk, food, biota, soil, and people.

At a fuel processing plant there is a potential for discharge of iodine 131, krypton 85, and tritium. Release of iodine 131 is minimized by storage of fuel elements at least 100 days before processing. Krypton release is greater than from a nuclear reactor and cannot as yet be effectively removed before release to a stack. Tritium is not yet removable and is discharged as a liquid or gaseous waste.

Careful surveillance is therefore necessary to ensure that discharges of these radionuclides are kept to a minimum and within acceptable limits. Sampling may include any combination of the following, with consideration of the facility monitored:

Air—around the installation and at some distance, as noted above.
Water—at plant outfalls, receiving stream and downstream; also groundwater from nearby wells.
Soil—immediately around the installation and at some distance, including stream bottom muds.
Biological specimens—fish, deer, rodents, plants—especially those used for food; also where available shellfish, ducks, plankton, and other plant life.
Milk—from dairy farms in vicinity.
Spaces—indoor spaces, containers, and conduits.

Table 7–10 gives suggested frequency of sampling and determinations.

Sampling stations are usually located on the site being monitored, in the immediate vicinity, and at some distance. Those onsite and in the immediate vicinity should preferably be fixed and be continuous monitors. Stations at a distance could be used for periodic grab sampling, selected if possible with reference to other stations maintained by the Weather Bureau, health department, a university, U.S. Geological Survey, radio station, or airport that could provide supporting data. Stream samples include water, mud, and biota. Tiles, stones, or slides are suitable for the collection of biological attachments.

A monitoring station might contain a continuous water sampler, a high-volume air sampler, silver nitrate filter, film badge, adhesive paper, silica gel, rain gauge, ionization chamber, and such other equipment or devices as may be indicated by a safety analysis and the nature of the facility being monitored. Table 7–10 shows the type of monitoring indicated around a nuclear reactor or fuels processing plant.[19] Table 7–11 shows some radioactivity concentrations that might be found in clean and contaminated surface waters, algae, and bottom muds.

The air monitoring around a boiling-water reactor is described by

[19] Sherwood Davies, "Environmental Radiation Surveillance at a Nuclear Fuel Reprocessing Plant," *Am. J. Public Health,* **58,** 2251–2260 (December 1968).

TABLE 7–10   SUGGESTED MONITORING AROUND A NUCLEAR REACTOR OR A FUEL REPROCESSING PLANT

Sample*	Frequency	Determination	Remarks†
Water intake and effluent	Daily where indicated	Gross alpha and beta of dissolved and suspended solids	Quarterly composite shall not exceed the standard established by USAEC. Make complete isotopic analyses if exceeded.
Receiving waters	Daily	Same as above	—
Fish, shellfish, mud, stream, water supplies, plankton	Quarterly‡ upstream and downstream from outfalls	Gross and gross beta activity	Identification of radionuclides if standard exceeded.
Milk, farms in vicinity	Weekly	Iodine 131, strontium 90, tritium, and cesium 137	Identification of radionuclides if standard exceeded.
Air around facility	Continuous monitors, daily	Iodine 131, krypton 85, gross beta on particulates	Shall not exceed standard established by USAEC. If exceeded or approached, monitor environment and eliminate source.
Vegetation, animals	Growing season‡	Gross beta	Same as for milk.
Fallout, downwind	Daily cumulation, biweekly	Gross beta	Relate to fallout from weapons or other testing.

*Include meteorological data and stream flows.

† See Table 7–9 and "Radiation Protection Guides for Water, Food, and Air, in this chapter for ranges and limits.

‡ Quarterly or annually for radionuclides of long half-life; for short-lived nuclides, no more than two or three half-life intervals.

Thomas.[20] Within a 1-mi radius distributed on a 500-ft grid will be 50 thermoluminescent dosimeters; 3 air monitoring stations to sample airborne particulates, radioiodine, heavy particulate fallout, and rainwater. Air filters will be scanned continuously by a beta-gamma-sensitive G-M tube, and readings will be reported at the control center. In addition

[20] Fred W. Thomas, "TVA's Air Quality Management Program," *J. Power Division,* ASCE, 131–143 (March 1969).

TABLE 7-11   SOME RADIOACTIVITY CONCENTRATIONS IN
STREAM SAMPLES

	Gross Alpha p Ci/l	Gross Beta p Ci/l
Stream sample	0 to 21	0 to 334
(relatively clean)	0 to 3	0 to 72
	0 to 220	7 to 370
	0 to 33	7 to 67
Stream sample	4,700	5,500
(contaminated)	300,000	25,000
Water supply—raw	0 to 25	3 to 440
treated	0 to 11	3 to 144
	p Ci/gm dry	p Ci/gm dry
Algae		
Clean water	0 to 25	41 to 322
Clean water	2 to 83	0 to 220
Contaminated water	to 2,500	to 10,000
Mud		
Clean water	13 to 30	18 to 71
Contaminated water	660	6,600

Note: Rainfall usually contains more radioactivity than surface water.

four air monitors about 10 mi away in an urban area will provide continuous reports to the control center. Weekly composite samples will also be collected from stations about 40 mi from the plant. The air monitoring is supplemented by the sampling of fish, mussels, algae, and other biota as well as milk, water, vegetation, and soil. State and Federal agencies also have responsibilities for public health protection.

Careful plans should also be made for immediate emergency monitoring and action in case of an accidental discharge to the environment. This is necessary to protect the public while an assessment is being made. Gross activity levels are useful for rapid measurements, as are samples from fixed air and water monitoring stations.

The Public Health Service, the Atomic Energy Commission, some state and local health departments, and other agencies routinely sample the air, water, and food supplies. Reports on the surveillance maintained are issued monthly by the Public Health Service and periodically by other agencies. Surveillance is also maintained over fallout.

Space and Personnel Monitoring

A radiation installation should be responsible for its own monitoring program. This should include the facility work and storage areas, sources,

TABLE 7-12   RADIATION DETECTING DEVICES

Detector	Types of Radiation Measured	Typical Full Scale Readings	Use	Minimum Energy Measured	Directional Dependence	Advantages	Possible Disadvantages
Scintillation counter	Beta, X, gamma, neutrons	0.02 mR/h to 20 mR/h	Survey	20 keV for X rays. Variable for betas.	Low for X or gamma	1. High sensitivity 2. Rapid response	1. Fragile 2. Relatively expensive
Geiger-Müller counter	Beta, X, gamma	0.2 to 20 mR/h or 800 to 80,000 counts/min	Survey	20 keV for X rays. 150 keV for betas.	Low for X or gamma	Rapid response	1. Strong energy dependence 2. Possible paralysis of response at high-count rates or exposure rates 3. Sensitive to microwave fields 4. May be affected by ultraviolet light
Ionization chamber (Cutie pie)	Beta, X, gamma	3 mR/h to 500 R/h	Survey	20 keV for X rays. Variable for betas.	Low for X or gamma	Low energy dependence	1. Relatively low sensitivity 2. May be slow to respond
Alpha counter	Alpha	100 to 10,000 alpha/min	Survey	Variable	High	Designed especially for alpha particles	1. Slow response 2. Fragile window
Film	Beta, X, gamma, neutrons	10 mR and up	Survey and monitoring	20 keV for X rays. 200 keV for betas.	Moderate	1. Inexpensive 2. Gives estimate of integrated dose 3. Provides permanent record	1. False readings produced by heat, certain vapors, and pressure 2. Great variations with film type and batch

534

Instrument	Radiation	Range	Use	Energy		Advantages	Disadvantages
Film (cont'd)							3. Strong energy dependence for low-energy X rays
Pocket ionization chamber and dosimeter	X, gamma	200 mR to 200 R	Survey and monitoring	50 keV.	Low	1. Relatively inexpensive 2. Gives estimate of integrated dose 3. Small size	1. Subject to accidental discharge
B F$_3$ Counter	Neutrons	0-100,000 c/min	Survey	Thermal		Designed especially for neutrons	

Source: *Radiation Protection in Educational Institutions*, Rep. No. 32, National Council on Radiation Protection and Measurements, Washington, D.C., July 1, 1966.

waste disposal, emergency procedures, and personnel monitoring. A competent health physics safety unit should be established and given responsibility for the monitoring and authority to take prompt corrective action whenever indicated.

Instruments for detecting and measuring radiation are necessary for an effective monitoring program, and these must be selected with consideration of their sensitivity and purpose to be served. A summary of some radiation-detecting devices giving their characteristics, advantages, and disadvantages is given in Table 7–12.

RADIATION PROTECTION PROGRAM

General

A state radiation protection program involves control of X-ray units, radioactive materials (medical and nonmedical), waste disposal, and environmental monitoring and surveillance. Particle accelerators and other radiation machines, certain radionuclides produced artificially and in small quantity, radium, and other naturally occuring radionuclides are included. In all cases the goal must be elimination of all unnecessary exposure to ionizing radiation.

The licensing and regulation of nuclear reactors and reactor-produced radionuclides is the sole responsibility of the Atomic Energy Commission. The control of reactor-produced radionuclides can be specifically delegated to a state government adopting control procedures acceptable to the AEC. The possession and use of certain reactor-produced radionuclides requires a general license for stated small quantities and a specific license for a particular use not coming under a general license.

A permit to construct or install a nuclear reactor must be obtained from the AEC, and then a separate operating license is needed before the fuel is inserted and the reactor is placed in operation. Preoperation and postoperation environmental surveillance and monitoring are usually conducted cooperatively by the AEC, Public Health Service, state health department, other state agencies concerned, and the industry involved. The preoperational survey should be started at least one year before the proposed date of operation and should cover an area within a 20-mi radius of the facility.

Planning

The planning of a state or local radiation program requires that the problem be defined and means for bringing it under control be developed and implemented.

The first step, then is to conduct a survey to determine:

1. The location, number, and types of radiation sources within the area of jurisdiction. State professional licensing agencies, professional societies, telephone-book yellow pages, equipment manufacturers, and supply houses may be helpful.
2. The legislative and regulatory needs.
3. The types of problems that can be anticipated and are to be resolved.
4. The training and education needs of staff.
5. The required workload, staff, equipment, and budget.

Reference to the following may be helpful in planning a local program, keeping in mind that it is offered only as a guide.[21]

> Manpower allocation
> > X-ray equipment—70 percent
> > Radioactive materials—30 percent
> Manpower allocation to X-ray equipment
> > Medical—60 percent
> > > Therapy—1 percent
> > > Fluoroscopic—24 percent
> > > Dental and radiographic—35 percent
> > Nonmedical—10 percent
> > > Fluoroscopic—5 percent
> > > Radiographic—5 percent
> Manpower allocation to radioactive materials
> > Medical—10 percent
> > > AEC—5 percent
> > > Non-AEC—5 percent (radium)
> > Nonmedical—20 percent
> > > AEC—15 percent
> > > Non-AEC—5 percent (radium)

The time required to make the initial complete survey of a particular installation has been estimated as follows:[22]

Dental X ray	10.5 man-hours
Chiropody X ray	10.5
Radiographic (chest X ray)	11.5
Medical fluoroscope	11.5
Miscellaneous isotope	8.0

These estimates include travel, inspection, data analysis, and report writing for a complete and comprehensive survey by a two-man team. The time can be cut in half by using one experienced radiological health specialist. A comprehensive reinspection, however, would generally require in all less than two man-hours.

---

[21] "The Responsibility of Local Health Agencies in the Control of Ionizing Radiation," report of the Committee on Local Control of Ionizing Radiation of the Conference of Municipal Public Health Engineers, November 1962.
[22] Ibid.

Control Program

A control program should include registration of all radiation sources; education of the inspectors and users, licensing or certification of X-ray technicians, and the dissemination of informational literature; inspection of all equipment and the procedures used; discussion of deficiencies observed and their correction with the operator and then confirmation of the recommendations in writing; follow-up inspections to ensure compliance and administrative enforcement when indicated; reinspection every two to three years. This would include medical and dental, veterinarian, industrial, laboratory, university, and hospital uses.

Coordination should be maintained with other agencies, such as with the building department (including plumbing and electrical), for referral of applications involving installation of X-ray equipment or construction of a laboratory planning to use radioactive materials. Liaison should also be maintained with the department of public works in connection with possible problems and need for monitoring at the sewage treatment plant, incinerator, or sanitary landfill.

In some instances when a large number of X-ray installations are to be controlled, consideration might be given to a requirement that the installation be routinely surveyed by a licensed individual. A report of satisfactory compliance with statutory regulations would be submitted annually to the regulatory agency. The agency would establish licensing requirements, make periodic evaluations on a statistical random-sampling basis, and take whatever enforcement action that would be indicated. This approach may not significantly reduce the size of the regulatory staff but would strengthen the program.

Coordination With Other Agencies

Federal, state, and local program activities should be carefully integrated to avoid duplication and ensure complete coverage. Each should perform, in a previously agreed on systematic manner, what each is best equipped to do by virtue of its legal responsibility and resources, its physical location, and its staffing.

The Public Health Service traditionally provides training opportunities to state and local health personnel. It supports research, provides leadership in program development, and sets certain standards. Educational institutions, research centers, the Federal Radiation Council (functions placed in the Environmental Protection Agency in 1970), Bureau of Standards, and the Atomic Energy Commission play an important role in education, surveillance, standards setting, and research. The Federal Office of Civil Defense also participates in case of a serious accident.

The state usually has a basic responsibility to promote local programs when indicated, to carry out training, to provide assistance, to provide laboratory support, to evaluate and make recommendations for program improvement, and to require that a competent program be conducted.

The local agency, if given the resources, should make routine inspections of X-ray equipment and other facilities if properly staffed, conduct education, carry out enforcement when necessary, and report the work done, problems, and accomplishments to the state. If a proper program cannot be carried out locally, the state agency having the overall responsibility will have to find a way to carry out the program.

OTHER RADIATION SOURCES

The electromagnetic spectrum, Figure 7–1, shows the relationship of the various types of radiation and their approximate wavelengths. Certain electronic products can emit harmful radiations if not properly designed, constructed, installed, or used. These include color television receivers, diathermy units, electron microscopes, voltage regulators, vacuum condensers, vacuum switches, rectifiers, microwave ovens, radar microwaves, devices using laser and maser beams, and other intense magnetic fields.

The Public Health Service was given responsibility in 1968 under the Radiation Control Act to control electronic product radiation. Electronic products such as mentioned above, including X-ray machines, manufactured after October 18, 1968 are required to comply with the regulations effective January 22, 1970.[23] The regulations spell out procedures for repairing, replacing or refunding the cost of radiation defective or nonstandard products and establish procedures to prevent importation of electronic products found to be in violation of the act.

Color television receivers operating at higher than design kilovoltage or with improper shielding can be potential hazards. Viewing at least 6 ft from the tube is advised in any case. The radiation should not exceed 0.5 mR/hr at any accessible point 5 cm (about 2 in.) from the surface.[24]

Microwaves

Microwave ovens with poor-fitting doors, which can operate with the door open or partly open, or with a safety interlock system that can be bypassed can cause serious harm to the user. Microwave ovens are

[23] Published in the *Federal Register,* Vol. 35, No. 15–Thursday, January 22, 1970.
[24] Recommended by the International Commission on Radiological Protection and the National Council on Radiation Protection. Also, limiting value in the United Kingdom.

used in the home, in mass-feeding facilities, and as an adjunct to quickly heat precooked meals from automatic vending machines and caterers. Dirt, grease or metal particles on door seals can permit microwave leakage, hence cleanliness is essential.[25]

Microwaves and laser and maser beams are forms of energy that produce heat when absorbed in tissue. The eye is particularly vulnerable. Direct exposure as well as reflected laser light beams from great distances can be harmful. Laser beams are in the ultraviolet and infrared range of the electromagnetic spectrum.

Microwaves are also used in airport radar installations, electrotherapy, and television relays. Microwaves pass through glass and other clear materials; they are reflected by metals but are absorbed by nonmetallic materials such as water, food, and human tissue (eyes, internal organs, other) with the production of heat. By the time burning or pain is felt, the damage is done. A densiometer or power density meter is used

TABLE 7-13   MICROWAVE STANDARDS

Maximum Exposure (in milliwatts per square centimeter (mW/cm^2)	Reference
10	Underwriter Laboratories, U.S.
10	U.S.A. Standards Institute
10 continuous	U.S. Army and Air Force
10 to 100 limited	U.S. Army and Air Force
10 continuous	Great Britain (post office regulation)
1.0 prolonged	Sweden
0.01 continuous	U.S.S.R.
0.1 2-hr maximum	U.S.S.R.
1.0 15 to 20 minutes*	U.S.S.R.
1.0†	U.S. Public Health Service
5.0‡	U.S. Public Health Service

*Protective goggles required.

† For microwave ovens at time of sale.

‡Not more than 1 mW/cm^2 prior to oven sale and not more than 5 mW/cm^2 throughout the useful life of a microwave oven measured 2 in. (5 cm) from the external oven surface.

to monitor radiation leakage. Some standards for the control of electronic equipment are given in Table 7–13.

Microwave ovens and other electromagnetic energy sources (diathermy units, some radio frequencies, neon lights, gasoline engines and their

[25] Robert L. Elder and Walter E. Gundaker, "Microwave Ovens and their Public Health Significance," *J. Milk and Food Technology,* **34,** 444–446, (September 1971).

ignition systems) may interfere with the proper functioning of cardiac pacemakers. Pacemaker users should be alerted to the potential interference possibilities from outside electromagnetic sources.

BIBLIOGRAPHY

Basic Radiation Protection Criteria, Rep. No. 39, National Council on Radiation Protection and Measurements, Washington, D.C., January 15, 1971.

Dental X-Ray Protection, Rep. No. 35, National Council on Radiation Protection and Measurements, Washington, D.C., March 9, 1970.

Ionizing Radiation, American Public Health Association, Inc., 1815 Eighteenth Street, N.W., Washington, D.C., 1966.

Medical X-Ray and Gamma-Ray Protection for Energies up to 10 MeV—Equipment Design and Use, Rep. No. 33, National Council on Radiation Protection and Measurements, Washington, D.C., February 1, 1968.

Medical X-Ray and Gamma-Ray Protection for Energies up to 10 MeV—Structural Shielding Design and Evaluation, Rep. No. 34, National Council on Radiation Protection and Measurements, Washington, D.C., March 2, 1970.

Radiation Levels: Air, Water and Food, Royal Society of Health, London, S.W. 1, England, 1964.

Radiation Protection in Educational Institutions, Rep. No. 32, National Council on Radiation Protection and Measurements, Washington, D.C., July 1, 1966.

Radiological Health Handbook, U.S. Public Health Service, Dept. of HEW, Government Printing Office, Washington, D.C., January 1970.

Routine Surveillance of Radioactivity Around Nuclear Facilities, U.S. Public Health Service Pub. No. 999–RH–23, Dept. of HEW, Government Printing Office, Washington, D.C., December 1966.

8

FOOD PROTECTION

The health and disease aspect of food and the investigation of food-borne illnesses were discussed in Chapter 1. This chapter concentrates on the technical and sanitation aspects.

DESIGN OF STRUCTURES

Locating and Planning

The essential elements of a preliminary investigation to determine the suitability of a site for a given purpose was discussed in Chapter 2. Time and money spent in study, before a property is purchased, is a good investment and sound planning. For example, a food processing plant that requires millions of gallons of cooling or processing water would not be located too distant from a lake or clear stream unless it was demonstrated that an unlimited supply of satisfactory well water or public water was available at a reasonable cost. An industry having as an integral part of its process a liquid, solid, or gaseous waste would not locate where an adequate dilution or disposal facility was not available unless an economical waste treatment process could be devised.

Such factors as topography, drainage, highways, railroads, watercourses, exposure, swamps, prevailing winds, dust, odors, insect and rodent prevalence, type of rock and soil, availability and adequacy of public utilities, the need for a separate power plant, and water, sewage, or waste treatment works should all be considered and evaluated before selecting a site for a particular use.

Structural and Architectural Details

The design of structures is properly an engineering and architectural function that should be delegated to individuals or firms that have become expert in such matters. Certain construction details will be shown and discussed here as they apply particularly to food establishments, although the basic principles will apply equally as well to other places.

The construction material used is dependent on the type of structure and operation. Some materials can be used to greater advantage because of location, availability of raw materials, labor costs, type of skilled labor available, local building codes, climate, and other factors.

Floors

There are many types of floor construction; a few of them will be briefly discussed. Floors in food processing plants, dairy plants, kitchens, toilet rooms, and similar places should be sloped $\frac{1}{8}$ to $\frac{1}{4}$ in./ft to a drain. A trapped floor drain is needed for every 400 ft^2 of floor area, with the length of travel to the drain not more than 15 ft.

Concrete floors. Wear causes dust and pitting. New floors require at least seven days for curing and must be kept wet during this time. Concrete can be colored. Special finishes include rubber resin enamel, concrete paint or sealer, chemical treatment, wax over paint, or sealer. Warm linseed oil treatment makes concrete acid-resistant. Wire-mesh reinforcement will reduce cracking, particularly in coolers and driers where large temperature variations are expected. Floors in receiving rooms should be armored. Trucks and dollys with rubber or composition wheels reduce concrete wear and noise.

Wood strip or block flooring must be close-grained, seasoned, and carefully laid. A new floor requires sanding, two coats of a penetrating-type sealer or primer, buffing, and a wax, resin, or plastic finish. Avoid use of water.

Asphalt or mastic tile, vinyl tile, and **rubber tile.** Use water sparingly. Strong soap, lye, varnish, or lacquer causes covering to shrink, curl, and crack. Use proper sealer, cleaner, and wax, usually water-based. Can be used below grade with waterproof adhesive.

Terrazzo and mosaic floors require six months for curing. Clean with warm water and treat with terrazzo seal. Solvents, oils, strong alkalis, and abrasives cause discoloration and destruction.

Vitreous tile, ceramic tile, and **packing-house brick tile** are hard, nonabsorbent, resist acid, and are not scratched by steel or sand. Cement joints should be full $\frac{1}{16}$ in. Ceramic or quarry tile properly laid is best for kitchens, dairies, breweries, bakeries, shower rooms, and similar places. Wash with hot water and neutral cleaner to keep floor in good condition. Soap and hard water cause discoloration.

Linoleum floors require wax treatment and damp mopping. Water, alkali, or oil compounds cause linoleum to break down. Dampness causes linoleum to raise. Linoleum provides insect harborage and is not recommended for use in food-preparation rooms.

Cork floors are treated with a filler and water-emulsion wax. Avoid use of water. Provide furniture rests. Do not use directly on basement concrete floors.

Oxychloride, magnesite, or **composition cement** flooring has been used for marine decking, industrial building flooring, and other purposes. A penetrating-type dressing about every six months and waxing is the usual maintenance. Excessive water causes deterioration.

New or renovated floors require a conditioning or primary treatment. This usually involves a preliminary cleaning, then sealing and finishing to prevent penetration of moisture, soil, fungi, and bacteria and also to improve the wearing qualities of the floor. Following the conditioning the floor should be cleaned periodically thereafter and properly maintained to keep down dust and the microorganism population, prevent accidents due to slippage, and remove fire hazards. Manufacturers of flooring can give specific advice for the sealing and care of their particular material.

A good floor cleaner is all-purpose and neutral, requiring no rinsing; it should dissolve completely, have a low surface tension so as to pick up soil, and be economical and easy to use. A proper maintainer has a nonoil base and does not harm the floor material, furniture, walls, or equipment. It must not soften the floor finish; it should be applied either as a spray or powder, reduce slippage, be noninflammable, be easily removed from the mop or applicator, impart a pleasant aroma, and have good spreading or coverage quality. Ensure proper ventilation when using chemical cleaners. Rope off wet floors until dry.

Walls and Juncture with Floor

The line or point of juncture between the wall and floor, and with built-in equipment, should form a tight, sanitary cove and a smooth and flush connection. Some typical base details are illustrated in Figure 8–1. Walls should be smooth, washable, and kept clean. Many materials have been used for interior walls and wall finishes. These include glass blocks, plaster, cement, clay tile and cement, concrete tile and block, marble, concrete, ceramic tile, glazed brick and tile, and architectural terra cotta wall blocks. Absorbent materials such as wallboard and rough plaster require careful treatment and maintenance where moisture or water is an integral part of a process. These surfacings are to be avoided in such cases if possible. The wall finish should be a light color in work and processing rooms. Windowsills that are sloped down at an angle of about 30 degrees simplify cleanliness.

Hollow walls and partitions, hung ceilings, and boxed-in pipes and equipment will provide a channel of communication, shelter, and protection for insects and rodents. They should be eliminated or, if this is not possible, special entrance openings or ducts should be provided to

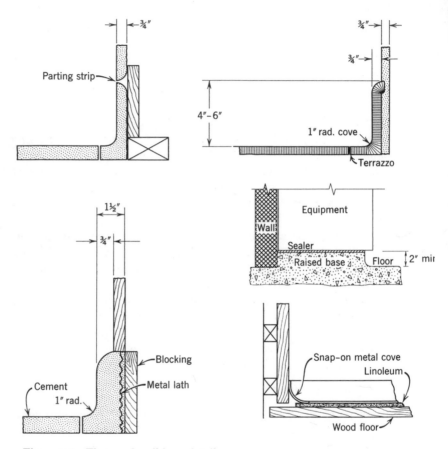

Figure 8–1   Floor and wall-base details.

make possible simple and direct fumigation. The use of gypsum blocks is not recommended where water or dampness might be encountered.

Condensation

A difficult problem in food plants and restaurants is the condensation of moisture on cold pipes or surfaces, with resultant dampness, annoying dripping of water, and possible contamination of food and food-preparation tables. Condensation is caused by warm, humid air in contact with cool pipes or surfaces below the dew-point temperature. Lowering the humidity of the air, cooling the air, or increasing the temperature of the surface above the dew point will prevent condensation. Dehumidification can be accomplished by mechanical means and indirectly by good ventilation, which also would cause a lowering of the air temperature. Increase of the temperature of the surface is generally accomplished by insulation; the type and thickness of insulation is determined by the particular air temperature and humidity.

Lighting

Adequate light, with consideration to interior surface finish and color, is essential to proper operation and the maintenance of cleanliness. A minimum of 30 ft-c and an average of 100 is recommended at work surfaces.

Washing Facilities, Toilets, and Locker Rooms

The provision of adequate, convenient, and attractive sanitary facilities including dressing rooms or locker rooms is not only required by most health, industrial hygiene, and labor departments, but also is a good investment. Contented employees make better workers, and the habitual practice of personal hygiene in food establishments especially is essential. Wall-hung fixtures and toilet partitions in washrooms and the proper selection of floor and wall materials substantially reduce cleaning costs.

The number of plumbing fixtures to provide depends on the type of establishment and the probable usage. Local standards are usually available; but in the absence of regulations, the suggestions given in Table 11–6 can be used as a guide. Washbasins and showers should be connected with both hot and cold running water or with tempered water. Soap or a suitable cleanser and sanitary towels are standard accessories. Extra washbasins with warm water, soap dispenser, and paper towels at convenient locations in all workrooms help promote cleanliness.

Water Supply

An improperly constructed and protected water supply is a common deficiency at food establishments and particularly at dairy farms. This condition prevails probably because the water supply is related to the production of milk and food only indirectly and hence is considered unimportant in the spread of disease or product quality. It is also possible that the average inspector, being inexperienced in water-supply sanitation, overlooks weaknesses in the development and protection of the water supply source.

A safe and adequate water supply under pressure is fundamental to the promotion of hygiene and sanitary practices at a dairy farm, pasteurization plant, and other types of food-processing plants. It is good practice to oversize the piping and provide extra tees at regular intervals to permit future connections. Where a contaminated or potentially contaminated water supply under pressure is also available, it must not be connected to the plant potable water system unless special approval is obtained from the health department and required backflow preventers are properly installed. The availability of an unsafe water supply in a plant is dangerous; it should be eliminated if at all possible. An inadequate water supply will hamper and in many instances make impractical good sanitary practices and procedures. Contaminated water can become a link in the chain of infection rather than a barrier and, in addition, lead to the production of poor-quality food. The development, protection, and location of wells and springs and the treatment of water are discussed in Chapter 3.

Liquid and Solid Waste Disposal

Where the disposal or treatment of industrial wastes is a problem, the sanitary sewer and waste drainage systems should normally be separate. Clear, unpolluted water should not as a rule be mixed with the sewage or industrial wastes and thus aggravate these problems, but they can be disposed of separately without treatment. The waste treatment problem can usually be reduced by good plant housekeeping. Solid wastes are collected and stored for separate disposal; liquid wastes are kept at a minimum by careful operation and supervision; and where possible wastes are salvaged. Information concerning the treatment and disposal of sewage and industrial wastes is given in Chapter 4. Solid waste management is discussed in Chapter 5.

FOOD HANDLING, QUALITY, AND STORAGE

Food Handling

Temperature control and clean practices should be the rule in kitchens and food-processing plants if contamination is to be kept to a minimum. All food contact surfaces and equipment used in preparation must be kept clean and in good repair. Frozen meat, poultry, and other bulk frozen foods should be thawed slowly under controlled refrigeration (36 to 38°F is ideal) and *not left to stand at room temperature overnight to thaw*. Frozen vegetables and chops need not be thawed but can be cooked directly. Prepared foods, especially protein types, should be served immediately, kept temporarily at a temperature of 45°F or less, or on a warming table maintained at a temperature above 140°F until served. If not to be served, the food should be refrigerated in shallow pans to a depth of 2 to 3 in. within 30 min at a temperature of 45°F. Bulk foods should be cut into smaller pieces and refrigerated within 30 min of preparation unless immediately served. Spoilage of hot foods is prevented by prompt refrigeration. Heat foods to at least 165°F.

Food handlers are expected to have hygienic habits. A washbasin located in the kitchen or workroom supplied with warm running water, soap, and individual paper towels is necessary and conducive to greater personal cleanliness. Standard instructions to food handlers should include the following precautions.

1. Keep perishable foods covered and in the refrigerator until used. Serve cold foods cold and hot foods hot!

2. Wash hands thoroughly after using the toilet. Use plenty of soap and warm water.

3. Do not pick up food with the fingers during food preparation unless absolutely necessary, and never when serving. Use a serving spoon, fork, spatula, or tongs.

4. Keep hands clean and fingernails short and clean. Keep fingers out of food.

5. Keep body and clothes clean and wear a head covering or hairnet.

6. Pick up cups, spoons, knives, and forks by the handles and keep the fingers out of glasses, cups, soup bowls, and dishes.

7. Cover nose and mouth with a paper tissue when sneezing or coughing, then discard tissue and wash hands thoroughly. Do not smoke where food is prepared.

8. Report to a doctor at the first sign of a cold, sore throat, boils, vomiting, running sores, fever, or loose bowels. Stay at home!

9. Help keep the entire premises clean. Store foods in a clean, dry place, protected from overhead drippage and animal, human, rodent, and insect contamination. Keep food preparation tables and utensils clean; avoid cross-contamination between food and unclean surfaces.

Food Inspection

Food for human consumption is expected to be clean, wholesome, and sanitary. Decayed, insect-infested, moldy or musty food is not acceptable. In any case, purchased food should be in full weight and measure and of the indicated grade. The best time to check this is on delivery against the order or invoice, before acceptance. Although special training is needed for expert determination, some general guides are summarized in Table 8–1.

Microbiological Standards

Natural microbial variations in different foods and the statistical aspects of sampling present considerable difficulties in the establishment of firm standards. In addition, because of normal errors inherent in

TABLE 8-1   TESTS FOR FOOD SPOILAGE OR POOR QUALITY

Canned food—Grade A, B or C
1. Swelled top and bottom.
2. Dents along side seam.
3. Off-odor.
4. Foam.
5. Milkiness of juice.

   This applies to canned vegetables, meats, fish, and poultry. Home-canned foods should be cooked thoroughly. Do not taste!

Fresh fish
1. Off-odor.
2. Gray or greenish gills.
3. Eyes sunken, dull; pupils gray.
4. Flesh easily pulled away from bones.
5. Mark of finger nail indentation remains in flesh.
6. Not rigid, scales dry and dull.
7. Oysters and clam shells not tightly closed. Shell gives dull sound on tapping.

Raw shrimp
1. Pink color on upper fins and near tail. Soft, dull, sticky.
2. Off-odor similar to ammonia.

   Some types of shrimp are naturally pink. Cooked shrimp are also pink. Both are wholesome if the odor is not abnormal.

Meat—Grade Prime, Choice, Good
1. Off-odor, sourness, taint.
2. Slimy to touch.
3. Beef soft, dark, coarse-grained; soft fat, yellow.
4. Lamb or veal flesh dark, fat yellow.

   Beef usually spoils first on the surface. Pork spoils first at meeting point of bone and flesh in the inner portions. To test for spoiled beef or pork use a pointed knife to reach the interior of the meat. An off-odor on the knife means spoilage.

Left-over food
1. Discoloration.
2. Off-color.
3. Mold.

   Any food that has not been refrigerated below 45°F may be considered slightly spoiled. The off-odor of spoiled food is not always apparent. Do not keep cooked food such as ground meats, hollandaise sauce, cream fillings, cream sauces, custards, or ham, chicken, egg, or fish salads.

   Bacterial spoilage of food begins as soon as it becomes warm. Refrigeration will delay this spoilage, but will not destroy toxins previously formed due to contamination and improper storage.

Salads and desserts

   Chicken salad, tuna, and other fish salads, nonacid potato salad, all types of custard-filled pastries, and some types of cold cuts must be kept refrigerated at all times. All have been touched with the hands during their manufacture and may be considered slightly contaminated.

   Refrigeration will keep contamination from increasing. Spoilage is often impossible to detect until foods are totally spoiled. Serve salads immediately after taking from refrigerator.

Frozen foods

   Frozen foods will spoil if kept out of the refrigerator. Spoilage is caused by growth of bacteria on the food. Thaw poultry and roasts in refrigerator for 2–3 days before use.

   Cook frozen vegetables thoroughly before serving to destroy any contamination that may be present. Do not overcook.

   Heat TV dinners thoroughly, 165°F, before use.

Fruits and vegetables—Fancy, No. 1, No. 2
1. White or grayish powder around stems of fruit and at juncture of leaves and stems of cabbage, cauliflower, celery, and lettuce.
2. Poorly developed leaves, dry or yellow, soft, loose, coarse, sprouted, slime, mold, insect infested, unclean.
3. Fruits soft, blemishes, molded, decayed, worms; citrus fruits soft, light. The powder indicates spray residues.

   Most of the chemicals used by growers are not dangerous but some may be. All fruits and vegetables must be washed before

Source: Adapted from Ohio Department of Health and other sources.

550

TABLE 8-1 TESTS FOR FOOD SPOILAGE OR POOR QUALITY (cont'd)

eaten or cooked. Cooking will not destroy the spray chemicals.

Dressed poultry—Grade A, B
1. Stickiness or rancidity under wing, at the point where legs and body join, and on upper surface of the tail.
2. Darkening of wing tips.
   Dressed poultry should be washed thoroughly before cooking. Wash your hands after handling.

Cereal
1. Insects in cereal.
2. Lumps, mustiness, mildew.
   Spread the cereal on brown paper. If insects are present they will be easily seen. If even one is observed destroy the entire batch of cereal.
   These insects are not dangerous, but neither are they appetizing.

Smoked meats
1. Pale color, soft, moist, flabby.
2. Sausage, bologna, frankfurters slimy, moldy, discolored spots, internal greening.

Cured fish
1. Offensive odor, soft flesh.

Eggs—Grade AA, A, B, C
1. Candling shows large air cell.
   Jumbo 28 oz/doz, extra-large 26 oz, large 24 oz, medium 21 oz, small 18 oz.
2. Cracked, dirty or leaky shell.

Butter—Grade AA, A, B
1. Off-flavor, fishy, stale, unclean, rancid or cheesy flavor—scored 83–87½.
2. Slightly objectionable flavor—scored 89–91½.
3. Made from unpasteurized cream.
   A score of 92 indicates a clean, sweet

butter lacking in rich creamy flavor. A score of 93–100 indicates a clean, sweet, creamy, excellent butter. Butter made from raw milk is dangerous.

Cheese
1. Evidence of contamination or infestation; mold not characteristic of the product.
2. Made from unpasteurized product and not stored at least 30 days above 35°F. Should be made from pasteurized product.

Chemical tests*
1. Solution of malachite dye mixed with ground meat will turn bright red if sodium sulfite added to preserve or mask decomposition of meat. Zinc plus dilute mineral acid gives off H_2S in presence of sulfite.
2. The filtrate, when alcohol is added to meat, will turn red or pink if artificial coloring has been added.
3. Shucked oysters with a pH of 5.8–5.4 suspicious, pH can be 5 in oysters from certain beds. Use a drop of oyster liquor on chlorophenol red test paper and compare with standard. If washed oyster meat turns persistent red when methyl red indicator solution used, oysters are spoiled. Teaspoonful of crab meat in water plus 0.5 ml Nessler Reagent turns deep yellow or brown if meat is spoiled.
4. Cadmium-plated utensils detected by rubbing a swab moistened with 10 percent nitric acid and placing on moistened filter paper impregnated with 20 percent solution of sodium sulfite. A canary yellow stain indicated cadmium.
5. Cyanide is detected by a special test paper that turns orange and then brick red in 5–10 minutes when suspended in air space of bottle containing suspected polish if cyanide exceeds 0.5 percent.

*Walter D. Tiedeman and Nicholas A. Milone, Laboratory Manual and Notes for E. H. 220, Sanitary Practice Laboratory, School of Public Health, University of Michigan, Ann Arbor, 1952, and as revised 1971. Field testing equipment available from La Motte Chemical Products Co., Chestertown, Md.

TABLE 8-1   TESTS FOR FOOD SPOILAGE OR POOR QUALITY (cont'd)

6. Rodent stains fluoresce in ultraviolet light. Cook's test paper placed on stain turns blue when moistened.
7. Arsenic is detected by a test similar to the sulfite determination.
8. The presence of fluoride is detected by a special test paper in the presence of citric acid.

Custard pastries

Custard-filled pies, puffs, eclairs, and similar pastries, to be considered safe, should be

1. Prepared with custard or cream filling that has been heated and maintained for at least 10 min at 190° F or 30 min at 150° F or
2. Rebaked for 20 min at an oven temperature of 425° F or 30 min at 375° F and
3. Cooled to 45° F within 1 hr after heating and maintained below 45° F until consumed.

Only properly pasteurized milk or cream should be used; filling equipment should be cleanable and sterilized before each use, and no cloth filling bags are to be used.

laboratory techniques, it is practical to allow some leeway in the standards proposed and refer to them as guidelines. One might use the geometric mean of, say, ten samples or allow one substandard sample out of four. Parameters used include total aerobic count, toxigenic molds, number of coliforms, number of *E. coli,* coagulase-positive staphylococci, salmonellas, Shigella, *Clostridium perfringens, Clostridium botulinum,* and beta-hemolytic streptococci as indicated.

Some of the guidelines (administrative standards) that have been used, which if exceeded show the need for investigative action, are given in Tables 8–2 and 8–3.

Dry Food Storage

Practically all foods, whether canned, pickled, dried, or chemically preserved, deteriorate on storage. They change in color and texture, develop off-odors or flavors, and lose nutritional value. Storage temperature is the single most important factor that affects the storage life of food items. A temperature of 40°F is good for the storage of canned and dehydrated foods 3 to 5 years. A temperature up to 70°F is acceptable for short-term storage; but the useful storage life of most products at 100°F is 6 months or less. Dried fruits are best stored at 32 to 40°F and 55 percent relative humidity. It is also of practical value to know in connection with food storage that insects become active at a temperature of around 48°F. There are no such things as nonperishable or semiperishable foods; they begin to deteriorate as soon as prepared or packaged. This was a conclusion of the Quartermaster Food and Container Institute for the Armed Forces, sponsored by the Department of the Army Office of the Quartermaster in March 1955. Refrigeration however will prolong storage life of canned food.

TABLE 8-2   SOME MICROBIOLOGICAL GUIDELINES FOR PREPARED FOODS*

Test	Total Count per gram
Standard plate count†	100,000
Coliform‡	20
Fecal streptococcus	1,000
Staphylococcus§	100
Clostridium perfringens	Negative
Salmonellae‖	Negative

*Results should be interpreted in the light of the sanitation inspection.

†Ham and meat sandwiches less than 25,000; salad sandwiches less than 40,000; shelled eggs in laboratory 0 to 3; after shelling at commercial egg-breaking plant prior to freezing liquid eggs at a good plant 24,000 to 2,800,000, average 1,600,000. [Charles Senn, "Microbiology—An Essential in Food Inspection," *J. Environmental Health* (September–October 1964.)]   Frozen cream-type pies not more than 50,000.

‡Count of 20/gram or less is attainable in cooked seafoods and cooked meals that have been frozen. (Committee on Food Microbiology and Hygiene of the International Association of Microbiological Societies—Conference, Montreal, Canada, August 16–18, 1962.) A count of less than 10 is attainable in custard-filled pastries. Foods should be negative for fecal coliform.

§Should be negative for coagulase-positive.

‖Analysis made on finished product when raw ingredient in frozen food is likely to contain *salmonella*.

Special note: Celery is the major source of bacterial contamination of salad. Immersion in boiling water for 30 sec, then chilling under cold tap water reduces the total bacterial count, coliform, enterococci, and staphylococcus coagulase-positive to acceptable levels without impairment of crispness or palatability. [Syed A. Shahid et al., "Celery Implicated in High Bacteria Count Studies," *J. Environmental Health* (May–June 1970.)]

Pasteurization of all liquid whole eggs and egg products is required by the Food and Drug Administration (FDA).

TABLE 8-3   ADMINISTRATIVE GUIDELINES (FDA) FOR CRAB CAKES — COOKED, FROZEN; CRABS — DEVILED, COOKED, FROZEN

1. Sample correlates with establishment inspection showing substantial insanitary conditions, and

2. Examination of a minimum of 10 subs shows any of the following:
   a. Coliform >3.6/g (MPN) in 20% or more of subs, or
   b. *E. coli* >3.6/g (MPN) in 20% or more of subs, or
   c. Coag. pos. staph. >3.6/g in 20% or more of subs, or
   d. Plate count >10,000/g as geometric avg. of subs.

Source: Robert Angelotti, "Catering Convenience Foods—Production and Distribution Problems and Microbiological Standards," *J. Milk and Food Technology* (May 1971), p. 231.

Food-storage rooms generally adjoin unloading platforms and connect as directly as possible with the food preparation areas of the kitchen. The storage room that provides a floor area of approximately 1 to 1½ ft² per meal served per day is usually adequate in size. Storerooms should be clean, cool, dry, dark, ventilated, protected against the entrance of insects and rodents, and organized. Artificial lighting should provide 10 to 20 ft-c of light. The floor should be well drained but without a floor drain, cleanable, and above any possible high-water or sewage flooding. Food must not be stored in basements that might be flooded. If sewer or waste pipes pass through storerooms, the pipes should be provided with drip pans to carry off possible leakage or condensation and food must not be stored directly beneath the pipes. Good ventilation is important. It can be obtained with screened windows opened at the top and bottom, an exhaust fan, a circulating fan, upper and lower wall and door louvered vents that do not open into a hot kitchen.

Storerooms should be furnished with shelves, dunnage, and covered containers. With shelves 10 to 20 in. deep and at least 18 in. away from the walls, so as to be accessible from all sides, and the bottom shelf 12 in. or more above the floor, stock inventory and cleanliness are simplified. Food-storage platforms should be movable, at least 12 in. off the floor and away from the wall. Bulk items are stored in the original packing on dunnage, and loose foods on shelves, in some predetermined order. Sugar, coffee, flour, rice, beans, and other dry stores are stored in metal or other waterproof containers with tight-fitting covers. Old stock should of course be used first; the dating of new supplies will help accomplish this objective. Provide a 3½-ft aisle space between shelving.

Some of the precautions to take in the care of fruits and vegetables are given in Table 8–9. See also "Refrigeration" this chapter, Chapters 1 and 10, and the folded insert in Chapter 1.

MILK SOURCE, TRANSPORTATION, PROCESSING, AND CONTROL TESTS

Milk Quality

The quality of milk reaching the ultimate consumer is largely determined at the farm where the milk is produced. The type of herd, feed, and health of the cows are important; but unless the raw milk is obtained, handled, and stored in a sanitary manner, from the cow to the processing plant and ultimate consumer, the final product will be mediocre or even unacceptable.

A healthy milk herd is expected to be mastitis-, tuberculosis-, and brucellosis[1]-free. Requirements for testing, accreditation, and disposition of reactors or segregation of unsound animals are established by the Bureau of Animal Industry, U.S. Department of Agriculture. Decisions of the federal agency are usually accepted by the state agency having jurisdiction. The milk produced should not be bloody, stringy, or otherwise abnormal, nor should the milk be used within 15 days before calving or 5 days after calving. Milk from cows treated with antibiotics should be withheld from the market until at least 3 days have elapsed.

The reservoir of bovine tuberculosis, although greatly reduced, has not been eliminated; hence the dairy farmer, department of agriculture, and health department must be constantly on the alert against the introduction of the disease in a clean herd. Other illnesses that can be transmitted to man via milk include brucellosis, Q fever, salmonellosis, shigellosis, infectious hepatitis, diphtheria, staphylococcic infections, and streptococcic infections. Calves can be vaccinated against brucellosis. Staphylococci, streptococci, and coliform microorganisms are associated with mastitis. Every person having anything to do with the processing of milk should be free of communicable diseases or running sores.

The production of a clean milk requires that abnormal milk be discarded; that milk be promptly cooled; and that the utensils and equipment used, the udders and teats of all milking cows, as well as the flanks, bellies, and tails, the milker's hands and clothing, and the milking area be clean. In addition, the milker's hands, milk utensils, and equipment, the udders and teats should be rinsed or wiped with an approved bactericidal solution. Although these elementary principles may be obvious, they require continual emphasis and reemphasis with some dairy farmers. If after repeated instruction a dairy farmer cannot be relied on to produce a clean milk, his milk should be excluded from the market.

Dairy Farm Sanitation

Routine inspections of dairy farms by representatives of industry, and occasionally by official agencies, should be made with the owner or manager so that unsatisfactory conditions and practices can be pointed out and discussed. Control programs usually include the collection of milk samples for bacterial examinations. A low-count milk can be produced at every farm. Experience shows that attention to the following procedures will result in a good-quality, low-count milk.

1. Keep milking equipment and utensils clean.
    a. Take all milking equipment apart immediately after each milking and rinse in lukewarm water. Replace equipment and utensils having open seams.
    b. Scrub all parts, including dairy utensils, in hot water with a brush and

---

[1] Also known as Bang's disease, cause of contagious abortion of cattle.

good washing compound, then disinfect. Hot water at a temperaure of 180–190°F is very good, or use a strong (200 mg/l) chlorine solution according to the manufacturer's directions to disinfect rubber, valves, and other parts. Store all equipment on a clean rack in the milkhouse.

c. Remove milk-stone deposits with an acid cleaner. One ounce of gluconic acid or similar milk-stone remover per gallon of disinfecting water will help prevent milk-stone and water-hardness deposits.

d. Replace all worn rubber parts. Two sets of inflations used every other week will last longer.

e. Boil rubber parts for 15 min once a week in a solution of 2 tbsp of lye in a gal of water, or soak rubber gaskets and teat cups, and other rubber pieces, for ½ hr in a solution of hot water containing 2 oz of lye per gallon. Use enamelware pot free of chips and cracks. Rinse the parts and store in a cool, dry, dark place. Disinfect before reuse.

2. Cool promptly all milk in a clean milkroom or milkhouse.

a. Maintain water level above milk level in can cooler, but do not submerge cans. Change water in cooler frequently. Maintain water temperature at 40°F or less and provide circulation of water in cooler: OR

b. Cool milk in a refrigerated farm bulk-milk tank and keep at 36 to 38°F.

3. Prevent external contamination.

a. Brush down cobwebs in barn and control dust.

b. Keep barn and cows clean. Clip and clean flanks, tails, and udders of milking herd. Whitewash barn walls and ceilings at least once a year.

c. Assure proper sewage, wastewater, and manure disposal. Control flies and keep yard drained.

d. Use a protected water supply that is safe and adequate. Contamination with psychrophilic bacteria can affect milk bacterial count and shorten raw-milk storage life.

e. Feed cows odorous feeds after milking.

4. Use good milking procedures.

a. Massage udders with a warm disinfecting solution just prior to milking.

b. Prohibit wet-hand milkings. Wash hands thoroughly; disinfect with a quaternary solution and dry before milking.

c. Use single-service pads for straining milk.

5. Keep the milking herd healthy.

a. Make monthly or quarterly chemical milk-screening tests on herd milk. Cows should be free from tuberculosis, brucellosis (Bang's disease), mastitis, and other diseases. Obtain veterinary diagnosis when mastitis is suspected and treat if necessary.

b. Use a strip cup, check for abnormal milk, and use proper milking procedures.

c. Provide ample, clean, dry bedding.

Detailed inspection report forms and explanations of regulations are available from the U.S. Public Health Service, state and local health departments, and in some instances agriculture departments. Some industries provide supplementary information.

Barn

The milking barn should have 20 ft-c of artificial light, with 50 ft-c at the cow's udder, as well as natural lighting. Adequate air space and ventilation are necessary to prevent overcrowding and condensation, excessive odors and dampness. Hinged windows tilting open at the top and other vertical or horizontal means of ventilation are suitable. A window area of 4 ft²/60 ft² of floor surface and a minimum of 500 ft³ of air space per animal are recommended. Concrete manure gutters and walks at least 3 ft wide with a 6 in. coved curb at walls integral with the floor are easy to keep clean. A hose bib connection (protected against freezing if necessary) in the milking barn, stable, or parlor, connected with a safe water under pressure, simplifies the dairy-cleaning operation. The water outlet should be equipped with a back-flow preventer. In any case, the milking area must be clean and walls, ceilings, and windows constructed of cleanable material that keeps out dirt, flies, and odors. Manure must be spread on the fields or otherwise disposed of at least every 4 days in the fly season. Temporarily, it may be stored in a drained area not closer than 25 ft to any building, although daily removal is preferred. Storage pits, where provided, must be flytight.

Pen Stabling

Pen stabling, also referred to as loafing barn, loose housing, and straw shed, is a method in which the cattle are permitted to roam at will within a bedded area and to an adjoining open space. A feeding and milking area in addition to the milkhouse are necessary adjuncts. The loafing area should be well drained and allow 70 ft² bedded area per animal, plus an additional 25 ft² per animal for a paved feeding area. Mixed manure and fresh bedding as needed to keep bedding dry may be permitted to accumulate for 3 to 6 months to a year and then removed. A more frequent cleaning interval and manure removal before the fly season are recommended. The feeding area usually requires weekly cleaning. The milking area is separated from the loafing area and should be so constructed as to be cleanable. Centrally located milking areas or centers are encouraged. It is claimed that pen stables reduce the risk of injury to cattle, are less expensive to build, save labor, and result in greater milk production since the cattle are free to browse, eat, and rest at will.

Pipeline Milker

A labor-saving device on the farm is the pipeline milker. Milk flows from the teat-cup assembly and hoses of the milking machine to a special glass or stainless steel pipeline terminating in a refrigerated bulk-milk storage tank or other container. This equipment requires special attention to assure it remains clean. Off flavors have been attributed to pipelines. A water heater of adequate capacity and equipment for circulating detergent solution mixed with air at a velocity of at least 5 fps are needed to clean the pipeline in place.[2] At this velocity a 10-gal can is filled in 14 sec by a 2-in. stainless steel line, in 12.5 sec by a 2-in. glass pipeline, and in 22 sec by a 1½-in. glass pipeline. Turbulent flows give better cleaning. Other parts, including the milking machine and teat-cup assemblies, the air lines, air chambers, valves, fittings, and dead ends, are disassembled and immediately brush-cleaned by hand. Pipelines can be cleaned as follows:

1. Prerinse with lukewarm water (100 to 120°F) immediately after use until water at outlet is clear. Do not recirculate.

2. Wash for 10 min by circulating an adequate volume of a proper detergent solution (130 to 140°F) through the system at a velocity that will give turbulent flow. Use an acid detergent solution once a week if alkaline detergent is normally used. The type of washing solution and cleaning procedure should take into consideration the characteristics of the water supply. Rinse out detergent with hot water.

3. Before use, rinse with 180°F water measured at outlet (a separate booster heater for this purpose is recommended), 160°F water containing 50 to 100 mg/l chlorine solution, or other approved disinfectant for at least 20 min; open valves momentarily to allow contact with interior of valve. Cap the pipelines when finished.

4. If the total length of piping is less than about 50 ft, it will probably be easier and more economical to take the piping apart rather than attempt in-place cleaning. The system should be self-draining or dismantled.

Milkhouse

A milkhouse of proper size and construction is a necessary part of a dairy farm. The provision of adequate facilities for the cleansing, disinfection, and storage of utensils and milking equipment and for the

TABLE 8-4   SUGGESTED MILKHOUSE AND BULK-MILK TANK SIZES

Dimensions Inside, (ft)	Size of Tank, (gal)
14 x 14	Up to 350
14 x 18	350 to 850
16 x 20	850 to 1,250
16 x 22	1,250 and longer

Source: R. W. Guest, W. W. Irish, and R. P. March, *Milkhouses for Bulk Tanks,* Agricultural Engineering Extension Bull. 326, Cornell University, Ithaca, N.Y., p. 6.

refrigeration of milk to a temperature of 38°F are basic essentials. Some recommendations for milkhouse and refrigerated bulk-milk tank sizes are given in Table 8-4. A milkhouse floor plan including details is also shown in Figure 8-2.

[2] See pages 613, 618–629 for hot water and heater requirement calculations.

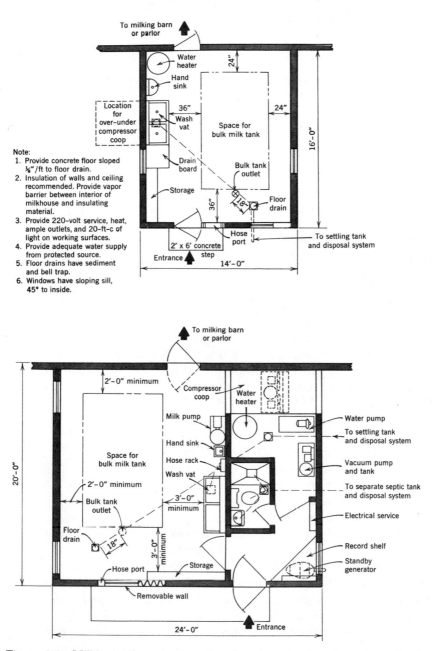

Figure 8-2 Milkhouse layouts show the arrangement of equipment on the floor plan for the small 14′ × 16′ milkhouse or large 20′ × 24′ milkroom with utility room. (Adapted from R. W. Guest, W. W. Irish, and R. P. March, *Milkhouse for Bulk Tanks*, Agricultural Engineering Extension, Bull. 326, Cornell University, Ithaca, N.Y., p. 5.)

Bulk Cooling and Storage

Bulk cooling and storage of milk on the farm reduces handling and the possibilities for contamination, results in rapid cooling of the milk, and requires less space than cans. The cooling is accomplished in a stainless-steel-lined unit that also serves as a storage tank. It is placed 2 to 3 ft from walls or fixtures. The tank is insulated, mechanically cooled, and equipped with a motor-driven agitator, thermometer, thermostat bulb, outlet valve, and measuring stick.

A direct-expansion cooling system would require about 1 hp (1¼ hp with air-cooled condenser) of compressor motor capacity per 50 gal of milk to be cooled at each milking. A chilled water or ice-bank system would require about ⅓ hp of compressor motor capacity for each 50 gal of milk to be cooled. A can cooler would require ⅛ hp per can for cooling both night and morning milk. In any case, farm-refrigerated milk-storage tanks should be designed to cool the milk to 45°F or less within 1 hr, and to 36 to 38°F in 2 hr and should comply with health department regulations.

The theoretical milk-cooling capacity of the cooling system = temperature reduction in degrees F × 8.6 × 0.93 = Btu/hr/gal. Add 15 percent to obtain recommended capacity. Milk with 3½ percent fat and 8.6 percent nonfat solids has a specific gravity at 68°F of 1.033 and weighs 8.60 lb/gal. Milk has a specific heat of 0.93 Btu/lb. The temperature reduction = 90°F − 37°F = 53°F; average = 27°F. The theoretical milk cooling capacity = 27 × 8.6 × 0.93 = 216 Btu/hr/gal. Recommended capacity = 216 + 32 = 248 Btu/hr/gal in a direct-expansion bulk-milk cooling system. In an ice-bank system, 1 lb of ice takes 144 Btu to melt. To cool a gallon of milk from 90°F to 37°F will require the removal of (53 × 8.6 × 0.93) = 424 Btu from the milk, that is, the melting of 3 lb of ice.

Milk is pumped out of a farm milk tank directly into a tank truck through a special hose carried by the hauler. The ends of the hose and outlet valve are sanitized by the hauler, who also rinses the farm tank thoroughly with lukewarm water after all milk is pumped out. Hot and cold water under pressure and a water hose in the milkhouse are practically essential to do a proper job. As soon as possible after rinsing, the agitator, thermometers, outlet valve plug, measuring stick, and strainers should be removed, brushed, and then washed in a warm chlorinated alkaline cleaning solution followed by a cold-water rinse. The tank's inside surfaces and covers require the same treatment. The parts removed are reassembled and the entire tank is sanitized before being placed in use again. Make a visual inspection with a strong light. Use an acid rinse if needed.

Transportation

Milk should be hauled in insulated trucks to the receiving station or milk-processing plant as soon as it is removed from the cooler. In this way milk that has been carefully produced will not deteriorate in quality while in transit.

The transportation of bulk milk is accomplished by insulated tank trucks designed and built to meet established standards. Equipment used to sample, fill, and empty the tank must also be of sanitary design and construction. This equipment and the transportation tank are required to be cleaned and sanitized immediately after being emptied in a room provided for this purpose. The room should be clean, heated, drained, and provided with facilities for washing the milk tank, including an adequate supply of hot and cold water under pressure. Water supply must be of satisfactory sanitary quality and drainage disposal in accordance with local requirements and should not cause a nuisance.

Pasteurization

Pasteurization is the process of heating every particle of milk or milk product to at least 145°F and holding at or above such temperature continuously for at least 30 min., or to at least 161°F and holding at such temperature continuously for at least 15 sec., in approved and properly operated equipment (Figure 8-3). A temperature of at least 150°F for 30 continuous min (155°F for 30 min in practice) or at least 166°F for 15 sec (175°F for 25 sec in practice) is needed for milk products that have a higher milk-fat content than milk or contain added sweeteners. Higher temperatures and shorter times are also being used. The heat treatment should be followed by prompt cooling.

The effectiveness of pasteurization in the prevention of illnesses that may be transmitted through milk-borne disease organisms has been demonstrated beyond any doubt. The continued sale and consumption of raw milk must therefore be attributed to ignorance of these facts. Pasteurization does not eliminate pesticide residues, anthrax spores, or toxins given off by certain staphylococci; but the production of toxins is nil when milk is properly refrigerated. The equipment used to pasteurize milk is of three general types: the holder, high-temperature short-time, and in-the-bottle pasteurizers.

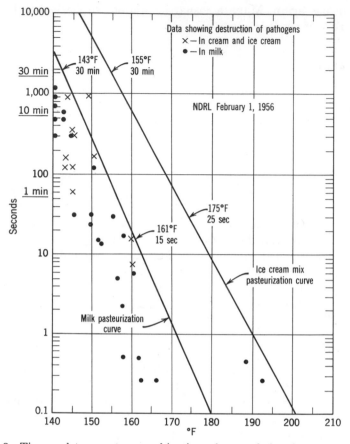

Figure 8–3 Time and temperature combinations that result in adequate pasteurization. Time and temperature combinations read from these curves will give bacterial destruction comparable to that obtained by presently accepted pasteurization treatments. (From DHEW-PHS-BSS-CDC, Atlanta, Ga., April 1959. Based on selected references.)

Holder Pasteurizer

Holder pasteurizers are referred to as batch, pocket, and continuous flow-type pasteurizers. They have a capacity of 100 to 500 gal or more. In the batch-type pasteurizer (Figure 8–4) the milk is heated in the pasteurizer vat or tank and held at 145°F for 30 min, and then the milk is run over a surface cooler and bottled. The milk may be precooled in the pasteurizer vat to about 115°F and then run over the surface cooler. The pocket, or multiple tank-type, pasteurizer is no longer in general use. In the continuous flow-type pasteurizer, which consists of a series of tubes, heated milk is pumped by a calibrated pump from the lowest tube and passed out at the top from the highest tube. All tubes are jacketed.

High-Temperature Short-Time Pasteurizer

The high-temperature short-time (HTST) pasteurizer uses accurate and dependable controls, including the flow-diversion valve, calibrated pump, and heat-exchanger equipment to assure proper pasteurization of large quantities of milk in a short time. In this type of pasteurizer a large number of plates are carefully clamped together, with water and milk on alternate sides. It is compact and makes possible heating, holding, regenerating, and cooling all in one unit. See Figures 8–4 and 8–5.

In-the-Bottle Pasteurizer

In-the-bottle pasteurization is a method for heating milk in the final bottle with hot water or steam to accomplish pasteurization as in the holder type, followed by cooling. Indicating and recording thermometers, circulation of the hot water, auxiliary heating of the air above the milk level, and holding of the bottle and milk inside the bottle at the proper temperature are frequently wanting, making this method of pasteurization questionable unless carefully supervised. In an emergency raw milk can be rendered safe to drink if heated in a water bath to a temperature of 165°F and then immediately cooled.

Vacuum Pasteurizer

A recent development is the vacuum pasteurizer in which every particle of milk or milk product is heated to at least 162°F or up to 200°F. The process involves passage through vacuum chambers that remove off-flavors and odors, as well as cooling, with adequate safety controls.

Ultra-High Temperature Pasteurizer

Ultra-high temperature (UHT) pasteurization of milk and milk products is heat treatment at a temperature of 190 to 270°F, with a holding time of 2 sec or less. The U.S. Public Health Service's *Grade "A" Pasteurized Milk ordinance* (1965) mentions holding times of 1, 0.5, 0.1, 0.05, and 0.01 sec for respective temperatures of 191°, 194°, 201°, 204°, and 212°F. At 270°F milk is sterilized rather than pasteurized. Timing in the holding tube and speed of response of the recorder-controller flow-diversion valve system are critical, requiring sensitive equipment and competent supervision. Testing equipment for regulatory agency use in measuring time-temperature response is not yet available.[3]

[3] Harold Irvin, "UHT Pasteurization," *J. Environmental Health,* **31,** No. 2, 128–131 (September–October, 1968).

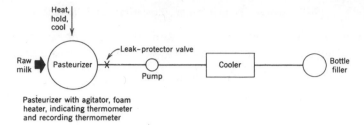

Batch-type pasteurizer

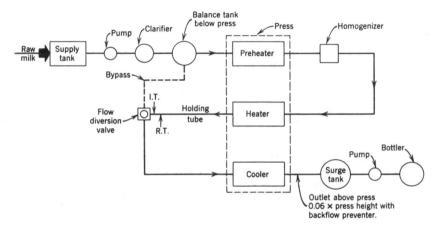

High-temperature short-time pasteurizer

Figure 8–4 Some simple pasteurizer flow diagrams.

Source: "Procedure for Testing Pasteurization Equipment," U.S. Public Health Service, Dept. of HEW, 1969; also Appendix I, *Grade "A" Pasteurized Milk Ordinance,* U.S. Public Health Service, Pub. No. 229, Dept. of HEW, 1965.

Note:

1. I.T. = Indicating thermometer. R.T. = Recording thermometer. Holding tube slopes up at least $\frac{1}{4}''$ per ft. Pump may be substituted for homogenizer. Flow-diversion valve is set to bypass milk when the temperature of the milk drops to $161\frac{1}{2}°F$, and to pasteurize milk when the temperature of the milk raises to $162°F$.

2. Capacity of holding tube $= \dfrac{\pi \times D^2}{4} \times L \times 7.48 = $ "A" gal.

   D = inside dia. of tube in ft; L = length of tube in ft.

3. Theoretical time to fill a 10-gal can and provide 16-sec hold $= x; \dfrac{16}{A} = \dfrac{x}{10}; x = \dfrac{160}{A}.$

4. Test holding time by means of salt conductivity or improved test for precision.

5. Holding time for milk $= \dfrac{1.032(TMw)}{Ww}$, in which:

   1.032 = specific gravity for milk with 4% fat at 68°F = 1.013 with 20% fat and 0.995 with 40% fat at 68°F.

   T = average holding time for water in sec.

   Mw = average time to deliver a measured weight of milk in sec.

   Ww = average time to deliver an equal weight of water in sec.

This process requires built-in safety factors to ensure adequate heat treatment. A pasteurization temperature of 194°F in the vacreator for a calculated hold of 1 sec for milk, and 200°F for 3 sec with frozen desserts in the Roswell Pasteurizer will meet USPHS standards. Because of variations of particle velocity in laminar flow, the holding time or

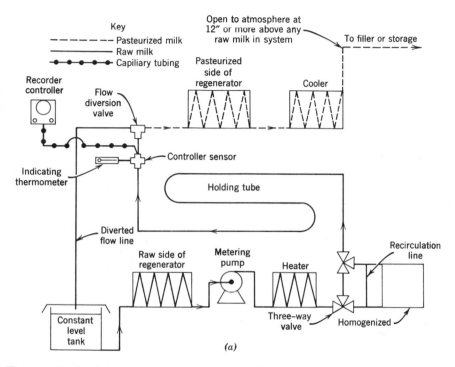

Figure 8–5a Milk-to-milk regeneration pasteurizers. Homogenizer upstream from holder.

tube length is made twice the calculated hold. This can be checked by determining the time to fill a can of known volume. See Figure 8–4.

Cooler

Following pasteurization the milk is promptly cooled below 45°F to keep the bacteria count from materially increasing. This is accomplished by a cooler. A common type is the surface- or external-tubular cooler. It may be a single series of tubes or a cabinet tube type. A surface cooler is usually divided into an upper and lower section. The upper section is generally cooled by water and sometimes chilled raw milk flowing inside the tubes. The lower section usually circulates brine, ammonia, ice water, or "sweet" water that has been cooled. The warm milk is distributed by means of a perforated pipe or trough over the top of the cooler and is cooled as it flows in a thin sheet over the surface of the cooler.

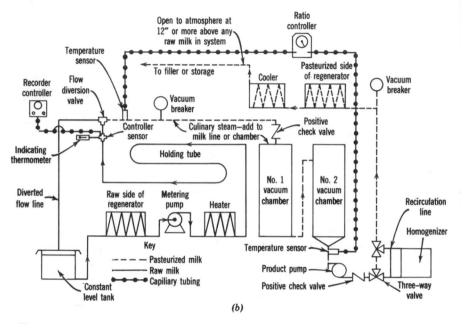

Figure 8–5b Milk-to-milk regeneration pasteurizers. Homogenizer and vacuum chambers downstream from flow dispersion valve. (From *Grade "A" Pasteurized Milk Ordinance,* recommendations of the U.S. Public Health Service, Pub. No. 229, Dept. of HEW, Washington, D.C., 1965.)

Another type of cooler is the internal-tubular cooler. It may be a single series of tubes or a cabinet tube type. In this method there are two concentric tubes; the milk flows in the center tube and the refrigerant flows in the outside tube.

When large volumes of milk are to be cooled, a plate heat exchanger or regenerator is generally used (Figures 8–4 and 8–5). Hot pasteurized milk on one side of the plate is cooled by cold raw milk, water, or brine on the opposite side of the plate.

The cooled milk is conducted to a bottle filler and capper, where the milk is bottled. Bottled milk and filled cartons are placed in cases that are stored in a cooler until delivered. The bottles are filled by means of a rotary-gravity or vacuum-type filler. Empty bottles are fed automatically from a conveyor to the bottle filler. In very small operations a hand-operated bottle filler and capper is used.

Bottle Washer

An important unit of a pasteurizing plant is the bottle washer, unless single-service paper or plastic cartons are used. Bottle washers are of various types, depending usually on the total number of bottles to be washed.

1. At small plants bottle washing is a manual operation. The bottles are soaked, washed by hand on a revolving brush, rinsed, and then sanitized by steam or a hot-water spray in the case.

2. In the larger plant a case washer is usually used. As the name implies, bottles placed in the case are individually subjected to water-pressure cleansing by a washing solution, a rinse, and then a final sanitizing rinse.

3. At large plants a soaker-type washer is used. Bottles are placed in the machine slot or pocket and subjected to a prerinse, soaking in alkali solution for several minutes, an inside and outside alkali spray, sometimes automatic brushing, then a final rinse. The highest temperature reached is about 160°F, which is gradually reduced. The final rinse is fresh cold water, which usually contains a chemical sanitizer.

4. The bottle-washing solution used frequently determines the effectiveness of the entire washing operation. A good washing compound dissolves dried milk and foreign matter, rinses easily, and is an effective disinfectant. Experience shows that sodium hydroxide (caustic soda) is the compound of choice for machine use. It is used as a 2 to 3 percent solution. In hard-water areas the water should first be softened or a suitable detergent used. Addition of ¾ lb of 76 percent commercial caustic soda to one ft^3 of water, or 7½ gal, will make a 1 percent solution of sodium hydroxide.

Cooler and Boiler Capacity

Adequate power plant and refrigeration capacities are necessary to the processing of milk. The advice of persons experienced in the design and operation of these units should be obtained when constructing a new plant or when making major modifications. A table of small boiler and refrigeration capacities for different size plants is given in the table on page 568 as a general guide.

Coolers for the storage of milk are of the walk-in type. Refrigeration is obtained by coils on walls circulating brine, by direct expansion of ammonia in coils on walls, and by unit coolers equipped with coils and a fan. The unit cooler keeps the room drier and provides better air circulation. Freon, carbon dioxide, methyl chloride, sulfur dioxide, and ammonia are common refrigerants.

Plant Capacity, gallons of milk	Steam Boiler Rated Horse-power Capacity	Refrigeration Unit Rated Capacity, tons*
150	15	5
300	27	11
750	42	20
1500	72	23
2500	92	32

* A ton equals 288,000 Btu in 24 hr. This is also the amount of refrigeration accomplished by the melting of 1 ton of ice.

Quality Control

Quality control involves herd health, milk handling, transportation, processing, and distribution. Field and laboratory testing coupled with inspection, supervision, education, surveillance, enforcement, and evaluation are the major methods used.

The tests used to control the quality of milk are explained in detail in *Standard Methods for the Examination of Dairy Products*.[4] The major tests are discussed here. Raw milk quality is determined by temperature; sediment; odor and flavor; appearance; reduction time; direct microscopic counts, including clumps of bacteria, leucocytes, and streptococci; standard plate counts, abnormal milk tests; freedom from antibiotics; and thermoduric determination. Tests for brucellosis and animal health are also made. Pasteurized milk quality is indicated by the standard plate count, direct microscopic count, phosphatase test, coliform test, taste and odor tests. Other common tests are for butterfat, total solids, and specific gravity. See Table 8–5.

Milk should be promptly cooled to 50°F or less within 2 hr after milking and maintained at that temperature until delivered to the receiving station or pasteurizing plant, unless delivered within 2 hr after milking is completed. The importance of proper cooling on the rate of bacterial growth is demonstrated by the results of studies at the Michigan Agricultural Station on freshly drawn milk. Milk having an initial count of about 4300 bacteria/ml decreased to 4100 at 40°F and increased to 14,000 at 50°F and to 1,600,000/ml at 60°F after 24-hr storage. It is apparent therefore that milk that reaches the pasteurizing plant or receiving station 24 hr after milking, at a temperature above 50 to 60°F,

[4] 12th ed., American Public Health Association, 1740 Broadway, New York, 1967.

TABLE 8-5 MAXIMUM ALLOWABLE TEMPERATURE, BACTERIOLOGICAL, AND CHEMICAL STANDARDS

Product	Temp-erature	Bacterial Limit	Coliform Limit	Anti-biotics	Phos-phatase
Raw milk for pasteurization	50°F	100,000† 300,000‡	–	None	Not applicable
Pasteurized milk and milk products§	45°F	20,000‖	10	None	<1 μg¶
Pasteurized cultured products	45°F	Not applicable	10	None	<1 μg¶
Sterilized milk and milk products	–	<1	<1	None	Not applicable

†Individual producer milk. ‡Commingled milk. §Except cultured products but including imitation milk. ‖If greater than 10,000 review handling prior to pasteurization. ¶By Scharer Rapid Method or equivalent as microgram of phenol per milliliter of milk.

1. Bacterial, coliform, and phosphatase results are reported as per milliliter. The bacterial limit may be a standard plate or direct microscopic count.

2. Rinse of empty milk bottle, carton, or other container shall be free of coliform organisms and contain less than 1 bacteria per milliliter. Swab counts from 5 different 8-in.^{2}-areas shall be free of coliforms and not exceed 250 colonies/40 in.^{2}.

3. Sediment. Sample from off-bottom unstirred 40-qt can of milk shall contain less than 2 mg of sediment when compared to photographs of standards.* Sample of stirred milk from a can, weight tank, farm bulk-milk tank, plant storage tank, or transportation tank shall not exceed the 1 mg standard.

4. Temperature of sample in transit and as received at an official or officially designated laboratory must be less than 50°F.

5. When one or two of the last four consecutive bacteria counts, coliform results, or cooling temperatures taken on separate days exceeds the standard, send written notice. When three of last five exceed standard, suspend permit and withdraw product.

6. When phosphatase standard is exceeded, immediately determine cause. Milk involved shall not be sold. With the Scharer Rapid Phosphatase Test, a value of one microgram or more of phenol per milliliter of milk, milk product, reconstituted milk product, or cheese extract would indicate improper pasteurization or contamination with unpasteurized product. See *Standard Methods for the Examination of Dairy Products* for other tests.

7. Pesticide residues of up to 0.05 ppm in whole milk and 1.25 ppm on a milk-fat basis in manufactured dairy products are permitted under FDA tolerance levels. The maximums apply to DDT and its chemical degradation products DDD and DDE or any combination.

*Obtainable from Photography Division, Office of Information, U.S. Dept. of Agriculture, Washington, D.C.

will probably have a high bacteria count. Although fresh raw milk has bacteriostatic properties that help retard bacterial growth, the duration of this factor is variable.

The sediment test shows the amount of extraneous material in milk but will not show dissolved material. A pint of milk from the bottom of an unstirred 40-qt can is strained through a Lintine or similar disc of absorbent cotton. When milk in a farm bulk-milk tank, plant storage tank, or transport tank is to be tested, a 1-gal well-mixed sample warmed to 90 to 100°F is collected. The color of the stain produced, from yellow to brownish black, and particles retained are a simple visual indication of the amount of dirt and abnormal substances in the milk. The test and its interpretation is given in *Standard Methods*. Failure to properly wash a cow's udder just prior to milking is the common cause of high-sediment test results. A clarified or strained milk will not show the dirt, hence the test may only encourage straining of the milk at the farm rather than cleaner milking procedures. However, other tests will also reveal insanitary practices. The test is sometimes used on processed milk.

Odor, flavor, and appearance are physical tests usually made on the receiving deck or at the bulk-milk tank at the same time sediment tests are made. They are referred to as platform tests. Experienced inspectors can make very rapid and accurate determinations by sight and smell on the quality of a milk. Sour milk, dirt, and odors associated with mastitis, improper cooling, dirty utensils, horses, consumption of leeks and other weeds, or a disinfectant are easily detected. Feed flavors can be prevented by the feeding of cows after milking or at least 5 hr before milking; removal of cows from a lush pasture 3 hr before milking; elimination from pastures of wild onions, leeks, skunk cabbage, and certain mustard weeds; and storage of silage or strong-smelling feed away from the cows. Barny or musty flavors and odors can be controlled by keeping the cows, milkers, and stables clean. Good ventilation and dry conditons without dust are also important. Salty flavors are due to the use of milk from cows infected with mastitis or from cows being dried off (strippers). This milk should be discarded. Rancid milk is caused by use of milk from stripper cows, milk from cows late in lactation, slow cooling of milk, and excessive agitation. A malty flavor is caused by high bacteria count, dirty utensils and cows, poor cooling, and slow cooling. Very high bacteria count, dirty utensils, and slow or poor cooling cause sour milk. An oxidized or cardboard flavor is caused by milk coming in contact with copper, copper alloys, or iron equipment; exposure of milk to sunlight; copper or iron in water supply; and possibly milk from special cows. Medicinal flavors are due to certain medications for the teats, creosote-base disinfectant barn sprays, and insect sprays used just before or during milking.

Reduction tests—when it is not possible to collect and examine a large number of samples for bacterial examinations, an indication of the sanitary condition of the milk can be obtained by means of the methylene blue reduction test or the resazurin test.

In the methylene blue reduction test 1 ml of fresh methylene blue thiocyanate solution is added to a 10-ml sample of milk and incubated at $36°C \pm 1°C$ (98.6°F). Observations for loss of color are made after 30 min and at hourly intervals thereafter, and results are recorded as MBRT in whole hours. A proposed USDA standard for manufacturing raw milk sets limits of $2\frac{1}{2}$ and $4\frac{1}{2}$ hr.

In the resazurin reduction test, resazurin dye is used in place of methylene blue. The reduction time to 5P 7/4 Munsell color end-point (triple reading test requiring 3 hr) should not be less than $2\frac{3}{4}$ hr for Grade A milk. A proposed USDA standard for manufacturing raw milk establishes reduction time limits of $1\frac{1}{2}$ and $2\frac{1}{4}$ hr. Advantages of the resazurin test are that it takes less time to perform; the presence of leucocytes is indicated, although not reliably; and color readings are more distinct.

Comparisons between the reduction time and microscopic and plate counts have been attempted, but exact agreement is not expected. In general the time for a sample to change from blue to white with the methylene blue test, or purple to pink with the resazurin test, is inversely proportional to the bacterial content of the sample when incubation starts. The resazurin test must be used on fresh milk, not over 24 hr old. The reduction tests are not especially suitable on low-count, well-cooled milk, but are effective in detecting poor-quality milk.

The direct microscopic count (DMC) tells the number of isolated bacteria and groups of bacteria in stained films of milk dried on glass slides, as determined with the aid of a compound microscope. It is a rapid test used on small batches of raw milk to show the types of bacteria present and the possible source as dirty utensils (bacterial clumps), infected udders, or poor cooling (bacteria in pairs) provided the source, age, and temperature of the milk are known. On small volumes of pasteurized milk it is possible to detect poor-quality raw milk, the presence of leucocytes and sometimes thermoduric, psychrophilic, thermophilic, and other types of bacteria that do not normally grow on standard agar plates, and contamination after pasteurization. Generally most dead bacteria disintegrate within several hours after pasteurization; those that remain do not stain well, and hence only a few of the dead bacteria are counted when the direct microscopic count is made. *Streptococcus lactis* in pairs, threes, or double pairs of oblong cocci generally indicate poor cooling or unclean cans. Masses of bacteria indicate unclean utensils, milking machines, or tubing.

This test is of limited value when made on mixed milk from bulk-milk tanks in view of the large dilution and generally effective cooling. The DMC method should not be used on very low-count milk.

The direct microscopic somatic cell count (DMSCC) is a more precise test than DMC. It gives the total leucocyte and epithelial cells in a sample of raw milk. A leucocyte cell count over 500,000/ml or a total cell count of 1,000,000/ml, together with long-chain streptococci, gives strong indication of infected udder and effects of mastitis.

Screening tests for the detection of abnormal milk.[5] Various tests are used, such as the California Mastitis Test, the Wisconsin Mastitis Test, the Modified Whiteside Test, and Catalase Test. However, only the DMSCC should be used as a basis for legal action and possibly the shutting off of a farm milk supply. Elimination of the cause includes milking hygiene, dipping of teats in 4 percent hypochlorite solution after milking, and penicillin treatment of all quarters at the time of drying off. Veterinary assistance may be needed to bring difficult problems under control.

The standard plate count (SPC) shows the approximate number of bacteria and clumps of bacteria that will grow in 48 hr on a standard medium held at a temperature of $32°C \pm 1°C$. Wide variations are common. The plate count is especially suited to products having low bacterial densities and as a measure of the bacterial quality of certified milk and pasteurized milk and milk products, except fermented milk products. It is also recommended for process sampling. High counts on freshly pasteurized milk suggest that thermoduric bacterial counts be made on producer samples to learn the source. Recent studies show that the SPC is superior to the DMC in detecting poor-quality raw milk and hence is the method of choice, particularly with bulk-tank milk. But excessive numbers of psychrophilic bacteria are not detected by the routine plate count. Incubation at a temperature of $7°C \pm 1°C$ $(44.6°F)$ for 10 days is suggested.

Coliform test—practically all raw milk will show the presence of the coliform group of organisms, usually less than 100/ml in high-quality milk. Where pipeline milkers and farm bulk-milk tanks are used, the coliform count is a measure of utensil sanitization, udder cleanliness, and milking hygiene. Properly pasteurized milk will usually show the

---

[5] See *Screening Tests for the Detection of Abnormal Milk,* U.S. Public Health Service Pub. No. 1306, Dept. of HEW, Washington, D.C., 1965.

absence of coliform organisms. A positive test for coliform organisms in pasteurized milk greater than 10/ml (should be less than 3) is an indication of improper processing or excessive contamination following pasteurization by improperly cleaned and disinfected equipment, utensils, or dripping of condensate into pasteurized milk. Dust, exposure of the pasteurized milk to the air, flies, or other insects, and poor storage may also contribute coliforms. Some authorities claim that strains of coliform organisms exist that are heat resistant and are not completely destroyed by pasteurization.

The phosphatase test shows whether milk has been properly pasteurized and whether it has been contaminated with raw milk after pasteurization. Phosphatase is an enzyme present in all raw cow's milk and is almost completely inactivated by proper pasteurization. The enzymes surviving pasteurization release phenol from disodium phosphate substrate. Upon the addition of 2, 6 dichloroquinonechloroimide (CQC), the liberated phenol reacts with CQC, producing an indophenol blue color, the intensity of which depends on the amount of enzyme present and hence the degree of pasteurization or raw milk added. In the Scharer rapid phosphatase test, a reading of 1 microgram (μg) or more of phenol per milliliter of milk, milk product, reconstituted milk product, or cheese extract is an indication of improper pasteurization or contamination with raw milk or milk product. Other methods described in *Standard Methods* are the modified spectrophotometric phosphatase method, the Cornell phosphatase test, and the dialysis phosphatase test. Sale of underpasteurized products should be prohibited.

A false positive phosphatase test may be obtained on HTST or UHT pasteurized milk or milk products such as chocolate milk, cream or other high-fat products, and old cream having a high bacteria count, particularly when not continuously or adequately refrigerated. Differentiation of reactivation, from residual phosphatase, is explained in *Standard Methods*. Sterile milk products that have been allowed to warm up may require microbiological tests to determine the quality.

An overpasteurized milk could conceivably have added to it a very small quantity of raw milk that would not be detected by the phosphatase test; but this would undoubtedly introduce coliform organisms that would be detected by the coliform test. Pathogenic bacteria likely to be found in raw milk are destroyed more rapidly than the phosphatase enzyme.

The phosphatase test is the best single indicator of the safety of milk. When combined with the coliform test and bacteria count, a fairly complete indication is obtained of the sanitary quality of milk.

Psychrophiles or *cryophiles* are a species of bacteria that can grow commonly within the range of 35 to 50°F. Raw milk and cream are particularly susceptible if stored for 2 or 3 or more days. Psycrophiles are introduced into the milk by organisms (Pseudomonas) on dirty equipment, unclean hands, or in a contaminated water supply used to rinse the equipment. Milk stored at a temperature below 40°F will retard the growth of psychrophiles; higher temperaures will favor their growth, depending on genera and species. Pasteurization will normally destroy most psychrophiles present in raw milk. Their presence in pasteurized milk is generally an indication of postcontamination, but may also be due to an excessive number of psychrophiles in the raw milk, inadequately cleaned and sanitized contact surfaces, or use of a contaminated water supply. They also cause flavors in milk and poor keeping quality. This becomes important in the holding of raw milk and every-other-day milk delivery. Coliform microorganisms grow well but slowly with time in refrigerated pasteurized milk. Proper cleaning and sanitizing of equipment in the pasteurization plant, protection of the equipment from contamination, use of a water supply of satisfactory sanitary quality, and milk storage below 40°F will control psychrophiles. Use of a chlorinated wash-water supply containing 5 to 10 mg/l available chlorine is effective in controlling these bacteria.[6]

Thermoduric bacteria withstand pasteurization at 145°F and 161°F. They grow best at a temperature of 70 to 98°F. Milk that is not properly refrigerated at the farm would permit the growth of thermodurics. The cow's udders, improperly cleaned utensils and milking machines, feed, manure, bedding, and dust are sources of thermoduric bacteria. Preheating equipment, pasteurizers, bottles, and piping that have not been properly washed and sanitized can harbor thermoduric bacteria.

When high standard plate counts are reported on pasteurized milk or cream, laboratory pasteurization of individual producer raw milk samples contributing to a batch will reveal the presence and producer source of thermoduric bacteria. *Most thermophilic bacteria* grow at a temperature of 113 to 158°F. Their optimum temperature for growth is 131°F but they can grow at 98.6°F or lower. The presence of thermophiles in large numbers may be due to repasteurization of milk or cream, prolonged holding of milk or cream in vats at pasteurization temperatures, stagnant milk in blind ends of piping at pasteurization temperatures, continuous use of preheaters, long-flow holders or vats for more than 2 to 5 hr without periodically flushing out equipment with hot water, pas-

---

[6] J. C. Olson, Jr., R. B. Parker, and W. S. Mueller, "The Nature, Significance, and Control of Psychrophilic Bacteria in Dairy Products," *J. Milk and Food Technol.*, **18**, No. 8, 200–203 (August 1955).

sage of hot milk through filter cloths for more than 1 to 2 hr without replacing cloth, residual foam on milk that remains in vats when emptied at end of each 30-min holding period, and growth of thermophiles in milk residues on surfaces of pasteurizing equipment.

Mesophilic bacteria grow at a temperature of 59 to 113°F. They prefer a temperature of 98.5°F. Pathogenic bacteria are in this group.

Agglutination-type tests including the milk ring test are used to determine the presence of brucellosis. The ring test is made on samples of mixed milk from a herd, usually as delivered to the receiving station. A positive milk ring test can then be followed up by individual blood testing of the infected herd to determine the responsible animals. This makes possible more effective use of the veterinarian's time. The milk ring test will detect one infected animal in a herd of 40 when a sample of the pooled milk is examined.

MILK PROGRAM ADMINISTRATION

The milk industry and the regulatory agencies have a joint responsibility in ensuring that all milk and milk products consistently meet the standards established for protection of the public health. Inspection duplication should be avoided; instead there should be a deliberate synergism of effort. With proper planning and cooperation the industry, local, state, and federal systems can actually strengthen the protection afforded the consumer. The role of industry and official agencies to accomplish the objective stated is described below.

Certified Industry Inspection

Industry quality-control inspectors are qualified by the official agency (usually health or agriculture) based on education, experience, and examination to make dairy-farm inspection pursuant to the milk code or ordinance. Certificates are issued for a stated period of time of 1 to 3 years, may be revoked for cause, are renewed based on a satisfactory work record, and may require participation in an annual refresher course. Copies of all inspections, field tests, veterinary examinations, and laboratory reports are promptly forwarded to the official agency or to an agreed on place and kept on file at least one year.

Cooperative State-Public Health Service Program for Certification of Interstate Milk Shippers (IMS)

The voluntary federal-state program is commonly referred to as the IMS Program. A state milk sanitation rating officer certified by the U.S. Public Health Service makes a rating of a milk supply. The name of the supply and rating is published quarterly by the USPHS. If the milk and milk products are produced and pasteurized under regulations that are substantially equivalent to the USPHS Milk Ordinance, and are given an acceptable milk sanitation compliance and enforcement rating, they may be shipped to another area of judrisdiction that is participating in the IMS Program.

The procedures for rating a milk supply are carefully designed with detailed instructions to be followed by the industry and the rating officer. Independent evaluations are made by USPHS rating officers to confirm ratings given, or changes since the last rating, and to ensure reproducibility of results.

Official Local Program Supervision and Inspection

The official agency makes regular review of the industry inspection files mentioned above, takes whatever action is indicated, and at least annually makes joint inspections with the industry inspector of a randomly selected significant number of dairy farms, including receiving stations. The quality of work done is reviewed, the need for special training is determined, and recommendation concerning certificate renewal is made to the permit-issuing official. A similar review is also made of the sample collection; transportation; and the procedures, equipment, and personnel in the laboratory making the routine milk and water examinations.

In addition the local regulatory agency collects official samples as required by the state milk code, advises the industry having jurisdiction of the results and corrective action required, participates in training sessions, and serves as the state agent in securing compliance with the state milk code. The authorized local city or county agency usually has responsibility for the routine inspection of processing plants, sample collection, and overall program supervision for compliance with the code. The agency sanitarian serves as a consultant to the industry in the resolution of the more difficult technical, operational, and laboratory problems. This whole procedure makes possible better use of the qualified industry inspector and the professionally trained sanitarian, with better direct supervision over dairy farms and pasteurization plants and more effective surveillance of milk quality. In some states the local activities are carried out in whole or in part by the state regulatory agency.

Official State Surveillance and Program Evaluation

The state department of health, and in some instances the state department of agriculture, share responsibility for milk sanitation, wholesomeness, and adulteration. The responsibility is usually given in state law, sanitary code, or milk code, and in rules and regulations promulgated pursuant to authority in the law. Most states have adopted the *Grade "A" Pasteurized Milk Ordinance*, 1965 Recommendations of the USPHS, or a code that is substantially the same. This makes possible a reasonable basis for uniformity in both interstate and intrastate regulations and interpretation. However, short-term and alleged economic factors frequently limit reciprocity and interstate movement of milk. Both industry and regulatory agencies need to cooperate in the elimination of milk codes as trade barriers. Health agencies usually have no objection to reciprocity where milk-quality compliance and enforcement is certified under the IMS Program. However, not all states or farms participate in the program.

The IMS Program also gives the state regulatory agency a valuable tool to objectively evaluate the effectiveness of the local routine inspection and enforcement activities. The types of additional training needed by the qualified industry inspector, the assistance the dairy farmer should have, and the supervisory training needed by the regulatory agency sanitarian become apparent. Changes in technology and practices are noted, and the need for clarification of regulations, laws, and policies are made known to the state agency.

A common function of a state regulatory agency is periodic in-depth evaluation of local milk programs. This includes the effectiveness with which the local unit is carrying out its delegated responsibilities as described above, quality of work done, staff competency and adequacy, reliability of the official laboratory work, record keeping, equipment and facilities available, number of inspections and reinspections made of dairy farms and processing plants, and their adequacy or excessive frequency. The state agency usually has responsibility for approving all equipment used in milk production and service from the cow to the consumer. The standards recommended by national organizations are generally used as a basis for the acceptance of equipment. See pages 580 and 581 and footnotes on pages 594 and 595.

HOSPITAL INFANT FORMULA

Diarrhea of the newborn has been responsible for deaths in hospital nurseries. *Escherichia coli* is commonly recovered from the stool. Enteroviruses and *Salmonella* are also recovered; *Shigella* rarely. Sanitary engineers and sanitarians have been effective in applying their knowledge and experience in the control of milk- and food-borne illness to hospital formula-room sanitation.

The control program should be presented as a consultation service furnished through the cooperation of the hospital administrator and the health department. The health department would make available its combined epidemiological, nursing, medical, sanitation, and laboratory resources. After the initial surveys and compliance with accepted standards, the hospital would be expected to carry on its own inspection program; but the health department team would be utilized whenever a problem appeared. A baby formula-inspection program should include the following:

1. Establish appropriate lines of communication between the hospital personnel and the health department to ensure continuing supervision and consultation service.

2. Assist in establishing procedures to minimize contamination of baby formulas, bottles, and nipples during the handling and bottling processes, and assist in the control of formula constituents.

3. Assist in establishing procedures designed to ensure that terminal sterilization is adequately carried out and that infections are not introduced into the nursery through poor hygiene and sanitation practices.

4. Assist in establishing procedures to ensure proper handling of the terminally sterilized product until time of consumption by the infants.

5. Arrange for routine weekly bacteriological testing of one bottle of each product out of every 50 prepared or purchased, the test specimen to be selected and collected by health department personnel for laboratory examination. Emphasis is to be placed on the sanitary survey and day-to-day surveillance, the bacteriological testing being only ancillary and confirmatory to the techniques followed and equipment provided.

The *Control of Communicable Diseases in Man,* gives additional information.[7] A total plate count of less than 10/ml in terminally heated formula is easily obtainable, and a count of 3 or less is practical.[8] Rinse samples of 8-oz bottles that have only been properly washed should show a plate count of less than 300/ml, with no coliform microorganisms present.

Precautions to follow include:
1. Discard scratched and chipped bottles as well as old and porous nipples.

2. Thoroughly rinse all bottles and nipples with lukewarm water immediately after use to simplify and make more effective the subsequent cleansing operation.

3. Each week treat all bottles and nipples by soaking in a 1-percent acid solution to prevent the buildup of mineral deposits, by the use of a milk-stone remover found effective in the dairy industry.

4. Use a suitable detergent, followed by a clear water rinse, based on the water characteristics, in washing bottles and nipples (never use green soap or bar soap).

[7] American Public Health Association, 1740 Broadway, New York, N.Y. 10019, 1970.
[8] J. A. Salvato and Louis Lanzillo, "Infant Formula Inspection Program as an Aid in the Prevention of Diarrhea of the Newborn," *J. Milk and Food Technol.,* **20,** 127–131 (May 1957).

5. After washing and rinsing nipples, boil in water for 5 min.

6. Use a maximum-registering thermometer in each batch of formulas terminally heated.

7. Assure that bottles used in the nursery are not mixed with baby bottles from other parts of the hospital.

8. See that daily records are kept showing the maximum temperature of the maximum-registering thermometer for each batch of formulas together with the pressure and temperature during terminal heating, the number of formulas and other fluid bottles heat-treated, and the temperature of the refrigerator in which formulas are stored.

9. Assure that materials for formula preparation are stored in a separate locked room.

Central preparation, sterilization, and distribution of formulas in single-service bottles simplify control. The infant formula-inspection and control activity is only one part of the program for the prevention of diarrhea of the newborn. It should be integrated with special medical and nursing preventive techniques as well as the sanitary engineering aspects of air sanitation, cross-connection, and back-siphonage studies.

RESTAURANTS, SLAUGHTERHOUSES, POULTRY DRESSING PLANTS, AND OTHER FOOD ESTABLISHMENTS

Supervision of food establishments such as restaurants, caterers, commissaries, pasteurizing plants, frozen-dessert plants, frozen prepared-food plants, vending-machine centers, slaughterhouses, poultry processing, bakeries, shellfish shucking and packing plants, and similar places is in the public interest. This responsibility is usually vested in the state and local health and agriculture departments and also in the Food and Drug Administration, the U.S. Department of Agriculture, the U.S. Public Health Service, and in the U.S. Department of Interior. The industry affected as a rule also recognizes its fundamental responsibility.

Basic Requirements

It is to be noted that certain basic sanitation requirements are common to all places where food is processed. McGlasson has proposed a set of standards under the following headings:[9]

1. Location, construction, facilities, and maintenance: (a) Grounds and premises. (b) Construction and maintenance. (c) Lighting. (d) Ventilation. (e) Dressing rooms and lockers.

2. Sanitary facilities and controls: (a) Water supply. (b) Sewage disposal.

[9] E. D. McGlasson, "Proposed Sanitation Standards for Food-Processing Establishments," Association of Food & Drug Officials of the United States, Vol. 31, No. 2, April 1967.

(c) Plumbing. (d) Restroom facilities. (e) Hand-washing facilities. (f) Food wastes and rubbish disposal. (g) Vermin control.

3. Food-product equipment and utensils: (a) Sanitary design, construction, and installation of equipment and utensils. (b) Cleanliness of equipment and utensils.

4. Food, food products, and ingredients: (a) Source of supply. (b) Protection of food, food products, and ingredients.

5. Personal: (a) Health and disease control. (b) Cleanliness.

A similar set of basic standards was published in the *Federal Register* in April 1969.[10] The above general sanitation requirements should be supplemented by specific regulations applicable to a particular establishment or operation. Excellent codes, compliance guides, and inspection report forms are available. Some additional valuable sources of information are:

1. *Grade "A" Pasteurized Milk Ordinance,* 1965 Recommendations of the USPHS, Pub. No. 229.

2. *Food Service Sanitation Manual,* 1962 Recommendations of the USPHS, Pub. No. 934.

3. *Frozen Desserts Ordinance and Code,* recommended by the USPHS, 1958.

4. *Poultry Ordinance,* 1955 ed., USPHS Pub. No. 444, *Recommendations Developed by the Public Health Service in Cooperation with Interested States and Federal Agencies and the Poultry Industry.*

5. *AFDOUS Frozen Food Code,* Association of Food & Drug Officials of the United States, Vol. XXVI, No. 1, January 1962. (Adopted June 22, 1961.)

6. *Recommendations and Requirements for Slaughtering Plants,* D. E. Brady, Merle L. Esmay, and John McCutchen, Missouri College of Agriculture, Bull. No. 634, September 1954.

7. *Manual of Recommended Practice for Sanitary Control of the Shellfish Industry,* U.S. Public Health Service, Part I, 1959, Part II, 1962.

8. *The Vending of Food and Beverages—A Recommended Ordinance and Code,* 1965 Recommendations of the USPHS.

9. *A Sanitary Standard for Manufactured Ice,* 1964 Recommendations of the Public Health Service Relating to Manufacture, Processing, Storage and Transportation. U.S. Dept. of HEW, Washington, D.C.

10. *Sanitation Standards for Smoked Fish Processing,* Part I, Fish Smoking Establishment 1967 Recommendations, U.S. Dept. of the Interior, Fish and Wildlife and Parks, and U.S. Dept. of HEW, Public Health Service, Washington, D.C.

11. *Voluntary Minimum Standards for Retail Food Store Refrigerators,* Health and Sanitation, (CRS–S1–67), Commercial Refrigerator Manufacturers Association, 111 West Washington St., Chicago, Ill.

12. *Sanitary Standards for Food-Processing Establishments,* 1969 Recommendations for study purposes only, Public Health Service, U.S. Dept. of HEW, Environmental Sanitation Program, NCUI, Cincinnati, Ohio.

13. *Bakery Establishments,* 1969 Recommendations, Addendum B for study pur-

---

[10] Title 21—Food and Drugs, Chapter 1—Food and Drug Administration, U.S. Dept. of HEW, Part 128—Human Foods; Current Good Manufacturing, Processing, Packing, or Holding, *Federal Register,* April 25, 1969.

poses only, Public Health Service, U.S. Dept. of HEW, Environmental Sanitation Program, NCUI, Cincinnati, Ohio.

14. *Beverage Establishments,* 1969 Recommendations, Addendum C for study purposes only, Public Health Service, U.S. Dept. of HEW, Environmental Sanitation Program, NCUI, Cincinnati, Ohio.

15. *Ordinance and Code Regulating the Processing of Eggs and Egg Products,* 1968 Recommendations, U.S. Dept. of HEW, Environmental Sanitation Program, Cincinnati, Ohio.

16. *Manual of Inspection Procedures of the U.S. Dept. of Agriculture,* Consumer and Marketing Service, Meat Inspection, July 1968, Washington, D.C.

Manuals and guides are not a substitute for the exercise of intelligent and mature judgment. They are, however, indispensable administrative aids that, with continual revision and supervision of field personnel, will help maintain uniform quality enforcement of a sanitary code. The reasonable and intelligent interpretation of the regulations and what represents good practice requires an understanding of the basic microbiological, engineering, and administrative principles involved. These are discussed below and throughout this text.

Inspection and Inspection Forms

Routine or frequent inspections of food-processing and food-service establishments alone will not be adequate to ensure the maintenance of proper levels of sanitation. This must be supplemented by education, motivation, persuasion, legal action, management supervision, qualified workers, laboratory sampling at critical points, and quality (including flavor) control. The primary responsibility for cleanliness and sanitation rests with management. Inspection frequency, enforcement, administration, and program evaluation are discussed in Chapter 12.

Inspection report forms are valuable tools to assure complete investigations and uniform policy procedures. See Figures 8–6 through 8–10. They can also be made available to the industry to help in self-policing and in greater cooperation between the industry and official agencies. Preparation of an inspection form that is self-explanatory in showing what is acceptable, supplemented by a satisfactory compliance guide, can have a very desirable educational effect. A report form checked off entirely in the "Yes" column would indicate the establishment is in substantial compliance with existing regulations. Items checked "No" would indicate deficiencies that should be followed up on subsequent reinspections. When these conditions are corrected, the third column, "CM," would be checked, indicating correction made of a previously reported deficiency. The date could be inserted in the CM square in place of a check mark to show when the correction was made. It is usually good practice to

Establishment Name _____ Location _____

Owner _____ Address _____ Date of Inspection _____

Type _____ Maximum No. Meals served daily _____ Seating Capacity _____

Item check 'Yes' satisfactory; 'No' unsatisfactory; 'CM' correction made. Explanation of defects numbered on back.

1. *Food-drink*	Yes	No	CM
(a) Storage—no contamination by overhead pipes; not on floor subject to flooding, dust, depredation by rodents			
(b) Protection—readily perishable foods stored at 45°F or less or 140°F or above; Water from refrig. equip. properly drained			
(c) Display—no open displays			
(d) Handling—minimum manual contact with food or drink			
(e) Milk—milk products from approved source: milk, etc., served in original bottles; dispensed from bulk dispenser; oysters, clams, mussels, app'd. source: if shucked, kept in original container			
2. *Utensils* (multiuse—single serv.)			
(a) Construction—all multiuse utensils easily cleanable, kept in good repair; no corrosion, open seams, chipped, or cracked utensils; no cadmium or lead utensils			
(b) Cleanliness—eating and drinking utensils thoroughly cleaned after each use; other utensils cleaned each day; single service utensils used once			
(c) Storage and handling—stored above floor in clean place protected from splash, dust, flies, etc., inverted or covered where practical; no handling of contact surfaces; single service utensils stored and handled properly; dispensing spoons, dippers kept clean in running water			
3. *Equipment*			
(a) Construction—easily cleanable, no corrosion, good repair			
(b) Cleanliness—clean cases, counters, shelves, tables, meat blocks, slicers, refrigerators, stoves, hoods			
(c) Location—properly located to facilitate cleaning around and beneath the equipment			
4. *Washing facilities*			
(a) Manual type—3 compartment sink and long-handled wire baskets: water for disinfection maintained at 180°F or higher for 2 minutes; satis'. chem. disinfection where hot water is not practical			
(b) Mechanical type—wash water maintained at 150-160°F; wash cycle not less than 40 sec.; rinse water maintained at 180°F or higher; rinse cycle not less than 10 sec; rinse water flow pressure controlled between 15 and 20 psi; wash sol. OK			
5. *Premises*			
(a) Floors—easily cleanable construction, smooth, good repair; clean			
(b) Walls and ceiling—all: clean, good repair, walls smooth and washable			
(c) Lighting and ventilation—all rooms in which food is prepared, stored, or in which utensils are washed shall be well lighted and ventilated			
(d) Clean, sanitary, orderly			

	Yes	No	CM
(e) Doors and windows—all openings into outer air shall be effectively screened and doors shall be self-closing			
(f) Insect and rodent control—flies, roaches, and rats under control; structure rat proofed; poisons stored in locked cabinet			
6. *Employees* (Male_____Female_____) Clean outer garments; hands, arms clean; no tabacco used where food is prepared			
7. *Toilet—lavatory facilities* Comply with plumbing code; adequate, conveniently located for employees: good repair, clean, warm and cold running water; well lighted; outside ventilation; self-closing door; soap and individual towel service			
8. *Garbage—refuse* Stored in tight nonabsorbent washable receptacles; tightly fitted covers provided; receptacles covered; proper disposal			
9. *Water supply* (Public___Private___) Running water accessible as required; supply adequate and of a safe sanitary quality; no submerged inlets or cross connections			
10. *Disposal facilities* (Public_____ Private___) Liquid wastes into public sewer or satis. private system; no interconnection with washing machines or sinks, kettles, warm tables, coolers			
11. *Disease control* No person at work with any communicable disease, sores, or infected wounds			
12. *Ice* Source, storage, handling satis.			

Water heater(s):

Storage in gal _____

Recovery capacity in Btu per hr _____

Therm. setting _____ °F

Adequate _____

Refrig. capacity:

Tons per 24 hr _____

Cubic feet walk-in _____

Cubic feet freezer _____

Cubic feet upright _____

Adequate _____

Dishwashing:

Manual—No. compartments _____

    —Disinfection by hot water _____

    "   "   Chemical _____

    —Thermometer or test kit provided _____

    —Type of booster heater & cap. _____

    —Name of disinfectant _____

    "   "   detergent _____

Machine—Mfg., No. type _____

    —No. tanks _____

    —Pres. gauge on rinse_____reading_____psi

    —Temp. gauge on wash_____rinse _____

    —Name of detergent_____feeder _____

Length of: soiled-dish counter_____ft

    drainboard_____ft

Figure 8-6   Eating place, inspection form.

Trade name.. Operator..

Address.. Veterinarian..

Personnel: M......F......Permit No..........Killed Annually: Beef......Hogs......Sheep......Other........

1. *Building*	Yes	No	CM
(a) Site drained, graded, clean			
(b) Structure in repair, location meets zoning restrictions			
(c) Rooms 12 ft high, space for storing meat and washing utensils; slaughtering separated from processing			
(d) Floors concrete or equal, coved at walls up 12″, graded to trapped drain, drip gutter under dressing rail, separate drain for bleeding and viscera hanging area			
(e) Walls and ceilings cleanable, concrete, tile or equal, wall impervious to 5 ft height, ceiling tight and cleanable			
(f) Pens have concrete floor, are drained, clean, 150 ft to corral			
(g) Insects and rodents absent, building ratproofed			
(h) Dressing rooms and lockers clean, separated from work			
2. *Lighting and Ventilation*			
(a) Adjustable windows, ducts, or fans; hoods over vats, no condensation or excessive odors, heating adequate			
(b) Windows 15% floor area, no glare, 10 ft candles of light min. and 25 at grading and insp. areas, light-colored walls, ceilings			
(c) Openings screened, flies control			
3. *Water supply*—100 gal per animal			
(a) Source protected, construction satisf. and adequate, rooms accessible with hose, no cross-connections, quality ok			
4. *Toilet facilities*			
(a) Location convenient, screened, ventilated, lighted, tissue, backflow preventers, clean			
(b) Lavatories with warm running water, soap, paper towels			

5. *Refuse*	Yes	No	CM
Flytight receptacles for offal manure, blood, condemned meat			
6. *Sewage and waste disposal*			
Sewage to sewer or disposal system, other liquid wastes separate, operation proper			
7. *Animals*			
Disease-free from ante-mortem veterinary inspection			
8. *Equipment*			
(a) Work tables, blocks, vats are cleanable and free from cracks			
(b) Chilling room stores carcasses 6 hr, air circulation in cooler; floors, walls, hooks, trays cleanable; thermometers			
(c) Washing and flushing (150°F) water under pressure, tanks, brushes, cleansing powders adequate			
(d) Transporting vehicles covered, cloths and waterproof paper			
9. *Handling, refrig., disinfection*			
(a) Veterinarian post-mortem sup, carcasses stamped, washed, protected from contamination			
(b) Processed meat heated to 170°F for 30 min., meat therm. used			
(c) Carcasses separated in refrig. supported off floor, placed immediately in chilling room at approx. 40° and stored below 34° after 24 hr, therm. used, no spoiled meat			
(d) Hot water, brushes and cleansing powders used, equip. clean			
(e) Utensils submerged 2 min. in 170° water or 200 ppm chlorine			
(f) Transportation vehicles clean used only for edible meats, meats wrapped and protected			
(g) Personnel clean, have washable outer garments, aprons, rubber boots, gloves; have clean habits and in good health			

Remarks and Explanation of Defects: (CM indicates correction made)..

....

....(over)

Inspected by.. Date..

Figure 8–7 Slaughterhouse inspection form.

	Yes	No	CM			Yes	No	CM
1. Plant rodentproof and free from rodents				12. Ice used for dressing birds clean and handled in clean manner				
2. Structure in repair, location meets zoning restrictions				13. Holding cages and batteries cleaned daily, washing facilities provided, all kept clean				
3. Structure screened, free of flies, roaches, other vermin				14. Clean scalding water provided (170–180°F) and scalders have water changed frequently to prevent accumulation, spray chambers have nozzles clear				
4. Light in workrooms 30–50 ft candles 30 in. above the floor				15. Crops and vents of birds cleaned and birds washed clean				
5. Ventilation prevents condensation and odors				16. Spray and scrubber type washers and coolers provided				
6. Water supply satisfactory sanitary quality and adequate				17. Superchlorinated water (10–20 ppm) available and used for water flushing of birds, for cavity and bird washing after eviscerating, giblet washing				
7. Work rooms accessible with hose supplied by hot and cold water and steam; hot water supply adequate				18. All tables, buckets, cans, shackles, hooks, trays, cookers and other equipment kept clean				
8. Floors in killing and dressing area smooth concrete or equal, coved at walls up 12″, graded to trapped drain				19. Convenient handwashing facilities for workers				
9. Walls and ceilings cleanable concrete, tile or equal; walls impervious to 5 ft height				20. Toilet, washroom, and dressing room facilities adequate, and clean				
10. Floors free of blood and vent and crop material; washed as often as necessary to prevent accumulations				21. Cutting, wrapping, weighing, packing, and cooler equipment clean and adequate				
11. Sewage, wastes, offal, and refuse properly disposed								

Remarks and Explanation of Defects (CM denotes correction made)...

....

....

....

....

....

See Poultry Plant Sanitation, Supplement No. 1, Institute of American Poultry Industries, 59 East Madison, Chicago 2, Illinois.

Inspected by... Date...........................

Figure 8–8 Poultry dressing plant inspection form.

Name _____ Operator _____

Address _____ Wholesale _____ Retail _____

Personnel: M _____ F _____ Products _____ Permit No. _____

	Yes	No	CM		Yes	No	CM
1. Location on or above ground level, in clean surroundings				11. Locker and dressing space, storage place for clean and soiled linen, uniform, and aprons; lighted and clean			
2. Structure in repair, rodent-proof, screened, window display enclosed, baked goods protected, lighting adequate				12. Personnel clean, head covering, clean habits, good health, supervised			
3. Floor finish smooth, cleanable, and nonabsorbent, sloped to drain that is trapped, kept clean and in repair				13. Design and construction of equip. cleanable and kept in repair (pans, kettles, molders and rollers, dough troughs, mixers, beaters, flour bins and chutes, conveyors, sieves), equipment cleaned at end of each day and kept clean*			
4. Walls and ceilings cleanable, kept clean and in repair							
5. Ventilation adequate to prevent condensation, remove smoke and odors. Hoods equipped with proper exhaust fans and ducts				14. Bakery and confectionery products free of vermin and other filth; flour, meal, farina, etc. stored in original pkg. at least 6″ off floor or in tight receptacles; raw ingredients from approved source; perishable products stored at 45°F or less and frozen products at 0°F or less			
6. Rats, mice, roaches, flies, silverfish, beetles, weevils, lice, mites, moths, ants, effectively controlled							
7. Extermination and cleaning program in effect with responsibility established in a sanitary supervisor who has authority to make corrections				15. Equip. and facilities adequate for cleaning and disinfection; includes wash sinks, automatic hot water, detergents, and disinfectants, scrapers, and brushes			
8. Water supply satisfactory, sanitary quality, easily accessible, adequate volume and pressure				16. Refrigeration adequate for perishables; custards and creams refrigerated, perishable products so labeled and marked kept below 45°F			
9. Wastes disposed of to proper sewer or private sewage disposal or treatment system				17. Disposition of leftovers satisfactory			
10. Toilets and washrooms convenient, have washbasins with warm water, soap, and individual towels (1 w.b. to 10 to 15); adequate waterclosets (1 w.c. to 10 to 15); handwashing sign; cleanable floors, walls, ceilings; adequate light and ventilation; tissue; kept clean				18. Adequate containers for storage of refuse, clean, no accumulations			
				19. Delivery trucks clean, baked goods protected, perishables refrigerated			

* Equipment meets standards of Bakery Industry Sanitation Committee, 3 "A" Standards Committee, National Sanitation Foundation.
(Ultraviolet light, "black light," shows urine stains. Microanalysis of food and ingredients reveals presence of rodent hairs, dirt, and fragments of pellets and insects.)

Remarks: _____

_____ (over)

Inspected by _____ Date _____

Figure 8–9 Bakery and confectionary inspection form.

1. *Water storage*	Yes	No	CM
(a) Shellfish stored in fill and draw or continuous flow tanks			
(b) Water meets U.S.P.H.S. drinking water standards and is of proper salinity			
(c) Fill and draw tanks emptied within 24 hours and water maintained of satisfactory sanitary quality by chlorination			
(d) Continuous flow tanks supplied with satisfactory chlorinated water at rate to replace water in not more than 24 hours			
(e) Tanks side walls extend 4 inches above surrounding floor and 1 foot above high water			
(f) Tanks maintained in a sanitary condition and employees wading in tanks wear clean rubber boots			

2. *Shellfish from approved source*			
(a) Shellfish packaged by person having permit. If in interstate traffic, person has permit from other state whose program is certified by the U.S.P.H.S.			
(b) Shellfish tagged with name, address, and certificate or permit number of shipper, the contents and source of shellfish, date of shipment, name and address of consignee			
(c) Receiver of shellfish keeps record of quantity and source and sale			
(d) Boats clean, bilge water disposed away from shellfish, watertight receptacles for excreta, receptacles emptied on shore in sanitary manner			

Remarks (Explain "No" items on back) (CM indicates correction made)

....

....

....

....

....

Inspected by.............................. Date ....

3. *Buildings*	Yes	No	CM
(a) All buildings kept clean, free of debris and unused material			
(b) Adequate toilet facilities			
(c) Toilets clean, separate from workrooms; sewage disposal proper			
(d) Washbasins with hot and cold water, soap and sanitary towels, convenient, wash hands signs posted			
(e) Buildings adequately lighted, ventilated, screened			
(f) All walls, ceilings, and floors cleanable and floors drained			
(g) Approved water supply			
(h) Cleanable shell stock storage bins and rooms			

4. *Shucking*			
(a) Shucking and packing rooms separated, cleanable, free of insects			
(b) Shucking benches nonabsorbent, cleanable, disinfected, and cleaned			
(c) Shucking blocks free of cracks, cleaned			
(d) Containers for waste material, and waste material disposed of in a sanitary manner			
(e) Shucked shellfish washed with cold water of satisfactory sanitary quality			
(f) All shucking equipment cleanable			
(g) Refrigeration for shucked shellfish above freezing and below 40°F			
(h) Shucked shellfish stored and shipped at temperature between 32° and 40°F not in contact with ice in clean and new metal or equal containers, adequately sealed			
(i) Containers embossed with certificate number of shipper preceded by state			

5. *General*			
(a) Personnel clean and have clean habits			
(b) No person in plant ill or has communicable disease			
(c) Shellfish handled and shipped in clean containers and at temperature to keep them alive			
(d) Cars and trucks in which shellfish shipped clean			

Figure 8–10 Shellfish control inspection form.

leave a copy of the original inspection report form with the owner, operator, or other responsible person. When subsequent reinspections fail to show improvement, it is customary to confirm the deficiencies in a letter, with reference to previous inspections and notifications and a scheduled date for final inspection before taking further action.

Regulatory agencies usually have standard inspection report forms covering activities under their supervision. However, some forms for the sanitary inspection of eating places, slaughterhouses, poultry processing plants, and bakeries are reproduced here for convenience.

Shellfish

Shellfish include the mollusks scallops, oysters, soft-shell and hard-shell (quahaug) clams, and mussels, as well as the crustacea lobsters, crabs, and shrimp. For control purposes oysters, clams, and mussels are given primary consideration. Since oysters are frequently eaten raw and in large quantities, they are more frequently involved in outbreaks. Mussels from the Pacific Coast, England, and Europe contain at certain times of the year a chemical poison that is not destroyed by cooking. Because of this, California has a quarantine on mussels from June to September. Shellfish can transmit typhoid fever, cholera, dysentery, and gastroenteritis. Lobsters, shrimp, crabs, and scallops are less likely to cause illness as they are usually thoroughly cooked before being served. Improper handling of these shellfish before service can, however, introduce contamination.

The pollution of shellfish waters with sewage and industrial wastes decreases the fish and shellfish population and introduces a disease-transmission hazard. Typhoid fever outbreaks traced to contaminated shellfish between 1900 and 1925 led to U.S. Public Health Service certification of dealers involved in interstate shipment and to control by state and local regulatory agencies. Infectious hepatitis is an added hazard.

The USPHS evaluates compliance with minimum standards by inspection of a representative number of handling and processing plants. The state regulating agencies should adopt adequate laws and regulations for the sanitary control of shellfish from the source to the consumer. The local industry is expected to comply with established procedures regarding identification information on each package of shellfish, record-keeping from the point of origin to the point of ultimate sale, and sanitary control of shellfish. Numbered certificates issued by the authorized state agencies are accepted by the USPHS for approved producers and dealers engaged in interstate commerce. The USPHS maintains a current list of approved shellfish shippers.

Records of all shellfish shipments must be kept; this includes reshipments. The name of the shipper and consignee, the date harvested, state and water source, the type of shellfish, and permit certificate number of the packer and reshipper must appear on each carton or lot.

Spawning takes place when the water is warm, usually between May and August. The shellfish seed or spat is released at this time, attaches itself to some hard surface, and grows. Oyster shells dumped back into shellfish beds provide a ready growing surface. The parent shellfish is not as meaty or of best flavor during the spawning season but is edible. Problems in refrigeration during warm weather mitigate against the production and distribution of shellfish.

Oysters grow best a pH of 6.2 to 6.8 and at a temperature of 41°F. The pH can be as low as 5.0. The best salinity of the water is 0.24 to 0.27 percent. A salinity greater than 0.31 percent is unsuitable.

Oyster-shell stock or shucked oysters sampled at the source should not show an MPN of 230 or more coliform organisms per 100 grams or more than 500,000 total plate-count per gram. An MPN of 2400/100 g may be occasionally permitted, provided needed corrections are promptly made. Shucked shellfish should be stored at a temperature of 45°F or less. Shellfish growing-area water should have a fecal coliform median MPN of less than 7.8/100 ml, with not more than 10 percent of the counts showing a fecal coliform MPN of greater than 33/100 ml.

Shellfish standards have been suggested to assist in sanitary control at the consumer level.[11] These are tabulated below.

Category	Coliform MPN	Standard Plate Count
Acceptable	Not more than 16,000 per 100 ml	Not more than 50,000 per ml
Acceptable on condition*	Less than 160,000 per 100 ml	Less than 1,000,000 per ml
Rejectable	160,000 or more per 100 ml	1,000,000 or more per ml

* Shipments will be reported to the shellfish control organization of the originating state for investigation and will not be rejected unless the report of the investigating authority is unsatisfactory.

For further reference see *Manual of Recommended Practice for Sanitary Control of the Shellfish Industry*, Part I—"Sanitation of Shellfish—Growing Areas," 1959, and Part II—"Sanitation of the Harvesting and Processing of Shellfish," 1962, U.S. Public Health Service, Pub. No. 33, Dept. of HEW, Washington, D.C.

[11] U.S. Public Health Service, Rep. No. 71, Dept. of HEW, Washington, D.C., October 1956, p. 1064.

To be representative, water samples should be taken during different seasons of the year, and at varying tides, depths, and watershed runoffs. Examinations must be made in accordance with the procedures described in *Recommended Procedures for the Examination of Sea Water and Shellfish.*[12] Sanitary surveys of shellfish growing areas and tributary watersheds, including chemical and bacterial examinations of the waters and inspections of shellfish plants, should be made periodically by the state shellfish regulatory agency. Shellfish concentrate radioactive materials and pesticides.

The quality of the water source of shellfish will determine the need for special treatment and cleansing period (24 to 48 hours) for shellfish before being placed on the market. Cleansing or purification processes involve the placing of the bivalves in tanks fed with clean seawater, under the direction of the control agency. Sanitary handling and processing must be assured to maintain a satisfactory product from the source to the consumer. This should include thorough washing at the source, clean packing, and storage at temperatures that will keep the shellfish alive and active without loss of shell liquor. Sand trapped in the shell will irritate and cause a loss of shell liquor, with general weakening of the bivalve. Oysters free themselves of contaminating viruses and bacteria within 12 to 24 hours of exposure in purified seawater.[13]

Burlap and jute sacks used to transport shellfish must be kept clean and dry because damp or wet sacks will cause bacteria to multiply rapidly. Clean barrels, boxes, or baskets are better shipping containers.

Shucked shellfish are shellfish that are taken from the shell and washed and packed in clean containers constructed of impervious material. The operation should be conducted under controlled sanitary conditions. The packer's certificate number and state, name and address of the distributor, and contents are permanently stamped on the container.

Fresh oysters can keep for two weeks or somewhat longer on ice. Frozen oysters can be kept six months or more. Soft-shell clams and mussels should not be kept more than 2 days. Shellfish shipped to market in the shell are required to be handled in a sanitary manner, stored at a temperature of 36 to 40°F, and kept alive. Repacking of fresh oysters as breaded frozen oysters or a breakdown in sanitary precautions introduces health hazards. An effective shellfish control program requires the active participation of state and local regulatory agencies. Shellfish should be obtained only from certified dealers.

[12] 4th ed., American Public Health Association, New York, 1970.
[13] *New York Times,* November 7, 1965, report on research done at the Gulf Coast Shellfish Research Center at Dauphin Island, Ala.

Typical Kitchen Floor Plans

Some floor plans are included here to suggest the extent of the problem and the approach that should be used when a new establishment is constructed or major alterations are proposed. Many details, equipment, facilities, and trades are involved, all of which require specialized knowledge. The preparation of scale drawings and plans by a competent person, such as a registered architect or engineer who has had the related experience, invariably saves money and anguishing frustrations. It is strongly recommended that preliminary plans be reviewed by the local building department, health department, or other agencies responsible for inspection and supervision of the particular operations, regardless whether this is a requirement of any sanitary code, regulation, or law. One should seek assistance from equipment manufacturers and others. Most official agencies are only too happy to be consulted in advance rather than require major corrections after a job is completed and business started.

DESIGN DETAILS

General Design Guides

Restaurant and equipment design, including equipment arrangement, is a specialized field. In many instances, however, basic sanitation problems are incorporated rather than eliminated in the design. Food handling, storage, preparation, and service should receive careful consideration. Some guides in planning that result in efficient, economical, and sanitary operations are summarized below and illustrated in Figures 8–11 and 8–12. Planning starts with study of food flow and type of service.

1. Aisle spaces should be not less than 36 in.

2. Unless built-in in a sanitary manner, equipment should be spaced 18 to 30 in. from other equipment or walls and 12 to 18 in. above the floor to make possible cleaning. A smaller space that is still completely accessible for cleaning may be permitted.

3. Built-in or stationary equipment that eliminates open spaces, crevices, rat and insect harborage, and provides a coved or smooth joint where equipment abuts other equipment, the wall, ceiling, or floor will simplify sanitary maintenance.

4. The floor finish or covering selected for different spaces should recognize the type of traffic and uses to which the spaces will be put. The same would apply to wall and ceiling finishes. Linoleum or asphalt tile that is not grease-resistant, for example, is not a suitable floor covering for restaurant kitchens; quarry tile is ideal. The limitations of various floor coverings are discussed under structural and architectural details.

5. Wooden shelving should be 12 to 18 in. above the floor, planed smooth, but not painted, shellacked, or varnished. Oilcloth or paper lining complicates the housekeeping job.

6. Bins, cutting and work boards, meat hooks, shelving in walk-in and reach-in refrigerators, and other equipment that comes into contact with food should be removable for cleaning.

7. Provision should be made in the kitchen for a washbasin or utility sink, connected with hot and cold running water, for the convenience of food-handlers.

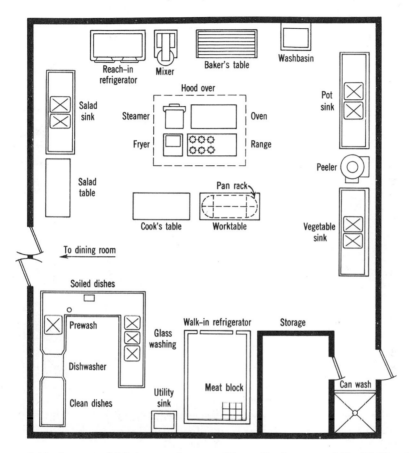

Figure 8–11 Layout of kitchen equipment. (From *Environmental Health Practice in Recreational Areas,* U.S. Public Health Service Pub. No. 1195, Dept. of HEW, Washington, D.C., 1965, p. 70.)

8. A garbage-can storage room is a practical necessity in the larger restaurants. This problem is discussed in greater detail in Chapter 5.

9. Garbage grinders are very desirable in the dishwashing area, in the vegetable preparation space, in the pot-washing area, and in the butcher shop.

10. Avoid submerged water inlets where possible; provide vacuum breaker and check valve on waterline if not possible. Inlet lines should be 2½-pipe diameters and not less than 1 in. above the overflow rim. Run wastelines from potato-peeling

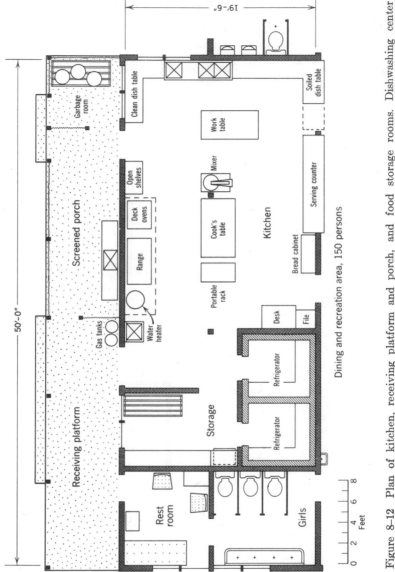

Figure 8-12 Plan of kitchen, receiving platform and porch, and food storage rooms. Dishwashing center along wall at extreme right. (From *Cornell Miscellaneous Bull.* No. 14, New York State Colleges of Agriculture and Home Economics, Ithaca, March 1953.)

machine, egg boiler, dishwashing machine, steam table, refrigerator, air conditioning equipment, and kettles to an open drain or to a special sink. See the section on plumbing for greater details.

11. Pipes passing through ceilings should have pipe sleeves extending 2 in. above the floor surface.

Lighting

Proper lighting is essential for cleanliness and accident prevention and to reduce fatigue. Proper lighting takes into consideration the task, spacing of light sources, elimination of shadows and glare, control of outside light, and lightness of working surfaces, walls, ceilings, fixtures, trim, and floors. Some desirable lighting standards are given in Table 8–6.

TABLE 8–6   LIGHTING GUIDELINES

Space	Footcandles	Space	Footcandles
Bakery	30 to 50	Restaurant	
Brewery	30 to 50	Dining room	3 to 30
Candy making	50 to 100	Kitchen	70
Canning		Dishwashing	30 to 50
General	50 to 100	Quick-service	50 to 100
Inspection	100	Food quality control	50 to 70
Dairy		Cooler, walk-in	30
Pasteurization	30	Storage rooms	20
Bottle washer	100	Toilet rooms	20 to 30
Milkhouse	20 to 30	Dressing rooms	20 to 30
Inspection	100	Offices	70 to 100
Laboratory	100	Corridors, stairways	20
Gages, on face	50	Other spaces, minimum	5
Milking, at cow's udder	50	Recreation rooms	20 to 30
Meat packing			
Slaughtering	30		
Cleaning, cutting, etc.	100		

Note: The lighting should be measured on the working surface, in walk-in coolers and storerooms at the floor, and in other areas 30 in. above the floor.

Typical color reflection factors are: white, 80 to 87%; cream, 70 to 80%; buff, 63 to 75%; aluminum, 60%; cement, 25%; brick, 13%.

Ventilation

Ventilation is considered adequate when cooking fumes and odors are readily removed and condensation is prevented. This will require sufficient windows that can be opened both from the top and bottom or ducts that will induce ventilation. Hoods with exhaust fans should be provided, extending 6½ ft above the floor and 2 ft beyond ranges, ovens, deep fryers, rendering vats, steam kettles, and over any other equipment that gives off large amounts of heat, steam, or soot. Exhaust fans for small hoods usually have a capacity of 75 cfm; large hoods have 150 cfm fans. Mechanical ventilation can be provided for kitchens with fans having the capacity of 15 to 25 air changes/hr, depending on the amount of cooking. It is advisable to draw fresh air in from a high point, as far away as possible from the exhaust fan, to prevent drafts and inefficient ventilation. Hood ventilation is given in Table 8–7.

TABLE 8–7   EQUIPMENT VENTILATION

Equipment	Rate of Ventilation (cfm/ft^2)
Range hoods 2 ft or more in height	100
Range hoods less than 2 ft in height	125
Hoods for steam tables	60
Domestic-type range used for warming only	none

Required exhaust fan = hood length × width × cfm in table.

Uniform Design Standards

There is a distinct need for reasonably uniform standards and criteria to guide the manufacturing industry as well as design architects and engineers. In this way food equipment, materials, and construction would meet the requirements of most regulatory agencies, and manufacturers would not need special equipment design for different parts of the country. Such standards and criteria have been established by joint committees representing industry and regulatory agencies, and many have received wide acceptance.[14] They are highly recommended for reference and design purposes. (See Figures 8–13, 8–15, and 8–17.)

Space Requirements

A common problem in existing restaurants is the inadequate space in the kitchen for the number of meals served. Inefficiency and the result-

[14] 1. *National Sanitation Foundation Standards,* National Sanitation Foundation, School of Public Health, Ann Arbor, Michigan.
    2. *3-A Sanitary Standards,* Committee on Sanitary Procedures, International Association of Milk, Food and Environmental Santitarians, Inc.; the Sanitary Stan-

ant confusion compound the problem. Some suggested average space requirements are listed in Table 8–8; however, complete scale drawings should be made to fit individual situations. See also pp. 590 to 593. How the space is used is as important as the space available.

Refrigeration

Refrigeration slows down bacterial activity and preserves food, including many microorganisms. It is frequently stated that food stored at a temperature of 50°F is adequately protected against spoilage. This generalization can be very misleading unless such factors as storage method, frequency of opening and closing of the refrigerator door, insulation, type of food stored, length of storage, box outside-temperature, compressor capacity, temperature and weight of food placed in the storage box, maintenance, and similar variables are taken into careful consideration. Actually, a temperature of 50°F will favor the growth of *Salmonella enteriditis*. Studies show that the growth of salmonellas and staphylococci in perishable foods is inhibited when the internal temperature is at or below 42°F.[15] Hence it is extremely important to distinguish between the air temperature in a refrigerator and the internal temperature of the food. A temperature of 38 to 40°F has been found more satisfactory in practice for the storage of most foods that are used within 24 or 48 hours. Frozen foods should be stored compactly below 0°F.

Low-temperature heat-treated hermetically packed foods, such as cured hams with salt, also require refrigeration. All canned food is not necessarily sterilized. In some instances the can merely serves as a container, for a meat or other food product, that has been subjected to sufficient heat to only inhibit the growth of spoilage organisms. Improper

dards Subcommittee, Dairy Industry Committee; and the U.S. Public Health Service.

3. *BISSC Sanitation Standards,* Baking Industry Sanitation Standards Committee.

4. Meat Inspection Division, U.S. Dept. of Agriculture, Washington, D.C.

5. Poultry Inspection Division, U.S. Dept. of Agriculture, Washington, D.C.

6. Bureau of Commercial Fisheries, U.S. Dept. of Interior, Washington, D.C.

7. *National Plumbing Code,* American Standards Association, ASA A40.8-1955, 29 West 39th St., New York.

8. *National Electrical Code,* National Fire Protection Association, 60 Batterymarch Street, Boston, Mass., 1971.

9. *National Building Code,* American Insurance Association, 85 John Street, New York, 1967.

10. See also state and local standards, regulations, and guidelines.

[15] Robert Angelotti, Milton J. Foter, and Keith H. Lewis, "Time-Temperature Effects on Salmonellae and Staphylococci in Foods," *Am. J. Public Health,* **51,** No. 1, 76–88 (January, 1961).

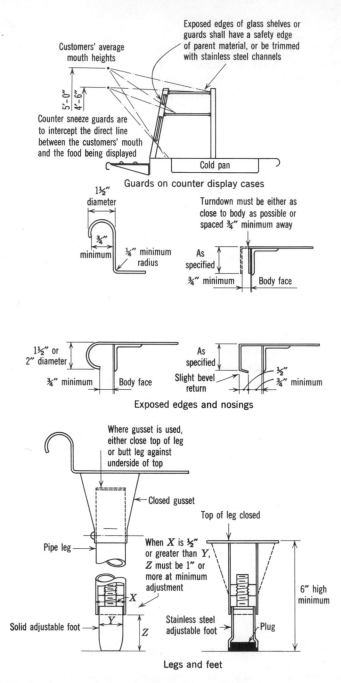

Figure 8-13 Equipment design details. (From *National Sanitation Foundation Standards,* Standard No. 1, "Soda Fountain and Luncheonette Equipment," April 1969, and Standard No. 2, "Food Service Equipment," April 1965, Ann Arbor, Mich.)

TABLE 8–8   SUGGESTED SPACE REQUIREMENTS FOR KITCHENS

Type of Food Service	Kitchen,* (ft² per seat)	Dining Room, (ft² per seat)
Children's camp	7 to 9	12
School lunchroom	5½ to 6½	10
Restaurant	7½ to 9½	14
Lunchroom (counter, service, and tables)	8 to 12	15 to 20†
Hospitals	20 to 30 per bed	12 to 15 per bed
Cafeteria	12	15†

*Includes storage, receiving, and dishwashing spaces, also employees' toilet rooms and locker space; deduct 4 to 5 ft² per seat if not included. Larger areas are suggested for the smaller places. The greater the turnover per hour the larger the kitchen per seat.

†Includes serving area. Experience shows that with experienced help a cafeteria can serve 300 people per hour in the line and 600 people per hour with a limited menu. With counter service, allowing 2 ft per stool, 4 persons can be served per hour. With table service, the turn-over is 3 to 4 timer per 2 hours. (Ned Greene, *How to Serve Food at the Fair,* Pyramid Books, New York, 1963.)

refrigeration of such foods may cause the growth of pathogenic organisms without showing swelling of the can.

In general, fresh meats should be stored at a temperature of 34 to 38°F; fresh fish at 34 to 38°F, poultry at 34 to 38°F; fresh fruits at 35 to 40°F; dairy products, including eggs and salads, at a temperature of 40 to 45°F; leafy green and yellow vegetables at 40°F; tomatoes, potatoes, onions, lemons, cucumbers, and melons at around 50°F; cured meats at 40°F in a dry, well-ventilated space; butter is best held at 0°F. A more exact breakdown of recommended storage temperatures with time limitations is given in Table 8–9.

Accurate thermometers located in the warmest compartments of refrigerators will show the adequacy of refrigeration. Sandwich and salad mixtures and cut or leftover food should be placed in shallow pans at a depth not greater than 3 in. to accelerate the rapid cooling of such foods. Hot foods are usually cooled down to about 140 or 150°F before being placed in the refrigerator, but in any case should be refrigerated within 30 min. In this way bacterial growth is retarded and the cooling capacity of the refrigerator is not exceeded. Foods in shallow pans placed in the freezing compartment will cool rapidly.

The storage of fish, ham, bacon, cabbage, onions, and similar odorous foods in the same cooler or refrigerator with milk, butter, eggs, salads, desserts, and fruit drinks that pick up odors and flavors should be avoided. Dairy products and meats are best stored in a separate refrigerator. Large pieces of meat should be hung so that they do not touch,

TABLE 8-9   STORAGE DATA

Product	Storage Temp.		Sp. Heat, Btu/lb/°F*	Approximate Storage Period	Common Measure
	Long Storage	Short Storage			
Apples	31 to 33	35 to 40	0.90	2 to 8 months	1 bushel = 45 to 50 lb 2½ bushel = barrel 5 ft³ per barrel
Asparagus	32 to 34	40 to 45	0.95	2 to 4 months	
Bananas	56 to 60	35 to 40	0.81	7 to 10 days	50 lb per bunch
Blackberries	32 to 34	35 to 40	0.89	7 to 10 days	
Beans	32 to 34	40 to 45	0.92	3 to 4 weeks	
Beef (fresh)	32 to 34	35 to 40	0.77	3 months	750 lb average beef
Butter	40 to 45	40 to 45	0.64	5 months	50 to 60 lb per tub
Cabbage	32 to 34	35 to 45	0.93	3 to 4 months	
Candy	60	65 to 70	0.93	3 months	
Carrots	32 to 34	40 to 45	0.87	2 to 4 months	
Cauliflower	32 to 34	35 to 40	0.90	2 to 3 weeks	
Celery	31 to 33	35 to 40	0.91	2 to 4 months	
Cheese	34 to 36	40 to 45	0.67	4 months	
Corn	32 to 34	35 to 40	0.80	1 to 2 weeks	
Cranberries	34 to 36	40 to 45	0.85	1 to 3 months	
Cucumbers	50 to 60	50 to 60	0.93	6 to 8 days	54 lb = bushel
Eggs (fresh)	31 to 33	40 to 45	0.76	6 months	30 doz = case = 50 lb case 12″ × 12″ × 25″
Fish (iced fresh)	30 to 32	34 to 38	0.82		
Grapes	30 to 32	35 to 40	0.90	3 to 5 months	
Lamb	32 to 34	34 to 38	0.67		75 lb avg. wt

598

Item					
Lard	32 to 34	45 to 50	0.52	1 to 4 months	
Lemons	50 to 60	50 to 60	0.92	2 to 7 days	
Lettuce	32 to 34	35 to 40	0.90	1 to 3 weeks	
Melons	50 to 55	50 to 55	0.91		
Nuts (dried)	32 to 34	35 to 40	0.30	8 to 12 months	bushel = 50 lb.
Onions	32 to 34	50 to 55	0.90	5 to 6 months	bag = 70 lb
Oranges	32 to 34	40 to 45	0.90	1 to 2 months	crate 15″ × 15″ × 30″
Oysters, liquid	32 to 38	40 to 43	0.90		
Peaches	32 to 34	35 to 40	0.92	1 to 2 weeks	bushel = 56 lb
Pears	32 to 34	35 to 40	0.91	1 to 4 months	box = 40 lb
Peas (green)	32 to 34	40 to 45	0.80	1 to 3 weeks	
Pork (fresh)	32 to 34	34 to 38	5*–0.7	1 month	Avg. wt of hog 250 lb
Potatoes	36 to 50	45 to 50	0.77	3 to 5 months	150 lb/bag, 60 lb/bushel
Poultry	28 to 34	34 to 38	0.80		
Sausage – fresh	32 to 34	35 to 40	0.89		
smoked	33 to 35	40 to 45	0.68		
Strawberries	32 to 34	35 to 40	0.89	7 to 10 days	24-qt case = 45 lb
Tomatoes	50 to 55	50 to 55	0.95	7 to 10 days	Bushel = 50 lb
Veal	31 to 33	34 to 38	0.71		

Source: *Kramer Engineering Data*, Catalog No. R-114, Kramer-Trenton Co., Trenton, N.J.

Note: A temperature of 36—42°F and humidity of 80—95 percent make an excellent environment for the growth of mold and yeast.

*The amount of heat required to lower the temperature of 1 lb of the product 1°F up to freezing temperature.

and all foods including roasts, chops, and steaks should be arranged so that air can circulate freely. Wrapping should be removed and the food kept loosely covered. Smoked meats may be kept tightly wrapped. If not stored separately, wrap butter, cheese, and fish tightly.

Refrigeration in a food establishment, camp, or institution is adequate when it is possible to store and maintain all perishable foods at a proper temperature, at least over weekends, without packing and crowding. The refrigerator should be relatively dry and permit air to circulate freely.

The size refrigerator to provide for a particular purpose is dependent on many variables, most of which can be controlled. The number and type of meals served, the delivery schedules, and the purchasing methods all affect the required storage space. As a guide, a refrigerator capacity of 0.15 to 0.3 ft^3 per meal served per day, with 0.25 as a good average, has been found adequate to prevent overloading over weekends and holidays. Where shopping is done twice a week, the refrigerator should provide, 2½ ft^3 per persons served three meals a day; 3 to 3⅓ ft^3 is preferred. In practice many foods are purchased or delivered daily, hence the required refrigerator volume should be adjusted accordingly. Additional storage space is needed for bottled-beverage cooling and for frozen foods. Allow 1 ft^3 for 50 to 75 half-pints of milk and 0.1 to 0.3 ft^3 per meal served per day for frozen food storage.

The storage racks in a refrigerator or walk-in cooler should be slotted to permit circulation of cold air in the box and should be removable for cleaning. If the box is provided with a drain, it must not connect directly to a sewer or waste line, but rather to an open sink or drain that is properly trapped, unless drainage is to a drip pan that is manually emptied. A ceramic tile floor with a coved base is very desirable.

If cans or boxes are stored in a cooler or refrigerator it is advisable to provide duckboards or racks on the floor. This makes possible better air circulation and helps keep the floor and supplies clean. In any case, refrigerators need daily cleaning and inspection for possible food spoilage. Washing the inside at frequent intervals with a warm mild detergent solution, followed by a warm-water rinse containing borax or baking soda and then drying will keep a box sweet and clean. Wipe the door gaskets daily. Do not use vinegar, caustic, or salt in cleaning solutions.

Many types of refrigerators on the market use different principles and refrigerants. When possible, provide entrance to a walk-in refrigerator through an air lock to prevent the cold air inside from rolling out and the outside warm air from pouring in. An efficient arrangement for refrigerators is a plan providing a group of walk-in coolers opening into a central room that can serve as a general cold storage room for certain foods that do not require a low temperature and as an access room from the unloading platform or kitchen. Another alternative would be two boxes in series with entrance to the freezer room through the cold room.

Refrigeration Design

The proper refrigeration of food is one of the most important precautions for the prevention of food poisonings and infections, as well as for the preservation of food. Refrigeration is the extraction of heat from a mass, thereby lowering its temperature. Ice has been used for many years to preserve food and is the standard that serves as a base for the measure of refrigeration. One pound of ice in melting will absorb 144 Btu of heat. A ton of refrigeration is 288,000 Btu/24 hr = 2000 lb of ice × 144 Btu.

The amount of heat required to lower one pound of a product one degree Fahrenheit is known as the specific heat. For example, the amount of heat to be extracted from 1 lb of water to lower its temperature 1 degree is 1 Btu, up to 32°F. To change the 1 lb of water to ice requires the extraction of 144 Btu from the water; this is known as its latent heat of fusion or the heat absorbed due to a change from the liquid to the solid state. But to lower the temperature of ice at 32°F requires the extraction 0.5 Btu per pound of ice per degree of drop in temperature. Similarly, each food product has its own specific heat, before freezing and after freezing, and its latent heat of fusion. For example, it is generally assumed for calculation purposes that the freezing point of most food products is about 28°F. Fresh pork has a specific heat of 0.6 Btu/lb before freezing, 0.38 Btu after freezing, and a latent heat of fusion of 66 Btu/lb. One cubic foot of air at a temperature of 90°F and 60 percent relative humidity that is cooled to 35°F will release 2.43 Btu.

Calculation of refrigeration compressor capacity takes into consideration the sources of heat through wall losses, air changes and leakage, pounds of products refrigerated, and miscellaneous losses such as electric light bulbs, men working, and heat released due to respiration of fruits and vegetables.

Heat transmission through walls, floors, and ceilings is by convection, conduction, and radiation. The amount of heat transferred depends largely on the overall coefficient of heat transmission of the wall materials in Btu per hour per square foot of surface per degree difference in temperature. Computation of heat transmission coefficients is explained in detail in *The American Society of Heating, Refrigeration, and Air-Conditioning Engineers Guide (ASHRAE Guide)*[16] and other related

[16] 345 East 47th Street, New York, N.Y., 1970.

standard texts. For example, 1-in. typical corkboard has a conductivity of 0.30 Btu per hour per square foot per degree Fahrenheit or 7.2 Btu/24 hr. Sawdust (dry) has a conductivity of 0.41 Btu per inch thickness or 9.84 Btu per 24 hours per square foot per degree. One inch of cement plaster has a conductivity of 8.0 Btu or 192 Btu/24 hr.

Air changes in a refrigerator each time the door is opened. Cold air rolls out and warm air enters, thereby lowering the temperature of the air in the refrigerator. A walk-in cooler having a volume of 300 ft³ will probably have about 35 air changes/24 hr due to door opening and infiltration.[17] This will amount to 10,500 ft³ of air. If the air has a temperature of 90°F and 60 percent relative humidity and is to be cooled to 35°F, it will represent 2.43 Btu/ft³ of air, or in the instance cited (10,500 × 2.43) = 25,500 Btu of heat per 24 hr to be removed by the refrigerating unit. In restaurants, where refrigerators normally receive heavy usage, it is good practice to estimate a 50 percent increase over normal in the number of air changes. Usage heat loss in Btu per 24 hr per ft³ of refrigerator capacity has been estimated by Kramer and others.

When food is placed in a refrigerator it is usually necessary to remove heat in the food in order to cool it to the desired temperature. The heat load is dependent on the weight of the product, its specific heat, the temperature reduction, and the speed with which the product is to be cooled. If 100 lb of fresh poultry is to be cooled from 65 to 35°F in 6 hr, the heat load will be:

$$\frac{100 \text{ lb} \times 0.80 \text{ sp heat} \times 30° \text{ temperature reduction} \times 24 \text{ hr}}{6 \text{ hr}}$$

$$= 9600 \text{ Btu/24 hr}$$

If the products refrigerated are fresh fruits or vegetables, adjustment must also be made for the heat released by these foods in storage. It is estimated for example that strawberries stored at 40°F give off about 3 Btu per 24 hr.

Miscellaneous losses include 3.42 Btu per hour per watt of light bulb, 3000 Btu per hour for each motor horsepower, and 750 Btu per hour per man working in a box. A safety factor of at least 10 percent is added to the total load, and compressor capacity is based on a 16-hr operating day to provide for a defrosting cycle. A horsepower hour is 2546 Btu or 0.7457 kilowatt hour. Keep motor and refrigerating unit free of lint and dirt to permit air circulation, and be sure door gaskets keep cold air from escaping.

[17] *Kramer Engineering Data,* Catalog No. R-114, Kramer-Tenton Co., Trenton, N.J.

An example will serve to illustrate the refrigeration design principles and data for the calculation of refrigeration load. Given: a box 10 ft × 10 ft × 8 ft outside, 3-in. cork insulation standard construction with two layers of insulation paper and sheathing inside and outside, room temperature 95°F, and box temperature 35°F, 100 lb of fresh poultry received at 65° and 150 lb of lean beef received at 65° cooled to 35° in 24 hr and 50 w electrical load.

Solution:
1. Box outside surface area: 520 ft²
2. Temperature reduction: 60°
3. Wall loss = 1.8 Btu per ft² per degree temperature reduction (see Table 8–10)
   = 1.8 × 520 × 60 = 56,200 Btu/24 hr
4. Air change = 9 × 9 × 7 ft (assume inside dimension 1 ft less than outside)
   = 567 ft³ × 23 air changes per 24 hr (see Table 8–13)
   = 13,040 ft³

TABLE 8–10   HEAT TRANSMISSION COEFFICIENTS

Insulation Material	Wall Heat Loss* per °F Room Minus Cooler Temperature
Corkboard, 1″	7.2
Single glass	27
Double glass	11
Corkboard paper and sheathing on both sides	3.6 for 1″-cork
	2.4 for 2″-cork
	1.8 for 3″-cork
	1.5 for 4″-cork

*Btu/24 hr/ft² of outside surface. See *ASHR & ACE Guide* for other coefficients.

5. Btu to cool 1 ft³ of air from 95 to 35° = 2.79 (see Table 8–11)
   Hence to cool total air change requires

$$13,040 \times 2.79 = 36,500 \text{ Btu/24 hr}$$

6. Product load of poultry and beef
   Specific heat of poultry = 0.80 Btu/lb (see Table 8–12)
   Specific heat of beef = 0.77 Btu/lb

$$\text{Product load} = \left[\frac{100 \times 0.80 \times (65-35°)}{24} + \frac{150 \times 0.77(65-35°)}{24} \right] 24$$

   = 2400 + 3460 = 5860 Btu/24 hr
7. Electrical load = 3.42 Btu per watt per hour. Assume 8-hr lighting
   = 3.42 × 50 × 8 = 1368 Btu/24 hr
   Total load = 56,200 (wall) + 36,500 (air change) + 5,860 (product)
   + 1368 (electric)
   = 99,928 Btu/24 hr plus 10 percent safety factor 0.1 (99,928)
   = 99,928 + 9993 = 109,921 Btu/24 hr

TABLE 8-11   BTU PER CUBIC FOOT OF AIR REMOVED IN COOLING TO STORAGE CONDITIONS*

Storage Room Temp., °F	Temperature of Air Outside, Assumed at 60 percent Relative Humidity			
	85°	90°	95°	100°
65	0.85	1.17	1.54	1.95
60	1.03	1.37	1.74	2.15
55	1.34	1.66	2.01	2.44
50	1.54	1.87	2.22	2.65
45	1.73	2.06	2.42	2.85
40	1.92	2.26	2.62	3.06
35	2.09	2.43	2.79	3.24

*Data for the calculation of refrigeration load adapted from *Kramer Engineering Data,* Catalog No. R-114, Kramer-Trenton Co., Trenton, N.J.

Compressor capacity, based on 16-hr operation,

$$\frac{109,921}{16} = 6875 \text{ Btu/hr}$$

Approximate calculation:
Usage heat loss = 1.52 Btu per ft³ per deg F per 24 hr (see Table 8–14)
$\qquad$ = 1.52 × 567 ft³ × 60° = 52,000 Btu/24 hr
$\quad$ Total = 52,000 (usage) + 56,200 (wall) = 108,200 Btu/24 hr
Compressor capacity, based on 16-hr operation,

$$\frac{108,200}{16} = 6760 \text{ Btu per hr}$$

Cleansing

The effectiveness of cleansing, including bactericidal treatment of equipment and utensils, is determined by the results. A standard measure is the bacteriological swab test.[18] One swab for each group of five or more similar utensils, cups, glasses, or five 8-in.² surface areas is used. A standard plate count, 32°C for 48 hr, of less than 100 per utensil or surface examined and the absence of coliform organisms indicates a satisfactory procedure. There are many field and laboratory methods to determine the effectiveness of cleansing materials and operations. Some are discussed here.[19]

[18] William G. Walter, Ed., *Standard Methods fo rthe Examination of Dairy Products,* 12th ed., American Public Health Association, New York, N.Y., 1967.
[19] Walter D. Tiedeman and Nicholas A. Milone, *Laboratory Manual and Notes for E. H. 220,* Sanitary Practice Laboratory, School of Public Health, University of Michigan, Ann Arbor, 1952.

TABLE 8-12   SPECIFIC HEAT, BTU PER POUND*

Product	Btu per lb per °F Before Freezing	Btu per lb per °F After Freezing	Latent Heat of Fusion Btu per lb
Asparagus	0.95	0.44	134
Berries, fresh	0.89	0.46	125
Beans, string	0.92	0.47	128
Cabbage	0.93	0.47	130
Carrots	0.87	0.45	120
Dried fruit	0.42	0.27	32
Peas, green	0.80	0.42	108
Potatoes	0.77	0.44	105
Sauerkraut	0.91	0.47	129
Tomatoes	0.95	0.49	135
Fish, fresh	0.82	0.41	105
Fish, dried	0.56	0.34	65
Oyster, shell	0.84	0.44	115
Bacon	0.55	0.31	30
Beef, lean	0.77	0.40	100
Beef, fat	0.60	0.35	79
Beef, dried	0.34	0.26	22
Liver, fresh	0.72	0.40	94
Mutton	0.81	0.39	96
Poultry	0.80	0.41	99
Pork, fresh	0.60	0.38	66
Veal	0.71	0.39	91
Eggs	0.76	0.40	98
Milk	0.90	0.46	124
Honey	0.35	0.26	26
Water	1.00	0.50	144

*Data for the calculation of refrigeration load adapted from *Kramer Engineering Data,* Catalog No. R-114, Kramer-Trenton Co., Trenton, N.J.

TABLE 8-13   AIR CHANGE FOR STORAGE ROOMS (32° TO 50°) CAUSED BY DOOR OPENING AND INFILTRATION*

Cubic Feet	Air Change per 24 hr	Cubic Feet	Air Change per 24 hr
250	38.0	800	20.0
300	34.5	1,000	17.5
400	29.5	1,500	14.0
500	26.0	2,000	12.0
600	23.0	4,000	8.2

*Data for the calculation of refrigeration load adapted from *Kramer Engineering Data,* Catalog No. R-114, Kramer-Trenton Co., Trenton, N.J.

TABLE 8-14   USAGE HEAT LOSS, INTERIOR CAPACITY PER °F DIFFERENCE BETWEEN ROOM AND BOX TEMPERATURE, PER 24 HR

Volume in ft^3	Heat Loss, Btu per ft^3—Service		Volume in ft^3	Heat Loss, Btu per ft^3—Service	
	Normal	Heavy		Normal	Heavy
15	2.70	3.35	400	1.62	2.38
50	2.42	3.10	600	1.52	2.28
100	2.12	2.85	800	1.48	2.22
200	1.85	2.60	1,000	1.42	2.15
300	1.70	2.45	1,200	1.38	2.10

1. The suitability of a detergent can be roughly determined by rinsing it off the hands (if not a caustic) in the available water. It should not leave a greasy or sticky feeling. A normal solution in warm water should completely dissolve without any precipitate. Dishes and utensils should show no spots or stains. A measure of rinsibility is the alcohol test. A drop of alcohol on the dry surface allowed to evaporate should leave no deposit. Another field test uses phenolphthalein. A drop of standard phenolphthalein is added to the wet surface of the utensil to be tested. If the water turns red, it is an indication of residual alkalinity and hence poor rinsing. If the pH of the tap water is above 8.3, the surface to be tested should be wetted with distilled water. The cleanliness of a glass can be observed by filling it with water and watching it drain. A clean glass would show a continuous unbroken water film; breaks in the water film would indicate unremoved soil. Metal parts, however, could have a uniform coating of soil that would not give a water break. Salt sprinkled over the utensil or equipment to be examined, while still wet, will not adhere to surfaces that have not been cleaned. A common field test is the pouring of plain soda water in a glass. Gas bubbles will cling to unclean areas. Other aids are a good flashlight, paper tissues, filter pads, a spatula, and studied individual observation.

2. A fluorochromatic method uses a nearly colorless fluorochrome such as brilliant yellow uranine, fluorescent violet 2G, ultraviolet light (black light), and others. Dishes, for example, would be preflushed, immersed a few seconds in a fluorochrome solution, and washed in the usual manner. The dishes on the clean-dish table are then examined under ultraviolet light. Any soil remaining would be fluorescent.

3. The efficiency of cleansing materials and operations can be determined by various methods with greater accuracy in a laboratory. Methods particularly suited for research use a standard soil, radioactive tracers, and fluorescent dyes.

4. Pots, pans, and other equipment require brushing in the detergent washing operation. Rancid, stale, and other off-flavors and odors in prepared foods can frequently be traced to incompletely removed food or mineral films on food preparation equipment. Three-compartment sinks, each 24″ × 24″ or 24″ × 30″, with 3-ft drainboards, and a stem or electric immersion heater or gas booster under the last two sinks are practical necessities. A pot and pan mechanical washer is on the market. Its effectiveness has not been proven.

5. A method for detecting residual grease and protein or starch films on china, melamine plastic, glass, or silverware is the use of a dry mixture of 80 percent talcum dust (not commercial powder) and 20 percent safranin by weight. The

mixture is placed in a saltshaker covered with cheesecloth to limit the flow and dusted over the surface of the utensil or article to be tested. It must be dry. Then hold the utensil under an open tap for at least 30 sec, tilt to drain off water, and examine for any red color. The intensity of the color is an indication of the amount of residual soil present.[20]

Proper cleansing involves the use of a suitable washing compound or detergent for the hardness, pH, and the available water should have a temperature that will result in the removal of fats, proteins, and sugars, usually referred to as "soil" or organic matter. The detergent must be used in adequate amounts. Too little does not accomplish cleansing; too much means greater expense and more care and time for satisfactory rinsing.

In some cases chemical disinfectants and combination cleanser–sanitizers are used because an adequate supply of hot water is not available in the first place. As a result, some operators of eating and drinking places resort to chemical compounds in an effort to obtain clean, disinfected dishes and utensils.

Chemical disinfection is less reliable than 180°F hot water and retards air drying. Chemical sanitizers are not effective on dishes or utensils that are not properly cleansed. When properly used, hypochlorite solutions can produce satisfactory results. A solution strength of 100 mg/l should be used. It can be prepared by mixing ¼ oz of 5¼ percent solution with one gallon of water. The solution should be replaced when the concentration falls to 50 mg/l or when the water cools below 75°F. The equivalent of a 3-compartment sink is essential to rinse the soil off between the wash and disinfection step, as organic matter and milk use up chlorine, thereby making it ineffective as a disinfectant. A 200 mg/l solution is recommended when the hypochlorite is used as a spray. At a pH of 7.0 a 250-mg/l solution for a 2-min contact period is necessary for effective disinfection. At a pH of 8.5 a 250-mg/l solution for a 20-min contact period is necessary for effective disinfection. It should also be pointed out that chlorine turns silver black.

Other chemical disinfectants, such as quaternary ammonium compounds, have definite limitations. Bicarbonates, sulphates, magnesium chloride, calcium chloride, and ferrous bicarbonate interfere with the bactericidal quality of quaternary ammonium compounds. Quaternaries should therefore only be used where their successful preformance has been demonstrated in the establishment concerned. The limits within which satisfactory results can be assured should be specified by the manufacturer of the sanitizer, as water chemical quality can be expected to vary from community to community.

The so-called detergent–sanitizer combinations have been reported to be effective under certain conditions. They are more likely to be weakened in cleansing and in disinfecting power by organic matter and by incompatibility under conditions of use. Iodine compounds may have application for hand dishwashing in warm

[20] E. H. Armbruster, "Method for Detecting Residual Soil on Cleaned Surfaces," *J. Am. Dietetic Assoc.*, **39**, 228–230 (1961).

water at a concentration of 25 mg/l. Iodine vaporizes at a temperature above 80°F, especially at 120°F.

If chemical disinfection is used, equipment should be provided and tests made by the operator to assure that the sanitizer is maintained at the proper strength. This objective is rarely achieved in practice.

The type of dishwashing method or machine that may be used depends on the number of dishes or pieces to be washed per hour, the money available, hot-water storage and heating facilities, individual preferences, and other factors. For example, if fewer than 400 pieces are to be washed per hour, a 3-compartment sink or single-tank dishwashing machine can be used. If 1200 pieces are to be washed per hour, then three 3-compartment sink arrangements or one single-tank dishwashing machine may be used. If 2400 pieces are to be washed per hour, a 2-tank dishwashing machine may be the best answer. It soon becomes apparent that the cost of labor alone becomes the factor to be weighed against the cost of a dishwashing machine. Manufacturers of approved-type dishwashing machines are in a position to make recommendations for specific installations.

Soiled silverware requires special handling to assure removal of soil. It has been found that silverware that has been allowed to soak at least 15 min at 130 to 140°F in a good detergent can be hand or machine washed with uniformly good results.

The relatively high humidity in most food establishments and the incompleteness of many routine cleaning operations favor the development of molds and odors. Molds grow in refrigerators, on foods, floors, walls, ceilings, and equipment. Since mold spores are carried by air currents, packing, and dirt, they are difficult to keep out; hence their control depends on cleaning, humidity reduction, and treatment. A recommended mold-control program includes cleansing of affected area with an alkaline detergent, spraying with a 5000-mg/l sodium hypochlorite solution with care to materials that might be harmed, drying of the hypochlorite, spraying with a 5000-mg/l quaternary ammonium compound, and then respraying of the entire section with a 500-mg/l quaternary solution every week or two.[21] Molds are killed by a solution of sodium pentachlorphenate or sodium orthophenylphenate. Practically all off-odors in food establishments are due to decomposing food, grease, or other organic material. Their control rests on proper design and routine cleanliness. The use of deodorants or special odorous disinfectants temporarily masks the odor, but cannot substitute for a warm detergent solution and "elbow grease."

[21] Thomas D. Laughlin, *Modern Sanitation Practices,* Institutions Div. of Klenzade Products, Inc., Beloit, Wis., 1954.

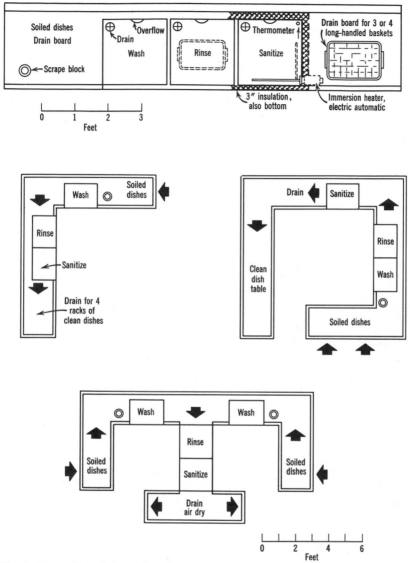

Figure 8–14 Hand dishwashing plans.

Hand Dishwashing, Floor Plans, and Designs

For hand washing, a 3-compartment sink and long-handled wire baskets are required. See Figure 8–14. The following steps are recommended:

1. *Scrape and flush* all dishes and utensils promptly—before the food soil has a chance to dry. A spray-type flusher delivering water under pressure at a tempera-

ture of 110°F, or a pan of hot water and brush, will remove most of the heavy soil.

2. *Wash* in the first compartment with water at a temperature of 110 to 120°F using an effective washing compound (detergent). The detergent solution strength must be maintained. Glasses will require brushing; wash, rinse, and sanitize separately from dishes.

3. *Rinse* in the second compartment by immersion in clear hot water. The dishes, cups, and utensils from the washing operation can be placed directly in the wire basket, previously submerged in this sink, to save on extra handling. Place cups and glasses in a venting position. A hot water spray is also suitable for rinsing if thoroughly used and the drain is kept open.

4. *Sanitize* in the third compartment by immersing the basket of dishes in water at a temperature of 180°F or higher for at least 30 sec, although 2 min is generally recommended. Steps 3 and 4 can also be accomplished by thorough spraying of carefully racked dishes with 180°F water in a properly designed and maintained single tank-stationary rack dishwashing machine or in an improvised spray-rinse cabinet.

5. *Air dry*—do not towel! Towelling is an expensive, time-consuming, and insanitary operation. Towels become moist, warm, and collectors and spreaders of germs. Store dishes, cups, and utensils in a clean dry place and handle them in a sanitary manner. Cups should face down; knives, forks, and spoons with handles up. Dishes and utensils properly sanitized with scalding hot water will dry in less than one minute. Plastic dishes will take longer unless a drying agent is used in the rinse water.

Glass washing becomes a special problem where facilities are inadequate. The glass, being transparent, is subject to critical visual inspection on being raised to one's lips, thereby creating demand for a glass that looks clean. In any case, glasses should receive cleansing and sanitization treatment equal to that given dishes and utensils. A procedure that has been found effective requires the washing of glasses separate from dishes and includes the following steps:

1. *Remove* cigarette butts, papers, ice, liquid, and straws. Then prerinse the glasses with warm water to remove milk and syrup. This will also warm up the glasses and prevent them from cracking when placed in hot water. If glasses cannot be prerinsed promptly, soak them under warm water to prevent the milk, syrup, and soil from hardening.

2. *Wash* glasses in warm, 110 to 120°F, water containing an organic acid detergent using a stiff bristle brush to remove lipstick and stubborn soil. Stationary and motor-driven cluster-type brushes that are placed in the bottom of wash tanks are available.

3. *Rinse* the glasses in clean warm water or in a hot-water spray. If a basket is used care should be taken to stack the glasses in a venting position. Use a special basket that holds glasses in place.

4. *Sanitize* the glasses as described under "Hand Dishwashing," 4, or in clear hot water containing ¼ oz of 5¼ percent hypochlorite solution per gallon (100 mg/l) of water in the sink. Make up a fresh tank of hot water and chlorine solution when the water cools. A water softener or a special solution added to

the final rinse tank may prevent the formation of spots. Drying the bottom of the glass will also help prevent the formation of spots. Quaternaries and iodophors may also be reliable disinfectants under controlled conditions.

5. *Air dry* the glasses by placing them bottoms up on a wire or plastic draining and drying rack after the disinfecting rinse. Do not towel the glasses dry.

Some establishments have found it more convenient to wash and sanitize glasses in the kitchen sinks or dishwasher instead of at the fountain or bar. Although this procedure will entail keeping a somewhat larger stock of glasses, it is offset by increased efficiency, less confusion at the counter, and improved service. Where a dishwasher is used, the glasses should be washed before the dishes. If this is not possible, the dishwasher should be thoroughly cleaned, new detergent solution made up for the wash tank, and fresh rinse water added in a 2-tank machine before the glasses are washed. Special hot-water glass washers and sanitizers that greatly simplify the entire cleansing routine are also available. In either case, glasses should first be washed in a sink with stationary or rotating brushes and washing solution to assure the consistent removal of all lipstick and stubborn soil. Glasses that have become coated or clouded can be made sparkling clean by soaking in an organic acid solution similar to that used in the dairy farm or milk plant for milkstone removal, or by the use of an organic acid detergent. Where equipment and personnel are not available to produce clean glasses, paper cups should be used. Paper service for drinking water, ice cream, milk, and other drinks will greatly reduce the dishwashing problem and frequently improve public relations.

A fundamental objective in the design of a dishwashing space is the continuous movement of dirty dishes through the cleansing operation to the point of distribution or storage without interruption or pile-up. This requires an efficient floor plan that eliminates confusion, provides adequate soiled-dish drainboard space, sufficient wash, rinse, and sanitizing sinks, drainboard space for air drying of long-handled baskets, an adequate number of long-handled baskets, sufficient personnel, and properly designed water-heating facilities. Some floor plans or flow diagrams are shown in Figures 8–14 and 8–15. In operation, an extra long-handled basket can be kept in the rinse compartment when the wash and rinse tanks are adjoining. Then as dishes are washed they can be placed directly in the basket to be rinsed, sanitized, and air-dried without further individual handling. If the wash and rinse tanks are not adjoining, the basket can be placed on the drainboard alongside the wash tank. In small establishments one man can perform all operations. During rush periods, or when the number of dishes is large, the tasks can be divided between two persons, extra wash tanks provided as illustrated, or duplicate 3-compartment sink arrangements can be installed. About 400 pieces can be washed per hour.

Hot Water Requirements

Maintenance of the water in the sanitizing compartment at the proper temperature requires auxiliary heat. This is also desirable in the wash and rinse compartments, unless the water is replaced frequently. A gas burner beneath an insulated sink or an electric immersion heater in the sink are the common methods of keeping the water in the third compartment at 180°F or higher. Other heat sources are gas- or oil-burning units, steam and hot-water jackets, "side arm" heaters, or the equivalent.

Thermostats that automatically control water temperatures, and tank insulation, are useful accessories that help maintain proper water temperatures and also reduce heat losses. A thermometer installed on the third compartment is necessary to show the dishwasher when the water is at disinfection temperature. Thermometers on the wash and rinse tanks are also useful. When an open flame is used beneath the sink to heat the water, a copper sheet placed against the sink bottom will prevent burning out the bottom of the sink. Dial thermometers and gages are frequently inaccurate. They should be checked against a reliable standard. If an adequate number of sinks is not available, a 20-gal pot of hot water can be kept on a stove and dishes immersed in boiling water with the aid of a special basket. The 3-can method used by the U.S. Army is very satisfactory for field use. The first can contains hot water with a detergent and brush, the second can boiling water for rinsing, and the third can boiling water for sterilization.

An example is given below showing calculation of the amount of energy required to heat water in sinks.

Given: A sink $18'' \times 24''$ containing 12 in. of water at a temperature of 120°F. The sink is open at the top and uninsulated. One cubic foot of water = 62.4 lb. Room temperature = 80°F. Water surface loss is approximately 1430 Btu for 100°F temperature difference per hour per square foot. For 80° temperature difference the loss is 940 Btu, and for 120° the loss is 2320 Btu. Failure to insulate the tank results in a heat loss of 215 Btu per ft^2 per hr when the temperature of the hot water is 180°F and the air temperature is 70°F, or about 1.95 Btu per hr per deg per ft^2.

Required: The heat energy to keep the water temperature at 180°F, with air temperature at 80°F, to disinfect 100 to 400 pieces per hour. Assume the temperature of the wash water is 100°F, and a dish, saucer, and cup weigh 2 to 3 lb—say 3 lb or an average of 1 lb per piece. The specific heat, or the amount of heat required to raise 1 lb of crockery 1 deg is 0.22 Btu.

1. The heat loss through the sides and bottom of the sink with no insulation, in
   Btu/hr

   = ft² of surface area × wall loss in Btu/per hr per deg × deg temperature difference between the water and room
   = 10 × 1.95 × 100
   = 1950 Btu/hr

2. The heat loss from an open water surface, in Btu/hr

   = ft² of water surface × surface loss in Btu for 100° temperature difference between water and room
   = 3 × 1430
   = 4290 Btu/hr

3. Heat absorbed by 100 dishes, in Btu/hr

   = weight of dishes in lb × specific heat of crockery in Btu × temperature rise in degs
   = 100 × 0.22 × 100
   = 2200 Btu/hr

4. The heat required to raise the temperature of water in a sink 18 × 24 in. containing 12 in. water from 120 to 180°F, in Btu/hr

   = lb of water × specific heat of water × total temperature rise (rather than average)
   = (1.5 × 2 × 1)62.4 × 1 × 60
   = 11,250 Btu/hr

Therefore: The total heat loss is the sum of the loss from the sink sidewalls and bottom, water surface, and heat absorbed by the crockery sanitized in 1 hr.

Total heat loss for
(after temperature of
180°F is reached)

—100 pieces = 1950 + 4290 + 2200 = 8440 Btu/hr
—200 pieces = 6240 + 2(2200) = 10,640 Btu/hr
—300 pieces = 6240 + 3(2200) = 12,840 Btu hr
—400 pieces = 6240 + 4(2200) = 15,040 Btu hr

The heater provided should be at least capable of conducting 11,250 Btu/hr to the water in the sink. This heat will be adequate to disinfect about 250 pieces/hr as noted above. If 400 pieces are to be disinfected/hr, the heater should be capable of conducting 15,040 Btu/hr. Hence an electric immersion heater would have a capacity $= \dfrac{15,040}{3412} = 4.5$ kW-hr. An open-gas flame heater, being only about 50 percent efficient overall would have a capacity $= \dfrac{15,040}{0.50} = 30,080$ Btu/hr.

NOTE: The wall heat loss of the sink can be reduced to approximately 0.24 Btu per degree temperature difference per hour with 3-in. insulation. The water surface heat loss of 4290 Btu during buildup of water temperature can be reduced to 429 Btu (1/10) by providing an insulating cover.

Electric immersion heaters for sinks require (1) a 115-volt circuit for the lower capacities of 1500 to 3000 (or 4000) W and a 230-V circuit for the higher capacities; (2) thermostat control; (3) variable or three-point heat switch; and (4) burn-out protection or low-water cutoff controls. Immersion heaters are manufactured by General Electric, Schenectady, N.Y., Electro-Therm Inc., Silver Springs, Md., and others. If sink bottom and wall insulation is omitted, radiation heat losses will be increased by about eight times and must be compensated for by increased wattage

Where a gas or oil burner is used beneath a sink, precautions must be taken to fire-retard the space around the burner. This space should be vented to the outer air and a flame guard provided. In addition, a copper sheet placed between the flame and bottom of stainless steel or other type of sink will prevent burning out of the bottom. The American Gas Association recommends, for furnishing 180°-water in sinks supplied with 140°-inlet water, gas burner inputs in Btu per hour for under sink burners of:

26,000 Btu/hr for 16 × 16 to 18 × 18 in. sinks
30,000 Btu for 18 × 20 to 20 × 20 in. sinks
34,000 Btu for 20 × 20 to 22 × 22 in. sinks
38,000 Btu for 22 × 24 to 24 × 24 in. sinks[22]

For example, a sink 24 × 24 in. with 12 in. of water will hold $(2 × 2 × 1)62.4 =$ 249.6 lb of water. If the water enters at 120°F and is to be heated to 180°F, the heat required $= 249.6 × 60 = 14,976.0$ Btu/hr. The heat loss through bottom and sides of tank without insulation = surface area × wall loss × degrees temperature difference between water and room $= 12 × 1.95 × 100 = 2,340$ Btu/hr. Heat loss from open water surface $= 1,430$ Btu/ft² for a 100° temperature difference, which for 4 ft² $= 5720$ Btu. The total heat loss, assuming a 1-hr temperature buildup $= 14,976 +$ $\dfrac{2340}{2} + \dfrac{5720}{2} = 19,006$ Btu. With an open-gas flame heater assumed to have an overall efficiency of 50 percent, the heat input $= 19,006/0.50 = 38,000$ Btu/hr.

Machine Dishwashing, Floor Plans, and Designs

When more than about 400 dishes and cups must be washed per hour, either a duplicate 3-compartment sink and drainboard installation is required or a mechanical dishwasher should be provided. Of course, a dishwashing machine can be provided regardless of the number of dishes washed. Small machines are available to serve the average household and up to 50, 125, 250, 400, or 3000 persons. In any case the equipment, including the hot-water supply, must always be proper and adequate to meet peak operating requirements. Since the average person does not know which dishwashing machines will produce satisfactory results, he

[22] J. Stanford Setchell, *Enough Hot Water—Hot Enough,* American Gas Association, 420 Lexington Avenue, New York 10017, N.Y.

should be guided by the recommendations of the National Sanitation Foundation and purchase a machine that contains their seal of approval.

For proper machine dishwashing the following steps are recommended to help assure consistently satisfactory results.

1. *Scrape* dishes clean of food and soil with the hand or a rubber scraper. This operation can also be accomplished by a pressure water spray or a large recirculated stream of water at a temperature of about 110°F. A garbage grinder or food-waste disposer is a very desirable unit to incorporate in this operation. The dishwashing layout should provide a soiled-dish table equipped with a removable strainer the width of the table, just ahead of the machine, to intercept liquid garbage before it enters the dishwashing machine.

2. *Rack dishes.* Place dishes in special wooden or wire racks or flat on the converyor belt. Do not stack or otherwise overcrowd the dishes. Spray the rack of dishes with a manual preflusher delivering water at a temperature of 140°F under pressure or preflush each dish by hand. See Figure 8–15.

 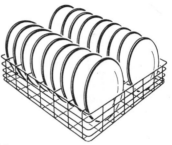

Typical Racks for Dishes

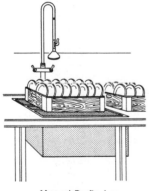

Manual Preflusher

Figure 8–15 Equipment details. (From *National Sanitation Foundation Standards,* Standard No. 1, "Soda Fountain and Luncheonette Equipment." July 1952, and Standard No. 2, "Food Service Equipment," October 1952, Ann Arbor, Mich.)

3. *Wash and sanitize.* Feed racked dishes into the dishwashing machine at the established speed for the machine. The time of the wash and rinse cycles, the volume and temperature of water required, and other details for each type of machine are given in Tables 8–15 and 8–16. This information is summarized from

TABLE 8-15   STATIONARY RACK DISHWASHING MACHINE, WASH AND RINSE WATER, VOLUME, TEMPERATURE, AND TIME

Rack	150 to 160°F, Wash Cycle per Rack*			Alternate Wash Cycle per Rack†			Rinse, 180° to 195°F‡		
	Water Volume	Time; not less than	Pump Capacity	Water Volume	Min. Pump Capacity	Time; not less than	Water Volume, in gal		
							15 psi	20 psi	30 psi
20″ × 20″	92 gal	40 sec	140 gpm	92 gal	75 gpm	10 sec	1.50	1.73	2.10
18″ × 18″	75	40	112	75	50	10	1.25	1.44	1.75
16″ × 16″	60	40	90	60	40	10	1.00	1.15	1.40
Other	0.23 × rack area in square inches	40	0.35 × rack area in square inches	0.23 × rack area in square inches	0.15 × rack area in square inches	10	0.43 gal/in.² of rack area		

Source: Adapted from National Sanitation Foundation Standard No. 3, *Spray-Type Dishwashing Machines*, April 1965.

*Wash water temperature thermostatically maintained at 150 to 160°F.

†Time of wash cycle increased beyond 40 sec to apply not less than the indicated volume of wash water per rack.

‡Rinse water flow pressure on line at machine controlled between 15 and 25 lb. (A pressure control valve adjusted to keep the line pressure at 20 psi at all times is strongly recommended. If pressure in line can fall below 15 psi a hydraulic analysis and correction is required.) In single temperature door-type machine, minimum rinse time shall be 30 sec and spray volume 23 gal for each 400 in.² of rack area and rinse water temperature 165°F; used rinse water is discharged to waste. Temperature not less than 165°F in single temperature door-type machine.

616

TABLE 8-16 CONVEYOR DISHWASHING MACHINE, WASH AND RINSE WATER, VOLUME, TEMPERATURE, AND TIME

Type Machine	Wash Cycle*			Recirculated Rinse, 170°F			Final Rinse, 180° to 195°F
	Volume of Water per Lineal Inch of Conveyor length	Minimum Period of Wash	Pump Capacity	Volume of Water per Lineal Inch of Conveyor length	Minimum Period of Rinse	Pump Capacity	Minimum Flow†
Single-tank conveyor (dishes prewashed or water scrapped).	3 gal per 20-in. width; 2.7 gal per 18-in. width; 3.3 gal per 22-in. width; and 3.6 gal per 24-in. width (160°F min.)	15 sec	140 gpm	None	—	—	6.94 gpm across conveyor to cover space of at least 6 in. in direction of travel measured 5 in. above conveyor, with rack or conveyor width of 20 in.
Multiple-tank conveyor with dishes in inclined position or in racks (dishes prewashed or water scrapped).	1.65 gal per 20-in. width; 1.48 gal per 18-in. width; 1.82 gal per 22-in. width; 1.98 gal per 24-in. width (150°F min.)	7 sec	125 gpm	1.65 gal per 20-in. width; 1.48 gal per 18-in. width; 1.82 gal per 22-in. width; 1.98 gal per 24-in. width	7 sec	125 gpm	4.62 gpm across conveyor to cover space of at least 3 in. in direction of travel measured 5 in. above conveyor with conveyor width of 20-in.

Source: Adapted from National Sanitation Foundation Standard No. 3, *Spray-Type Dishwashing Machines*, April 1965. See Standard No. 5 for hot-water supply and equipment.

*Each lineal inch sprayed from above and below.

†Rinse-water flow pressure on line at machine controlled between 15 and 25 lb. (A pressure control valve adjusted to keep the line pressure at 20 psi at all times is strongly recommended. If pressure in line can fall below 15 psi, a hydraulic analysis and correction is required.)

Standard No. 3, *Spray-Type Dishwashing Machines,* published by the National Sanitation Foundation in April 1965. For satisfactory results, the proper use of a good washing compound or detergent that is suitable for the water hardness is necessary. The concentration of detergent should be maintained at 0.2 to 0.3 percent. This can be obtained at the beginning with about 1 lb or 1⅔ cups of washing powder to 50 gal of fresh wash water. But the detergent must be supplemented during the dishwashing operation to compensate for the soil, rinse water, and make-up water entering the wash-water tank and diluting the detergent concentration. This is accomplished by automatic detergent feeders or by the hand feeding of detergent. Follow the manufacturers' directions. Studies show that the detergent concentration becomes ineffectual after 10 to 15 racks of dishes are washed. An automatic detergent feeder requires cleaning after each dishwashing period, just like a dishwashing machine. In spite of this care, many difficulties have been encountered with "automatic" feeders. In hand feeding it is preferable to apportion the detergent over each rack of dishes, say at the rate of a level tablespoonful for each rack of dishes entering the machine, beginning at the start of the dishwashing operation, after the machine has been charged. Another problem concerns the wash- and rinse-water tempratures. Provision for heating the water in the dishwashing machine wash tank or tanks is often made by the manufacturer, but this should be checked under operating conditions. The connection of an adequate volume of 180°F water to the machine final-spray rinse line is usually under a separate contract and is often neglected, unless the machine representative, designing architect or engineer, health official, or plumbing contractor calls attention to the need. This problem and its solutions are discussed separately under water heaters.

4. *Air dry* the dishes while in the racks. If rinsed at the proper temperature the china dishes will dry in less than 1 min. Plastic cups, dishes, and trays dry slowly. Plastic is hydrophobic, causing water to form in droplets that take longer to evaporate than a film. Drying agents are now available that can be fed into the rinse water, thereby overcoming this difficulty. Sufficient space on an airy dish table is necessary for the air drying to take place promptly. Consideration should be given to adequate ventilation to prevent condensation.

An efficient floor plan adapted to each establishment is necessary to realize the advantages associated with the use of a dishwashing plan. Some plans that show the essential elements are illustrated in Figures 8–11, and 8–16.

Hot-water Requirements[23]

To meet the hot-water demands of a dishwashing machine it is necessary to have a separate heater and storage tank, a two-stage heater, a separate instantaneous heater, or a booster heater. Thermostatic hot-water controls, including a hot-water circulation pump, will help assure a dependable water supply at proper temperature. The required volume

[23] For design details see *ASHRAE Guide and Data Book Systems 1970,* American Society of Heating Refrigerating and Air-Conditioning Engineers, Inc., 345 East 47th Street, New York, pp. 543–572.

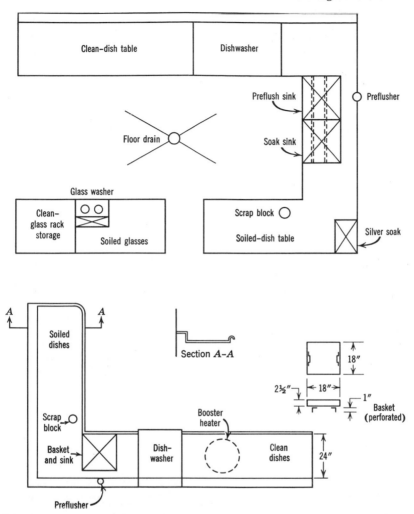

Figure 8–16 Dishwashing arrangements—machine.

of water is indicated in Tables 8–15, 8–16, and 8–17. To this should be added the amount of water required to fill the wash and recirculating rinse tanks.

1. Fluctuations in water pressure due to large demands and hydraulic friction losses will cause variations in the rate of flow. This is apparent from the basic hydraulic formula $Q = VA$, in which Q is the rate of flow in cubic feet per second (cfs), V is the velocity of the water in feet per second (fps), and A is the area of the pipe in square feet. But $V = \sqrt{2gh}$ and $h = p/w$, in which $g = 32.2$ feet

TABLE 8–17   APPROXIMATE VOLUME OF 180°F WATER
USED WITH TIMED RINSE

Dishwashing Machine Types	Rack Size or Width	Mechanical Capacity per hour, maximum	Gallons of 180°F Rinse Water Required*		
			per hour, maximum*	per rack	per minute
Single-tank, stationary rack.	20″ × 20″	60 racks	104	1.73	10.4
	18″ × 18″		87	1.44	8.67
	16″ × 16″		69	1.15	6.94
Single-tank rack conveyor.		240 racks	416	1.73	6.94
Multiple-tank conveyor. Dishes inclined.		continuous	277†	−	4.62
Multiple-tank belt conveyor. Dishes flat.		continuous	347†	−	5.78

*This assumption provides for some wasted rinse water due to inefficiency in operation; in practice machines are not operated without interruption, nor are racks always fully loaded. Some designers say the machine is actually in operation only about two-thirds of any hour, but then add 50 percent to the hourly flow to compensate for wasted 180°F-rinse water. Check actual rinse-water usage with manufacturer and actual water flow under operating conditions. Flow is based on 20 psi at machine.

†Add 2.31 gpm if make-up water for multiple tank machines is drawn from final rinse water line. Assure that *National Sanitation Standards* are met. Do not use push-through-type machine as the amount of rinse water depends on the "human element"; require automatic wash and rinse timers. Assume a 20 × 20-in. rack holds 24 pieces, an 18 × 18-in. rack 20 pieces, and a 16 × 16-in. rack 16 pieces. A common estimate is that an average complete restaurant meal results in 8 dishes or pieces, a luncheon in 6 pieces, a cafeteria meal in 4 pieces, and an elaborate dinner 10 pieces.

per second per second, h = head of water in feet, p = water pressure in pounds per square foot, and w = 62.4 pound for water. It can be seen therefore that a change in velocity, pipe diameter, or water pressure will cause a change in flow, and similarly a change in water flow will cause a change in velocity or pressure. It should be cautioned that many waters are hard or corrosive. Most hard waters contain minerals that form a scale on the inside of pipe when heated and corrosive waters eat away the inside of pipe, both of which conditions increase the friction and hence reduce the flow of water to the dishwashing machine. The removal of hardness and control of corrosion are discussed in Chapter 3.

2. The provision of a pressure-reducing valve on the 180°F rinse line at the machine will assure that a larger volume of 180°F hot water will not be used by the machine than that for which the water heater was designed, or is needed, for proper functioning of the machine. The pressure-reducing valve should be set to maintain the water pressure at the machine preferably at 20 psi. If a larger volume of hot water is permitted to be used by the machine, due to a higher water pressure or manipulation of valves, the water heater will be overloaded beyond the design capacity, resulting in a higher water consumption and hence lower than 180°F-water being delivered to the dishwashing machine.

3. On the other hand, if the water pressure in the rinse line is less than 15 psi under conditions of maximum use, there will be an inadequate volume of 180°F rinse water delivered to the dishwashing machine to do the intended rinsing and disinfection job. This would call for a hydraulic analysis and perhaps a booster pump. A pressure in excess of 25 psi will result in atomization or a fine mist spray at the nozzles, which is not effective in rinsing. By maintaining the water pressure within the established limits, the hot-water rinse line of a known length, and the diameter usually not less than 1 in., it is possible to assure the required rate of hot rinse-water flow. Hence, failure to deliver 180°F rinse water will immediately indicate an inadequate heater or heater and storage tank, or, in an old installation, excessive friction in the water lines. There should be no shutoff valves or drawoffs between the heater and dishwashing machine. A partial closure of a valve on the line will result in a higher hot-water temperature or inadequate flow of rinse water. Drawoffs between the heater and dishwashing machine will reduce the water pressure in the line and the available volume of 180°F-water to the machine.

The required size or capacity of the water heater to serve a dishwashing machine, kitchen fixtures, hand washing sinks, and miscellaneous outlets can be estimated. Two design temperatures are used, 140 to 150°F for general utility purposes and 180 to 200°F for the dishwashing machine. The heating system must be designed to meet the estimated peak hot-water demands for its duration as well as normal hot-water needs. Several methods for computing hot-water requirements are given here.

1. One method suggested by a manufacturer of heating equipment is based on the experience that the amount of hot water required per meal including the peak hot-water demand will average 1.8 gal.[24] This is divided as follows:

28 percent for food preparation, pot and pan washing, etc., lasting 2 to 2½ hr.
55 percent for dishwashing and related activities carried on at same time lasting 1 hr or more.
17 percent for utensil and equipment cleaning and general cleanup lasting 1 to 1½ hr.

An average usage of 1.5 gal per person per meal with a very similar distribution is suggested in the *Cornell Miscellaneous Bulletin* 14.[25] With a 1-hr dishwashing period, it is apparent that the maximum demand to be designed for is 55 percent of 1.8 gal or 1.0 gal per hour per meal served during the heaviest 1-hr feeding period. If the dishwashing period for the same number of dishes is extended over a 2-hr period, the rinse water requirement would be at the rate of 0.5 gal per hour per meal.

[24] *Hot Water Requirements for Food Service Establishments,* Bull. No. HDH, 987A, A. O. Smith Corporation, Kankakee, Ill.
[25] "A Central Camp Building for Administration and Food Service," *Cornell Miscellaneous Bulletin* 14, New York State Colleges of Agriculture and Home Economics, Cornell University, Ithaca, N.Y., March 1953.

At restaurants serving light meals, the peak flow is estimated at 0.7 gal per hour per meal; at school lunchrooms and cafeterias 0.7 to 0.8 gal; and 1.2 gal where elaborate meals are served.

2. Another method of estimating the hourly hot-water needs is to count the maximum number of dish racks to be washed and rinsed in 1 hr. Knowing the amount of rinse water used per rack from Table 8–15 or the manufacturer, it is a simple matter to compute the total amount of 180°F rinse water used per hour. Add to this the amount of water required to fill the dishwasher tank or tanks. For example, if 1.25 gal of 180°F rinse water is used per rack 18 × 18 in. and a rack is washed every 50 sec, which is the maximum possible in a timed single-tank machine set to wash 40 sec and rinse 10 sec, then a maximum of 90 gal will be used in 1 hr. If the wash-water tank holds 20 gal, the total amount of water used in a maximum hour would be 110 gal. In practice, it is sometimes assumed that racks can be fed at a maximum rate of 1/min in a single-tank stationary-rack dishwasher, or 60 racks/hr, which would mean a water consumption of 75 gph, with 18 × 18 in. racks, rather than 90 gal.

3. The American Gas Association suggests that 0.8 gal of 180°F-water is used per meal in a single-tank dishwashing machine and 0.5 gal per meal in a two-tank dishwashing machine, in addition to 1.2 gal of 140°F-water per meal. It is assumed that eight dishes are used per meal.

4. *Design calculations.* The required heater capacity in Btu can be calculated from the formula:

$$\text{Btu} = \frac{\text{gph} \times 8.3 \times \text{temperature rise}}{\text{overall efficiency}}$$

where Btu = required input rating of the heater

gph = gallons per hour of 180°F-water required, usually heated to 190°F when the heater is located some distance from the dishwasher

Temperature rise = temperature of water as it leaves the heater minus the temperature of the water entering the heater

Overall efficiency[26] = the overall heating-system efficiency of gas may be taken at 60–70 percent provided the heater and storage tank, if provided, are insulated. Water heaters hold between 2 to 5 and 10 gal in the coils, or surrounding heating tubes; this water

[26] NOTE. The thermal efficiency of electric heaters may be taken at 100 percent, gas heaters at 60–70 percent, steam boilers at 80 percent, oil heaters at 60 percent, and coal fire heaters at 50 percent. One ft^3 of natural gas = 1050 Btu. See local utility company for other ratings.

is available to meet a momentary demand. The heat loss from long runs of pipe can be appreciable. Table 8–18 gives the heat loss from bare and insulated pipe.

5. *Example:* A restaurant serves a maximum of 450 meals. Dishes are to be washed over a period of 1½ hr. What type of dishwasher should be used? Several types of dishwashers are shown in Figures 8–17a,

TABLE 8-18   HEAT LOSS FROM PIPE CARRYING HOT WATER

| Nominal Pipe Size in inches | Btu per Hour Heat Loss per Linear Foot of Horizontal Pipe in Still 70°F Air Carrying Hot Water at Temperature Stated* | | | | | |
| | Bare Iron Pipe | | | Insulated Pipe 85% Magnesia 1 in. | | |
	120°F	150°F	180°F	120°F	150°F	180°F
½	27.2	45.8	66.6	8.2	13.3	18.5
¾	33.0	55.2	80.2	9.3	15.0	20.9
1	39.5	66.3	96.6	10.5	17.1	23.8
1¼	48.9	81.8	119.5	12.2	19.8	27.6
1½	54.5	92.0	134.1	13.3	21.5	30.0

*Does not include Btu needed to reheat water in pipes. For example, 10 ft of 1-in. pipe holds 1 gal and will require 415 Btu to raise its temperature 50°F, or wasting of water until 180°F water is obtained at the machine or outlet.

b, c, and d. What heater and storage-tank capacities are needed if the temperature of the incoming water to the heater is 40°F?

If an average meal uses 6 pieces and a 20×20 in. rack is used that holds 24 pieces, each rack will hold the dishes from 4 meals. With dishes from 450 meals to be washed in 1½ hr or 300 meals in 1 hr, a total of $300/4 = 75$ racks are to be washed per hour. If 18×18 in. racks that hold 20 pieces are used, each rack will hold the dishes from 20/6 or 3.33 meals. With dishes from 300 meals to be washed in 1 hr, a total of $300/3.33 = 90$ racks are to be washed per hour. The dishwashing machine that meets these requirements should be selected.

The machine selected may be a single-tank rack-conveyor dishwasher or a multiple-tank rack-conveyor dishwasher. Assume that a single-tank 18×18 in. rack-conveyor dishwasher is selected. Such a machine uses 1.73 gal of 180°F rinse water per rack of dishes. With 90 racks to be washed per hour, the hot rinse-water requirement would be $90 \times 1.73 = 156$ gph. (At the mechanical capacity this machine would use 416 gph and process 240 racks.) To this should be added the water to fill the dishwashing machine wash tank, say 25 gal, making a total of 181 gph. If no storage is provided,

the required gas heater $= \dfrac{181 \times 8.3 \times (190 - 40)}{0.60} = 376,000$ Btu/hr[27]

[27] Ibid.

If an electric heater were designed for, its capacity would be

$$= \frac{181 \times 8.3 \times (190 - 40)}{1.0} = 225,000 \text{ Btu/hr,}$$

or since 1 kW-hr = 3,412 Btu, the electric heater would have a capacity of $\frac{225,000}{3,412}$ = 66.0 kW-hr. It may be more practical to use an electric heater as a booster adjacent to the dishwashing machine to raise the rinse water from 140 to 180°F.

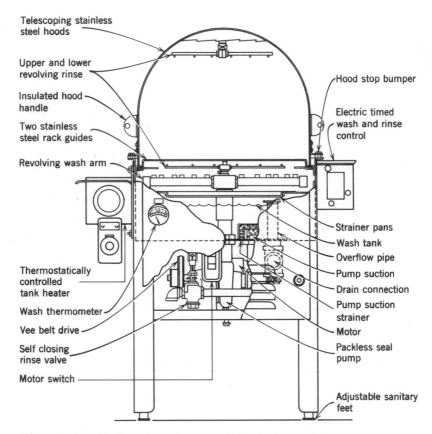

Figure 8–17a   Single-tank stationary rack hood-type machine.

If a 120-gal storage tank is provided and the interval between meal periods is 4 hr, a lesser capacity heater would be required, as follows: A 120-gal storage tank has an available storage of about 75 percent or 90 gal, leaving (160 — 90) or 70 gph to be met instantaneously by the heater. The required gas-heater capacity (adequate for 1½-hr dishwashing) under these circumstances would be:

$$\frac{(90/4 + 70) \times 8.3 \times (190 - 40)}{0.60} = 193,000 \text{ Btu/hr}$$

If a multiple-tank rack-conveyor dishwasher is used, say, 0.5 gal of 180°F rinse water is used per rack of dishes or equivalent. With 90 racks of dishes to be washed per hr, the hot rinse-water requirement would be 90 × 0.5 = 45 gph. Adding the water needed to fill the two dishwasher tanks of 60 gal (see manufacturer's

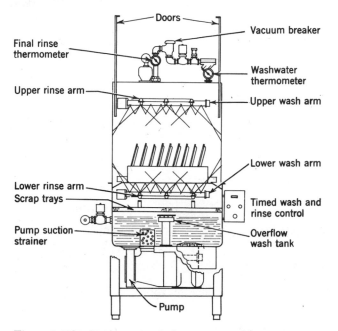

Figure 8–17b   Stationary rack door-type machine.

catalog), a total of 105 gal must be designed for. If no storage tank is provided the

$$\text{required heater} = \frac{105 \times 8.3 \times (190 - 40)}{0.60} = 218,000 \text{ Btu/hr}$$

If a 120-gal storage tank is provided (90 gal available) and the interval between meals is 4 hr, the

$$\text{required heater} = \frac{[\frac{90}{4} + (105 - 90)] \times 8.3 \times (190 - 40)}{0.60} = 78,000 \text{ Btu/hr}$$

This is adequate for a dishwashing period of 1⅓ hr, assuming 180°-water is used to fill the two dishwasher tanks. If it is considered that the 120-gal storage tank must be heated to obtain 90 gal, the

$$\text{required heater} = \frac{(\frac{120}{4} + 105 - 90) \times 8.3 \times (190 - 40)}{0.60} = 93,500 \text{ Btu/hr}$$

This would be adequate for a 2-hr dishwashing period.

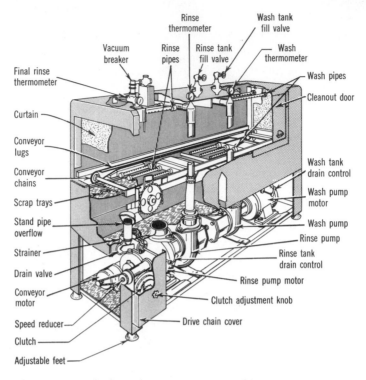

Figure 8–17c   Single-tank conveyor-type machine.

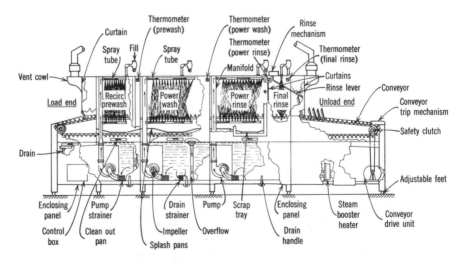

Figure 8–17d   Multiple-tank rackless conveyor-type machine. (From *National Sanitation Foundation Standards,* Standard No. 3, "Commercial Spray-type Dishwashing Machines," Ann Arbor, Mich., April 1965.)

6. Some designers add a factor of safety of 50 to 100 percent to the required heater capacity or preferably design for the mechanical capacity of the dishwashing machine. A common practice is the provision of a storage tank for the storage of all water at a temperature of 140°F. This tank supplies the 140°F needs and feeds the heater or booster supplying 180°F-water. Several water-heater flow diagrams are given in Figure 8–18.

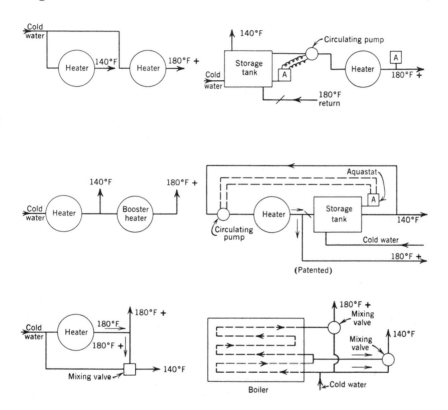

Figure 8–18   Water heater flow diagrams.
NOTE: 1. The heater capacity must equal the requirements for 180°F water plus 140°F water over and above the available storage. A return circulating line is necessary on a heater located at a distance from a dishwashing machine. 2. Thermometers, control valves, and relief valves omitted from drawing. 3. To supply 180°F water to a dishwashing machine requires heating the water to about 190°F.

7. For the problem given, in which dishes from 300 meals are washed, the method 1 would call for 300 gal of 180°F-water/hr; method 5 would require 181 gal; and method 3 would require 240 gal with a single-tank

machine and 150 gal with a two-tank machine, plus water to fill the one or two tanks. In view of the possible variations, it is important that the most probable hot-water requirements peculiar to the establishment under consideration be carefully studied and the best possible estimate made for design purposes. Such factors as type of meals, number of dishes, equipment available, possibilities for hot-water waste, type of personnel and supervision, dishes to be washed per hour, water pressure, and type of diswashing machine all have an effect on the hot-water requirements.

Hot Water for General Utility Purposes

The hot-water requirements for general utility purposes include the needs for the hand washing of dishes, pots, pans, and utensils; for wall, floor, and equipment cleanup; for personal hygiene and toilet rooms; for food preparation; and for waste. One estimate is 30 gal per sink per wash-up, with the dishwashing requirements broken down to: (1) prewash—0.25 to 0.5 gal/meal served; (2) detergent washing—0.75 to 1.0 gal/meal served; and (3) final rinsing—0.5 to 1.5 gal/meal served.[28] Total usage adds up to 1.5 to 3.0 gal/meal served. A factor of safety addition of 50 to 100 percent is suggested, depending on the type of restaurant and seasonal demands.

If it is assumed that 2 gal of 140°F-water are used per meal, that 50 meals are served, and that hand dishwashing is completed in 1 hr, then the total hot-water demand per hour will be 100 gal, or 150 gal if a 50 percent factor of safety is added. This can be obtained from a water heater having a recovery rate to produce 150 gal of 40°F-water in 1 hr. (A separate under-sink booster is also required.) If a storage tank is provided capable of supplying the total demand, and 3 hr are available to heat the water, then the heater recovery rate would be

$$\frac{150}{0.75} \times \frac{1}{3} = 67 \text{ gph}$$

The factor 0.75 represents 75 percent of the storage-tank capacity that may be withdrawn before the incoming cold water cools the remaining stored hot water below the desired temperature.

Another example might be the hot-water requirements of 300 gph in a kitchen, plus that of three washbasins and two showers. The washbasins are used by 30 persons in 1 hr and the two showers by four persons in 1 hr. Say that it takes a person 3 min to wash at a basin and 10 min to take a shower. A basin faucet will flow at the rate of 3 gpm, of which 2 gpm can be taken as hot water. A shower head will flow at the rate of 5 gpm, of which ¾ or 3½ gpm can be taken as hot water. Large variation can be expected depending on the type of head and water

[28] R. M. McLaughlin, paper presented at the Metropolitan Gas Heating and Air Conditioning Council, New York, December 14, 1950.

pressure. The required heater and storage-tank capacities can be obtained as follows:

$$3 \text{ basins} = 30 \text{ persons} \times 3 \text{ min} \times 2 \text{ gpm} = 180 \text{ gal}$$
$$2 \text{ showers} = 4 \text{ persons} \times 10 \text{ min} \times 3\tfrac{1}{2} \text{ gpm} = 140 \text{ gal}$$
$$\text{Kitchen usage} = 300 \text{ gal}$$
$$\text{Total} = 620 \text{ gph}$$

If a storage tank is provided to supply 620 gal of hot water it would have a capacity of $\dfrac{620}{0.75} = 827$ gallons. With a heat-up period of 3 hr, the required heater would have a recovery rate of $\dfrac{827}{3} = 276$ gph.

If a storage tank has a 450-gal capacity, of which $450 \times 0.75 = 337$ gal is available, the required heater would have a recovery capacity of $620 - 337 = 283$ gph. At this recovery rate the storage tank would be reheated in $450/283 = 1.6$ hr.

In practice the nearest standard-size storage tank and heater would be selected.

BIBLIOGRAPHY

Angelotti, Robert, Foter, Milton J., and Lewis, Keith H., *Time-Temperature Effects on Salmonellae and Staphylococci in Foods,* U.S. Public Health Service, Technical Report F60-5, Dept. of HEW, Cincinnati, Ohio, 1960.

West, Bessie Brooks, Wood, Levelle, and Harger, Virginia F., *Food Service in Institutions,* John Wiley & Sons, Inc., New York, 1967.

Bryan, Frank L., *Diseases Transmitted by Foods* (A Classification and Summary), CDC, U.S. Public Health Service, Dept. of HEW, Atlanta, Ga., 1971.

Donovan, Anne Claire, and Ives, Orville B., *Hospital Dietary Services,* U.S. Public Health Service Pub. No. 930-C-11, Dept. of HEW, Washington, D.C., 1966.

Establishing and Operating a Restaurant, Industrial (Small Business) Ser. No. 39, U.S. Government Printing Office, Washington, D.C., 1946.

Food Service Sanitation Manual, 1962 Recommendations of the U.S. Public Health Service Pub. No. 934, Dept. of HEW, Washington, D.C.

Grade "A" Pasteurized Milk Ordinance, 1965 Recommendations of the U.S. Public Health Service, Dept. of HEW, Public Health Service Pub. No. 229, Washington, D.C.

Joint FAO/WHO Expert Committee on Milk Hygiene, WHO Tech. Rep. Ser., No. 453, 1970.

Kotschevar, Lendal H., and Terrell, Margaret E., *Food Service Planning,* John Wiley & Sons, Inc., New York, 1963.

Milk Hygiene, WHO Mon. Ser., No. 48, Geneva, 1962.

Procedure for the Investigation of Foodborne Disease Outbreaks, 2nd ed., International Association of Milk, Food, and Environmental Sanitarians, Inc., Shelbyville, Ind., 1966.

Ramsey, Charles George, and Sleeper, Harold Reeve, *Architectural Graphic Standards,* John Wiley & Sons, Inc., New York, 1970.

Taylor, Joan, ed., *Bacterial Food Poisoning,* Royal Society of Health, London, England, 1970.

Walter, William G., ed., *Standard Methods for the Examination of Dairy Products,* 12th ed., American Public Health Association, Inc., 1740 Broadway, New York, 1967.

9

RECREATION AREAS AND
TEMPORARY RESIDENCES

BEACH AND POOL STANDARDS AND REGULATIONS

The sanitation of bathing places is dictated by health and aesthetic standards. Few people would knowingly swim or water-ski in polluted water, and insanitary surroundings are not conducive to the enjoyment of "a day at the beach." People are demanding more and cleaner beaches and pools, and a camp, motel, club, or resort without a pool or beach is not nearly as popular as one so equipped.

Health Considerations

From our knowledge of disease transmission, it is known that certain illnesses can be transmitted by improperly operated or located swimming pools and beaches through contact and ingestion of polluted water. Among these are typhoid fever, dysentery, and other gastrointestinal illnesses; conjunctivitis, trachoma, leptospirosis, ringworm infections, schistosomiasis, or swimmer's itch; upper respiratory tract diseases such as sinus infection, septic sore throat, and middle-ear infection. The repeated flushing of the mucous coatings of the eyes, ears, and throat and the excessive use of alum or lack of pH control expose the unprotected surfaces to possible inflammation, irritation, and infection. Contraction of the skin on immersion in water may make possible the direct entrance of contaminated water into the nose and eyes.

Stevenson reported that "an appreciably higher over-all illness incidence may be expected in the swimming group over that in the nonswimming group regardless of the bathing water quality."[1] It was further stated in his studies that "eye, ear, nose, and throat ailments may be expected to represent more than half of the over-all illness incidence, gastrointestinal disturbances up to one-fifth, and skin irritations and other illnesses the balance." Although based on limited data, swimming in lake water with an average coliform content of 2300/ml caused "a significant increase in illness incidence . . ." and swimming in river water "having a median coliform density of 2700/100 ml appears to have caused a significant increase in such (gastrointestinal) illness." The study also showed the greatest amount of swimming was done by persons 5 to 19 years of age.

Diesch and McCulloch summarized incidences of leptospirosis in persons swimming in waters contaminated by discharges of domestic and wild animals, including cattle, swine, foxes, racoons, muskrats, and mice.[2] Pathogenic leptospires were isolated from natural waters, confirming the inadvisability of swimming in streams and farm ponds receiving drainage from cattle or swine pastures.

Joyce and Weiser report that enteroviruses that are found in human or animal excreta, if introduced into a farm pond by drainage or direct flow, can constitute a serious public health hazard if used for recreation, drinking, or other domestic purposes.[3]

Many other studies have been made to relate bathing water bacterial quality to disease transmission with inconclusive or negative results.[4]

British investigators have drawn the following general conclusions:[5]

"(i) That bathing in sewage-polluted sea water carries only a negli-

---

[1] Albert H. Stevenson, "Studies of Bathing Water Quality and Health." Presented before the second session of the Engineering Section of the American Public Health Association at the 80th Annual Meeting in Cleveland, Ohio, October 23, 1952.

[2] Stanley L. Diesch and William F. McCulloch, "Isolation of Pathogenic Leptospires from Waters Used for Recreation," *Public Health Reports,* **81**, No. 4, 299–304 (April 1966).

[3] Gayle Joyce and H. H. Weiser, "Survival of Enteroviruses and Bacteriophage, in Farm Pond Waters," *J. Am. Water Works Assoc.,* **59**, No. 4, 491–501 (April 1967).

[4] "Sewage Contamination of Coastal Bathing Waters in England and Wales," Committee on Bathing Beach Contamination of the Public Health Laboratory Service, *J. Hygiene,* **57**, No. 4, 435–472 (December 1959); John M. Henderson, "Enteric Disease Criteria for Recreational Waters," *J. Sanitary Engineering Division,* ASCE, SA6, Proc. Paper 6320, 1253–1276 (December 1968); "Coliform Standards for Recreational Waters," Committee Report, *J. Sanitary Eng. Div.,* ASCE, SA4, Proc. Paper 3617, 57–94. (August 1963).

[5] "Sewage Contamination," ibid.

gible risk to health, even on beaches that are aesthetically very unsatisfactory.

"(ii) That the minimal risk attending such bathing is probably associated with chance contact with intact aggregates of faecal material that happen to have come from infected persons.

"(iii) That the isolation of pathogenic organisms from sewage-contaminated sea water is more important as evidence of an existing hazard in the populations from which the sewage is derived than as evidence of a further risk of infection in bathers.

"(iv) That, since a serious risk of contracting disease through bathing in sewage-polluted sea water is probably not incurred unless the water is so fouled as to be aesthetically revolting, public health requirements would seem to be reasonably met by a general policy of improving grossly insanitary bathing waters and of preventing so far as possible the pollution of bathing beaches with undisintegrated faecal matter during the bathing season."

The findings of the Public Health Activities Committee of the ASCE Sanitary Engineering Division are summarized in the following abstract:[6]

"Coliform standards are a major public health factor in judging the sanitary quality of recreational waters. There is little, if any, conclusive proof that disease hazards are directly associated with large numbers of coliform organisms. Comprehensive research is recommended to provide data for establishing sanitary standards for recreational waters on a more rational or sound public health basis. British investigations show that even finding typhoid organisms and other pathogens in recreational waters in not indicative of a health hazard to bathers but is only indicative of the presence of these diseases in the population producing the sewage. The Committee recommends that beaches not be closed and other decisive action not be taken because current microbiological standards are not met except when evidence of fresh sewage or epidemiological data would support such action."

In view of the available information, emphasis should be placed on elimination of sources of pollution (sewage, storm water, land drainage), effective disinfection of treated waste waters, and the proper interpretation of bacterial examinations of samples collected from representative locations.

In swimming pools there is a possibility of direct transmission of infection from one bather to another if the water does not have an

[6] "Coliform Standards," op. cit.

active disinfectant such as free available chlorine. Proper operation and treatment therefore are of prime importance.

Regulations and Standards

Health departments generally require that swimming pools and bathing beaches be operated under permit. Usually swimming pools are not to be constructed or altered until plans and specifications prepared by a licensed professional engineer or registered architect are submitted and approved by the commissioner of health. Regulations and standards cover water quality, design, construction, operation, maintenance, sanitary facilities, and related factors.

Beaches

Regulations governing the use of bathing beaches or natural partly artificial pools, and evaluation of their suitability, are based on (1) sanitary survey of the drainage area to the beach or pool, (2) the water quality including meteorological factors, (3) epidemiological data linking illnesses to the bathing area, and (4) water circulation and dilution.

The sanitary survey takes into consideration geographic factors and probable sources of pollution on the watershed tributary to the bathing beach. This includes sewage and industrial wastewater discharges, storm-water overflows, bird and animal populations, commercial and agricultural drainage, and their relationship to the beach. The location and volume of the pollution and its chemical, bacterial, and physical characteristics, the volume and quality of the diluting water, water depth, water surface area, tides, time of day and year, thermal and salinity stratification, confluence of tributaries, water currents, and prevailing winds are all evaluated in determining the suitability of the water for bathing. Also evaluated are safety hazards such as fast currents, submerged objects, beach slope and sharp drop-offs, condition and stability of the beach bottom in the wading area, and the water depth in the diving area. The New York State Public Health Law, for example, prohibits the maintenance of a bathing beach within 500 ft of a sewer outfall.

Bathing-beach water quality standards vary throughout the country.[7] They range from no standard to a permissible total coliform count of 50/100 ml to 2400/100 ml or higher; the preponderant number relates results to a sanitary survey of the bathing area, epidemiological data, and judgment. Interpretation of bacteriological results should take into consideration possibly associated illnesses, the sanitary survey, and

[7] Ibid.

meteorological factors such as rainfall, tidal currents, and prevailing winds as mentioned above.

Bathing-beach standards usually refer to the most probable number (MPN) of the coliform group of organisms per 100 ml of sample. This group includes the *Escherichia coli*, which usually inhabit human and animal intestines, and the *Aerobacter aerogenes* and the *Aerobacter cloacae*, which are frequently found in many types of vegetation and in pipe joint material, pipelines, and valves. The intermediate-aero-genes-cloacae (IAC) subgroups are sometimes found in fecal discharges, but usually in smaller numbers than the *E. coli*. *A. aerogenes* and inter-mediate types of organisms are reported to be commonly present in soil and in waters polluted sometime in the past. It is apparent therefore that all surface waters can be expected to be polluted to a greater or lesser extent and that bathing in "unpolluted" water is a practical im-possibility. It is also apparent that the total coliform test does not dis-tinguish between the more hazardous recent human and animal pollution and the soil, vegetation, and old or past human and animal pollution.

The coliform group of organisms includes the fecal coliforms (*Escherichia coli* at 44.5°C). The presence of fecal coliforms is thought to be more indicative of recent human and animal pollution and hence the presence of pathogenic organisms. The fecal coliforms may average 15 to 20 percent of the total coliforms in stream samples. See also Chap-ter 3, page 115. A recommended standard for water used for wading, swimming, water-skiing and surfing states:[8]

"Fecal coliforms should be used as the indicator organism for evaluat-ing the microbiological suitability of recreation waters. As determined by the multiple-tube fermentation or membrane filter procedures and based on a minimum of not less than five samples taken over not more than a 30-day period, the fecal coliform content of primary contact recreation waters shall not exceed a log mean of 200/100 ml, nor shall more than 10 percent of total samples during any 30-day period exceed 400/100 ml.

". . . the pH should be within the range of 6.5–8.3 except when due to natural causes and in no case shall be less than 5.0 nor more than 9.0.

". . . the clarity should be such that a Secchi disc (20 cm in diameter divided into four quadrants painted alternating black and white) is visi-ble at a minimum depth of 4 feet. In 'learn to swim' areas the clarity

[8] *Report of the Committee on Water Quality Criteria*, Federal Water Pollution Control Administration, U.S. Dept. of the Interior, Washington, D.C., April 1, 1968, pp. 8–14.

should be such that a Secchi disc on the bottom is visible. In diving areas the clarity shall equal the minimum required by safety standards, depending on the height of the diving platform or board."

Swimming Pools

The sanitary quality of swimming-pool water is determined by certain bacteriological, chemical, and physical tests. Examinations are made in accordance with the analytical procedures described in *Standard Methods* by competent laboratory personnel.[9]

The Joint Committee on Swimming Pools recommends:[10]

pH	7.2 to 8.2.[11]
Alkalinity	at least 50 mg/1.
Clarity	6-in. black disc on a white field readily visible at deepest point.[12]
Plate count (Agar, 35°C)	not more than 15% of samples over a considerable period of time contain more than 200 bacteria per ml.
Coliform organisms (dechlorinated sample)	not more than 15% of samples over a considerable period of time show positive (confirmed test) in any of five 10-ml portions or more than 1.0 coliform organisms per 50 ml in membrane filter test.

Other Recreation Waters

For surface waters for general recreational use, not involving significant risk of ingestion and in the absence of local epidemiological evidence to the contrary, a standard of "an average not to exceed 2000 fecal coliforms per 100 ml and a maximum of 4000 per 100 ml, except in specified mixing zones adjacent to outfalls" is suggested.[13]

For waters where the probability of ingesting appreciable quantities is minimal, "the fecal coliform content, as determined by either the multiple-tube fermentation or membrane filter technique, should not exceed a log mean of 1000/100 ml, nor equal or exceed 2000/100 ml in more than 10 percent of the samples."[14]

---

[9] *Standard Methods for the Examination of Water and Wastewater,* American Public Health Association, 1015 Eighteenth Street, N.W., Washington, D.C. 1971.
[10] *Suggested Ordinance and Regulations Covering Public Swimming Pools,* American Public Health Association, 1015 Eighteenth Street, N.W., Washington, D.C. 1964.
[11] See Swimming Pool Operation this Chapter.
[12] The National Swimming Pool Institute recommends that a 2-in. diameter disc with red and black alternate quadrants be clearly visible and the colors discernible in 15 ft of water.
[13] *Report of the Committee on Water Quality Criteria,* op. cit.
[14] Ibid.

Sample Collection

Water samples should be collected in wide-mouth bottles by plunging the bottle downward and then forward until filled, while the bathing area or swimming pool is in use. The sampling points should be in the bathing area of a beach or near the pool outlet or outlets and at such representative points as will indicate the quality of the bathing water. A sample per 300 ft of beach in about a 2-ft depth of water is suggested. Samples of chlorinated water must be collected in sodium-thiosulfate-treated bottles so as to dechlorinate the sample and thus give a true indication of the quality of the water at the time of collection. Sodium thiosulfate should not be flushed out. In any case, the sample should be returned to the laboratory for examination as soon as possible after collection.

Accident Prevention

The elimination of tripping or slipping hazards in pools, runways, and decks and controlled area bathing with adequate lifeguard supervision will do much to prevent accidents and drownings. Attention to accident hazards and protection of life are musts at all public beaches and pools, including those at resorts, schools, clubs, associations, and so forth. Failure to do so will place the management in an untenable position in case of lawsuits.

The American Red Cross suggests that one guard may be sufficient for each 100 bathers provided they are in a confined area.[15] On surf bathing beaches one guard in a tower for every 100 yd of beach plus a guard in a boat every 200 yd in the swimming area is recommended. Double this number may be needed on weekends and holidays. One lifeguard director for every 75 swimmers plus one trained "life saver" for every 10 persons in the water has also been recommended at recreational camps and similar places. Bathing areas and pools should be marked for swimmers and nonswimmers. Lifeboats with anchors and bamboo poles 10–15 ft long, surfboards, torpedo buoys, heaving lines of $\frac{3}{16}$-in. manila line made up in coils containing 75 ft of line attached to 15-in. diameter ring buoys, and grappling irons are standard equipment for bathing beaches.

Life-saving equipment per 2000 ft² of pool-water area should include a bamboo pole 10 to 15 ft long or shepherd's crooks and a ring buoy not more than 15 in. in diameter, with 60 ft of $\frac{3}{16}$-in. manila line

---

[15] *Life Saving and Water Safety*, American Red Cross, Blakiston Co., Philadelphia, 1937; also, Bull. No. 27, New York State Department of Health, Albany, 1962.

attached and a ¼-in. heaving line equal to 1½ times the pool width. An elevated lifeguard tower should be required for a pool 2250 to 4000 ft², and additional towers for larger pools. Where a lifeguard is not on duty, a sign in plain view should state in clear letters, at least 4 in. high, "Warning—No Lifeguard on Duty." Another sign should state, "Children Should Not Use Pool Without An Adult in Attendance.[16] Check state and local laws; many prohibit bathing at public beaches and pools unless a lifeguard is on duty. All pools should be fenced and kept locked when not under supervision.

Where the water is not crystal clear at all times, as in natural ponds and dammed-up streams, it is especially important that bathing be carefully organized. A system of checking such as "roll call" or "tag board" to determine who is in the water and who comes out, and a "cap" system to distinguish nonswimmers, beginners, and swimmers, or a "buddy" system whereby bathers are paired according to their abilities are accepted and highly recommended safety practices.

A telephone should be readily accessible and the telephone numbers for respirators, ambulances, doctors, and hospitals conspicuously posted for use in case of emergency. Also needed are a 24-unit first-aid kit, cot and blankets, and a trained first-aid man.

Artificial pools, lakes, coastal beaches, and ponds may have holes, steep sloping bottoms, sudden drops, large rocks, stumps, heavy weed growths, tin cans, bedsprings, broken glass, and miscellaneous debris that can cause injury or drowning. These should of course be removed or, if not practical, carefully marked. The bottom should be gently sloping, clean, clear, and firm at least out to a depth of 4 or 5 ft. Silt, muck, or quicksand bottoms are not suitable. Where it is not feasible to remove the silt, muck, or mud and replace it with sand, a polyethylene sheet covered with 12 in. of sand or a paved bottom might be substituted at least in the wading area. This will help keep the water clear and reduce the drowning hazard.

SWIMMING POOL TYPES AND DESIGN

Swimming pools are generally classified as artificial pools and partly artificial pools. Artificial pools are usually constructed of concrete or steel. Plastic or composition materials are also used. Pools are designed and operated as recirculating pools with filtration and disinfection, or sometimes with disinfection only, and as fill- and-draw pools or as flow-through pools.

[16] *Suggested Ordinance and Regulations,* op. cit.

Recirculating Swimming Pool

In the typical recirculating-type swimming pool a pump takes water out of the pool, passes it through a filter, chlorine is added, and the water is returned to the pool. Water lost by evaporation, splashing, and backwashing of the filters is replaced by fresh water. This type of pool permits use by a maximum number of bathers; a minimum amount of water is wasted, and fuel for heating the water, where desired, is saved because the filtered pool water is reused. If the filters are omitted the organic and dirt load cumulate, residual chlorine control is difficult, and frequent change of water and cleaning is necessary. Filtration is provided by means of gravity rapid-sand filters, pressure sand, anthrafilt or diatomaceous earth filters, and by vacuum filters. The pressure filters are most commonly used on swimming pools.

Fill-and-Draw Pool

The fill-and-draw pool is filled with fresh water, used for some period of time, then emptied, cleaned, and refilled. Such pools are nothing more than common bathtubs in which the pollution introduced is circulated among the bathers. Their use is generally prohibited.

Flow-Through Pool

The flow-through pool is fed by a continual supply of fresh water, without treatment, which causes an equal volume of pool water to overflow to waste. Although the bacterial pollution is reduced, it is not completely flushed out of the pool, as shown in Table 9–1. For example, a pool 60′ × 20′, holding 55,000 gal and provided with 166,500 gal of water per day to displace the pool water in 8 hr, can, according to the formula $Q = 6.25T^2$, accommodate 17 bathers per hour. $Q =$ quantity of fresh water (no disinfectant) required per bather, which in this case is 400 gal, and $T =$ the turnover period in hours, 8 hr in this example. On the other hand, this same pool would accommodate 48 bathers at any one time if the water were continuously chlorinated and recirculated, the restriction being a desired pool area of 25 ft^2 per bather in the pool, maintenance of 0.6 mg/l free available chlorine, and water clarity.

Partly Artificial Pool

The partly artificial pool is made by damming a stream, causing water to back up to form a small lake. If the size of the artificial pool is small, the permissible number of bathers per hour would be governed largely by the dilution or volume of water entering the pool. But if a large lake is formed such as to permit the natural laws of purification

TABLE 9-1   EFFECTIVENESS OF CONTINUOUSLY FLOWING FRESH
DILUTING WATER IN REMOVING POLLUTION FROM A SWIMMING
POOL IN THE ABSENCE OF DISINFECTION

1 Number of Times Each Day of 24 hr Water is Displaced by Fresh Water	2 Displacement Period, in Hours T	3 Computed Quantity in Fresh Water per Bather Q	4 Percent of Pollution of Water Remaining in Pool	5 Number of Days for Pollution to Reach Uniform Values Shown in Column 4
1	24	3,600	58	9
2	12	900	16	4
3	8	400	5	3
4	6	225	2	2

Source: New York State Health Dept., Bull. 31, Albany, 1950.
Note: Values given in columns 4 and 5 are in accordance with the "dilution law."
$Q = 6.25T^2$.
Assumption: Daily increment of pollution equals that initially present.

to operate, the permissible number of bathers would be governed by
the results of bacteriological examination of a series of samples of water
collected during the bathing period and the sanitary survey.

Summary of Pool Design

Site Selection and Layout

1. Accessible to users; space for parking, recreation, picnicking, bathhouse,
and purification equipment.
2. Adequate and satisfactory water supply.
3. Adequate sanitary sewer or separate disposal system.
4. Pool drainage and wastewater disposal proper. High water table relief.
5. Site drained, gently sloping, 100 yd or more from roads, railroads, factories,
and wooded areas.
6. Pool area enclosed with high wall or fence.

Size of Pool

1. Provide for diving, swimming, and nonswimming areas. A minimum length
of 60 ft and water depth of at least 6 ft at the deep end is suggested for
public pools. A depth of 8½ ft is required for a 0.0- to 2.0-meter diving
board, 10 ft for a 2.1- to 3.0-meter board, 11½ ft for a 3.1 or more meter
board, with 20- to 30-ft minimum pool width.
2. Use Table 9-2 as guide in determining pool size.
3. Allow at least 10 ft² of deck space per bather and include 5-ft wide strip

TABLE 9-2   SWIMMING POOL DESIGN BATHING LOAD LIMITS

Authority	Diving	Area (ft^2) Swimming	Nonswimmers
APHA Bathers expected at time of maximum load	300 ft^2 around each diving board or platform	24	10
New York State Bathers actually in water	—	25	25
Bathers present in pool	—	12½	12½
Illinois Bathers actually in water	—	30	15
Iowa State College	For cities under 30,000; maximum day = 5 to 10 percent of population and maximum day at any one time = one-third maximum day.		
National Recreation Assn.	Provide swimming and bathing space 3 percent of population. Allow 15 ft^2 of water area per person.		
Tile Council of America	600 ft^2 per 1,000 for communities of 4,000 people or less to 320 ft^2 per 1,000 for communities up to 90,000 population.		
Anonymous	Poor area required = $\frac{1}{5}$ × total population served (for communities greater than 5,000 population).		

Note: Adjust required pool area for class use such as in schools, YMCA, and for special local conditions. Pool costs estimated in 1965 at $12.50 to $13.50/ft^2 of water area without bathhouse and $16.50 to $18.00 including bathhouse; recirculation equipment, fencing, filters, piping, etc., included. Pools greater than 2,000 to 4,000 ft^2 cost less per square foot.

around pool. Provide floor drain for each 100 ft^2. Pave deck space; do not use grass or sand. Deck space 3 to 4 times pool-water area is preferred.

Source of Fresh Water

1. Use municipal source if available.
2. Use an existing stream or lake if clean to fill pool through the filters.
3. Develop a well or spring if necessary.

Recirculation System

1. Design to replace water in 6 to 8 hr. For a private pool 12 hr may be acceptable.
2. Provide inlets on four sides, at least 12 in. below water surface, not more than 15 ft apart or 10 ft from sidewalls; use directional inlets with gate valve or similar control. Maximum pipe velocity 6 fps.

3. Eliminate dead areas.

4. Pool drain(s) permits pool to be emptied in 4 hr or less. No direct connection to sewer.

5. Outlet opening or grating area is at least 4 times drain pipe. Use multiple outlets if pool is more than 30 ft wide. Grate velocity 1½ fps max.

6. Design for head loss of 5 to 7 lb/in.^{2} in sand filter, 30 to 50 lb in diatomite filter, and 20 in. of mercury with vacuum diatomite filter.

7. Select correct pump for type of filter and head loss in recirculation piping.

Accessories

1. Water heater with automatic thermal control for indoor and some outdoor pools.

2. Locate water heater, if provided, on water flowing from filters to pool. Use cold water for filter-washing. Temperature of water entering pool not greater than 110°F.

3. Hair and lint catcher. Provide spare on bypass, with valves. Area of strainer openings at least 10 times area of main recirculating line.

4. Vacuum cleaner connected to portable pump or recirculating pump, with proper connections in pool sides at least 8 in. below the pool water surface.

5. Space heaters and ventilation for indoor pools.

6. Residual chlorine and pH testing kits. Chlorine range 0.3 to 2.0; pH range 6.8 to 8.0 or 7.0 to 8.2.

7. Spare parts for chlorinator, including ammonia bottle and gas mask where gas machine used.

8. Hose bibs provided to hose down walks and floors.

9. Steps and ladders for pools more than 30 ft long.

Disinfection

1. Chlorination is most common. Bromine, iodine, and chlorine dioxide are also used.

2. Chlorinator capacity adequate to dose indoor pool water at 5 mg/l and outdoor pool at 10 mg/l. Manufacturers recommend gas chlorinator capacity of 15 lb for first 100,000 gal plus 10 lb for each additional 100,000 gal of pool volume.

3. Gas chlorinator housed in separate room; can be mechanically ventilated without danger to attendants and bathers. See page 652 and Figure 9–7.

4. Compare first cost, operation, safety, and maintenance of positive feed hypochlorinator versus solution feed gas chlorinator.

Clarification of Swimming-Pool Water

1. Pools usually require filtration equipment. See Table 9–3.

2. Provision made for batch treatment with alum or other chemical by means of solution feed pump, fresh water or surge tank, or "alum pot." Assure that several minutes' coagulation can take place before water reaches filters.

TABLE 9-3   SWIMMING-POOL FILTRATION AND CAPACITY DATA

	Vertical Pressure Filters		
Diameter (in.)	Surface Area (ft²)	Inlet (in.)	Waste to Sewer (in.)
30	4.9	1½	2
36	7.1	2	2½
42	9.6	2	2½
48	12.6	2½	3
54	15.9	2½	3
60	19.6	3	4
66	23.8	3	4
72	28.3	4	5
78	33.2	4	5
84	38.5	4	5
90	44.2	4	5
96	50.3	5	6
102	56.8	5	6
108	63.6	5	6
114	70.9	5	6
120	78.5	5	6

Diatomite Filter Septa

Diameter (in.)	Cylindrical Area (ft² per ft of length)
2	0.524
3	0.785
4	1.047
5	1.309
6	1.571

Filter rate (Lower rate is recommended)	-2.0 to 3.0 gpm/ft² with sand or anthrafilt.
	-1.0 to 2.0 gpm/ft² with diatomite.
Wash-water rate	-9 or 12 gpm/ft², but 15 gpm is preferred. The lower rate is very satisfactory when using anthrafilt since it has about one-half the density of sand.
Recirculating pump max. capacity, gpm	$=$ total filter area in ft² $\times$ filter rate in gpm/ft²
Hours for one pool turnover	$= \dfrac{\text{pool capacity in gal}}{\text{recirculating pump capacity in gpm}} \times \dfrac{1}{60}$
	$=$ 6 to 8 hr for public pool; 12 hr for private pool.
Quantity of filtered water provided	$=$ at least 50 gal per person during period individual is in pool.
Recirculating pump capacity	$=$ sum of flows from each recirculation inlet, which is made proportional to the volume of water in that part of the pool.

3. Alkalinity adjustment by means of soda ash, added by dry chemical feeder, solution feeder, solution pot, or soda-ash briquets. Calcite or calcium carbonate graded sand may be used in filter. Other compounds are available.

4. Filter sand 30 in. deep, clean and sharp, effective size 0.4 to 0.55 mm, uniformity coefficient not greater than 1.75. Anthracite 0.6 to 0.8 mm size. Filters provide 18 in. freeboard between top of sand and overflow.

5. Rate of filtration 2.0 to 3.0 gpm/ft² with sand or anthrafilt, and 1.0 to 2.0 gpm with diatomite. The lower rates are recommended. Up to 5.0 gpm permitted for private residential pools. Higher rates are used.

6. Wash-water rate 15 gpm/ft² desirable; 9 to 12 gpm is acceptable. Use 9 gpm with anthracite.

7. Continuous feed of diatomite said to result in less diatomite use.

8. Provide on filter system a rate of flow indicator for recirculation and backwash measurement, air-release valve on each filter shell at top, pressure gauges on influent and effluent lines and on either side of hair catcher, two baskets for hair catcher, sight glass on waste-discharge line.

Overflow Gutters

1. Overflow gutters provided completely around pool; a 12- to 18-in. flat gutter sloping away from pool also used. Overflows easily accessible for cleaning and inspection.

2. Outlet drains not less than 2½ in., 15 ft on centers. Drainage to sewer through air gap or to recirculating pump or balancing tank.

3. Pools less than 30 ft wide may be provided with skimmers built into the side and corners of the pool to take the place of gutters if acceptable to the health department. A skimmer provided every 30 ft and an overflow rate of 30 gpm/ft of skimmer has been suggested.[17] See Figure 9–1.

4. At least 50% of recirculation from gutters or skimmers.

No Connection with Potable Water Supply

1. Design does not provide opportunity for pool water to enter drinking-water system.

2. Introduce fresh water into the suction side of the recirculation pump, preferably through a balancing tank. A loop at least 35 ft above the pool walk can also be used provided no back pressures can occur. Other arrangements are illustrated in the text, Figures 9–2 and 9–3.

Bathhouse

1. Facilities include separate dressing rooms, showers, toilets, and wash basins for each sex in proportion shown in Table 9–4. Showers have at least 90°F water. See Figure 9–4.

[17] H. F. Eich and William M. Tibbett, "Pool Skimmers vs. Overflow Gutters," *Swimming Pool Age,* **22,** 30 (November 1956). Skimmers and filters should comply with National Sanitation Foundation Standards or equal.

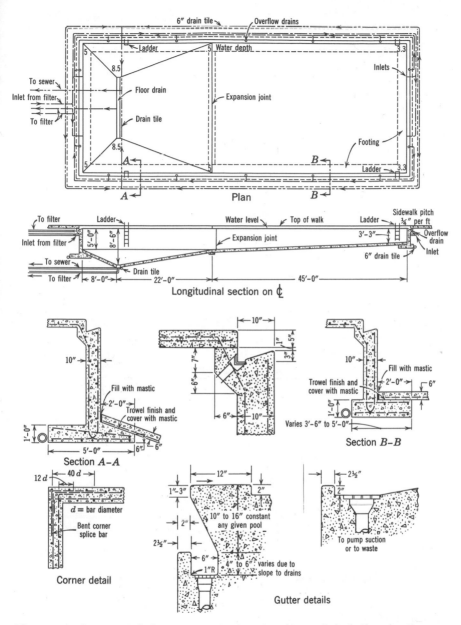

Figure 9–1 Suggested design for 30 × 75-ft swimming pool, including details.

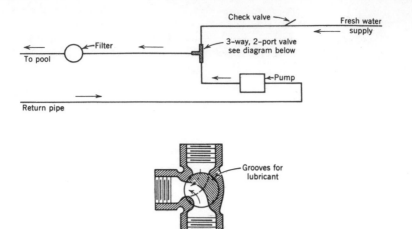

Diagram of 3-way 2-port valve

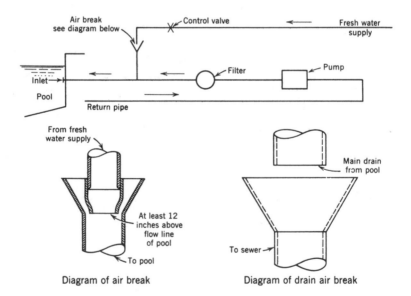

Figure 9–2 Acceptable means for adding fresh water to a pool. (From New York State Health Dept., Bull. 31, Albany 1950).

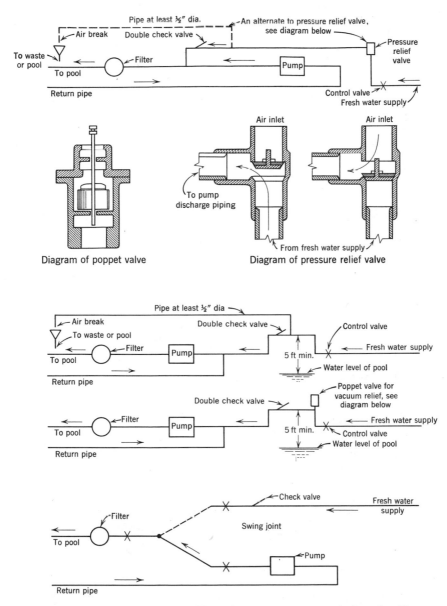

Figure 9-3 Acceptable means for adding fresh water to a pool. See also Figures 9-6 and 9-7. (From New York State Health Dept., Bull. 31, Albany 1950).

TABLE 9-4   PLUMBING FIXTURES RECOMMENDED AT SWIMMING POOLS

Fixture	Number Men Served				Number Women Served			
	*	†	‡	§	*	†	‡	§
1 water closet	75	60	15	60	50	40	10	40
1 urinal	75	60	20	60	–	–	–	–
1 washbasin	100	60	15	60	100	60	10	60
1 shower	50	40	4	40	50	40	4	40
1 service sink	At least one				At least one			

Note: For classes, the number of showers equals one-third the number of pupils in the maximum class. Provide at least one drinking fountain in pool area.

* American Public Health Association, *Suggested Ordinance and Regulations Covering Public Swimming Pools,* 1964.

† New York State Health Department, Bull. 27, Albany, 1962.

‡ Class in school, YMCA, and similar pool for 1-hr class. Multiply number of persons by 2½ if class is of 2-hr duration.

§ *Swimming Pools,* U.S. Public Health Service Training Manual, Dept. of HEW, Washington, D.C., 1959.

2. Dressing rooms provide 7 ft²/female and 3.5 ft²/male patron expected at maximum periods. Ventilated.

3. Floor drains not over 25 ft apart and floor slopes between ⅛ and ½ in./ft.

4. Entrance and exit at shallow end of pool.

5. Bathers from dressing room must pass toilets and go through shower room before entering pool.

6. All electrical work, indoor and outdoor, shall be installed, comply with, and be maintained in accordance with the Standards of the National Electrical Code.

A Small Pool Design

To illustrate some of the design principles an example is given below. Figure 9–5 shows a cross-section of the pool. Figure 9–6 shows a typical plan.

Size: 60′ × 30′ = 1,800 ft²

No. persons: $\dfrac{1800}{25}$ = 72 people at any one time; 144 in pool enclosure

Capacity: $\left[\dfrac{(3.5 + 5.5)}{2} 30 + \dfrac{(5.5 + 9)}{2} 20 + 9 \times 10 \right] 30 \times 7.5 = 83{,}250$ gal

Recirculation period: assume 8 hr

Required capacity of recirculating pump $= \dfrac{83{,}250}{8 \times 60} = 173.5$. Say 175 gpm

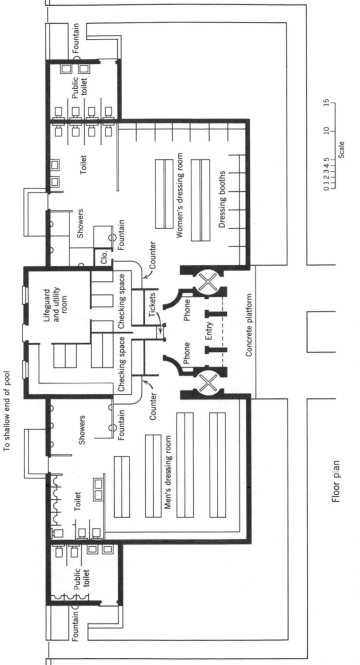

To shallow end of pool

Fountain

Public toilet

Toilet

Showers

Clo.

Fountain

Counter

Lifeguard and utility room

Checking space

Checking space

Tickets

Phone

Phone

Entry

Women's dressing room

Dressing booths

Concrete platform

Counter

Fountain

Showers

Men's dressing room

Toilet

Public toilet

Fountain

0 1 2 3 4 5   10   15

Scale

Floor plan

Figure 9-4 Suggested layout for a bathhouse with a capacity of 340 persons corresponding to a 40 × 100-ft pool.

619

Required electric motor (1,750 rpm, with magnetic starter and under voltage protection).

$$= \frac{\text{gpm} \times \text{total head in ft}}{3,960 \times \text{pump and motor efficiency}}$$

$$= \frac{175 \times (16' \text{ in filters} + 16' \text{ in recir. system})}{3960 \times 0.60 \times 0.80}$$

$$= 2.95$$

Say, 3 hp with pressure sand filters.

$$= \frac{175 \times (69' \text{ in filter} + 16' \text{ in recir. system})}{3960 \times 0.60 \times 0.85}$$

$$= 7.4, \text{ say } 7\tfrac{1}{2} \text{ hp with pressure diatomite filters (minimum)}$$

Piping: Main lines to and from recirculating pump = 4 in.; lines to pool inlets = 2 in.; provide minimum of 4 pool inlets, more preferred. Locate inlets 18 to 24 in. below water surface. Provide 2-in. vacuum cleaner connections and flow measurement.

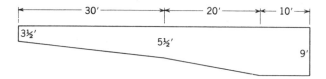

Figure 9–5   Typical summer resort swimming-pool design.

Hair catcher: 4-in.; coagulant feeder: one; soda ash feeder: one; one vacuum cleaner, and one flow indicator.

Chlorinator type: hypochlorinator, positive feed.

Filters: required area = $175\tfrac{3}{3}$ = 58 ft^2; 3 5-ft dia. filters = 3×19.6 = 59.8 ft^2. Provide 3 5-ft dia. pressure sand filters; *or* one diatomite filter with total septum area of 88 ft^2, with air release at top of each filter shell consisting of $\tfrac{1}{2}$-in. line with globe valve, and pressure gauges on inlet and outlet lines.

Pool water heating:

H = heater size in Btu/hr
Q = volume of water in pool in gal
T = degrees F water temperature increased
8.33 = weight of 1 gal of water in lb

Then:

$$H = Q \times 8.3 \times T$$

A pool 30×60 ft in area and with a depth of $3\tfrac{1}{2}$ ft to 5 ft in one-half the area and up to 9 ft in the remaining area holds approximately **85,000** gal of water (actually **83,250** gal). The temperature of the pool water is cooled to 58°F, and it is required that the temperature of this water be raised to 78°F. Determine required heater output.

$$H = 85,000 \times 8.3 \times 20$$
$$H = 14,110,000 \text{ Btu/hr}$$
$$H = 705,500 \text{ Btu/hr with a 20-hr pool heat-up time}$$

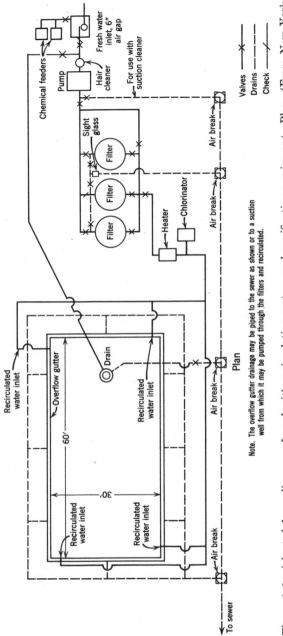

Figure 9-6 Adapted from diagram of pool with recirculating system and purification equipment—Plan. (From New York State Health Dept., Bull. 31, Albany 1950).

651

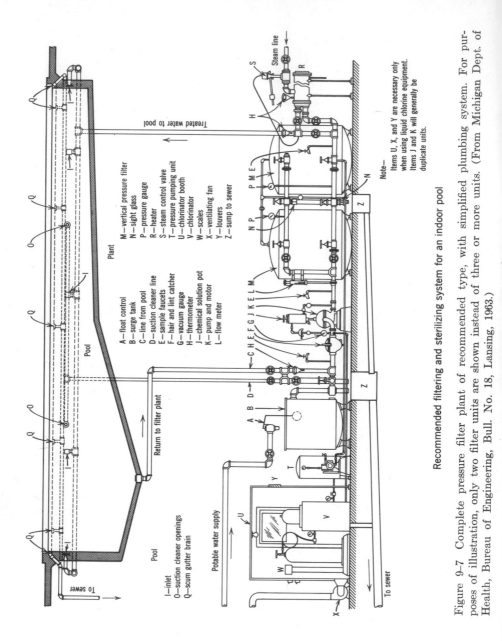

Plant

A—float control
B—surge tank
C—line from pool
D—suction cleaner line
E—sample faucets
F—hair and lint catcher
G—vacuum gauge
H—thermometer
J—chemical solution pot
K—pump and motor
L—flow meter

M—vertical pressure filter
N—sight glass
P—pressure gauge
R—heater
S—steam control valve
T—pressure pumping unit
U—chlorinator booth
V—chlorinator
W—scales
X—ventilating fan
Y—louvers
Z—sump to sewer

Pool

I—inlet
O—suction cleaner openings
Q—scum gutter drain

Note—
Items U, X, and Y are necessary only
when using liquid chlorine equipment.
Items J and K will generally be
duplicate units.

Recommended filtering and sterilizing system for an indoor pool

Figure 9–7 Complete pressure filter plant of recommended type, with simplified plumbing system. For purposes of illustration, only two filter units are shown instead of three or more units. (From Michigan Dept. of
Health, Bureau of Engineering, Bull. No. 18, Lansing, 1963.)

If the heat loss from the water surface is 12 Btu per square foot per hour per degree temperature difference, and the average air temperature is 58°F, the water surface heat loss is:

$$(30 \times 60 \times 12 \times (78 - 58) = 432,000 \text{ Btu/hr}$$

Since the heater required will have an output of 705,500 Btu/hr and exceeds the water surface heat loss of 432,000 Btu/hr, it will be adequate for the example given.

The heater output capacity can also be estimated by assuming a heat requirement of 7 to 10 Btu per hour per gallon of water recirculated in an indoor pool.

A recommended filtering and disinfecting system for an indoor pool including surge tank for make-up water, steam heating, and other appurtenances is illustrated in Figure 9–7.

SWIMMING POOL OPERATION

Bacterial, Chemical, and Physical Water Quality

The bacterial standards for artificial swimming pool water approach the Public Health Service Drinking Water Standards. Maintenance of such water quality will be affected by many factors, many of which can be controlled by the pool personnel. These include:

1. The operation and maintenance of purification equipment provided, such as strainers, pump, filters, chemical feed apparatus, and chlorinator.

2. The bathing load, or number of persons permitted to be in the pool at any one time and in any one day, based on the capacity of the purification equipment and pool area.

3. Pollution introduced by the bathers and enforcement of warm water and soap cleansing showers before entering or reentering the pool enclosure.

4. Supervision of pool personnel and bathers.

5. Maintenance of pool decks, general cleanliness, and separation of recreation and picnic areas from the pool. Prohibition of spectators from pool enclosure.

The accepted and effective method of maintaining satisfactory pool-water quality in a properly designed pool is by continuous recirculation, chlorination, and filtration over a 24-hr period. Attempts to economize by intermittent operation invariably lead to ineffective or nonuniform chlorination, algal growths, and reduced water clarity. This is particularly true at outdoor pools and at pools subject to periodic heavy bathing loads. Proper water level must be maintained in a pool if the scum gutters and/or skimmers are to function as intended.

It is good practice to collect at least two water samples in thiosulfate-treated bottles from a pool, one at each end, during the bathing period of maximum usage. Where adequate laboratory facilities are available, samples should be collected at least once a week. The bacterial analyses should be made at an approved laboratory near the pool, if possible, to reduce the time lapse between collection and examination of the samples. The tests made may include the coliform test, which indicates the presence of intestinal-type organisms; the total bacteria count, which indicates the concentration of bacteria in the pool water; and the streptococci test, which indicates the presence of organisms associated with respiratory illnesses. Some interest has also been shown in staphylococci, fecal coliforms, fecal streptococci, and *Pseudomonas*. Of the tests mentioned, the coliform test is most valuable for general use, although the others give additional information. Streptococcal bacteria found in pools are more resistant to chlorine than coliform bacteria.

As in other activities, the results of bacterial analyses must be interpreted in the light of a sanitary survey of the facility. Representative samples are an indicator or check on the effectiveness of pool operation at times of peak usage. Deficiencies in equipment or operation usually become apparent before they become too serious. Hence unsatisfactory results should be immediately investigated to eliminate their cause before any harm is done.

Disinfection

Chlorine, bromine, iodine, chlorinated cyanurates, and ultraviolet ray lamps have been used to disinfect swimming pool water. Chlorine and to some extent bromine are the chemicals of choice. Chlorinated cyanurates and ultraviolet ray lamps have distinct limitations. Iodine has not been used to any extent but is reported to be a satisfactory pool disinfectant and more stable than chlorine in outdoor pools.

The maintenance of at least 0.6 mg/l free available chlorine and a pH of 7.2 to 7.6 in a pool water will produce consistently satisfactory bacteriological results. Free available chlorine is the sum of chlorine as HOCl and OCl$^-$. The HOCl component is the markedly superior disinfectant, but the proportion is largely dependent on the pH of the water. For example, at a pH of 7.2, 62 percent of the available chlorine is in the form of HOCl; at pH 7.4, 52 percent is available as HOCl; at pH 7.6, 42 percent is available as HOCl; and at pH 7.8, 32 percent is available as HOCl, all at a temperature of 68°F. See Table 3–7 and pages 158 and 657.

Eye irritation is said to be caused by prolonged swimming in water with a pH about or below 7.4. A pH of 6.5 to 8.3 can be tolerated for a limited period of time because of the high buffering capacity of the lacrimal fluid.[18]

If ammonium alum or free ammonia is used, it will be impractical to maintain a free available chlorine in the pool water because the ammonia would combine with the chlorine to form chloramines or combined chlorine. Combined residual chlorine is a slow-acting disinfectant and very ineffective as a bactericide in swimming pools, where it is important to immediately neutralize any contamination introduced by the bathers. With continuous recirculation and chlorination, this pollution can be neutralized by adding sufficient chlorine to maintain an adequate free available chlorine in the pool water.

The chlorination of water containing iron in solution will cause discoloration of the water and staining of pool walls. See page 193.

When bromine is used as the disinfectant, to find the bromine residual with a chlorine comparator multiply the residual chlorine reading by 2.25 to convert it to bromine. A residual of 2.0 mg/l should be maintained. Pure bromine at room temperature is a deep red color with a strong odor. It fumes readily and is extremely corrosive. It is a very satisfactory disinfectant.

Control of pH

The chemical quality of the pool water is generally measured by tests for the pH value and chlorine residuals. Occasional laboratory determination of alkalinity, free carbon dioxide, and so forth, is desirable, but pH and residual chlorine tests should be made at least three times a day, before and during peak loads. The maintenance of a proper pH and alkalinity ratio is important for other than possible corrosion reasons. Water coagulation, disinfection with chlorine, and eye irritation are effected by the acidity–alkalinity balance. Sand grains may become coated with resultant plugging and ineffective filtration if the water is hard and too much soda ash is used. In practice, however, the pH of the pool water will drop rapidly, particularly where chlorine gas is used. Chlorine as hypochlorous acid and hypochloric acid reacts with the alkalinity in the water at a rate of 1 part chlorine to 1.2 parts alkalinity. This can be compensated for by adding 1.2 mg/l alkalinity as $CaCO_3$ for each part of chlorine applied to maintain a balance.

Alum, which is used as a coagulant, tends to lower the pH and alkalinity of a water. It is necessary to add an alkali to offset the lowering

[18] Eric W. Mood, "The Role of Some Physico-Chemical Properties of Water as Causative Agents of Eye Irritation of Swimmers," *Report of Committee on Water Quality Criteria*, Federal Water Pollution Control Administration, U.S. Dept. of the Interior, Washington, D.C., April 1, 1968. p. 15.

effect of the alum. This is done by adding a pH adjuster such as sodium carbonate, also known as soda ash or washing soda, to the pool water by means of a chemical feed machine that introduces the chemical, usually into the recirculating pump suction line, or by the placing of soda ash briquets near the pool recirculation outlet. Caustic soda or sodium hydroxide will also produce the desired results but this material, also known as lye, can cause severe burns; hence its use is not recommended unless competent and reasonably intelligent personnel are available. If the pH gets too high, it should be lowered by diluting with fresh water. Acid should not be used; although sodium bisulfate is safer to use than hydrochloric or sulfuric acid. Chlorination by means of sodium or calcium hypochlorite will add alkalinity to pool water.

Clarity

The physical quality of the water—its appearance and clarity, is determined by sight tests previously described. The clarity of an artificial swimming pool water is maintained by continual filtration. Where sand or anthrafilt filters are used, alum is the chemical usually used to trap and coagulate the suspended matter and color in the water. For best results, in addition to pH control previously discussed, the alum must be fed in small controlled quantities, be well mixed with the recirculated water, and then be allowed to coagulate before it reaches the filter. One way in which this is done is to add filter alum (aluminum sulfate) into a reaction or make-up tank and then introduce the alum into the suction side of the pump, where it is thoroughly mixed with the pool water in passing through the pump and coagulated before reaching the filters. A dosage of 3 to 4 lb of alum per 10,000 gal of pool water is normally used. At some pools, excellent results are obtained by slowly adding a small quantity of alum immediately following backwash of the filters, to replace the flocculant mat on top of the sand bed, following which the chemical feed is stopped.

Swimming Pool Water Temperature

At indoor pools, it is desirable to control the temperature of the pool water and air. An air temperature about 5°F warmer than the temperature of the water in the pool is recommended. The maximum water temperature suggested for general use is 80°F; 74 to 76°F is comfortable. Outdoor pools may require water replacement or the addition of large quantities of ice to keep the water from getting too warm. The water temperature should not exceed 85°F (30°C). Swimming or bathing in water above this temperature for any length of time has a temporary debilitating effect. Spray aeration of recirculated water to the pool can lower the water temperature 5 to 10°F. Swimming in 40°F water causes fatigue and breathing difficulty in a short time—10 min or less.

Testing for Free Available Chlorine and pH

The residual chlorine test and the test for pH of swimming pool water require a special procedure and technique to give better accuracy.[19] Combined chlorine is of little value as a disinfectant in a swimming pool water. Therefore the reporting of residual chlorine as free chlorine that is actually combined chlorine can produce a false sense of security.

If the test for residual chlorine is not carefully made, it may appear to show the presence of free available chlorine, whereas the chlorine present may in fact be in the combined form. This occurs when the water temperature is above 35°F, especially when the water temperature is above 68°F. At a temperature below 35°F, free available chlorine will be readily detected, as orthotolidine reacts immediately with free available chlorine to produce the characteristic color independent of the temperature. However, combined chlorine, if present, reacts slowly with orthotolidine at this low temperature. Therefore, to determine the free available chlorine present in a pool water, it is first necessary to cool the sample to below 35°F, before the reagents are added, as explained for water supply practice. Because free available chlorine reacts readily with soil or organic matter, it becomes extremely important to use clean glassware and a clean technique when making the test.

The test for combined chlorine should be made at a temperature of 68°F and the color reading taken at the end of 5 min. The maximum color develops in about 5 min. If the pool water temperature is much above 68°F, the maximum color is produced sooner and also fades more rapidly, thereby making a proper color reading difficult. If the temperature of the water sample is much below 68°F, the time for maximum color development after the addition of orthotolidine reagent is considerably extended. See page 158, Chapter 3.

A new field test for free available chlorine reported to be accurate in waters at a temperature of 46°F to 95°F has been developed. Leuco crystal violet is used as the indicator.[20]

---

[19] W. A. Sanderson, "Control of High Residual Chlorination in Swimming Pools," Division of Laboratories and Research, New York State Health Department, Bull. 33A, Albany, November 1954; A. P. Black, R. N. Kinman, M. A. Keirn, J. J. Smith, and W. E. Harlan, "The Disinfection of Swimming Pool Water," Parts I and II, *Am. J. Public Health,* **60,** No. 3 and 4, 535–545 and 740–750, (March and April 1970).

[20] A. P. Black and G. P. Whittle, "New Methods for the Colorimetric Determination of Halogen Residuals. Part II. Free and Total Chlorine," *J. Am. Water Works Assoc.,* **59,** 607 (May 1967).

When high free available chlorine is maintained in a pool water, the chlorine will bleach the pH color indicator, giving an improper reading. If the free available chlorine is first removed, a proper reading can be obtained. This can be done by adding one drop of ¼ percent sodium thiosulfate solution to the pool water sample. This will destroy the free available chlorine present without appreciable effect on the pH value. The glassware used must be clean and the sample must not be contaminated by lime, alum, or other soil.

Algae Control

Algae development in a pool causes a slimy growth on the walls and bottom, reduced water clarity, increased chlorine consumption, and a rapid rise in the pH of the pool water in one day. The wall and bottom growths penetrate into cracks and crevices, making them difficult to remove once they become attached. The suspended or floating type are more easily treated. However, the best and simplest control method is to prevent the algae from developing, and this can be accomplished by maintaining a free chlorine residual of at least 0.6 mg/l in the pool water at all times. This cannot be accomplished by intermittent recirculation and chlorination.

If algae difficulties are anticipated or experienced, several control measures can be tried. These include heavy chlorination, copper sulfate treatment, quaternary ammonium treatment, emptying the pool and scrubbing the walls and bottom with a strong chlorine, caustic soda, or copper sulfate solution, and combinations of the methods mentioned. Caustic soda can cause severe burns; its use by a lay person is not recommended.

Heavy chlorination, or superchlorination, is the preferred treatment. Dosage of the pool water so as to maintain a free chlorine residual of 0.6 to 2 mg/l, together with recirculation and filtration, should be effective in preventing and destroying the algae growths normally found in a pool. Hand application of 2½ to 5 gal of 5 percent sodium hypochlorite to a 100,000-gal pool will speed up the treatment. Other forms of chlorine, such as 25 or 70 percent calcium hypochlorite, can of course also be made up into 1 to 5 percent solution and the clear supernatant evenly distributed in the pool water. Bathers should not be permitted to use the pool until the residual chlorine drops to less than 4 mg/l. Iron, manganese, and also hydrogen sulfide if present will precipitate out of solution in the presence of free chlorine.

Copper sulfate has long been known for its ability to control algal growths. Most algae are destroyed by a dosage of 5 lb per million gallons; but practically all forms are killed at a dosage of 2 mg/l or 16.6

lb of copper sulfate per million gallons of water. The copper sulfate crystals can be easily dissolved in the pool water by dragging the required amount, placed in a burlap bag, around the pool. There are dangers in the use of copper sulfate with certain waters. If the pool water has a high alkalinity, a milky precipitate will be formed that also interferes with the action of the copper ion, and waters high in sulfur or hydrogen sulfide will react and produce a black coloration upon the addition of copper sulfate. If in doubt, try dosing a batch of the water in a barrel or pail. Too high a concentration of copper sulfate will tend to discolor one's hair or bathing suit; hence pool water that has been overdosed should be diluted.

Quaternary ammonium compounds have been suggested for algae control and some satisfactory results have been reported, but free residual chlorination is made difficult. Care must be used not to overdose because foaming will be produced.

If the methods mentioned above are not effective, it will be necessary to drain the pool and scrub the bottom and walls with a 5 percent solution of copper sulfate, or spray and scrub with a 1 percent solution of calcium or sodium hypochlorite, followed by rinsing with a hose after about 15 min.

Prevention of Ringworm Infections

The prevention and control of ringworm infections are discussed in Chapter 1.

At one time, the provision of a footbath containing a strong solution of chlorine or other fungicide at the entrances to a pool and in gym shower rooms was required by health officials. Experience has shown, however, that footbath solutions were rarely properly maintained, with the result that they became in fact incubators and spreaders of the infection rather than inhibitors. The contact time between the feet and fungicide solution was too brief to do much good. The fungus on the feet is usually so imbedded in the skin that the solution could not penetrate to the spores.

Where footbaths are built into the floor, they can serve the very valuable purpose of preventing the tracking of dust, sand, and dirt into the pool. They must, however, be provided with a continuously running spray and an open drain.

Towels and bathing suits used at public pools can also be the means whereby ringworm and other infections are spread. It is therefore essential that towels and suits be carefully sterilized after each use. This can probably be done best by a public laundry. If sterilization is done at the pool, equipment and supervision must be provided to assure that towels and suits will be washed in hot water and suitable soap or detergent, rinsed in clean water, and then dried by artificial heat at a temperature above 175°F, which will destroy bacteria and the fungus spores. Chemical disinfection may also be effective, although the heat treatment is believed to be more reliable. Woolens, silks, nylons, and elasticized materials will require special handling. Cold-water disinfection can be accomplished by soaking suits 5 min in a 1:1000 dilution of alkyl-dimethyl-benzyl-ammonium chloride or other quaternary ammonium compound. This is reported to show "no adverse effect on dyes or materials commonly used in swimming suits." Temperatures above 110°F will shrink most woolens.

Personnel

As in most other instances, the designation of one competent person who is responsible for satisfactory operation of the entire swimming pool is a basic necessity. Only in this way can safe, clean, and economic operation be assured. In large pools this person would be the manager, and he should have under his supervision the bathhouse attendants, lifeguards, and water purification plant operator. A competent manager should be a good administrator who is familiar with all phases of pool operation, including the water-treatment plant, collection of water samples, making of routine control tests, and keeping of operation reports. At summer resorts, camps, and similar places, the senior lifeguard may be given the overall responsibility for pool operation for reasons of economy. This does not usually prove to be satisfactory unless the lifeguard can be trained before the summer season in the fundamentals of pool operation.

Most states have laws or sanitary codes regulating the operation of swimming pools and bathing beaches. The owner has a legal responsibility to comply with these laws and rules. His failure to do so would make him liable to prosecution for being negligent. It is to the owner's interest therefore, and for his protection, that he keep such records as will show proper operation of the swimming pool purification equipment, depth of the pool water, lifeguards on duty, elimination of accident hazards, and any other measures taken to protect the users of the pool. Swimming pool operation report forms are available from most health departments. To be of value these reports must be carefully and accurately filled out each day. Analysis of reports will indicate where failure is occurring and need for new equipment. The operator should be provided with a residual chlorine test kit having a range of 0.2 to 3.0 mg/l and a pH test kit having a range between 6.8 and 8 or 7 and 8.2.

Pool Regulations

Regulations should be as logical and concise as possible. The following are samples:

1. Urinating, spitting, or blowing the nose in the pool can spread disease to other bathers. Use the overflow gutter for expectoration and toilet facilities to help keep your pool clean.

2. Persons having skin disease, running sores, a cold, or other infection endanger the health of their fellow bathers; hence they cannot be permitted to use the pool until well.

3. Spectators carry dirt on their feet that would be tracked into the pool. This lowers the quality of the pool water; therefore spectators are not permitted inside the pool enclosure.

4. A pool is a common bathtub in which the water is kept clean by continuous purification and replacement of the water. Dirt and bacteria may build up in the pool faster than the purification equipment can take it out if bathers do not remove perspiration, dirt, and dust from their bodies. Therefore, take a warm water and soap shower in the nude before entering the pool and after using the toilet.

5. The maintenance of order is necessary to prevent accidents and drownings and permit maximum enjoyment of the facilities. Obey the lifeguard and attendants promptly. There shall be no swimming in the absence of an attendant.

SWIMMING POOL MAINTENANCE

Recirculating Pump

1. Inspect and service regularly.

2. Check motor bearings for lateral and vertical wear. Brushes and commutators should wear evenly; clean commutator with fine sandpaper when placing in operation and turn down commutator if uneven.

3. Pump impeller has little end play or slippage; check suction and pressure produced by pump, lubrication, and stability of pump mounting. No entrance of air in suction. Slight leakage from packing gland.

4. Check pump capacity against operating head.

5. Compute backwashing, recirculating rate, and pool turnover against design. A new pump may be needed.

6. Determine water clarity, uniformity of chlorination, and bacteriological quality of pool water.

7. If repairs are needed call in competent pump man or electrician at beginning of season or during holiday periods.

8. Drain pump casing, cover pump and motor with canvas, and remove fuses from switch boxes at end of season. If subject to flooding, remove.

Hair Catcher

1. Maintain a regular daily cleaning schedule.

2. Keep a replacement basket strainer on hand. A hair catcher in parallel with proper valves is very desirable.

Filters

1. Open air valve at top of each filter shell to release air. If problem recurs frequently find and eliminate cause.

2. At end of each season after backwashing, drain each filter, open manhole, and inspect sand. The sand should be level, without dirt, hair, holes, or cracks. Open inlet valve to filter to slowly admit water. If sand comes up unevenly or boils in spots, sand may have to be cleaned or replaced or the underdrain system may be partially clogged.

3. Sample of sand at depth of 6 to 8 in. should be clean.

4. If backwash rate is not adequate to clean sand bed, anthracite may be substituted, which requires only about two-thirds the backwash rate for sand.

5. Deep raking of sand surface may permit washing out of mud and dirt.

6. Sodium hydroxide treatment may be used to clean sand bed. Add 1 to 2 lb per square foot of filter area to 12 in. of water over the sand. Lower the water level to just above the sand surface; let stand 12 hr, drain out solution, and repeat. Backwash and observe condition of sand as explained in (2). A strong solution of chlorine, sulfur dioxide (sulfuric acid), salt, or detergent may also be used to remove organic matter.

7. Excessive use of soda ash and alum with a hard water may cause cementation of sand grains and improper water filtration. Break up incrustations or replace sand. Use of a metaphosphate with a hypochlorinator may prevent this difficulty.

8. Failure to use a filter aid with a diatomite filter or pretreatment of water high in iron or manganese will cause clogging of the filter tubes. Check head loss every few weeks; remove and clean tubes whenever necessary, soak in 8% solution of sodium hexametaphosphate for 2 hr and scrub; replace damaged tubes; use a new gasket for filter shell head when it is opened.

9. Backwash and drain filters at end of summer season to prevent freezing. Check for needed repairs.

10. See that valves operate easily and properly. Grease stems and lubricate moving parts. Check packing and repack if necessary. Drain and inspect all piping and fittings at end of each season.

Chemical Feed Equipment

1. Dismantle at least once a year; inspect and replace worn parts. Special attention should be given to gaskets, check valves, shutoff valves and seats, tubing and piping, diaphragms, pistons, and special gears.

2. Rinse with clear water and drain before storing.

3. Coat metal parts subject to corrosion with petroleum jelly, plastic, or paint.

4. Check oil level and lubricant cups in accordance with manufacturer's instructions.

Pool Structure

1. Overflow troughs, drains, walls, floors, walks, lockers, shower rooms, toilets, and fixtures kept clean and in good repair.

2. At least one person is delegated responsibility for twice a day cleanup. Proper equipment and cleaning compounds provided.

3. Depressions in walks leveled; broken or cracked floors repaired.

4. Vacuum cleaning equipment provided and used. Pool walls and floors clean, without growths, smooth but not slippery.

5. Winter protection of a pool includes drainage of all equipment, piping, tanks, floor drains, and fixture traps. Trapped floor drains should be sealed watertight. If pool is designed to withstand ice pressure due to freezing water, the pool need not be drained. Pool walls with an outward batter of 1 to 1½ in. per foot of height will permit ice to expand and raise.

WADING POOLS

Definitions of wading pools vary. All are meant to be shallow, up to 24 in. maximum depth, and specifically for the use of children. They are usually included with the design of a swimming pool as an independent structure.

Public health and recreational agencies have recognized for some time that wading pools can actually be the cause of more illness than swimming pools. The irresponsible child may relieve himself in the pool and drink the same water with innocent abandon. Dirt, sand, grass, food, and other debris are carried into the wading pool in such quantity as to make continued purification of the wading pool water impractical.

Types

Wading pools have been constructed and operated as spray showers with open drain so as not to collect water, as flow-through type, as recirculation and filtration with chlorination type, and as fill-and-draw pools. The fill-and-draw wading pool cumulates pollution and is rarely kept clean; it should not be used at public places. The recirculation, with continuous chlorination, type of pool may be suitable if the pollution load is not too great and careful supervision can be given to the operation. The recirculation should include filtration and addition of sufficient chlorine to maintain at least 0.6 mg/l free residual chlorine in the pool water. Because of the heavy pollution, the wading pool should not normally be connected with the swimming pool recirculation system. The flow-through-type wading pool may give acceptable results if drained and cleaned preferably twice a day, but in any case at least daily, and an adequate continuous flow of clean water into the pool is maintained to replace the water in the wading pool in 3 hr or less. With this flow of fresh water a water of good quality can be maintained, provided the flow is continuous at least during the period of use, which may average 10 to 12 hr. Of all the types of wading pools mentioned, the spray type with an open drain is the safest and cleanest. Children seem to enjoy the sprays immensely; the danger of drowning and infec-

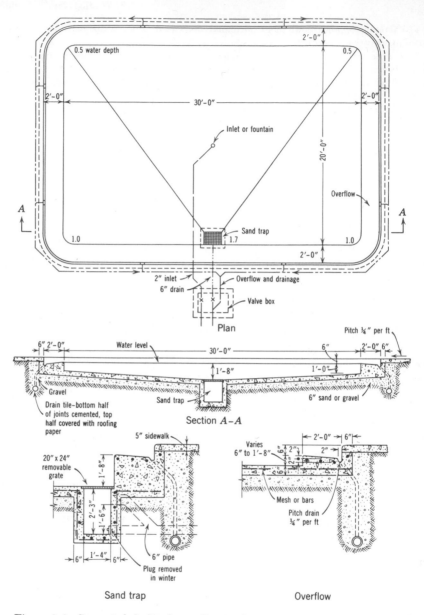

Figure 9–8 Suggested design for wading pool.

tion from polluted water is eliminated. Existing wading pools can easily be converted to spray-type pools in which there is no standing water.

The design details of some typical wading pools are shown in Figure 9–8. A gently sloping floor, say 6 in. in 10 ft, a drain provided with a sand trap, and a drained concrete apron at least 10 ft wide completely around the pool are elements common to all wading pools. A 6-in. drain is adequate for most wading pools; it must be so graded as to make the backing up of sewage impossible. Where a sand play area is provided, and this is very desirable, constant maintenance will be needed. This would include frequent raking to remove foreign matter and turnover of the sand to prevent the development of odors, particularly where the sand tends to remain wet and is in the shade. An underdrain system in such cases is necessary. Disinfection of the sand by sprinkling with a chlorine solution, followed by a fresh-water spraying from a garden hose may be found periodically desirable. Toilet facilities should be convenient to the wading pool users to help reduce pollution of the wading pool and sand play area.

BATHING BEACHES

The location of bathing beaches, sanitary surveys, bacterial standards, health considerations, and accident prevention were previously discussed. A suggested camp waterfront layout which includes numerous safety features is shown in Figure 9–9.

Disinfection and Water Quality

Unsatisfactory laboratory reports on a bathing-beach water call for a sanitary survey and evaluation to determine their significance. If a beach is being excessively polluted by sewage or other wastes, the obvious solution is reduction of the pollution at the source to the point at which it does not adversely affect the water quality. The importance of a sanitary survey and its intelligent interpretation in such cases is of major significance. When the remaining pollution reaching the bathing area or introduced by the bathers is small, or when the dilution provided (as determined by the dilution formula $Q = 6.25T^2$, in which Q = quantity of water per bather per day in gallons and T = the replacement period in hours) is not adequate to meet bacteriological standards, disinfection of the water may be possible.

The method for applying chlorine is determined by such factors as the type of beach, water current, size, and facilities available. If a sewer outfall is the culprit, the simplest emergency procedure would be heavy

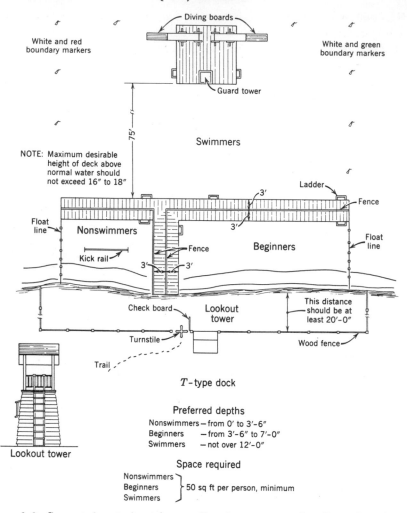

Figure 9–9 Suggested waterfront layout. Drawing not to scale. (From Boy Scouts of America, Division of Program, Engineering Service.)

chlorination, thorough mixing, and 15 to 30-min retention of the disinfected sewage before discharge to the receiving body of water. Chlorine can also be added to the bathing area at one time to treat a given volume of water several times a day; it can be added directly and continuously to the bathing water by means of submerged orifices, or the chlorine can be added in high concentration to water recirculated back to the bathing area. Each method has its limitations and requires careful planning and supervision.

The batch treatment or treatment of a given volume of water is a special problem in each case. The chlorine solution, up to 10 or 14 percent, can be added before each bathing period from a motorboat or rowboat as a spray or by means of a belt-driven solution feeder so as to traverse the bathing area. Since a large quantity of organic matter will be present to absorb the chlorine, a sufficient quantity of chlorine solution will have to be added to satisfy, at least partially, the initial chlorine demand. One might start with a dosage of 5 mg/l and be guided by the results of residual chlorine and bacteriological tests. It will be exceedingly difficult to maintain a residual chlorine for any extended period of time.

The continuous addition of chlorine, using a gas solution feed machine, through an underwater distribution system with orifices has been tried with partial success. Rubber or plastic piping should be used. If there is a current, the perforated chlorine distribution piping system would of course be located upstream. Where a small stream is dammed up and there is a measurable flow through the bathing area, chlorine can be added by means of one or more improvised drip chlorinators strategically located.

Good results can be obtained by continual chlorination and recirculation of the water in the bathing area. The capacity of the pump or pumps, number and location of inlets and outlets, and capacity of the chlorinator must be determined to fit individual cases.

The effectiveness of chlorination will be readily apparent by observing the clarity of the water and presence or absence of algae. The algae may appear as a green coloring in the water resembling pea soup, as a green scum, or as a dark flaky deposit. Experience has repeatedly shown that when a free residual chlorine is maintained in the water at all times, the bathing water is relatively clear of algal growths and attractive. In all cases technical control over the treatment, including the maintenance of adequate daily operation reports, is necessary.

Control of Algae

The control of algae is discussed in some detail in Chapter 3.

Under average conditions in recreation areas, a copper sulfate dosage of 5 lb per million gallons evenly applied at intervals of 2 to 4 weeks will prevent the development of most microorganisms, but may result in some fish kill. In the eastern part of the United States, the treatment should start in April or, if microscopic examinations are made of the water, when more than 300 organisms, or areal standard units, per milliliter of sample are reported.

Chlorination treatment can also be effective in a partly artificial pool, such as that formed by damming up a small stream. However, to be effective, it must be under good technical control and at least 0.6 mg/l free residual chlorine maintained in the water at all times. This will not only prevent the growth of most algae, but will also help keep the water cleaner looking and relatively clear.

Control of Aquatic Weeds

Aquatic weeds are objectionable in bathing areas. They become entwined around the legs and arms of bathers, interfere with swimming, and provide harborage for snails.

One of the simplest ways of controlling the growth of aquatic weeds, where possible, is lowering the water level sufficiently to cause the plants to dry out. If this is followed by burning and removal of the remaining debris, reasonably good control will result. The physical cutting and removal of the weeds is also temporarily effective. Where aquatic weeds grow above the water surface chemical control by the use of weed killers is possible.

Sodium arsenite as As_2O_3 at a rate of 4 to 7.5 mg/l has been very effective to control submerged rooted or anchored aquatic weeds, but its use is no longer advised. The chemical is very toxic to man and dangerous to the applicator. Other chemicals such as Endothall, Acrolein, Diquat, and 2,4-D granules or pellets may be used *if approved* by the control agency. Diquat, 2,4-D low-volatile esters, Silvex, or Amitrole-T may be permitted for floating weeds. The manufacturer's directions should be carefully followed, and when the water is located on a water supply watershed permission must first be obtained from the water-supply officials. Sodium arsenite is not usually allowed.

The health, agriculture, and conservation departments should also be consulted because some of the chemicals available are toxic to humans and fish. In many instances knowledge regarding persistence and effect on the ecology may be limited. A permit to apply chemicals to public waters is usually required. Only approved chemicals should be used and only when necessary. See also Chapter 10.

Control of Swimmer's Itch

Swimmer's itch is a form of schistosomiasis. The larvae of certain schistosomes are found in lakes and salt water beaches in many parts of the world including North America. Migratory fowl (and rodents) carry and distribute the nonhuman schistosomes. Bird droppings containing the parasites fall in water. The eggs hatch and the larvae (mircadia) must enter an appropriate species of snail[21] within 36 hr or die. The organism develops in the snail and is released in the water in the fork-tailed larval form (cercariae). The larvae bore into the bather's skin where they die, causing a rash and severe itching that is known as swimmer's itch.

Elimination of the snail, which is necessary to the life cycle of schisto-somes, will break the life chain and prevent schistosome dermatitis. The control of debris and aquatic growths in shallow water, to which snails become attached, will reduce snail harborage. But actual destruction of the snail offers the best protection. The application of copper sulfate, sodium pentachlorophenate, copper pentachlorophenate, or copper car-bonate so as to provide a dosage of 10 mg/l to the water will kill snails in the water, if the water can be impounded for 48 hr, as well as the free-swimming cercariae. Ehlers and Steel report that the "application of 2 pounds of copper sulfate and 1 pounds of copper carbonate for 1000 lineal feet of shoreline will kill the snails for a season or longer."[22] The solution is applied to the bottom being treated by means of a hose dragged behind a boat. A dosage of 3 lb of copper sulfate or copper carbonate per 1000 ft² of water surface to be treated is also recommended to kill the snail.[23] Other chemicals are reported effective in killing snails, but it must be remembered that the dosages mentioned here will kill fish life and perhaps bleach swimming suits; hence special precautions should be taken. Under certain circumstances acidity control of the water might be possible. In such cases maintenance of a pH of 7.0 or less will discourage snail growth. If snail control is not possible, swimming in endemic areas should be restricted.

TEMPORARY RESIDENCES

Operation of Temporary Residences

Camp, motel, hotel, and resort management is a very specialized field, requiring more than just the ability to provide an entertaining program for children and adults. There are more than 10,000 children's camps and an unknown number of summer family camp colonies, dude ranches, hostels, boarding houses, motels, hotels, and camp grounds scattered

---

[21] *Stagnicola emarginata* and *Physa parkeri* beach snails, and *Lymnaea stagnalis* and *Stagnicola palustris* swamp snails.
[22] Victor M. Ehlers and Ernest W. Steel, *Municipal and Rural Sanitation*, McGraw-Hill Book Co., New York, 1965, pp. 412.
[23] *Sanitary Control of Army Swimming Pools and Swimming Areas*, TB Med **163**, War Dept., Washington, D.C., May 1945.

throughout the United States. In many cases these places are communities unto themselves, with their own utilities and services. The operation problems approach those of small villages; some have a capacity of more than 1000 persons.

Camp leadership training courses usually include study of programming, nature lore, arts and crafts, counselor leadership, camping and woodcraft, fishing, aquatics, music, and related activities. The provision of basic sanitary facilities and an environment conducive to good health is frequently overlooked or taken for granted. Every camp director should have the basic knowledge to understand the significance of the multiple health and sanitation problems that may arise. He should know the resources available to him and be able to decide promptly the proper steps that may have to be taken to satisfactorily solve the problems. The health departments are anxious to help. They too are desirous of seeing that the health of the campers and guests is protected. Consult with the local health department and seek its advice in health and sanitation matters.

The fundamentals and problems associated with disease transmission; location and planning; water supply; sewage and solid waste disposal; swimming pools and bathing beaches; food, including milk; insects; rodents; noxious weeds; and housing are described elsewhere in this text and should be referred to for more detailed information.

Camps and Resort Hotels

It has been estimated that more than 5,000,000 children in the United States have the opportunity of going to summer camp each year. Probably an equal number of adults visit hotels and camps. The importance of maintaining minimum standards of sanitation, health, and safety at these establishments is recognized by state and local health department laws.

The development of children and the extension of their education through planned and supervised activities are major objectives of many camps. The camping philosophy followed should be understood and interpreted in determining the satisfactoriness of camp operation, maintenance, and supplied facilities.

A typical inspection report form is shown in Figure 9–10. Many variations and more detailed versions are used to help determine if an establishment is in compliance with the intent of existing regulations. Actually, trained judgment is necessary to determine what is adequate or satisfactory under the particular conditions of operation. A more detailed explanation of the items can be found in the text under the appropriate headings. Each permitting authority has specific regulations

NAME _____ TYPE _____

LOCATION _____ T.V.C. _____

Person interviewed _____ Title _____

Capacity _____ Occupants: Adults _____ Children _____ Illness _____

No.	Item	Yes	No	CM	No.	Item	Yes	No	CM
	General					*Water supply*			
1.	Permit and code posted				23.	Apparently safe			
2.	Competent sanitation supervision				24.	Adequate quantity, hot and cold			
3.	Medical and nursing care adequate				25.	Operation reports satisfactory			
					26.	Sources—protected			
4.	Adult for child care available				27.	No cross- or inter-connections			
5.	Soil drainage satisfactory				28.	Storage protected			
6.	General conditions sanitary; insects, rodents, weeds controlled				29.	Common cup prohibited			
					30.	Drinking fountain satisfactory			
	Housing					*Toilets and wastewater disposal*			
7.	Structurally safe				31.	Separate toilet facilities			
8.	Adequate size				32.	Toilets convenient and adequate			
9.	Easy to keep clean				33.	Privy location, construction, maintenance, satisfactory			
10.	Watertight roof and sides								
11.	Lean-tos exclude rain				34.	Disposal systems satisfactory			
12.	Cleanable floors provided				35.	Location satisfactory			
13.	Stoves fire protected and vented					*Other sanitation*			
14.	Buildings, grounds, etc., clean				36.	Pasteurized milk used			
15.	Fire escapes and exits provided				37.	Refrigeration adequate			
16.	Sleeping quarters adequate				38.	Food storage satisfactory			
17.	Buildings adequately ventilated				39.	Food handling satisfactory			
	Kitchen and dining rooms				40.	Dishes washed, cleansed, sanitized			
18.	Separate from dormitory and toilet				41.	Garbage storage satisfactory			
19.	Screening adequate				42.	Garbage disposal satisfactory			
20.	Equip. adequate for food prep.				43.	Shower facilities provided			
21.	Floors, walls, ceilings clean				44.	Pools, etc., conform to sanitary code			
22.	Ventilation adequate				45.	Communicable diseases reported			

NOTE: CM indicates correction made.

EXPLANATION OF OBSERVED DEFECTS BY NUMBER

_____
_____
_____
_____
_____
_____
_____

To the operator of the above described property: You are notified that the items checked in the "No" column are in violation of the requirements of the State Sanitary Code. Immediate elimination of any or all violations is required.

Received by _____ Inspected by: _____

Title _____

Date _____ Title _____

Figure 9–10 Reinspection form for camps and resorts. (See text for intepretation of each numbered item.)

to be complied with. The more common items numbered as in Figure 9–10 are briefly explained here to guide and help obtain better understanding of the intent of the laws.

Compliance Guide

1. *Permit.* A current permit signed by the health officer and displayed in a prominent place would indicate compliance with the sanitary code at the time of issuance. The person to whom the permit is issued is required to comply with the provisions of the permit.

2. *Competent sanitation supervision.* At least one competent person who is very familiar with all sanitary facilities at the camp and who is charged with the responsibility of keeping the grounds and buildings clean, the water supply, swimming pool, and sewerage systems working properly, and the hot-water systems, refrigerators, and dishwashing facilities functioning should be available at all times the property is open for use. At hotels and motels he should also keep a record of guests' names, addresses, and car license numbers.

3. *Medical and nursing supervision.* There should be a registered professional resident nurse employed at all camps where children are not physically normal or where the total number of persons is at any time greater than 75. In smaller camps prior arrangements should be made for the employment of a registered nurse in case of need. At all camps there should be a licensed physician on call, a person at the camp trained in first aid, an equipped first-aid cabinet, and telephone. An infirmary including hot and cold running water, examining room, isolation and convalescent space, and bathroom with flush toilets are practically essential at the larger camps. In the smaller camps a tent with a fly and mosquito netting and floor are the minimum facilities. See Figure 9–11.

4. *Adult supervision.* A competent camp director, or his assistant, should be available at all times at camps accommodating children under 16 years of age, in addition to such adult supervision as may be needed. It is common to have one counselor for every 6 or 8 campers, plus a service and activities staff. The camping philosophy will govern. In any case, the director or his deputy should assume the personal responsibility to make daily inspections of all environmental factors, including the cleanliness of food handlers, food servers, refrigeration, water-supply treatment, sewage and refuse collection and disposal, sleeping quarters, swimming or bathing areas, fire-fighting facilities, and accident hazards. The director's competence should include practical and educational experience. The counselors should be at least 18 years old.

Where a rifle or archery range is established, be sure that its use is permitted only under competent supervision and where established precautions can be taken.

5. *Surface drainage.* If surface water is not carried away or readily absorbed by the soil, or if groundwater, rock, or clay is close to the surface, making the subsurface disposal of sewage and other liquid wastes impractical, surface drainage would be considered unsatisfactory unless special provisions are made. The grounds shall be kept reasonably dry.

6. *General conditions sanitary.* Where papers, tin cans, bottles, garbage, or other refuse is strewn about in buildings or on the ground, where numerous flies are present, where building interiors or exteriors are not clean looking, or where there is sewage or other liquid-waste overflow, conditions are not considered to be sanitary. This is also interpreted to mean that poison ivy, ragweed, mosquitoes, ticks,

chiggers, mice, rats, or other insects and rodents that may transmit disease or cause injury should be effectively controlled or eliminated. The control of insects, rodents, and noxious weeds is discussed in Chapter 10.

7. *Structurally safe.* A building is considered structurally safe when it is capable of supporting 2½ to 4 times the loads and stresses to which it is or may be

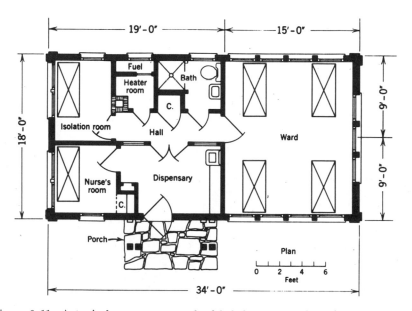

Figure 9–11 A typical summer camp, health lodge. (From *Organized Camp Facilities,* National Park Service, U.S. Dept. of the Interior, reprint from *Park and Recreation Structures.*)

subjected. The piers, foundations, girders, beams, flooring, and stairways must be sound and sturdy. Stairways and balconies should include handrails and balusters. The floors should not give significantly when stamped on. A licensed professional engineer or registered architect is qualified to determine the structural safety of a building.

8. *Adequate size.* Adequacy depends on the use to which the building is put. For example, sleeping quarters should provide at least 50 ft² of floor area per bed. Kitchens should provide sufficient space for storage and to avoid overcrowding and confusion. Dining rooms should allow 1½ to 3 ft around tables for serving and free movement of chairs or benches. The provision of 20 to 25 ft² per person in recreation and assembly rooms, 12 to 15 ft² per person in mess halls, and 7 to 9 ft² per person in kitchen areas, including storage and dishwashing, have been found adequate, with the smaller areas being more suitable for camps accommodating more than 100 and the larger areas for camps of 25 or less. See Figures 9–12 and 9–13.

9. *Easy to keep clean.* Smooth, washable finishes that are light colored, readily drained, and well lighted are easier to keep clean. More detailed information

on floor and wall construction and lighting is given in Chapter 8 and in the following paragraphs.

10. *Watertight roof and sides.* Dormitories, cabins, tents, dining halls, and similar structures should have watertight roofs and sides protected by overhanging eaves,

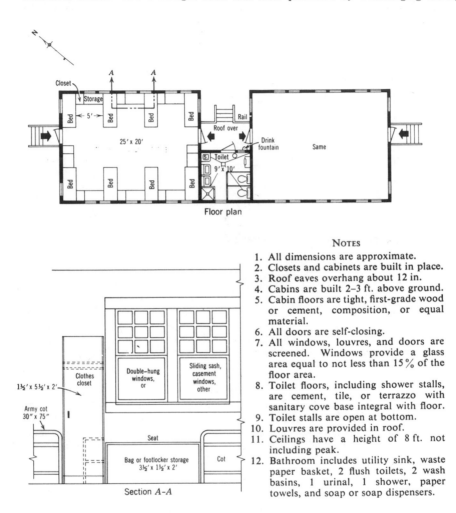

NOTES

1. All dimensions are approximate.
2. Closets and cabinets are built in place.
3. Roof eaves overhang about 12 in.
4. Cabins are built 2–3 ft. above ground.
5. Cabin floors are tight, first-grade wood or cement, composition, or equal material.
6. All doors are self-closing.
7. All windows, louvres, and doors are screened. Windows provide a glass area equal to not less than 15% of the floor area.
8. Toilet floors, including shower stalls, are cement, tile, or terrazzo with sanitary cove base integral with floor.
9. Toilet stalls are open at bottom.
10. Louvres are provided in roof.
11. Ceilings have a height of 8 ft. not including peak.
12. Bathroom includes utility sink, waste paper basket, 2 flush toilets, 2 wash basins, 1 urinal, 1 shower, paper towels, and soap or soap dispensers.

Figure 9–12 A double-cabin housing plan with connecting toilet.

storm canopies, or windows to keep out the rain. Floors should be so constructed as to remain dry and free of dampness. To accomplish this objective the floors, including joists, are raised above the ground with the space beneath ventilated. Masonry floors built on the ground require damp-proofing.

11. *Lean-tos exclude rain.* Lean-tos are short-term shelters; nevertheless where

provided they should be constructed so as to keep regularly used spaces relatively dry.

12. *Cleanable floors provided.* Where floors are frequently wet, hard acid brick or quarry tile laid with full $\frac{1}{16}$-in. Portland cement or equal hard joints, sloped to a drain, make good floors. Where floors are wet infrequently, top-grade maple

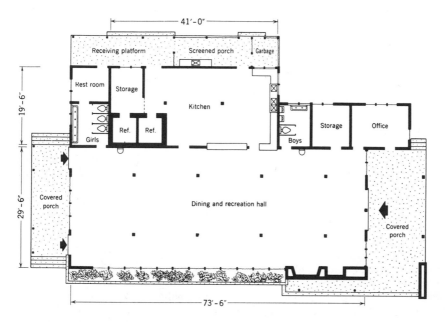

Figure 9–13   Plan of Cornell central camp building. [From *Cornell Miscellaneous Bull.* 14, New York State Colleges of Agriculture and Home Economics, Cornell University, Ithaca (March 1953).]

wood floors, well laid, are preferred. Other types of floor materials and coverings that do not entrain dust, grease, or other soil and that can be kept clean are satisfactory. All corners should be rounded and a sanitary flush cove base provided around the floor at the walls. See Chapter 8.

13. *Stoves fire protected and vented.* Heating stoves and baking ovens should rest on a fireproof or fire-resistant floor. In existing buildings the equipment can be placed on a built-up fireproof floor consisting of sheet metal, $\frac{1}{8}$-in. layer of asbestos, and 4 in. of masonry, all extending about 24 in. in all directions beyond the stove. Portable stoves should be prohibited. All cook stoves, water heaters, and space heaters should be properly vented to the outer air and provision made for adequate dilution air and back-draft protection.

Hoods over kitchen ranges should provide at least a 9-in. space between the hood and combustible material. Vents should extend to a properly constructed lined masonry chimney or an exterior stack extending at least 2 ft out from the building and 2 ft above the roof in such a manner that no danger can occur if grease in the hood and vent should catch fire. Horizontal runs should be not longer than 20 ft and slope up at least ¼ in. per foot, be provided with a back-draft diverter, and the connection made to the chimney should terminate flush with the interior of the chimney lining. Asbestos cement pipe is not suitable as a flue unless completely enclosed in masonry. Terra cotta clay pipe is standard. Hoods, exhaust ducts, or vents, fans, and filters require regular cleaning. Stoves burning wood, coal, gas, or oil must be vented to the outside air. A spark screen should be installed on chimney caps where a solid fuel is used.

Suitable fire extinguishers should be readily available at all times. A foam or carbon dioxide type of extinguisher is recommended for gasoline, paint, oil, or grease fires. The loaded stream or dry chemical type may also be used. A carbon dioxide extinguisher is recommended for electrical fires. The dry chemical type is also suitable. A soda–acid or water-type extinguisher is suitable for wood, paper, textile, and rubbish fires. Water extinguishers can be protected from freezing by adding calcium chloride containing a corrosion inhibitor. Five pounds to 9 qt of water will protect to 10°F. Eight pounds six ounces of calcium chloride to 2 gal of water will protect to −20°F, and 10 lb will protect to −40°F.

One 2½-gal or two 1½-gal fire extinguishers per 1000 ft^2 of floor space accessible within 100 ft is recommended. Provide one additional 2½-gal or two 1½-gal extinguishers for each additional 2000 feet^2. For each transformer or power generator provide one 4-lb carbon dioxide extinguisher; and in each kitchen or other areas where flammable grease or liquids are used, stored, or dispensed, provide one 12-lb carbon dioxide or dry chemical extinguisher. In any case, be guided by local fire and building department regulations where they exist.

At small camps, it is also good practice to have available 50 to 100 ft of 1-in. garden hose and convenient hose-bib water outlets within reach of all structures. Adequate water volume and pressure are of course essential. At large camps, a community-type water system as described in Chapter 3 should be provided. All extinguishers require annual inspection. The soda and acid type and foam type should be discharged, refilled, and tagged. The water-tank type should be operated to assure it is not clogged and refilled. Other type extinguishers should be weighed, refilled if necessary, and tagged. Do not overlook the usefulness of such simple devices as covered water buckets in tent and cabin areas, fire shovel, ground-fire beater, stirrup pump near streams, and back pump.

14. *Buildings and grounds clean.* Trash, weeds, brush, poison ivy, or manure should not be permitted to accumulate. Barns, stables, picket lines, and so on should be kept clean. Structures existing should be kept in repair or torn down.

15. *Fire escapes and exits provided.* In general, there should be more than one exit from a building. When sleeping quarters or assembly rooms are provided above the ground floor, at least two separated exists should be indicated that are accessible from all rooms through approaches 3 ft or more in width; an additional exit should be provided for each 20 persons or fraction in excess of 40. Stair halls, stairs, or passages connecting the entrance or first story with the second story should be separated by means of a self-closing fire-resistant door. Exits should be plainly marked. Buildings with more than two floors should not be used unless special fire protection and escape facilities are provided, including outside iron stairways or fire-resistant stairways with self-closing doors, sprinklers, and fire extinguishers. Also provide fire alarms, have a fire plan for emergency (including drill), and determine and comply with local regulations.

16. *Sleeping quarters adequate.* Beds that are spaced 5 ft apart, with an allowance

of at least 50 ft² of floor space per bed, prevent overcrowding and provide for clothing storage.[24] Where this is not possible, at least 40 ft² per person and head-to-foot sleeping should be used. If a minimum spacing between beds of 3 ft cannot be obtained, temporary housing such as under canvas should be furnished to alleviate overcrowding. Sleeping quarters should provide at least 400 to 600 ft³ of air space per bed and a window area equal to not less than 10 to 15 percent of the floor area. In special cases a lesser volume than 400 ft³ may be permitted. Double-decker beds are considered two beds; they are not recommended since an accident hazard is introduced and the spread of upper-respiratory diseases is facilitated.

17. *Buildings adequately ventilated.* This is accomplished when the accumulation of body odors is prevented, usually when an air change of 10 ft³/min is provided per person. The necessary air change is frequently secured in the summer by opening windows and by air leakage or seepage through the walls, ceilings, floor, windows, and doors. An area of window glass equal to 10 to 15 percent of the floor area preferably divided between opposite walls, with one-half of the area openable, is desirable. Toilet rooms should have openable screened windows (which are used) or separate forced-air or gravity ventilation providing 4 to 6 air changes per hour.

18. *Kitchens and dining rooms separate from dormitory and toilet.* This is meant to prohibit sleeping in the kitchen and dining room and to provide a separation between the kitchen and toilet room. A hand-washing vestibule ahead of the toilet compartment accomplishes this objective.

19. *Screening adequate.* Rooms in which food is prepared or served should have all usable windows and other openings equipped with full-length 16-mesh wire screening during the fly season. Sleeping rooms, dining rooms, recreation halls, and theaters are frequently screened for protection against mosquitoes and other insect pests. In endemic insect-borne disease areas screening is necessary. The number of flies and mosquitoes actually present, the control measures used, and the effectiveness of residual insecticides applied also determine the need for screening.

20. *Equipment adequate for food preparation.* Utensils, tables, cutlery, dispensers, sinks, stoves, and so forth used in the preparation of food should be sufficient in number and constructed smooth and seamless or with flush seams. Stainless steel is highly recommended. Tables should have nonabsorbent tops; spatulas, scoops, butter forks, tongs, relish jars, mustard jars, and ketchup bottles should be adequate and easily cleanable. See Chapter 8 and National Sanitation Foundation Standards. Stoves and ovens should be large enough to prepare the food for a usual meal.

21. *Floors, walls, ceilings clean.* The walls, floors, and ceilings of kitchens, dining rooms, and workrooms should be so constructed as to permit them to be readily cleaned, and they should be kept clean and in good repair. Light colors, tightness, and smoothness facilitate cleanliness. A concrete, tile, or composition floor that is badly cracked; a wooden floor that is slivered, warped, or poorly laid so as to leave spaces between strips where dirt and water could collect; or a floor covering that is cracked and with holes worn through would not be satisfactory.

---

[24] *Military Sanitation,* FM21-10, AFM160-46, Dept. of the Army and the Air Force, May 1957, states in reference to barracks, "There should be a space allowance of from 60 to 72 square feet per person."

The floor should be well drained, even, smooth (but not slippery), tight, and water-repellent without cracks, holes, or slivers. Walls should be washable to splash height.

Equipment placed 18 in. away from the walls or other stationary equipment and 8 in. above the floor will simplify cleaning.

Proper lighting simplifies the working and cleaning operations and reduces accident hazards. Lighting should meet the recommendations tabulated in Chapter 8. A more efficient design results when considerations are given to the spacing of light sources, elimination of glare, shielding, control of outside light, lightness of working surfaces, walls, ceilings, fixtures, trim, and floors.

22. *Ventilation adequate.* The ventilation is adequate when cooking fumes and odors are readily removed and condensation is prevented. This will require windows that can be opened both from the top and bottom and, in most cases, also ducts that will induce ventilation. Hoods with exhaust fans should be provided over ranges, ovens, deep-fat fryers, rendering vats, steam kettles, and over any other equipment that will give off large amounts of steam or soot.

23. *Water supply apparently safe.* A water supply is considered safe when a complete sanitary survey of the water system shows that the water is obtained from a source adequately purified by natural agencies or adequately protected by artificial treatment. The water supply system should also be free from sanitary defects and health hazards. In addition, water samples collected from representative points on the distribution system and examined by an approved water laboratory should consistently show the absence of coliform microorganisms. The physical and chemical characteristics of the water should not be objectionable.

The explanations and sketches given in Chapter 3 describe in greater detail what is satisfactory and what is good construction.

If ice is used for beverages, only that which is obtained from an approved source and which is distributed in a clean manner should be used. Tongs or scoops and not the hands should be used for the serving of ice.

Water available at any public establishment at any tap, faucet, or hose is assumed to be safe to drink and should be so considered unless it is inaccessible.

The provision made to secure water on hikes or camping trips should be investigated. Iodine or chlorine water-purification tablets should be available and their use understood.

24. *Adequate quantity of hot and cold water.* The quantity of water should be ample for drinking, handwashing, cooking, and dishwashing, flush toilets if provided, and bathing and laundry purposes. This total will probably vary from 30 to 75 or more gal per person per day, depending on the number of plumbing fixtures, water pressure, type of establishment and guests, location, water usage, and other variables. A water outlet should be available within 100 ft of any structure or recreational area.

Computation of the probable total water demand, storage, pipe sizes, and pump sizes is discussed in Chapter 3. The design of the hot-water supply and demand for kitchen and general utility purposes is discussed in Chapter 8. It should be remembered that peak water demands may be 5 to 10 times average demands and that pneumatic water-storage tanks can make available only about 20 percent of the tank volume.

Some suggested hot-water storage tank and heater sizes suitable for Boy Scout-type camps are shown in Table 9-5 for comparison and reference purposes. Various rule-of-thumb guides are given in providing hot water. At camps a minimum

TABLE 9-5   HOT-WATER STORAGE TANK AND HEATER SIZES

Facility	Capacity, persons	Approximate Tank Capacity* (gal)	Heater Size (Btu per hr)
Central kitchen	125 to 150	225	150,000†
	200 to 300	500	190,000‡
Central showers	125 to 150	250	60,000§
	200 to 300	400	90,000§
Health lodge	—	30	Home size

Source: From *Camp Sites and Facilities*, Boy Scouts of America, New Brunswick, N.J., 1950.

* Use nearest commercial size; specify minimum working pressure of 85 psi for galvanized steel tank, with standard tappings. Insulate tank with 2-in. asbestos cement on 1-in. mesh chicken wire or expanded metal wrapped around tank. Also insulate hot-water delivery lines.

† Rated to raise 140 gal/hr 90° for hand-type dishwashing. Add booster heater to deliver 180°-water to disinfecting rinse sink, and an immersion on under-sink heater to keep water at about 200°F.

‡ Rated to raise 200 gal/hr 90° for general purposes, and as a booster to furnish 180–200°F-water to spray-type dishwashing machine with circulator.

§ Provide mixing valve, inaccessible to campers, to limit water temperature at shower heads to 110°F.

of 3 to 5 gal per person per day is suggested for showers alone. Another value used is 5 to 7 gal/person for all purposes with 40 to 70 percent in storage and about 30 percent provided by the heater per hour. See also Table 11–9.

25. *Operation reports satisfactory.* Where a water supply is treated, accurate and complete daily reports on the operation of the treatment plant must be kept and submitted monthly to the health department. A special form that includes the desired specific information is usually provided by the health department. The report will show whether the resort owner or manager is doing his job properly and also serve as evidence that reasonable care has been taken to treat the water. To be of value, however, all entries must be accurate. The same principles apply to sewage treatment and swimming pool operation reports.

26. *Sources of water protected.* See Chapter 3.

27. *No cross-connections or interconnections.* This subject is illustrated and discussed in greater detail in Chapter 3 and in Chapter 11 under plumbing.

28. *Water storage protected.* Proper construction is illustrated in Chapter 3.

29. *Common drinking cup prohibited.* A glass or cup should not be left near a water faucet whereby anyone may use it. Provide paper cups in sanitary dispensers or sanitary drinking fountains in the ratio of one to 50 to 75 persons.

30. *Drinking fountains satisfactory.* A satisfactory drinking fountain is one having a protected nozzle extending above the rim of the bowl in which the nozzle throws an inclined, uniform jet of water. Waste water should be drained away to a disposal system or seepage pit to prevent the collection of pools of water. Use ASA standard Z4.2 for sanitary drinking fountain.

31. *Separate toilet facilities provided.* Separate toilet facilities for campers and visitors are required for each sex. These may be flush toilets or sanitary privies.

TABLE 9-6   PLUMBING FIXTURES RECOMMENDED FOR CAMPS, TRAILER PARKS, AND PICNIC AREAS

| Fixture | Number of Persons per Fixture | | | | | | | |
| | Children's Camps | | Trailer Parks* | | Campgrounds* | | Picnic Areas* | |
	Male	Female	Male	Female	Male	Female	Male	Female
Water closet or privy seat	12 to 15	10 to 15	90	45	60	30	120	60
Urinal†	15 to 20	–	135	–	60	–	120	–
Shower	12 to 15	12 to 15	135	135	–	–	–	–
Washbasin	10 to 15	10 to 15	45	45	60	60	120	120

*In comfort stations located within 300 ft of users. Minimum of 2 water closets, 2 washbasins, 1 urinal for males, and 3 water closets and 2 washbasins for females recommended. Assume 3 persons per parking site. Provide one slop sink or service sink per comfort station. Where individual laundry is done at camp, add one washing machine, dryer, and laundry tub or tray for each 60 persons.

†Three ft of urinal trough may be considered equivalent to 2 urinals.

32. *Toilets convenient and adequate.* Toilets provided should be near the sleeping quarters and centers of activities and not further away than 150 ft. Toilets for food handlers and kitchen help should adjoin the kitchen. Flush toilets are recommended for kitchen personnel. Washbasins with warm running water, soap, and paper or other type of sanitary towels are essential for food handlers, preferably in or adjoining the kitchen.

Table 9–6 shows the minimum recommended number of plumbing fixtures to provide at trailer parks, picnic areas and camps, with hot and cold running water at showers, washbasins, service sinks, and laundry tubs.

The height of water closets and washbasins should be adjusted to the age of the children served. Suggested heights are given below.

	Height to Rim of Fixture	
Age	Washbasin	Water Closet
Up to 6	1' 9''	10''
7 to 9	2' 1''	10''
9 to 12	2' 3''	1' 0''
Over 12	2' 6''	1' 2''

The walks and toilet rooms or latrines should be clear, well marked, and lighted at night. See Figures 9–14 and 9–15.

33. *Privy location, construction, and maintenance satisfactory.* Privies should be located above groundwater, at least 100 ft from places where food is prepared or served, and convenient to the users—within 150 ft. Other details concerning location, construction, and maintenance are given in Chapter 4. Privies that are properly constructed, located, and kept clean are relatively inoffensive and satisfactory sanitary devices. Modern recirculating toilets may be used in place of privies.

34. *Disposal systems satisfactory.* Sewage disposal systems are considered to be satisfactory when the overflow or exposure of inadequately treated sewage, including bath, sink, and laundry wastes on the surface of the ground is prevented. When there is an overflow or when an overflow is imminent, immediate steps should be taken to correct this dangerous condition.

To protect the owner and also comply with the requirements of most sanitary codes, a licensed professional engineer or architect competent in matters of sewage disposal and treatment should be engaged to prepare plans for an adequate system. The plans should be comprehensive, with construction details of existing and proposed work, including structures served, water system, sewerage system, bathing area, roads, typography, soil profiles, and any other features necessary for proper design and review of the plans. Following examination and approval of the plans by the health department sanitary engineer, the job can be let out for competitive bids. Construction of the sewage disposal or treatment system should be under the supervision of the designing engineer or architect to assure compliance with the approved plans.

Most camp sewerage systems include one or more septic tanks, which should be preceded by a grease trap if designed to receive kitchen wastes. Septic tanks should be inspected each year before the camping season and cleaned when the sludge or scum approaches the depth given in Table 4–8. Grease traps require

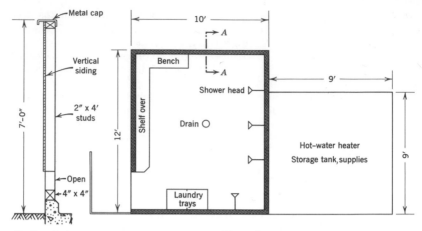

Section *A-A*   Shower house

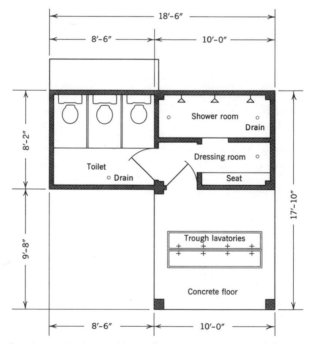

Figure 9–14 Latrine and shower-house floor plans. (Unit latrine from *Organized Camp Facilities,* U.S. Dept. of the Interior, Washington, D.C., 1938.)

frequent cleaning, depending on the size and type. Under-sink grease traps are not advised. Outside septic-tank-type grease traps should be cleaned at least once a month. More frequent inspections should be made to determine the cleaning schedule for each establishment if the carry-over of grease and clogging of sewers is to be prevented. Where other more elaborate treatment systems are provided, each unit should be maintained in proper working order so as not to lower the overall efficiency of the treatment process and to prolong the useful life of the system.

35. *Location of sewage disposal or treatment system satisfactory.* Sewage disposal systems are satisfactorily located when they are at a lower elevation, at least 100 and preferably 200 ft, from sources of water supply and 50 to 100 ft from lakes, streams, or swamps. Many other factors, as explained in Chapters 3 and 4, must be taken into consideration.

36. *Pasteurized milk used.* Only pasteurized milk, cream, cottage cheese, or dairy products obtained from an inspected and approved source should be used.

37. *Refrigeration adequate.* Refrigeration is adequate when it is possible to store and maintain all perishable foods at the proper temperature over weekends, without packing or crowding. The box should be relatively dry and permit air to circulate freely. There is never too much refrigeration space provided at camps.

Refrigerator sizes, temperatures, food storage, and maintenance are discussed in Chapter 8. Camps should provide at least $2\frac{1}{2}$ ft³ of refrigeration space per person accommodated; 3 ft³ is better.

38. *Food storage satisfactory.* If perishable fresh or prepared foods are left standing outside of refrigerators, food storage is unsatisfactory. If food is packed or disorganized in refrigerators, stored on the floor, or uncovered, the storage is not satisfactory. (Foods are packed in freezers.) Storerooms and storage spaces should be provided that are rodent-, insect-, and damp-proof, or food and dry stores should be kept in containers providing this protection. Food and drink should be wholesome. Details are given in Chapter 8.

39. *Food handling satisfactory.* Tongs, ladles, spatulas, or forks should be used to handle food in place of the hands whenever possible, and clean practices must be the rule if food poisonings and infections are to be prevented. Frozen meat, poultry, and other frozen foods, except vegetables and chops, should be thawed slowly in the refrigerator and not thawed by letting them stand overnight in a warm kitchen. Prepared foods should be either served immediately, kept on a warming table maintained at a temperature above 140°F, or refrigerated in pans to a depth of 2 to 3 in. at or below 45°F.

Food handlers are expected to have clean hygienic habits, to wash their hands thoroughly with soap and warm water after soiling them and after visiting the latrine, to wear clean clothing, to bathe daily, and to be free from communicable diseases.

Milk is one food that is particularly susceptible to contamination. Satisfactory handling and protection of milk can be obtained by serving it in quart, pint, or half-pint bottles or paper containers. Bulk-milk dispensers in which a special can with sanitary outlet is cleansed, disinfected, filled, and sealed at the pasteurizing plant that eliminates the objections to the use of the dipper and bulk milk are available. A prior understanding is necessary before the pasteurizing plant will agree to handle this equipment. The dispenser must be an approved type. Leftover milk returned from the dining room must not be reused for drinking. It should be thrown away.

Other precautions to prevent food poisoning and infection are given in Chapters 1 and 8. Pesticides should be carefully stored and used. See Chapter 10.

40. *Dishes washed, cleansed, and disinfected.* The dishwashing operation and equipment are discussed in considerable detail in Chapter 8.

41. *Garbage storage satisfactory.* There should be available an adequate number of covered metal receptacles to store all the garbage and other soiled refuse from one meal. See Chapter 5.

All the liquid wastes should be discharged to a properly designed disposal system through a grease trap.

42. *Garbage disposal satisfactory.* Probably the most satisfactory method for the disposal of garbage is to make arrangements for the use of a municipal incinerator or sanitary fill. When this is not practical, all refuse, including garbage, kitchen wastes, tin cans, bottles, ashes, and rubbish should be disposed of in a camp- or resort-operated sanitary fill. A description of the sanitary fill operation is given in Chapter 5.

An open dump invariably encourages the breeding of rats and flies. It is always unsatisfactory. Selling garbage for hog feeding encourages the spread of trichinosis. It should be discontinued in the interest of public health unless provision is made by the collector to first boil all garbage 30 min.

43. *Shower facilities provided.* Bathing facilities consisting of showers supplied with hot and cold running water or with tempered water are satisfactory. Floors should be nonskid, graded to a trapped drain, and the walls constructed of impervious material to splash height. Shower facilities convenient to kitchen personnel are a major consideration to help assure cleanliness. The availability of a swimming pool or bathing beach is not a substitute for warm water and soap cleansing showers, but is further reason for providing warm water showers to prevent greater pollution of the swimming pool. In tick-infested areas daily showers should be encouraged. Common wash pans should be discarded as they are just as much disease spreaders as common drinking cups or common bath and dish towels. See Figures 9–14 and 9–15.

44. *Bathing areas conform with sanitary code.* Pools and bathing beaches are operated and maintained in conformity with sanitary-code regulations. No bathing should be permitted at any time unless under the supervision of a competent person tained in life-saving procedures.[25] Construction of pools and bathing beaches, the life-saving procedures established, records kept, operation, and maintenance are covered earlier in this chapter. The importance of assuring that effective safety and life-saving plans are actually being practiced cannot be sufficiently emphasized. Pool should be fenced in. See this chapter for details.

45. *Reports of communicable diseases.* Persons in charge of a camp must report cases of diseases that are presumably communicable. It is the duty of the person in charge of a camp and the doctor who may be in attendance to report immediately to the health officer the name and address of any individual in the camp known to have or suspected of having a communicable disease. Strict isolation shall be maintained until official action has been taken. The method of isolation should be approved by the health officer. The person in charge should not allow such individual to leave or be removed without permission of the health officer.

Whenever an outbreak of suspected food poisoning or an unusual prevalence

---

[25] Should be over 21, have American Red Cross water-safety instructor's certificate or equivalent, and be assisted by a certified senior lifesaver for each 25 swimmers.

of any illness in which fever, diarrhea, sore throat, vomiting, or jaundice is a prominent symptom, it is the obligation of the person in charge of a camp to report immediately by telephone, telegram, or in person the existence of such illness to the health officer having jurisdiction. Health officers are required to investigate such outbreaks or the unusual prevalence of disease to determine the cause and prevent its repetition or spread. See also pages 9 to 25 and 48 to 58.

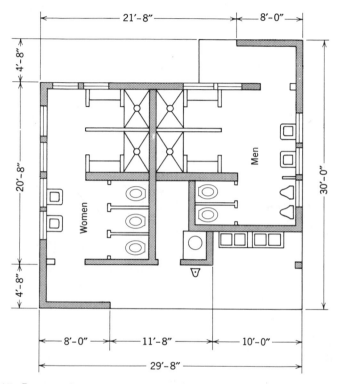

Figure 9–15 Layout of a permanent type of comfort station. (From *Environmental Health Practice in Recreational Areas,* U.S. Public Health Service Pub. No. 1195, Dept. of HEW, Washington, D.C., 1965.)

46. *Duty to enforce regulations.* It is the responsibility of the person operating a camp or the person to whom a permit is issued to see that all regulations are faithfully observed at all times.

Travel-Trailer Parks

A travel trailer is "a vehicular, portable structure built on a chassis, designed as a temporary dwelling for travel, recreation and vacation, having body width not exceeding 8 feet and its body length does not exceed 32 feet."[26] Larger units are referred to as mobile homes. Trailers include pickup coaches, motor homes, and camping trailers.

There has been a great increase in vacation travel by trailer in the United States and elsewhere. The Mobile Home Manufacturers Association estimated that in 1964 there were about 400,000 travel trailers being used in addition to several hundred thousand camping trailers and pickup coaches. The more than 2000 privately owned parks and 1200 national and state parks and campgrounds are frequently overtaxed during the vacationing season. Proper planning, designing, operation, and maintenance of trailer parks and campgrounds are essential. Most states recognize the potential hazards involved where insanitary conditions are permitted to exist and have adopted regulations to protect the users. Accepted standards to promote good sanitary practices are summarized below for easy reference.

Site. Adequate size, well drained, no water accumulations or breeding places for insects or rodents, relatively free of dust, smoke, soot, noise, and odors. If possible select site accessible to public water and sewerage systems, main highways, and service areas. Comply with zoning, building, health department, and other regulations. See Chapter 2.

Roads and parking areas. Roads should be at least 18 ft wide, 12 ft wide if serving less than 25 trailers, continuous, and dust controlled. Make roads 24 ft wide if two-way and 34 ft wide if cars parked on both sides. Provide space for off-street parking and maneuvering. See Figure 9–16.

Trailer space. Not more than 25 trailer spaces/acre; at least 10 ft from other trailer or structure. Add recreational area of not less than 8 percent of the gross ar a of the site, not less than 2500 ft².

Plans. Submit plans to health and zoning or planning agencies for approval. Show to scale the location and area of the trailer park, topography, the location of coach spaces, roads, service buildings, and other structures, water lines, fire hydrants, sanitary sewers, storm-water drains, catch basins, refuse storage racks, private wells, and sewage disposal or treatment systems. Include service building floor plan and fixtures, details of the water system and sewage disposal or treatment works, trailer water, electric and sewer connections, cold- and hot-water storage, and heater capacity.

Service building. Provide at least one. Access walks surfaced and lighted with at least 5 ft-c of illumination. Building within 300 ft of spaces served. Construction cleanable; floors drained to sewerage system; maintained clean, screened, ventilated, windows as high as practicable, providing at least 10 percent of floor area; laundry room and mirrored areas have 40 ft-c of illumination, heated to 70°F in cold

---

[26] *Environmental Health Guide for Travel Trailer Parking Areas,"* prepared by U.S. Public Health Service, Dept. of HEW, Washington, D.C.; published by Mobile Home Manufacturers Association, 6650 North NW Highway, Chicago, Ill., January 1966. Also, *Environmental Health Guide for Mobile Home Parks,* (same source as above) and *Mobile Home Court Development Guide,* U.S. Dept. of Housing and Urban Development, Washington, D.C., January 1970.

weather; separate men's and women's toilet rooms; hot and cold water for every lavatory, sink, tub, shower, and clothes washer. A floor plan is shown in Figure 9–17.

Minimum plumbing facilities are discussed earlier in this chapter. The number may be reduced if trailer-park usage is limited to only self-contained trailers.

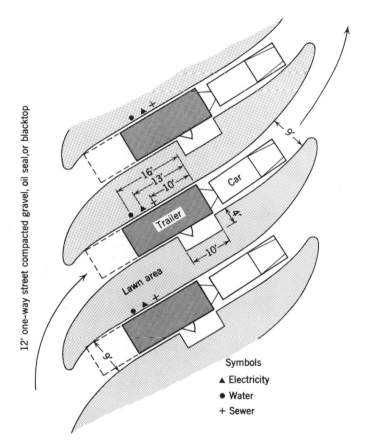

Figure 9–16 Typical drive-through parking spaces. (From *Environmental Health Guide for Travel Trailer Parking Areas,* prepared by U.S. Public Health Service, published by Mobile Home Manufacturers Association, 6650 North Northwest Highway, Chicago, Ill., January 1966.)

Water supply. At each trailer space and service building, connect a public water supply if available or an approved private supply that provides an adequate volume of water. The source should provide at least 100 gal per trailer space per day for individual connections and 50 gal per space per day without connections. The distribution system, including storage and pumping equipment, should be designed for a maximum momentary water demand based on Table 9–7.

A special water-supply outlet in the ratio of one to each 100 spaces is needed for filling trailer water-storage tanks. A flexible hose, shutoff valve, and backflow preventer are provided.

A trailer water-supply connection extending at least 4 in. above the ground with a ¾-in. valved outlet is needed at each space capable of supplying water

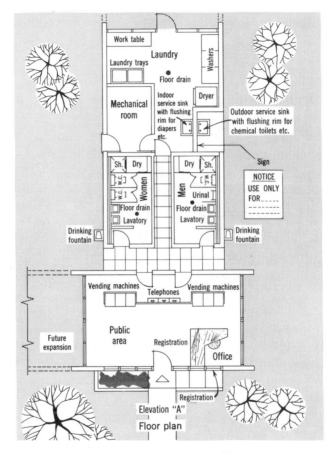

Figure 9–17 Service building. (From *Environmental Health Guide for Travel Trailer Parking Areas.*)

at a minimum pressure of 20 lb/in.² See Figure 9–18. Sanitary-type drinking fountains should be provided at service building and recreation areas. Surface water or questionable groundwater supplies should not be used unless use and treatment have been approved by health authorities. Daily records are required.

See Chapter 3 for water-supply design, storage, construction, and protection details. *Sewerage system.* Plumbing complies with state and local regulations. Special attention is given to possibility of back-siphonage; no connection permitted where possibility exists. Provide a 4-in. sewer connection extending 4 in. above ground surface and trapped below frost for each trailer space as illustrated in Figure 9–19. No wastewater shall discharge to the ground surface.

TABLE 9–7   ESTIMATED MAXIMUM WATER DEMAND FOR TRAVEL-TRAILER AND MOBILE-HOME PARKS

Number of Spaces	Demand Load (gpm)	
	Travel*	Home†
25	42	65
50	65	115
75	85	155
100	115	180
150	150	235
200	180	285
250	215	325
300	230	370

*Travel-trailer park at 4 fixture units per space.
†Mobile-home park at 8 fixture units per provided space.

Sewer lines are 10 ft from water lines and are designed for 2 fps velocity when flowing full. See Chapter 4. Joints are watertight, infiltration minimum, and surface water excluded. Manholes not more than 400 ft apart, at juncture of 2 or more sewers and at change in direction. Capped cleanouts with plugs to grade are permitted on 4- and 6-in. lines in place of manholes if at least 100 ft apart. Vents are needed on airtight system. Minimum 6-in. sewers are recommended.

Connect to public sewer. If this is not possible, design the treatment plant for a minimum flow of 100 gal per trailer space per day and so as not to cause a health hazard or nuisance. Design of plant must meet state and local requirements. See Chapter 4 for design and construction details. Separate provision should be made for storm and surface water drainage.

Refuse handling. Tightly covered containers providing 30-gal capacity per trailer space should be adequate. Bulk containers can also be used. Can storage rack is cleanable and located within 150 ft of any trailer. See Chapter 5 for acceptable disposal methods if municipal or contract service is not available.

Insect and rodent control. Mosquitoes, flies, roaches, rats, mice, and so forth are eliminated or kept under control. See Chapter 10.

Miscellaneous. Electrical wiring and apparatus complies with local codes and National Electrical Code.[27] Fuel oil and bottled gas tanks are protected and securely connected to stoves or heaters by means of metallic tubing.[28] Take fire prevention

---

[27] *National Electrical Code,* C1-1962, American Standards Association, 10 East 40th Street, New York, 1962. Also National Fire Protection Association, Boston, 1971.
[28] *Standard for Fire Protection in Trailer Courts,* NFPA No. 501A, National Fire Protection Association, 60 Batterymarch Street, Boston, Mass., 1964.

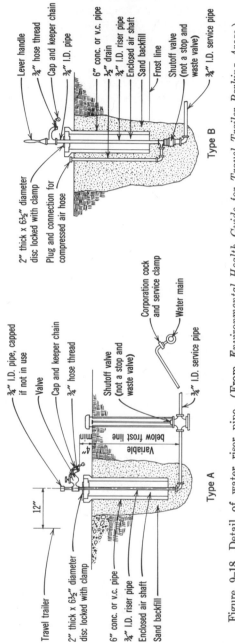

Lever handle
¾" hose thread
Cap and keeper chain
¾" I.D. pipe
6" conc. or v.c. pipe
½" drain
¾" I.D. riser pipe
Enclosed air shaft
Sand backfill
Frost line
Shutoff valve (not a stop and waste valve)
¾" I.D. service pipe

2" thick x 6½" diameter disc locked with clamp
Plug and connection for compressed air hose

Type B

¾" I.D. pipe, capped if not in use
Valve
Cap and keeper chain
¾" hose thread
Shutoff valve (not a stop and waste valve)
¾" I.D. service pipe

Variable below frost line
4" min

Corporation cock and service clamp
Water main

Travel trailer

2" thick x 6½" diameter disc locked with clamp
6" conc. or v.c. pipe
¾" I.D. riser pipe
Enclosed air shaft
Sand backfill

12"

Type A

Figure 9–18 Detail of water riser pipe. (From *Environmental Health Guide for Travel Trailer Parking Areas.*)

and protection precautions.[29] Make alterations, repairs, and additions in conformance with good practices. Keep pets under control and conditions sanitary. Maintain a register of occupants and vehicles. Eating and drinking establishments are constructed and maintained in compliance with Public Health Service, state, and local ordinances. See Chapter 8.

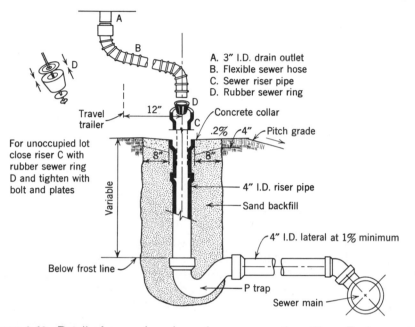

A. 3″ I.D. drain outlet
B. Flexible sewer hose
C. Sewer riser pipe
D. Rubber sewer ring

Concrete collar

Travel trailer

For unoccupied lot close riser C with rubber sewer ring D and tighten with bolt and plates

12″

.2%   4″   Pitch grade

8″   8″

4″ I.D. riser pipe

Sand backfill

4″ I.D. lateral at 1% minimum

Variable

Below frost line

P trap

Sewer main

Figure 9–19   Detail of sewer riser pipe and sewer connection. (From *Environmental Health Guide for Travel Trailer Parking Areas.*)

Migrant-Labor Camps

The problems associated with recruitment, labor relations, minimum wages, housing, unemployment compensation, health and welfare, child labor, and education at migrant-labor camps have received nationwide study. Reports and publicity have been generally unfavorable. The problems are difficult but not impossible to ameliorate. This area of work will tax the ingenuity and patience of the most dedicated people in public health, as well as those in other official and voluntary agencies having an interest in this challenging field.

An indication of the attitude of some enforcement officials and oper-

[29] *Standard for Fire Protection in Mobile Homes and Travel Trailers,* NFPA No. 501B, ibid. Also Standards for Mobile Home Parks, NFPA No. 501A, 1971.

ators of farm-labor camps is obtained from the following comments relating to environmental sanitation.

If bathing facilities are provided at farm-labor camps the people will not use them; the tubs will be used to store coal and collect trash.

Do not provide showers—the soil is too tight. Sewage will overflow onto the ground surface and create a more dangerous public health hazard.

The farm laborer does not expect bathing facilities. He never had any where he came from anyway.

A galvanized iron tube is provided which is perfectly adequate for taking a sponge bath. That's what we used when we were kids.

A washbasin and pitcher of water is all you need to keep clean.

The water supply is inadequate now—how can we provide laundry tubs and showers with running water?

The health department should first prove that our well-water supply is adequate before asking us to put in showers.

Showers are too expensive; we cannot afford to put them in.

The growers will not go along with the health department.

Political influence will be used to prevent the health department from carrying on a progressive farm-labor program.

The local health department will be voted out of existence.

You will not get the support of your own staff.

There is no health hazard involved in the failure to provide showers.

The opposition expressed by these statements is really no different from that which one would expect when embarking on any new environmental sanitation program. The half-truths must be attacked with education, painstaking engineering investigation, conferences, sound advice, and patience. When this is coordinated with a long-range plan to improve the environmental sanitation conditions and the work of other agencies, a successful program will result. See Chapter 12 and Table 9–8. The ultimate solution of the problems rests on elimination of the migrant labor camp system.

Specific items relating to health and sanitation at farm-labor camps are similar to those discussed previously under "Temporary Residences" and under the various headings in the remainder of the text. Inspection forms are shown in Figure 9–20. Good information is also available from most state health departments, extension services, U.S. Public Health Service, and Department of Labor. Some construction and facilities guides are given in Table 9–8.

A five-year enforcement program is graphically illustrated in Figure 12–2.

Mass Gatherings

The assemblage of large numbers of people in a limited area requires that certain minimum facilities be provided for the protection of the health, safety, and welfare of the people. The gatherings can vary from several thousand to several hundred thousand with and without overnight

TABLE 9–8   SOME CONSTRUCTION GUIDES AND FACILITIES AT FARM-LABOR CAMPS

Use	Area, Number, or Volume Recommended
Kitchen area	8 ft² /person including service counter, storage shelves, refrigerator, sink with hot and cold water.
Dining area	10 to 12 ft² /person.
Sleeping area	50 ft² /person plus storage space.
Showers and washbasins, with 90 to 120° F water	3 to 5 gal/person; 30% min. provided per hour by heater and 2 gal/person in storage.
Kitchen—hot water	3/4 to 1 gal/person per hour.
Water closet or privy seat	1/15 men, and 1/15 women.
Urinal	1/20 men.
Shower	1/15 persons.
Washbasin	1/15 persons with hot and cold water.
Laundry tub or machine	1/40 to 50 persons; min. of one tub.

accommodations and facilities. The types of events may include fairs, jamborees, auto races, music festivals, carnivals, and similar "happenings" held in open areas. The magnitude of such events can completely overwhelm all but the largest communities unless the highways, space, communication, sanitation, and related facilities are designed, operated, and maintained to handle the mass of people likely to descend upon the site. Such events should require a permit from the health department or other designated authority to indicate that adequate preparations have been made.

Some guidelines to assist in preparation for mass gatherings follow:[30]

1. Estimated maximum attendance and area reached by advertising.
2. Expected opening and closing date.
3. Name, owner, and location of property.
4. Name of operating "person."
5. Statement from state department of transportation giving capacity of connecting highways.

[30] Adapted from "Guidelines for Preparation of Engineering Report to Accompany Application for Permit to Conduct a Mass Gathering in New York State," New York State Department of Health, Albany, June 30, 1970; and "A Resolution to License and Regulate Persons Engaged Within the Boundaries of the County, but Outside the Limits of the City, Villages and Incorporated Towns, in the Business of Providing Entertainment or Recreation, or Providing for the Lodging of Transients; adopted September 13, 1949, as amended," County Board of Supervisors of Jackson County, Ill., February 1970.

Name and Address				Town or Village	County	Capacity	No. Occupants		
							M	F	C

							Location Code & Camp Number						Date	
1						8	9				14			

Remarks (List by numbers)
(Explain violations)

Housing	
Adequate surface drainage, grounds clean	15
Structures easy to keep clean	16
Structurally safe, sound and in good repair	17
Structures weathertight and watertight	18
Structures rodentproof	19
Floors sound, in good repair; smooth and readily cleanable	20
Ceiling height—80% at least 7 feet	21
Three sq ft. of storage area provided per person	22
Sleeping quarters for parents separated from children	23
Sleeping quarters for groups of opposite sex separate, except families. Walls, floor to ceiling or 10 feet	24
Minimum 27 inch clear space above bed surface	25
Natural light for sleeping quarters, kitchens, eating spaces	26
Natural ventilation at least 40% of window area	27
Adequate light facilities in sleeping rooms, kitchens, eating spaces, bathroom, washroom, toilet room, public spaces	28
Wall type electric convenience outlets installed in sleeping rooms, bathrooms, washrooms, cooking and eating spaces	29
Electric wiring properly installed and maintained	30
Stoves fire protected and vented as required	31
Automatic hot water and heating equipment provided with safety controls	32
Electric stoves and heaters approved by Underwriter Laboratory; gas heaters and stoves approved by American Gas Association	33
At least one outside exit above second floor. If window, not more than 14 ft. above ground	34
At least two doors from sleeping rooms for 10 or more persons	35
Enclosed stairs above second floor with self closing door top and bottom, swing in direction of travel to exit	36
Fire resistant construction provided in housing of 15 or more	37

Water	
Readily available—within 100 feet	38
Source protected	39
No cross or interconnections	40
Storage protected	41
Drinking fountains of satisfactory type	42

Toilets, privies and urinals	
One seat per 15 occupants	43
One urinal per 15 men	44
Located within 200 ft; privy within 50- 200 ft	45
Separate facilities for each sex	46

Common cooking and dining	
Separate from toilet and sleeping rooms	47
One cook stove or hot plate per 10 occupants or 1 per 2 families	48
Food and utensils storage shelves and counter space	49
Non-absorbent easily cleaned wall surfaces where required	50

Cooking in individual quarters	
Separate area of 10 sq. ft. per occupant	51
Stove or hot plate with minimum of 2 burners	52
Food and utensil storage shelves and food counter	53
Space for dishwashing	54
Nonabsorbent easily cleaned wall surfaces where required	55

Shower facilities	
One showerhead per 15 occupants	56
Separate facilities provided for each sex	57
Dry dressing space adequate	58

Laundry facilities	
One laundry tray or washtub per 25 occupants	59
One mechanical washer per 50 occupants	60
One laundry tray or tub per 100 occupants provided with mechanical washers	61

Hand washing facilities	
One unit per 15 occupants	62

STATUS CODE EXPLANATION

1. Compliance with Code requirement
2. Noncompliance with Code requirement
3. Not applicable
4. Not applicable because variance granted

Any items either in noncompliance with the Code or not applicable because of a variance granted shall be explained in the remarks section.

Person Interviewed	
Where Interviewed	Date Interviewed
Inspected by (signature)	Date Inspected
Rep. Rec'd by (sig.)	Date Rec'd

Permittee Note   1. Refer also to Part 15 of the State Sanitary Code for specific requirements.
2. The permittee is notified that items coded #2 are in violation of the requirements of the State Sanitary Code. Immediate elimination of any or all violations is required.

Figure 9-20a New York State Department of Health Migrant Labor Camp Structural Inspection Report (Part I).

Figure 9–20b MIGRANT LABOR CAMP *OPERATION AND MAINTENANCE* INSPECTION REPORT (Part II)

Name and Address	Town or Village	County	Capacity	No. Occupants		
				M	F	C

	Location Code & Camp Number		Date	
1	8	9	14	

General

Valid permit to operate	15
Grounds and common use spaces of buildings clean	16
Competent person in charge of maintenance	17
Adult supervisor for children under 16 years of age	18
Arrangements for medical care and first aid kit	19
Notice of construction, enlargement, conversion and permit as required	20

Housing

Fifty sq. ft. of floor area per person in sleeping rooms	21
Required beds and mattresses provided and in good condition	22
Adequate electric light outlets provided and working	23
Elec. and gas stoves and heaters of approved type and elec. type properly grounded	24
Heating facilities properly vented and installed with safety devices	25
Temperature at least 68° in all occupied or used rooms Oct. 1–May 1 or as required	26
Screens provided for all windows and exterior opening of living quarters	27

Water

Satisfactory quality	28
Adequate quantity of both hot and cold water	29
Hot water temperature at least 110° to 120°	30
Chlorine residual adequate if treated	31
Operation reports on water treatment satisfactory	32
No common cup	33

Toilets, privies and urinals

Properly maintained	34
Adequate ventilation provided	35

Sewerage

Proper disposal of all sewage and other liquid wastes	36
No inadequately treated liquid or solid waste on ground surface	37

Commissary and sale of food

Milk, milk products, and food properly stored	38
Adequate mechanical refrigeration	39
Only pasteurized milk sold or furnished	40

Central kitchen and dining rooms

Requirements of Part 14 met	41

Common cooking and dining

Properly screened	42
Sink with hot and cold running water	43
Adequate mechanical refrigeration	44
Adequate seating facilities	45

Cooking in individual quarters

Sink with hot and cold running water in new housing after March 1, 1968	46
Adequate mechanical refrigeration	47
Adequate seating facilities	48

Garbage and other refuse

Adequate containers provided	49
Covers available and in use	50
Garbage and other refuse storage satisfactory	51
Garbage and other refuse satisfactorily disposed of as needed	52

Shower facilities

Hot and cold or tempered water	53

Laundry facilities

Adequate hot and cold running water	54

Hand washing facilities

Hot and cold running water	55

Swimming pools and bathing beaches

Requirements of Part 6 met	56

Responsibilities of occupants

Dwelling units maintained in a clean and sanitary condition	57
Garbage and refuse properly disposed of	58
Insects, rodents or other pests controlled	59
Plumbing fixtures and housing properly used and cared for	60
Illegal heaters not in use	61
Cook stoves or burners not used for heating	62

Note to Key Punch Operator: Autoskip to column 78

REMARKS (List by numbers) (Explain violations)

Person Interviewed	

Where Interviewed	Date Interviewed

Inspected by (signature)	Date Inspected

Rep. Rec'd by (sig.)	Date Rec'd

	4
78	80

San. 102 A (rev. 7/68)

Figure 9–20b New York State Department of Health Migrant Labor Camp Operation and Maintenance Inspection Report (Part II).

6. Statement from state and local police that traffic control plan is adequate to serve 20,000 persons per hour.

7. Statement from local civil defense director approving plans for the assemblage and its evacuation.

8. Statement from local fire authority approving the fire-protection plan.

9. Emergency medical plan, including medical personnel, first-aid, hospital, and ambulance arrangements, and emergency evacuation facilities such as a helicopter.

10. Plan showing area of site and location in relation to adjoining towns within 20 mi.

11. Statement certifying all required construction, facilities, services, utilities, and operational plans are functional and approved 48 hr before assemblage time.

12. Security plan acceptable to state police ensuring crowd control, security enforcement, narcotics and drug control.

13. Plan of site showing boundaries, roads, toilet facilities, water supply, assemblage areas, first-aid stations, food service areas, refuse storage and disposal, sewage and wastewater disposal, water storage tanks, lakes, ponds, streams, wells, electric service, telephones, radio communications, emergency access and egress roads, parking areas, lighting, and other services and facilities necessary for operation.

14. Space—50 ft^2 per person.

15. Water supply—1 pt/person/hr; source, distribution, quality, and quantity comply with state health department standards.

16. Toilet facilities, for each sex—one seat/100 persons and one lavatory/100 persons; 40 percent of required seats for men should be urinals. Where portable toilets are used, obtain adequate cleanup and servicing schedule, including permit for waste disposal.

17. Refuse storage and disposal—provide in areas of assemblage 25 ft^3 of storage receptacle per 100 persons per day (24 hr), including policing and approval of disposal site.

18. Food service—provide detailed plan for food service including food sources, menu, mandatory use of single-service dishes and utensils, refrigeration facilities, food handling.

19. Vector control—where mosquito and biting fly population are in excess of 15 specimens per trap per night or other potential disease vectors are found, vector populations shall be reduced to a satisfactory level 48 hr before people assemble and shall be maintained for duration of the event.

20. Noxious weeds such as poison ivy or poison oak shall be removed from accessible areas.

21. Signs—show to the extent needed location of all facilities including roads, toilet facilities, first-aid stations, fire-fighting equipment, parking areas, camping sites, eating places, exits.

22. Drinking fountains—one sanitary-type drinking fountain including waste disposal facility per 500 persons.

23. Sewage and other waste water disposal—complies with health department standards and permit obtained for collection system and disposal.

24. Bathing areas—show location and safety measures; permit from health department.

25. Lighting—illuminate public areas of site at all times with no reflection beyond boundary of site.

26. Noise control—sound level not to exceed 70 decibels on the A scale of standard sound-level meter at perimeter of site.

27. Parking—provide usable space at rate of 100 passenger cars per acre and 30 buses per acre located off public roadways.

28. Public liability and property damage insurance as needed and reasonable in relation to the potential risks and hazards involved.

29. Performance bond—provide security sufficient to execute plans submitted and cover reasonable contingencies.

30. Communication center—at least one for operator and one for permit-issuing official.

31. Operation and maintenance—adequate staff, equipment, facilities, and spare parts, as well as communications and security.

There is a special need for the permit-issuing official to ensure co-ordination and cooperation between state and local agencies. The agencies having a responsibility and role in any mass gathering include the local government, state and local police, health, fire, transportation, and civil defense. Each should clearly understand in advance its particular responsibilities and plan accordingly. One individual should be in charge and serve as liaison officer, with the permit-issuing official having overall responsibility.

BIBLIOGRAPHY

American Camping Association Standards, Bradford Woods, Martinsville, Ind.

Black, A. P., Keirn, M. A., Smith, J. J., Dykes, G. M., and Harlan, W. E., "The Disinfection of Swimming Pool Waters," Part II, *Am. J. Public Health,* **60,** No. 4, 740–750 (April 1970).

Black, A. P., Kinman, R. N., Keirn, M. A., Smith, J. J., and Harlan, W. E., "The Disinfection of Swimming Pool Waters," Part I, *Am. J. Public Health,* **60,** No. 3, 535–544 (March 1970).

Environmental Health Guide for Travel Trailer Parking Areas, prepared by U.S. Public Health Service; Dept. of HEW, Washington, D.C.; published by the Mobile Home Manufacturers Association, 6650 North Northwest Highway, Chicago, Ill., 1966.

Environmental Health Practice in Recreational Areas, U.S. Public Health Service Pub. No. 1195, Dept. of HEW, Washington, D.C., 1968.

Minimum Property Standards for Swimming Pools, FHA 4516.1, U.S. Dept. of Housing and Urban Development, Washington, D.C., June 1970.

Resort Owner's Guide, N.Y. State Department of Health, Albany, 1965.

Salomon, Julian Harris, *Camp Site Development,* Girl Scouts of America, 830 Third Avenue, New York, 1959.

Suggested Ordinance and Regulations Covering Public Swimming Pools, American Public Health Association, 1015 Eighteenth Street, N.W., Washington, D.C., 1964.

Swimming Pool Operation, Engineering Bull. No. 18, Michigan Department of Health, Lansing, 1963.

Swimming Pools, CDC Training Manual, U.S. Dept. of Health, Education, and Welfare, Atlanta, Ga., 1959.

Mallman, W. L., "Public Health Aspects of Beaches and Pools," *Critical Reviews in Environmental Control,* Chemical Rubber Co., Cleveland, Ohio, June 1970, pp. 221–255.

10

VECTOR AND WEED CONTROL
AND PESTICIDE USE

As a general principle, the control of arthropods, rodents, and weeds where needed should take into consideration the life cycle and the conditions favorable to growth. For example, in the development of a multi-purpose reservoir or recreation area, the results of entomological, biological, and engineering studies should identify potential mosquito, aquatic weed, schistosome, or other related problems and the best means for prevention or control. The full potential of naturalistic and source-reduction measures should be applied before considering chemical means.

In the control of houseflies, roaches, and rats, the primary emphasis should be on basic community environmental sanitation to eliminate the conditions that make possible their survival and reproduction. After this has been conscientiously done, and if the problem has not been brought under adequate control, then the judicious use of the proper pesticide may be considered. There are also situations where chemosterilants, synthetic attractants (Mediterranean fruit-fly control), synthetic repellants, parasites and predators, release of sterile male flies (screwworm fly from pupae treated with gamma radiation), and quarantine have application.

PESTICIDES

Pesticide Use

There has been much public pressure to eliminate the use of persistent pesticides. This has led to controls in the United States, Great Britain, Sweden, Canada, Italy, and possibly other countries to restrict the use of certain pesticides and completely ban others (DDT and 2,4,5-T). An exception is usually made to permit DDT and other chemicals needed for the prevention or control of human, plant, and animal diseases and other essential uses for which no alternative is available. For example, the U.S. Public Health Service states that "in barrier woodland treatments for the control of *Culiseta melanura* to protect persons from infection with eastern equine encephalitis, DDT, at present, is the only insecticide considered adequate for this purpose."[1] The World Health Organization considers DDT irreplaceable in public health at the present time for the control of some of the most important vector-borne diseases of man. In addition, DDT is considered to be essential in the eradication of malaria and in the control of trypanosomiasis, onchocerciasis, louse-borne typhus, and bubonic plague. When a complete ban is imposed, there is a danger of promoting the use of a pesticide that has a higher acute toxicity, with a greater direct risk to man, wildlife, and livestock. In such instances a persistent pesticide that does not leave undesirable residues on a crop or elsewhere in the environment or cause concentration to harmful levels through the food chains may be preferable. Table 10–1 indicates the relative oral and dermal toxicity of selected pesticides. The decision to use any pesticide, whether persistent or not, should weigh the expected good against the undesirable effects. This requires comprehensive knowledge. Some of the factors to consider are:[2]

1. The development of strains of pests that are resistant to pesticides.
2. The need for repeat treatments.
3. The possible flourishing of other pests because their natural enemies have been destroyed.
4. The undesirable side effects on fish, birds, and other wildlife; parasites and predators; honey bees and other beneficial insects; man and domestic animals; and food crops.
5. The hazard of residual pesticides on food crops.
6. The direct hazard to the workers involved in pesticide application.
7. The effect of reducing and simplifying the biota; the advantages of preserving a diverse stable natural environment as compared to controlled manipulation of the environment.

The objective in pest control should be the use of a combination of biological, physical, chemical, and educational measures that will give

[1] "Public Health Pesticides," National Communicable Disease Center, U.S. Public Health Service, Dept. of HEW, Atlanta, Ga., 1970.
[2] Ray F. Smith, "Pesticides: Their Use and Limitations in Pest Management," talk presented at a research and training conference on "Concepts of Pest Management," March 27–29, 1970, in Raleigh, N.C. The conference was sponsored by the North Carolina State University and the Entomological Society of America.

TABLE 10-1   ACUTE ORAL AND DERMAL LD^{50} TOXICITY VALUES OF SELECTED PESTICIDES TO FEMALE RATS (mg/kg).*

Chemical Class	Pesticide	Oral	Dermal
Organochlorine	chlordane	430	690
	DDT	118	2,510
	dieldrin	46	. 60
	lindane	91	900
	methoxychlor	6,000†	>6,000
Organophosphorus	Abate	13,000	>4,000
	Diazinon	285	455
	dichlorvos	56	75
	dimethoate	245	610
	Dursban	82	202†
	fenthion	245	330
	Gardona	1,125	>4,000
	malathion	1,000	>4,444
	naled	250‡	800‡
	parathion (ethyl)	3.6	6.8
	parathion (methyl)	24	67
Carbamate	propoxur (Baygon)	86	>2,400
	carbaryl	500	>4,000
Arsenical	Paris green	100	2,400
Botanical	pyrethrum	263§	—
Miscellaneous	diphacinone	1.9‡	—
	pindone	280‡	—
	warfarin	3‡	—

*All figures taken from tests performed by the Atlanta Toxicology Branch, Division of Pesticides, Food and Drug Administration.
†Sex not specified.
‡Data for male specimens.
§Unpublished data.
Source: "Public Health Pesticides," National Communicable Disease Center, U.S. Public Health Service, Dept. of HEW, Savannah, Ga., 1970.

the best results with the least amount of a selective chemical poison and a minimum hazard to man and wildlife. One should carefully consider the possible spread of chemical contamination via the air, water, food, and soil, the household, transportation and handling, accidents, and occupational exposure.

If control cannot be achieved without the use of chemicals, a chemical should be selected that is specific and has the least effect on nontarget species. Only the area affected should be treated, and care should be taken to avoid streams and lakes. The treatment should be timed to coincide with the presence of the pest, and only the amount of chemical needed and type needed to achieve the result in the least time should be used. Select the formulation that will limit the range of the chemical, that is, granules for concentrated toxic action, emulsions within bodies of water, and solutions or suspension in oil on surfaces. Consider the use of repellants or noisemakers to move nontarget species from the area during the effective period of the chemical.[3] Consider also the possible insecticide resistance.

The distribution and use of pesticides are controlled by federal laws. In many instances, state and local laws and regulations parallel the federal laws. The Federal Insecticide, Fungicide, and Rodenticide Act requires that any pesticide shipped in interstate commerce be first registered with the U.S. Department of Agriculture. For a pesticide to be permitted for application on a raw agricultural food or feed, the residue cannot exceed the tolerance established for the product by the Department of Health, Education, and Welfare under the conditions of use. A product exceeding the tolerance may be considered adulterated. In addition, a pesticide added intentionally or incidentally to a processed food is considered an additive under the Food, Drug, and Cosmetic Act, Food Additive Regulations, which require, among other conditions, that no additive be used that can cause disease, such as cancer, under the conditions of use. These regulations, including those relating to adulterated and misbranded food, are enforced by the Food and Drug Administration, except that the Department of Agriculture enforces the law and regulations for meat and poultry products.

In New York State the purchase, distribution, sale, use, and possession of pesticides is restricted.[4] Certain pesticides may be used, only upon issuance of a commercial or purchase permit, as listed on the approved label as registered with the New York State Department of Environmental Conservation. The use of the following pesticides will not be allowed:

1. Bandane
2. BHC
3. DDD, TDE
4. DDT
5. Endrin

[3] J. Henry Wills, "Pesticides—A Balanced View," *Health News,* Vol. 44, New York State Department of Health, Albany, April 1967.
[4] Order Establishing Rules and Regulations of the Department of Environmental Conservation Relating to Restricted-Use Pesticides, Department of Environmental Conservation, Albany, N.Y., January 1, 1971.

6. Mercury compounds
7. Selenites and selenates
8. Sodium fluoroacetate
9. Strobane
10. Toxaphene

Pesticide Groupings

A pesticide may be defined as any substance or mixture of substances intended for eliminating or reducing local populations of, or for preventing or decreasing nuisance from, any insect, rodent, fungus, weed, or other form of plant or animal life or viruses, except viruses, microorganisms, or other parasites on or in living man or animals; and any substance or mixture intended for use as a regulator of plant growth and development, a defoliant or a desiccant, but not materials intended primarily for use as plant nutrients, trace elements, nutritional chemicals, plant inoculants and soil amendments.

A grouping of pesticides suggested by Scott is given below for convenient reference.[5] There are thousands of formulations. Repellents, attractants, impregnants, and auxiliaries are also mentioned.

Inorganic contact poisons
    Sulfur group—sulfur, lime-sulfur
    Mercury group—colomel, corrosive sublimate
Organic contact poisons
    Synthetic
        Nonsubstituted hydrocarbons—formaldehyde, ferbam, naphthalene, benzyl-benzoate, xanthone
        Chlorinated hydrocarbons—pentachlorophenol, BHC, lindane, methoxychlor, DDT, DDD, heptachlor, chlordane, aldrin, dieldrin, toxaphene, endrin, dilan, dimite
        Organic phosphorus group—malathion, diazinon, parathion, TEPP, EPN, HETP, DDVP (dichlorvos),* systox, potasan, schradan, ronnel, dicapthon, dimethoate
        Organic nitrogen group—dinitrophenol, DNOC, diphenylamine, prolan, dinitronaphthol
        Organic sulfur group—aramite, sulfenone, pyrolan, sevin, isolan, ovex
        Organic thiocyanate group—lethane, thanite, loro
    Natural
        Petroleum—fuel oil, kerosene
        Alkaloids—nicotine, anabasine, hellebore, ryania
        Esters—pyrethrum, pyrethrins, allethrin, scarbrin, rhododendron, mammein

[5] Harold G. Scott, "The Chemical Control of Insects," U.S. Public Health Service Pub. No. 772, Dept. of HEW, Washington, D.C., 1962.

NOTE: State or local regulations may impose certain restrictions on the use of these compounds; therefore the individual should consult local or state authorities on the accepted-use practice.

* Vapor dispensers of these pesticides should not be used where food is prepared, in the home, institutions, or other continuously occupied spaces.

Rotenoids—rotenone, elliptone, sumatrol, malaccol, degulin, milletia

Resins—turpentine, sage oil, basil oil, croton oil, amorphin, phellodendrin

Stomach poisons

Arsenic group—arsenic trioxide, lead arsenate, copper arsenate, Paris green

Fluorine group—sodium fluoride, 1080, cryolite, sodium fluosilicate

Other groups—phosphorus paste, borax, tartar emetic, thalium sulfate, SALP

Fumigants

Naphthalene, dichlorobenzene

Hydrogen cyanide, methyl bromide, carbon tetrachloride, ethide, sulfur dioxide

Repellents

Diethyltoluamide, indalone, Rutgers 612, dimethylphthalate, dibutylphthalate

Attractant

Geranoil, eugenol, ammonia, saccharine, anethol

Impregnants

Benzil, benzlbenzoate, eulan, M-1960, benzocaine, mitin, eulan, boconizer

Auxiliaries

Synergists, inhibitors, solvents, emulsifiers, wetting agents, spreaders, perfumes, dilutents.

Pesticide Disposal

The methods now available for the disposal of surplus pesticides and containers are not completely satisfactory for all materials. They include thermal decomposition, chemical neutralization, burial, and biological degradation.

Thermal decomposition requires 900 to 1000°C for varying periods of time. With a very few exceptions this method appears adequate to degrade 99 percent or more of most commercial pesticidal formulations.[6] Care must be used to hold the pesticides at this temperature long enough to decompose them and to remove air pollutants before the gases are discharged to the atmosphere. A wet scrubber and filtration through a porous clay bed and carbon filter, with lagoon treatment of the wastewater, are suggested. Satisfactory methods have not yet been developed. Mercury, arsenic, lead, and similar toxic compounds should not be incinerated unless special residue handling and disposal facilities are available.

Chemical neutralization must be tailored to each specific material. It is feasible with most of the organophosphate and carbamate insecticides, but not with the chlorinated hydrocarbons. Some pesticides are destroyed by nitric acid or sulfuric acid, whereas others are destroyed by sodium hydroxide or ammonium hydroxide. Still others are destroyed by chlorine compounds, peroxides, or other types of active chemicals.

---

[6] M. V. Kennedy, B. V. Stojanovic, and F. L. Shuman, Jr., "Chemical and Thermal Methods for Disposal of Pesticides," Mississippi Agricultural Experiment Station, Mississippi State University, State College, Miss., 1969.

Calcium hypochlorite seems to have the broadest application. Strong acid or alkaline hydrolysis does not give complete treatment.[7]

Burial introduces possible contamination of surface waters and groundwaters from leaching and runoff of the pesticides. Careful shallow burial with 18 in. of earth of *small* quantities of pesticides in a clay soil is generally acceptable until a better method is found. The location should be well above groundwater, downgrade, and several hundred feet from any sources of water supply and should be isolated to protect animals and children. The site should have a permanent marker with the date, amount, and name of pesticide stated. *Do not* bury pesticides in sandy soils because the risk of leaching the pesticide into surface water or groundwater is too high.

Biological or natural degradation of some short-lived materials is satisfactory, while for other more persistent materials it is too slow. The persistence of some pesticides in soils is shown in Table 10-2. The per-

TABLE 10-2   PERSISTENCE OF PESTICIDES IN SOILS

Pesticide	Half-life $(T^{1/2})$ (yr)*
Lead, copper, arsenic	10 to 30
Dieldrin, BHC, DDT insecticides	2 to 4
Triazine Herbicides	1 to 2
Benzoic acid herbicides	0.2 to 1
Urea herbicides	0.3 to 0.8
2,4-D; 2,4,5-T Herbicides	0.1 to 0.4
Organophosphorus insecticides	0.02 to 0.2
Carbamate (carbaryl) insecticide	0.02

Source: James N. Pitts, Jr., and Robert L. Metcalf, eds., *Advances in Environmental Sciences and Technology,* Vol. 1 Wiley-Interscience, New York, 1970.
*Estimates from PSAC Report (1965), Lichtenstein (1966).

sistence is variable, as can be seen, and depends on the particular chemical's reactivity, water solubility, and susceptibility to biochemical degradation.

Where possible and where facilities are available, surplus pesticides and containers should be disposed of as described above. Small quantities of combustible containers or packaging may be incinerated or buried where permitted. Containers should first be rinsed at least twice and the rinse returned to the spray tank before disposal. Where these methods

[7] Ibid.

are impractical, pesticides and large containers, such as 1-gal and larger jugs and 55-gal drums, should be returned to the manufacturer or formulator. The alternative is to store these materials in a locked building or room until an acceptable method of disposal becomes available.

THE HOUSEFLY

The housefly is the most abundant and familiar nonbiting insect about human habitations during the warm months of the year. The female lays its eggs, as many as 2700 in 30 days, in almost any moist, rotting, or fermenting material within 4 to 12 days after it is full grown. Horse, hog, chicken, and cow manure, human excrement and sewage, garbage, poorly digested sewage sludge and scum, rotting fruits and vegetables, decaying grains, soiled rags and papers serve as good breeding places. In warm weather the larvae, or maggots, hatch out from the eggs in 10 to 24 hr, become pupae in about 4 to 10 days, and adult insects in 3 to 6 more. Cold or lack of food and adverse moisture conditions will prolong this cycle from about 12 to 30 days. See Figure 10–1.

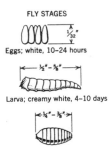

FLY STAGES

Eggs; white, 10–24 hours

Larva; creamy white, 4–10 days

Pupa; redish brown, 3–6 days

Figure 10–1 Fly stages.

Although warmth, moisture, and suitable food are required for growing maggots, relatively cool and dry materials are preferred during the pupal stage. The optimum temperature for breeding is 80 to 90°F. Temperatures of 115°F or above will kill eggs and larvae; below 44°F the fly is inactive. Larvae and emerging adults crawl through loose debris toward light. The full-grown flies are attracted by odors. Although "tagged" houseflies have been found 7 to 20 mi from breeding places, a significant number travel only 1 to 2 mi. Blowflies have been found 15 to 28 mi from breeding places, but are not numerous beyond 3 to 4 mi away.[8] Flies are most numerous in the late summer and early

[8] Herbert F. Schoof, "How Far Do Flies Fly?," *Pest Control* (April 1959).

fall. In warm climates and in heated buildings, breeding can take place throughout the year.

Spread of Diseases

The fly has a hairy body and sticky padded feet. A single fly may carry as many as 6,500,000 microorganisms. Bacteria may be carried in the digestive system of the fly for as long as 4 weeks; the bacteria can be transmitted to succeeding generations. The fly's instinct for survival and attraction to odors leads it to open privies, garbage dumps, overflowing sewage-disposal units, or grossly sewage-soiled banks of streams in search of the partially digested or decaying food that is to its liking. In feeding on the filthy material, the fly covers its legs, body, and wings with germs. After the main food course, rich in broken-down proteins, the fly is ready for its dessert. For this course the fly may alight on the milk bottle, cup, pudding, fruit, cake, bread, the baby's bottle, face, or mouth. Since the fly is on a liquid diet, it is necessary for it to transform the cake frosting or pudding, for example, into a liquid. It is well-equipped to do this by regurgitating some of the liquids already swallowed, such as sewage and microbe-laden saliva, until the relatively solid material is softened sufficiently to be swallowed. In so doing, only part of the regurgitated liquid and saliva can be taken in again. Hence the fly leaves behind part of its vomit, germs from its legs, and incidentally, its excretions. If undisturbed, the vomit and feces dry to form the familiar black specks frequently seen on some restaurant kitchen walls.

The housefly can be an important agent or mechanical carrier in the spread of such diseases as typhoid fever, paratyphoid and other salmonella infections, bacillary and amebic dysentery, cholera, typhus, anthrax, diarrhea or gastroenteritis, conjunctivitis, trachoma, possibly pulmonary tuberculosis, and poliomyelitis. The role played by houseflies in the transmission of many disease microorganisms and parasitic worms has been demonstrated in the laboratory and by epidemiological circumstantial evidence. A community-wide fly-control program in Hidalgo County, Texas, resulted in a significant reduction in the incidence of shigellosis and salmonellosis in the population. The extent to which the housefly is involved in the spread of other diseases remains to be determined and investigated.

Fly Control

Fly control requires the elimination of fly-breeding places, destruction of larvae and adults, and making food inaccessible. The application of good sanitary practices in an area throughout the year will be most effective in the prevention of fly breeding and in making a control program successful. Fly population surveys are sometimes made to give an indication of the problem and as a means to evaluate the effectiveness of control measures. This can be accomplished by means of fly-trap surveys using baited traps, flypaper strips and cone traps, or fly grills. Fly grills give quick and valid results. See Fig. 10–2.

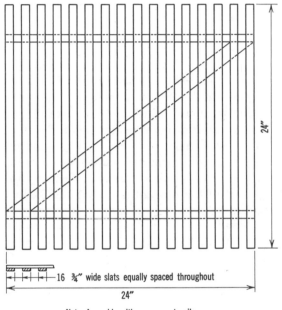

Note: Assemble with <u>screws</u>, not nails

Figure 10–2 Fly-counting grill.

The elimination of fly breeding depends primarily on environmental sanitation and secondarily on the proper use of insecticides, poisoned baits, electric grids, and traps. This involves cleanliness, good housekeeping, and the application of recommended sanitary control measures. Garbage should be wrapped in several layers of newspaper and placed in covered, washable, and cleaned containers. The receptacles should be stored in a cool or airy place and be of sufficient number and size to prevent spillage. In a warm climate, or if freezing weather is expected, a separate room kept at a temperature of 40 to 45°F should be provided at commercial establishments. A room with concrete or equal sidewalls and floor, equipped with a floor drain and hot water, promotes cleanliness. Garbage cans should be cleansed with hot water and a cleaning compound such as lye at least daily to keep down odors and prevent fly-breeding. Soiling of the garbage can will be reduced by wrapping drained garbage in paper or by using a special water-resistant lining. Spilled garbage and indiscriminately discarded food or trash should of course be collected promptly. Clean animal shelters and remove droppings from lawns.

Garbage and trash may be completely burned in a properly designed incinerator or disposed of in a sanitary fill as explained in Chapter 5. Garbage grinders in kitchen sinks and as part of the dish-scraping tables in restaurants are highly recommended where local conditions permit.

Privies, to be sanitary, must be constructed so that flies and rodents cannot gain access to the pit. Sewage and waste water should be discharged into a sewerage system so that exposure of inadequately treated sewage on the surface of the ground or into streams is prevented. Incompletely digested sludge and septic tank or cesspool cleanings are usually required to be buried at least 200 to 500 ft from any watercourse, swamp, or lake in trenches with at least 2 ft of earth cover. Thorough spraying with kerosene, fuel oil, or other larvicide or dusting with borax before and while covering will make the breeding of flies less likely.

Although the elimination of fly-breeding and the destruction of fly larvae will reduce greatly the number of adult flies, it is not usually possible to eliminate flies completely. Those flies that do develop or come from adjoining places can be killed on alighting by spraying an approved residual insecticide on screens, walls, and resting places just before the fly season. In addition to this residual spray, all buildings, particularly kitchens, nurseries, infirmaries, and those places where food is prepared or served should be completely screened. Do not spray food, utensils, pets, stoves, or open fires. Fly-repellent-type fans may be used in place of screening. Screen doors should open outward so as not to drive flies into a room when entering.

Insecticides suitable for use in fly control in food-processing plants, dairies, poultry houses, and for outdoor space spraying are summarized in Table 10–3.

Filth encourages the presence of flies. Complete dependence for killing flies cannot be placed on insecticides alone, as strains of houseflies are developing that are resistant to insecticides. Careful attention to the basic sanitation principles outlined here and good housekeeping will ensure success in controlling these disease-carrying insect pests. Store insec-

TABLE 10–3   ORGANOPHOSPHORUS INSECTICIDES FOR USE IN FLY CONTROL*

(State regulations may impose certain restrictions on the use of these toxicants in dairies or at other specified sites; therefore the individual should be certain that his usage conforms with local restrictions.)

Type Application	Toxicant	Formulation	Remarks
Residual		For 50 gal of finished spray, add water to:	
	Diazinon	2 gal 25% EC or 16# 25% WP	—Maximum strength permitted 1%. Labeled for use in dairy barns, milk rooms, and food-handling establishments, but not poultry houses.
	dimethoate	1 gal 50% EC	—Maximum strength permitted 1%. Can be used in dairy barns (except milk rooms), meat-processing plants, and poultry houses.
	Gardona	8# 50% WP or 6# 75% WP or 2 gal 2#/gal EC	—Maximum strength permitted 2%. Labeled for use in dairy barns but not poultry houses.
	malathion	2-4.5 gal 55% EC or 32-64# 25% WP	—Maximum strength permitted 5%. Labeled for use in dairy barns, poultry houses, meat-packing plants; premium grade material accepted for use in milk rooms and food-handling plants.
	naled	1 gal 50% EC	—Maximum strength permitted 1%. For use in dairy barns except milk rooms, in food-handling establishments† and in poultry houses.
	ronnel	2 gal 25% EC or 16# 25% WP	—Maximum strength permitted 1%. For use in dairy barns, milk rooms, food processing plants, and poultry houses.
	fenthion	0.7-1.3 gal 93% EC	—Maximum strength permitted 1.5%. Not to be used in dairy barns, poultry houses, or food processing plants.

AVOID CONTAMINATION OF HUMAN AND ANIMAL FOOD, WATER CONTAINERS. DO NOT TREAT MILK ROOMS OR FOOD-PROCESSING AREAS WHILE IN OPERATION. REMOVE ANIMALS FROM STRUCTURE DURING SPRAY OPERATION WHEN LABEL SO ADVISES.

Source: "Public Health Pesticides," National Communicable Disease Center, U.S. Public Health Service, Dept. of HEW, Savannah, Ga., 1970.

TABLE 10–3 (cont'd)

Type Application	Toxicant	Formulation	Remarks
Impregnated Cord	Diazinon and parathion	To be prepared by experienced formulators only.	Install at rate of 30 lin ft of cord/100 ft² of floor area. Accepted for use in dairies and food-processing plants† but not in poultry houses. Handle and install cords per manufacturer's instructions.
Bait	Diazinon	1# 25% WP plus 24# sugar; 2 fl oz 25% EC plus 3# sugar in 3 gal of water.	Apply 3-4 oz (dry) or 1 to 3 gal (wet)/1000 ft² in areas of high fly concentration. Repeat 1 to 6 times per week as required. Avoid application of bait to dirt or litter.
	dichlorvos	3-6 fl oz 10% EC plus 3# sugar in 3 gal water.	The use of permanent bait stations will prolong the efficacy of each treatment.
	malathion	2# 25% WP plus 23# sugar.	These toxicants are available as commercial baits labeled for use in dairies and in food processing plants.† None of these baits should be used inside homes. Neither should Diazinon nor trichlorfon be used in poultry houses.
	naled	1 fl oz 50% EC plus 2.5# sugar in 2.5 gal water.	
	ronnel	2 pt 25% EC plus 3# sugar in 3 gal water.	
	trichlorfon	1# 50% SP plus 4# sugar in 4 gal water.	DO NOT CONTAMINATE FEED OR WATERING TROUGHS.
Outdoor Space Spray‡	Diazinon§ fenthion	11 gal 25% EC in 34 gal water	—Apply 15 gal/mi.
	dichlorvos	6 gal 50% EC in 44 gal water	—Apply 15 gal/mi.
	dimethoate§	3 or 6 gal 50% EC in 50 gal water	—Apply 20 or 10 gal/mi.
	malathion	5 gal 55% EC in 41 gal water	—Apply 20 gal/mi.
	naled	1.5 gal 65% EC in 50 gal water	—Apply 15-20 gal/mi.
Larvicide	Diazinon	1 fl oz 25% EC to 1 gal of water	Apply 7-14 gal/1000 ft² as a coarse spray. Repeat as necessary, usually every 10 days or less. For chicken droppings, use only where birds are caged. Diazinon is not labeled for use in poultry houses.
	dichlorvos	2 fl oz 10% EC to 1 gal of water	
	dimethoate	0.5 pt 43% EC to 2.5 gal of water	

TABLE 10-3   (cont'd)

Type Application	Toxicant	Formulation	Remarks
Larvicide (cont'd)	malathion	5 fl oz 55% EC to 3 gal of water	AVOID CONTAMINATION OF FEED OR WATER AND DRIFT OF SPRAY ON ANIMALS.
	ronnel	1 pt 25% EC to 3 gal of water	

EC—Emulisifiable Concentrate   WP—Wettable Powder   SP—Soluble Powder
* For information on chemicals to be used against livestock and crop pests and for their residue tolerances on crops, consult Agriculture Handbook No. 331 (1968), U.S. Department of Agriculture, or your State Agricultural Experiment Station or Extension Service.
† Includes dairies, milk rooms, restaurants, canneries, food stores, warehouses, and similar establishments.
‡ Based on swath width of 200 ft.
§ Not specifically labeled for outdoor space applications.

ticides out of the kitchen and where they cannot be mistaken for flour, baking powder, syrup, or other food. Use with care.

MOSQUITO CONTROL

Disease Hazard

The more important diseases transmitted by mosquitoes are dengue or breakbone fever, encephalitis, filariasis, malaria, Rift Valley fever, and yellow fever. Only malaria and the encephalitis diseases are found in the United States. A summary of these diseases including etiologic agent, reservoir, transmission, incubation period, and control is given in Table 1-5.

In many parts of the world mosquito-borne diseases are a major cause of debility and death. In others the mosquito is a pest, cause for considerable annoyance and discomfort. If an intensive program of source elimination, chemical treatment, disease detection and control, and education is interrupted *before eradication or control is accomplished,* or if surveillance and control of treated areas is abandoned, the mosquito vector and the disease will soon (within 5 years) become reestablished.

Culiseta melanura is probably the most important vector of eastern equine encephalitis in the United States. *Anopheles quadrimaculatus* (Figure 10-3) is the principal malaria vector in the United States, although it is pretty much under control since the advent of DDT.

Anopheles albimanus is the principal vector in tropical America. The disease is spread when the female Anopheles mosquito bites a person or animal infected with the malaria parasites. The parasites go through changes in the body of the female mosquito until its cycle is completed. Then, when the mosquito bites a susceptible well person, the parasites are injected to develop the disease in the individual, who in turn becomes an additional reservoir of the disease until cured. It therefore becomes

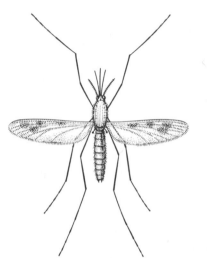

Figure 10-3 Anopheles mosquito (*A. Quad-rimaculatus*) much enlarged.

important in areas where the disease is prevalent to prevent the mosquito from gaining access to man. This is accomplished by spraying living quarters, screening, and the use of mosquito nets, gloves, and repellents until residual insecticides applied over large areas brings the anophelines under control. Where the anophelines or culicines are present, infected persons should be protected by screens or bed nets to prevent the mosquito from becoming infected. Suppressive drugs are very valuable in malarious areas.

Life Cycle and Characteristics

Mosquitoes breed in water. Weeds, tall grass, and shrubbery offer harborage to the adults. Eliminate standing water and the mosquito will be prevented from reproducing. Actually this is very difficult to accomplish because the mosquito will breed on the edges of ponds and streams, in salt marshes, holes in stumps or trees, open cesspools, open septic tanks, overflowing sewage, water barrels, street catch basins, clogged eaves troughs, empty tin cans and jars, and similar places.

Mosquito eggs hatch into the larva or wiggler stage in a matter of hours or to 2 days. After 6 to 8 days of growth they enter the pupal or tumbler stage, from which the adult mosquito emerges in about 2 days.

The female mosquito feeds only on the blood of man and animals. The male mosquito feeds on the nectar from flowers. If mosquitoes are very numerous the chances are they are breeding close by, although they may travel several thousand yards in search of a meal. Of course airplanes, trains, boats, and automobiles can and do carry mosquitoes long distances.

Mosquitoes have different biological characteristics and breeding habits. The *Aedes aegypti,* for example, have a flight range of less than $\frac{1}{2}$ mi, whereas *Aedes sollicitans* and *taeniorhynchus* may range for many miles. Eggs of some *Aedes* laid in damp or even dry soil are known to survive for months without water and then hatch out when flooded. Because of variations such as these, control measures should be based on the characteristics and life cycle of the particular mosquito species identified in the area. See Figure 10–4.

Control Program

A mosquito-control program should start with a sanitary survey. This would locate and map existing and potential breeding places, permit the collection and identification of larvae and adult mosquitoes, and make possible appraisal of the problem and planning of a control program. Light traps at key locations make possible collection of adults and an estimation of mosquito density. Supplemental surveys and surveillance are required throughout the year to determine sites suitable for early spring mosquito breeding, likely sites favorable for later breeding, and discovery of mosquito species not previously identified that might require special control. Available reports and studies by local universities and state and county agencies may also identify and locate species of mosquitoes present, concentration indices, and diseases reported in the past. Continual surveillance of the various vector species and evaluation of results should provide the basis for program direction and control. Water management is an extremely important aspect of mosquito control.

The field surveys will identify the problem areas that require primary attention. A plan can then be developed for the timing and application of the most appropriate control measure, that is, source reduction, biological, naturalistic, or chemical methods. A combination of methods may be best in some instances.

Physical source reduction is simplified by the use of topographical maps. Survey data spotted on the map will show the relative location of problem areas and their relationship to habitation and proposed new realty subdivisions or other developments. The need for and feasibility of tide gates, draining, diking, and pumping; filling, surface grading, and diversion of water; the relocation of streams and ditches; subsoil drainage; and brush clearing becomes readily apparent by studying the map and making field investigations.

Drainage involves the making of a survey to determine the breeding places, preparation of a map, determination of volume of water runoff, and type, cross-sectional area, and slope of main ditches. Main ditches should be staked out to approximate line and grade. Laterals may be constructed in the field by eye or line level as needed. Some typical ditch and covered drain cross-sections are illustrated in Figure 10–5. It should be remembered that drainage plans, excavations, and construction must include regular maintenance to remain effective and keep the cost of mosquito control within reasonable limits.

When a pond, lake, or reservoir is involved, the fluctuations in water level and resultant shallow areas should be located. Places conducive to mosquito breeding should be cleared, deepened where possible, and exposed to wave action. Areas subject to seasonal flooding or water backup should be graded to stone-lined channels draining to the pond, lake, or reservoir.

Where the filling of breeding places is possible, a sanitary landfill can combine source elimination and economical solid waste disposal with land reclamation. Hydraulic fill can also be used to deepen channels and fill low areas.

Biological measures include the stocking of ponds with *Gambusia affinis*, goldfish, or other "top feeders" that feed on larvae. The removal or encouragement of shade, depending on the specie of mosquito to be controlled, is also useful. For tidal marshes the stocking of ditches with the killifish, a salt-water minnow, and maintenance of water circulation are effective against larvae.

Naturalistic measures include rendering the water unsuitable for mosquito breeding by changing the chemical or physical characteristics of the water. For example, acid waters, a changing salinity of water, muddying water, changing water level, and agitation of water surface do not favor the growth of mosquito larvae.

Chemical control should be reserved for situations in which the other measures are not feasible. One should keep in mind the need for repeat

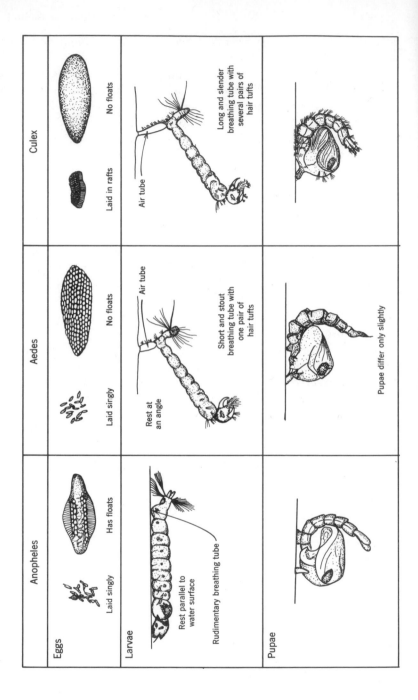

	Anopheles	Aedes	Culex
Eggs	Laid singly Has floats	Laid singly No floats	Laid in rafts No floats
Larvae	Rest parallel to water surface Rudimentary breathing tube	Air tube Rest at an angle Short and stout breathing tube with one pair of hair tufts	Air tube Long and slender breathing tube with several pairs of hair tufts
Pupae		Pupae differ only slightly	

716

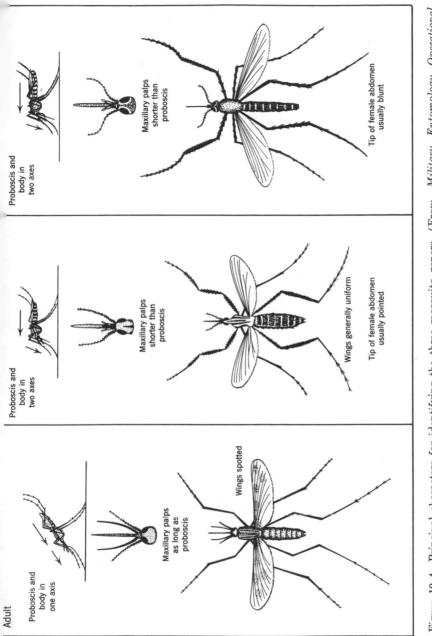

Figure 10-4 Principal characters for identifying the three mosquito genera. (From *Military Entomology Operational Handbook*, TM 5-632. Dept. of the Army, Washington, D.C., June 1965.)

Typical ditch cross sections

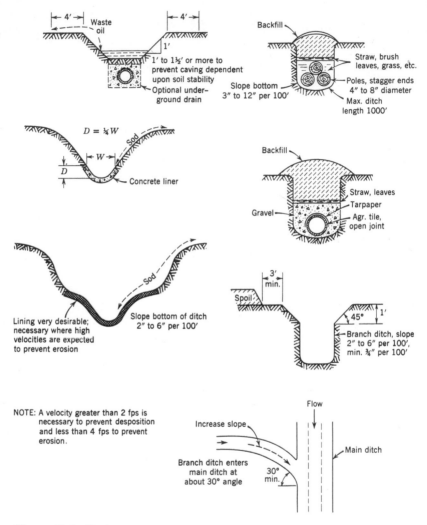

NOTE: A velocity greater than 2 fps is necessary to prevent desposition and less than 4 fps to prevent erosion.

Figure 10–5 Drainage measures.

treatment, the cost, and the limitations and hazards of pesticide use as explained on pages 699 to 702. Pesticides used in mosquito control are shown in Table 10–4.

Preflood or prehatch larvicide treatment of areas identified in surveys known to produce an early generation of mosquitoes should be treated late in winter or early spring if the site cannot be eliminated.

For larvicide spraying of small water surfaces, a compression garden-type sprayer is adequate. A power-spray unit of the orchard type is preferable for larger water surfaces. Space-spraying is accomplished by hand or power sprayers, fogging apparatus, smoke or aerosol generators, and airplanes. In marshes airboats, helicopters, and special tractors and vehicles are used.

Other larvicides, in addition to No. 2 fuel oil or Paris green, include borax and soap emulsions of pyrethrum extract in kerosene. Paris green (copper-aceto-arsenite) is especially suited for control of malaria-carrying species. Borax at the rate of 2 oz/gal of water in fire buckets will prevent breeding. Drip cans are satisfactory in slow-moving streams. A pyrethrum emulsion can be prepared using 8 oz of 40 percent liquid soap to 1 gal of water and 2 gal of kerosene pyrethrum extract. For effectiveness against larvae of the *Culex* and *Aedes* species, which are bottom feeders in contrast to the Anopheles larvae, which are surface feeders, add Paris green to moist sand and scatter the sand in breeding pools if Paris green pellets are not available.

Domestic Mosquito Control

In many cases mosquitoes are home grown and hence can be controlled in a community through combined individual action based on the life cycle of the particular species. Fundamental is the elimination of standing water in depressions, mud flats, flood plains, barrels, tin cans, drain-clogged or sagged roof gutters, and flat roofs. Inspect septic tanks, cesspools, and sewage drain fields for overflow and for openings or leaks and repair if defective. Stack containers that can accumulate water upside down and cut up or have collected old tires and cans. Treat sewer catch basins and areas that cannot be drained or filled by weekly spraying with fuel oil, kerosene, or other larvicide.

The spraying or fogging of shrubbery and tall grass around habitation in the early morning or evening will give temporary relief from adult mosquitoes for a week or two. The application of pesticide should be based on knowledge of the species involved, the source, and seasonal prevalence. A pyrethrin or dichlorvos spray beneath and around lawn furniture will help keep down mosquitoes for several hours, Dichlorvos plastic strips and pyrethrin space sprays are useful for indoor control.

For detailed information on pesticides for control of adult mosquitoes and larvae see Table 10–4. Use care and follow directions when applying pesticides.

Temporary relief from mosquito biting can be obtained by the use of repellents applied to the exposed skin and to the clothing. Common repellents are diethyltoluamide, indalone, dimethylphthalate, and Repellent 612.

TABLE 10-4   PESTICIDES CURRENTLY EMPLOYED IN MOSQUITO CONTROL*

(State or local regulations may impose certain restrictions on the use of these compounds; therefore the individual should consult local or state authorities on the accepted use practices.)

Type Application	Toxicant†	Dosage	Remarks
Residual Spray		(mg/ft²)	For use in United States as an interior house treatment.
	malathion	100 or 200	—Particularly persistent on wood surfaces and remains effective for 3 to 5 months.
	BHC	25 or 50	FOR USE IN OVERSEAS ZONES AS A STANDARD APPLICATION FOR TREATING THE INTERIOR OF HOMES IN MALARIOUS AREAS. A suspension formulation is most effective. Dosage and —cycle of retreatment depend on the vector, geographic area, and
	DDT	100 or 200	transmission period. DDT and dieldrin are effective for 6 to 12 months, BHC for 3 months. When
	dieldrin	25 or 50	the vectors are resistant to these organochlorine compounds, malathion should be used. Its efficacy is 2.5 to 3 months.
Residual fumigant	dichlorvos	1 dispenser/ 1,000 ft³	—Formulated in resin. Suspend from ceiling or roof supports. Provides 2½ to 3½ months of satisfactory kills of adult mosquitoes. Do not use where infants, ill or aged persons are confined.
		1 dispenser/ catch basin.	—Suspend dispenser 12″ below catch-basin cover.
Outdoor ground-applied space spray	carbaryl	(lb/acre) 0.2-1.0	Dosage based on estimated swath width of 300 ft. Apply as mist or fog from dusk to dawn. Mists are usually dispersed at rates of 7 to 25 gal/mi at a vehicle speed of 5
	fenthion‡	0.01-0.1	mph. Fogs are applied at a rate of —40 gal/hr dispersed from a vehicle moving at this speed; occasionally at much higher rates and greater
	malathion	0.075-0.2	speeds. Finished formulations contain from 0.5 to 8 oz/gal actual insecticide in oil or, in the case of the nonthermal fog generator, in a
	naled	0.02-0.1	water emulsion. Dusts can also be used.

TABLE 10-4   (cont'd)

Type Application	Toxicant†	Dosage	Remarks
Larvicide		Lb/acre	Apply by ground equipment or
	Abate	0.05-0.1	airplane at rates up to 10 qt of
			formulation/acre depending on
	Dursban‡ §	0.0125-0.05	concentration used. Use oil or
			water emulsion formulation in
	fenthion‡ § ‖	0.02-0.1	areas with minimum vegetative
			cover. Where vegetative cover is
	malathion	0.2-0.5	heavy, use granular formulations.
			—DO NOT APPLY PARATHION
	methoxychlor	0.05-0.2	IN URBAN AREAS. For prehatch
			treatment on an area basis, use
	parathion‡	0.1	methoxychlor (1 to 5 lb/acre).
	or methyl‡		Organophosphorus compounds
	parathion		such as Dursban and fenthion pro-
			vide prolonged effectiveness in con-
			taminated water at dosages 5 to 10
			times those listed.
	Paris green	0.75	Apply Paris green pellets (5%) at
			—rate of 15 lb/acre with ground ma-
			chines or airplane.
			Apply to cover water surface in
			catch basins or at a rate of 15 to
	fuel oil	2 to 20 gal/acre	—20 gal/acre in open water courses.
			With a spreading agent at a rate of
			0.5%, the volume can be reduced
			to 2 to 3 gal/acre.

Source: "Public Health Pesticides," National Communicable Disease Center, U.S. Public Health Service, Dept. of HEW, Savannah, Ga., 1970.

Note: The purchase and use of DDT and other persistent pesticides is restricted in many states. Check with department of agriculture, health, or environmental control.

* When insecticides are to be applied to crop lands, pasture, range land, or uncultivated lands, consult agricultural authorities as to acceptable compounds and application procedures.

† Other compounds, such as Thanite, Lethane 384, and propoxur (Baygon) may have uses in certain of the categories mentioned. If so, follow the directions on the label.

‡ For use by trained mosquito control personnel only.

§ Not to be applied to waters containing valuable fish, crabs, or shrimp.

‖ Label requires a 3-wk interval between applications except for fog treatments.

CONTROL OF MISCELLANEOUS ARTHROPODS

Bedbugs

Bedbugs (Figure 10–6) disturb the rest and comfort of individuals, and their continued presence in sleeping quarters is an indication of

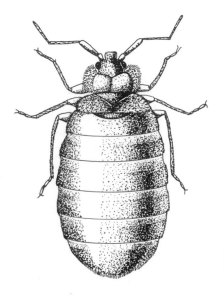

Figure 10–6   Bedbug (dorsal view). (From *Military Entomology Operational Handbook,* TM 5-632, Dept. of the Army, Washington, D.C., June 1965.)

poor housekeeping. Bedbugs are nocturnal in their habits. They are transported and distributed by clothing, laundry, baggage, bedding, and old furniture. The following extermination procedure is recommended:

Determine the source. Find where the bedbugs are hiding. Habitual hiding places are usually made evident by black or brown irregular spots, which indicate that bedbugs are or have been present in that place. These spots are usually close to the victim's sleeping quarters and are easily detected. Bedbugs live and lay their eggs in the crevices and cracks of floors, in the seams and folds of infested mattresses and bedclothes, and in the walls of buildings not far from the individual who supplies the blood meal needed for their reproduction. Favorite hiding places also are cracks in posts and bed frames, the interior of springs, and crevices of bedsteads.

Destroy eggs, larvae, and adults. Many liquid insecticides and dusts are effective in killing bedbugs. Malathion, dichlorvos, ronnel, and lindane sprays are some of the insecticides commonly available. A residual oil solution or emulsion is preferred until the bedbugs develop resistance, in which case lindane is effective. Walls, floors, bed frames, and bedding should be throughly sprayed. Pyrethrin sprays help drive bedbugs out from cracks and crevices. It may be necessary to repeat the application two or three times to kill eggs hatching out after initial spraying. Care should be used in selecting an insecticide that does not stain. Bedding and mattresses that have been sprayed should be aired 4 to 8 hr before use.

Clothing and bedding should be exposed daily, if possible, to fresh air and sunlight. Hand-picking, brushing with a stiff bristled brush, and the shaking of blankets is also recommended, particularly if insecticides are not readily available. Steam, scalding hot water, and the use of a blow torch are effective when they can be applied with safety.

Ticks

Infected wood ticks and dog ticks (Figure 10-7) are carriers of Rocky Mountain spotted fever, Q fever, relapsing fever and tularemia. The adult

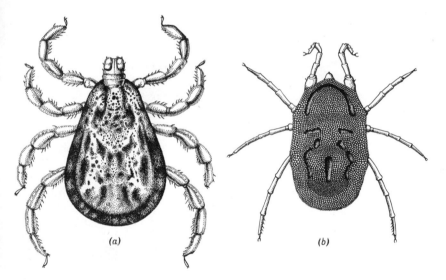

Figure 10-7 (a) Common wood tick; (b) relapsing fever tick.

wood tick appears during the spring and early summer in the northwestern states, and the dog tick appears throughout the summer in the eastern and southern states. The period of greatest prevalence is April to September. Rabbits, squirrels, woodchucks, dogs, horses, sheep, and cattle are frequently infested with ticks. The disease-causing organism is conveyed to man through the bite of the tick or by contact with crushed tick blood or feces through a scratch or wound.

Identification. Mature ticks are reddish brown in color and may have white markings on the back. They are usually one-quarter of an inch long, are oblong or seed-shaped, and have eight prominent legs.

Tick-bite prevention. One of the best ways of preventing infection is to stay away from tick-infested areas during the tick season. If that is not possible, wear clothing that completely protects the legs, arms, and hands. Inspect the body and clothing when in the field during rest periods and immediately remove any ticks found, being careful not to crush them. On returning to camp from an infested area undress, thoroughly inspect the body and clothing, and immediately remove and destroy any ticks. Inspection of the neck, back, head, and so on by another person is helpful because no tick should be overlooked. It is most important that ticks be removed within a few hours. One does not usually feel it when the tick is in the act of biting, and those attached to the body can only be found by a complete and careful inspection.

Removal of ticks. Proper method of removal and destroying of ticks is important. Use fine-pointed tweezers or a penknife, if available, for removal by insertion under the tick. *Do not crush* the tick on your body or between the fingers. Apply gentle but firm traction on the tick, being careful not to leave the mouth parts in the skin. Do not use force; a slow steady pull is required. The tick will soon let go. Or apply a few drops of oil, turpentine, kerosene, or gasoline between the skin and the undersurface of the tick. This application will kill the tick in a few minutes. A tick may also loosen its hold if a lighted cigarette is applied. The use of heat or oil is not recommended by some authorities because the irritation is said to cause the tick to regurgitate into the wound. The hands should be thoroughly washed with soap and warm water after handling ticks, and an antiseptic such as iodine, Mercurochrome, or merthiolate should be applied to the bites. Area control can be obtained with 1 to 3 lb of chlordane or toxaphene per acre. Other pesticides are also used. If found indoors, apply carbaryl dust, diazinon, malathion, or lindane into cracks around windows, baseboards, and floors. Infested dogs and kennels may be dusted with lindane, carbaryl, coumaphos, or malathion. Do not use on cats; use 1 percent rotenone.

Chiggers

The chigger is a small, bright-red mite, almost invisible. The chigger mite larva has 6 legs; the nymph and adult, 8. The chigger is the larval stage of the mite and is known as the "red bug" or "red spider." American mites do not cause disease, although the severe itching from chigger bites causes great annoyance and results in loss of sleep and fatigue. The oriental mites found in the South Pacific, however, transmit a serious disease called scrub typhus.

The adult chigger lays its eggs in the spring of the year when warm weather comes. The adult then dies; but the eggs hatch and chiggers emerge. Since the larvae have but one ambition, that is to find a host, they soon attach themselves to some passing or resting animal. The larva sinks its mouth parts into the skin, injects an irritating fluid, and sucks up the broken-down skin with some blood. After feeding for 3 or more days, if not scratched off, the larva drops off and enters the nymph stage, from which the adult develops. The nymph and adult feed on organic matter; the larva only feeds on humans and other animals, including reptiles and birds.

When possible, mite-infested areas should be marked off and avoided or cleared. Since this is sometimes impractical, certain precautions can be taken when hiking or camping in infested areas. Individual protection can be obtained by wearing clothing that fits snugly around the wrists, ankles, and neck. Clothing impregnated with 2 g of diphenyl carbonate or benzil per ft² of cloth remains effective in repelling chiggers even after several washings. Lindane, 5 percent powders of benzil, dimethylphthalate, sulphur, pyrethrum, and derris powder applied to clothing also kill or repel chiggers. A shower with soap and warm water will lessen the effect of chigger bites, and hanging infested clothing in the sun will cause chiggers to fall off or die. Ammonia, rubbing alcohol, vaseline, iodine, and a weak lysol solution will relieve the irritation from chigger bites.

Many chemicals are available to control chiggers on the ground, lawns, and brush. Wettable sulphur applied as 50 percent suspension in water at the rate of 2½ lb/1000 ft², or as a powder at the rate of 1 lb/1000 ft², controls chiggers for about 2 weeks. Lindane at the rate of 0.5 lb/acre and toxaphene and chlordane at the rate of 1 to 2 lb/acre, thoroughly applied to foliage, are other recommended insecticide sprays. Dieldrin has been found very effective against the scrub typhus chigger. The use of some of these pesticides may be restricted or prohibited by federal and state agencies. Additional control can be obtained by cutting grass and brush to the ground and either hauling away or burning the brush when dry. Burning over the camp area with a power oil burner is especially desirable when possible.

Lice

The presence of lice on the body or in the clothing is pediculosis, or lousiness. Such infestation may be due to the head louse, body louse, or pubic louse, also known as the crab louse (Figure 10–8). Lice require human blood to live; they obtain it by puncturing the skin, thereby causing itching and irritation. The resulting scratching may also cause secondary infections such as impetigo.

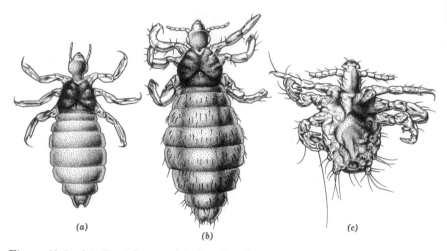

(a)

(b)

(c)

Figure 10–8 (a) Head louse, adult female; (b) body louse, adult female; (c) crab louse, adult male. (From *Insect and Rodent Control,* TM 5-632, Dept. of the Army, Washington, D.C., October, 1947).

The body louse and head louse develop in about 17 days and the crab louse in 22 to 25 days. Body lice are found in clothing, along seams, and occasionally on body hairs. Head lice prefer hairy parts of the body, such as the hair behind the ears. Crab lice infest hair around the groin, armpits, and eyebrows. The adult female body louse begins to lay eggs 4 days after maturity, at the rate of from 5 to 10 eggs a day. This will continue for 30 days under favorable conditions.

Lousiness is spread by contact with infested persons, blankets, clothing, common burshes, combs, hats, and possibly by toilet seats. Crowding, insufficient bathing and laundry facilities, and low standards of personal hygiene, chiefly infrequent bathing, favor the spread of lice.

The control of pediculosis begins with detection of the sources of infestation and treatment of those affected. Routine medical inspection of children should include scrutiny of the head under a strong light. If lice are found, a careful search should be made of the clothing and body of all children and adults living, training, working, or playing together. The U.S. Army recommends that all persons in a unit be deloused if inspection reveals 5 percent or more are infested.

Delousing may be accomplished by the proper application of powder or liquid insecticides prepared for the purpose to hairy parts of the body, clothing, and bedding. Repeat the application if necessary after 1 week and after 2 weeks, or until no further lice are found. Most of the insecticide louse powders contain 10 percent DDT or 1 percent lindane or malathion as the active ingredient. Abate dust is also reported to be very effective. When insecticides are not available, shaving or close clipping of the infested areas, followed by repeated scrubbing with soap and hot water will be effective. When the head is involved, clipping of the hair short, the application of vinegar and kerosene to the scalp if free of cuts or sores, and combing with a fine-toothed comb to remove the nits has been found effective. However, as previously stated, use of an insecticide is preferred. Clothing and bedding may be fumigated with methyl bromide if competent supervision is available, and also by dry heat or steam. Storage for at least 30 days will cause lice to die of starvation. Pressing of clothing, especially the seams, with a hot iron is also effective.

In any case, attention should be directed to personal hygiene and home cleanliness as the best preventive measures. Frequent bathing with hot water and soap, clean clothing, and clean surroundings will discourage the infestation and survival of lice.

Fleas

Fleas are associated with house pets, such as cats and dogs, and rats. Fleas can transmit disease, as explained in Chapter 1, although this is unlikely in the United States. The adult fleas attach themselves to the animals and feed on their blood. They breed in the fur or hair of pets, and the larvae hatch and live in dust in cracks, under carpets, and in cat or dog bedding. Figure 10–9 shows the human flea.

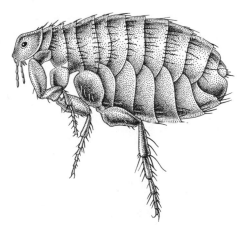

Figure 10–9 The human flea (much enlarged).

Where rats are present, a 5 percent DDT dust applied to rat runs and harborage areas will be picked up by the rat on its fur and be carried to the flea. Sometimes a 10 percent DDT dust is preferred as it will be more effective against other ectoparasites that may be present. A 0.5 percent diazinon, 1 percent lindane, 2 percent malathion, or 1 percent ronnel spray is effective against soil and house infestation by cat and dog fleas. Dusts of malathion, lindane, and carbaryl can also be used outdoors.

Oil solutions must not be applied to domestic animals as the poisons will be absorbed by the skin. Infested dogs and cats can be dusted with 1 percent pyrethrum, 1 percent rotenone, or 5 percent malathion.

Roaches

The continued presence of cockroaches in any establishment is due to poor housekeeping in the premises or in adjoining premises. These pests can be controlled by the maintenance of scrupulous cleanliness in kitchens, dining rooms, and food-storage spaces, and by the elimination of breeding places, the removal of sources of food for the insects, and the periodic application of dusts and sprays. Cockroaches are illustrated in Figure 10–10. The sanitary measures to apply are:

Clean premises and close openings. Remove temporarily all food and equipment from the place to be treated. Scrub the entire room clean, starting from the rafters or ceiling and ending with the floor. Include shelves, window sills, tops and bottoms of tables, corners, and floors under refrigerators, vegetable bins, ice boxes, sinks, and other equipment. Close all holes in floors, walls, ceilings, and around pipes and plumbing.

Treatment. Insecticides used in roach control are given in Table 10–5. Treat walls, starting from the ceiling and working down to the floor. The cracks around doors, windows, pipes; the space around kitchen cabinets, refrigerators, stoves, sinks, wood framing, baseboards, floor covering; the bracing under tables and benches should be thoroughly covered. A pyrethrum spray applied before a residual treatment will drive roaches out of cracks and crevices and give a quick kill where heavy infestations are encountered.

Ants

A 2 percent chlordane kerosene (deodorized) solution or emulsion is very toxic to ants. In buildings it can be applied with a sprayer or paintbrush to places where ants are seen crawling and at points of entry. A 0.5 percent chlordane emulsion or wettable powder is effective in controlling ants in lawns and ant mounds. Where the use of chlordane is not permitted, malathion, dichlorvos, or diazinon may be used although

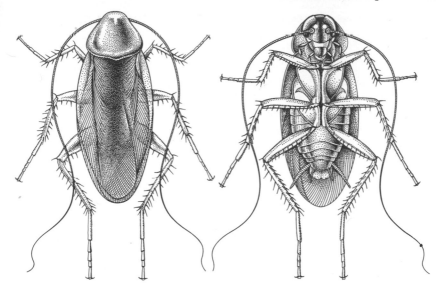

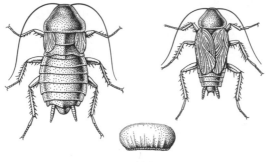

Figure 10-10   American cockroach: (a) view from above; (b) view from beneath; (c) Oriental cockroach, female and male, about natural size. (From *Insect and Rodent Control,* TM 5-632, Dept. of the Army, Washington, D.C., October 1947.)

they are less effective. Dusts and Paris green bait, 5 percent in brown sugar, are also effective.

Punkies and Sand Flies

The *Culicoides* species (punkie) apparently breeds in moist soil and is found along coastal areas, stream and pond shores, and in tree holes. Members of this species are called punkies or biting midges.

TABLE 10-5   INSECTICIDES USED IN COCKROACH CONTROL

Insecticide	Formulation	Percent Concentration*
propoxur (Baygon)	Spray	1.0
	Bait	2.0
Diazinon	Spray	0.5†
	Dust	1.0†
dichlorvos	Spray	0.5
	Bait	1.9
Dursban‡	Spray	0.5
fenthion‡	Spray	3.0
Kepone	Bait	0.125
malathion	Spray or dust	5.0

*Maximum allowable.
† 1% spray and 2 to 5% dust for pest control operators only.
‡Pest control operators only.

Salt-marsh sand flies (*Phlebotomus*), shown in Figure 10-11, can be controlled by use of fly spray where people gather. Parks, camping areas, and living quarters can be treated with a residual spray. Ordinary screens do not keep out these sand flies, but painting the screens with a 5 percent methoxychlor or malathion emulsion will give temporary protection.

The *Phlebotomus* species is a disease problem in southern Asia, the Mediterranean region, and in the high altitudes of South America. Pappataci fever, kala azar, verruga peruana, and oriental sore can be spread. Some species are found in the United States, but they pose no threat. Residual spraying of buildings, stone walls, and other resting places gives long-term control.

Blackflies

Blackflies (Figure 10-12) are blood-sucking insects belonging to the *Simuliidae* species, and are commonly called buffalo gnats. Only a few of the species are annoying biters. They are a disease vector for onchocerciasis in Mexico, South America, and Africa; they are pests in Alaska, nothern United States, and Canada.

Although some temporary relief can be obtained from mosquito-type spraying, better control is possible by killing the larvae. Individual control is impractical; area-wide control is necessary. The larvae of blackflies are found attached to rocks and vegetation in flowing streams. Application of DDT oil solutions, emulsions, or suspensions to a stream by

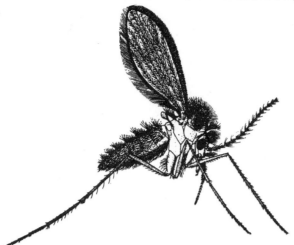

Figure 10–11 Sand fly (*Phlebotomus*). (From *Military Entomology Operational Handbook,* TM 5-632, Dept. of the Army, Washington, D.C., June 1965.)

means of a drip can or bottle has been found very effective, but the use of DDT is no longer advised because of its persistence and effect on nontarget species.

Airplane treatment with methoxychlor oil solution plus 0.5 to 0.75 percent emulsifier such as Triton X 100 is effective when applied over large areas. A spray mixture of 1 gal of 15 percent methoxychlor per

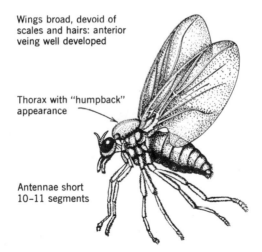

Figure 10–12 Adult blackfly (*Simulium spp*). (From *Military Entomology Operational Handbook,* TM 5-632, Dept. of the Army, Washington, D.C., June 1965.)

flight mile applied in approximately 100-ft swaths spaced ¼ mi on centers, to cross streams at right angles, is used.

Termites

There are two major groups of termites in the United States, the subterranean and the above ground. Subterranean termites travel from underground colonies to wood above ground by means of the connecting tunnels that they build. The above-ground termites may be the powderpost, dry-wood, damp-wood, or rotten-wood varieties. Termites, to survive, need food such as wood or other cellulose materials, moisture such as moist soil, and particularly protection from ants, which can enter termite tunnels and galleries.

Control requires elimination of buried wood from ground and prevention of contact of wood structural members, skirting, or siding with the ground (6-in. clearance). Good construction practices should be followed, including space ventilation and drainage; solid masonry construction, such as reinforced concrete; solid or filled hollow-block construction; and use of treated wood posts, piers, and steps. Screening (18 × 18 mesh) of windows, louvers, eaves, ventilators, and doorways and caulking of cracks, joints, and other openings will help keep out the above-ground termites.

The soil around buildings can also be treated where needed to provide a barrier next to foundations, piers, and walls and under concrete slabs. The chemicals used include 1 percent chlordane, 0.5 percent dieldrin, 0.5 percent aldrin, and 0.5 percent heptachlor, normally as water emulsions; but they can also be used as oil emulsions under professional direction.[9] The application rate should follow recommendations on the label, but for foundation walls with crawl space the dose will run around 2 gal/5 lin ft/foot width and depth. Children and pets should be kept away from the treated soil, and the chemicals should not be applied in the vicinity of well-water supplies, streams, or lakes.

CONTROL OF RATS AND MICE

Rat Control Program

Rats are carriers of disease germs, fleas, lice, mites, and intestinal parasites. They destroy food and cause an estimated damage of $900,000,000 per year in the United States. They breed prolifically; the gestation period averages about 22 days. A female rat becomes sexually mature in 2 to 3 months, produces about 20 surviving offspring per

[9] Dept. of the Navy, the Army, and the Air Force, *Military Entomology Operational Handbook,* TM 5-632 Washington, D.C., June 1965.

year, and has a life expectancy of 1 to 3 years, with an average of 1 year.[10] A rat drops 25 to 150 pellets of feces, 10 to 20 ml of urine, and several hundred hairs per day. Rats migrate when there is a lack of food, water, and shelter; when there are floods or crop failures; when dumps are abandoned; when buildings are vacated; and when warm weather and then cold weather come. Their normal range is 100 to 150 ft.

A rat control program should recognize that mouse and rat resistance to anticoagulant bait has been reported in Scotland, Denmark, England, and Wales and that it may be developing in a local work area.[11]

Rats can be controlled by community and coordinated premises sanitation through removal of food, water, and harborage. The number of rats is largely determined by the condition or capacity of an existing environment to support a prevailing population. Rat control is an integral part of a housing program. Public cooperation and individual hygienic practices on a continual basis are essential. Start with a community plan and program for action to include the following:

1. *Community survey.* Identify areas infested and degree of infestation. Determine condition of premises' sanitation, refuse storage and collection frequency, residential and business structure, dilapidation and deterioration (census data). Record data on inspection forms, maps, coded cards; analyze data and make charts, summaries, and graphs. Use these data as a basis for program planning, education and operation, followed by evaluation and program redirection. Establish a priority plan, administrative organization, and trained staff. Attack the rat infestation on a contiguous block-by-block, neighborhood-by-neighborhood basis. The survey and effectiveness evaluation should always be made by the regulatory agency; the control work may be done by the agency or by contract. Premises and block survey forms are very useful in identifying and analyzing the problem.[12]

2. *Public participation.* Solicit assistance of local neighborhood organizations, residents, and property owners in pointing out infested areas. Determine planning priorities and the extent of the participation of local people in eliminating the causes of the rats.

3. *Education and information.* Inform local officials, the mass media, and the community. Employment of local residents on a part-time or full-time basis to explain to their neighbors what brings rats and how to eliminate them can be one of the most effective measures that can be taken, particularly in run-down neighborhoods. These people have been designated as health guides, health educator aides, sanitation aides, or by similar titles. They are given special training in

[10] A female rat averages 4 to 7 litters per year, with 8 to 12 young per litter; but normally only about 20 survive.

[11] E. W. Bentley, "Developments in Methods for the Control of Rodents," *Royal Society of Health J.,* 129 (May–June 1970). Resistance found in No. Ca. 1971 and N.Y. State 1972.

[12] K. S. Littig, B. F. Bjornson, H. D. Pratt, and C. F. Fehn, *Urban Rat Surveys,* Environmental Control Program, U.S. Public Health Service, Dept. of HEW, Washington, D.C., February 1969; Joseph A. Salvato, Jr., *Control of Rats and Mice,* New York State Department of Health, Albany, October 1969.

rodent control and general environmental sanitation and work under the direction of a professionally trained supervisor. This should be supplemented by community education through the schools, news media, demonstrations, and other means to help the people learn how to help themselves. Public information examples are shown in Figures 10–13 and 10–14.

4. *Building, housing, and sanitary code enforcement.* Make sure that modern

HERE HE IS . . . THE RAT

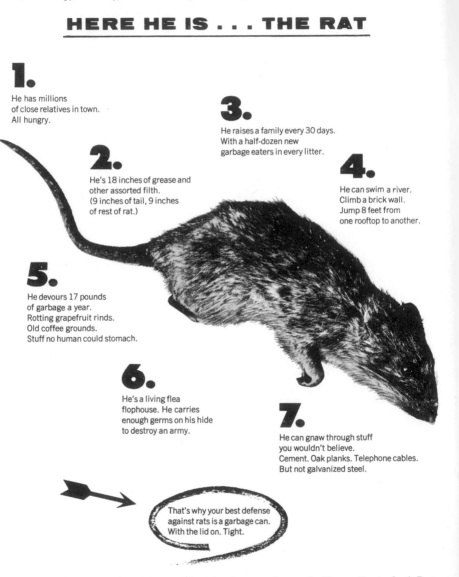

1.
He has millions of close relatives in town. All hungry.

2.
He's 18 inches of grease and other assorted filth. (9 inches of tail, 9 inches of rest of rat.)

3.
He raises a family every 30 days. With a half-dozen new garbage eaters in every litter.

4.
He can swim a river. Climb a brick wall. Jump 8 feet from one rooftop to another.

5.
He devours 17 pounds of garbage a year. Rotting grapefruit rinds. Old coffee grounds. Stuff no human could stomach.

6.
He's a living flea flophouse. He carries enough germs on his hide to destroy an army.

7.
He can gnaw through stuff you wouldn't believe. Cement. Oak planks. Telephone cables. But not galvanized steel.

That's why your best defense against rats is a garbage can. With the lid on. Tight.

Figure 10–13   Educational item, "Here he is . . . the rat." (From *Control of Rats and Mice*, New York State Dept. of Health, Albany, 1969.)

• Rats live on garbage. Wherever they can get it. But no rat ever chewed through galvanized steel.

• That's why your best defense against rats is a garbage can. With the lid on. Tight.

The directions are simple. Keep the lid on your garbage can, tight. With your garbage inside. Always.

Rats are smart. If a hungry rat finds there's no loose garbage on your premises, he won't hang around for long. Neither will his brother rats. They'll get the message fast.

Each rat consumes some 17 pounds of garbage a year. To get at it, he'll climb brick walls, swim half a mile underwater, gnaw through cement and swing from exposed ceiling beams. In his spare time he creates more rats, with a single pair producing 880 descendants a year.

To fight the rat, attack the loose garbage problem on two fronts.

One is block-by-block. The other is a campaign, urging every household to stow garbage in the can, with the lid on, at all times.

Figure 10–14   Educational item, "Starve a rat today." (From *Control of Rats and Mice,* New York State Dept. of Health, Albany, 1969.)

codes have been adopted and that they include sound, rodent-proof construction and maintenance. Competent and adequate staffs to ensure proper construction in the first instance and then systematic inspection to ensure that use and maintenance of structures and premises are in conformance with housing and sanitary codes are essential. This includes sheds, garages, alleys, backyards, empty lots, and vacant buildings. Abandoned structures should be demolished and removed.

5. *Rat eradication and harborage removal.* The killing of rats should be planned on a block-by-block and neighborhood basis just before permanent control measures are applied. The need for concurrent ectoparasite control must be determined. Rat eradication is only a temporary expedient if not accompanied by the permanent elimination of food supply and harborage. Vacated buildings must be treated

promptly to prevent the migration of rats in search of food to adjacent structures. This may also be necessary prior to demolition if reinfestation occurs. Local pest-control operators can play an important role in this activity.

6. *Adequate municipal refuse-collection service.* All communities should have an adequate municipal or contract collection service. This should include public, business, and residential areas, and at least daily collection in business districts and weekly collection in residential areas. Twice-a-week collection during the fly-breeding season and in problem neighborhoods is necessary. Proper storage and disposal are essential parts of a satisfactory refuse-collection service.

7. *Community sanitation and good housekeeping.* It should be recognized that without the maintenance of community sanitation and good housekeeping, a rat poisoning and killing program will produce only a temporary improvement. When the poisoning and killing are suspended, the rat population will again multiply as high as the available food supply and environment will support. Adequate refuse storage, collection, and disposal; rat-stoppage; housing code enforcement; and good housekeeping practices, including proper food and refuse storage and environmental hygiene, must be emphasized. Education of local residents in rat control measures is necessary so that they do not contribute to and become part of the problem.

8. *Control rats in sewers, waterfront, dumps.* Arrangements should be made for a sewer inspection and maintenance program, waterfront treatment and cleanup, and conversion of dumps to sanitary landfills preceded by poisoning. These aspects will be discussed further.

9. *Administration and interagency coordination.* All existing local agencies should cooperate and coordinate their activities to eradicate rats. One agency, such as the health department, could serve as the coordinator to ensure a planned comprehensive approach and complementary effort. Agencies involved might include, in addition to the health department, the housing code enforcement, urban renewal, sanitation, public works, buildings, fire, education, social services, and extension service agencies, as well as federal and state agencies having related programs and responsibilities.

10. *Training and employment of residents of substandard housing areas.* It has been found that residents of problem neighborhoods can communicate more effectively with the local people and obtain their cooperation in the application of rat control techniques. Such individuals can be given special instructions in basic environmental sanitation, lead poisoning control, and in the availability of community services to give them competency to serve as health guides or sanitation aides in the neighborhood. They can also assist in obtaining utilization of health services, correcting housing code deficiencies, and giving advice on better and more nutritious diets. The experience gained by the health guide or sanitation aide in rodent control and in community education can lead to other employment and advancement.

11. *Evaluation.* Evaluation involves the setting of objectives, ways of measuring achievement of the objectives, periodic measurement of the success obtained, and redirection of effort as indicated by the results. Evaluation of rodent control programs is essential if expenditure of money and effort is not to go on endlessly.[13]

[13] Clyde Fehn and Bayard F. Bjornson, "Developmental Evaluation Techniques for Community Rat Control Programs," Environmental Control Administration, Consumer Protection and Environmental Health Service, U.S. Public Health Service, Washington, D.C., November 1, 1968.

This activity should be carried out by the heatlh department or other designated control agency. The evaluation should include the measurement of recurring rat signs through resurvey, complaint investigation, bait stations, sanitation maintenance, and the possible development of bait resistance. A field-oriented laboratory support program is essential.

Rat Control Techniques

A brief outline of some specific rat control measures is given below:

Starve Rats

Starve rats by eliminating their supply of food and water. An adult rat eats about 1 oz of food and drinks 1 oz of water per day.

1. Store garbage in tightly covered metal containers.
2. Store foods in tightly covered containers of metal or glass. Sweep floors and stairways free of bits of food.
3. Store fruits and vegetables in refrigerator or in rat-proof room.
4. Keep laundry soap, candles, wax paper, and so on where rats cannot get at them.
5. Keep garbage cans, storage areas, and picnic areas neat and free of spilled food and litter.
6. Ensure adequate refuse storage, collection, and disposal and cleanliness of streets, alleys, empty lots, and premises.
7. Eliminate standing water. Fill in depressions and drain water to sewer, street, or soakage pit.

Build Rats Out

Rat-proof buildings and remove rat harborages. See Figure 10–15.

1. Find the holes rats use to enter buildings. Seal all holes in basement walls and floors with good cement or concrete. See that all doors and windows fit tight, and screen all windows that are kept open. Make all doors self-closing and outward-opening and see that they are kept closed.
2. Tear out infested wooden basement floors and replace with concrete.
3. Pile boards or lumber at least 18 in. above the ground. Remove old sheds and storage bins.
4. Keep premises, including yards and alleys, clean and free of accumulations of junk and debris.

Kill Rats by Using Poisons

Poisoning is not a substitue for basic environmental sanitation, only a supplement. Poisoning alone will only provide a temporary reduction in the rat population. Some common methods follow:

1. Mix fortified red squill or an anticoagulant rodenticide such as Warfarin, Pival, Diphacinone, Fumarin, or ANTU poison with fish, cereal, grain, or meat. Other poisons are arsenious oxide (arsenic trioxide and sodium arsenite), yellow phosphorus (hazardous, fire), and calcium cyanide (releases hydrocyanic acid on

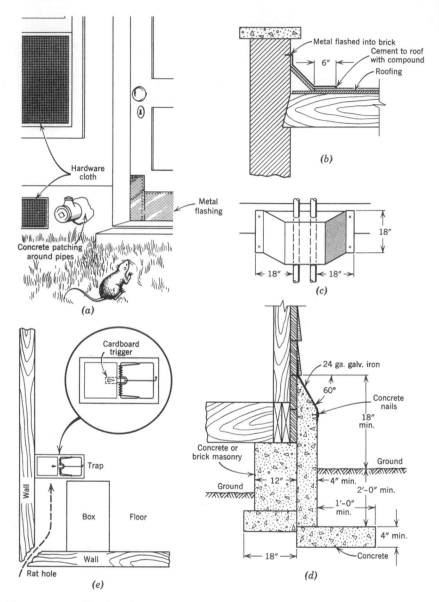

Figure 10–15 Rat-Proofing and trapping: (a) rat-proofing; (b) roof protection; (c) cable guard; (d) curtain wall; (e) trapping. (Parts a and e from "Feed People Not Rats," U.S. Dept. of Interior, Fish and Wildlife Service.)

exposure to moist air). These poisons should be used only by experienced people and should not be used in populated areas. Vary the type of fresh bait. Red squill, ANTU, and the anticoagulants are relatively harmless in the concentrations used. ANTU is quick-acting (48 hrs) against the Norway rat: its use more than once a year is not advised. ANTU is quite toxic to dogs and hogs. Place anticoagulants and ANTU in bait boxes so as not to be accessible to cats and dogs. ANTU and red squill are not effective against roof rats or house mice.

Anticoagulants are the preferred rat poisons for most purposes, but continuous feedings are required for 3 to 10 days. They cause capillaries to break down, resulting in internal bleeding, and they also prevent normal coagulation of the blood. Bait should be available for at least 2 weeks. Rodenticides and bait concentrations to poison rats—*Rattus rattus* and *Rattus norvegicus*—are given in Tables 10-6 and 10-7. *Rattus norvegicus* is also referred to as the Norway rat, sewer

TABLE 10-6   MULTIPLE-DOSE RODENTICIDES USED AGAINST MICE, ROOF RATS, AND NORWAY RATS

Rodenticide	Percent Concentration		
	Mice	Roof Rat	Norway Rat
diphacinone*	0.012-0.025	0.005-0.01	0.005-0.01
Fumarin pindone*	0.025-0.05	0.025	0.025
warfarin*	0.025-0.05	0.025	0.005-0.025

*May be used in dry or liquid bait.
Dilution factors:

    0.05%   (500 ppm) = 1 part of 0.5% concentrate to   9 parts of bait.
    0.025% (250 ppm) = 1 part of 0.5% concentrate to 19 parts of bait.
    0.01%   (100 ppm) = 1 part of 0.5% concentrate to 49 parts of bait.
    0.005% (50 ppm) = 1 part of 0.5% concentrate to 99 parts of bait.

rat, house rat, barn rat, wharf rat, and brown rat; *Rattus rattus* is also called the roof rat and black rat. The common terms may not accurately identify the actual species.

Sodium fluoroacetate, also known as "1080," is the most effective fast-acting rodenticide. Fluoracetamide (1081) is in the same group. It is very toxic and should be used only by professional exterminators under carefully controlled conditions.[14] Norbormide has a toxicity approaching 1080 to rats, whereas most other species of animals and birds are reported to survive oral doses of 1000 or more mg/kg.[15] Zinc phosphide is also very toxic and should not be used around habitation.

2. Place poisoned baits every 3 ft along rat runways. Use closed bait boxes at floor level indoors in restaurants and homes. Use bait boxes or burrow-baiting outdoors but, if exposed to rain, anticoagulants should be mixed in paraffin. CAUTION: Handle With Care. Protect children and pets. Place bait in pipes, underneath buildings, in basements and other relatively inaccessible places.

[14] The use of 1080 or 1081 is not permitted in some states.
[15] D. Glen Crabtree, W. H. Robison, and V. A. Perry, "New Concept in Rodent Control," *Pest Control* (May 1964).

TABLE 10-7   SINGLE-DOSE RODENTICIDES USED AGAINST RATS AND MICE

Rodenticide	Concentration (%) in Baits		
	Rattus norvegicus	*Rattus rattus*	*Mus musculus*
ANTU	1.0-3.0	*	*
arsenic trioxide	1.0-1.5	1.0-1.5	*
fluoroacetamide (1081)	2.0	2.0	2.0
norbormide	1.0	*	*
red squill (fortified)	5.0-10.0†	*	*
strychnine	*	*	0.3-0.5
sodium fluoroacetate (1080)	0.2-0.3	0.2-0.3	0.2-0.3
zinc phosphide	0.5-1.0	0.5-1.0	0.5-1.0

Source: "Public Health Pesticides," National Communicable Disease Center, U.S. Public Health Service, Dept. of HEW, Savannah, Ga., 1970.

*Not used against these species.

†The percent concentration used depends on the toxicity of the fortified red squill, i.e., a formulation with an oral LD_{50} to rats of 500 mg/kg would be prepared at the 10% level.

3. Pick up and destroy poisoned baits, except anticoagulants, after 24 hr of exposure.

4. Consult a reliable pest-control operator in difficult situations. State and local health departments, the state extension service, the U.S. Fish and Wildlife Service, and the U.S. Department of Agriculture can also provide guidance.

5. Use only poison products that note on the package the killing ingredients, the antidote, the word "POISON," the skull and crossbones, and the name and address of the manufacturer. Confirm actual quality of bait used by laboratory tests.

6. Some recommended bait formulations for large-scale control are given below.

Traps Rats

1. Use common wooden-base or steel snap traps and plenty of them. Provide at least 10 at a time. Tie down traps.

2. Set the traps at a right angle to and against the wall. Use fresh baits and change them daily, except for anticoagulants. See Figure 10-15. Try bacon strips, doughnuts, apple, fish, or bacon-scented oatmeal.

3. Enlarge the trigger of the common snap trap by fastening a 2-in. square of cardboard over the trigger using scotch tape. No bait is needed.

Maintain Constant Vigilance

1. Rat control measures must be continuous to be effective.

2. Make periodic surveys of buildings and neighborhoods, keeping in mind sources of rat food and water, probable harborage and nests, and what can be done to eliminate rats and prevent reinfestation. Maintain and check bait stations.

3. Public and individual cooperation in the maintenance of a clean community and home are necessary. If there is a relaxation in sanitation, renew educational campaign.

4. Ensure that a modern housing code, including rodent control, is adopted and enforced. See Chapter 11.

Rat Control at Open Dumps

Rats will travel to nearby farms and residential areas if the food supply at an open dump is reduced or cut off. This would occur when a site is abandoned or converted to a sanitary landfill. A poison-bait application program should be started ten days before a site is abandoned or converted and before earth-moving equipment is brought to the site. Post the entrance and warn nearby residents to keep children and pets at home. Baiting should be continued for 3 to 10 days after the open dump is completely converted to a sanitary landfill or closed and covered with at least 2 ft of compacted earth. This will prevent the migration and breeding of rats. A properly operated sanitary landfill will not support rats.

How to Proceed

Make a sketch of the open dump. Place bait under boxes along rat runs on the top surface, face and sides of the dump, and in all burrow openings using a long-handled spoon. Mark bait locations on the sketch and check every two or three days; replace bait eaten as needed. Apply bait at the rate of $\frac{1}{4}$ lb/yd² of open dump area including dump face. Treat the dump when there is the least activity; late Friday afternoon followed by a quiet weekend should give good results. Do not apply bait during or before rain or snow is forecast.[16]

Recommended Bait Formulations

Experienced operators have found that in general the initial use of zinc phosphide rat poison (with antimony-potassium tartrate added to cause vomiting if accidentally consumed by other animals) followed by an anticoagulant is an effective baiting sequence at dumps. However, red squill and ANTU are also satisfactory. In any case, Compound 1080 or thallium sulphate should not be used because of its extreme toxicity and associated hazards. The following formulation is recommended:

1. Lean ground horse meat, ground fresh fish, or canned cat food; poultry mash during freezing weather—70 lb.

---

[16] G. C. Oderkirk, E. M. Mills, and M. Caroline, "Rat Control on Public Dumps," *Pest Control* (August 1955); "Controlling Rats on Dumps," Fish and Wildlife Service, U.S. Dept. of the Interior, May 1960.

2. Rolled oats (acts as an extender, but may be omitted; if so, increase meat or fish by 20 lb)—20 lb.

3. Flour if using ANTU or zinc phosphide—5 lb.

4. Water—add enough to make bait quite moist. A large amount is needed if rolled oats are used; corn or peanut oil (about 2 gal) should be used as a binder during freezing weather with poultry mash instead of the meat or fish.

5. Add *one* of the following poisons to the above bait:

    a. Fortified red squill (minimum toxicity 500 mg/kg, LD 50)—10 lb.

    b. ANTU, use once a year, (for brown rats only)—1½ lb.

    c. Zinc phosphide (63%) rat poison (with antimonypotassium tartrate)—1½ lb.

    d. Anticoagulant (0.5 percent concentrate), Warfarin, Pival, Fumarin,—5 lb—or

    e. Diaphacinone—2 lb.

The anticoagulants are more expensive and need to have an edible oil added to prevent drying out. They must be eaten over a period of 3 to 10 days (requiring a bait exposure of at least 2 weeks) to cause death. Commercially prepared baits and wax-treated bait blocks may be purchased.

Canned dog or cat food in the ratio of 90 lb of meat to 10 lb of corn meal is a simple and rapid bait base. Prebaiting will determine the location of heavy infestation, suggest the amount of poisoned bait to use, and increase percentage kill. See also Tables 10–6 and 10–7.

Rats may be imported with refuse dumped at a refuse disposal site after it is converted to a sanitary landfill. However, proper daily operation, including prompt spreading and compaction of the refuse as dumped, daily 6-in. earth cover, and a final cover of 2 ft of compacted earth will eliminate the rats brought in.

Control of Rats in Sewers

Sanitary and combined sewers provide food, water, and harborage, permitting sewer-rat populations to reproduce until the environment can no longer support the growth. Population pressure then force the rats to move out through burrows, broken drains and sewers, manholes, catch basins, and house laterals to streets, yards, plumbing, and buildings. Rats can also return to sewers by the same routes when the cold weather sets in. Hence, sewer rats are a constant reservoir for community reinfestation unless controlled. Poison baits are normally placed on manhole ledges or suspended from manholes or catch basins. Zinc phosphide and anticoagulant baits are used, but spoil in a few days unless coated or mixed with paraffin. At least three applications are needed, followed by a monitoring system.[17] Baits should be checked at least twice during the winter months.

[17] W. C. Hickling and J. W. Peterson, "Rat Control as a Public Works Problem," *Public Works* (August 1968).

Control of Rats in Open Areas

Waterfronts, streams in urban and suburban areas, railroad yards, stockyards, granaries, and farms may also provide food, water, and harborage to support a rat population. These places also require surveillance and control, as described above.

Control of Mice

In general, the control of mice *Mus musculus Linnaeus* is based on the same techniques as those used in the control of rats. House-mice populations may increase as rat populations are reduced. Trapping, unless the infestation is heavy, will often be sufficient. Trap baits include peanut butter, fried bacon, cereal, gumdrops, and rolled oats. Mice can pass through a ½-in. diameter hole. Zinc phosphide bait containing antimonypotassium tartrate emetic, anticoagulants, and strychnine alkaloid impregnated grain give satisfactory results. DDT as a 50-percent "micronized" powder applied to runways and harborages also gives good results. Strychnine is highly toxic and its use around habitation should be avoided. See Tables 10–6 and 10–7 for rodenticides and bait concentrations to poison mice. Again, it must be emphasized that care must be observed in the proper handling of the materials, that a sufficient quantity of the material be used to ensure satisfactory results, and that sanitary measures be taken to eliminate the conditions that permit the rodents to exist.

Rat Control by Use of Chemosterilants

Various antifertility agents have been proposed to prevent the reproduction of rats. This technique has promise and can be applied as an *adjunct* to basic environmental sanitation and rat poisoning, particulary in those places where it is not practical to remove food, water, and harborage. Infested combined sewers, sanitary sewers, streams, wharfs, and farms are examples. But first an effective bait formulation that rats will accept must be developed. Studies with the use of a male chemosterilant show that even with 90-percent sterility, the remaining sterile males are quite adequate to impregnate the females with no resultant reduction in the rat population. Female chemosterilants seem to offer the most promise.[18]

In practice, it is rare to find community sanitation that is 100 percent effective, and complete dependence on area-wide poisoning for rat control

[18] Joe E. Brooks and Allan Bowerman, staff reports, New York State Department of Health, Albany, 1968–1971.

is expensive and in many cases impractical. Figure 10–16 illustrates how a combination of methods, including chemosterilants, might be applied to eradicate a rat population.

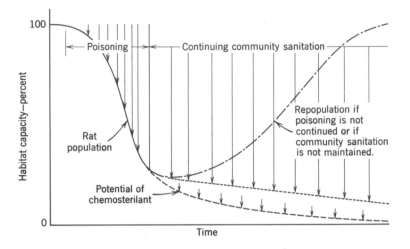

Figure 10–16   Rat population reduction and eradication using different techniques.

PIGEON CONTROL

The Health Hazard

Although the public health significance of pigeons has yet to be fully determined,[19] they do pose economic and aesthetic problems and are a potential disease threat to man.[20] The hazard is greatest to workers engaged in the removal of pigeon manure and nests; suitable dust respirators and protective clothing should be used. The diseases that may be spread to man by pigeon droppings and dust are ornithosis, histoplasmosis, cryptococcosis, and others; but the probability is remote except where there are large pigeon populations in close proximity to man.

Control Measures

Pigeon control should start with the removal of food, water, and harborage. Emphasis should be placed on sanitation and pigeon stoppage.

[19] E. S. McDonough, Ann L. Lewis, and L. A. Penn, "Relationship of Cryptococcus neoformans to Pigeons in Milwaukee, Wisconsin," *Public Health Reports,* **81,** No. 12 (December 1966).
[20] Harold George Scott, "Pigeon–Borne Disease Control through Sanitation and Pigeon Stoppage," *Pest Control,* **32,** No. 9 (September 1964).

The following measures have been used where permitted by state and local laws and where there is no serious public opposition.

1. Eliminate sources of food. Prohibit feeding. Remove animal wastes and keep animal shelters clean. Clean up spilled grain and other food sources. Properly store and collect refuse. Convert open dumps to sanitary landfills. Clean empty lots of weeds and debris.

2. Trap and humanely dispose of pigeons. This is a slow and expensive operation.

3. Locate and enclose nesting places. Close openings in buildings, sheds, attics, eaves, steeples. Screen with 1½ in. durable mesh hardware cloth. To keep out sparrows use ¾-in. mesh; use ½-in. mesh to keep out rats and ¼-in. mesh to keep out mice and bats. Slope ledges and windowsills on angle of 45 deg. or more.

4. Place repellents on roosting places such as ledges, signs, ridges, roof gutters, and cornices. Special chemicals and naphthalene, calcium chloride, and lye have been used but they need frequent replacing.

5. Use professional sharpshooters.

6. Use poisoned bait (strychnine-treated corn).

7. Use treated bait to temporarily anesthetize the pigeon. Then pick up and remove; release songbirds and humanely dispose of pigeons. Chemosterilants are still in the experimental stage.

8. Cats discourage pigeon roosting. Owls, hawks, mice, and rats prey on pigeons.

9. Mist sprays (5 percent ammonia in water with a detergent) penetrate the feathers and wet the body, causing freezing.

10. Grounded electric wires or screens can be used. Noisemakers and explosives are of temporary effectiveness.

A control program to eliminate nesting places should be preceded by ectoparasite control. It is necessary to destroy the mites, ticks, lice, fleas, and other bugs in the nests so they do not migrate when the pigeons fail to return. Infested areas can be treated with malathion emulsion or dust. These principles can also be applied to the control of nuisance sparrows and starlings.

CONTROL OF POISON IVY, POISON OAK, AND POISON SUMAC

The number of persons poisoned in the United States each year has been estimated at at least 500,000. In the many suburban communities where poison ivy or poison oak grows unchecked, it is a continual source of parental concern. The best preventive is recognition of the plant and its avoidance. See Figure 10–17.

The poison of poison ivy, oak, and sumac is an oleoresin (urushoil) that is found in the leaves, bark, flowers, roots, and fruit, but not in the wood. It is effective even after months of drying. The oily poison is nonvolatile; it is not found in pollen. Birds, livestock, pets, tennis balls, golf clubs and balls, shoes, gloves, auto tires, and tools that have come into contact with the plant serve to spread the poison.

Control measures for poison ivy are generally applicable to poison oak and poison sumac.

Eradication

Grubbing the plant out by the roots is possible although there is danger of infection. Part of the root system and debris are often left behind. It is an expensive method and is impractical to remove completely.

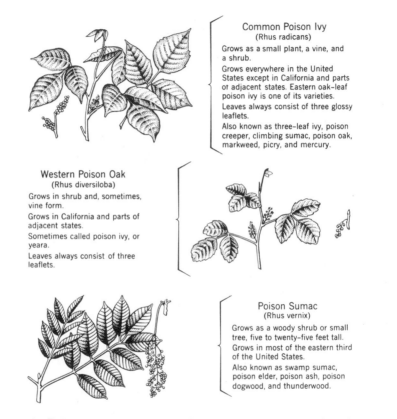

Common Poison Ivy
(Rhus radicans)

Grows as a small plant, a vine, and a shrub.

Grows everywhere in the United States except in California and parts of adjacent states. Eastern oak-leaf poison ivy is one of its varieties.

Leaves always consist of three glossy leaflets.

Also known as three-leaf ivy, poison creeper, climbing sumac, poison oak, markweed, picry, and mercury.

Western Poison Oak
(Rhus diversiloba)

Grows in shrub and, sometimes, vine form.

Grows in California and parts of adjacent states.

Sometimes called poison ivy, or yeara.

Leaves always consist of three leaflets.

Poison Sumac
(Rhus vernix)

Grows as a woody shrub or small tree, five to twenty-five feet tall.

Grows in most of the eastern third of the United States.

Also known as swamp sumac, poison elder, poison ash, poison dogwood, and thunderwood.

Figure 10–17 Poison ivy, poison oak, poison sumac. (From American Red Cross, *First Aid Textbook,* Doubleday & Co., Inc., Garden City, N.Y., 1957). NOTE: These plants belong to the same genus: *Rhus.* Since they contain the same poisonous substance, treatment for poisoning caused by them is the same for all. *Artist: Jane Roller.*

Burning off the plant is not advisable because smoke will carry particles long distances and spread the infection.

Chemical treatment is the recommended method of controlling these noxious weeds, but the chemical must not be applied where it can drain into a reservoir, stream, lake, or other body of water. Check with health and conservation agencies. Carefully follow the manufacturers' recommendations. Some of the better known weed killers are listed.

1. Ammonium sulfamate, known commercially as Ammate, does a good job; it will kill all vegetation but is somewhat corrosive to spray equipment. Application of a solution of 1 lb of Ammate crystals to 1 gal of water gives excellent control in 48 hr.

2. Borax is also effective but will damage surrounding plants if not carefully applied. The recommended dosage is 4 lb/100 ft^2.

3. The weed killer, 2,4-D, dichlorophenoxyacetic acid, takes 2 to 3 weeks. Here again care must be used not to damage other broad-leaf plants. It is not too effective in shaded areas.

4. A refinement of 2,4-D is 2,4,5-T (trichlorophenoxyacetic acid). It is a hormone-type selective herbicide that can be applied to stems or leaves during the growing season. The spray must do a thorough wetting job to be effective. A 2,4,5-T Diesel-oil solution sprayed thoroughly around the base of a plant in late winter or early spring or summer does a good job. Equal amounts of 2,4-D and 2,4,5-T make a good overall herbicide. Use of 2,4,5-T is restricted in some places.

5. A solution consisting of 3 lb of salt in 1 gal of water applied at the rate of 1 gal to 800 ft^2 does a good sterilizing job.

6. Calcium chlorate made up by 1 lb in 1 to 2 gal water is also effective. However, it is highly inflammable and will cause burns.

7. Repeated ploughing and cultivation will cause vines to gradually die out.

8. Sodium arsenite formulated with 1 lb of chemical in 5 gal of water is very effective but must be used with care as it is very poisonous to man and fish.

9. Amino triazole (Amitrole) is a very selective weed killer for poison ivy. It is available as a liquid concentrate, a water soluble powder, or in a pressurized aerosol can. The chemical kills the leaves and roots in 10 to 14 days.

Remedies

Various remedies have been suggested in the past; the best is recognition and avoidance of the plant. Better treatment methods can be developed. Some of the remedies used are listed below.

1. An old remedy that has been found effective if done *immediately* after exposure is repeated thorough scrubbing with soap and water. A good detergent should be equally effective in removing the oily irritant.

2. Aluminum acetate and lead acetate. A solution of a zirconium oxide combined with the antihistamine drug, phenyltoloxamine dihydrogen citrate, applied to exposed areas of the skin within 8 hr is reported to alleviate or prevent the symptoms. This should be applied under medical supervision.

3. A 10 percent solution of tannic acid dissolved in alcohol repeated at 6-hr intervals has been recommended. Do not use on face or genitals.

4. A 5 percent solution of ferrous sulfate or copperas is a better preventive treatment.

5. Calamine lotion with about 2 percent phenol relieves itching.

6. Jewelweed plant juice swabbed on blisters is found effective by some persons.

7. A medication injected as a liquid that is said to provide immunity against poison ivy and poison oak for up to 12 months is available.

The disease tends to disappear in 10 days to 3 weeks, and in uncomplicated cases in 5 to 7 days. The symptoms may appear after a few hours to 7 days. Only mild cases of dermatitis should be treated by lay persons; refer problem cases to a physician.

RAGWEED AND NOXIOUS-WEED CONTROL

In general, there are three kinds of weeds. "Annuals" live only one year, but are propagated by their seeds if not destroyed. Ragweed and crabgrass are examples of annuals. "Biennials" grow slowly the first year but develop a taproot and growth of leaves close to the ground. The second year rising stems and seeds are produced. Formation of seeds should be prevented to control this group. Examples are burdock and wild carrot. "Perennials" live three years or longer, each year producing runners, underground stems, roots, or bulbs. Poison ivy, Canada thistle, milkweed, bindweed, wild onions, and dandelions are examples of perennials. To control perennials requires complete destruction of all roots.

The control of weeds is accomplished by preventing spread of weeds into new areas, destruction of top weeds and underground parts of weeds, and destruction of weed seeds in the soil. Chemicals offer one of the most effective means of controlling weeds. Some are selective in killing weeds but not crops or grasses.

Weeds take over when land is neglected. Some of the conditions that encourage the growth of weeds are (1) soil abuse, (2) overgrazing, (3) erosion, uncontrolled drainage, and flooding, (4) overcultivation, (5) deforestation, (6) same kind of trees, (7) abandoned farms. It may take 10 to 15 years before plant growth progresses from the weed stage, by plant succession, to a stabilized shrub and tree phase; however, a tremendous amount of damage may be done in the interim.

Reasons for Control

It was reported in 1963 that 12 million Americans suffered from asthma and hay fever.[21] Another report states that approximately 5 percent of the persons living in the northeastern United States have pollen hay fever and that 80 to 90 percent are sensitive to ragweed pollen.[22]

[21] Bulletin of the Research and Statistical Division, National Tuberculosis Association, October 1963.
[22] A. H. Fletcher, "Report of the Research Coordinating Committee of the Northeastern Weed Control Conference for 1953."

Although mild cases do not cause great inconvenience, it can cause complications with asthma. Asthma occurs in about two-thirds of the persons who have suffered from hay fever for 25 or more seasons.[23] Hay fever can also lead to more serious illnesses. Thousands of persons are incapacitated annually during the pollen season.

Weeds disfigure the landscape, introduce driving hazards by reducing visibility, detract from recreational areas, clog ditches, and encourage mosquito breeding.

It is also estimated that weeds cause a loss to agriculture of about $1.00/acre. Another estimate places the loss from weeds at $5,000,000,000 annually to the farmer.[24]

Hay Fever Symptoms and Pollen Sampling

Hay fever causes sneezing, itchy runny eyes, swelling of the nasal passages, and an accompanying water discharge. Inhalation of air laden with pollen causes the symptoms to a greater or lesser degree depending on type, susceptibility, and apparently heredity. Relief may be obtained by acquired immunity, moving to a pollen-free area, air conditioning, and weed control.

A mature ragweed plant, for example, can produce up to one billion pollen grains in one season. Concentrations of less than 25 pollen grains/yd^3 of air in 24 hr usually do not produce allergic reactions. One method of making pollen counts is to read a light oil- or clear petroleum jelly-coated slide (glass $1'' \times 3''$ microscope slide with one end frosted), kept protected from the rain, after 24-hr exposure. The pollen grains on a 1.8 sq cm area are counted. A pollen count of 25 or more equals 1 point; the slide having the maximum count for a day is read, and 1 point is allowed for each 100 count; the total seasonal count is divided by 200 to give additional points. The sum of all these points is the pollen index for the area. An index of less than 5 is considered practically free of ragweed pollen contamination. An index of 5 to 15 is moderately free, and an index greater than 15 is indicative of a heavy concentration of pollen. Another generally accepted standard uses a 2 cm^2 area (2cm $\times$ 1cm) for counting. In this method a ragweed pollen grain count of 7/cm^2 is sufficient to show the symptoms of pollinosis or hay fever. Readings of 25 or more signify heavy contamination. There is need for improved methods, equipment, and techniques to give more precise results.

[23] Alfred H. Fletcher, "Procedures of Promoting and Operating Ragweed Control Program," *Public Health News,* **36**, 170–175 (May 1955), N.J. Dept. of Health.
[24] Gilbert H. Ahlgren, "A Cooperative Weed Control Program for New Jersey," *Public Health News,* **34**, 243 (June 1953), N.J. Dept. of Health.

Problems

1. Some pollens are carried long distances but are diluted in the atmosphere. Most ragweed pollen grains settle to the ground within about 200 ft of the source.
2. Responsibility for control may be vague.
3. Legal support may be lacking.
4. Many weeds, grasses, and trees contribute pollen that causes hay fever.
5. Measuring devices to measure the amount of pollen in air are not accurate.
6. Concentration and type of pollen causing specific allergic reactions is not well established.

A Control Program

1. The most practical way to control hay fever is to treat the environment rather than the patient. Therefore plan a broad program and attack on ragweed, the principal culprit.
2. Make a survey to locate ragweed. Use police, Boy Scouts and Girl Scouts, older boys, civic clubs, schools. Locate ragweed areas on a map, scale 1 in. = 400 to 600 ft. Fifty square miles can be mapped by two men in a car in a few days.
3. Have adopted an ordinance prohibiting ragweed. Solicit the support of hay-fever sufferers in key position.
4. Educate the public in possible control methods. Pull out before flowering; spray, and so forth. Give cost estimates.
5. Study results of the survey, publicize program, and obtain budget allowance for a long-range program.
6. Enlist the cooperation of adjoining communities and conservation and highway departments.
7. Use existing equipment such as mosquito spray equipment and purchase additional equipment within budget limitations.
8. Start the spray program in early summer. Proceed with demonstrations, publicity—use newspapers, posters, circulars, schools, civic organizations, official support, and private agency support. Ragweed appears in early spring. Give periodic reports to the press. See Figure 10–18.
9. Seed with grass, Japanese honeysuckle, and so on, to replace ragweed. Issue leaflets dealing with identification and control of plants detrimental to health.
10. Spraying in early or middle summer will prevent pollenation. A spraying program using 2,4-D centrally directed and administered has proven effective in urban areas. Conduct training program for field and supervisory personnel.
11. NOTE! Weeds will not grow where the ground is occupied with grass or other acceptable vegetation. Grass that is kept cut or grazed will not cause appreciable hay fever, provided it is not permitted to grow and get out of control. Spraying with 2,4-D kills ragweed, prevents seeds from maturing the following year, and encourages grasses and some plants to grow, thus discouraging weed growths. Integrate the program with tree spraying, caterpillar control, and work of other departments. Start cutting weeds and grass on streets and highways.
12. Start the program in area where most people will benefit.
13. Control ragweed along highways, particularly where shoulders are not seeded to grass. Ragweed is the principal pollen-producing offender.
14. Program should be under supervision of the health department, for it is conducted for the health of the citizens. A coordinating committee should contain

a health officer, agricultural specialist, highway superintendent, conservationist, and a public health engineer or sanitarian.

15. Intelligent field supervision is necessary to prevent damage to valuable plants.

16. Evaluate control measures by determining ragweed area reduction as result of spraying. Areas not sprayed will determine work areas in the next year. Pollen

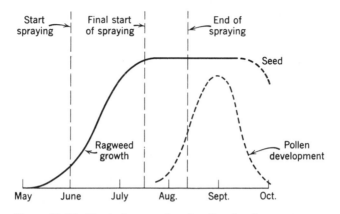

Figure 10–18   Typical ragweed and pollen development.

reduction, as measured by pollen sampling stations over wide areas and at different atmospheric levels, correlated with time of year, day, weather, and wind direction, will also indicate the progress being made if sufficient readings to be representative are taken.

Eradication

Pull out by the roots before flowering.

Prohibit by ordinance the growing of noxious weeds on private property and provide for spraying of any private property where the owner has permitted the noxious weeds to grow.

Use a special burner or flame thrower for nonselective control such as on reservoir shore and on ditch or stream banks where chemicals cannot be used.

Chemical treatment, centrally directed and administered, is effective. A good weed killer should be selective, readily absorbed by plants, nontoxic to humans, and effective in minute concentrations; 2,4-D is the chemical of choice. Weed killers are listed below.

1. 2,4-D was discovered in 1944.[25] It is available as a powder, liquid, or tablet. It is a selective weed killer. 2,4-D will control annual, biennial, and perennial weeds. Several general types are available. The type used depends on cost, weed,

[25] Synthetic hormone. Use of 2,4,5-T and possibly others may not be permitted.

and equipment. Broad-leaf weeds and bent grasses, including most decorative plants, are killed, thereby encouraging the growth of grasses and other plants that in turn compete with and discourage the growth of weeds. The chemical works on water plants too. It is more effective in the light than in the shade. See Table 10–8.

TABLE 10–8   PLANTS SUSCEPTIBLE TO HORMONE WEED KILLERS

Chemical	Plants Susceptible
2,4-D	In general most species of broad-leaved annual weeds. Beans, grapes, tobacco, tomatoes, some flowers, cotton, sugar beets, sunflowers, and cultivated legumes such as alfalfa, clovers, cowpeas, soybeans, and vetches, and broadleaf vegetables such as cabbage, rape, peppers, and cucumbers. (Also chickweed, henbit, pigweed, purslane, spurge.) Dandelion, plantain, field bindweed (½ to 1 lb/acre), Canada thistle (repeated applications of ½ to 2 lb/acre). Water hyacinth (8 lb/acre), cattails (a minimum of 3 lb acid equivalent/acre). No adverse effect reported on small fishes and water fauna.
MCPA	Similar to 2,4-D. Creeping buttercup, water hyacinth, and other waterweeds at rate of ½ lb/acre.
TCA	Bermuda grass, quackgrass, Johnson grass, prickly pear and other cacti, Kentucky bluegrass, bromegrass, para grass, and many annual grasses.
2,4,5-T*	Woody shrubs or trees usually require repeated application. Osage orange, brambles, poison ivy, hawthorn, gum, currant, hickory, wild roses, oaks, maple, alder, locust, poison oak. (Also black medic, chickweed, clover, knotweed, spurge.)
Note: (a)	2,4-D and MCPA are harmless to animals and man in the recommended dosages. Certain impurities in the compounds could prove harmful.
(b)	Ten pounds of 10 percent acid equivalent hormone weed killer will dose weeds at rate of 1 lb/acre.
(c)	The effect of weed killers is usually gone by the following season.

Source: Lee Ling, *Weed Control by Growth-Regulating Substances,* FAO Agricultural Studies, No. 13, Food and Agriculture Organization of the United Nations, Rome, Italy, January 1951.

*Can cause deformation of the offspring when fed to rats. Its use as a defoliant and ingestion in large quantities through water and food is suspected by WHO as possible cause of birth defects in children. (Brewer, *New York Times,* December 7, 1969). Its use may not be permitted in some states.

2. Sodium chlorate is a good patch killer; but the chemical presents a fire hazard.[26] Use 200 to 300 lb of dry sodium chlorate per acre, or a solution of 1½ oz in 1 gal water for each 100 ft^2.

3. Sodium arsenite is very toxic to humans and animals.

[26] Soil sterilant.

4. Kerosene and other petroleum products.[27]

5. Borax applied at the rate of about 2000 lb/acre will persist for 2 years.[28] Also other boron compounds.

6. Ammonium sulfamate (known commercially as AMS or Ammate).

7. 2,4,5-T, MCPA, 2,4-DB, 2,4-DEP, TCA, IPC.[29]

8. Also Ametryne, Atrazine, Benefin, Bensulfide,[30] Paraquat, DCPA, Dichlobenil, Diuron, DSMA, Fenuron, Monuron,[31] Sesone, and many others.

9. Education and promotion are essential to assure continuation of appropriations and public support. Exhibits, news releases, radio programs, demonstrations, and other health education measures should be used on a planned basis.

Equipment for Chemical Treatment

Use a mounted 100 to 1000-gal tank, a 15 to 30-gpm pump, capable of producing 50 to 100 psi; two 100-ft lengths of ½- or ¾-in. pressure hose. A nozzle diameter of about 0.08 in. is satisfactory. A man to handle each hose, one man to help, and one driver will be needed. One of the men should be responsible for keeping written reports of the work done each day. Converted street flushers have been used.

A knapsack sprayer of 3- or 5-gal capacity, sprinkling cans, garden sprayer, and hand carts equipped with perforated pipe distributor are suitable for patch treatment.

The use of a fogging machine or aerial sprayer is not recommended except for large isolated areas since it cannot be controlled for spot treatment. A jeep with tank trailer and pressure sprayer is very useful.

Chemical Dosages

In the early season, when the weeds are 4 to 6 in. high, use a 0.1 percent 2,4-D solution applied at the rate of 150 gal/acre or 1 gal to about 300 ft². Use 300 gal/acre when the weeds are in full leaf. One pound of 2,4-D to 100 gal of water makes a 0.12-percent solution. In all cases, the weeds should be thoroughly wetted to produce satisfactory results.

One four-man crew can spray 2 to 6 acres per day depending on size and proximity of plots. As much as 30 acres can be covered in open country.

Control of Aquatic Weeds

The control of algae and other aquatic organisms and aquatic weeds is discussed in Chapter 3. Aquatic weed control as it relates to bathing beaches is discussed in Chapter 9.

[27] Ibid.
[28] Ibid.
[29] All synthetic hormones.
[30] Soil sterilant.
[31] Ibid.

BIBLIOGRAPHY

Agricultural Chemicals, Manufacturing Chemists' Association, Inc., 1825 Connecticut Avenue, N.W., Washington, D.C., 1963.

Bjornson, Bayard F., Pratt, Harry D., and Littig, Kent S., *Control of Domestic Rats & Mice,* Environmental Control Administration, U.S. Public Health Service, Dept. of HEW, Rockville, Md., 1969.

Cleaning our Environment—the Chemical Basis for Action, a report by the Subcommittee on Environmental Improvement, Committee on Chemistry and Public Affairs, American Chemical Society, Washington, D.C., 1969, pp. 195–244.

DeVaney, Thomas E., *Chemical Vegetation Control Manual for Fish and Wildlife Management Programs,* U.S. Dept. of the Interior, Washington, D.C., January 1968.

Hayes, Wayland J., Jr., *Clinical Handbook on Economic Poisons,* U.S. Public Health Service Pub. No. 476, Dept. of HEW, Washington, D.C., 1963.

Headquarters, Dept. of the Army, *Military Entomology Operational Handbook,* TM 5-632, Washington, D.C., 1965.

Newell, Edward Lee, Jr., "The Influence of Rodents in Solid Waste Practice," *Waste Age* (November–December 1970).

Pratt, Harry D., and Littig, Kent S., *Insecticides for the Control of Insects of Public Health Importance,* U.S. Public Health Service Pub. No. 772, Dept. of HEW, Washington, D.C., 1962.

Report of the Secretary's Commission on Pesticides and Their Relationship to Environmental Health, U.S. Dept. of Health, Education, and Welfare, Washington, D.C., December 1969.

Safe Use of Pesticides, American Public Health Association, 1015 Eighteenth Street, N.W., Washington, D.C., 1967.

Suggested Guide for Weed Control 1967, Agricultural Handbook No. 332, U.S. Dept. of Agriculture, Washington, D.C.

11

HOUSING AND THE RESIDENTIAL ENVIRONMENT

The WHO Expert Committee on the Public Health Aspects of Housing defined housing (residential environment) as "the physical structure that man uses for shelter and the environs of that structure including all necessary services, facilities, equipment and devices needed or desired for the physical and mental health and social well-being of the family and individual."[1]

Every family and individual has a basic right to a *decent home* and a *suitable living environment.* However, large segments of population in urban and rural areas throughout the world do not enjoy one or both of these fundamental needs. Housing therefore must be considered within the context of and relative to the total environment in which it is situated, together with the structure, supplied facilities and services, and conditions of occupancy.

The realization of a decent home in a suitable living environment requires clean air, pure water and food, adequate shelter, and unpolluted land. Also required are freedom from excessive noise and odors, adequate recreation and neighborhood facilities, and convenient services in an environment that provides safety, comfort, and privacy. These objectives are not achieved by accident but require careful planning of new communities and conservation, maintenance, and replanning of existing communities to ensure that the public does not inherit conditions that are impossible or very costly to correct. In so doing that which is good or sound, be it a structure or a natural condition, should be retained, restored, and reused.

[1] *WHO Tech. Rep. Ser.*, 1961, No. 225, p. 6.

SUBSTANDARD HOUSING AND ITS EFFECTS

Growth of the Problem

Practically all urban and rural areas contain substandard, slum, and blighted areas.[2] The causes are numerous; they are not easily detected in the early stages or for that matter easily controlled.

There has been in the United States a rapid growth in population and in the size of urban areas, with most of the growth taking place in the suburbs. Many of the older cities that have not enlarged their boundaries are no longer experiencing a population increase. The city of Buffalo and Erie County, New York, population trend graphs (Figure 11-1) are typical.

Between 1800 and 1910 a very rapid growth took place in the cities of the United States due to the mass movement of people from rural to urban areas and as a result of heavy immigration. Many of the newer immigrants sought out their relatives and compatriots, who usually lived in cities, thereby straining the available housing resources.

With the movement of large numbers of people to cities, urban areas became congested; desirable housing became unobtainable. Inadequate facilities for transporting people rapidly and cheaply to and from work made it necessary for many people to accept less desirable housing in the cities, close to their work. The inability of the ordinary wage earner to economically afford satisfactory housing left him little choice but to accept what housing was available. Some landlords and speculators took advantage of the situation by breaking up large apartments into smaller dwelling units and by constructing cheap housing.

Unfortunately, assistance or leadership from local governing units to control potential problems is slow; there is usually a lag between the creation of housing evils and the enactment of suitable corrective legislation and enforcement. For example, the first Tenement House Law ap-

[2] Substandard housing is said to exist when there are 1.51 or more persons per room in a dwelling unit, when the dwelling unit has no private bath or is dilapidated, or when the dwelling unit has no running water. Other bases used are described under "Appraisal of Quality of Living."

A slum is "a highly congested, usually urban, residential area characterized by deteriorated unsanitary buildings, poverty, and social disorganization." (*Webster's Third International Dictionary*, 1966). A slum is a neighborhood in which dwellings lack private inside toilet and bath facilities, hot and cold running water, adequate light, heat, ventilation, quiet, clean air, and space for the number of persons housed. It is also a heavily populated area in which housing and other living conditions are extremely poor.

To blight is to "prevent the growth and fertility of; hence to ruin; frustrate" (Webster). A blighted area is an area of no growth in which buildings are permitted to deteriorate.

plicable to New York City was passed by the New York State legislature in 1867. However, subsequent amendments and discretionary powers vested in the board of health resulted in nullifying the law to a large extent. A new Tenement House Law of 1901 was made mandatory for New York City and Buffalo. But many amendments were made to

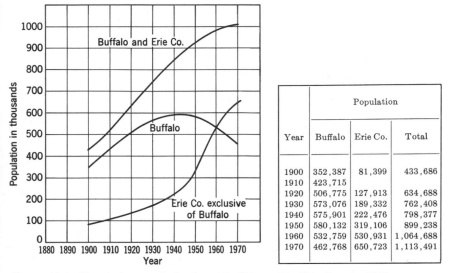

| | Population | | |
Year	Buffalo	Erie Co.	Total
1900	352,387	81,399	433,686
1910	423,715		
1920	506,775	127,913	634,688
1930	573,076	189,332	762,408
1940	575,901	222,476	798,377
1950	580,132	319,106	899,238
1960	532,759	530,931	1,064,688
1970	462,768	650,723	1,113,491

Figure 11–1   Population trends in city of Buffalo and in Erie County, N.Y.

the law within the next 10 years, as a result of powerful pressure from property owners, which practically defeated the original intent of the law. The experience in New York City and Buffalo shows the wisdom of having an informed public opinion to support legislation for the public good and to combat the pressures of vested interests. It also shows the practical dangers of discretionary powers. In 1912 the legislature strengthened the law. The Tenement House Law applied only to dwellings with three or more families who did their cooking on the premises. A Multiple Dwelling Law with wider coverage became effective in New York City in 1929, replacing the Tenement House Law. It became effective in Buffalo in 1949. It is to be noted that between 1867 and 1912 the existent laws were of doubtful effectiveness, yet in the 32-year period between 1880 and 1912 the New York City population increased from about 2 million to 5 million and that of Buffalo from 155,000 to 425,000.

Obsolescence is another factor that causes the growth of slums. Land and property used for purposes for which they may have been well suited in the first place may no longer be suitable for that purpose.

An example is the slum frequently found on the rim of a central business area, originally a good residential district convenient to business. This may start with the expansion or spill-over of business into the contiguous residential areas, thereby making the housing less desirable. People next door or in the same building, desiring quiet and privacy, move. Owners are hesitant to continue maintenance work, causing buildings to deteriorate. The landlord, to maintain his income, must either convert the entire building to commercial use or lower rents to attract lower-income tenants. If he converts, more people leave. If he lowers the rents, maintenance of the building is reduced still further and overcrowding of apartments frequently follows. The progressive degradation from blight to slum is almost inevitable. As blight spreads so does crime, delinquency, disease, fires, housing decay, and welfare payments.

Then there are the areas that are slums from the start. The absence of or failure to enforce suitable zoning, building, sanitary, and health regulations leads to the development of "shanty towns" or poor-housing areas. Add to this small lots, cheap, new, and converted dwellings and tenements that are poorly located, designed, and constructed, just barely meeting what minimum requirements may exist, and a basis for future slums is assured.

An indifferent or uninformed public can permit the slums to develop. The absence of immediate and long-range planning and zoning, lack of parks and playgrounds, poor street layout, weak laws, inadequate trained personnel to enforce laws, pressure groups, lack of leadership from public officials and local key citizens, and poor support from the courts and press make the development of slums and other social problems only a matter of time.

In recent years the population growth has been taking place outside the major cities, but the rate of housing construction and rehabilitation has not kept pace with the needs of population growth. The loss of housing due to obsolescence, abandonment, decay, and demolition further compound the problem.

According to a recent report, the United States will have to provide, through its building construction industry, 2.4 million new and 200,000 rehabilitated housing units per year for 10 years to properly house its population.[3] It has been estimated that there are between 6.7 and 12 million substandard dwelling units, not including those that are overcrowded and those falling into decay each year. Stevenson estimates that some 34 million Americans live in 11 million dwelling units that

[3] President's Committee on Urban Housing, *A Decent Home,* December 1968, U.S. Government Printing Office, Washingon, D.C.

are either overcrowded or have structural or plumbing deficiencies.[4] But the building industry has only been averaging (1964–1970) between 1.2 and 1.5 million dwelling units a year.[5] If a family cannot afford to spend more than 20 percent of its gross earnings on housing, an estimated 7.8 million American families cannot afford to own a new home.[6]

It is apparent that unless the rate of new-home construction is accelerated, the rehabilitation of sound substandard dwelling units strongly encouraged, and the conservation and maintenance of existing housing required, that a decent home for every American family will never be realized. Added to this is the need to provide public housing and financial assistance to the low-income family.

There comes a time, in the ownership of income property, when the return begins to drop off. This may be due to obsolescence and reduced rents or increase in the cost of operation and maintenance. At this point, the property may be sold (unloaded), repairs may be made to prevent further deterioration of the property, the property may be sold and demolished for a more appropriate use, or a minimum of repairs is made consistent with a maximum return. This is a critical time and will determine the subsequent character of a neighborhood. In situations where repairs are not made or where a property is sold and repairs are not made, the annual rental from substandard property may equal or exceed the assessed valuation of the property. A complete return on one's investment in five to seven years is not considered unusual in view of the so-called risks involved. Because of this, housing ordinances should be diligently enforced and require owners to reinvest a reasonable part of the income from a property in its conservation and rehabilitation, *at the first signs of deterioration.* This would tend to prevent the "milking" of a property and nonpayment of taxes. The burden on the community to acquire for nonpayment of taxes and demolish a worthless structure, or maintain an eyesore and fire and accident hazard, would be lessened. Cause for further property devaluation and extension of the blighting influence would also be reduced. See Figure 11–2.

---

[4] Albert H. Stevenson, "Home and Community Health: An Environmental Problem," Proceedings of the First Invitational Conference on Health Research in Housing and It's Environment, Airlie House, Warrenton, Va., March 17–19, 1970, p. 49.

[5] *U.S. News & World Report,* November 9, 1970. Source: 1964–1969, U.S. Dept. of Commerce; 1970–1971 estimate USN & WR Economic Unit.

[6] The U.S. Census Bureau reports that in 1969 there were 24.3 million persons with incomes below the poverty level. The poverty level for a family of four living in a city was set at $3,745 per year and for a farm family of four at $3,197.

Figure 11–2 Run-down, filthy, vermin-infested backyards present many real health hazards. (Buffalo Municipal Housing Authority.)

It is an unfortunate practical fact, because of the complexity of the problem, that effective city code enforcement and rehabilitation are in many places not being accomplished. Efforts suffer from frustration and lack of support, aided and abetted by governmental sympathy and financial assistance (urban renewal, welfare). A greater return on time and money invested can be realized by giving greater assistance to those communities and property owners demonstrating a sincere desire to conserve and renew basically sound areas. Evidence of actual maintenance and improvement of the existing housing supply, code enforcement, encouragement of private building, low-interest mortgages, and provision of low- and middle-income housing are some of the facts that should guide the extent and amount of financial assistance a community receives.

Health, Economic, and Social Effects

The effect housing has on the health of the people of a community is not subject to exact statistical analysis. For example poverty, malnutrition, and lack of education and medical care also have important effects on health. These may mean long hours of work with resultant fatigue, improper food, and lack of knowledge relating to disease prevention or sanitation and personal hygiene. The problem is compounded by the slum itself in the feelings of inferiority and resentment of the residents against others who are in a better position. In addition, slums are characterized as having high delinquency, prostitution, broken homes, and other social problems. Then (who can say?) are some people sick because they are poor, or are they poor because they are sick? Many studies show that as a matter of practical fact bad housing is profoundly detrimental to the life, health, and welfare of a community. The results of a few studies are summarized in Table 11–1.

The higher morbidity and mortality rates and the lower life expectancy associated with bad housing are also believed to be the cumulative effect or result of continual pressures on the human body. Dubos points out, in a related discussion, "Many of man's medical problems have their origin in the biological and mental adaptive responses that allowed him earlier in life to cope with environmental threats."[7] He adds, "The delayed results of tolerance to air pollutants symbolize the indirect dangers inherent in many forms of adaptation, encompassing adaptation to toxic substances, microbial pathogens, the various forms of malnutrition, noise or other excessive stimuli, crowding or isolation, the tensions of competitive life, the disturbances of physiological cycles, and all other uncontrolled deleterious agencies typical of urbanized and technicized societies. Under normal circumstances, the modern environment rarely destroys human life, but frequently it spoils its later years."

Emphasis must be on preventive sanitation, medicine, and engineering to avoid some of the contributory causes of early and late disability and premature death. This avoidance includes the insidious, cumulative, long-term insults to the human body and spirit, as well as maintenance and improvement of those factors in the environment that enhance the well-being and aspirations of people.

Although this discussion concentrates on the environmental health aspects of housing, it is extremely important that concurrent emphasis be placed on the elimination of poverty and on improved education for those living in poor housing and neighborhoods. It is essential that the causes of poverty and low income be attacked at the source, that usable skills be taught, and that educational levels be raised. In this

---

[7] Rene Dubos, paper delivered at the Smithsonian Institution Annual Symposium, February 16–18, 1967; *The Fitness of Man's Environment*, Smithsonian Institution Press, Washington, D.C., 1968.

TABLE 11-1 EFFECTS OF SUBSTANDARD HOUSING ON HEALTH AND CITY COSTS

Communicable Diseases (CD)	Tuberculosis	Health and Other	Police	City Costs
CD rate 65% higher, VD rate 13 times higher, CD death rate as high as 50 years ago.*	Half of cases from ¼ of population.*	Source of 40% of mentally ill in state institutions.*	Juvenile delinquency twice as high.*	20% of area in city brings in 6% of real-estate tax.¶
Intestinal disease rate 100% higher in homes lacking priv. flush toilet.†	TB rate 8 times higher.†	Infant death rate 5 times higher.*	20% of area in city accounts for 50% of arrests, 45% of major crimes, 50% of juvenile delinquency.¶	Contributes 5½% of real-estate tax but takes 53% of city services.**
	Secondary attack rate 200%.§	Infant mortality as high as 50 years ago.*		Slums cost $88 more per person than they yield; good areas yield $108.**
Meningococcis rate 5½ times higher‡	Death rate 8.6 times higher.‡	Life expectancy 6.7 years less.‖	2.6 times more arrests, 1.9 times more police calls, 2.9 times more criminal cases, 3.7 times more juvenile delinquents.††	20% of area in city accounts for 45% of city service costs, 35% of fires.¶
Infective and parasitic disease death rate 6.6 times higher.‖	20% of area in city accounts for 60% of cases.¶	64% of out-of-wedlock cases.§§		1.5 times more fires, 15.7 times more families on welfare, 4 times more nursing visits.††
Pneumonia and influenza death rate 2 times higher.§		Interest rate for mortgages higher in blighted areas.	50% of murders, 60% of manslaughters, 49% of robberies.§§	Account for most of the welfare benefits.
		Fire insurance rates higher.		Slums yielded ½ cost of services required.‡‡
		Accident death rate 2.3 times higher.‖		
		Injuries, burns, and accidental poisonings 5–8 times the national average.‖		

* Release by Dr. Leonard Scheele, Surgeon General, U. S. Public Health Service Pub. No. 27, 1949, regarding 6 cities having slums.
† National Health Survey by U.S. Health Service, 1935-1936.
‡ Dr. Bernard Blum, *J. Am. Public Health Assoc.*, 39, 1571-1577 (December 1949).
§ Paper on housing and progress in public health by W. P. Dearing, M.D., presented at the University of North Carolina, April 16, 1950.
‖ Erie County Health Dept, N.Y., 1953 Annual Report.
¶ Miscellaneous city studies.
** Raymond M. Foley, "To Eradicate our Vast Slum Blight," *New York Times*, Magazine section, November 27, 1949.
†† Report from city of Louisville, Ky.
‡‡ John P. Callahan, "Local Units Fight Problem of Slums," *New York Times*, July 22, 1956.
§§ The Twenty-seventh Annual Report of the Buffalo Urban League, Inc., N.Y.
‖‖ Albert H. Stevenson, Airlie House Conference, U.S. Dept. of HEW, Washington, D.C., March 17-19, 1970.

way more individuals can become more productive and self-sufficient and develop greater pride in themselves and in their communities.

APPRAISAL OF QUALITY OF LIVING

APHA Appraisal Method

The American Public Health Association appraisal method for measuring the quality of housing was developed by the Committee on the Hygiene of Housing between 1944 and 1950. This method attempts to eliminate or minimize individual opinion so as to arrive at a numerical value of the quality of housing that may be compared with results in other cities and may be reproduced in the same city by different evaluators using the same system. It is also of value to measure the quality of housing in a selected area, say at 5-year intervals, to evaluate the effects of an enforcement program or lack of an enforcement program. The appraisal method measures the quality of the dwellings and dwelling units as well as the environment in which they are located.

The items included in the APHA dwelling appraisal, Tables 11–2 and 11–3, are grouped under "Facilities," "Maintenance," and "Occupancy." Additional information obtained includes rent, income of family, number of lodgers, race, type of structure, number of dwelling units, and commercial or business use. The environmental survey reflects the proximity and effects of industry, heavy traffic, recreational facilities, schools, churches, business and shopping centers, smoke, noise, dust, and other factors that determine the suitability of an area for residential use.

The rating of housing quality is based on a penalty scoring system, shown in Table 11–2. A theoretical maximum penalty score is 600. The practical maximum is 300; the median is around 75. A score of zero would indicate all standards are met. An interpretation of the dwelling and environmental scores is shown in Table 11–5. It is apparent therefore that according to this scoring system, either the sum of dwelling and environmental scores or a dwelling or environmental score of 200 or greater would classify the housing as unfit.

Application of the APHA appraisal method requires trained personnel and experienced supervision.[8] The survey staff should be divorced from other routine work so as to concentrate on the job at hand and produce information that can be put to use before it becomes out-of-date. In

[8] A trained sanitarian can inspect about ten dwelling units per day. For every four inspectors there should be one trained field supervisor, three office clerks, and one office supervisor.

TABLE 11-2 APPRAISAL ITEMS AND MAXIMUM STANDARD PENALTY SCORES (APHA)

Item	Maximum Score	Item	Maximum Score
A. *Facilities*		17. Rooms lacking window*	30
		18. Rooms lacking closet	8
Structure:		19. Rooms of substandard area	10
1. Main access	6	20. Combined room facilities‡	
2. Water supply* (source)	25		
3. Sewer connection*	25	B. *Maintenance*	
4. Daylight obstruction	20	21. Toilet condition index	12
5. Stairs and fire escapes	30	22. Deterioration index* (struc-	
6. Public hall lighting	18	ture, unit)§	50
		23. Infestation index (structure,	
Unit:		unit)§	15
7. Location in structure	8	24. Sanitary index (structure,	
8. Kitchen facilities	24	unit)§	30
9. Toilet* (location, type,		25. Basement condition index	13
sharing)†	45		
10. Bath* (location, type,		C. *Occupancy*	
sharing)†	20	26. Room crowding: persons	
11. Water supply* (location and		per room*	30
type)	15	27. Room crowding: persons	
12. Washing facilities	.8	per sleeping room*	25
13. Dual egress*	30	28. Area crowding: sleeping	
14. Electric lighting*	15	area per person*	30
15. Central heating	3	29. Area crowding: nonsleeping	
16. Rooms lacking installed		area per person	25
heat*	20	30. Doubling of basic families	10

Source: *An Appraisal Method for Measuring the Quality of Housing,* Part II, "Appraisal of Dwelling Conditions," American Public Health Association, Washington, D.C., 1946.

Note: 1. Maximum theoretical total dwelling score is 600, broken down as:

Facilities 360   Maintenance 120   Occupancy 120

2. Housing total = dwelling total + environmental total.

* Condition constituting a basic deficiency.

† Item score is total of subscores for location, type, and sharing of toilet or bath facilities.

‡ Item score is total of scores for items 16-19 inclusive. This duplicate score is not included in the total for a dwelling but is recorded for analysis.

§ Item score is total of subscores for structure and unit.

764

TABLE 11-3   BASIC DEFICIENCIES OF DWELLINGS (APHA)

Item*	Condition Constituting a Basic Deficiency†
A. Facilities	
2.	Source of water supply specifically disapproved by local health department.
3.	Means of sewage disposal specifically disapproved by local health department.
9.	Toilet shared with other dwelling unit, outside structure or of disapproved type (flush hopper or nonstandard privy).
10.	Installed bath lacking, shared with other dwelling unit or outside structure.
11.	Water supply outside dwelling unit.
13.	Dual egress from unit lacking.
14.	No electric lighting installed in unit.
16.	Three-fourths or more of rooms in unit lacking installed heater.‡
17.	Outside window lacking in any room of unit.‡
B. Maintenance	
22.	Deterioration of class 2 or 3 (penalty score, by composite index, of 15 points or over).
C. Occupancy	
26.	Room crowding: over 1.5 persons/room.
27.	Room crowding: number of occupants equals or exceeds 2 times the number of sleeping rooms plus 2.
28.	Area crowding: less than 40 ft^2 of sleeping area/person.

Note: Some authorities include as a basic deficiency unvented gas space heater, unvented gas hot-water heater, open gas burner for heating, lack of hot and cold running water.

*Numbers refer to items in Table 11-2, "Appraisal Items and Maximum Standard Penalty Scores."

†Of the 13 defects that can be designated basic deficiencies, 11 are so classified when the item penalty score equals or exceeds 10 points. Bath (item 10) becomes a basic deficiency at 8 points for reasons involving comparability to the U.S. Housing Census; deterioration (item 22) at 15 points for reasons internal to that item.

‡The criterion of basic deficiency for this item is adjusted for number of rooms in the unit.

practice, it is found desirable to select a limited area or areas for pilot study. The information thus obtained can be used as a basis for determining need for extension of the survey, need for new or revised minimum housing standards, extent of the housing problem, development of coordination between existing official and voluntary agencies, the part private enterprise and public works can play, public information needs, and so forth.

Census Data

Much valuable information is collected in the U.S. Census of Population and Housing. Information summarized in the 1970 Census includes

TABLE 11-4 ENVIRONMENTAL SURVEY—STANDARD PENALTY SCORES (APHA)

Item	Maximum* Penalty Score
A. Land crowding	
1. Coverage by structures—70% or more of block area covered.	24
2. Residential building density—ratio of residential floor area to total = 4 or more.	20
3. Population density—gross residential floor area per person 150 ft² or less.	10
4. Residential yard areas—less than 20 ft wide and 625 ft² in 70% of residences.	16
B. Nonresidential land areas	
5. Areal incidence of nonresidential land use—50% or more non-residential.	13
6. Linear incidence of nonresidential land use—50% or more non-residential.	13
7. Specific nonresidential nuisances and hazards—noise and vibration, objectionable odors, fire or explosion, vermin, rodents, insects, smoke or dust, night glare, dilapidated structure, insanitary lot.	30
8. Hazards to morals and the public peace—poolrooms, gambling places, bars, prostitution, liquor stores, nightclubs.	10
9. Smoke incidence—industries, docks, railroad yards, soft-coal use.	6
C. Hazards and nuisances from transportation system	
10. Street traffic—type of traffic, dwelling setback, width of streets.	20
11. Railroads or switchyards—amount of noise, vibration, smoke, trains.	24
12. Airports or airlines—location of dwelling with respect to runways and approaches.	20
D. Hazards and nuisances from natural causes	
13. Surface flooding—rivers, streams, tide, groundwater, drainage annual or more.	20
14. Swamps or marshes—within 1000 yd, malarial mosquitoes.	24
15. Topography—pits, rock outcrops, steep slopes, slides.	16
E. Inadequate utilities and sanitation	
16. Sanitary sewerage system—available (within 300 ft), adequate.	24
17. Public water supply—available, adequate pressure and quantity.	20
18. Streets and walks—grade, pavement, curbs, grass, sidewalks.	10
F. Inadequate basic community facilities	
19. Elementary public schools—beyond $\frac{2}{3}$ mi, 3 or more dangerous crossings.	10
20. Public playgrounds—less than 0.75 acres/1000 persons.	8
21. Public playfields—less than 1.25 acres/1000 persons.	4
22. Other public parks—less than 1.00 acres/1000 persons.	8
23. Public transportation—beyond $\frac{2}{3}$ mi, less than 2 buses/hr.	12
24. Food stores—dairy, vegetable, meat, grocery, bread, more than $\frac{1}{3}$ mi.	6

Source: *An Appraisal Method for Measuring the Quality of Housing,* Part III, "Appraisal of Neighborhood Environment," American Public Health Association, Washington, D.C., 1950.

*Maximum environment total = 368.

TABLE 11-5   HOUSING QUALITY SCORES (APHA)

Factor	A—Good	B—Accept-able	C—Border-line	D—Sub-standard	E—Unfit
Dwelling score	0 to 29	30 to 59	60 to 89	90 to 119	120 or greater
Environmental score	0 to 19	20 to 39	40 to 59	60 to 79	80 or greater
Sum of dwelling and environmental scores	0 to 49	50 to 99	100 to 149	150 to 199	200 or greater

number of dwelling (housing) units; the population per owner- and renter-occupied unit; the number vacant; the number of dwelling units with private bath, including hot and cold piped water as well as flush toilet and bathtub or shower, and the number lacking some or all these facilities; the number of dwelling units occupied by whites and Negroes; the number of dwelling units having 1.00 or fewer persons per room, the number with 1.01 to 1.50, and the number with 1.51 or more; the monthly rental; and the value or sale price of owner-occupied one-family homes. Other statistics on selected population characteristics for areas with 2,500 or more inhabitants and for counties are available. This information can be used as additional criteria to supplement reasons for specific program planning. Plotting the data on maps or overlays will show concentrations sometimes not discernible by other means.

The accuracy of census data for measuring housing quality has been questioned and hence should be checked, particularly if it is to be used for appraisal or redevelopment purposes. It is nevertheless a good tool in the absence of a better one.

Health, Economic, and Social Factors

It is frequently possible to obtain morbidity and mortality data for specific diseases or causes by census tracts or selected areas. Also available may be the location of cases of juvenile delinquency, public and private assistance, and probation cases. Sources of fires and rodent infestation and areas of social unrest give additional information. Health, fire, police, and welfare departments and social agencies should have this information readily available. See Table 11–1. Tabulation and plotting of these data will be useful in establishing priorities for action programs.

Planning

The location of existing and proposed recreational areas, business districts, shopping centers, churches, schools, parkways and throughways, housing projects, residential areas, zoning restrictions, redevelopment areas, railroads, industries, lakes, rivers, and other natural boundaries help to determine the best usage of property. Where planning agencies are established and are active, maps giving this as well as additional information are usually well developed. Analysis and comprehensive planning on a continuing basis are essential to the proper development of cities, villages, towns, counties, and metropolitan areas. The availability of state and federal aid for community-wide planning should be investigated. Plans for urban renewal, housing-code enforcement, rehabilitation, and conservation should be carefully integrated with other community and state plans before decisions are made. These subjects are discussed in greater detail in Chapter 2.

Environmental Sanitation and Hygiene Indices

Most modern city and county health departments can carry on a housing-inspection program based on minimum housing standards, provided competent personnel is assigned. Where housing inspections are made on a routine basis and records are kept on a punch-card system, visible card file, or by both methods, problem streets and areas can be detected with little difficulty. A survey and follow-up form based on a modern housing ordinance is shown in Figure 11-3. It should list recommendations for correction based on what is practical from field observation so as to serve as the basis for an accurate confirmatory letter to the owner. The type of deficiencies found, such as the lack of a private bath and toilet, dwelling in disrepair, or lack of hot and cold running water, can give a wealth of information to guide program planning. To this can usually be added the origin of complaints, which if plotted on a map will give a visual picture of the heavy work-load areas. Reliable cost estimates for complete rehabilitation of selected buildings and a simple foot survey should be made to confirm administrative judgment and decisions before any major action is taken. Many apparently well-thought-out plans have fallen down under this simple test.

Other Survey Methods

Other techniques are being refined for making a rapid appraisal of the physical environment. These include aerial surveys, external ground-level surveys, and the Public Health Service Neighborhood Environmental Evaluation and Decision System (NEEDS).[9] The selection of

[9] "NEEDS" Neighborhood Environmental Evaluation and Decision System, Bureau of Community Environmental Management, U.S. Public Health Service, Dept. of HEW, Washington, D.C., 1970.

Address _____ Dates inspected _____

Type of Structure and Occupancy _____

(frame, stucco, brick veneer, solid brick; residential, factory, store)

Owner and Address _____

No.	Item	Yes	No	CM	No.	Item	Yes	No	CM
1.	Water supply in each apt. satisfact. quality and quantity (no x-conn).				11.	Occupant keeps dwelling unit and fixtures clean and sanitary.			
2.	Private in each apt.				12.	Space and water heaters adequate, properly connected and vented to outer air; back-draft guard.			
	(a) water closet								
	(b) washbasin								
	(c) shower/tub				13.	Premises free of rodent and vermin infestation, rodent-proof.			
	(d) kitchen sink								
	(e) cabinets and counter				14.	Refuse, garbage, and ashes storage proper and adequate.			
	(f) refrig. and stove								
3.	Piped hot water for				15.	One or more apartments above 2nd floor have 2 means of egress			
	(a) washbasin								
	(b) shower/tub				16.	Public halls and stairs lighted, daylight and artificial in MD.			
	(c) kitchen								
4.	Plumbing, heating, electricity, and fixtures properly installed and maintained.				17.	Property and dwelling properly drained and sewered.			
5.	Water-repellent floor and base in toilet room and bathroom.				18.	Owner keeps public areas of building and premises clean.			
6.	Window 1/10 floor area in every room; openable, adequate light and air or induced ventilation for bath.				19.	Living in cellar prohibited.			
					20.	Dwelling in good repair, safe, sanitary, and weatherproof (handrails, stairs, walls, wiring, floors, siding, doors, frames, plaster, porch, eaves, roof, foundation beams firm and sound).			
7.	Dwelling unit provides 150 ft^2 for one and 100 ft^2 area for each additional occupant.								
8.	Dwelling can be heated to 68°F.				21.	Lodging house has one washbasin, shower or tub, and water closet per 6 persons.			
9.	Sleeping rooms provide 70 ft^2 for one person and 50 ft^2 for each additional person.				22.	Lodging house supplies clean linen and towels prior to letting and weekly.			
10.	Every habitable room has 2 electric outlets; bathroom, w.c. stall, laundry, and hall have a light fixture, min. 3w/ft^2.				23.	Cooking in lodging house done in approved and lawful kitchen or kitchenette only.			

NOTE: Explain each "No" item on back by item number and follow with recommendation for correction. "CM" is checked or dated when correction is made. "MD" denotes three or more dwelling units.

Floor	Apt.	Total Hab. Area	Total Hab. Rooms	Bedrooms No.	Bedrooms Area	Persons	Shelter monthly rental

Remarks: (tenant names, agent, change in ownership)

_____

_____

_____

_____

Inspected by: _____

EH-49,5

Figure 11-3 Dwelling survey.

a small number of significant environmental variables can also provide a basis for a rapid survey if checked against a more comprehensive survey system.

NEEDS

This is a five-staged systems technique designed to provide a rapid and reliable measure of neighborhood environmental quality. The data collected are adapted for electronic data processing to reduce the time lapse between data collection, analysis, program planning, and implementation.

In Stage I an exterior sidewalk survey is made of 10 to 20 blocks in each problem neighborhood of a city to determine which are in greatest need of upgrading. The conditions analyzed include structural overcrowding, population crowding, premises conditions, structural condition of housing and other buildings, environmental stresses, condition of streets and utilities, natural deficiencies, public transportation, natural hazards and deficiencies, shopping facilities, parks and playgrounds, and airport noise. Time: one man-hour per block.)

In Stage II the neighborhood(s) selected are surveyed in some depth to determine the physical and social environmental problems facing the residents. About 300 dwelling units and families are sampled in the study area. This phase includes interviews to determine demographic characteristics, health problems, health services, interior housing conditions, and resident attitudes.

In Stage III the data collected in Stage II are computer processed and analyzed with local government and community leaders. Problems are identified and priorities for action are established.

In Stage IV programs are developed to carry out the decisions made in Stage III. The community participates and is kept aware of the study results and action proposed.

In Stage V the decisions made and programs developed are implemented. Federal fiscal assistance is also solicited, and the information and support made possible by this technique are used to strengthen applications for grants.[10]

APHA Criteria

The American Public Health Association Committee on the Hygiene of Housing has listed the criteria to be met for the promotion of physical, mental, and social health on the farm as well as in the city dwelling.

[10] Ibid.

Thirty basic principles, with specific requirements and suggested methods of attainment for each are reported in *Basic Principles of Healthful Housing,* published by the American Public Health Association in 1938. The principles stated are grouped under four headings, as shown below:

Fundamental Physiological Needs.

1. Maintenance of a thermal environment that will avoid undue heat loss from the human body.

2. Maintenance of a thermal environment that will permit adequate heat loss from the human body.

3. Provision of an atmosphere of reasonable chemical purity.

4. Provision of adequate daylight illumination and avoidance of undue daylight glare.

5. Provision for admission of direct sunlight.

6. Provision of adequate artificial illumination and avoidance of glare.

7. Protection against excessive noise (and radiation).

8. Provision of adequate space for exercise and for the play of children.

Fundamental Psychological Needs.

1. Provision of adequate privacy for the individual.

2. Provision of opportunities for normal family life.

3. Provision of opportunities for normal community life.

4. Provision of facilities that make possible the performance of the tasks of the household without undue physical and mental fatigue.

5. Provision of facilities for maintenance of cleanliness of the dwelling and of the person.

6. Provision of possibilities for esthetic satisfaction in the home and its surroundings.

7. Concordance with prevailing social standards of the local community.

Protection Against Contagion.

1. Provision of a water supply of safe sanitary quality, available to the dwelling.

2. Protection of the water supply system against pollution within the dwelling.

3. Provision of toilet facilities of such a character as to minimize the danger of transmitting disease.

4. Protection against sewage contamination of the interior surfaces of the dwelling.

5. Avoidance of insanitary conditions in the vicinity of the dwelling.

6. Exclusion from the dwelling of vermin that may play a part in the transmission of disease.

7. Provision of facilities for keeping milk and food undecomposed.

8. Provision of sufficient space in sleeping rooms to minimize the danger of contact infection.

Protection Against Accidents.

1. Erection of the dwelling with such materials and methods of construction as to minimize danger of accidents due to collapse of any part of the structure.

2. Control of conditions likely to cause fires or to promote their spread.

3. Provision of adequate facilities for escape in case of fire.

4. Protection against danger of electrical shocks and burns.

5. Protection against gas poisonings.

6. Protection against falls and other mechanical injuries in the home.

7. Protection of the neighborhood against the hazards of automobile traffic.

These "basic principles" have been expanded to reflect progress made and present-day aspirations of people to help achieve total health goals as defined by the World Health Organizaticn. The American Public Health Association Program Area Committee on Housing and Health, in 1968, prepared what is undoubtedly the best comprehensive statement of principles available to guide public policy and goal formulation.[11] These can also serve as a basis for performance standards to replace specification standards for building construction, living conditions, and community development. The major headings, or objectives, of the committee report are quoted below. The reader is referred to the committee report for further explanation of the items listed.

BASIC HEALTH PRINCIPLES OF HOUSING AND ITS ENVIRONMENT

I. *Living Unit and Structure.*

"Housing" includes the living unit for man and his family, the immediate surroundings, and the related community services and facilities; the total is referred to as the "residential environment." The following are basic health principles for the residential environment, together with summaries of factors that relate to the importance and applicability of each principle.

A. Human Factors.
1. Shelter against the elements.
2. Maintenance of a thermal environment that will avoid undue but permit adequate heat loss from the human body.
3. Indoor air of acceptable quality.
4. Daylight, sunlight, and artificial illumination.
5. In family units, facilities for sanitary storage, refrigeration, preparation, and service of nutritional and satisfactory foods and meals.
6. Adequate space, privacy, and facilities for the individual and arrangement and separation for normal family living.
7. Opportunities and facilities for home recreation and social life.
8. Protection from noise from without, other units, and certain other rooms and control of reverberation noises within housing structures.
9. Design, materials, and equipment that facilitate performance of household tasks and functions without undue physical and mental fatigue.
10. Design, facilities, surroundings, and maintenance to produce a sense of mental well-being.
11. Control of health aspects of materials.

B. Sanitation and Maintenance.
1. Design, materials, and equipment to facilitate clean, orderly, and sanitary maintenance of the dwelling and personal hygiene of the occupants.

[11] "Basic Health Principles of Housing and Its Environment," *Am. J. Public Health* **59**, No. 5, 841–852 (May 1969). See also *Housing: Basic Health Principles & Recommended Ordinance,* American Public Health Association, 1015 Eighteenth Street, N.W., Washington, D.C., 20036 (1971).

2. Water piping of approved, safe materials with installed and supplied fixtures that avoid introducing contamination.

3. Adequate private sanitary toilet facilities within family units.

4. Plumbing and drainage system designed, installed, and maintained so as to protect against leakage, stoppage, or overflow and escape of odors.

5. Facilities for sanitary disposal of food waste, storage of refuse, and sanitary maintenance of premises to reduce the hazard of vermin and nuisances.

6. Design and arrangement to properly drain roofs, yards, and premises and conduct such drainage from the buildings and premises.

7. Design and maintenance to exclude and facilitate control of rodents and insects.

8. Facilities for the suitable storage of belongings.

9. Program to assure maintenance of the structure, facilities, and premises in good repair and in a safe and sanitary condition.

C. Safety and Injury Prevention.

1. Construction, design, and materials of a quality necessary to withstand all anticipated forces that affect structural stability.

2. Construction, installation materials, arrangement, facilities, and maintenance to minimize danger of explosions and fires or their spread.

3. Design, arrangement, and maintenance to facilitate ready escape in case of fire or other emergency.

4. Protection against all electrical hazards, including shocks and burns.

5. Design, installation, and maintenance of fuel-burning and heating equipment to minimize exposure to hazardous or undesirable products of combustion, fires, or explosions and to protect persons against being burned.

6. Design, maintenance, and arrangement of facilities, including lighting, to minimize hazards of falls, slipping, and tripping.

7. Facilities for safe and proper storage of drugs, insecticides, poisons, detergents, and deleterious substances.

8. Facilities and arrangements to promote security of the person and belongings.

II. *Residential Environment.*

The community facilities and services and the environment in which the living unit is located are essential elements in healthful housing and are part of the total residential environment:

A. Community or Individual Facilities.

1. An approved community water supply or, where not possible, an approved individual water supply system.

2. An approved sanitary sewerage system or, where not possible, an approved individual sewage disposal system.

3. An approved community refuse collection and disposal system or, where not possible, arrangements for its sanitary storage and disposal.

4. Avoidance of building on land subject to periodic flooding and adequate provision for surface drainage to protect against flooding and prevent mosquito breeding.

5. Provision for vehicular and pedestrian circulation for freedom of movement and contact with community residents while adequately separating pedestrian from vehicular traffic.

6. Street and through-highway location and traffic arrangements to minimize accidents, noise, and air pollution.
7. Provision of such other services and facilities as may be applicable to the particular area, including public transportation, schools, police, fire protection, and electric power, health, community and emergency services.
8. Community housekeeping and maintenance services, like street cleaning, tree and parkway maintenance, weed and rubbish control, and other services requisite to a clean and aesthetically satisfactory environment.

B. Quality of the Environment.
1. Development controls and incentives to protect and enhance the residential environment.
2. Arrangement, orientation, and spacing of buildings to provide for adequate light, ventilation, and admission of sunlight.
3. Provision of conveniently located space and facilities for off-street storage of vehicles.
4. Provision of useful, well-designed, properly located space for play, relaxation, and community activities for daytime and evening use in all seasons.
5. Provision for grass and trees.
6. Improved streets, gutters, walks, and access ways.
7. Suitable lighting facilities for streets, walks, and public areas.

C. Environmental Control Programs.
To promote maintenance of a heathful environment necessitates an educational and enforcement program to accomplish the following:
1. Control sources of air and water pollution and local sources of ionizing radiation.
2. Control rodent and insect propagation, pests, domestic animals, and livestock.
3. Inspect, educate, and enforce so that premises and structures are maintained in such condition and appearance as not to be a blighting influence on the neighborhood.
4. Community noise control and abatement.
5. Building and development regulations.

Minimum Standards Housing Ordinance

Building divisions of local governments have traditional responsibility over the construction of new buildings and their structural, fire, and other safety provisions as specified in a building code. Fire departments have responsibility for fire safety. The health department and such other agencies as have an interest and responsibility are concerned with the supplied utilities and facilities, their maintenance, and the occupancy of dwellings and dwelling units for more healthful living. This latter responsibility is best carried out by the adoption and enforcement of a housing ordinance. See Chapter 2 for code definitions, also this chapter.

In an enforcement program, major problems of structural safety or alterations involving structural changes for which plans are required would be referred to the building division. Serious problems of fire safety would be referred to the fire department. Interdepartmental agreements and understanding can be mutually beneficial and make possible the best use of the available expertise. This requires competent staffing and day-to-day cooperation.

The *APHA-PHS Recommended Housing Maintenance and Occupancy Ordinance* has been prepared for local adoption.[12] It updates and replaces the APHA's *A Proposed Housing Ordinance* of 1952 but retains most of the fundamental principles and adds certain important administrative and enforcement features. The ordinance should apply to *all* existing, altered, and new housing. Other model ordinances are also available.

The essential features of the housing ordinance are summarized here for easy reference. These are minimum standards and in many instances should be exceeded.

Summary of APHA-PHS Recommended Housing Maintenance and Occupancy Ordinance[13]

Minimum Standards for Basic Equipment and Facilities.
1. Kitchen sink properly connected and operating.
2. Flush toilet and washbasin properly operating and connected.
3. Tub or shower properly operating and connected.
4. Kitchen cabinets, shelving, and counters adequate for permissible occupancy.
5. Proper cook stove and refrigerator in operation or adequate space for them if provided by occupant.
6. Suitable facilities for safe storage of drugs and poisonous household substances.
7. Exterior doors of dwelling unit have safe locks.
8. Kitchen sink, washbasin, and tub or shower connected with adequate hot- and cold-water lines.
9. Water heater properly installed and supplies adequate water at 120°F or higher.
10. Sound handrails for steps with five or more risers and porches 3 ft or higher.
11. Every dwelling unit has dual safe means of egress leading to ground level or as required by local law.
Minimum Standards for Light and Ventilation.
12. Every habitable room, bathroom, and water closet compartment has window to the outdoors 10 percent of floor area, or
    a. Skylight on top floor, 10 percent of floor area, as substitute for window.
    b. Mechanical ventilation in kitchen, bathroom, or water closet compartment as substitute for window.
13. Rooms have openable windows equal to at least 45 percent of minimum window area or other approved ventilation.

---

[12] Prepared by Subcommittee on Housing Regulations and Standards, Program Area Committee on Housing and Health, American Public Health Association, in collaboration with Bureau of Community Environmental Management, U.S. Public Health Service Publ. No. 1935, Dept. of HEW, Washington, D.C. 1969.
[13] See *APHA-PHS Recommended Housing Maintenance and Occupancy Ordinance* for definition of terms and complete explanation of all regulations.

14. Where electric service is within 300 ft, habitable rooms provide at least 3 watts/ft^2 of floor area and have at least two separate electric outlets; water closet compartment, bathroom, laundry room, furnace room, and kitchen have at least one light fixture, all properly connected and maintained. Switches conveniently located.

15. Public hall and stairway adequately lighted and provide at least 10 ft-c of light on tread, in three-family or more dwelling at all times; if less than three families, light switches acceptable.

16. Door and window openings screened if needed.

17. Windows near ground used for ventilation screened to prevent entrance of rodents.

Minimum Thermal Standards.

18. Dwelling can be heated to temperature of 68°F measured 18 in. above floor.

19. No space heater using a flame unless properly vented.

General Requirements for Safe and Sanitary Maintenance of Dwelling and Dwelling Unit.

20. Structure is sound, weather-tight, damp-free, watertight, rodent-proof, affords privacy, and is in good repair. Premises drained, clean, safe, and sanitary.

21. Every window, exterior door, and entranceway is weather-tight, rodent-proof, and in working condition and good repair.

22. Stairways, porch, etc., safe, sound, and in good repair.

23. Plumbing fixtures, water and waste pipes properly installed and in good condition. Fixtures supplied with safe water.

24. Water closet and bathroom floor reasonably impervious and cleanable.

25. Service, facility, equipment, or utility is constructed, installed, maintained satisfactorily.

26. Service, facility, equipment, or utility not discontinued except for repair.

27. Exterior walls, foundations, basements, roofs, pipes, porches, wires, skirting, lattice, doors, grilles, windows below or within 48″ of grade, and other openings or accessory structures are rat-proofed.

28. All construction and materials, ways and means of egress, conform with local fire regulations.

Maximum Density, Minimum Space, Use and Location Requirements.

29. Dwelling unit provides at least 150 ft^2 floor space for first occupant and 100 ft^2 per additional occupant. Maximum occupancy is 2 times number of habitable rooms.

30. No more than one family plus two boarders per dwelling unit unless under permit.

31. Dwelling unit has at least 4 ft^2 of floor-to-ceiling closet space or equivalent additional space.

32. Dwelling unit of two or more rooms, every room for sleeping provides at least 70 ft^2 for first person and 50 ft^2 for each additional.

33. Access to two or more sleeping rooms, a bathroom, or water closet compartment, that is not through bedroom or bathroom or water closet compartment.

34. At least one-half of floor area 7 ft high; floor area where less than 5 ft not part of total floor area.

35. Cellar space not used as habitable room or dwelling unit.

36. Basement space used as a habitable room or dwelling unit has:

    a. Dry floor and walls.

b. Total window area in each room equal to 10 percent floor area and entirely above grade or equivalent.

c. Openable window area equal to 45 percent unless adequate ventilation provided.

d. No pipes, ducts, etc., less than 6'8" above floor.

Responsibilities of Owners and Occupants.

37. Dwelling unit not occupied or let unless it is clean, sanitary, fit for human occupancy, and complies with state and local requirements.

38. Owner of dwelling containing two or more units keeps public areas of dwelling and premises clean and sanitary.

39. Occupant of dwelling or dwelling unit keeps parts he occupies and controls clean and sanitary.

40. No rat harborage maintained in or about any dwelling or dwelling unit.

41. Occupant of dwelling or dwelling unit stores and disposes of all rubbish and garbage in a clean, safe, and sanitary manner. Garbage container is watertight and vermin-proof.

42. Owner provides containers or facilities for garbage and refuse disposal in dwelling containing more than two units. Occupant provides container when two or fewer units.

43. Owner provides and hangs screens and storm doors and windows.

44. Occupant of single-dwelling unit keeps unit free of vermin, and occupant of dwelling in which one of several units infested keeps his own unit free; owner responsible when two or more units infested or when maintenance is needed.

45. Occupant keeps all plumbing fixtures clean, sanitary, and operable.

46. Supplied heat maintains at least 68°F from (date) to (date) and from (time) to (time) of day.

Rooming House, Dormitory Rooms, and Rooming Units.[14]

47. Annual permit issued; may be suspended on inspection.

48. Sanitary facilities provided in ratio of:

   a. One flush water closet to six roomers. (Flush urinal may equal one-half of water closets in rooming house for males.)

   b. One washbasin to six roomers.

   c. One bath or shower to six roomers.

   d. Fixtures properly connected and operating.

   e. Facilities are in dwelling and accessible from common passageway.

49. Washbasin, tub, or shower supplied with hot and cold water.

50. No flush water closet, washbasin, bath, or shower in basement unless approved by health officer.

51. Bed linen changed once a week and prior to letting to new occupant.

52. Bedding clean and sanitary.

53. Rooming unit provides 110 ft^2 for one person and 4 ft^2 of closet space additional.

54. No cooking in rooms; communal cooking only if approved.

55. Access doors to rooming unit have proper locks.

56. Safe dual means of egress to ground level as required by local law, without passing through any other room.

57. Walls, floors, ceilings, premises maintained sanitary by operator.

---

[14] Requirements are in addition to other applicable items.

Additional sections in the ordinance deal with the adoption of plans of inspection (for the inspection of dwellings): powers and duties; licensing of multiple dwellings and rooming houses; authority to adopt rules and regulations; notice of violation; penalties; repairs, demolition, and revolving fund; collection and dissemination of information; conferences and hearings; emergencies; conflict of ordinances and partial invalidity; effective date; and legal notices.

The administrative section of a housing code should permit the phased improvement of dwellings and dwelling units first in selected areas having identifiable grossly substandard living conditions. The expected level of compliance would be consistent with the intent of the housing code regulations and the achievable environmental quality of the neighborhood. Low-interest loans, technical guidance, limited grants, craftsmen and labor assistance, tax relief, and other inducements are essential to encourage rehabilitation, stem the tide of deterioration and blight, and make possible compliance with the housing code. Concurrently, community services and facilities would need to be improved and maintained at an adequate level.

HOUSING PROGRAM

Approach

Housing can be a complex human, social, and economic problem that awakens the emotions and interests of a multitude of agencies and people within a community. Governmental agencies must decide in what way they can most effectively produce action; that is, whether in addition to their own efforts it is necessary to give leadership, encouragement, and support to other agencies that also have a job to do in housing.

Resolution of the housing problem involves just about all official and nonofficial groups or agencies.[15] Organization begins with the housing coordinator, as representative of the mayor or community executive officer. The groups involved include an interagency coordinating committee; citizens advisory committee; health, building, public works, law, and fire departments; the urban renewal agency; the housing authority; financing clinic; planning board; air pollution control agency; office of community relations; public library; rent-control office; universities and technical institutes; welfare department; council of social agencies;

[15] Emil A. Tiboni, *Philadelphia's Plan of Action for Housing and Neighborhood Improvement*, January 1955.

neighborhood and community improvement associations; banking institutions; real-estate groups; consulting engineers and architects; general contractors; builders and subcontractors; the press; and service organizations.

Components for a Good Housing Program

Good housing does not just happen but is the result of far-sighted thinking by individuals in many walks of life. In some cases a few houses are built here and there and areas grow with no apparent thought being given to the future pattern being established. This is typical of small subdivisions of land and developments where there is no planning. In other instances, typical of large-scale developments, construction proceeds in accordance with a well-defined plan. The growth of existing communities, the quality of services and facilities provided, and the condition in which they are maintained usually depend on the leadership and coordination provided by the chief executive officer and the controls or guides followed. The more common support activities are described below.

Planning Board

A planning board can define the area under control and, by means of maps, locate existing facilities and utilities. Included are highways, railroads, streams, recreational areas, schools, churches, shopping centers, residential areas, commercial areas, industrial areas, water lines, sewer lines, and so forth. In addition, plans are made for future revisions or expansion for maximum benefit to the community. A zoning plan is needed to delineate and enforce the use to which land shall be put, such as for residential, farm, industrial, and commercial purposes. This also provides for protection of land values. Subdivision regulations defining the minimum size of lots, width and grading of roads, drainage, and utilities to be provided may also be adopted. New roads or developments would not be accepted unless in compliance with the subdivision standards. The planning board, if supported, can guide community changes and, if established early, can direct the growth of a community, all in the best interest of the people.

Department of Public Works

A public works department usually has jurisdiction over building, water, sewers, streets, refuse, and, if not a separate department, fire prevention.

The building division's traditional responsibility is regulation of all types of building construction through the enforcement of a building code. A building code includes regulations pertaining to structural and architectural features, plumbing, heating, ventilation and air conditioning, electricity, elevators, sprinklers, and related items. Fire structural regulations are usually incorporated in a building code, although a separate or supplementary fire-prevention code may also be prepared. Certain health and sanitary regulations such as minimum room sizes, plumbing fixtures, and hot- and cold-water connections should of course be an integral part of a building code so that new construction and alterations will comply with the health department's housing-code minimum standards.

The water and sewer division would have the fundamental responsibility of making available and maintaining public water supply and sewerage services where these sanitary facilities are accessible. In new subdivisions of land the provision of these facilities, particularly when located outside the corporate limits of a city or village, is usually the responsibility of the developer.

The division of streets would maintain streets, including snowplowing, cleaning streets, rebuilding old roads, and assuring the proper drainage of surface water. In unincorporated communities, the highway department assumes these functions. New roads would not be maintained unless dedicated to the city, village, or township and of acceptable design and construction.

Refuse collection would include garbage, rubbish, and ashes. This function may be handled by the municipality, by contract, or by the individual.

Public Housing and Urban Renewal

In existing urban communities the housing authority has responsibility for the construction and operation of dwellings for low-income families and for the rehousing of families displaced by slum clearance. The urban renewal agency assembles and clears land for reuse in the best interest of the community and rehouses persons displaced as a result of its actions. In basically good neighborhoods, unsalvable housing is demolished, good housing is protected, and sound substandard housing is rehabilitated.

Health Department

State and local health departments have the fundamental responsibility of protecting the life, health, and welfare of the people. Although most cities have health departments, many areas outside of cities do not have the services of a completely staffed county or city-county health department. However, where provided, the health department responsibilities are given in a public health law and sanitary code. In addition to communicable disease control, maternal and child health care, clinics, nursing services, and environmental sanitation, the health department should have supervision over housing occupancy, maintenance and facilities, food sanitation, water supply, sewage and solid waste disposal, pollution abatement, air pollution control, recreation, sanitation, control over radiation hazards, and the sanitary engineering phases of land subdivision. The environmental sanitation activities, being related to the planning, public works, housing and redevelopment activities, should be integrated with the other municipal functions. For maximum effectiveness and in the interest of the people, it is also equally proper that the services and talents available in the modern health department be consulted and utilized by the other municipal and private agencies.

Health departments have the ideal opportunity to redirect and guide nuisance and complaint investigations; lead-poisoning elimination; carbon monoxide poisoning prevention; insect, rodent, and refuse control activities into a planned and systematic community sanitation and housing hygiene program. By coordination with nursing, medical care, and epidemiological activities, as well as those of other agencies, the department is in a position to constructively participate in elimination of the causes that contribute to the conditions associated with slums and substandard housing. The health department should also assist in the training of building, sanitation, welfare, and fire inspectors and others who have a responsibility or interest in the maintenance and improvement of living conditions.

Private Construction

New construction and rehabilitation of housing in accordance with a good building and housing code are essential to meet the normal needs of the people. Private construction and rehabilitation of sound structures should be encouraged.

Loan Insurance

Mortgage loan insurance by the government to stimulate homeownership has been fundamental to the construction and preservation of good housing. However, private financial institutions also have a major function and obligation.

Outline of a Housing Program

Several approaches have been suggested and used in the development of a housing program. They all have several things in common and generally include most of the following steps:[16]

1. Establishment of a committee or committees with representation from official agencies, voluntary groups, the business community, and outstanding individuals as previously described.

2. Identification of the problem—the physical, social, and economic aspects and development of a plan to attack each.

3. Informing the community of survey results, housing needs, and the recommended action.

4. Designation of a board or commission to coordinate and delegate specific functions to be carried out by the appropriate agency, for example, urban renewal, redevelopment, public housing, code enforcement, rehousing, rehabilitation, refinancing.

5. Appropriation of adequate funds to support staffing and training of personnel.

6. Appraisal of housing and neighborhoods and designation of urban, suburban, and rural areas for (a) clearance and redevelopment, (b) rehabilitation, (c) conservation and maintenance. This involves identification of structures and sites selected for preservation, interim code enforcement, rehabilitation of basically sound structures not up to code standards, spot clearance, provision or upgrading of public facilities and services, and land-use control.

7. Preparation and adoption of an enforceable housing code and other regulations that will upgrade the living conditions and provide a decent home in a suitable living environment.

8. Institution of a systematic and planned code-enforcement program, including education of tenants and landlords, *prompt* encouragement and *requirement* of housing maintenance, improvement and rehabilitation where indicated.

9. Concurrent provision and upgrading of public facilities and services where needed.

10. Provision of new and rehabilitated housing units through public housing, private enterprise, nonprofit organizations, individual owners, and other means.[17]

11. Aid in securing financial and technical assistance for homeowners.

12. Liaison with federal, state, and local housing agencies, associations, and organizations.

[16] Joseph G. Molner and Morton S. Hilbert, "Responsibilities of Public Health Administrators in the Field of Housing," and Charles L. Senn, "Planning of Housing Programmes," *Housing Programmes: The Role of Public Health Agencies,* Public Health Papers No. 25, WHO, Geneva, 1964; "Expert Committee on the Public Health Aspects of Housing," *WHO Tech. Rep. Ser.,* 1961, **225**; "Appraisal of the Hygienic Quality of Housing and its Environment," *WHO Tech. Rep. Ser.,* 1967, **353**; Subcommittee of the Program Area Committee on Housing and Health, *Guide for Health Administrators in Housing Hygiene,* American Public Health Association, New York, 1967.

[17] Kenneth G. MacIntosh, "Lower-Income Home Ownership Through Urban Rehabilitation—a new venture for private business," National Gypsum Co., Dept. of Urban Rehabilitation, Buffalo, N.Y.

13. Requirement that payments to welfare recipients not be used to subsidize housing that does not meet minimum healthful standards.

14. Evaluation of progress made and continual adjustment of methods and techniques as may be indicated to achieve the housing program goals and objectives.

These efforts need to be supplemented by control of new building construction, land subdivision, and mobile home parks. Also to be included are migrant-labor camps, camp and resort housing, and housing for the aged, chronically ill, handicapped, and those on public assistance.

Solutions to the Problem

The more obvious solutions to the housing problem are production of new housing, redevelopment, slum clearance, and public housing for low-income families. Increasing emphasis is being placed on cooperative ownership and the conservation and rehabilitation of existing sound housing to prevent or slow down blight. It is probably the most economical way of providing additional, more healthful dwelling units and at the same time protect the existing surrounding supply of good housing. Privately financed housing, redevelopment, and public housing cannot reach their full usefulness unless the neighborhood in which they are carried out is also brought up to a satisfactory minimum standard and is protected against degradation.

Federal, state, and local governments have an essential role in providing leadership and support. The Federal Housing Act of 1949 and subsequent amendments make assistance available to finance housing for low- and moderate-income families, for urban renewal programs including rehabilitation and conservation, and for redevelopment of urban communities. The Model Cities amendment (1966) authorizes federal funds to help coordinate and finance the rebuilding, both physically and socially, of sections of cities. A local legally constituted agency is necessary to represent the municipality in its dealings under the act.

Eligibility for urban renewal funds requires formulation of a workable program "for effectively dealing with the problem of urban slums and blight within the community and for the establishment and preservation of a well-planned community with well-organized residential neighborhoods of decent homes and suitable living environment for family life."[18] The workable program also makes a municipality eligible for assistance for concentrated code enforcement, special help to blighted areas, demoli-

[18] *How Localities Can Develop a Workable Program for Urban Renewal*, October 1, 1954, and *Workable Programs for Small Communities and Rural Non-Farm Areas*, November 1956. Housing and Home Finance Agency, Government Printing Office, Washington, D.C. See also pages 69 and 70.

tion grants, rehabilitation grants and loans, neighborhood facilities improvement and development. The workable program requires:[19]

1. Adoption of adequate minimum standards of health, sanitation, and safety through a comprehensive system of codes and ordinances effectively enforced.

2. Formulation of a "comprehensive community plan" or a "general plan"—implying long-range concepts—and including land use, thoroughfare, and community facilities plans; a public improvement program; zoning and subdivision regulations. See Chapter 2.

3. Identification of blighted neighborhoods and analysis for extent and intensity of blight and causes of deterioration to aid in delineation for clearance or other remedial action.

4. Setting up an adequate administrative organization, including legal authority, to carry on the urban renewal program.

5. Development of means for meeting the financial obligations and requirements for carrying out the program.

6. Provision of decent, sanitary housing for all families displaced by urban renewal or other governmental activities.

7. Development of active citizen support and understanding of the urban renewal program.[20]

The restoration of a substandard housing area, in an otherwise sound and healthy neighborhood, must include removal of the blighting influences to the extent feasible and where indicated. These include heavy traffic, air pollution, poor streets, lack of parks and trees, poor lighting, dirty streets and spaces, inadequate refuse storage and collection, inadequate water supply and sewerage, unpainted buildings, noisy businesses and industries, and tax increases for housing maintenance and improvement.

Further explanations and specifics to carry out some of the above program functions are given in the sections that follow.

Selection of Work Areas

A housing program must keep in proper perspective community short- and long-term plans, special surveys and reports, area studies, pilot block enforcement, and the routine inspection work to enforce a minimum standards housing ordinance. Continual evaluation of the program is necessary to assure that the control of blight and deterioration of buildings and neighborhoods, as well as the rehabilitation of substandard dwellings, is being carried out where encountered. When indicated and possible, such action should be carried out on a planned block or area basis. This makes accomplishments more apparent and also awakens community pride, which can become "infectious." A blighted block does not always have a neon sign at the head of the street. The very insidious-

[19] Ibid.
[20] Ibid.

ness of blight makes it difficult to recognize and will have to be deliberately sought out and attacked. It may be only one or two houses in a block, which are recognized by a dilapidated outward appearance, by the inspector's intimate knowledge of the neighborhood and people, by office records showing a history of violations, or by routine survey reports. Signs of deterioration and dilapidation should be attacked immediately before decay and blight take over and incentive to make and support repairs becomes an almost impossible task.

The selection of conservation, rehabilitation, clearance, and redevelopment areas should take into consideration and adapt the following criteria:

1. The grading of socioeconomic areas using a weighted composite consisting of overcrowding, lack of or dilapidated private bath, lack of running water, and other measurable factors reported by the Bureau of the Census, including income and education.

2. The grading of areas using health indices and mortality and morbidity data.

3. The grading of areas on the basis of social problems, juvenile delinquency, welfare and private agency case load, adult probation, early venereal diseases, social unrest, and so forth.

4. The areas having the high, average, and low assessed valuation, dilapidation, and overcrowding.

5. Grading of neighborhoods using the APHA or equal appraisal system, health and building department surveys, and plotted data. A sampling of 20 percent of the dwellings can give fairly good information.

6. Results of surveys and plans made by planning boards, housing authorities, and redevelopment boards.

7. Selection of work areas with reference to these criteria as well as existing and planned housing projects, parks, redevelopment areas, parkways or throughways, railroads, and industrial or commercial areas; also existing barriers such as streams or lakes, good housing areas, swamps, and mountains.

8. Health department environmental sanitation and nursing division office records and personal knowledge of staff.

9. Visual foot surveys and combinations of criteria 1–8.

Enforcement Program

Enforcement of a housing conservation and rehabilitation ordinance involves use of the same procedures and techniques that have been effective in carrying out other environmental sanitation programs. The development of a proper attitude and philosophy of the intent of the law and its fair enforcement should be the fundamental theme in a continuing in-service training program. A housing enforcement program can proceed along the following lines.

1. Inspection of a pilot area to develop and perfect inspection techniques and learn of problems and practical solutions to obtain rehabilitation of substandard areas.

2. Routine inspection of all hotels and rooming and boarding houses for compliance with minimum-standards housing ordinance. Satisfactory report may be made condition to the issuance of a permit and a license to operate.

3. Routine inspection of multiple dwellings. An initial inspection on a two- or three-year plan will reveal places requiring reinspection.

4. Inspections throughout a city, village, or town on an individual structure, block, and area basis to obtain rehabilitation and conservation of sound housing or demolition of unsalvable structures.

5. Concentration of inspection in and around salvable areas to spread the border of improved housing so that it will merge with a satisfactory area.

6. Redirection of complaint inspections to complete housing surveys when practical. This can make available a large reservoir of personnel for more productive work.

7. Continuing in-service training with emphasis on law enforcement through education and persuasion, alteration and reconstruction, letter reporting, and financing to obtain substantial rehabilitation—not patchwork.

8. Close liaison with all city departments, especially city planning, building division, urban renewal agency, welfare, and the courts.

9. Continuing public education, including involvement of community and neighborhood organizations, to support the housing program and help or guide owners to conserve and rehabilitate their homes.

The enforcement program should take into consideration the problems, attitudes, and behavior of the people living in the dwelling units and the changes that need to be effected with the help of other agencies. The enforcement program must also recognize that, to be effective, subsidies and other forms of assistance to low-income families will be necessary to support needed alterations and rehabilitation.

Staffing Patterns

Various staffing patterns have been suggested to administer a housing-code enforcement program. One basis is a program director supported by one assistant to supervise up to eight inspectors, each making an estimated 1000 inspections per year. Slavet and Levin suggest the following ratios:[21]

1. One inspector per 10,000 population or one inspector per 1000 substandard dwelling units. This assumes inspection and reinspection to secure compliance at an average of 200 substandard dwelling units per year over a 5-year period.

2. One inspector per 3000 standard dwelling units, assuming inspection of 600 units per year over a 5-year period, in addition to the staff needed to handle complaints.

3. One financial specialist for every three or four housing inspectors.

4. One community relations specialist for every three or four housing inspectors.

5. One rehabilitation specialist for every two or three inspectors.

[21] Joseph S. Slavet and Melvin R. Levin, *New Approaches to Housing Code Administration*, National Commission on Urban Problems, Research Rep. No. 17, Washington, D.C., 1969.

6. One clerk for every three or four inspectors.
7. One supervisor for every six to eight inspectors.

Some Contradictions

An enforcement program will reveal many contradictory facts relating to human nature. These only serve to emphasize that the housing problem is a complex one. The social scientist and anthropologist could study the problems and assist in their solution provided he works in the field with the sanitarians and engineers.

1. For example, it has been found that a person or family living in a substandard housing area may be very reluctant to move away from friends and relatives to a good housing area where the customs, religion, race, and very environment are different.

2. A housing survey may show that a structure is not worth repairing, and the expert construction engineer or architect may be able to easily prove that conclusion. But the owner of a one-, two- or three-family building will rarely agree. He will go on to make certain minimum repairs or even extensive repairs to approach the standards established in the housing ordinance, leaving you wondering if there really is any such thing in fact as a nonsalvable dwelling.

3. Then there is the not infrequent situation where two old structures are located on the same lot, one facing the street and the other on the rear-lot line. If one structure is demolished, another cannot be built on the same lot because zoning ordinances usually prohibit such intensive lot usage by new structures. The owner, therefore, rather than lose the vested right due to prior existence of the structure, with respect to the zoning ordinance, may choose to practically rebuild the dwelling at great cost rather than tear it down.

4. Another common occurrence is the tendency for some landlords to seize upon health department letters recommending improvements to bring the building into compliance with the housing ordinance as the opportunity to evict tenants and charge higher rentals. This is not to say that the owner is not entitled to a fair return on his investment; but the intent is that no one should profit from human misery. A fair return on one's investment has been given as 10 percent of the assessed valuation or purchase price of the property. Since the health department's objective is to improve the living conditions of the people, it should not become a party to such actions. As a matter of fact, it is the unusual situation where needed housing improvements cannot be made with the tenant living in the dwelling unit, even though some temporary inconvenience may result. Hence in such situations it is proper for the health department to state that it is its opinion that the needed repairs can be made without evicting the tenants, thereby leaving the final determination in the exceptional cases with the owner and the courts.

5. A rather unexpected development may be the situation where an owner agrees to make repairs and improvements, such as new kitchen sink and provision of a three-piece bath with hot as well as cold water under pressure, only to be refused admittance by the tenant. For the tenant knows that the rent may be increased by $20 to $30 per month and suddenly decides that a private bathroom is a luxury he can get along without. In such case the courts have granted eviction orders and have sanctioned the reasonable increase in rent.

Fiscal Aspects

The financing of improvements is sometimes an insurmountable obstacle to the lay person. As an aid to the people affected by a housing enforcement program, a "financing clinic" can be formed to advise home-owners in difficult cases. The clinic may be a committee consisting of a banker, an architect, a general contractor, a plumbing contractor, and a representative of the building division, redevelopment board, federal agencies, and community. The department representative would help the owner present the problem. If the homeowner is instructed to submit estimates of cost from two or three reliable contractors, a sounder basis for assistance is established.

An enforcement program must recognize that many investors use the criterion that a dwelling is not worth purchasing unless it yields a gross annual income $\frac{1}{4}$ to $\frac{1}{5}$ of the selling price. Another rule of thumb, for rehabilitation of a dwelling, is that the cost of alteration plus selling or purchase price shall not be more than five times the gross annual income.

Getting Started

The report that appeared in the *New York Times*, October 6, 1957, is an excellent example of a sound approach to housing conservation and rehabilitation. It is quoted below.

> Home owners whose properties are in Milwaukee's first "conservation" area are not likely to complain to the Mayor when their houses are inspected in the door-to-door canvass being made by the city. Mayor Frank Zeidler's home is in the area, too.
>
> The inspections are planned to find out whether the district's middle-aged homes meet housing code standards. Although the courts can enforce compliance with the code, the drive is aimed primarily at informing owners of dangerous deterioration, with the expectation that they will correct the conditions within a reasonable time.
>
> Mayor Zeidler told his neighbors that "this is in the nature of getting free advice." According to the National Association of Housing and Redevelopment Officials, the drive can save a neighborhood from becoming a slum which, experience shows, costs taxpayers more than it repays in tax revenues.

Enforcement Procedures

The fair and reasonable enforcement of the *intent* of a minimum standards housing ordinance requires the exercise of a great deal of trained judgment by the administrators, supervisors, and field inspectors. Continuing in-service training of the housing staff is essential to carry out the purpose of the housing ordinance in an effective manner. A trained educator can offer valuable assistance and guidance in planning the training sessions and in interpreting the program to the public and legislators. Health department sanitarians are admirably suited to carry out the housing program because of their broad knowledge in the basic sciences and their ability to deal effectively with the public.

The enforcement measures and procedures can be summarized as follows:

1. Preparation of a housing operating manual giving a background of the housing program, guides for conduct, inspection report form based on the housing ordinance, interpretation of sections of the ordinance needing clarification for field application, suggested form paragraphs for routine letters, including typical violations and recommendations for their correction, follow-up form letters, building construction details, and related information. Typical form letters, a few form paragraphs, and architectural details are included in the text to suggest the type of material that can be included in a housing manual to obtain a certain amount of uniformity and organization in a housing enforcement program.

2. Complete inspection of all types of premises and dwelling units occupied as living units. A comprehensive checklist or survey form based on the housing ordinance similar to Figure 11-3 could be used.

3. Preparation of a complete rough-draft letter by the sanitarian based on his field inspection, listing what was found wrong, together with recommendations for correction.

4. Review of the inspection report and suggested letter by a supervising sanitarian for completeness and accuracy. The letter is signed by the supervisor and a tickler date confirmed for a follow-up inspection. See Figure 11-4.

5. Maintenance of a visible card-index file on each multiple dwelling, boarding house, rooming house, and hotel, in addition to a file on each dwelling, including one- and two-family structures.

6. A reinspection letter based on the findings reported in the first letter and remaining uncorrected. This may take the form of an abridged original letter listing what remains to be done, a letter calling for an informal hearing, a letter giving 30 days in which to show substantial progress, or a special letter. See Figure 11-5.

7. In some instances substantial improvements are requested that justify a reasonable rent adjustment. This should be brought out at the time of informal hearings when financial hardship is pleaded. Informal administrative hearings are very useful.

8. Office consultation with owners to investigate long-range improvement programs within the limits of the law and the federal, state and private assistance available.

9. Maintenance of a list of competent contractors who perform alteration work.

10. Review of cost estimates with the owner for completeness.

11. Arranging for expert consultation to finance improvements, including a hearing before a "financing and construction clinic," which should have representation from the lending institutions as well as the contractors, architects, and engineers.

12. In recalcitrant cases, a letter from the chief city judge or justice providing for a pretrial hearing to show cause why a summons should not be issued, followed by the hearing. A similar procedure could be followed by the corporation counsel or county attorney.

13. Issuance of a summons by the corporation counsel at the request of the

ERIE COUNTY DEPARTMENT OF HEALTH

Division of Environmental Sanitation

601 City Hall, Buffalo 2, N.Y.

MOhawk 2800
Extension 28

Re:

C. Buffalo, New York

Dear

The City of Buffalo recently adopted a minimum standards housing ordinance to assist in the conservation and rehabilitation of existing housing. In this connection Mr.   of this Department made a survey of dwellings in your neighborhood on   Listed below are the conditions found at your building which were in need of improvement or correction.

We shall be glad to discuss any questions or difficulties you may have in complying with the minimum housing standards. Would you please advise this office in writing within 10 days of the action you propose to take?*

Your cooperation will help prevent the deterioration of property values in your neighborhood and help make Buffalo a more healthful place in which to live.

Very truly yours,

J. A. Salvato, Jr., P.E.,

Chief, Bureau of General Sanitation

By...

* This letter is not to be construed as reason for removal of tenants, unless specifically stated therein.

† Alterations or additions must be made in accordance with all applicable laws. Consult with the Division of Buildings, Room 325, City Hall for a permit to perform building, plumbing or electrical work required here. A building unlawfully occupied or with an increased number of living units must be brought into compliance, or if this is not permitted by law, it must revert to its original use.

JAS/er (1) 9/54:1000

Figure 11–4 Form letter to confirm findings at time of first inspection.

COUNTY OF ERIE
HEALTH DEPARTMENT
601 CITY HALL
BUFFALO 2, N.Y.

Re:

C. Buffalo, New York

Dear

This department notified you by letter on   of the
housing deficiencies existing on your premises and the need for making
corrections.   However, a reinspection showed that no substantial progress had
been made, and to date we have received no reply from you.

Inasmuch as the conditions enumerated in our letters to you are violations
of the Buffalo Minimum Housing Standards Ordinance, it is imperative that
the needed improvements be made.   Your failure to reply must be interpreted
as an indication of lack of good faith or misunderstanding of the intent of our
letters.

Under the circumstances, you are requested to appear in Room 605, City
Hall, at   to show cause why legal action should not be
started.   It is hoped that you will come prepared with an itemized report,
indicating when the needed improvements will be completed, so as to make
unnecessary further action by this department.

We wish to emphasize that it is our intention and obligation to follow up on
violations of the Minimum Housing Standards Ordinance until such time as all
work is done.   It is only in this way that existing housing in the City of Buffalo
can be conserved and rehabilitated.

Very truly yours,

J. A. Salvato, Jr., P. E.,

Chief, Bureau of General Sanitation

By...

JAS/er(3)
9/54:500

Figure 11–5   Informal hearing letter.

enforcement agency when other measures have failed. A record of appearances
and results should be kept on forms such as those in Figures 11–6 and 11–7.

14. Issuance of a summons by an authorized member of the enforcement agency
in emergencies.

15. Cooperation with the press with a view toward obtaining occasional special
reports, feature articles, and community support.

BUREAU OF GENERAL SANITATION

LEGAL ACTION SUMMARY

Location .... Structure use ....

Name .... Address ....

Name .... Address ....

Name .... Address ....

CASE HISTORY

Inspection dates	Inspector	Deficiencies Found	Date of letter (L) or personal notice (PN)	Section Violated

Person cited for informal hearing—Date and results..

....

Legal action approved by..

Case No........................ Referred to C. C. .... Date of summons .... Issued by........................

Summons returnable... Court .... Time ....

Letter of (withdrawal) to Corporation Council—Date........................... by...........................
      (dismissal)

Reason ....

Figure 11-6 Legal action summary.

Date of Appearance	Presiding Judge	Disposition (Fine—Dismissal—Adjournment)	Adjourned to

REMARKS (Reason for indicated disposition)..

....

....

Inspector's Signature:

Date of Appearance	Presiding Judge	Disposition (Fine—Dismissal—Adjournment)	Adjourned to

REMARKS (Reason for indicated disposition)..

....

....

Inspector's Signature:

Date of Appearance	Presiding Judge	Disposition (Fine—Dismissal—Adjournment)	Adjourned to

REMARKS (Reason for indicated disposition)..

....

....

Inspector's Signature:

Date of Appearance	Presiding Judge	Disposition (Fine—Dismissal—Adjournment)	Adjourned to

REMARKS (Reason for indicated disposition)..

....

....

Inspector's Signature:

GS–H–20

Figure 11–7 Court history.

16. Cooperation and liaison with all private and public agencies having an interest in housing.

17. Notification of welfare department, when public assistance is being given for rental, because payment for housing should be made only when the facilities and conditions of maintenance and occupancy meet minimum standards.

18. Education of the legal department so as to obtain an aggressive, trained, and competent attorney who can become experienced in housing problems and be able to guide the legal enforcement phase of housing work.

19. Stimulate resurgence of neighborhood pride in the rehabilitation and conservation of all buildings on a block or area basis. Information releases and bulletins explaining what community groups and individuals can do to improve their housing is an excellent approach.[22]

20. Discourage profit at the expense of human misery by making it possible through state law to file a lien in the county clerk's office against a property that is in violation of health ordinances, for the information of all interested and affected persons.

21. Notification of the mortgage holders and insurance carriers of existing conditions. Establish an escrow fund (2 percent a year) for rehabilitation only.

The enforcement procedure includes news releases and information bulletins, inspection, notification, reinspection, second notification, third notification, informal hearings, formal hearings, pretrial hearings, and summonses. In brief, therefore, the enforcement procedure consists of education and persuasion and legal action only as a last resort.

Legal Conduct of Sanitarians[23]

1. Must be aware at all times that they are public servants, that they represent the health officer, commissioner, and the health department; that their misconduct will result in loss of prestige for department.

2. As witnesses for the people in court, and in all their official activities, they should show a fair, impartial, courteous, and objective attitude.

3. To indicate the efficiency of their department in court, they must be thoroughly prepared to testify. Such preparation depends, basically, on the thoroughness of the inspection and the records made of such inspection. Every inspection should be made with the thought that it may form the basis for a court trial (memorandum to refresh recollection).

4. Must be punctual in court and present in courtroom when case is called. Papers should be reviewed in morning before court session begins to make certain that they are in proper form and signed and to refresh recollection about case.

5. Avoid talking to defendant or his attorney in court unless instructed to do so by the judge. If they have any questions refer them to the corporation counsel or prosecutor for the people.

6. Above all, avoid showing any personal interest or vindictiveness or officious attitude. The court is inclined to take your word for any situation you describe in an expert, official manner. But as soon as you show that you have a personal feeling, the court feels that you have a grudge against the defendant.

[22] *Neighborhood Conservation Information Bulletin,* Detroit City Plan Commission, Committee for Neighborhood Conservation and Improved Housing, Detroit, Mich.
[23] From New York State Health Dept. Field Training Center Notes, prepared by Irving Witlin.

7. You have the burden of proving your case beyond a reasonable doubt.

8. It is a common court tactic, as in sports, to get your opponent riled so that he will lose his head. The defendant's attorney may try that one on you.

9. Do not pretend to be an expert on everything. Use ordinary, simple language—remember that the judge is not a sanitarian.

10. In preparing a complaint, carefully read the section or regulation applicable and as succinctly as possible indicate how that section was violated. Remember that every word in a law has been carefully thought of to give a certain common and usual meaning. Do not try to give it your own personal interpretation.

HOUSING FORM PARAGRAPHS FOR LETTERS

It is necessary to have some reasonable uniformity and accuracy when writing letters or reports to housing-code violators. This becomes particularly important when a housing-code enforcement unit has a large staff and when court action is taken. By using a dwelling-inspection form such as Figure 11–3, a form letter such as Figure 11–4, and the form paragraphs listed below, it should be possible for the inspector to readily prepare a draft of a suitable letter for review by his supervisor. The form paragraphs can and should be modified as needed to reflect more precisely the unsatisfactory conditions actually observed and practical suggestions for their correction. Reference to pages 796, 797, 800, 802, 804, 818–820, and 823 will be helpful is describing violations.

Dilapidation

Dilapidation of a structure is usually quite obvious on inspection, but may be difficult to describe in a letter so as to convey fully the observed deterioration. Professional engineers use as a guide that a building to be considered safe must be able to support $2\frac{1}{2}$ to 4 times the loads and stresses to which it is or may be subjected.

Certain conditions that may be deemed dangerous or unsafe need explanation. For example, a 12-in. beam that has sagged or slanted more than one-quarter out of the horizontal plane of the depth of floor structural members in any 10-ft distance would be more than 3 in. out of level in 10 ft and hence unsafe. An interior wall consisting of $2'' \times 4''$ studs or $4''$ terra-cotta tile blocks more than one-half out of the vertical plane of the thickness of those members between any two floors would be more than 2 in. out of plumb and hence unsafe.

A stairs, stairway, or approach is safe to use when it is free of holes, grooves, and cracks that are large enough to constitute a possible accident hazard. Rails and balustrades are expected to be firmly fastened and maintained in good condition. Stairs or approaches should not have rotting or deteriorating supports, and stairs that have settled more than 1 in. or pulled away from the supporting or adjacent structure may be dangerous. Stair treads must be of uniform height, sound, and securely

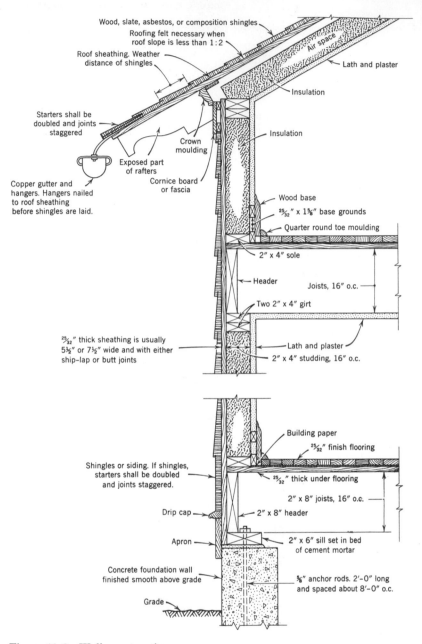

Wood, slate, asbestos, or composition shingles

Roofing felt necessary when roof slope is less than 1 : 2

Roof sheathing. Weather distance of shingles

Air space

Lath and plaster

Insulation

Starters shall be doubled and joints staggered

Insulation

Crown moulding

Exposed part of rafters

Cornice board or fascia

Copper gutter and hangers. Hangers nailed to roof sheathing before shingles are laid.

Wood base

$\frac{25}{32}$″ x 1⅝″ base grounds

Quarter round toe moulding

2″ x 4″ sole

Header

Joists, 16″ o.c.

Two 2″ x 4″ girt

$\frac{25}{32}$″ thick sheathing is usually 5½″ or 7½″ wide and with either ship-lap or butt joints

Lath and plaster

2″ x 4″ studding, 16″ o.c.

Building paper

$\frac{25}{32}$″ finish flooring

Shingles or siding. If shingles, starters shall be doubled and joints staggered.

$\frac{25}{32}$″ thick under flooring

2″ x 8″ joists, 16″ o.c.

Drip cap

2″ x 8″ header

Apron

2″ x 6″ sill set in bed of cement mortar

Concrete foundation wall finished smooth above grade

$\frac{5}{8}$″ anchor rods. 2′-0″ long and spaced about 8′-0″ o.c.

Grade

Figure 11–8 Wall construction.

796

fastened in position. Every approach should have a sound floor and every tread should be strong enough to bear a concentrated load of at least 400 lb without danger of breaking through. See Figure 11–9.

Incomplete Bathroom

The (first floor rear apartment) did not have a complete bathroom. A complete bathroom including a water closet, tub or shower, and wash-

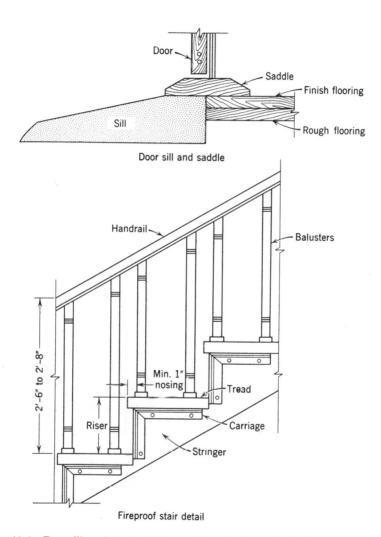

Figure 11–9 Doorsill and stair details. (Adapted from C. G. Ramsey and H. R. Sleeper, *Architectural Graphic Standards,* John Wiley & Sons, New York, 1970.)

Figure 11–10 Alteration for shower and washbasin addition.

basin connected with hot and cold running water is required to serve each family. See Figures 11–10 and 11–11. The bathroom shall have a window not less than 3 ft² in area, providing adequate light and ventilation. A water-repellent floor with a sanitary cove base or equivalent is necessary.

No Hot Water

There was no piped hot water in the (kitchen of the first-floor front apartment). This apartment shall be provided with hot water or water-

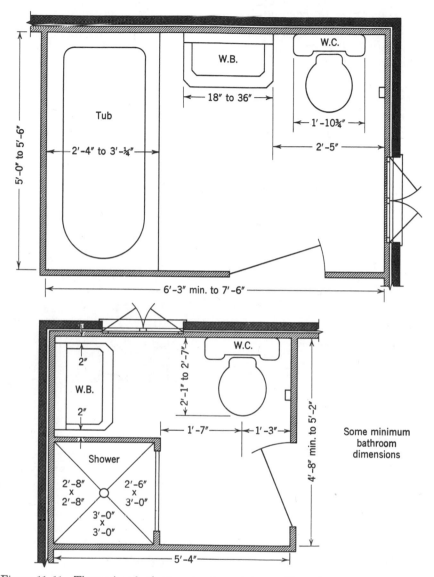

Some minimum
bathroom
dimensions

Figure 11–11   Three-piece bathroom showing minimum dimensions.

heating facilities of adequate capacity, properly installed and vented. The heater shall be capable of heating water so as to permit water to be drawn at every required kitchen sink, lavatory basin, bathtub, or shower at a temperature of 120°F.

Leaking Water Closet

The water-closet bowl in the (describe location) apartment was (loose) (leaking) at the floor. (When the toilet is flushed the water drains onto the floor and seeps through the ceiling to the lower apartment.) The water closet should of course be securely fastened to the floor, floor flange, and soil pipe so that it will be firm and not leak when flushed. (It will also be necessary to repair the loose ceiling plaster in the lower floor and repaint or repaper as needed.) See Figure 11–12.

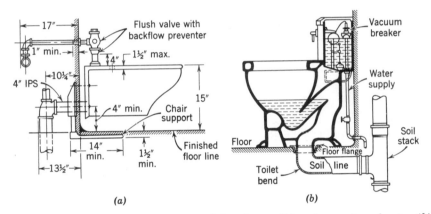

(a)　　　　　　　　　　　　　　　　　　*(b)*

Figure 11–12 Plumbing details: (*a*) siphon-jet wall-hanging water closet; (*b*) integral water closet and tank combination. (From Svend Plum, *Plumbing Practice and Design,* Vol. 1, John Wiley & Sons, New York, 1943.)

Floors Not Water Repellent

The (bathroom) (water-closet compartment) (floor covering) was (worn through) (broken) (bare wood with open joints). The floor should be (repaired) (made reasonably watertight), and the floor covering should extend about 6 in. up the wall to provide a sanitary cove base. Satisfactory material for the floor covering is inlaid linoleum, rubber or asphalt tile, smooth cement concrete, tile, terrazzo, and dense wood with tightly fitted joints covered with varnish, lacquer, or other similar coating providing a surface that is reasonably impervious to water and easily cleaned.

Exterior Paint Needed

The exterior paint has (peeled, worn off) exposing the bare wood, rusting the nails, causing splitting and warping of the siding. This will lead to the entrance of rainwater, rotting of the siding, sheathing, and studs, as well as inside dampness and falling of the plaster! You are urged to immediately investigate this condition and make the necessary repairs, including painting of the house, to prevent major repairs and expenses in the future.

Rotted and Missing Siding

The (shingles, siding, apron, cornice, exposed rafters) (on the north side of the house at the second-floor windows and foundation) was (were) (rotted and missing). Decayed material should of course all be removed, the sheathing repaired wherever necessary and the (shingles, siding, etc.) replaced. Following the carpentry work, all unpainted or unprotected material exposed to the elements should be painted to prolong its life.

Sagging Wall

The (door frames and window frames) in the (location) were out of level, making complete closure of the doors and windows impossible. Outside light could be easily seen through the openings around the (window rails and door jambs). The supporting (beams, girders, posts, and studs) should be carefully inspected as there was evidence that some of these members were rotted, causing the outside wall to sag. The building should be shored and made level wherever necessary. The unsound material should be replaced, and the improperly fitting (doors, windows, and framing) repaired so as to fit and open properly. See Figures 11-8 and 11-13.

Loose Plaster

The plaster was (loose) (and buckled) (and had fallen) from the living-room ceiling and walls) in the (name apartment or other location) over an area of approximately (10 ft²). All loose plaster should be removed and the wall replastered; following curing it should be painted or papered to produce a cleanable, smooth, and tight surface.

Leaking Roof

There was evidence of the roof leaking over or near the (kitchen, living room, etc., in the tenant apartment). The (paint, paper, was stained and peeling). It is essential that the leak be found and repaired, not only to prevent the entrance of water and moisture in the apartment but also to prevent loosening of the plaster, rotting of the timbers, and extension of the damage to your property.

No Gutters or Rain Leaders

There were no gutters and rain leaders on this building. Gutters and rain leaders should be placed around the entire building and connected to (the sewer if permitted) to ensure proper drainage of rainwater. This

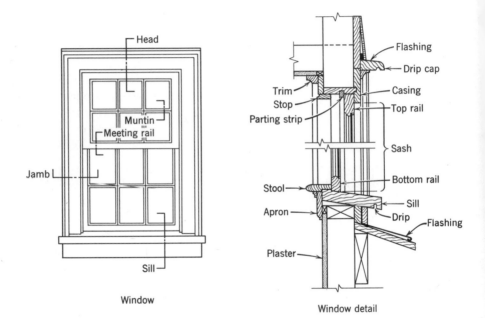

Window

Window detail

Figure 11–13   Window details.

will also make less likely rotting and seepage of water through siding and window frames and entrance of water into the (cellar) (basement).

No Handrails

There were no handrails in the stairway between the (first and second floor at the rear). This is a common cause of preventable serious accidents. Handrails should be provided and securely fastened at a height of 30 to 32 in., measured above the stair tread.

Refuse in Attic

There were (rags, refuse, paper, and trash) in attic. These materials are a fire hazard and provide harborage for mice and other vermin. All rags, paper, and trash must be removed from the attic and the attic maintained in a clean and sanitary condition at all times.

Water-Closet Flush Tank Not Operating Correctly

The (water ran continuously) in the water-closet flush tank in the (John Jones's apartment); OR the water closet in the (John Jones's apartment) could not be flushed. The (broken, worn, missing) (ball-cock valve, ball-cock float, flush-valve ball, flush lever, flush handle) should be repaired or replaced to permit proper flushing of the water closet. See Figure 11–14.

Garbage Stored in Paper Box or Bag

Garbage is being stored in (open, uncovered baskets) (paper bags) (paper boxes) (in the rear yard). This encourages rodent, fly, and vermin breeding. All garbage should be drained, wrapped, and properly stored in tightly covered containers. It will be necessary for you to procure needed receptacles for the proper storage of all garbage until collected.

Dilapidated Garbage Shed

The garbage shed in the (specify location) was in a dilapidated, rotted, and insanitary condition. Garbage sheds tend to accumulate garbage and encourage rodent, fly, and vermin breeding. This dilapidated garbage shed should be removed and the premises cleaned. Store the garbage cans on an elevated rack or concrete platform. (Enclose pamphlet showing some suggested storage racks.) (See pages 395 and 396.)

Debris in Yard or Vacant Lot

The vacant lot located at (specify location) was found littered with (broken concrete, asphalt, and rubbish). This is unsightly and may serve as a rat harborage and as an invitation to dump on the property. It is requested that you make a personal investigation of the conditions reported and arrange to have the lot cleared and cleaned. It is also recommended that you post a "No Dumping" sign to discourage future littering of the property.

Dirty Apartment

The apartment on the (second floor) is in a very insanitary condition. (Describe.) Every occupant is expected to keep his apartment and the

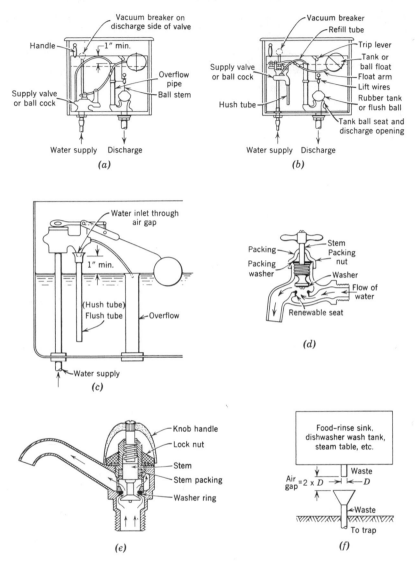

Figure 11–14 Plumbing details: (a) submerged type of ball cock; (b) elevated type of ball cock; (c) venturi-type vacuum breaker; (d) compression type faucet; (e) crane "Dial-e-ze" faucet; (f) air-gap waste receptor. [(a) (b) (d) (e) adapted from Forrest B. Wright, *Rural Water Supply and Sanitation,* John Wiley & Sons, New York, 1956.]

premises he controls in a clean and sanitary condition at all times. (Copy of letter to tenant.)

Overcrowded Sleeping Room

The bedroom(s) in the (specify location) were overcrowded. There were (three) persons in a room having an area of (80) ft^2 and (four) persons in a room (85) ft^2. Every room occupied for sleeping purposes shall contain at least 70 ft^2 of floor space for one person and 50 ft^2 for each additional person. (Suggest correction.) This apartment should not be rerented for occupancy by more persons than can be accommodated in accordance with this standard.

No Window in Habitable Room

There was no window to the outside air provided in the (living room, kitchen, bedroom). Every room used or intended to be used as a (living room, kitchen, bedroom) is required to have a total unobstructed window area of at least 10 percent of the floor area. Consideration should be given to the possibility of (cutting in a new window) (providing a skylight) if the room is to be continued in use as a (bedroom, living room, or kitchen).

Unlawful Third-Floor Occupancy

The third floor of this building had been converted and was occupied for living purposes. The conversion or alteration of a third floor or attic in a frame building for living or sleeping purposes is prohibited by Chapter X of the city ordinances. Discontinue the use of the third floor immediately. (Refer copy of letter to division of buildings.)

Unlawful Cellar Occupancy

The cellar was being used for (sleeping, living, purposes). A cellar may not be used for living purposes, hence this space must be permanently vacated. (The housing ordinance defines a cellar as "an enclosed space having more than one-half of its height below curb level. A cellar shall not be counted as a story.")

Clogged Sewer

The soil stack, building drain, or sewer was apparently clogged, for sewage from the upper apartment(s) backed into the (kitchen sink, water closet) in the (first-floor front apartment). The clogged sewer must be cleared and, if necessary, repaired to eliminate cause for future complaint.

Unvented Heater

The gas water-heater(s) in the (name room or space and locate) was (were) not vented. Unvented heaters in bathrooms and sleeping rooms have been the cause of asphyxiation and death. These heaters must be properly vented to the chimney or outside air.

Furnace Flue Defective

The furnace flue had rusted through in several places (and the connection to the chimney was loose), causing waste gases to escape into the basement. Since such gases rise and seep into the upper apartment(s) and have been known to cause asphyxiation, it is imperative that the flue be repaired and the collar sealed to prevent leakage of any waste gases. This should also improve the efficiency of the furnace.

Rubber-Hose Gas Connection

The gas heater(s) in the (tenant apartment) has (have) (plastic pipe, rubber hose) connection(s). Such materials eventually leak and may cause death in the household. It will be necessary to replace all plastic and rubber-hose connections with rigid, metal pipe.

Rat Infestation

There was evidence of a very bad rat condition existing in this building as indicated by (explain condition). All holes in the foundation (floors) should be sealed with cement mortar and openings around wood framing closed with metal flashing or with cement mortar where possible. Traps and repeated use of a rodenticide such as Warfarin are suggested to kill rats inside the building. All sources of food and harborages must be eliminated. Such control measures should be continuous for at least two or three weeks to be effective. (Enclose pamphlet giving additional details dealing with accepted control measures.)

Roach Infestation

The apartment was apparently infested with roaches as indicated by the roachy odor, roaches observed hiding under the sink, baseboard, moldings (stains in the kitchen cabinet, pellets of excrement in the dish cabinet). Roaches are sometimes brought in with boxes of food, baskets, or bags; dirt and filth encourage their reproduction in large numbers. Thorough cleaning, filling of cracks around frames with plaster or plastic wood, followed by the proper application of an insecticide in selected places and in accordance with the manufacturer's directions should bring the problem under control. (The enclosed pamphlet gives additional detailed information . . . use only in severe cases or when several apartments are infested.)

Overflowing Sewage Disposal System

The sewage disposal system serving your dwelling was (seeping out onto the surface) (discharging into the ditch in front of your home). This is not only a health hazard to you but also to your neighbors, children, and pets. Immediate steps should be taken to determine the cause and make corrections. (See Chapter 4 for more details.)

Improperly Protected Well Water Supply

A sanitary survey of your well water supply shows it to be subject to contamination. The well (is uncapped) (has a hole around the casing where surface water can drain down and into the well) (does not have a tight seal at the point where the pump line(s) pass into the casing as noted by drippage observable from looking into the well). The necessary repairs should be made to prevent contamination of your water supply, and then the well should be disinfected as explained in the enclosed instructions. (See Chapter 3 for more details.)

Major Repairs

In view of the major repairs and improvements needed, only some of which having been reported above, plans prepared by a registered architect should be submitted showing the existing conditions and all proposed alterations, for approval by this Department and the Division of Buildings, before any work is done. This procedure makes possible the receipt of comparable bids from several contractors and usually results in more orderly prosecution of the work at a minimum cost.

Minor Repairs

In view of the repairs and improvements needed, a sketch drawn to scale should be prepared showing existing and proposed work, to assure that the work can be done as intended. The sketch should be submitted to and be approved by the Division of Buildings and this Department before any work is done. This procedure makes possible the receipt of comparable bids from several contractors and usually results in more orderly prosecution of the work at a minimum cost.

Obtain at Least Three Estimates

We urge you to obtain at least three estimates from reputable contractors before having any work done. Written bids should be requested and assurance obtained from the contractor that the estimate is all inclusive.

PLUMBING

Plumbing Code

Sanitary plumbing principles that are based on the latest scientific studies should be fundamentally similar, but will be varied in application depending on the local conditions. Many plumbing designs and standards currently in existence are based on an unsound old rule of thumb or prejudice. They could be reviewed with profit in the light of present-day knowledge.

The *National Plumbing Code*, ASA A40.8–1955,[24] and Public Health Service Pub. No. 1038[25] are comprehensive standard codes of minimum requirements for use throughout the country. The interested person would do well to have copies in his reference file. The sizing of water supply, drainage, vent, and storm drain piping is concisely covered.

A complete set of definitions of plumbing terms is given in ASA A40.8–1955 and in PHS Pub. No. 1038. One term used frequently is "fixture." This means "installed receptacles, devices, or appliances either supplied with water or receive on discharge liquids or liquid-borne wastes or both." The bathtub, sink, water closet, dishwasher, and drinking fountain are examples of plumbing fixtures.

Health departments can accomplish more in the interest of public health by seeing that proper standards of plumbing exist and are enforced than by actually doing the plumbing inspection.

For example, the health department can see to it that plumbing and building codes prohibit dangerous cross-connections and interconnections and require a private three-piece bathroom and kitchen sink served by hot and cold water in every new dwelling unit. This is fundamental to the prevention of disease, the promotion of personal hygiene, and sanitation.

Plumbing codes should specify an adequate number of fixtures for private, public, and industrial use, all properly supplied, trapped, vented, and sewered as noted in Tables 11–6 to 11–9. Water connections with unsafe or questionable water supplies would be prevented, and connections or conditions whereby used or unsafe water could flow back into the

---

[24] *National Plumbing Code*, ASA A40.8–1955. The American Society of Mechanical Engineers, 29 West 39th Street, New York, N.Y. (1955).
[25] *Report of Public Health Service Technical Committee on Plumbing Standards*, U.S. Public Health Service, Pub. No. 1038, Dept. of HEW, Washington, D.C. (September 1962).

potable water system would be prohibited. A safe water supply and proper sewage disposal should of course be assured.

Housing codes would be expected to make reference to a modern plumbing code. Housing codes would be of little value unless they were applicable to all new, altered, and existing one-, two-, or more family dwellings, hotels, boarding houses, and rooming houses.

The health department should serve as a consultant to the building, plumbing, water, and sewer divisions. Any new or revised codes or regulations should first be reviewed and approved by the health department before being considered for adoption to assure that fundamental principles of public health and sanitary engineering are not violated.

Some plumbing details, with particular emphasis on backflow prevention, recommended minimum number of plumbing fixtures, the application of indirect waste piping, and other details are given in the Tables 11–6 to 11–9 and are discussed below.

Backflow Prevention[26]

The backflow of pollution into a water-distribution piping system is a very real possibility. The best way to eliminate the danger is to prohibit direct connection between the water system and any other system or tank containing polluted or questionable water. This can be accomplished by terminating the water supply inlet or faucet a safe distance above the flood-level rim of the fixture. The distance, referred to as the air gap, is 1 in. for a ½-in. or smaller diameter faucet or inlet pipe, 1½-in. for a ¾-in. diameter faucet, 2 in. for a 1-in. diameter faucet, and twice the effective opening (cross-sectional area at point of water supply discharge) when its diameter is greater than 1 in. When the inside edge of the faucet or pipe is close to a wall, that is, within 3 or 4 times the diameter of the effective opening, the air gap should be increased by 50 percent.

Sometimes, however, it is not possible or practical to provide an air gap. Under such circumstances an approved type of backflow preventer, such as that shown in Figure 11–15, may be used if permitted by the local regulations. The backflow preventer should be installed on the outlet side of the control valve, a distance not less than 4 times the nominal diameter of the inlet, measured from the control valve to the flood-level rim of the fixture, and in no cases less than 4 in. This is illustrated in the sketch of the siphon-jet wall-hanging water closet, Figure 11–12(a).

---

[26] See *Water Supply and Plumbing Cross-Connections,* Public Health Service **Pub.** No. 957, U.S. Dept. of HEW, Washington, D.C., 1963.

TABLE 11-6 MINIMUM NUMBER OF PLUMBING FIXTURES

Type of Building Occupancy	Type of Fixture					
	Water Closets	Urinals	Lavatories	Bathtubs or Showers	Drinking Fountains	Other Fixtures
Assembly—places of worship.	*Number of persons* / *Number of fixtures* 150 women — 1 300 men — 1	*Number of persons* / *Number of fixtures* 300 men* — 1	1	—	1	
Assembly—other than places of worship (auditoriums, theaters, convention halls).	*Number of persons* / *Number of fixtures* 1-100 — 1 101-200 — 2 201-400 — 3 Over 400, add 1 fixture for each additional 500 men and 1 for each 300 women.	*Number of persons* / *Number of fixtures* 1-200 — 1 201-400 — 2 401-600 — 3 Over 600, add 1 fixture for each 300 men.*	*Number of persons* / *Number of fixtures* 1-200 — 1 201-400 — 2 401-750 — 3 Over 750, add 1 fixture for each 500 persons.	—	1 for each 300 persons.	1 slop sink.
Dormitories—school or labor, also institutional.	Men: 1 for each 10 persons. Women: 1 for each 8 persons.	1 for each 25 men. Over 150, add 1 fixture for each 50 men.*	1 for each 12 persons. (Separate dental lavatories should be provided in community toilet rooms. A ratio of 1 dental lavatory to each 50 persons is recommended.)	1 for each 8 persons. For women's dormitories, additional bathtubs should be installed at the ratio of 1 for for each 30 women. Over 150 persons add 1 fixture for each 20 persons.	1 for each 75 persons.	Laundry trays, 1 for each 50 persons. Slop sinks, 1 for each 100 persons.

Type of building or occupancy	Water closets	Urinals	Lavatories	Bathtubs or showers	Drinking fountains	Kitchen sink, laundry, etc.
Dwellings—one and two-family.	1 for each dwelling unit.	—	1 for each dwelling unit.	1 for each dwelling unit.	—	1 for each dwelling unit.
Dwellings—multiple or apartment.	1 for each dwelling unit or apartment.	—	1 for each dwelling unit or apartment.	1 for each dwelling unit or apartment.	—	Kitchen sink, 1 for each dwelling unit or apartment. For apartments or multiple dwelling units in excess of 10 apartments or units, 1 double laundry tray for each 10 units or 1 automatic laundry washing machine for each 20 units.
Industrial—factories, warehouses, foundries, and similar establishments.	*Number of each sex* / *Number of fixtures* 1-10 → 1 11-25 → 2 26-50 → 3 51-75 → 4 76-100 → 5 1 fixture for each additional 30 employees.	Where more than 10 men are employed: *Number of men* / *Number of urinals* 11-30 → 1 31-80 → 2 81-160 → 3 161-240 → 4	*Number of persons* / *Number of fixtures* 1-100 persons → 1 to 10 Over 100 persons → 1 to 15	1 shower for each 15 persons exposed to excessive heat or to occupational hazard from poisonous, infectious, or irritating material.	1 for each 75 persons.	—
Institutional—other than hospitals or penal institutions (on each occupied story.)	1 for each 25 men. 1 for each 20 women.	1 for each 50 men.*	1 for each 10 persons.		1 for each 50 persons.	

TABLE 11-6   MINIMUM NUMBER OF PLUMBING FIXTURES (cont'd)

Type of Building Occupancy	Type of Fixture					
	Water Closets	Urinals	Lavatories	Bathtubs or Showers	Drinking Fountains	Other Fixtures
Hospitals		—	1	1	—	1
Individual room wards	1 for each 8 patients.	—	1 for each 10 patients.	1 for each 20 patients.	1 for each 100 patients.	slop sink/ floor.
Waiting rooms						
Employees	Same as public.	Same as public.	Same as public.	—	Same as public.	
Penal institutions	1 in each cell.	1 in each exercise room.	1 in each cell.	1 on each cell block floor.	1 on each cell block floor.	1 slop sink/ floor.
Prisoners	1 in each exercise room.	—	1 in each exercise room.	⋯	1 in each exercise area.	
Employees	Same as public.	Same as public.	Same as public.	—	Same as public.	
Public buildings, offices; business; mercantile, storage, and institutional employees.	*Number of each sex* / *Number of fixtures*: 1-15 → 1; 16-35 → 2; 36-55 → 3; 56-80 → 4; 81-110 → 5; 111-150 → 6; 1 fixture for each additional 40 employees.	Urinals may be provided in men's* toilet rooms in lieu of water closets but for not more than 1/3 of the required number of water closets.	*Number of employees* / *Number of fixtures*: 1-15 → 1; 16-35 → 2; 36-60 → 3; 61-90 → 4; 91-125 → 5; 1 fixture for each additional 45 persons.	—	1 for each 75 persons.	1 slop sink/ floor.
Schools	*Boys* / *Girls*					
Elementary	1/40 / 1/35	1/30 boys.	1/50 pupils.	In gym or pool showerrooms, 1/5 pupils of a class.	1/100 pupils but at least 1 per	Slop sinks, 1 on each floor.
Secondary	1/75 / 1/45	1/30 boys.	1/50 pupils.			

| Working men, temporary facilities | 1/30 working men. | 1/30 working men. | 1/30 working men. | — | 1 fixture or equivalent for each 100 working men. |

Source: *Report of Public Health Service Technical Committee on Plumbing Standards*, September 1962, U.S. Public Health Service Pub. No. 1038, U.S. Dept. of HEW, Washington, D.C., 1963.
*Where urinals are provided for the women, the same number shall be provided as for men.

TABLE 11-7   SIZE OF FIXTURE SUPPLY AND DRAIN

Type of Fixture or Device	Pipe size (in.) Supply	Drain	Type of Fixture or Device	Pipe size (in.) Supply	Drain
Bathtubs	$\frac{1}{2}$	$1\frac{1}{2}$	Sink, service, P trap	$\frac{1}{2}$	2
Combination sink and			Sink, service, floor trap	$\frac{3}{4}$	3
tray	$\frac{1}{2}$	$1\frac{1}{2}$	Urinal, flush tank, wall	$\frac{1}{2}$	$1\frac{1}{2}$
Drinking fountain	$\frac{3}{8}$	1	Urinal, direct,		
Dishwasher, domestic	$\frac{1}{2}$	$1\frac{1}{2}$	flush valve	$\frac{3}{4}$	2
Kitchen sink,			pedestal, flush valve	1	3
residential	$\frac{1}{2}$	$1\frac{1}{2}$	Water closet,		
commercial	$\frac{3}{4}$	$1\frac{1}{2}$	tank type	$\frac{3}{8}$	3
Lavatory	$\frac{3}{8}$	$1\frac{1}{4}$	Water closet,		
Laundry tray, 1 or 2			flush-valve type	1	3
compartments	$\frac{1}{2}$	$1\frac{1}{2}$	Hose bibbs	$\frac{1}{2}$	–
Shower, single head	$\frac{1}{2}$	$1\frac{1}{2}$	Wall hydrant	$\frac{1}{2}$	–

The minimum water pressure at the outlet, at times of maximum demand, shall not be less than 8 psi, except for direct flush valves, where 15 psi is required, and where special equipment requires other pressure. Adapted from U.S. Public Health Service Pub. No. 1038, Dept. of HEW, Washington, D.C. See also Chapter 4 for trap sizes and fixture unit ratings.

TABLE 11-8   DISTANCE OF FIXTURE TRAP FROM VENT

Size of Fixture Drain (in.)	Distance, (ft) Maximum
1¼	2½
1½	3½
2	5
3	6
4	10

Where it is not practical to connect the backflow preventer above the flood-level rim of the fixture, it may be acceptable to terminate the end of the inlet pipe a distance 4 times the diameter of the pipe and not less than 4 in. above the maximum overflow level. This is permissible provided there is a free discharge to the atmosphere or floor, should the waste pipe receiving the overflow become clogged.

Indirect Waste Piping

Waste pipes from fixtures or units in which food or drink is stored, prepared, served, or processed must not connect directly to a sewer or

TABLE 11-9 HOT-WATER DEMANDS AND USE FOR VARIOUS TYPES OF BUILDINGS

Type of Building	Maximum Hour	Maximum Day	Average Day
Men's dormitories	3.8 gal/student	22.0 gal/student	13.1 gal/student
Women's dormitories	5.0 gal/student	26.5 gal/student	12.3 gal/student†
Motels: number of units†			
20 or less	6.0 gal/unit	35.0 gal/unit	20.0 gal/unit
60	5.0 gal/unit	25.0 gal/unit	14.0 gal/unit
100 or more	4.0 gal/unit	15.0 gal/unit	10.0 gal/unit
Nursing homes	4.5 gal/bed	30.0 gal/bed	18.4 gal/bed
Office buildings	0.4 gal/person	2.0 gal/person	1.0 gal/person
Food service establishments:			
Type A—full-meal restaurants and cafeterias	1.5 gal/max. meals/hr	11.0 gal/max. meals/hr	2.4 gal/avg. meals/day*
Type B—drive-ins, grilles, luncheonettes, sandwich and snack shops	0.7 gal/max. meals/hr	6.0 gal/max. meals/hr	0.7 gal/avg. meals/day*
Apartment houses: number of apartments			
20 or less	12.0 gal/apt.	80.0 gal/apt.	42.0 gal/apt.
50	10.0 gal/apt.	73.0 gal/apt.	40.0 gal/apt.
75	8.5 gal/apt.	66.0 gal/apt.	38.0 gal/apt.
100	7.0 gal/apt.	60.0 gal/apt.	37.0 gal/apt.
130 or more	5.0 gal/apt.	50.0 gal/apt.	35.0 gal/apt.
Elementary schools	0.6 gal/student	1.5 gal/student	0.6 gal/student*
Junior and senior high schools	1.0 gal/student	3.6 gal/student	1.8 gal/student*

Source: Copyright by the American Society of Heating, Refrigerating and Air Conditioning Engineers, Inc. Reprinted by permission from *ASHRAE Guide And Data Book*, New York, 1970.

*Per day of operation.

†Interpolate for intermediate values.

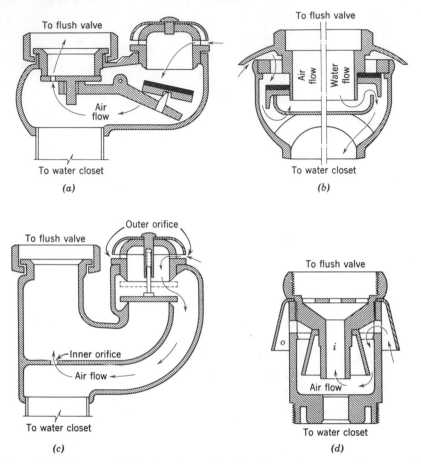

Figure 11–15 Types of siphon-breakers: (*a*), (*b*), (*c*) moving parts; (*d*) nonmoving part. [From Roy B. Hunter, Gene E. Golden, and Herbert N. Eaton, "Cross-connections in Plumbing Systems," Research Paper RP 1086, *J. Research National Bureau of Standards,* **20** (April 1938).]

drain. Stoppage in the receiving sewer or drain would permit polluted water to back up into the fixture or unit. Waste piping from refrigerators, iceboxes, food rinse sinks, cooling or refrigerating coils, laundry washers, extractors, steam tables, egg boilers, steam kettles, coffee urns, dishwashing machines, sterilizers, stills, and similar units should discharge to an open water-supplied sink so that the end of the waste pipe terminates at least 2 in. above the rim of the sink.

A commercial dishwashing machine waste pipe may be connected to the sewer side of a floor drain trap when the floor drain is located next to the dishwashing machine.

An alternate to the installation of a water-supplied sink waste receptor is the provision of an air gap in the fixture waste line, at least twice the effective diameter of the drain served, located between the fixture and the trap. This is illustrated in Figure 11–14(f).

The water-supplied sink or air-gap waste receptor should be in an accessible and ventilated space and not in a toilet room.

Plumbing Details

A few typical details and principles are illustrated for convenient reference and as guides to good practice. Many variations are to be found, dependent on local conditions and regulations. See Figures 11–16 and 11–17.

Other

See Tables 4–3, 4–4, and 11–7 for fixture unit ratings for sewage flow computations, trap sizes, drain sizes, special fixture values, and fixture unit load to building drains and sewers. Figure 3–19 gives curves for estimating probable water demand of a sum of fixture units. Table 3–13 lists weights of fixtures, and Table 3–14 gives the size of fixture supply pipe, rate of flow, and required pressure during flow for different fixtures.

VENTILATION

Spread of Respiratory Diseases

The danger of spreading respiratory diseases can be reduced by preventing overcrowding, by applying dust-control measures, and by preventing accumulation of the microorganism concentration through recirculation of the same air. This, of course, is in addition to good personal hygiene, reduction of contact, disinfection of eating and drinking utensils, sanitary disposal of mouth and nose discharges, and maintenance of natural resistance. Air-borne infections are discussed in Chapter 1.

Thermal Requirements

Good ventilation requires that air contain a suitable amount of moisture and that it be in gentle motion, cool and free from offensive body odors, poisonous and offensive fumes, and large amounts of dust. Comfort zones for certain conditions of temperature, humidity, and air movement are given by the American Society of Heating, Refrigerating and Air Conditioning Engineers, Inc.[27] The optimum condition is given as 77.5°F for winter and summer, with a 1.2°F difference. Relative humidities

[27] *Handbook of Fundamentals,* ASHRAE, 345 East 47th Street, New York, 1967, p. 122.

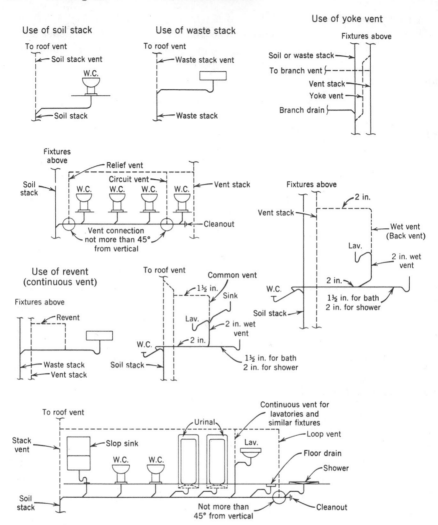

Figure 11–16 Some fixture plumbing details. (From N.Y. State Building Construction Code—Plumbing, April 1, 1957.)

below 70 percent are reported to have no influence on human confort. Air movement, radiant heat, the individual, and tasks being performed must also be taken into consideration.

There is no one temperature and humidity at which everyone is comfortable. People's sensations, health, sex, activity, and age all enter into the comfort standard. McNall recommends a temperature range between 73 and 77°F and humidity between 20 and 60 percent for lightly clothed

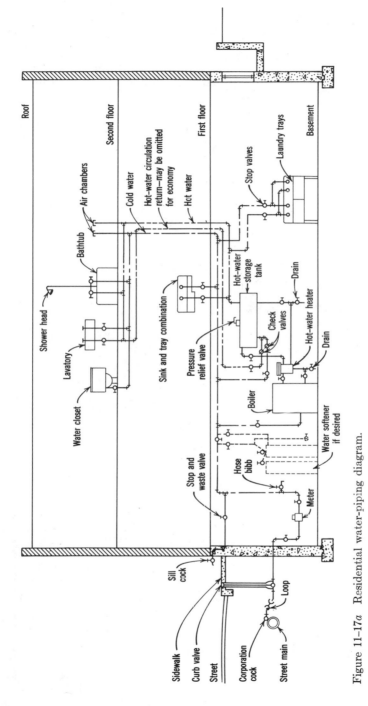

Figure 11-17a   Residential water-piping diagram.

819

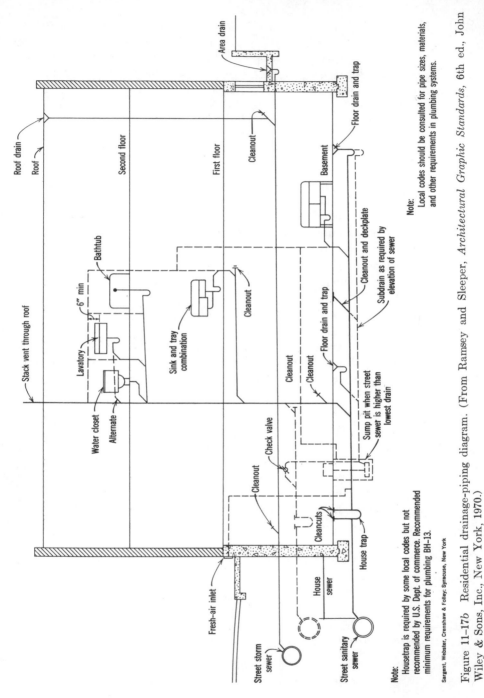

Figure 11–17b Residential drainage-piping diagram. (From Ramsey and Sleeper, *Architectural Graphic Standards*, 6th ed., John Wiley & Sons, Inc., New York, 1970.)

Note:
Housetrap is required by some local codes but not recommended by U.S. Dept. of commerce. Recommended minimum requirements for plumbing BH-13.

Sargent, Webster, Crenshaw & Folley; Syracuse, New York

Note:
Local codes should be consulted for pipe sizes, materials, and other requirements in plumbing systems.

adults engaging in sedentary activities in residences.[28] Lubart gives as the comfort level a range of 68°F at inside relative humidity of 50 percent to 76°F with 10 percent relative humidity.[29] Indoor relative humidity of 60 percent or higher would cause excessive condensation and greater mildew, corrosion, and decay. Ordinarily, however, only a temperature and ventilation control is used in the home, with no attempt being made to measure or control the relative humidity.

Space Ventilation and Heating

In schools, separate venting of each classroom to the outside is preferred. Good standards specify that the ventilating system provide a minimum air change of 10 to 15 ft^3 per minute per pupil to remove carbon monoxide and odors, without drafts. The air movement should not exceed 25 ft per minute, and the vertical temperature gradient should not vary more than 5°F in the space within 5 ft of the floor. Air conditioning or electric fans may be required to meet special requirements. Temperature control should be automatic.

In cold weather, when forced ventilation is combined with heating by means of ducts and registers or openings, uncomfortable drafts and large temperature variation should be avoided in the design of the ventilating systems. There is also danger of spreading air-borne infections by recirculating dust and pathogenic bacteria or viruses in dry used air. Recirculation will cause an increase in the concentration of these organisms due to the cumulative effect of recirculating the same used contaminated air. This will occur unless sufficient clean fresh air is admitted and part of the used air is exhausted to waste. During cold months of the year greater fuel consumption can be expected. The usual method of air purification by washing and filtration is relatively inefficient in removing bacteria or viruses from used air, although it can be effective in removing dust. Heating appliances producing or using dangerous gases must be vented adequately to the outside air through tight flue connections.

A minimum of one to two air changes per hour can often be secured by normal traffic and leakage through walls, ceilings, floors, and through or around doors and windows. Under ordinary circumstances, proper ventilation can be obtained by natural means with properly designed

---

[28] Preston E. McNall, Jr., report in *Proceedings of the First Invitational Conference on Health Research in Housing and Its Environment,* Airline House, Warrenton, Va., March 17–19, 1970, U.S. Public Health Service, Dept. of HEW, Washington, D.C., p. 27.

[29] Joseph Lubart, "Winter Heating an Enemy Within," *Medical Science,* 33–37 (March 1967).

windows, both in the winter and summer. Openable windows, louvers, or doors are needed to ventilate and keep relatively dry attics, basements, pipe spaces, and cellars. Tops of windows should extend as close to the ceiling as possible, with consideration to roof overhang, to permit a greater portion of the room being exposed to controlled sunlight. The net glass space should at least equal 10 to 15 percent of the floor area and the openable window area at least 45 percent.

In recreation halls, theaters, churches, meeting rooms, and other places of temporary assembly, the requirement of at least 10 ft³ of fresh clean air per minute per person cannot ordinarily be met without some system of mechanical or induced ventilation. Any system of ventilation used should prevent short-circuiting and uncomfortable drafts.

Toilet Ventilation

Bathroom and toilet-room ventilation is usually accomplished by means of windows or, in fireproof dwellings, ventilating ducts. The common specification is that the window area shall be at least 10 percent of the floor area and not less than 3 ft², of which 45 percent shall be openable. A system of mechanical exhaust ventilation maintained and operated to provide at least four changes per hour, of the air volume of the bathroom or toilet room, during the hours of probable use, is usually specified for ventilation where windows are not relied on or are not available for ventilation. Fans that are activated by the opening and closing of doors would not provide satisfactory ventilation.

Venting of Heating Units

Proper venting is the removal of all the products of combustion through a designated channel or flue to the outside air with maximum efficiency and safety. Gravity-type venting relies largely on having the vent gases inside the vent hotter (thus lighter) than the surrounding air. The hotter the vent gases, the lighter they are and the greater their movement up through the vent. Thus, in order to keep the vent gases hot so that they may work at maximum efficiency, proper installation and insulation are necessary.

Factors that prevent proper venting are abrupt turns; downhill runs; common vents to small, uninsulated vent pipe; conditions that cause back drafts; and unlined masonry chimneys. Stained and loose paper or falling plaster around a chimney is due to poor construction. A masonry chimney will absorb a great deal of the heat given off by the vent gases, thus causing the temperature in the chimney to fall below the dew point. The high moisture in vent gases condenses inside the chimney, forming sulfuric acid. This acid attacks the lime in the mortar, leaching it out and creating leaks and eventual destruction of the chimney. It is therefore necessary to line a masonry chimney with an insulating pipe, preferably terra cotta flue lining.

Figure 11–18 (a-b) shows chimney conditions apt to result in back drafts. The flue or vent should extend high enough above the building or other neighboring obstructions so that the wind from any direction

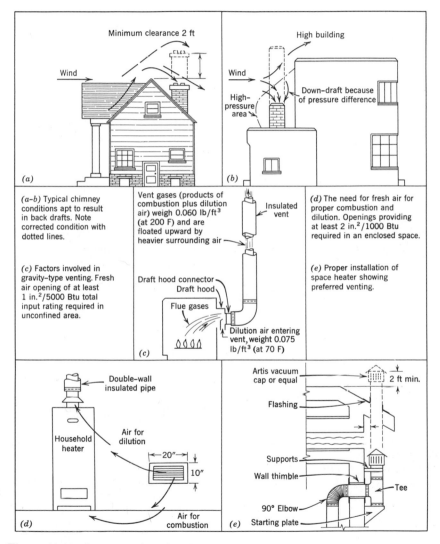

Figure 11-18 Some venting details. (Drawings are typical and not necessary in full accordance with any code.)

will not strike the flue or vent from an angle above horizontal. Unless the obstruction is within 30 ft or unusually large, a flue or vent extended at least 3 ft above flat roofs or 2 ft above the highest part of wall parapets and peaked roof ridges will be reasonably free from downdrafts. If it is not possible to carry the vent up, a special backdraft guard such as illustrated in Figure 11–18(e) will afford a measure of protection.

To ensure proper venting as well as proper combustion, sufficient amounts of fresh air are required, as shown in Figure 11–18. An opening of 100 to 200 in.^{2} will usually provide sufficient fresh air under ordinary household conditions; this opening is needed to float the flue gases upward and ensure proper combustion in the fire box. Proper venting and an adequate supply of fresh air are also necessary for the prevention of carbon monoxide poisoning or asphyxiation.

Before making any vent installations or installing any gas- or oil-fired appliances, consult the building code and the local gas or utility company. Standards for chimneys, fireplaces, and venting systems, including heating appliances and incinerators, are given in the National Fire Protection Association's Pub. No. 211.[30]

MOBILE HOME PARKS

Mobile homes are defined as "transportable, single-family dwelling units suitable for year-round occupancy and containing the same water supply, waste disposal, and electrical conveniences as immobile housing."[31] A mobile home is also defined as a "manufactured relocatable living unit."[32] Mobile homes produced since 1954 are 10 to 12 ft wide and up to 60 or more ft long. Wider units 20 to 24 ft wide are assembled at the home site by combining 10 and 12 ft wide sections.

It is estimated that in 1969 there were about 2 million mobile homes housing 5 million people in the United States and that 400,000 are being manufactured each year.[33] About half of these mobile homes are in 22,000 mobile home parks. A survey in 1968 revealed that 81 percent of mobile home families have no school-age children; 43 percent of the owners

[30] *Chimneys, Fireplaces, Venting Systems,* NFPA, 60 Batterymarch Street, Boston, Mass. 02110, 1970.
[31] *Environmental Health Guide for Mobile Home Parks,* prepared by U.S. Public Health Service, Dept. of HEW; published by Mobile Home Manufacturers Association, 6650 North Northwest Highway, Chicago, January 1966, p. 3.
[32] *Environmental Health Guide for Mobilehome Communities,* prepared by U.S. Public Health Service, Dept. of HEW; published by Mobile Home Manufacturers Association, 6650 North Northwest Highway, Chicago, Ill. Revised January 1971, p. 30.
[33] *Environmental Health Guide for Mobilehome Communities,* ibid.

are 34 years old or older, and 25 percent are over 55. The typical home is 12 ft wide and 60 ft long.

The parks established have all of the potential environmental sanitation problems of a small community. Because of this, standards have been prepared to guide mobile home manufacturers, operators, owners, and regulatory agencies to help promote good, uniform sanitary practices. Compliance with these standards is facilitated by reference to the pertinent chapters in this text; to the guides cited in footnotes 31 and 32, which include a recommended ordinance; and to the guide published by the U.S. Department of Housing and Urban Development.[34] (See also Chapter 9 for design and related details.)

INSTITUTION SANITATION

Definition

An institution, is a complete property, its buildings, facilities, and services having a social, education, or religious purpose. It includes schools; colleges or universities; hospitals; nursing homes; homes for the aged; prisons; reformatories; and the various types of federal, state, city, and county welfare, mental, and detention homes or facilities.

Institutions as Small Communities

Most institutions are communities unto themselves. They have certain basic characteristics in common that require careful planning, design, construction, operation, and maintenance. These include (1) site selection, planning and development for the proposed use, including subsoil investigation and accessibility and relationship to sources of noise and air pollution; (2) a safe, adequate, and suitable water supply for fire protection as well as for institutional use; (3) sewers and a wastewater disposal system; (4) roads and a storm water drainage system; (5) facilities for the storage, collection, and disposal of all solid wastes generated in the institution; (6) boilers and incinerators with equipment and devices to control air pollution; (7) food preparation and service facilities; (8) housing and facilities for the resident population; and (9) insect, rodent, and noxious weed control. In addition, depending on the particular institution, they might have recreation facilities, such as a swimming pool or bathing beach. A hospital or educational institution might have its own laundry. A state training school or institution might have a dairy farm or produce farm, pasteurization plant, and food-processing plant. The environmental engineering and sanitation con-

[34] *Mobile Home Court Development Guide,* January 1970.

cerns at all these places are in many instances quite extensive and complex. The reader is referred to the appropriate subject matter throughout this text.

The material that follows will highlight sanitation at various types of institutions.

Hospitals and Nursing Homes

The hospital is expected to provide an environment that will expedite the recovery and speedy release of the patient. Carelessness can introduce contaminants and infections that delay recovery and may overburden the weakened patient, thereby endangering his survival. It has been estimated that 1.5 million patients, out of some 30 million, incur infections in hospitals annually.[35] Basic to the prevention of complications are hygienic medical and nursing practices and air, water, food, and equipment sanitation.

An extension of the hospital is the extended-care facility, where a patient can recuperate under medical and nursing supervision. This is supplemented by the skilled nursing home, where full-time nursing care and needed medications are provided under medical direction and supervision of a registered professional nurse. A third type of facility is the intermediate-care home, which provides physical assistance and care for people who cannot take complete care of themselves.

In recent years a great deal of special emphasis has been placed on hospital and nursing home construction, equipment, and inspection or survey of operation and services. Federal and state standards are required to be met with respect to fitness and adequacy of the premises, equipment, personnel, rules and bylaws, standards of medical care, and hospital services. Plans for new structures and for additions or modifications of existing facilities are also reviewed for compliance with federal and state requirements to help ensure the best possible facilities for medical care.

The survey or inspection of hospitals and nursing homes takes into consideration the administration, fire prevention, medical, nursing, environmental sanitation, nutrition, and related matters, as shown in Figures 11–19 and 11–20. A proper initial survey of these diverse matters calls for a professionally trained team consisting of a physician, sanitarian, engineer, nurse, nutritionist, and hospital administrator. The proper interpretation of the environmental health survey form items requires a well-rounded educational background and specialized training, with sup-

[35] Robert L. Elston, "The Role of Hands in the Spread of Diseases," *Am. J. Public Health*, 2211 (November 1970).

Name		Address			T.V.C.	
Operator		Persons interviewed				
Inst. No.	No. Cert. Beds	Inspected by		Date		

Item	S	U	Item	S	U
Structure and grounds			*Fire safety*		
1. Location			46. Fire-resistive construction		
2. Buildings and grounds			47. Interior finishes		
3. Accessible by emergency vehicles			48. Smoke and fire doors, open, ducts		
4. Service entrances			49. Fire-resistant enclosures—chutes,		
5. Elevators			shafts, stairs, kitchen, boiler and in-		
Water supply			cinerator rooms		
6. Supply and pressure, hospital and fire			50. Exit doors, access, stair, hall, signs		
7. Treatment, physical, biological,			51. Flame-retardant fabrics, drapes		
chemical			52. Flammable liquids		
8. Quality—hospital, protect. and sur-			53. Flammable anesthetics		
veillance			54. Oxygen and nitrous oxides		
9. Quality—special, medical, and lab.			55. Fire hydrants, hoses, standpipes		
10. Hot water			56. Fire extinguishers, portable		
Liquid and solid wastes			57. Automatic sprinklers in chutes,		
11. Sewage piping and disposal			soiled linen, trash, and storage rooms		
12. Biological wastes collection, storage,			58. Automatic extinguish. in kitchen		
disposal			hoods, shops, storage attics		
13. Solid wastes collection, storage dis-			59. Fire detection systems in boiler,		
posal			kitchen, labs, laundry, pantry, ga-		
Plumbing			rage		
14. Toilets, lavatories, tubs, showers,			60. Fire-alarm system, internal		
sinks			61. Alarms connected to fire department		
15. Cross-connection and backflow con-			or station		
trol			62. Electrical hazards controlled		
16. Drinking fountains			63. Anesthesia areas—electrical safe-		
Emergency power and light			guards		
17. Power to vital services			64. No smoking—signs, supervision		
18. Lighting to vital areas			65. Fire plans—posting, drills, training		
Ventilation, heat, and air conditioning			*Accident safety*		
19. Air flows, rates, pressure, differential			66. Handrails and grab bars—corridors,		
20. Air filtration			stairs, ramps, toilets, bath		
21. Air temperature			67. No obstacles—corridors, ramps,		
22. Air humidity			stairs		
23. Intakes and exhausts			68. Floors—nonslip and nontrip		
Laundry			69. Burn protection—heaters, hot water		
24. Soiled linen handling and transpor-			70. Patient furnishings and equipment		
tation			71. Electrical safety hazards		
25. Laundering			72. Accident reporting, records		
26. Clean linen handling and transpor-			73. Lighting levels		
tation			*Disinfection and sterilization*		
Dietary services			74. Infection control committee		
27. Food sources			75. Sterilization facilities		
28. Refrigerated food storage			76. Sterilization—procedure, surveill-		
29. Dry food storage			ance, records		
30. Food preparation			77. Disinfection of anesthesia apparatus		
31. Food serving, including water			78. Disinfection of patient-used articles		
32. Food utensils			79. Disinfection of operating, delivery,		
33. Food equipment			isolating, rms., nursery		
34. Ice making and handling			80. Disinfection of inhalation therapy		
35. Vending machines			apparatus		
Insect and rodent control			*Supporting services*		
36. Physical controls			81. Housekeeping		
37. Chemical controls			82. Plant maintenance		
Infant formula			*Refrigeration (nondietary)*		
38. Equipment, supplies, technicians,			83. Pharmaceutical and blood storage		
records			84. Morgue		
39. Approved source and handling			*Pools (therapy or swimming)*		
Space provisions			85. Recirculation and filtration		
40. Patient rooms			86. Disinfection		
41. Isolation rooms			*Radiations*		
42. Bath and toilet rooms			87. Diagnostic X-ray units		
43. Nursing service areas			88. Therapeutic X-ray units		
44. Other services—rooms, spaces			89. Teletherapy units		
45. Central and general storage			90. Radioact. materials—storage, use		
			91. Microwave ovens		

NOTE:
S means satisfactory equipment, construction, operation, maintenance.
U means unsatisfactory.
Contract Services _____ Housekeeping _____ Laundry _____ Dietary _____ Other _____ . Give name.

Figure 11–19 Hospital environmental health survey form.

Name _____ Address _____ T.V.C. _____
Operator _____ Person interviewed _____
Capacity _____Bed patients _____ Amb. patients _____ Personnel: RN _____Pract. nurses _____Other _____

Housing	Yes	No	CM		Yes	No	CM
1. Building, floors, stairs structurally sound				26. Separate toilet and shower for help convenient, washbasin with sanitary towels			
2. Rooms, kitchen, cellar, attic, yard clean				*Food and nutrition*			
3. Rooms, including kitchen, well lighted				27. Equipment and facilities sanitary and adequate for food storage, preparation, and serving			
4. Rooms, including kitchen, well ventilated				28. Dishes, trays, utensils clean and disinfected			
5. Rooms, basement, closets free of debris				29. Foodhandlers clean, have lockers			
6. Beds 4 ft apart and 1 ft from walls				30. Refrigeration adequate			
7. Windows, doors, openings screened				31. Satisfactory food storeroom			
8. Free of rat and vermin infestation				32. Pasteurized bottled milk served			
9. Bedding clean, storage satisfactory				33. Refuse storage and disposal satisfactory			
10. Heating suitable and adequate				34. No poisons stored in kitchen			
11. Frame building has above 1st floor:*				35. Menu planned week in advance			
(a) Outside iron stairs 4 ft wide				36. Meals nutritionally balanced			
(b) Exits with 3 ft 8 in. door				37. Food service attractive			
(c) Approved, inspected fire equipment on each floor including 1st floor, kitchen, boiler room				38. Leftovers used promptly			
(d) Helpless patients on 1st floor				*Medical and nursing service*			
12. Open stairwells fire protected				39. Mental and TB cases prohibited			
(a) Fire drill for employees monthly				40. Patient accidents and CD reported			
13. Home location satisfactory				41. RN always in attendance			
				42. Adequate competent help available			
Water supply				43. Employees free of CD			
14. Water supply approved; fire hydrant within 500 ft				44. Helpless patients get special attention			
15. Adequate supply hot and cold				45. Records kept giving:			
16. Sanitary drinking fountain or paper cups				(a) Name of MD for patient			
17. Cross-connection and interconnection prohibited				(b) Name and address of patient, sex, color, age, marital status, occupation, place of birth, admission, diagnosis, discharge, relative			
18. Backflow preventers on WC, flushers—bedpan				(c) Clinical history and treatment, medications, signed MD's orders			
Sewage and toilet facilities				(d) Qualification of nurses and attendants			
19. Sewage disposal satisfactory				(e) Narcotics on hand, dispensed			
20. Waterclosets—1 for 8 beds				46. Narcotics stored in locked cabinet			
21. Washbasins—1 for 8 beds				47. Drugs properly labelled in cabinet			
22. Shower or tub—1 for 12 beds				48. Adequate dressings and supplies			
23. Slop sink—each floor, and 1 for 24 beds				49. Recreation facilities for patients			
24. Bedpan sterilizer—1 for 24 bed patients				50. Occupational therapy provided			
25. Toilet room on each floor							

* Fire resistive construction generally required.   NOTE: CM indicates correction made.

REMARKS_____

_____

Date inspected _____. By _____

Figure 11–20 Nursing home inspection form.

Name _____ Location _____ T.V.C. _____

Principal _____ Supt. of Schools _____ Date _____

Grades ____ No. classrooms ____ No. boys ____ No. girls ____ Public ____ Private ____ Parochial ____ Boarding ____

Item	Yes	No	CM	Item	Yes	No	CM
1. Water system approved by HD				18. Thermometer at seat level, provided, which reads in winter at:			
2. One sanitary drinking fountain for every 75 children, OR				(a) 68–72°F in classrooms, auditoriums, offices, cafeterias			
3. One sanitary paper-cup dispenser for every 75 children				(b) 66–70°F in corridors, stairways, shops, laboratories, kitchen			
4. Water stored and dispensed in a sanitary manner				(c) 60–70°F in gymnasium			
5. One washbasin with warm and cold water for every 50 children				(d) 76–80°F in locker and shower room			
6. Soap, paper towels, and mirrors provided				(e) 80–86°F in swimming pool			
7. One toilet including tissue and partition for every 35–45 girls, and one toilet including tissue and one urinal for every 60 boys; separate				19. In nonheating season, when outdoor temperature reads 80°, 90°, 95°F, inside temperature reads 75°, 78°, 80°F, respectively			
8. Toilet and lavatory rooms convenient, clean, free of odors, ventilated, floors impervious and drained				20. Ventilation and heating satisfactory (10–30 ft³ per min per person), drafts and excessive heat prevented			
9. Sewage and excreta disposal satisfactory				21. Place provided to store clothes, lunch boxes, rubbers			
10. Buildings two or more stories, of fire-resistive construction				22. A separate adjustable and movable seat and desk available for each child			
11. Corridors, stairs, exits, doors marked and provide safe and ready escape from building in case of fire				23. Class room provides 20–25 ft²; per pupil			
12. Flammables stored in metal cans				24. Buildings, windows, rooms, chalk boards, lights, fixtures, corridors, walls, ceilings, etc., clean; grounds free of litter, insects, rodents, weeds, pools of water			
13. Fire extinguishers, sprinkler heads, fire hydrants, fire hose, fire alarm, fire escapes, panic bolts operable and tested every 6 mo				25. Swimming pool operated and maintained in conformance with sanitary code requirements			
14. Fire drills conducted, each floor emptied in 2 min				26. Lunch room or cafeteria operated and maintained in conformance with sanitary code requirements			
15. Window shades eliminate glare				27. Safety precautions taken in shops, laboratories, play area			
16. Seats in classrooms face away from window or lighting sources				28. Solid wastes properly stored and disposal satisfactory			
17. Natural and artificial light provide:				29. Air pollution prevented			
(a) 20 ft-c in classrooms, libraries, offices, shops, laboratories, gymnasium, pool				30. A competent person assigned responsibility to see that all environmental hygiene precautions are observed by teachers and students. (Incorporate in curricula.)			
(b) 30 ft-c in sewing, drafting, and arts and crafts rooms							
(c) 40 ft-c in sightsavings classrooms				NOTE: CM indicates correction made.			
(d) 10 ft-c in auditorium, assembly rooms, cafeterias							
(e) 5 ft-c in locker rooms, corridors, stairs, toilet rooms							

REMARKS _____

_____

Date inspected _____ By _____

Figure 11–21 School sanitation inspection form.

Item	S	U		S	U
Water supply			*Dietary*		
1. Quality meets drinking water standards			23. Food sources approved		
2. Quantity—yield, storage, pressure adequate			24. Refrigerator storage temperature, space, clean		
3. Operation, maintenance, and reports satisfactory			25. Dry storage clean, dry, space		
4. Qualified operator			26. Food preparation, handling, cooling proper		
5. On routine sampling schedule			27. Food service temperature and protection satisfactory		
Sewerage			28. Utensils and equipment type, condition, satisfactory		
6. Surface discharge approved			29. Dishwashing—dishes, utensils		
7. Groundwater discharge approved			30. Handwashing facil. adequate, convenient		
8. Collection system adequate					
9. Industrial wastewater separate treatment			*Milk production*		
10. Treatment meets stream standards			31. Dairy-farm operation meets USPHS code		
11. Operation, maintenance, and reports satisfactory			32. Milking procedures proper		
12. Qualified operator			33. Cooling and storage satisfactory		
			34. Pasteurization proper		
Air pollution control			35. Bottling, storage, distribution		
13. Incinerator emissions meet standards			*Ionizing radiation*		
14. Boiler emissions meet standards			36. Diagnostic X-ray units satisfactory		
15. Process emissions meet standards			37. Therapeutic X-ray units satisfactory		
16. Fuel composition and use acceptable			38. Teletherapy units satisfactory		
			39. Radioactive materials properly stored, used		
Solid wastes					
17. Garbage storage and collection satisfactory			*Housing*		
18. Refuse storage and collection satisfactory			40. Rooms clean, lighted, ventilated		
19. Disposal satisfactory			41. Fire escape from rooms		
			42. Adequate space for occupancy		
Swimming pool and bathing beach			43. Insect and rodent control		
20. Life-saving equipment and lifeguards adequate			44. Clean bedding		
21. Adequate clarity			45. Heating safe and adequate for intended use		
22. Adequate treatment			46. Fire protection adequate		
			47. Fire-resistive construction		

S means satisfactory equipment, construction, operation, maintenance.
U means unsatisfactory. Use available codes, rules, and regulations for compliance. Mark items NA if not applicable.
Supplied services _____ Water Supply _____ Sewerage _____ Refuse Collection _____ Dietary _____ Other _____.
Give name of contractor or supplier.

Figure 11-22 Abbreviated institution environmental health inspection form. (For comprehensive interpretations see pertinent sections of this text.)

port from consultants when indicated. It is good procedure to coordinate the inspection program with the work of other agencies and develop continuing liaison with the county medical society, accreditation groups, local nursing home association, local hospitals, fire department, building department, social welfare services, and others who may be involved. This can strengthen compliance and avoid embarrassment resulting from conflicting recommendations.

Special attention should be given to non-fire-resistive hospitals and nursing homes. Until such places can be replaced with fire-resistive structures, they should be protected against possible fires. This would include automatic sprinkler and alarm; horizontal and vertical fire-stopping of partitions; enclosure and protection of the boiler room; outside fire escape; fire doors in passageways, vertical openings and stairways; fire extinguishers; a fire-evacuation plan and drills; and the housing of nonambulatory patients on the ground floor.

Schools

Schools house the children of a community during their formative years. The health, welfare, safety, and morals of the children must be guarded. Available resources in health and education departments should be used fully in the best interest of the children. A school sanitation inspection report form is shown in Figure 11–21 to suggest the broad areas to be considered when making an inspection. Guidance as to what is considered satisfactory compliance can be found in this text under the appropriate headings and also in federal, state, and local publications.

Other Institutions

Other institutions would include colleges and universities, boarding schools, reformatories, prisons, welfare homes, mental institutions, and detention homes. The occupancy and supplied facilities and services should comply with basic environmental engineering and sanitation practice as given in this text. An abbreviated general institution inspection form is shown in Figure 11–22.

BIBLIOGRAPHY

APHA Program Area Committee on Housing and Health, *APHA-PHS Recommended Housing Maintenance and Occupancy Ordinance,* Subcommittee on Housing Regulations and Standards, U.S. Public Health Service Pub. No. 1935, Dept. of HEW, Washington, D.C., 1969.
APHA Program Area Committee on Housing and Health, "Basic Health Principles

of Housing and Its Environment," *Am. J. Public Health*, **59**, No. 5, 841–852 (May 1969).

ASHRAE Guide and Data Book, 1970 Systems Vol., 1969 Equipment Vol., 1968 Applications Vol., 1967 Handbook of Fundamentals, American Society of Heating, Refrigerating and Air-Conditioning Engineers, Inc., 345 East 47th Street, New York.

An Appraisal Method for Measuring the Quality of Housing, Part I. "Nature and Uses of the Method," 1945; Part II. "Appraisal of Dwelling Conditions," Vols. A, B, C, 1946; Part III. "Appraisal of Neighborhood Environment," 1950, American Public Health Association, 1015 Eighteenth Street, N.W., Washington, D.C.

Andrzejewski, A., Berjusov, K. G., Ganewatte, P., Hilbert, M. S., Karumaratne, W. A., Molner, J. G., Perockaja, A. S., and Senn, C. L., *Housing Programmes: The Role of Public Health Agencies*, Public Health Papers No. 25, WHO, 1964.

Appraisal of the Hygienic Quality of Housing and its Environment, WHO Tech. Rep. Ser., 353, Geneva, 1967.

Basic Housing Inspection, U.S. Public Health Service, Pub. No. 2123, Dept. of HEW, Washington, D.C., 1970.

Basic Principles of Healthful Housing, American Public Health Association, Committee on the Hygiene of Housing, Washington, D.C., 1946.

Environmental Engineering for the School, U.S. Public Health Service Bull. No. 856 and Office of Education Pub. No. OE 21014, Dept. of HEW, Washington, D.C., 1961.

Environmental Health Guide for Mobilehome Communities, prepared by U.S. Public Health Service, Dept. of HEW; published by Mobile Homes Manufacturers Association, 6650 North Northwest Highway, Chicago, Ill. Revised January 1971.

Environmental Health Guide for Mobilehome Communities, prepared by U.S. Public Health Service, Dept. of HEW; published by Mobile Home Manufacturers Association, 6650 North Northwest Highway, Chicago, Ill., January 1966. Revised January 1971.

Expert Committee on the Public Health Aspects of Housing, *WHO Tech. Rep. Ser.*, 225, Geneva, 1961.

General Standards of Construction and Equipment for Hospitals and Medical Facilities, U.S. Public Health Service Pub. No. 930-A-7, Dept. of HEW, Washington, D.C., December 1967.

Goromosov, M. S., *The Physiological Basis of Health Standards for Dwellings*, Public Health Papers No. 33, WHO, Geneva, 1968.

Housing an Aging Population, Committee on the Hygiene of Housing, American Public Health Association, Washington, D.C., 1953.

Housing Policy Report, Supplement to *House & Home* (January 1954).

Knittel, Robert E., *Organization of Community Groups in Support of the Planning Process and Code Enforcement Administration*, U.S. Public Health Service, Dept. of HEW, Washington, D.C., 1970.

Krasnowiecki, Jan, *Model State Housing Societies Law*, U.S. Public Health Service, Pub. No. 2025, Dept. of HEW, Washington, D.C., 1970.

Mobile Home Court Development Guide, U.S. Dept. of Housing and Urban Development, Washington, D.C., January, 1970.

National Electrical Code, ASA, C1-1962, 10 East 40th Street, New York, 1962.

National Plumbing Code, ASA A40.8-1955, American Society of Mechanical Engineers, 29 West 39th Street, New York, 1955.

"Neighborhood Environment," *Ohio's Health* (November–December 1969).

New Approaches to Housing Code Administration, National Commission on Urban Problems, Research Rep. No. 17, Washington, D.C., 1969.

Nursing Homes: Environmental Health Factors—A Syllabus, U.S. Public Health Service, Dept. of HEW, Washington, D.C., March 1963.

Parratt, Spencer, *Housing Code Administration and Enforcement,* U.S. Public Health Service, Pub. No. 1999, Dept. of HEW, Washington, D.C., 1970.

Principles for Healthful Rural Housing, American Public Health Association, Committee on the Hygiene of Housing, Washington, D.C., 1957.

Proceedings of The First Invitational Conference on Health Research and Its Environment, Airlie House, Warrenton, Va., March 17–19, 1970; sponsored by the American Public Health Association and the U.S. Public Health Service, Dept. of HEW, Washington, D.C.

Ramsey, C. G., and H. R. Sleeper, *Architectural Graphic Standards,* John Wiley & Sons, Inc., New York, 1970.

Rehabilitation Guide for Residential Properties, U.S. Dept. of Housing and Urban Development, Washington, D.C., January 1968.

Subcommittee of the Program Area Committee on Housing and Health, *Guide for Health Administrators in Housing Hygiene,* American Public Health Association, Washington, D.C., 1967.

Tenants' Rights: Legal Tools for Better Housing, Dept. of Housing and Urban Development, Washington, D.C., 1967.

Uniform Building Code, International Conference of Building Officials, Pasadena, Calif., 1961.

U.S. Senate, Subcommittee on Housing and Urban Affairs, Committee on Banking and Currency, *Congress and American Housing,* Washington, D.C., 1968.

Walton, Graham, *Institutional Sanitation,* U.S. Bureau of Prisons, Washington, D.C., June 1965.

Woodward, H. C. A., "The 1969 Housing Act, the Qualification Certificate," pp. 198–201, and W. Combey, "The Rehabilitation of Houses," pp. 202–207, *Royal Society of Health J.,* **90**, No. 4 (July/August 1970).

12

ADMINISTRATION

ORGANIZATION

Introduction

The administrative structure on the federal, state, and local levels of government for the provision of environmental control services to the people has been changing.[1] This has been brought about by public recognition of the environmental pollution problems that have not been resolved and the desire for a higher quality of life.

The development and implementation of an effective environmental protection program will require consideration of the scientific, social, political, and economic constraints involved. A balanced appraisal and adjudication of competing objectives are then necessary. The planning process briefly described below and in Chapter 2 will be found useful to gather and analyze the facts, make plans, establish priorities, and make decisions for action programs. A recent report of a World Health Organization Expert Committee will also be found informative.[2] It reviews present trends in environmental health, planning, organization and administration, international collaboration, and technical needs and problems requiring future action. The U.S. Public Health Service has

---

[1] "We tend to meet any new situation by reorganizing, and a wonderful method it can be for creating the illusion of progress while producing confusion, inefficiency and demoralization." Petronius Arbiter, 54 A.D. *Newsletter,* Sanitary Engineering Division, American Society of Civil Engineers, January 1971. Actually the major changes relate to governmental, fiscal, and enforcement policy, priority, and support.

[2] "National Environmental Health Programmes: Their Planning, Organization, and Administration," *WHO Tech. Rep. Ser.,* No. 439, Geneva, 1970.

also prepared a document to guide environmental health administrators and planners in the development of appropriate and effective programs.[3]

No one organizational structure is necessarily the best for a country, state, or local unit of government because the problems existing, purposes to be served, type of government, and methods considered acceptable will vary. However, the national structure will usually influence the state and local structure, and the state structure will influence local structure.

A distinction should be made between the actual provision and operation of a service, such as refuse collection or water supply, and the control function to ensure that the service is satisfactory and meets established laws, rules, regulations, and standards for the protection of the general health, safety, and welfare. The environmental control functions may be carried out on a national, state, or local basis or on some combined or shared basis, depending on the source of a potential problem and on how widespread its effects. For example, controls for food and drugs, surface waters, and air that move across state boundaries, would be federal functions; but if such movement is confined within a state, the controls could be state functions. A sharing of responsibilities coupled with uniform standards would eliminate duplication and make for more efficient use of manpower.

The federal, state, and local control agencies are not in competition with one another, but rather are complementary. Proper utilization of the available resources at each governmental level can make possible the provision of a more complete service to the people, usually at less cost. Failure to realize administrative understanding among official agencies, or between federal and state or state and local agencies, is probably due to lack of cooperation, understanding, confidence, and two-way communication.

State and Local Programs

The activities conducted by a state or local environmental control agency are determined by the responsibilities delegated by the state and by the local governing body. This is usually done through the public health law, environmental protection law, sanitary code, and local ordinances.

A state, city, or country department or division of environmental control can provide basic environmental protection services through an organizational structure of bureaus, sections, and units, depending on the

[3] *Environmental Health Planning,* U.S. Public Health Service Pub. No. 2120, Dept. of HEW, Washington, D.C., 1971.

population served. The bureau director would serve as program administrator, with assistance from section chiefs as consultants or activity specialists. He would plan, organize, direct, and supervise the activities for which he is responsible, see that the required job is done, and that quality is maintained. In a small department or division, the director and assistant director might share responsibility for a combination of programs and activities.

Where district or regional officies are maintained by a state or a large local organization, the director should be a broadly trained person. He would be expected to have administrative and technical ability and be qualified to oversee the review and approval of architectural and engineering plans. Where qualified people are not available locally, the technical, planning, engineering, and architectural problems would be referred to a central office staff.

The organizational plan should generally promote decentralization of responsibility to the most competent level capable of delivering generalized services directly to the people as the need and acceptance develop. Variations and modifications are necessary, depending on the local situation. It is known, however, that a good generalized program of inspection and supervision requires competent personnel, including a number of specialists, depending on the program complexities. The retention of competent personnel requires adequate salaries, a dynamic and challenging program, recognition of the dignity of the individual, and pleasant, stimulating working conditions.

Manpower

The work in environmental control is carried out by a diverse group in and outside of government. Those who have been carrying the major burden are the sanitary or environmental engineers, sanitarians, radiological health specialists, sanitary chemists, veterinarians, entomologists, public health inspectors, environmental health technicians, and various other specialists. This group is being supported by biologists, attorneys, economists, administrators, and many others having a professional or lay interest in conservation and in improvement of the quality of the environment.

Job descriptions and educational requirements for some of the personnel engaged in environmental engineering and sanitation activities may be useful and are quoted below.

Environmental or Sanitary Engineer

"The environmental engineer applies engineering principles to the prevention, control, and management of environmental factors that influence man's physical, mental, and social health and well-being."[4] The American Academy of Environmental Engineers expands on this by saying, "Environmental Engineering means the application of engineering principles and the practices to one or more elements of the environment for the purpose of protecting or improving man's health and well-being. It includes the control of the air, land and water resources, man's personal and working environment in relation to his health, social and economic well-being, and the design and maintenance of systems to support life in alien and hostile environments."[5]

Sanitarian

"The sanitarian applies his knowledge of the principles of physical, biological and social sciences to the improvement, control, and management of man's environment."[6]

The U.S. Department of Labor goes further in this description by adding, "Plans develops, and executes environmental health practices for schools and other groups. Determines and sets health and sanitation standards and enforces regulations concerned with food processing and serving, collection and disposal of solid wastes, sewage treatment and disposal, plumbing, vector control, recreational areas, hospitals, and other institutions, noise, ventilation, air pollution, radiation and other areas. Confers with government, community, industrial, civil defense, and private organizations to interpret and promote environmental health programs. Collaborates with other health personnel in epidemiological investigations and control. Advises civic and other officials in development of environmental health laws and regulations."[7]

Program specialists in environmental engineering and sanitation include, among others, personnel engaged in industrial hygiene, radiation protection, air pollution control, solid wastes management, water pollution control, water supply, hospital engineering and sanitation, vector control, milk and food technlogy, research and related activities.

The minimum educational requirement for environmental engineers, sanitarians, and other environmental specialists is the baccalaureate degree. However, the trend is toward a requirement of graduate education in one of the basic disciplines or in an area of categorical program

[4] *Health Resources Statistics—Health Manpower and Facilities, 1969.* National Center for Health Statistics, U.S. Dept. of HEW, Rockville, Md. 20852, May 1970.
[5] *The Diplomate,* Environmental Engineering Intersociety Board, Inc., Washington, D.C. 20016, December 1967.
[6] *Health Resources Statistics,* op. cit.
[7] *Dictionary of Occupational Titles,* Vol. 1, 3rd ed.—Professional 079.118, U.S. Dept. of Labor, Washington, D.C., 1965.

specialization. In several basic disciplines the qualifying professional degree is the doctorate. The sanitarian's baccalaureate degree is usually with a major in environmental health or in the physical or biological sciences.[8]

Environmental Technician

"The environmental technician assists professional personnel in carrying out various elements of prevention and control programs, including inspections, surveys, investigations, and evaluations to determine compliance with laws and regulations. He obtains appropriate samples of air, food and water, and assists in performing tests to determine the quality of these samples. He assists in operating water purification, waste water treatment plants, and solid waste disposal facilities."[9]

Environmental Aide

"The environmental aide assists professional personnel and technicians in carrying out various elements of prevention, control and service programs. He performs routine tasks under supervision."[10]

The minimum appropriate educational requirement for the environmental technician is an associate degree in environmental health, radiologic technology, or related areas of specialization. A number of junior colleges and technical institutes offer technical training in environmental health or similar areas. The environmental aide normally is a high school graduate with varying amounts of appropriate short-course training in specialized subjects.[11]

Salaries and Salary Surveys

Competent direction is essential to a divisional program's effectiveness. As has been well established in private business, it is false economy to pay inadequate or marginal salaries. The money spent for an entire program is in jeopardy of being totally misdirected and hence wasted when placed in the hands of a mediocre person. If such a person must be "led by the hand" or continually checked, other more valuable time is being misdirected. It is fundamental that high salaries over a period of time attract the most competent people.

Surveys of professional positions by "experts" for the purpose of establishing equitable salaries are fraught with danger. In some cases it has been found politically expedient to employ a consultant to make a salary

[8] *Health Resources Statistics,* op. cit.
[9] Ibid.
[10] Ibid.
[11] Ibid.

survey and thus shift the blame for questionable decisions to an outsider. Since the greatest number of people are employed in the lower-salaried nonprofessional positions, which also have counterparts in private industry and which also represent a larger number of votes, the work done is more clearly appreciated and more likely to be acknowledged by upward salary adjustment. But the professional administrative positions are relatively few; the value of the work done is not understood unless the expert is well acquainted with environmental engineering and sanitation practices. It is more probable therefore that professional people will suffer rather than gain from a salary survey by experts. Every effort should therefore be made to separate salary surveys into professional and nonprofessional groups. Make certain that professional salaries are surveyed by people expert in environmental control programs.

ENVIRONMENTAL CONTROL PROGRAM PLANNING

In order to plan properly for an environmental control program it is important to know what the people want, what they need, what they can afford, and what is possible. The answers to these questions can evolve out of the planning process, an example of which is outlined below.

Statement of Goals and Objectives[12]

This statement should recognize the environmental quality aspirations of the community's. These might include decent housing, an adequate and safe water supply, clean air, proper sewage collection and disposal, proper solid waste collection and disposal, adequate and safe parks and recreation facilities, clean food service establishments, elimination of mosquitoes and rats, and other environmental quality matters of pressing concern to the people.

Data Collection, Studies, and Analyses

This phase involves the systematic collection of basic demographic, land-use, and economic data and their projection; research, study, and analysis of the data; the community institutions and government. Included would be sociological factors, characteristics and needs of the people, their expectations and attitudes, The political structure, laws, codes and ordinances, the tax structure, departmental budgets, and special background reports that may be pertinent should also be considered.

Basic data collection should identify and generally define both favor-

---

[12] See Chapter 2.

able and unfavorable natural and man-made environmental conditions and factors. These would provide background for the more in-depth environmental health and engineering study to follow and would include such matters as (1) weather and climate; (2) topography, hydrology, flooding, tidal effects, seismology, soils characteristics, and drainage; (3) natural and man-made air, land and water pollution; (4) flora and fauna of the area; (5) noise, vibrations, and unsightly conditions; and (6) condition of housing, community facilities, and utilities.

The data collection and analyses should also take into consideration factors that may influence the plan being prepared. This step would include evaluation of the effects of scientific and technological development; the probable increase in liquid, solid, and gaseous wastes; federal and state grants; governmental and private planning and construction programs; and availability of manpower.

This part of the study should probe in some depth those environmental factors that have or may have a direct or indirect effect on man's physical, mental, and social well-being both as an individual and as a member of a community. Epidemiological surveys and investigation of the adequacy of water supply, sewerage, refuse collection and disposal, recreation facilities, housing, and the many other environmental factors listed in Chapter 2 would be included.

Upon completion of the studies and analyses, the goals and objectives previously established should be reevaluated and revised. Feasible objectives should be more clearly defined within the broad goals established. It is also important to recognize and coordinate with other federal, state, local, and agency program plans and make indicated adjustments.

Environmental Control Program Plan

The information obtained above will usually reveal many weaknesses in need of attention through the implementation of an environmental control plan. Such a plan would identify those factors subject to environmental control and those factors requiring construction or rehabilitation of a facility, such as a water system, in order to meet a desired objective. This calls for the presentation and study of alternatives and consideration of the advantages and disadvantages of each. The establishment of priorities and committment of resources to achieve the agreed upon ends require political confirmation and support. At the same time essential services and facilities are maintained, recommendations are made, and support is given to the agencies responsible for the design, construction, and operation of needed environmental control facilities and services.

The factors or functions subject to control include all or most of those mentioned in Chapter 2 and those proposed by a WHO Expert Committee and quoted below.[13] The scope of an environmental control program to carry out the established goals and objectives would encompass:

1. Water supplies, with special reference to the provision of adequate quantities of safe water that are readily accessible to the user, and to the planning, design, management, and sanitary surveillance of community water supplies, giving due consideration to other essential uses of water resources.

2. Wastewater treatment and water pollution control, including the collection, treatment, and disposal of domestic sewage and other water-borne wastes, and the control of the quality of surface water (including the sea) and groundwater.

3 Solid waste management, including sanitary handling and disposal.

4. Vector control, including the control of arthropods, mollusks, rodents, and other alternative hosts of disease.

5. Prevention or control of soil pollution by human excreta and by substances detrimental to human, animal, or plant life.

6. Food hygiene, including milk hygiene.

7. Control of air pollution.

8. Radiation control.

9. Occupational health, in particular the control of physical, chemical, and biological hazards.

10. Noise control.

11. Housing and its immediate environment, in particular the public health aspects of residential, public, and institutional buildings.

12. Urban and regional planning.

13. Environmental health aspects of air, sea, or land transport.

14. Accident prevention.

15. Public recreation and tourism, in particular the environmental health aspects of public beaches, swimming pools, camping sites, and so forth.

16. Sanitation measures associated with epidemics, emergencies, disasters, and migrations of populations.

17. Preventive measures required to ensure that the general environment is free from risk to health.

The American Public Health Association Committee on Environment proposed the following program areas and also the planning considerations and methods to implement programs.[14]

I. *Environmental Program Area*
Wastes
  a. Air
  b. Sewage and liquid
  c. Solid
Water supply
Housing and residential environment

---

[13] "National Environmental Health Programmes," op. cit.
[14] "Environmental Factors in Health Planning," *Am. J. Public Health,* **58,** No. 2, 358–361 (February 1968).

Food and drugs
Radiation
Noise
Accidents
Occupational and institutional hazards
Vectors
Recreation
II. *Planning Considerations*
Health
Economic
Demographic and land use
Social
Asthetic
Resource conservation
(Also manpower, facilities and services)
III. *Methods and Techniques*
Research
Demonstration
Education
Standards
Legislation
Inspection
Enforcement
Planning
Evaluation
Incentives
Systems Analysis

A program plan should then be developed for each of the programs. The plan should list for each program the following:

1. *Need for the program*—as determined by analyses of the data collected, trends in technology and social development.

2. *Objectives*—long term (5 to 20 years), and short-term (1 to 5 years), with the specific tasks or change to be accomplished each year for the first five years.

3. *Law*—applicable public health law, sanitary code, rules and regulations. Authority and responsibility.

4. *Work load and methods to meet objectives*—program size and activities to be performed by surveys, inspections, technical consultation, sampling, engineering and architectural plan review, certification, permit, education, legal action, surveillance, conference, evaluation. Develop program size indicators. Indicate industry participation.

5. *Resource requirements*—manpower and data needs, coding and electronic data processing support, standards development, registry categorizing types of places, consultant services, support from other agencies, laboratory support, research and development.

6. *Evaluation*—effectiveness measures, that is, percent of places, facilities and samples meeting objectives, regulations or standards; reduction in violations, unsatisfactory conditions, complaints; reduction in morbidity and mortality; other tangible and intangible accomplishments.

7. *Financing*—estimated funding requirements, sources of funds (federal, state, local, grants, other). Capital construction needs (regular, first instance).

Administration

The implementation of a control program requires a management organization to administer a total environmental control program. This would include professional direction and management support, such as fiscal, legal, personnel, planning, education and public information, research, overall monitoring, surveillance, and evaluation. Administration should also have the competence and authority to develop staff support to make changes and redirect program efforts as needed.

PROGRAM SUPERVISION

Evaluation

Objective evaluation of the environmental engineering and sanitation program conducted by state and local agencies serves many purposes. It can provide the basis for integrating, adjusting, and balancing the program; it can be used to demonstrate the need for obtaining and retaining competent personnel; it aids the administrator of the program, and the supervisor of one or a group of activities, to determine whether available personnel are being utilized to do the work considered most important. In addition, it can provide facts for supporting program recommendations and policy determinations, and it has been found effective in showing which programs need more inspection time, which are receiving too much inspection time, and which are not producing results.[15]

To be of value evaluation studies should consider work load, work done, quality of the work, and its effectiveness. The data assembled must of course be reliable, and they must be interpreted in the light of the thoroughness and competence of the inspections. Evaluation should start in the planning stage, before a program gets started. Goals are set, plans are made to reach feasible objectives within the goals, yardsticks or measurement are established, and analyses or studies are made to see how close one has come to the goals and what changes may be indicated. Figures 12–1 and 12–2 illustrate program evaluation techniques.

Reporting

The reporting of work done is a problem that must be realistically faced by every organization. Field personnel usually consider this an

[15] J. A. Salvato, Jr., "Evaluation of Sanitation Programs in a City–County Health Department," *Public Health Rep.*, **68**, No. 6, 595–599 (June 1953).

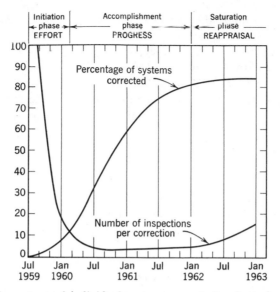

Figure 12–1 Percentage of individual systems corrected and number of inspections per correction in a community sewerage survey.

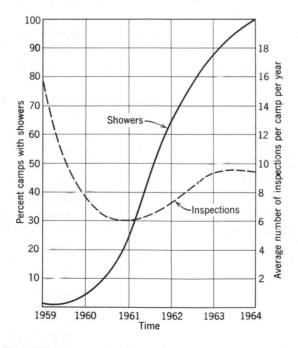

Figure 12–2 Farm-labor camps provided with showers, a planned program.

unnecessary chore, probably because the reason for keeping the records is not understood. It is important therefore that field personnel fully appreciate that their status, salary, and very existence are intimately tied up with accurate reporting of field visits and other work accomplished. The organization's monthly, quarterly, or annual report can be no more accurate than the reporting used. It is essential, therefore, that the system in effect be as simple as possible to give the desired information.

Time Sheet

A time sheet can serve practically all the reporting needs of an organization for statistical purposes. It can be greatly simplified by using a coding system, thereby adapting it for a mechanical record-keeping system. A form the time sheet can take is shown in Figure 12–3.

Coding

Personnel should all be given a code number such as 01, 02, 03, and so on. In setting up a code system it is suggested that numbers be allotted for future personnel. The date, hours worked, time in office and time in field, on vacation, on sick leave are self-explanatory. Under "Field" the field man writes the name of each premise visited. The time spent on each assignment including travel, is also noted.

Location is coded by giving a code number to each township, village, city, or district in the department jurisdiction. With this information in hand the field man simply puts down the code 01, 12, or whatever the case may be.

The reason for a field visit is of interest in determining how much of the work done is as a result of planning to carry out the sanitary code responsibility. This is a routine inspection. How much time is spent as the result of requests by the public for service, such as requests for assistance in the design of a private sewage disposal system; for examination of a sample of water from a school, private home, or camp; or for help at a pasteurizing plant or restaurant would be service requests. Then there is the fieldwork that is necessary because of a garbage complaint, overflowing cesspool, dirty lot, standing water, barking dog, pigeons, and so on. These inspections would be made by reason of a reported nuisance or complaint. Therefore the "Reason" for a field visit and its coding would be as follows:

Reason	Code
Routine inspection	01
Service request	02
Complaint or nuisance	03

Name _____ Code _____ Date _____

Hours worked: Office ____ Field ____ On vacation ____ On sick leave ____ In train. ____

| Name of Place | Field Activity | | | | | | Time | Samples | |
	Location	Reason	Type	Kind of Visit	Finding	Action	Hours	Number	Type
FIELD									

OFFICE　　　　　　　　*REMARKS* (Names, subject, training, activity)

Telephone Conferences _____

Office Conferences _____

Plans reviewed _____

Plans approved _____

Figure 12–3 Division of environmental control—time report.

The next items shown on the time report form are listed under "Field Activity." The "Type" of activities carried on depends on the extent of the division program. A rather complete listing is given in Table 12–1, together with a suggested coding. A more extensive breakdown may be desired.

Following "Type" on the time report is given "Kind of Visit." This

TABLE 12-1 A DIVISION OF ENVIRONMENTAL CONTROL ACTIVITY CODING

Activity Type	Code	Activity Type	Code
Air sanitation, public	01	Refuse—garbage, rubbish, ashes	47
Industrial	02	Vacant lot—weeds, drainage	48
Private	03	Insanitary housing—premises	49
Bathing—beaches	04	Other	50
Swimming pools	05	Parks and recreation—state	51
Carnivals and fairgrounds	06	Municipal	52
Emergency sanitation, civil		Private	53
defense	07	Places of public assembly	54
Food—eating places	08	Rabies control	55
Processing establishments*	09	Realty subdivisions—private water	
Distributing and vending†	10	supply and sewerage	56
Abattoirs—meat	11	Public water	57
Abattoirs—poultry	12	Public sewer	58
Other, specify	13	Public water and sewerage	59
Hospitals	14	Refuse—incinerators	60
Nursing homes	15	Sanitary fills	61
Convalescent homes	16	Open dumps	62
Other medical care facilities	17	Scavengers, swill, vehicles	63
Housing—multiple family	18	Other	64
Hotels, rooming houses	19	Schools—public	65
One- and two-family	20	Private	66
Mobile homes	21	Sewage and wastes—municipal	67
Other, specify	22	Industrial	68
Industrial and business places	23	Private dwelling	69
Insect, rodent, and vermin		Other	70
control	24	Stream pollution	71
Institutions—state	25	Temporary residences—day camps	72
County	26	Children's camps	73
City	27	Motels, hotels, tourist courts	74
Private	28	Trailer Parks	75
Ionizing radiation—medical‡	29	Hotels, lodging, boarding houses	76
Dental	30	Mobile-home parks	77
Chiropractic, veterinarian	31	Migrant-labor camps	78
Research, radioactive materials	32	Training, given	79
Milk—dairy farms, can	33	Received	80
Dairy farms, vat	34	Water supply—public, municipal	81
Dairy farms, bulk, machine	35	Quasi-public	82
Dairy farms, bulk, pipeline	36	Cross-connection	83
Transportation tank truck	37	Watering points—land, sea, air	84
Pasteurizing plants, HTST	38	Factory, industrial plant	85
Pasteurizing plants, vat	39	Private dwelling	86
Milk processing, other	40	Bottled water	87
Quality control supervision	41	Public springs or wells	88
Nuisances—water supply-pressure,		Weed control	89
taste, odor, appearance	42	Algae control	90
Sewage—plumbing, leaks,		Special survey, specify	91
clogging	43	Non-ionizing radiation	92
Air pollution, noise, odors	44		93
Animals—dogs, cats, pigeons	45		94
Vermin—insects and rodents	46		95

*Includes bakery, delicatessen, caterer, frozen meal preparation.
†Includes vending machine, supermarket.
‡Includes radiographic, fluoroscopic, therapeutic, and teletherapeutic machines.

is meant to show the kind of visit made, which can be tabulated as follows:

Kind of visit	Code
Initial inspection of year	01
Reinspection	02
Inspection during construction	03
Conference in field	04
Person not at home	05
Collection of sample only	06

"Persons not at home" incidencts can be reduced by making telephone appointments.

The listing under "Finding" indicates the status of the place, which can be reported as shown below. However, if the visit was actually a field conference or if no one was at home, there would be no finding.

Finding[16]	Code
Substantially satisfactory	01
Unsatisfactory	02
Continuing unsatisfactory	03
All corrections substantially made	04
System installed (private sewerage)	05

The listing under "Action" would indicate the steps taken in the field or in the office to implement the inspection. If no one was at home, there would normally be no action reported. The coding under this heading is:

Action[17]	Code
No cause for action	01
Written notice issued by inspector	02
Design prepared (private sewerage)	03
Excluded (dairy farm)	04
Reinstated (dairy farm)	05
Written notice by office	06
(Rough-draft letter attached to time report)	

The number of samples collected and plans reviewed are coded by type, as shown below. The listings can be reduced or added to depending on local conditions and the desired information.

[16] Every effort should be made to cause an inspection and report to be made while the man is in the field, rather than have a field conference which usually results in a lesser progress. On the other hand, some field conferences can be very effective.
[17] Ibid.

Sample Type	Code
Water supply—municipal, public	01
private dwellings	02
camps, restaurants, hotels, schools, bottled, milk plants, etc.	03
Sewage	04
Industrial wastes	05
Stream pollution	06
Swabs, restaurants	07
Milk—pasteurizing plants	08
street, route	09
farms, can	10
farms, bulk	11
deck test, temperature	12
deck test, MWT, CMT[18]	13
deck test, sediment	14
baby formulae	15
Cream—pasteurized	16
Rinses—cans, bottles	17
Swimming pools or beaches	18
Air—high volume	19
fallout	20
pollen	21
stack	22
materials' degradation	23
Radiologic—water	24
air	25
food	26
Other, specify	27
Plan Type	Code
Subdivision	01
Water supply and treatment	02
Sewerage and treatment	03
Swimming pool	04
Housing	05
Restaurant	06
Other, specify	07

Record Keeping and Data Processing

The time report should be completed and submitted daily to the supervisor or director. Attached should be an inspection report form, rough-draft letter, or memorandum covering each field visit. This material is reviewed each day by a supervisor for completeness and indicated quality of the visits. The time report is then referred to the clerical section for posting and statistical recording. Daily posting consists of marking on a visible card, for each establishment or place under routine

[18] MWT and CMT are chemical screening tests.

supervision, the date of inspection and the finding as satisfactory or unsatisfactory, by means of a green tab for satisfactory or red tab for unsatisfactory. Other records kept on a daily basis would include the results of water and milk examinations and such other information as may be indicated or desired for special study purposes.

Records for statistical recording can be kept by a manual or mechanical system, depending on the volume and resources available. Since each field visit requires the processing of one card or form, unless a hand tally is maintained, the number by the end of each reporting period can become quite large and will determine the need for a mechanical punching and sorting system. Local representatives of large data processing corporations are available who can explain adaptation of the coding system to their particular operation. A marginal punch card is shown in Figure 12–4. Special forms can be developed to fit the system

0 0 0 0 units	0 0 0 0 units	0 0 tens	0 0 0 0 units	0 0 0 units	0 0 0 0 tens	0 0 0 0 units	0 0 0	
1 2 4 7	1 2 4 7	1 2	1 2 4 7	1 2 4	1 2 4 7	1 2 4 7	1 2 4	
Personnel	Month		Location	Reason		Type	Visit	

	Samples		Time in			
Finding	Action	Number	Type	Office	Field	Office activity
1 2 4	1 2 4	1 2 4	1 2 4 7	1 2	1 2 4 7	1 2 4 7	1 2 4 7	1 2 4 7
units 0 0 0	units 0 0 0	tens 0 0 0	units 0 0 0 0	tens 0 0	units 0 0 0 0	units 0 0 0 0	units 0 0 0 0	units 0 0 0 0

Figure 12–4 Punch card for statistical recording. (Reduced.)

used, and the number of entries can be adapted to satisfy particular needs.

It will soon become apparent to the person using a punch-card system that a tremendous amount of information is available. The combinations possible, from the basic items punched, can give more data than can possibly be used. IBM and Remington Rand are mechanical systems, McBee is manual. The forms can be adapted as inspection reports if desired. The McBee system is suitable when the maximum number of cards per sorting does not exceed 500. When large numbers of inspections are involved, electronic data processing should be used for data recording and evaluation. Data collection forms and types should be designed with the help of specialists.

A visible card file, or its equivalent, is an indispensable administrative aid. The system can be as detailed or as simple as desired. If a file is kept on each establishment or facility under supervision, the visible card file can be very simple. If information is being assembled for a special study, a detailed breakdown may be justified. *It is essential to keep in mind the use to which the information is to be put* before asking the record-processing unit and clerical staff to spend the many tedious hours required to keep records month after month or year after year. A few accurate figures are more useful than a mass of figures of questionable accuracy.

A simple visible card file, which should be kept current on an annual basis, three-year basis, or such interval as is practical depending on the personnel, number of places, and time available to carry out the activities comprising the total program, is illustrated in Figure 12–5.

Name of place__________ Owner__________

Address__________ Person in charge__________

Capacity__________ Special features__________

(Laboratory results over or in premises file)

Dates of inspection__________

(5″ x 8″ card)

No.
CTV* Name | Permit | J | F | M | A | M | J | J | A | S | O | N | D

*Street number, district, city, village or town.

Figure 12–5 Visible file card. (Reduced.)

The city, village, town, district, or street number is placed in the lower left-hand corner, followed by the name of the place. The card does not leave the office.

A colored tab is placed over "Permit" to show that a permit has been issued, if required. Different colors can be used to indicate a temporary or conditional permit. Another colored tab is placed over the month an inspection or reinspection is made; a standard color such as green can be used to designate a place found in substantial compliance and a color such as red can be used for one found deficient. Attempts to list degrees of deficiency become involved, time consuming, and confusing. If several hundred or thousand places are being supervised under an activity, it is possible, by pulling out a visible card-file drawer, to see at a glance places that have not yet been inspected, places operating without any permit or with a conditional permit, the month a place was last inspected, and the places that have substantial deficiencies. In this way it is possible in a very short time to analyze all or several activities from an administrative or supervisory point of view and determine where time and effort should be directed.

Where a data processing system is used, a program can be prepared to print out the desired information, provided it has been coded and recorded.

Statistical Report

The preparation of a monthly, quarterly, or annual statistical report to summarize the work of a division or department is a common requirement. It is ideal if all the desired information can be obtained automatically at the end of each reporting period. This goal can be approached with the adoption of a mechanical, manual, or combination record-keeping system, depending on the resources available. Since it is not possible to reduce adequately all the work done to figures, and since such information usually needs interpretation, it is customary to accompany statistical data with a narrative report. Statistical report forms that have been used are shown in Tables 12-2 and 12-3.

Work Load and Program Supervision

The workload of a division or department is the total program and miscellaneous activities for which it is responsible. The program must take into consideration the responsibilities legally delegated, the demands of public officials and the people, and the public health needs and promotional work to be done.

The sum of the "places on record" (Table 12-3) for each activity is an indication of the work load in connection with places for which the department has a routine responsibility. The total work load is determined by adding to this total the number of special program services provided such as talks given, requests for service, sampling programs, plans reviewed and approved, conferences, educational efforts, planning, evaluation, and so forth.

The "Inspected" places show what part of the work has been done

TABLE 12-2   DIVISION OF ENVIRONMENTAL CONTROL— ACTIVITIES SUMMARY

Name of unit _____ From _____ To_____

Item	Number
1 Places on record	_____
2 Places inspected	_____
3 Number of inspections	_____
4 Number of field conferences	_____
5 No one at home	_____
6 Samples only collected in field	_____
7 Number of office conferences	_____
8 Number of telephone conferences	_____
9 Visits, routine	_____
10 Visits, service requests	_____
11 Visits, complaints, nuisances	_____
12 Number of inspections, substantially satisfactory	_____
13 Number of inspections, unsatisfactory	_____
14 Number of inspections, continuing unsatisfactory	_____
15 Number of inspections, corrections substantially made	_____
16 Number of private sewage-disposal systems installed	_____
17 No cause for action	_____
18 Notices issued by inspector	_____
19 Designs prepared, private sewerage	_____
20 Written notices or letters by office	_____
21 Samples collected and interpreted, total	_____
22 Water—municipal, public	_____
23 Private dwellings	_____
24 Camps, restaurants, hotels, schools, bottled, milk plants, parks	_____
25 Sewage	_____
26 Industrial wastes	_____
27 Stream	_____
28 Swabs	_____
29 Pasteurizing plants	_____
30 Street, route	_____
31 Milk, can	_____
32 Bulk	_____
33 Deck temperature	_____
34 Deck sediment	_____
35 Deck, MWT, CMT	_____
36 Cream	_____
37 Baby formula	_____
38 Rinses	_____
39 Swimming pool	_____
40 Bathing beaches	_____
41 Radiologic—water	_____
42 Air	_____

TABLE 12–2   (cont'd)

Item		Numbers
43 Food		_____
44 Air–high volume		_____
45 Fallout		_____
46 Pollen		_____
47 Stack		_____
48 Materials degradation		_____
49 Subdivision plans reviewed _____	approved	_____
50 Water-supply plans reviewed _____	approved	_____
51 Sewerage plans reviewed _____	approved	_____
52 Swimming-pool plans reviewed _____	approved	_____
53 Housing plans reviewed _____	approved	_____
54 Restaurant plans reviewed _____	approved	_____
55 Other plans reviewed _____	approved	_____
56 Engineer man-days in office _____	field	_____
57 Sanitarian man-days in office _____	field	_____
58 Sanitary inspector man-days in office _____	field	_____
59 Total man-days at work, corrected for vacation, sick leave, training, etc. _____		
60 Total man-days time in field _____	percent	_____
61 Total man-days time in office _____	percent	_____
62 Total weighted units of work performed _____		
63 Weighted units of work per man-day _____		

and what part remains to be done if each place is to receive an annual inspection. If an activity includes a large number of premises, such as multiple dwellings, restaurants, or dairy farms, and personnel are not available or not desired to make annual inspections, this activity may be set up as a 2- or 3-year goal. This information is very valuable in planning future work. For example, the multiple dwelling tabulation in Table 12–4 shows that less than half the structures on record were inspected by the end of the year. If all structures are to be visited annually, more emphasis would have to be placed on the inspection of this activity. Actually, because of the limited staff available and because divisions of environmental control rarely have the necessary staff to do all of the work that needs to be done, this particular activity was planned as a 3-year program. This had to be done in spite of the legislated responsibility that required that all multiple dwellings be inspected three times a year, if it were done at all.

The "Total No. Inspections Made" (Table 12–3) tells where inspection time is being spent and can be used to determine the average number of inspections each place has received. This information may indicate the need for redirecting inspection time or for greater field supervision.

TABLE 12-3   DIVISION OF ENVIRONMENTAL CONTROL PROGRAM EVALUATION

(See Table 12–1 for complete activity coding.)

Program Evaluation

From _____ To _____

Code	Activity	Number of Places								
		On Record	Inspected	Total Number Inspections Made	Satisfactory	Unsatisfactory	Continuing Unsatisfactory	Substantially Corrected	Field Conference
01	Air sanitation, public								
02	Industrial								
03	Private								
04	Bathing—beaches								
05	Swimming pools								
06	Carnivals and fairgrounds								
07	Emergency sanitation								
08	Food—eating places								
09	Processing establishments								
10	Distributing and vending								

TABLE 12-4   SAMPLE TABULATIONS FROM THE STATISTICAL
SUMMARY FORM USED FOR RECORDING INSPECTION DATA

Activity	Fourth Quarter	Total, Year
1. Multiple dwellings—3 or more families:		
(a) Number of places on record	7,083	7,083
(b) Number inspected of total on record—original for year	319	3,180
(c) Number of inspections made, original and reinspection	2,284	9,391
(d) Number of places unsatisfactory	316	1,653
(e) Number of places substantially corrected	350	1,079
2. Camps—recreational, trailer, tourist:		
(a) Number of places on record	129	129
(b) Number inspected of total on record—original for year	10	129
(c) Number of inspections made, original and reinspection	93	715
(d) Number of places unsatisfactory	19	106
(e) Number of places substantially corrected	2	83
(f) Number of permits issued	9	98

It must of course be considered in relation to the problems, the needs, and the results produced. In the camp tabulation (Table 12–4), for example, all places were inspected during the year about 5½ times. This figure might be considered high, but the high percentage of places eliminating all deficiencies or "substantially corrected" tends to confirm that the time was well spent. This could be investigated further to determine specifically what type of corrections were made. In addition, by looking at the visible card file, if one is maintained, one can direct subsequent inspections the following year first to those places having deficiencies and thus reduce the total number of inspections.

The number of places "Satisfactory," "Unsatisfactory," "Continuing Unsatisfactory," and "Substantially Corrected" (Table 12–3) show the progress being made and the condition of the places under supervision. If a place removes all deficiencies and on subsequent inspection is found to have slipped back, it would again be listed as "Unsatisfactory." When progress is not being made, additional inspection, supervision, or review of the program may be indicated. The key may be lack of direct supervision, poor quality of supervision, departmental policy, lack of staff promotional opportunities, need for in-service training or additional public health education, or poor morale. Interpretation or definition of unsatisfactory deficiencies, of course, may vary with the individual. The variations can be minimized with adequate in-service training, by the employment and retention of competent personnel, and by the development and use of compliance guides. See pages 672 to 685.

The "Number of permits issued" (Table 12–4) may be added to show those places under supervision that, on inspection, were found in compliance with existing regulations. This assumes that annual permits are issued only after an inspection report of satisfactory compliance is received.

When places "substantially corrected" are compared with the number of "inspections," an indication is obtained of the number of inspections per place corrected. The point of diminishing returns can thus be seen, as illustrated in Figures 12–1 and 12–2. It then becomes an administrative decision to determine whether a new approach should be used and the effect of this approach on public relations and on the future program.

What is an Inspection?

There appears to be a need to clarify what is meant by an "inspection" to ensure that inspections are meaningful, effective and adequate. In general it can be said that there are three types of inspections: (1) initial or complete inspections, (2) routine inspections (3) reinspections. Each is discussed separately below.

Initial or Complete Inspection

This is a detailed type of inspection, which should be made when a property or establishment is first brought under department control to determine compliance with the law, code, rules and regulations. The inspection is very comprehensive and includes the collection of basic data and information concerning ownership, operation, physical conditions, and equipment and facilities at the establishment. Sketch plot plans, floor plans, flow diagrams, equipment, and structure layout are included. See Figures 4–4, 4–29 to 4–32, 8–11, 9–6, and 9–9. Inasmuch as conditions do not remain static, it is advisable to have an initial or complete inspection made every three to five years and when major changes are made.

Routine Inspection

Routine inspections are usually complaint investigations or annual inspections of establishments based on a field inspection form to determine compliance with the law, state or local sanitary code or rules and regulations. All items on the inspection form should be completed and deficiencies explained at the bottom of the form or on the back under "remarks." Experience shows that if the written explanation of the violation is followed by a recommendation as to how the violation can be corrected, while the inspector is at the site, more practical recommendations are made. This can also serve as the basis for informal discussions with the operator of the establishment, thereby setting the basis for immediate correction of the violation. In any case, significant violations noted should be confirmed by letter with a timetable for correction as previously agreed upon with the operator at the time of the inspection. The report and letter involving important violations and recommendations should be reviewed by the program director or supervisor before the letter is sent out. Figures 8–6 to 8–10, 9–10, 9–20, 11–3, and 11–19 to 11–22 are examples of routine inspection forms.

Reinspection

A reinspection would be made as a follow-up on a routine inspection, particularly when significant violations are noted, to determine whether the needed corrections have been made. When a reinspection is made, the inspection report form should be completed in its entirety. In other words, all items should be inspected and not only those that were found unsatisfactory at the time of a previous inspection. A partial reinspection might be indicated when only one or two specific items are to be re-checked and other conditions are not likely to change since the last reinspection. It is emphasized, however, that in general all items on the inspection report form should be reviewed and not only those that were found unsatisfactory. This will take relatively little additional time and will ensure that deficiencies overlooked at the time of the routine inspection are observed. It will also leave the impression that a thorough inspection has been made.

There may be exceptions to the above procedures based on special circumstances. However, a uniform policy should be followed by the field offices to help ensure that thorough, competent inspections are being made and that the requirements of the law, sanitary code, and rules and regulations are being interpreted in a reasonably uniform manner.

Right of Inspection

Decisions of the U.S. Supreme Court have reaffirmed the right under Article IV of the Constitution of the United States, that "The right of the people to be secure in their persons, houses, papers, and effects, against unreasonable searches and seizures shall not be violated, and no warrants shall issue but upon reasonable cause, supported by oath or affirmation, and particularly describing the place to be searched and the person or things to be seized."[19] The conduct of sanitation, housing,

---

[19] *Camara* v. *Muncipal Court of the City and County of San Francisco*, 387 U.S. 541 (1967), and *See* v. *City of Seattle*, 387 U.S. 541 (1967).

and safety inspections of nonpublic areas can be made by consent specifically granted for the intended purpose. If not granted, an inspection can only be forced under an issued search or inspection warrant or equivalent court order. Any conditions observed without specific authority cannot be used as a basis for the swearing out of a search warrant. States lacking the legal authority for the issuance of a search warrant, other than in connection with criminal proceedings, need to amend state law to make possible inspections required by law and still safeguard the fundamental right of the people.

Frequency of Inspection

In the interest of professional program direction, no arbitrary frequency of inspection should be followed unless required by law or special agreement. The emphasis should be on getting proper operation rather than on the number of inspections per se.

The effectiveness of a program is largely a function of the technical competence of the assigned personnel, the quality of work done, and the direction it receives, rather than the number of inspections made per establishment. The success of a program will depend largely on the extent to which the owner, operator, manager, and employees accept their responsibilities under the law. Effort should be concentrated on places of greatest risk of illness or safety and should, among other factors, consider the volume of services and number of persons exposed. The frequency of inspection should vary according to circumstances. In some instances an establishment may be inspected frequently until either compliance is achieved or legal evidence is accumulated for legal action.

To specify an arbitrary frequency of inspection for a program after an initial inventory has been made fails to recognize that one place may not need an inspection, whereas other places in the same program may be sorely in need of help and inspection. For some programs, an *average* of three inspections per year is needed to maintain a satisfactory level of sanitation. Other programs may require on the average only one or two inspections. An average can also mean that some places receive one inspection in 2 years and that some places may receive six inspections in a year. A program, for various reasons, may be deliberately planned to be covered on a 2- or 3-year basis. Where annual permits are issued, a minimum of one inspection per year may be necessary. In other instances an annual inspection may be desirable if adequate personnel are available.

It should be remembered that there are other, perhaps more effective, ways of obtaining compliance with good sanitation practices. Continual evaluation, including comparison of effort versus accomplishment, hazard to the public health, results of surveillance sampling, and special problems should guide inspection frequency. To do otherwise is to proceed blindly.

Inspection Supervision

Review of inspection reports by trained, experienced, and competent sanitary engineers or sanitarians directly responsible for the supervision of specific activities will aid in evaluating the fieldwork from day to day. Incomplete reports, a high percentage of "No Violations," sketchy explanation of deficiencies observed, and no recommendations or vague and nonspecific ones readily become apparent. An unusually large or small number of inspections made in a day may indicate whether an inspector is trying to do a good job. This is discussed further under "Efficiency."

The ratio of supervising engineers or sanitarians to field personnel will depend on such factors as the difficulty of the work, its newness, and the degree of progress already made in obtaining satisfactory compliance. In a going housing or restaurant program, the ratio may be 1 supervisor to 6 or 8 inspectors; in a migrant-labor camp or recreation camp inspection program it may be 1 to 4; a special housing appraisal or stream pollution survey may require a ratio of 1 to 2 or 3. Since environmental engineering and sanitation activities are of a wide range of difficulty, broad generalizations need to be adjusted in individual situations.

Every effort should be made to encourage establishments under supervision to appoint a sanitation supervisor. His responsibility would include self-inspection and compliance with legal requirements as well as company policies.

The program activities mentioned above need not all be carried out with equal intensity at the same time. This is rarely possible or for that matter necessary. Gibson in fact pointed out the necessity of sanitation program planning so as to assure that greater effort is deliberately directed to those activities that are actual or potential problems.[20] For example, use of raw milk, lack of community water supply or sewerage, presence of malaria or rabies, radiation hazards, and a large number of public requests for a particular service are all areas of work that should receive priority attention. Traditional areas of work such as

---

[20] William C. Gibson, "Sanitation Planning," paper presented before the Engineering Section of the American Public Health Association, 81st Annual Meeting, New York, November 12, 1953.

restaurant supervision and complaint investigation may receive inspections out of proportion to the relative need, simply because they are usually well established and easier to perform. However, it is not enough to determine the key problems for emphasis. Those who are directly concerned, such as the enforcing agencies, the local elected officials, and operators of establishments, must recognize that a problem exists and accept responsibility for taking the action that may be indicated. It is therefore necessary to enlist the participation of those individuals or occupation groups directly involved with the carrying out of good sanitary practices so as to help make a sanitation program effective. These groups would include food and milk handlers; water, swimming pool, and sewage treatment plant operators; local plumbing and building inspectors; well drillers and plumbers; resort operators, school administrators, pest control operators, equipment manufacturers, architects, engineers, land planners, and surveyors; and others. Not to be forgotten is the importance of keeping the public and community leaders informed so as to obtain individual and organized community support when needed.

Production

Information useful in the planning and management of an operating program can be obtained from a minimum of data usually available in all departments. The number of inspections made is a record commonly kept. The number of man-days used in a program can be determined from payroll and attendance records. The number of productive hours per work day may be obtained by subtracting the hours spent in "nonproductive" work from the total working hours. A typical day might show the following nonproductive work: 1 hr in writing reports, discussing special problems, making appointments, receiving in-service training, and so forth; ½ hr in getting to the first assignment; and ½ hr in maintaining good public relations. Therefore, of a 7-hr workday (exclusive of the lunch hour), 5 hr were spent in "productive" work. From the data given it is possible to determine:

1. $\dfrac{\text{Average number of}}{\text{inspections per man-day}} = \dfrac{\text{Number of inspections}}{\text{Number of man-days}}$

2. $\dfrac{\text{Average number of hours}}{\text{per inspection}} = \dfrac{\text{Number of productive hours per day}}{\text{Number of inspections per man-day}}$

An example will illustrate the procedures for the determinations made above. During a one-year period 25,000 inspections were made in a city environmental sanitation program and 10,000 in a county suburban and rural program. There were 5067 man-days on duty in the city and 2777 in the suburban and rural areas. The net average annual work year corrected for holidays, vacation, and sick leave was 220 days. Thus:

1. Average number of city inspections per man-day $= \dfrac{25,000}{5067} = 4.9$

2. Average number of county inspections per man-day $= \dfrac{10,000}{2777} = 3.6$

3. Average number of hours per inspection in city $= \dfrac{5}{4.9} = 1.02$

4. Average number of hours per inspection in county $= \dfrac{5}{3.6} = 1.39$

These figures, which may be determined for a total program or for one activity, have many uses. Annual comparison of this information on an overall activity basis will indicate trends and may show where special attention should be directed. The figures can also be used to determine the approximate number of personnel needed to carry out an existing or new inspection responsibility. An adjustment should be made if an activity is of more or less than "average" difficulty.

For example, an analysis of a city environmental sanitation program showed that it had an annual workload of 17,685 places, and that 11,271 of them were inspected during the year a total of 35,145 times. Each place was therefore inspected an average of 3.1 times. The places not inspected amounted to 17,685 — 11,271, or 6414. If an average of 3.1 inspections per place was required for reasonable control, 6414 × 3.1, or 19,883, additional inspections would appear to be needed. Since this work was in a city where experience showed that 4.9 inspections could be made per man-day, 19,883 ÷ 4.9, or 4058, additional man-days or 4058 ÷ 220 = 20 men would be needed in addition to supplementary supervisory staff. This assumes that one man-year equals 220 man-days.

If, of the 6414 places not inspected, 2164, or one-third, represent an activity that is of minor public health importance, it may be possible to deemphasize inspections in this category to release men for other work. It is also probable that some licensing agency may be making inspections of the activity, thereby providing good reason for avoiding duplication and making only special inspections. Hence the number of men needed might be reduced by one-third.

If another activity has been brought under satisfactory control, the reduced number of inspections necessary may release two or three men. It may also be decided that the inspection of some activities could be conducted on a 2- or 3-year planned inspection interval rather than on an annual basis, thereby reducing the needed personnel still further. Under the circumstances, therefore, instead of asking for 18 additional men it may be possible to do an adequate job with four or six additional men properly used. This is not to say that one should always try to carry out a comprehensive environmental program with a fraction of the required staff. But it is meant to combat a tendency to keep asking for additional people without first thoroughly revaluating the uses being made of available personnel. It may be found that what is needed is not only additional personnel, but policy changes in the light of environmental engineering and sanitation progress made and the unfilled needs.

Efficiency

The inspection data may also be used in evaluating the efficiency of the sanitation program. It is first necessary to have determined by an impartial expert the average amount of time that should be required to make each type of inspection. One need not make an elaborate time study; an informed estimate with field checks is adequate. What is more important is retention of the figure without change. Multiplying the number of inspections actually made by the average time required for each type of inspection will give the total time that should have been spent to do the work reported. The ratio of the time required to the time actually spent will give the percentage efficiency. It is not inconceivable that an efficiency of 150 or 200 percent may be found when quantity rather than quality has been emphasized. Annual comparison of efficiencies can tell a very interesting story over a period of years. The numerical figure obtained is not nearly as important as the relative change from year to year.

Another measure of effectiveness is graphically illustrated in Figures 12–1 and 12–2. By plotting on an annual basis inspections versus corrections or improvements, trends are readily apparent. The point of diminishing returns can be seen and the need for revising inspectional procedures or administrative philosophy is shown.

Performance

Another way of evaluating a program or activity is by determining its cost from year to year and the services received. The amount of money required to perform a service is something everybody can understand. If services are related to cost, it can be said that an environmental program costs, say, A dollars per year, the amount provided and spent in the budget. If a division performs a total of B weighted services[21] per year, each service costs $\dfrac{A}{B}$ dollars.

$$\text{Cost per service} = \frac{\text{Budget in dollars } (A)}{\text{Weighted services performed } (B)}$$

[21] See Table 12–5, "Performance Report Based on Weighted Work Units."

Hence, if a budget is cut X dollars, and the cost per service is known approximately, one can say that the number of weighted services that can be performed will be reduced by an amount that can be computed.

For example, an environmental control budget is $200,000 and the weighted services performed = 12,500. The cost per weighted service = $\dfrac{\$200,000}{12,500}$ = $16.00. If the budget is reduced to $180,000, the weighted services anticipated will be reduced to $\dfrac{\$180,000}{\$16.00}$ = 11,250, or by 1250. This can be carried further by saying that in the previous year, for example, the restaurant inspection activity required 1000 weighted services, the private water supply survey and sampling activity required 1500 weighted services, the dairy-farm inspection activity required 2500 weighted services, the milk sampling program required 500 weighted services, and so forth. If the budget is cut $20,000, the weighted services that can be performed will be reduced by 1250 as previously computed. Hence, either the restaurant activity can be eliminated, or the private water supply survey and sampling activity practically eliminated, or the dairy-farm inspection activity cut in half, or the milk-sampling control and restaurant-inspection activity drastically reduced. If the budget director, finance committee, or others involved are confronted with the necessary reduction in services accompanying a budget cut, a better understanding of the services is obtained. To be effective, however, the officials must understand fully the principles involved; otherwise this will only be an administrative tool.

Using the same reasoning, one can also say that an increase in the budget by $20,000 will result in a corresponding increase in the services performed, if needed personnel can be employed. For example, with a budget of $220,000 and a weighted service cost of $16.00, the number of weighted services rendered will be $\dfrac{\$220,000}{\$16.00}$ = 13,750, or an increase of 1250. This might mean a new service, such as ionizing radiation control, or strengthening of existing services.

A performance-type of evaluation based on cost requires the reduction of all tangible services to comparable units. This would be a weighted unit of service. For example, a bathing beach inspection could be given a weight of one. The collection of a milk sample could be given a weight of 0.2. Other activities can be similarly weighted, as shown in Table 12–5. This summation can take the place of a statistical report. These are average weights and recognize the fact that the time to make inspections within each activity will vary but also tend to balance one another when a large number of inspections are made or samples collected. This list has been abbreviated; it can be greatly expanded to include a breakdown of weighted original inspections, reinspections, and partial inspections, as well as weights for different types of food establishments, temporary residences, nuisances, and so forth. This exactness may be desirable but is not always required. There is so much inherent variation within the particular inspection activity that a greater degree of accuracy in the final figure would be of no value. In other words the final accuracy cannot and need not be any greater than the components.

TABLE 12-5 PERFORMANCE REPORT BASED ON WEIGHTED WORK UNITS

Type of Service	Unit Weight	Actual Number Units	Actual Weighted Units	Forecast Weighted Units
Bathing beach, inspection	1.0	37	37	36
Swimming pool, inspection	1.7	38	65	38
Food establishment, inspection	1.2	900	1,080	1,320
Fairground or carnival, inspection	4.0	6	24	24
Hospital—maternity, inspection	2.0	196	392	150
Housing, inspection	1.2	296	355	450
Industrial or business place, inspection	1.0	24	24	20
Institution, inspection	2.0	2	4	4
Ionizing radiation, inspection	1.0	160	160	50
Milk—dairy farm, inspection	1.0	885	885	515
Milk—processing plant, inspection	2.0	234	468	244
Nuisance, investigation	1.0	247	247	500
Picnic ground, inspection	1.2	32	38	24
Rabies, investigation	1.0	137	137	186
Realty subdivision—rural, investigation	4.0	50	200	160
Realty subdivision—urban, investigation	1.0	12	12	8
Refuse disposal facility, inspection	1.0	52	52	40
Rodent or insect control, investigation	3.0	10	30	18
School—central, inspection	2.3	122	261	138
School—rural, inspection	1.0	53	53	80
Sewage—municipal, inspection	2.0	36	72	48
Sewage, wastes—industrial, inspection	2.0	6	12	10
Sewage—private, inspection	1.0	988	988	1,000
State park, inspection	2.0	2	4	4
Stream pollution, investigation	3.0	19	57	36
Temporary residence—camp, inspection	3.0	76	228	270
Temporary residence—other, inspection	1.0	246	246	180
Water supply—municipal, inspection	2.0	224	448	400
Water supply—quasi-public, inspection	1.0	63	63	60
Water supply—carriers, inspection	1.0	6	6	6
Water supply—private, inspection	1.0	901	901	1,000
Special survey, investigation	3.0	6	18	15
Field conference, visit	1.0	2,330	2,330	2,000
Not at home, visit	0.3	476	143	120
Unclassified	1.0	59	59	100
Collection of sample—milk, water, other	0.2	10,198	2,040	1,532
Plan review—subdivision, pool, sewerage, water	2.0	258	516	375
Total	—	—	12,655	11,086

Recapitulation

Estimated weighted units of work to be performed (at beginning of year)	11,086
Estimated budget expenditures—total costs	$205,998
(variable costs = $182,272.00, fixed costs = $23,726.00)	
Estimated total cost per weighted work unit	$19.56
Total actual weighted units of work performed	12,655
Total man-days at work, corrected for vacation, sickness, training, etc.	2,954
Average time in field	65%
Average time in office	35%
Average weighted units of work per man-day	4.3
Actual budget expenditures—total costs	$195,654
(variable costs = $171,928.00, fixed costs = $23,726.00)	
Actual total cost per weighted work unit $15.44.	$15.44

Cost of each activity = weighted unit $\times$ $15.44. For example, the milk control
activity cost is farm and plant inspections plus sampling = [885 + 468 + 480
(wt. samples)] $\times$ $15.44   = $28,300

Note: Fixed costs are those costs that are directly related to the initial operation, i.e.,
rent, light, heat, commissioner, program directors and secretaries, cleaning, etc. Variable
costs are primarily for staff and their maintenance. Additions should be made for retirement
benefits, hospitalization, workmen's compensation, and social security.

The weighted forecast shown in Table 12–5 would represent the goal established at the beginning of a reporting period, such as a year. By completing this summary each quarter, semiannually, and annually, it is possible to determine whether the goal is being achieved and, at the end of the year, if it has been achieved. The administrator, however, should be prepared to change his plans at the end of a quarter, if conditions warrant, so that his overall program reflects the needs as they may develop and as he interprets them.

It is apparent that an artificial or paper increase in units of work reported that is not based on a complete and competent service or inspection can easily produce a misleading lowered cost per weighted unit of work. As previously stated, quality inspection must be emphasized and assured through the employment of competent personnel, by training, and by supervision. The most important element, therefore, in obtaining the greatest value for the taxpayers' dollar is high-quality administration. In a health department this means the best health commissioner and program directors money can buy. And money alone cannot buy top administrators for long unless it is accompanied by pleasant and satisfying working conditions. This staff nucleus, plus essential auxiliary personnel, equipment, and the plant or building is therefore a basic or "fixed" cost in the establishment of a health department. It largely determines the quality of the health service. Other costs are considered "variable" costs.

A state health department parcelling out state aid to local health departments might do well to help guarantee adequate salaries for the "fixed" cost item. For nothing will better assure best usage of state allotted funds than the people in responsible charge. Of course other things must be considered, but this question of adequate salaries for dedicated professional career administrators is frequently overlooked. The establishment by law or sanitary code of minimum qualifications to be met for the employment of health department personnel is a step in the proper direction.

As previously pointed out, the cost per weighted service at the end of a year can be easily determined. The amount of money spent can be readily obtained. The number of weighted services rendered is known. It is therefore a simple matter to determine the actual cost of a weighted service. Comparison of the cost of a weighted service from year to year is an additional evaluation criteria. One should be cautioned against trying to compare programs of different departments on a cost basis. There are many, many variables that make such comparisons misleading and incorrect. Such variables as community needs, type of program, number of years a program is in effect, progress made, adequacy and competence of staff, quality of administrative direction, morale, extent of political infitration, quality of work and supervision, and many other variables would first have to be adjusted before valid comparisons can be made. On the other hand, comparison of a program within the same department from year to year is reasonable.

The principles outlined under performance budgeting have direct application to industry. It is also useful in determining the cost of carrying out an existing activity, a new activity, or a new environmental program.

ENFORCEMENT

Enforcement Philosophy

The administrator is constantly looking for ways to make more effective the limited time and personnel available to carry out the department or division program. Frequently doubts are raised as to whether the time and effort spent to obtain improvements in accordance with law or good practice are worth the results. In this connection education, persuasion, and motivation are fundamental to accomplish the bulk of the objectives. In some cases legal action is resorted to when all other means have failed in the time period considered reasonable. It is sometimes said that to resort to legal action is to admit failure. Perhaps it would be fairer to say that a program based *solely* on legal action is doomed to failure.

Hollis stated that "the most primitive approach by the sanitarian was to carry a big stick. A more sophisticated approach was to speak softly and carry the big stick in a velvet glove. With the advent of epidemiology, it proved effective to speak cogently and to carry a slide rule. . . . The pressure to comply with approved sanitation practice now rises less from a fear of epidemics or of legal sanctions and more from a desire for good living and common realization of mutual interest."[22] Stated another way, the primitive approach to sanitary regulation is police enforcement; the civilized approach is environmental health education; the sophisticated approach to law enforcement is self-inspection under regulatory surveillance and control, with effectiveness evaluation by industry and the regulatory agency.

Shattuck in his classic report stated that "local Boards of Health endeavor to carry into effect all their orders and regulations in a conciliatory manner; and they resort to compulsory process only when the

[22] Mark Hollis, "Public, Professional, Industrial Allies in Sanitation," *Public Health Reports,* **68,** No. 8, 805–810 (August 1953).

public good requires."[23] Sir Edwin Chadwick, a pioneer sanitarian (1854), was removed from office because he tried to force his ideas of cleanliness on the people of London.

Enforcement techniques are described below and in pages 785 to 795.

Performance Objectives and Specification Standards

There has been considerable discussion among administrators regarding the advantages and disadvantages of performance and specification standards to carry out environmental control responsibilities. Some of this discussion has been useful in clarifying the objectives of a program and how inspection and enforcement is to be carried out.

From a legal standpoint, the specification type of standard is easier to enforce than the performance standard. However, the objective of a program is not strictly legal enforcement, but is directed toward obtaining compliance (performance) with the *intent* of the law by elevating the level of operation through attitudinal changes, education, and persuasion. The interpretation of a standard should in most instances vary with the type of facility, the equipment available, the people employed, geographic location, construction, and many other variables. In most cases it is difficult to establish a definite standard that can be applied uniformly without causing unnecessary expense and undue hardship. On the other hand, where it is possible to establish an enforceable standard that has broad general application, this should be done.

For example, a standard for dishwashing could state that the dishes and utensils shall be so cleansed and bactericidally treated as to have a total bacteria count of not more than 100 per utensil surface examined, as determined by a test made in a laboratory approved for the purpose by the state commissioner of health. This would actually be a performance objective. On the other hand if an attempt were to be made to explain how dishes shall be washed, rinsed, and bactericidally treated, using a two-compartment sink, a three-compartment sink, a single-tank door-type dishwashing machine, a single-tank conveyor-type dishwashing machine, a two-tank conveyor-type dishwashing machine, for example, very detailed specifications would have to be written and continually updated. The same analogy can be used in determining whether a water supply is satisfactory. We can use the statement that a water supply shall be of safe sanitary quality, of adequate quantity and pressure. What is adequate and satisfactory requires detailed explanations, depending on the type of establishment, source of water, type of treatment,

[23] Lemuel Shattuck, *Report of the Sanitary Commission of Massachusetts, 1850,* Harvard University Press, Cambridge, Mass., 1948.

and many many other variables, all of which require professional judgment. A very common method of handling this problem is to include in the law the performance objective that the responsible person is expected to achieve and then make reference to certain publications that have official or semiofficial status, such as the *Public Health Service Drinking Water Standards, 1962,* or *Recommended Standards for Water Works.*[24] Stream pollution control standards may require that wastewater treatment plants meet certain effluent standards and freshwater or saltwater classifications.

It is extremely important to avoid the establishment of rigid specification standards unless some flexibility can be built in because a rigid specification is sometimes inadequate, and under certain circumstances, quite ridiculous. An example might be a requirement of 35 gal or even 100 gal of water per person per day at a camp or institution without regard to the type of operation, number and type of plumbing fixtures, peak demands, storage, and capacity of the water source.

For many years efforts have been made to get away from the old-fashioned "inspector" who simply completes an inspection checklist, finds what is wrong, and issues a summons without making any attempt to interpret the regulations, their intent, and the facilities provided. An investigation followed by explanation of the findings with the responsible person with a view toward getting at the source of difficulty so that the cause can be eliminated and the establishment operated in compliance with the objectives of the law would appear to be a more effective way to operate on a routine basis.

Specification-type standards are extremely valuable when they have universal application. In such cases they should be written into law. However, when a specification standard cannot be established because of the nature of the operation or lack of a reliable measurement tool, it is better to rely on the performance objective. It would then be up to the individual responsible to determine how to comply with the objective established. He might be referred to certain publications or educational materials for guidance or perhaps in some instances to a consultant. This does not prevent the enforcing agency from recommending or even requiring, when this is indicated, that certain definite procedures be followed when dealing with some recalcitrant and willful violators.

Correspondence

Letter and report writing can be one of the most effective means of obtaining improvements or correction of deficiencies. Continuity of

[24] See Chapter 3.

effort can be maintained; differences of opinion or misunderstanding can be reduced; and other interested people can be kept abreast by copies of letters and thus lend support to sound objectives. It is also a means of acquainting local officials with the diversity of problems they are actually being relieved of, with resultant greater appreciation of the services being rendered.

In many cases the only contact people have with the control agency is through a letter. The impression left may therefore be good or bad, depending on the tone of the letter, its appearance, the choice of words, completeness, and other factors. The letter can also indirectly serve as an educational piece and as a base for obtaining cooperation. If improperly worded or if inaccurate, the letter can be a source of embarrassment.

Form paragraphs for different activities have been found to be of great assistance in letter writing to confirm the results of large numbers of inspections. If letters are individually written to fit specific situations, and the form paragraphs are varied to apply accurately, a good letter can result. It is then possible to have the field inspector prepare a draft of the letter he feels should be sent based on the conditions actually observed. The letter would then reflect department policy and leave to supervisory personnel the task of reviewing the proposed letter and then signing the letter after it is typed. Form paragraphs should be periodically reviewed and improved. Suggested form paragraphs and letters were given in Chapter 11 (as applied to housing).

There will of course be instances when a letter is unnecessary. Many departments follow the practice of having the owner or operator sign the inspection report checklist to acknowledge receipt of a copy of the report. This also serves to officially advise the responsible person of any deficiencies and the need for making prompt corrections. A confirmatory letter could follow only when serious deficiencies were in evidence.

Compliance Guides

Satisfactory compliance guides explaining in some detail and with examples what is "adequate," "satisfactory," "proper," "good," "safe," and so on, can be very helpful to field personnel. Many parts of this text can serve this need. The sections on milk, camps, and resort hotels can be used as compliance guides. The Public Health Service has recommended that milk and restaurant ordinances incorporate the public health reasons for each regulation and what would constitute satisfactory compliance.

It is possible under such circumstances to obtain some of the best thinking and reasonably uniform interpretation of regulations. This is essential to prevent chaotic enforcement. On the other hand it is also essential that such compliance guides be reviewed periodically and revised as the need exists.

Compliance guides can be developed with the assistance of personnel responsible for a new activity. By so doing, a better understanding of the governing regulations is obtained by the personnel actually involved in the enforcement. It then also becomes an opportunity for in-service training applied to a specific problem. Such experience is usually much more effective and of lasting benefit. Representatives of the industry involved should also be requested to participate to help obtain acceptance and application of the principles involved.

In-Service Training

Every division having environmental control responsibilities should maintain a continuing program of in-service staff training. In a large division it could be the full-time duty of one person who is qualified to carry out this very important function. In a small division it would have to be an additional duty of the director or a member of his staff. Regardless of division size or work load, to neglect in-service training is to grind to a standstill. Diligence on the part of administrative personnel cannot produce effective results unless the entire staff, and particularly field personnel, are kept abreast of new developments and new or changing policies. Opportunities for discussion, attendance at short courses, conferences, and conventions should be encouraged for their stimulating effects and for improved individual performance.

In-service or on-the-job training of new personnel is a very necessary extension of a student's formal education. Unless provided as an integral planned part of a college curriculum, the young graduate is at a considerable disadvantage when on the job. It will probably take 2 or 3 years before he can function effectively with little direct supervision. Opportunities for field application and learning should be provided the student by the college or university during summer vacations, long holidays, by term papers, and by project-oriented theses. Cooperative college-industry programs have also shown the advantages of this type of training. State and full-time local environmental control agencies should assist in this approach by providing opportunities for the field training of engineers, biologists, veterinarians, sanitarians, technicians, and sanitary inspectors, even though this means extra work for those in charge. Colleges and universities should, in fairness to their students, seek out opportunities to give them the field experience they need.

Not to be forgotten is the value of technical or occupational group training of those persons directly responsible for carrying out good sanitary practices. Supervisors of food handlers; resort managers; operators of swimming pools, water and sewage treatment plants, and milk plants should also be given special in-service training. Attorneys, realtors, architects, consulting engineers, and others can also benefit from and be helpful in implementing special programs through the conduct of workshops or special conferences. The cooperation of a local educational institution can be of great assistance. See also Inspection Supervision this chapter.

Education

The educator and public information specialist are important members of a control agency, voluntary agency, and other related organizations. Their services as consultants can be extremely valuable in making more effective a department or division program for training, community organization, interpretaton to the community and legislators, release of educational material, and special reports. Other services can be rendered depending on the competency of the control agency staff and organization. In the final analysis, the division director and his staff will have to provide most of the basic technical information for the guidance and assistance of the educator and information specialist. This is as it should be, for the professional educator or information specialist is not expected to be an expert in stream pollution abatement or bonding, for example; but his techniques can be applied to make the specialized knowledge of the division staff more effective in obtaining public approval of a comprehensive plan and a bond issue for construction of an environmental facility improvement.

Education should start with planned intensive in-service training of all field and office personnel. The extent of the training would depend on the competency of the personnel; but it should nevertheless be a continuing activity.

The field engineer, sanitarian, or inspector is, in addition to his primary duty, an educator. Every time he talks to a person he is acting in the role of an educator. He explains what is wrong or what can be improved and discusses how corrections and improvements can be made. The occasion may frequently arise when other department programs and services can be mentioned and explained. Every inspection or field conference is an opportunity to point out the services rendered and enhance the prestige of the department, thereby making it easier to put across the total program for the benefit of the people.

The preparation of suitable literature, including leaflets, pamphlets, and media releases adapted to the local conditions, can do much to obtain acceptance of good environmental sanitation practices. It also helps to avoid confusion in the interpretation of recommendations. For example, a leaflet giving precautions to be taken before buying a home in a rural area can emphasize the superiority of public sewers and water supply or of the alternative of having to carefully look into the well-water supply quality and quantity and the construction of an approved septic-tank leaching system. A bulletin on rural water supply can illustrate and explain good well and spring location, construction, and protection. A leaflet on food poisoning gives opportunity to bring out the necessity of refrigeration and hygienic food-handling practices. A release on septic tanks can point out the importance of inspecting and cleaning them. Typical construction details of septic tanks, distribution boxes, tile fields, and leaching pits will help the contractor installing the system to do the job in accordance with the best practices so that approval can be readily given. The many illustrations in this book are adaptable for educational use. Most state and local agencies have pamphlets and leaflets available to meet the needs that exist.

Another important aspect of education and public relations involves the neutralizing of unfounded or emotional opposition and inaccurate information. In this way support can be developed for a bond issue to construct sewers and treatment plant or to obtain acceptance of a sanitary landfill site or incinerator. Public information activity should start in the planning phase and be continuous, culminating at a critical time, such as before voting on approval of a bond issue.

Other education opportunities are presented when asked to give talks to students, visitors, and civic and professional groups and when participating in meetings and conferences with representatives of other organizations. Attendance at short courses and scientific conferences should be encouraged, not only for the new information gained, but also for the stimulating effect on the individual and enthusiasm transmitted to others. Every situation should be used to point out the services available and how the individual can help himself, and how as a member of the community he can help improve the environment in which he and his family are living.

Persuasion and Motivation

Although many people are ready to grasp new ideas that will improve their standard of living, there are perhaps just as many who resist change. This requires the application of a little more effort to convince them and get them to do what is indicated. Here the personal contact can do much. The benefits to be derived, the reason for a particular regulation, and the economic, aesthetic, and disease dangers involved can be explained in some detail. Examples of how others have solved the problem and a visit to demonstrate the solution can be powerful motivating influences.

A written report to the responsible person formalizing a violation is always a good persuasive step. Compliance is encouraged if the letter is accompanied with sound recommendations and, if possible, illustrations. It is usually helpful if copies of letters are sent to interested persons such as the local health officer, town official, official agencies, the mortgage holder, and so forth. Letters should be clear, direct but polite, and persuasive.

In many cases a permit for operation, that incidentally promotes compliance and inspection supervision, is required. If an annual permit is needed its issuance can and should be withheld, with notification of the applicant, including reasons, until all legal requirements are complied with. If a continuing permit is issued it can be suspended and revoked after a hearing. It is well known that operation without a permit is a readily enforceable regulation. Most operators will want to avoid the threat of prosecution. As a rule, operators of public places are reasonable and will cooperate when confronted with letters and accumulated evidence by the enforcing agency. Sometimes the cost of legal services or effect on public relations is the determinant.

Another technique, when a formal permit or license is not required by law, is an informal certification system. In this process the health officer or other official certifies that a place meets certain established standards or perhaps higher than minimum standards. The certification is withdrawn when the standards are not met. The periodic publication of "certified places" can provide a competitive incentive and voluntarily raise industry and commercial facility and operation standards. This is made particularly effective when many of the operators subscribe to the system and when governmental agencies, corporations, and individuals require the certification as a prerequisite to doing business. The Voluntary Interstate Milk Shippers Program and certain camp, hotel, and motel association ratings are examples of the system.

Reinspections and additional field conferences are sometimes indicated to educate and persuade the offending party. The approach, attitude, patience, and ability of the inspector is frequently the deciding factor. In some cases a new inspector might be able to accomplish the desired objectives. A joint inspection with a supervisor might also have the desired effect. Certainly these approaches should be given a trial before deciding on other measures.

It cannot be denied that there will be some situations where solution of the problem appears impossible. A more forceful approach may be indicated in such cases. A registered letter signed by another person in higher authority sometimes brings the desired result. If this fails, an office conference may be called to give the violator an opportunity to explain his problems. Perhaps a step-by-step program of improvements can be agreed upon, with immediate attention being given to the dangerous conditions. Another approach would be the calling of an administrative or formal hearing. Typical notices using these procedures are included in Chapter 11 and in this section.

Legal Action

In general, the advice and assistance of the legal department should be enlisted before considering legal action. If any compliance sluggishness or resistance is detected, an educational program would be indicated. It should not be expected that the corporation counsel, county attorney, or other public prosecutor would have the same understanding of the problems and their effect on the public as the department head or the professional engineer or sanitarian. If this is recognized, it would be well to keep the legal department advised of any new legal and policy decisions in the field and to seek informal assistance and interpretations. Joint field visits provide an excellent opportunity to get out of the office and to get acquainted firsthand with the broad environmental control program—its objectives and problems.

If a division is well organized, with sound administrative procedures in effect, the file on a particular establishment should be replete with factual information. The educational and persuasive steps previously taken, including inspection reports and letters, will show the reasonableness of the division's actions. If this can be supplemented with office conferences or informal hearings that are confirmed in writing, the accumulated evidence alone is usually sufficient to convince the violator or his attorney to come to a prompt agreement. This information is also invaluable to the department counsel and makes possible a strong clear-cut case that is relatively easy to prosecute.

The administrator of a division or department must use a fine sense of judgment, statesmanship, and diplomacy in bringing up cases for legal action. Such actions should be kept at a minimum if one is not to be overwhelmed with the preparation of cases for legal action to the detriment of the total program.

An approach found effective, when other methods have failed, is the holding of a formal hearing before the board of health, commissioner, or preferably before an authorized hearing officer. This would be a civil action leading possibly to a fine and would thus overcome the objections to a criminal proceeding, when a violation is called a misdemeaner. The department attorney should be present at that time to guide and present the department's case. The outcome of this and previous persuasive attempts would determine the need for other court action. Once a case is turned over to the legal department the administrator should refer all further communication and requests for conferences, meetings, compromises, and so on, to the department attorney. He should not, however, forget about the action. He should maintain close liaison with the attorney assigned until the case is settled.

The procedures mentioned above are suggested for use in connection with routine legal enforcement. The administrator may find from experience that some can be omitted. Most departments have or can obtain hearing, affidavit of service, permit withheld, permit denied, order of board of health, and similar forms or letters to expedite compliance with sanitary code and law requirements. Some of these forms are shown in Figures 12–6, 12–7, 12–8, 12–9, and 12–10.

A system of issuing a departmental summons in accordance with a procedure previously agreed upon by the court may be possible in some areas. This privilege, however, should not be abused as it will only result in a greater number of adjournments and suspended sentences. The net result will be loss of prestige, a great deal of time lost in court, poor legal representation, and only a passing success at best.

Another standard approach in rural and suburban areas is the filing of information before the local justice of the peace. In some cases this works well, *particularly when the aggrieved local people can be present.* However, punishment of a violation as a misdemeanor is sometimes not effective for, as previously pointed out, there is a reluctance on the part of the courts to brand one found guilty of a misdemeanor as a criminal. Civil action would be preferable in such situations. Other types of penalties and sanctions should be explored with the agency counsel. For example, a violation could be declared an offense, and requirement of a license or permit to conduct a business can discourage operation in violation of the law. Injunction proceedings to close or cease an operation or a mandatory injunction to require a certain action or improvement are extremely effective techniques if permitted under existing law and allowed by the court.

The press and the courts can be valuable allies to a department that is sincerely trying to do a good job. It takes time to develop this prestige and acceptance. The department must demonstrate beyond any doubt that it is carrying on an intelligent, planned enforcement program, and not a day-to-day panicky program based solely on court actions.

There will always be situations where the slow educational and persuasive approach is not practical. Such cases may arise when a really dangerous condition exists and cooperation is nonexistent. When confronted with such situations the administrator must act summarily, in concert with his legal advisor, to eliminate with dispatch the dangerous condition as provided for in the existing laws.

DEPARTMENT OF HEALTH*

DENIAL OF PERMIT

Date Served.......................................

To...

Address of Business...

Type of Business...

Your health permit is hereby denied because of failure to comply with items contained in the notice issued on ....

Unless, by the date shown below, you discontinue operating a business for which a health permit is required, or comply with requirements for issuance of a permit, the matter will be referred to the County Attorney for legal action.

Effective Date...

Commissioner of Health

.... M.D.

By...

NOTE: Regulations provide "Whenever an application for permit is denied by the Health Officer the applicant may file with the Secretary of the Board an application for a hearing by the Board. Such application shall be filed within 7 days after receipt of notice of denial."

Via Certified Mail, Return Receipt Requested.

Figure 12–6 *Adapted from City of Los Angeles Health Department Form SH-151. Denial of permit form.

DEPARTMENT OF HEALTH*

NOTICE OF SUSPENSION OF PERMIT
TO CONDUCT A.....................................
AND OF TIME OF HEARING THEREON.

To...

....

NOTICE IS HEREBY GIVEN that the permit heretofore issued to you by the Health Department of the .... to operate a............................
.... at .... Street, in the................................,
is hereby suspended in accordance with the provisions of Section .... of the .... Sanitary Code.

Such suspension is made upon the grounds and for the reason that said place is operated and conducted in an extremely unsanitary manner and condition, thereby endangering the health of the citizens of the ....

YOU ARE FURTHER NOTIFIED that on the .... day of ...., 19......., at the hour of ....M., there will be conducted before the Board of Health in the board room of said Board, on the .... floor at .... Street in the city of ...., a hearing upon the question of revocation or further suspension of the above named permit on the grounds and for the reason above set out, at which time and place you may appear and show cause, if any you have, why such proposed action should not be taken by said Board.

YOU ARE FURTHER NOTIFIED that you may not operate said ....
..............................hereafter until the time of said hearing unless, because of the abatement of the conditions causing the suspension of your permit, the foregoing order of suspension is cancelled and proceeding before the Board is dismissed.

Commissioner of Health

.... M.D.

Effective ....M., ...19....

By...

Served this .... day of

................................, 19 ...

by leaving with...

Title ....

Figure 12–7 *Adapted from City of Los Angeles Health Department Form SH-55. Notice of suspension of permit.

BEFORE THE BOARD OF HEALTH

of ....

State of..

In the Matter of an alleged nuisance
detrimental to health existing upon the
property of, or maintained by...................

....

   Information having been presented to this Board that an alleged nuisance,
detrimental to health, exists on the premises owned or occupied by (or maintained
by).........................at...........................in the.........................of.........................:
namely ....

....

....

Notice is hereby given that a hearing upon the above mentioned alleged nuisance
will be given by the Board of Health of the..
of..................County, N. Y., at...................on the...................day of..................
19........ at .... o'clock; and you are hereby cited to appear before said Board
to show cause why said condition should not be declared a nuisance detrimental
to health and such nuisance abated, as provided by the Public Health Law (or
Sanitary Code of ....).

By the Order of the Board of Health

(Sgd) ....

Dated at...............................N. Y.

this...............day of...............19.....

Mailed to...

Via Certified Mail, Return Receipt Requested.

Figure 12–8   Hearing before board of health.

BEFORE THE BOARD OF HEALTII

of...
State of..

In the matter of an alleged nuisance
detrimental to health existing upon the
property of, or maintained by...................
....

Information having been presented to this Board that a nuisance, detrimental
to health, exists on the premises owned or occupied by (or maintained by)........
......................................at..in the...................................
of .... namely ....
....
....

And the said .... having been cited to show cause why such
premises, or the condition thereon, should not be condemned as a nuisance
detrimental to health and the said nuisance abated; and the said.........................
having appeared (failed to appear); NOW, upon proof of the facts showing the
existence of such nuisance, no sufficient proof to the contrary being shown,
IT IS ORDERED...
....
....

BY ORDER OF THE BOARD OF HEALTH

(Sgd)..

Dated at.............................N. Y.,
this .... day
of...
19...

Via Certified Mail, Return Receipt Requested.

Figure 12–9 Order of board of health.

Emergency Sanitation

Emergencies and natural disasters may be associated with war, displaced persons, epidemics, floods, hurricanes, earthquakes, volcanic eruptions, fires, major accidents, extended interruption of utilities and services, or other crises. On such occasions large masses of people may

---

AFFIDAVIT OF SERVICE OF NOTICE OR CITATION

State of................................⎫

County of...........................⎬ss:

.... of ....⎭

...., being duly sworn, says that he is over twenty-one years of age; that he served the annexed notice (or order) upon...........................

..............................and that he knew the person so served to be the same person mentioned and described in said notice (or order).

....

Sworn to before me this

.... day of .... 19........ .

....

....

Figure 12–10   Affidavit of service of notice or citation.

be at the mercy of nature. Their survival in an uncontrolled environment is dependent upon how soon relief can be mustered and a favorable or controlled environment restored. The speed with which action can be taken is largely dependent on the prior plans made to cope with emergencies. This must include the delegation of specific authority and responsibilities, the assignment of selected manpower resources, stockpiling and inventory of supplies and equipment.

National civil defense and local governmental organizations (including the Army, Navy, and Air Force), and the Red Cross are usually involved in major disasters. International agencies that have provided assistance are the League of Red Cross Societies and the United Nations including the World Health Organization (WHO), United Nations Children's Fund (UNICEF), Food and Agriculture Organization of the United Nations (FAO), and the World Food Programme.

The environmental engineeering and sanitation aspects of emergency sanitation are only touched upon in this text. Some guidelines are given on prevention of communicable diseases (Chapter 1), emergency water supply (Chapter 3), excreta disposal (Chapter 4), solid waste storage and disposal by modified sanitary landfill (Chapter 5), food protection (Chapter 6), temporary residences and mass gatherings (Chapter 9), and vector control (Chapter 10). A recent comprehensive publication has been prepared on emergency sanitation by the World Health Organization. It is recommended as a reference.[25] Numerous other technical manuals, bulletins, and guides are available through the national or local civil defense and emergency planning organizations.

BIBLIOGRAPHY

Changing Environmental Hazards, American Public Health Association, Washington, D.C., 1967.

Classification Code for Local Health Department Activities, Engineering Section Project, American Public Health Association, Washington, D.C., November 1950.

Ehlers, Victor M., and Steel, Ernest W., *Muncipal and Rural Sanitation,* McGraw-Hill Book Co., New York, 1965.

"Environmental Health in Community Growth—An Administrative Guide to Metropolitan Area Sanitation Practices," *Am. J. Public Health,* **53,** No. 5, 802–822 (May 1963).

Environmental Health Planning, U.S. Public Health Service Pub. No. 2120, Dept. of HEW, Washington, D.C., 1971.

Fleck, Andrew C., *Performance Budgeting,* Rensselaer County Health Dept., Troy, N.Y., April 1956.

Freedman, Ben, *Sanitarian's Handbook,* Peerless Publishing Co., New Orleans, La., 1970.

---

[25] M. Assar, *Guide to Sanitation in Natural Disasters,* World Health Organization, Geneva, 1971. Also *Red Cross Disaster Relief Handbook,* League of Red Cross Societies, Geneva, 1970; *An Outline Guide Covering Sanitation Aspects of Mass Evacuation,* Public Health Service Pub. No. 498, U.S. Dept. of HEW, Washington, D.C., 1956; *Military Sanitation,* FM 21-10, AFM 160-46, Departments of the Army and the Air Force, Washington, D.C., May 1957; *Emergency Water Services and Environmental Sanitation,* Department of National Health and Welfare, Ottawa, 1965.

Grad, Frank P., *Public Health Law Manual,* American Public Health Association, Washington, D.C., 1970.

Hanlon, John J., *Principles of Public Health Practice,* C. V. Mosby Co., Saint Louis, Mo., 1969.

Herman, Harold, and McKay, Mary Elisabeth, *Community Health Services,* Institute for Training in Municipal Administration, International City Managers' Association, 1140 Connecticut Aveune, N.W., Washington, D.C., 1968.

Manual for Public Health Inspectors, Canadian Public Health Association, 1255 Yonge Street, Toronto, 1961.

Maxcy-Rosenau, *Preventive Medicine and Public Health,* edited by Philip E. Sartwell, Appleton-Century-Crofts, New York, 1965.

"National Environmental Health Programmes: Their Planning, Organization, and Administration," Report of a WHO Expert Committee, *WHO Tech. Rep. Ser.* No. 439, Geneva, 1970.

Oviatt, Vinson R., *Proposed Instruction Manual for Sanitarians Using the McBee Marginal Punch Card System of Recording,* Michigan Dept. of Health, Lansing, November 1955.

Public Relations, WPCF Pub. No. 12, Water Pollution Control Federation, Washington, D.C. 20016, 1965.

Salvato, J. A., Jr., "Evaluation of Sanitation Programs in a City-County Health Department," *Public Health Reports,* **68,** No. 6 (June 1953).

APPENDIX I

MISCELLANEOUS DATA

Degrees $C° = \frac{5}{9}(F - 32°)$; Degrees $F° = \frac{9}{5}C° + 32$

Circumference of circle	$= 2\pi r = \pi D;$
	$\pi = 3.1416; r = \text{radius}; D = \text{diameter}$
Area of circle	$= \dfrac{\pi D^2}{4} = 0.7854 D^2$
Area of sphere	$= \pi D^2$
Volume of sphere	$= \dfrac{\pi D^3}{6} = 0.5236 D^3$
Volume of cone	$= \text{area of base} \times \frac{1}{3} \text{ altitude}$
Volume of pyramid	$= \text{area of base} \times \frac{1}{3} \text{ altitude}$
Area of triangle	$= \frac{1}{2} \text{ altitude} \times \text{base}$
Area of trapezoid	$= \frac{1}{2} \text{ (sum of parallel sides)} \times \text{altitude}$
1 acre	$= 43,560 \text{ ft}^2$
1 square mile (mi²)	$= 640 \text{ acres}$
1 inch (in.)	$= 2.54 \text{ centimeters (cm)}$
1 ppm (in water)	$= 1 \text{ mg per liter} = 8.34 \text{ lb per mil gal}$
1 cubic foot (ft³)	$= 7.48 \text{ gal} = 62.4 \text{ lb}$
1 U.S. gallon (gal)	$= 8.345 \text{ lb} = 3.785 \text{ liters} = 231 \text{ in.}^3$
	$= 0.833 \text{ British Imperial gallons}$
1 pound (lb)	$= 7000 \text{ grains} = 0.4536 \text{ kg} = 16 \text{ oz}$
1 grain per gallon	$= 17.1 \text{ parts per million}$
1 pound per square inch (psi)	$= 2.31 \text{ ft vertical head of water}$
	$= 2.04 \text{ in. of mercury}$
1 gallon per minute (gpm)	$= 1440 \text{ gal per day}$
	$= 0.133 \text{ cubic feet per minute}$
1 cubic foot per second (cfs)	$= 0.646 \text{ million gallons per day}$
1 million gallons per day (mgd)	$= 1.547 \text{ cfs}$
1 million gallons per day (mgd)	$= 1.547 \text{ cfs}$
1 horsepower hour (hp-hr)	$= 0.746 \text{ kW-hr} = 2546 \text{ Btu}$
	$= (33,000 \times 60) \text{ ft-lb}$
	$= 0.065 \text{ gal of diesel oil (approx.)}$
	$= 0.110 \text{ gal of gasoline (approx.)}$
	$= 10\text{–}20 \text{ ft}^3 \text{ of gas.}$
1 British thermal unit (Btu)	$= \text{quantity of heat required to raise the temperature of one pound of water one degree F}$
1 Btu per minute	$= 17.57 \text{ watts}$
1 Btu	$= 778 \text{ ft-lb}$
	$= 0.000393 \text{ hp-hr}$

1 ton of refrigeration	= 228,000 Btu per 24 hr
	= 2000 lb of ice $\times$ 144 Btu; 1 lb of ice absorbs 144.3 Btu of heat in melting
1 pound of water evaporated from 212°F	= 0.284 kW-hr = 970.3 Btu
1 kilowatt-hour (kW-hr)	= 3412 Btu per hour
1 boiler horsepower	= 33,479 Btu per hour

DIRECT CURRENT

Ohm's Law (dc)

$$I = \frac{V}{R};$$

I = strength of the current (amperes)

V = applied potential difference (volts)

R = resistance of the conductor (ohms)

Resistors in Series

$R = R_1 + R_2 + R_3 \ldots$ etc.

R = combined resistance

R_1 = resistance of first conductor or resistor

R_2, R_3, etc. = resistance of 2nd, 3rd, etc.

Resistors in Parallel

$$\frac{1}{R} = \frac{1}{R_1} + \frac{1}{R_2} + \frac{1}{R^3} \cdots \text{etc.}$$

Electric Power (dc)

$P = I \times V;$ P = power in watts;

1000 watts = 1.341 hp

Heat Developed by Conductor

$Q = 0.24I^2 \times R \times t;$ Q = calories

t = time in seconds

Transformer (dc)

$$\frac{V_s}{V_p} = \frac{n_s}{n_p};$$ V_s = voltage in secondary coil; I_s = amperage

n_s = number of turns in secondary coil

$$\frac{I_s}{I_p} = \frac{n_s}{n_p};$$ V_p = voltage in primary coil; I_p = voltage

n_p = number of turns in primary coil

Power

Watts = I^2R

     = IV ·

ALTERNATING CURRENT

$V = ZI;$ V = volts

$\qquad\qquad\qquad Z$ = ohms impedance of circuit

$P = VI\ Cos\ \theta;$ I = amperes

   and $\qquad\qquad P$ = power in watts

$P = I^2R;$ $Cos\ \theta$ = power factor; θ becomes zero when alternating current

(when $Cos\ \theta = 1$) $\qquad$ circuit contains resistance only, i.e., no capacity or inductance and Cos zero = 1

$Cos\ \theta = \dfrac{R}{Z};$ R = resistance (ohms)

Motor Loads (ac)

System	Power		Max. volts to ground
Single-phase (2-wire)	$P = VI\ Cos\ \theta;$	$V = 120$	120
Single-phase (3-wire)	$P = VI\ Cos\ \theta;$	$V = 240$	120
Two-phase (4-wire)	$P = 2VI\ Cos\ \theta;$	$V = 120$	120
Three-phase (delta)	$P = \sqrt{3}VI\ Cos\ \theta;$	$V = 240$	207.8, 120 lights
Three-phase (Y)	$P = \sqrt{3}VI\ Cos\ \theta;$	$V = 207.8$	120

Motor Speeds (ac)

$N = \dfrac{120f}{P};$ N = RPM of synchronous motor (squirrel cage)

$\qquad\qquad f$ = frequency, cycles per second

$\qquad\qquad P$ = number of poles

Note:

1. Synchronous motor runs at constant speed in contrast to induction motor, which lessens as load increases. Induction motor requires relatively small starting current, producing 200 or more percent of full-load torque.
2. Resistance in a line, due to travel of current some distance in a wire, will cause a drop in voltage, or line drop, and a loss in voltage and power. The power loss shows up as heat and varies as the square of the current. To prevent melting of the wire and possible fire when excessive current is drawn through a wire, a fuse of proper size or a circuit breaker is placed on the house line and on each branch circuit leaving the service panel.
3. See Table A, Page 888, for motor wiring.

Pump Efficiencies:

Deep-well displacement pumps have an efficiency of 35–40 percent. Small centrifugal pumps (10–40 gpm) have an efficiency of 20–50 percent. Larger centrifugal pumps (50–500 gpm) have an efficiency of 50–80 percent.

Small duplex, triplex, and reciprocating pumps in general have efficiencies of 30–60 percent; but large pumps have efficiencies of 60–80 percent.

TABLE A. MOTOR WIRING FOR VARIOUS HORSEPOWERS, MOTOR TYPES, AND CURRENTS

Horsepower	3-Phase Squirrel-Cage Induction Motors 220 Volts			3-Phase Squirrel-Cage Induction Motors 440 Volts			Single-Phase Induction Motors 115 Volts			Single-Phase Induction Motors 230 Volts			Direct Current Motors 115 Volts			Direct Current Motors 230 Volts		
	Approximate Full-Load Amperes	Wire Size, min AWG	Branch Circuit Fuse, Amperes	Approximate Full-Load Amperes	Wire Size, min AWG	Branch Circuit Fuse, Amperes	Approximate Full-Load Amperes	Wire Size, min AWG	Branch Circuit Fuse, Amperes	Approximate Full-Load Amperes	Wire Size, min AWG	Branch Circuit Fuse, Amperes	Approximate Full-Load Amperes	Wire Size, min AWG	Branch Circuit Fuse, Amperes	Approximate Full-Load Amperes	Wire Size, min AWG	Branch Circuit Fuse, Amperes
1/3							7.4	14	25	3.7	14	15						
1/2							10.2	14	35	5.1	14	15						
3/4							13	12	40	6.5	14	20						
1	3.5	14	15	1.8	14	15	18.4	10	60	9.2	14	30	8.6	14	15	4.3	14	15
1 1/2	5	14	15	2.5	14	15	24	10	80	12	14	40	12.6	12	20	6.3	14	15
2	6.5	14	20	3.3	14	15	34	6*	110	17	10	60	16.4	10	25	8.2	14	15
3	9	14	30	4.5	14	15				28	8*	90	24	6	40	12	14	20
5	15	12	45	7.5	14	25							40		60	20	10	30
7 1/2	22	10	70	11	14	35							58	3*	90	29	8	45
10	27	8*	80	14	12	45							76	2*	125	38	6	60
15	40	6	125	20	10	60							112	00*	175	56	4	90
20	52	4*	175	26	8*	80							148	4/0*	225	74	2*	125
25	64	3*	200	32	8	100												
30	78	1*	250	39	6	125												
40	104	00*	350	52	4*	175												
50	125	000*	400	63	3*	200												
60	150	4/0*	450	75	2*	250												

Source: Adapted from General Electric Co. 1955 Diary. See *National Electrical Code* for additional details.
Note: Values given are for not more than three conductors in a cable. Single-phase and direct current motors use two conductors.
*Next smaller size may be used with RH-type insulation.

Horsepower to Pump Water

Horsepower to pump water:

$$= \frac{\begin{array}{c}\text{gal of water}\\\text{pumped per min}\end{array} \times \begin{array}{c}\text{total head pumped}\\\text{against in ft}\end{array} \times \begin{array}{c}\text{weight of one}\\\text{gal of water in lb}\end{array}}{33,000 \text{ ft-lb per minute} \times \text{pump efficiency as a decimal}}$$

$$\text{Horsepower of the motor drive} = \frac{\text{gal per min} \times \text{total head in ft}}{3,960 \times \text{pump efficiency} \times \text{motor efficiency}}$$

Power input to a motor in watthours

$$= \frac{3,600 \times \text{number meter disk revolutions} \times \text{disk constant}}{\text{number seconds}}$$

Power input to a motor in watthours

$$= \text{voltage} \times \text{amperage} \times \text{power factor} \times 3,600 \times \left(\begin{array}{c}1 \text{ for single-phase}\\1.41 \text{ for two-phase}\\1.73 \text{ for three-phase}\end{array}\right).$$

NOTE: One rotation of meter disk = watts marked on disk.

Waterfall

Power of waterfall = $0.114 \times \text{cfs} \times \text{head in ft} \times \text{eff. (eff.} = \pm 88\%)$.

Channel Flow:

$$V = \frac{1.486 R^{2/3} S^{1/2}}{n}$$

where $R = \dfrac{\text{cross-sectional area of water (ft}^2)}{\text{perimeter wet (ft)}}$

$S = $ slope of water surface (ft per ft)
$n = 0.015 \pm$ (See hydraulic texts)
$Q = VA$; in which, $Q = $ quantity of water, in cubic feet per second (cfs)
$V = $ average velocity of flow, in feet per second (fps)
$A = $ area of pipe or conduit, in square feet.

For a stream having a trapezoidal section:

$$A = \tfrac{1}{2}(a + b)h$$

$V_{(average)} = 0.85 \times$ surface velocity. Surface velocity is time for a floating object to travel a known distance in a fairly straight section of stream.

Weir Discharge:

Discharge over 90° weir = $Q = 2.53 H^{5/2}$

Discharge over rectangular weir $= Q = 3.33(b - 0.2H)H^{3/2}$
(with end contractions)

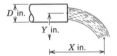

With no end contractions $Q = 3.33bH^{3/2}$; $b =$ at least $3H$
NOTE: Place hook gage at least 3 ft back of weir.
No or low velocity of approach.

Water Pressure

$$h = \frac{p}{w}$$

where $p =$ pounds per ft²
$w =$ pounds per ft³ $= 62.4$ for water

$$h = \text{head of water (ft)} = \frac{p \times 144}{62.4} = 2.3p, \text{ where } p \text{ is in psi}$$

$$h = \frac{V^2}{2g}$$

where $g = 32.2$ ft per sec per sec and $V =$ velocity in feet per second.

Flow of Water in Pipe:

$$h = \frac{flV^2}{d2g}$$

where $l =$ ft length
$d =$ ft diameter
$f =$ friction loss $= 0.02$ for cast iron   (see hydraulic texts).
Bernoulli's equation:

$$\frac{p_1}{w} + Z_1 + \frac{V_1^2}{2g} = \frac{p_2}{w} + Z_2 + \frac{V_2^2}{2g} + \text{loss head (ft)}$$

where $Z =$ feet available potential energy.

Flow from full-flowing horizontal pipe, in gpm $= \dfrac{2.83D^2X}{\sqrt{Y}}$;

FINANCE OR COST COMPARISONS

Compound Interest:

$$A = P(1 + i)^n;$$

where $A =$ amount or sum of principal and interest
$P =$ principal or amount borrowed
$i =$ rate of annual interest, or value of money
$n =$ number of years money is drawing interest.

Present Worth:

$$P_w = \frac{A}{(1 + i)^n};$$

where P_w = present worth of a sum of money A due in n years at compound interest rate i.

Sinking Fund (annuity):

$$S = \left[\frac{(1 + i)^n - 1}{i}\right] d;$$

where S = sinking fund or an amount that will accumulate in n years by making equal annual installments d, at interest rate i

or
$$d = \frac{Si}{(1 + i)^n - 1}$$

Bond Issue:

$$d = B\left[\frac{i}{(1 + i)^n - 1} + i\right];$$

where B = bond issue or present sum of money, to be paid off in equal annual installments d in n years at interest rate i. The bond issue includes principal, legal and engineering fees, and interest.

Capitalized Cost—Economic Comparison:

$$S_C = C + \frac{0}{i} + \frac{C - \text{salvage}}{(1 + i)^n - 1} + \frac{R}{(1 + i)^x - 1};$$

where S_C = capitalized cost of a project having a useful life of n years. It includes renewal or first cost C, annual cost of maintenance and operation $\frac{0}{i}$, and depreciation $\frac{C - \text{salvage}}{(1 + i)^n - 1}$ at interest rate i, and major repairs $\frac{R}{(1 + i)^x - 1}$ every x years. Where the salvage value of project or major repairs are nil, omit.

Total Annual Cost—Economic Comparison:

$$S_C i = C i + 0 + \frac{i(C - \text{salvage})}{(1 + i)^n - 1} + \frac{Ri}{(1 + i)^x - 1};$$

where $S_C i$ = total annual cost and $\frac{Ci}{(1 + i)^n - 1}$ represents the annual rate of depreciation or money set aside each year at compound interest i to equal the first cost C at end of n years.

Capital Recovery:

$$R = P \left[\frac{i(1 + i)^n}{(1 + i)^n - 1} \right]$$

where R is the annual rate of capital recovery; P is principal or first cost of installation; n is capital recovery period or useful life; i is interest rate.

CONVERSION FACTORS

Units of Weight

1 kilogram (kg) = 1,000 grams (gm)
1 milligram (mg) = 10^{-3} gm
1 microgram (μg) = 10^{-6} gm
1 nanogram (ng) = 10^{-9} gm
1 picogram (pg) = 10^{-12} gm

Units of Concentration—for Solids

1 part per million (ppm) = 1 mg/kg = 1 μg/gm
1 part per billion (ppb) = 1 μg/kg = 1 ng/gm
1 part per trillion (ppt) = 1 ng/kg = 1 pg/gm

Units of Concentration—for Water (on a weight-to-weight basis)

1 milligram per liter = 1 mg/l = approx. 1 ppm
1 microgram per liter = 1 μg/l = approx. 1 ppb
1 nanogram per liter = 1 ng/l = approx. 0.001 ppb

Units of Concentration—for Air (on a weight-to-volume basis corrected for temperature and pressure)

1 microgram per cubic meter = 1 μg per cu m = 1 μg/m³
1 nanogram per cubic meter = 1 ng per cu m = 1 ng/m³

APPENDIX II

METRIC EQUIVALENTS OF COMMONLY USED ENGLISH UNITS OF MEASUREMENT

Conversion Factors

English Unit	Multiplier	Metric Unit
acre	0.405	ha
acre-ft	1,233.5	m³
BTU	0.252	kg-cal
BTU/lb	0.555	kg-cal/kg
bu	35.24	l
bu	0.03524	m³
cfm	0.028	m³/min
cfs	1.7	m³/min
cfs/acre	4.2	m³/min/ha
cfs/mi²	0.657	m³/min/sq km
ft³	0.028	m³
ft³	28.32	l
in.³	16.39	cm³
yd³	0.75	m³
°F	0.555 (°F − 32)	°C
fathom	1.8	m
ft	0.3048	m
ft-c	10.764	lumen/m²
gal	3.785	l
gpd/acre	0.00935	m³/day/ha
gpd/ft	0.0124	m³/day/m
gpd/ft²	0.0408	m³/day/m²
gpm	0.0631	l/sec
gpm/ft²	40.7	l/min/m²
hp	0.7457	kw
in.	2.54	cm
lb	0.454	kg
lb/day/acre-ft	3.68	g/day/m³
lb/1,000 ft³	16.0	g/m³
lb/acre/day	0.112	g/day/m²
lb/day/ft³	16	kg/day/m³
lb/day/ft²	4,880	g/day/m²
lb/mil gal	0.12	g/m³
mgd	3,785	m³/day
mgd/acre	9,360	m³/day/ha
mile	1.61	km
ppb	10⁻³	mg/l
pcf	16.02	kg/m³
psf	4.88	kg/m²
psi	0.0703	kg/cm²
sq ft	0.0929	m²
sq ft/ft³	3.29	m²/m³
sq in.	6.452	cm²
sq miles	2.590	km²
tons (short)	907	kg
tons (short)	0.907	metric tons

SOURCE: *J. Water Pollution Control Fed.*, Part 2, 136–137 (March 1969).

APPENDIX III

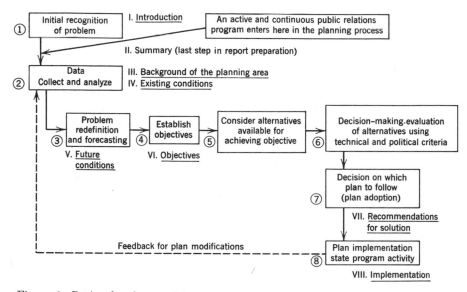

Figure 1 Basic planning model. (From Richard O. Toftner, *Developing A State Solid Waste Management Plan,* U.S. Public Health Service Pub. No. 2031, Dept. of HEW, Washington, D.C., 1970.)

The first step (1) in the planning process is awareness that a problem exists and needs to be solved. The second step (2) is to collect and analyze data relating to the problem. Such analysis makes possible a redefinition of the problem and a forecasting of future situation (3). Problem definition for both the present and future situation helps to suggest objectives (4) that if achieved would serve to solve the problems. Two or more alternatives (5) might be available for solving the problem and achieving objectives. The feasible alternative or alternatives (6) are selected by considering technical, political, social, and other factors. Once this decision has been made a plan for solution of the problems (7) can be adopted. Actual action for carrying out the plan (8) then follows. Effectiveness of the plan is measured during its implementation. These data are fed back into the continuing planning process to guide plan modifications if needed.

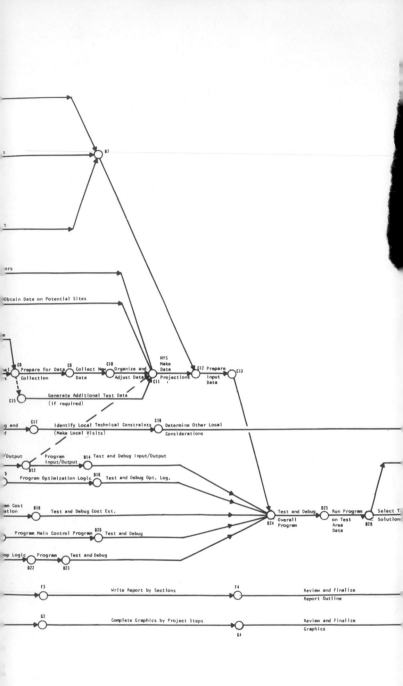

t

B7

t

ors

Obtain Data on Potential Sites

Prepare for Data
Collection
C8

Collect New
Data
C9

Organize and
Adjust Data
C10

NYS
Make
Data
Projections
C11

C12 Prepare
Input
Data

C13

Generate Additional Test Data
(if required)
C15

Identify Local Technical Constraints
(Make Local Visits)
C17

Determine Other Local
Considerations
C18

Program
Input/Output
B13

Test and Debug Input/Output
B14

Program Optimization Logic
B15

Test and Debug Opt. Log.
B16

am Cost
ation
B18

Test and Debug Cost Est.

Program Main Control Program
B20

Test and Debug

op Logic
B22

Program
B23

Test and Debug

Test and Debug
Overall
Program
B24

Run Program
on Test
Area
Data
B25

Select T
Solution
B28

F3

Write Report by Sections
F4

Review and Finalize
Report Outline

G3

Complete Graphics by Project Steps

Review and Finalize
Graphics
G4

lanagement study.

NEW YORK STATE
STATEWIDE SOLID WASTE MANAGEMENT STUDY
PHASE 2 - SYSTEMS MODEL

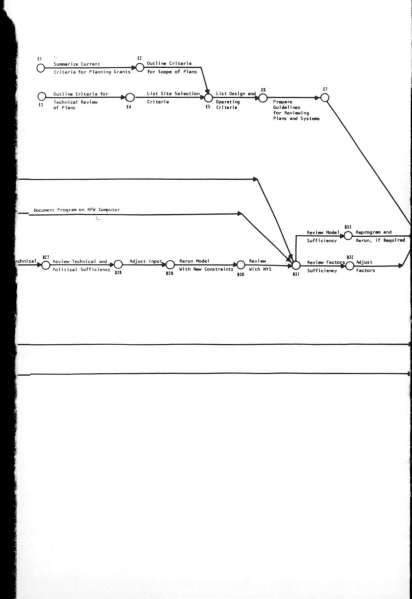

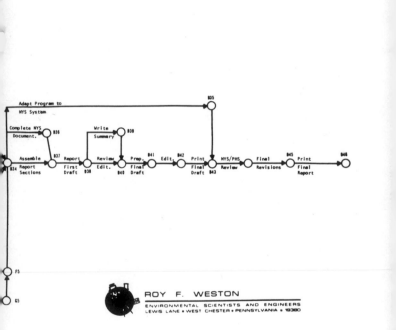

Adapt Program to
NYS System

Complete NYS   B36
Document.

B35

Assemble   B37   Report   Review   Prep.   B41   Edit.   B42   Print   NYS/PHS   Final   B45   Print   B46
Report   First   Edit.   Final   Final   Review   Revisions   Final
Sections   Draft B38   B40   Draft   B43   Report
B34

Write   B39
Summary

F5

G5

ROY F. WESTON
ENVIRONMENTAL SCIENTISTS AND ENGINEERS
LEWIS LANE • WEST CHESTER • PENNSYLVANIA • 19380

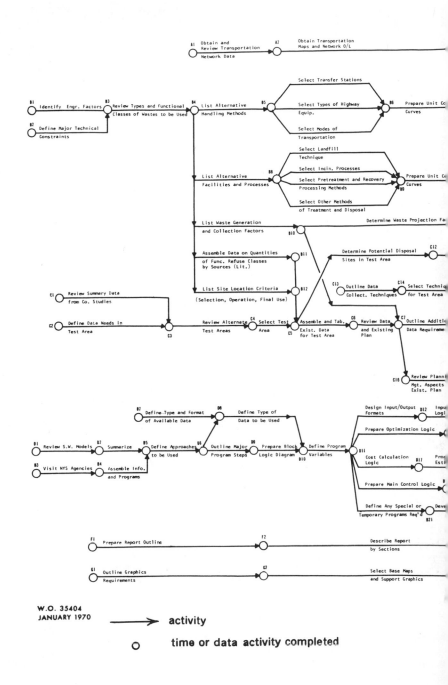

→ activity

○ time or data activity completed

Figure 2   A network diagram as applied to the preparation of a state solid waste

INDEX